# Healthcare Delivery in the Information Age

**Series Editor**
Nilmini Wickramasinghe
Epworth HealthCare
Richmond, VIC
Australia

Swinburne University of Technology
Australia

The healthcare industry is uniquely structured so that the receiver of the services (the patient) often isn't the predominant payer for those services (the insurance company). Healthcare interventions are often complex and typically involve multiple players including providers, payers, patients and regulators. This leads to economic dilemmas such as moral hazard, information asymmetry, and tangential considerations of cost versus quality creating obstacles on the road to delivering efficient and effective healthcare. Relevant data, pertinent information, and germane knowledge play a vital role in relieving these problems and can be most effectively obtained via prudently structured and well designed healthcare technology. Some of the major challenges facing today's healthcare organizations include demographic (longer life expectancy and an aging population), technology (incorporating advances that keep people healthier), and financial (escalating costs technological innovation) problems. In order to realize technology's full potential it is imperative to understand the healthcare-technology paradigm, develop sustainability models for the effective use of technology in a specific context, then successfully design and implement patient-centric technology solutions. Many of the problems with technology are connected to the platform-centric nature of these systems which cannot support seamless transfer of data and information, leading to inferior healthcare delivery. This new series focuses on designing effective and efficient technologically enabled healthcare processes to support the delivery of superior healthcare and provide better access, quality and value. It's main goal will be to identify the barriers and facilitators in moving from idea generation to concept realization and will navigate the key challenges in the field: bringing readers solutions and recommendations while identifying key factors in developing technology-enabled healthcare solutions.

More information about this series at http://www.springer.com/series/8783

Nilmini Wickramasinghe • Freimut Bodendorf
Editors

# Delivering Superior Health and Wellness Management with IoT and Analytics

Springer

*Editors*
Nilmini Wickramasinghe
Epworth HealthCare
Richmond, VIC
Australia

Swinburne University of Technology
Hawthorn, VIC, Australia

Freimut Bodendorf
Friedrich-Alexander-University
Erlangen-Nuremberg
Erlangen, Germany

ISSN 2191-5946 ISSN 2191-5954 (electronic)
Healthcare Delivery in the Information Age
ISBN 978-3-030-17346-3 ISBN 978-3-030-17347-0 (eBook)
https://doi.org/10.1007/978-3-030-17347-0



This Springer imprint is published by the registered company Springer Nature Switzerland AG.
The registered company address is: Gewerbestrasse 11, 6330 Cham, Switzerland

*For Our Families*

*This Series is dedicated to Leo Cussen: Learned scholar, colleague extraordinaire, and good friend.*

# Foreword

The first mHealth Summit in Washington, DC, drew a mere 500 attendees in 2010, but it grew nearly tenfold after only 4 years. The initial products typically featured basic, consumer-facing health and wellness monitoring sensors, along with some of the earliest and simplest wrist-wearable activity sensors like the Fitbit.

The arrival of ubiquitous mHealth about a decade ago dovetails with unexpected and exceptional adoption of mobile communication technologies. One might easily forget that the iPhone and apps were only introduced in 2008, followed by the blockbuster introduction of the iPad in late 2009. Right behind the Apple products came a growing wave of Google Android-based cell phones and tablets, which greatly expanded the market. Simultaneously, broadband cellular and wi-fi services were also progressing rapidly, covering much broader geographic footprints with growing reliability and capacity.

As cell phones, apps, and cellular data services were growing quickly, IoT, though described in the research literature, had not yet become a generic product reality. Many/most mHealth products were, at that point, tethered to apps, and the phone/tablet devices became the store-and-forward tools. Mass manufacturing of low-cost IP-addressable wi-fi and cellular chips needed for phones and apps opened a new product opportunity, though: inexpensive stand-alone or clustered mHealth sensors that could be used to collect personal health data with or without smartphones or tablets. Sporadic and isolated islands of proprietary IP-addressable consumer-facing products were showcased in the 2015 Consumer Electronics Show, but they exploded into widespread deployment in 2017 after the Amazon Alexa and Google Assistant ecosystems began to consolidate the market.

In January 2019, the US Consumer Electronics Show showcased thousands of IoT-ready mHealth sensors, monitors, and health and wellness monitoring and management ecosystems. Many of the products displayed are also FDA-approved for life-critical medical care, not just wellness and fitness monitoring.

Despite all of the underlying technology progress, commercial adoption of IoT-based mHealth and eHealth technologies is still very slow. Yes, consumers are indeed purchasing a large variety of mHealth devices for personal and family use, but those devices are rarely used to communicate directly with physicians.

This book provides a diverse selection of chapters that shed light on many important theory, research, and practice that must be understood and resolved in order for the true opportunities of improved patient care access, better patient care coordination, higher patient and clinician adoption and satisfaction, and, ultimately, improved quality, safety, efficiency, efficacy, and cost-effectiveness of patient care and wellness management.

There are four overarching themes that organize this book's sections: mobile- and sensor-based solutions, opportunities to incorporate critical aspects of analytics to provide superior insights and thus support better decision-making, critical issues around aspects of IoT in healthcare contexts, and applications of portals in healthcare contexts.

The chapters are introduced well in the Preface, but a few of them stand out because they illustrate some of the interesting opportunities and challenges of IoT in healthcare applications.

The vast majority of chapters deal with the first mobile- and sensor-based solutions chapter, because the healthcare field is reaping the benefit of decades of sensor miniaturization and pervasive mobile connectivity research and product development from other industrial and consumer markets, e.g. automobiles, aircraft, drones, computer games, and smartphones have all been leveraging low-power microscopic gyroscope technologies for at least two decades, same with RFID chips, barcode readers, and location-based marketing which began to impact retail and manufacturing processes at about the same time.

The flood of new, lightweight, wearable, and wireless sensors is creating a rich choice of discrete patient data that only existed inside high-acuity hospital settings in the 1990s. This health data is being created 24 hours a day, year round, in settings ranging from home, school, allied health, physical fitness, sports, and vacation destinations. One need not look further than the latest Apple Watch Series 4 release in 2018, which offered FDA-approved atrial fibrillation ECG monitoring as an example.

In this book, "Towards a Tricorder . . . " and "Towards a Better Life for Diabetic Patients," chapters describe the aspirational goals of modern mobile health technology inventors. Other chapters illustrate novel ways for clinicians to collaborate more closely ("Piloting a Mobile Tele-Simulation Unit . . . " and "Mobile Nursing..."). Also, chapters like "SmartCoping . . . ," "Changing Behavior of Kids with Obesity . . . ," and "Precision wellness . . . " address the growing global interest to empower patients to manage their own wellness and healthcare with personalized tools that suit their health and lifestyle most effectively.

The section "Opportunities to incorporate critical aspects of analytics to provide superior insights and thus support better decision making" presents a number of novel ideas about translating the rapidly growing flood of mobile health data into information and knowledge for better decision, action, and outcomes. Notable chapters include these three: "Knowledge Acquisition of Consumer Medication Adherence," because medication errors are believed to cause the majority of preventable medical errors, harm, and expenses; "Data Quality in Healthcare," because ongoing flaws in mHealth data accuracy has greatly slowed widespread adoption and use of the technologies; and "Enabling Value-Based Health Care with Business Analytics and Intelligence" because of the global shift by government agencies away from financial incentives and profits as the primary lever to manage healthcare expenses.

The third section highlights some addition, "Critical Issues Around Aspects of IoT in Healthcare Contexts." The chapter on "Implementing Lean Principles in the Healthcare Industry…" points out novel constraints of healthcare, such as healthcare's "failures" being measured in lives lost, a far less forgiving metric than lost profits. The "AR/VR In Healthcare" chapter identified significant gaps in healthcare-related research versus applications in entertainment and other industries, and the "Data Disparity Denial" chapter delves into cost-, decision-, and responsibility-shifting that pervasive adoption of "best practices" can create for aging populations around the world who are exhibiting similar co-morbidities like obesity and diabetes which may greatly complicate or inhibit access to necessary therapeutic treatment like hip, knee, or cardiac valve replacement.

The fourth and final section sheds new light on "Applications of portals in healthcare contexts," because the emerging global model of health and wellness includes a very strong emphasis on "patient empowerment." The chapter on "Older Adults Empowerment Through Training and Support…" provides an excellent example, because it addresses the global phenomenon, challenge, and opportunity presented by rapidly growing elder populations. Similarly, the "Toward Actionable Knowledge…" chapter illustrates the opportunities that can emerge when patient self-care is integrated with physician/clinician tracking and intervention for improved patient-centric care. The "Determining missing key elements in OIS (Oncologic Information System)…" chapter does a very good job exposing information, communication, and collaboration gaps between patients, caregivers, and physicians when making decisions about cancer treatment pathways.

In summary, the rapidly expanding quantity of data becoming available from mobile IoT health and wellness devices is now beginning to lend itself to advanced analytic tools. These tools and techniques are often adopted and adapted from consumer and industry applications. For example, newly emerging artificial intelligence (AI) and machine learning (ML) tools are being used to strengthen proven generations of Business Intelligence and Analytics software from other industries, too. There are promising applications of these analytic tools to assist patients and clinicians for a number of chronic diseases like diabetes, where treatment pathways and care coordination practices have begun to converge. Such convergence improves the researchers', patients', managers', and clinicians' ability to derive actionable information from large population data sets. In addition, individual patients and clinicians can fine-tune the tools to better match specific and unique patient circumstances.

Today, the path forward remains unclear, and somewhat obscure, however. There are several factors that make mobile IoT healthcare applications quite novel and challenging. First, the healthcare industry is not homogeneous across specialties, nations, or societal subsets.

Healthcare practice does not follow rules like baseball or computer games. In fact, modern healthcare practices historically developed from clinical subspecialties that evolved along very disparate pathways over dozens, sometimes hundreds or thousands, of years. Second, diseases and co-morbidities (simultaneous diseases in the same patient) are rapidly increasing because human longevity has increased so rapidly in the past century. These co-morbidities can be very unique to the patient, based on history, heredity, and/or environmental factors. Third, the progression of one or multiple diseases, and the actual responsiveness to different medicines or therapy, varies from patient to patient, and the variation may evolve uniquely during daily circadian body rhythms and over the course of time. Fourth, and somewhat perversely, the underlying diseases are actively antagonistic and opportunistic, and many are found to effectively evolve, morph, relocate, and/or rapidly create defence mechanisms in response to medical interventions (e.g., some staphylococcus bacteria have evolved to the point where they are immune to virtually all modern antibiotics!).

The final issue, which has become sort of "the elephant in the room," is balancing personal health privacy and confidentiality against cost, convenience, and societal needs. To date, few IoT technologies have been developed with robust encryption and security in mind. Most commercial products rely more heavily on vendor's proprietary communication, storage, and analysis techniques. Not only aren't those systems inherently secured with typical SSL or encryption techniques, but such systems also limit government or hospital data collection, aggregation, and analysis (see the HITSP.org Technical Note 905 for a longer detailed discussion about underlying issues).

The original US government "consumer empowerment" electronic medical data standards were intentionally separated from the unfortunately named "biosurveillance" standards (see www.HITSP.org). "Biosurveillance" was intended as a catch-all umbrella for the societal benefit of tracking, identification, and remediation of high-risk health problems like infectious diseases, but the term also carries the stigma of potential government tracking of its own citizens and possible privacy intrusion threats to citizen safety. Because of the risk of governmental (or employer) access and misuse of personal health data, the noble goal of helping consumers access and control their own health data (consumer empowerment) added the shadow of doubt and privacy concerns by other HITSP standards regarding biosurveillance and newborn baby screening.

The ongoing pathetic failures of managing privacy via fines and criminal prosecution (https://ocrportal.hhs.gov/ocr/breach/breach_report.jsf) provide a painful reminder that such issues *must* be addressed in order to gain enduring traction and accelerated adoption of mobile IoT technologies.

Because privacy and security solutions are rapidly being explored and implemented in finance, retail, and other industries, the privacy and security of the global healthcare infrastructure should improve in the near future. Once that is solved, patient and citizen's data can be "de-identified," allowing much more acceptable integration with other population health data.

As the privacy, quality, and quantity of heterogeneous mobile IoT health data improves, the ideas, innovations, and opportunities that are described in this book's chapters will all contribute to the improvement, personalization, affordability, equity, and transformation of healthcare. Kudos to the authors and editors for assembling this timely and inspiring collection!

Villanova University
USA

Elliot B. Sloane

# Preface

The *Internet of Things (IoT)* is made up of a network of devices all embedded with electronics, software, sensors, and connectivity to enable them to connect, interconnect, and exchange data (Brown 2016a, b; Internet of Things Global Standards Initiative; Hendricks 2015). This then creates further opportunities for direct integration of the physical world into computer-based systems which in turn results in efficiency gains, economic benefits, more effective operations, and typically a reduction in human labour (Vermesan and Friess 2013; Santucci 2016; Mattern and Floerkemeier 2016; Lindner 2015). In 2017, the number of IoT devices was over 8.4 billion (http://www.faz.net/aktuell/wirtschaft/diginomics/grosse-internationale-allianz-gegen-cyber-attacken-15451953-p2.html?printPagedArticle=true#pageIndex_1), and this is estimated to increase to 30 billion devices by 2020 (Nordrum 2016), while the global market value of IoT is anticipated to reach $7.1 trillion at this time (Hsu and Lin 2016).

IoT involves extending Internet connectivity beyond standard devices, such as desktops, laptops, smartphones and tablets, to any range of traditionally *dumb* or non-Internet-enabled physical devices and everyday objects (Hamid et al. 2019; Wigmore 2014; Internet of Things (IoT)). Embedded with technology, these devices can communicate and interact over the Internet, and they can be remotely monitored and controlled (The "Only" Coke Machine on the Internet; Mattern and Floerkemeier2010; Weiser 1991; Magrassi and Berg 2002). Healthcare to date has been a laggard in embracing technology in general and the full extent of the potential of IoT specifically; however, this cannot continue if healthcare is to deliver superior patient-centred high-value care (Wickramasinghe and Schaffer 2010; Wickramasinghe et al. 2017).

Healthcare delivery in the twenty-first century is currently facing the triple challenge of exponentially increasing costs, aging populations, and the rise of chronic care (Wickramasinghe et al. 2017). This is leading most countries around the world to look at technology-enabled healthcare reform that provides better quality, better access, and better value care in general and leveraging the opportunities and benefits afforded by the technologies of the IoT for the healthcare domain (Wickramasinghe et al. 2017).

Our book serves to present a miscellany of papers which focus on critical aspects around embracing the technologies of the IoT to enable and support superior healthcare delivery and wellness management. This is still a very nascent domain and many issues around health literacy, policy, privacy, and security, not to mention the direct and subtle as well as far-reaching implications for the various stakeholders (patients, clinicians, healthcare organizations, regulators, payers, and the community at large) which have yet to be fully understood or identified.

Our goal is to simply compile the chapters to make up this work. We wanted to share with you, the reader, some of the critical touch points and help to enrich discussions and discourse and inspire further research into this critical domain, a domain that touches all of us, and thus, all of us should form considered opinions about its future directions. It is our belief that through the judicious use of the technologies of the IoT, it will indeed be possible to enable and sustain an environment where digital technologies support better monitoring, better data, better communications so that we have better access, better quality, and a high value of healthcare delivery and wellness management for all of us, when and how they need it.

The chapters making up this book have been arranged into four main sections as follows: (1) mobile- and sensor-based solutions, (2) opportunities to incorporate critical aspects of analytics to provide superior insights and thus support better decision-making, (3) critical issues around aspects of IoT in healthcare contexts, and (4) applications of portals in healthcare contexts.

Specifically, they are as follows:

Part I: Mobile and Sensor Based Solutions

Chapter "Towards a Medical Tricorder – A 3D-Map to Categorize Diseases for Self-Care with Mobile Technology" by Hamper et al which describes an innovation challenge around designing a mobile IoT monitoring device to treat lifestyle cardiovascular and metabolic diseases and to enhance patient self-care.

Chapter "Piloting a Mobile Tele-Simulation Unit to Train Rural and Remote Emergency Healthcare Providers" by Jewer et al. which discusses the development and piloting of a Mobile Tele-Simulation Unit (MTU) prototype to address the challenges of emergency healthcare training in rural and remote settings.

Chapter "Drone Delivery for Medical Emergencies" by Scott, J and Scott C which provides a general overview of drone technology, and then explores ways drones can potentially improve emergency medical care by speeding delivery of medicines, devices, blood, vaccines, or even organs for transplant.

Chapter "Converting Disability into Ability Using IT/IS and Smart Textiles" by Shaukat et al which focuses on utilizing IS/IT (Information Systems/Information technology) and smart textiles to empower individuals suffering from physical disability due to nerve/ neuron damage due to brain injury.

Chapter "A Mobile Nursing Solution" by Kou et al. which discusses the design and development of a mobile nursing solution for medication management for a hospital in China.

Chapter "SmartCoping: A Mobile Solution for Recognizing Stress and Coping With IT" by Reimer et al. which describes a mobile biofeedback app, SmartCoping, designed to recognize personal stress based on calibration to a person's unique heart rate variability (HRV).

Chapter "Changing Behaviour Of Kids With Obesity With Gamification Wearables" by Schultz et al, which focuses on how gamification can be harnessed to assist with addressing childhood obesity.

Chapter "Precision Wellness: An Optimization Model" by Cooper and Wickramasinghe which serves to differentiate between precision medicine, precision wellness and related terms and the role for the tools and technologies of IoT (Internet of Things).

Chapter "The Development of a Wearable for an Automated Documentation and an Improved Staff Planning in Outpatient Care" by Ma and Weissenbaeck which presents a sensor-based solution to assist with assessing staff physical and psychological stress levels to assist with planning activities accordingly.

Chapter "Toward a Better Life for Diabetic Patients – Developing and Integrating a Non-invasive Self-Management Support Tool Within a Smart Digital Companion" by Nguyen et al. which examines the potential for a non-invasive solution to assess blood glucose levels of individuals with diabetes.

Part II: Opportunities to Incorporate Critical Aspects of Analytics to Provide Superior Insights and Thus Support Better Decision Making

Chapter "Intelligent Risk Detection in Health Care: Integrating Social and Technical Factors to Manage Health Outcomes" by Moghimi et al. which presents an integrative framework to support superior intelligent risk detection for various healthcare contexts.

Chapter "A Literature Review on Predicting Unplanned Patient Readmissions" by Eigner and Cooney in which the authors examined the literature between 2005 and 2017 and found 44 relevant articles regarding predictive models for determining patient readmission factors.

Chapter "Using Knowledge Management To Develop Superior Online Health Decision Support Solutions: The Case of Allergy Care" by Wickramasinghe presents a research in progress study that focusses on designing and developing a longitudinal knowledge base to track the progress and development of paediatric allergies.

Chapter "Opportunities for Using Blockchain Technology in Ehealth: E-prescribing in Germany" by Seitz and Wickramasinghe which examines the potential for the incorporation of blockchain technology to support successful e-prescribing in Germany.

Chapter "Knowledge Acquisition of Consumer Medication Adherence" by Vlahu-Gjorgievska et al which examines various analytical methods to assist with support better medical adherence.

Chapter "Addressing Data Accuracy and Information Integrity in Mhealth Solutions Using Machine Learning Algorithms" by Sako et al which present the potential of designing suitable machine learning algorithms to assist in identifying data accuracy levels in mhealth solutions.

Chapter "Enabling Value-Based Health Care with Business Analytics and Intelligence" by Wickramasinghe which presents the opportunities for leveraging BA (business analytics) and BI( business intelligence) capabilities into healthcare contexts for providing high value, patient centred quality care.

Part III: Critical Issues Around Aspects of IoT in Healthcare Contexts

Chapter "A review of Mixed Reality in Health Care" by John and Wickramasinghe which presents opportunities for Augmented reality, mixed reality and virtual reality solutions to be incorporated into various healthcare contexts to support a better patient experience and/or assist clinicians and clinical training.

Chapter "Implementing Lean Principles in the Healthcare Industry: A Theoretical and Practical Perspective" by Pakdil et al which presents the case for the benefits of incorporating lean principles into various healthcare technology implementations to enable the full realisation of the benefits afforded by the technology solutions.

Chapter "Data, Denial, and Disparity: Is This a New Digital Divide" by Wickramasinghe et al which examines the potential negative impacts of a technology enabled value-based care imitative that fails to recognise challenges and realities for vulnerable groups in the community.

Chapter "The Enabling Role for Technology in the Support of Care Coordination in Health Care" by Gibbings and Wickramasinghe which presents the merits of adopting a technology enabled care co-ordination vision for delivery superior patient-centred care.

Chapter "Managing the Risks of Emerging IoT Devices" by Paxton and Branca which provides insights on how to ensure suitable cybersecurity safe guards are in place when using IoT devices in healthcare contexts.

Chapter "Mosquitoes and Public Health: Improving Data Validation of Citizen Science Contributions Using Computer Vision" by Muñoz et al. which highlights important public health impacts and how IoT can potentially assist.

Part IV: Applications of Portals in Healthcare Contexts

Chapter "Using Responsive Web Design to Enhance the User Experience of Chronic Disease Management Portals for Clinical Uses" by Gunawardane and Wickramasinghe which highlights the need to design suitable web interfaces so that solutions load equally gracefully on computers, handheld devices and tablets for an optimal user experience.

Chapter "Older Adults Empowerment Through Training and Support and Its Implication on Proactive Self-Monitoring, Patient Engagement, and Connected Health" by Bozan and Mooney which discusses the benefits of training and support especially for older adults in regards to technology solutions and their effective adoption and use.

Chapter "An Evaluation of a Point-of-Care (PoC) System Implementation and Adoption in a Multi-Campus Private Hospital in Melbourne" by Muhammad and Wickramasinghe which examines the adoption and implementation of the OneVeiw Poitn-of-C-are systems into a private not for profit tertiary healthcare system in Melbourne, Australia.

Chapter "Leveraging the IOT to Enable the Guided Self-Determination Method" by Wickramasinghe et al. which examines the repurposing of the Guided Self-Determination (GSD) method into the Australian healthcare context.

Chapter "Determining Key Elements in Oncology Information System to Improve the Patient Experience and Clinical Care" by Shaukat et al. which examines several OIS (oncology information system) solutions to assess their strengths, weaknesses and key barriers and facilitators.

Chapter "Toward Actionable Knowledge: A Systematic Analysis of Mobile Patient Portals" by Noteboom and Abdel-Rahman which investigates and then identifies critical issues around designing suitable patient portals.

Chapter "A Lazy User Perspective to Patient Adoption and Use of Personal Health Records" by Kunene which examines critical issues around personal health records and patients portals.

Chapter "The Australian PCEHR OR My Health Record – The Journey Around a Large Scale Nationwide Digital Health Solution" by Wickramasinghe and Zelcer which traces key design and development and deployment of the national eHealth solution adopted by Australia.

No book can ever present in one volume a comprehensive collection covering all areas of IoT for healthcare; however, we hope this miscellany of chapters we present will challenge our readers and be thought provoking. We also hope that you have as much fun reading our book as we have had in compiling and writing it. In closing, we trust that on the completion of this book, researchers, scholars, practitioners, consultants, and the general public will all have a better understanding of how the technologies of the IoT can be harnessed to provide superior healthcare delivery and wellness management and will rise to the challenge of starting to build a better health and wellness environment for tomorrow and today.

East Kew, VIC, Australia<br>Nürnberg, Germany<br>September 2019

Nilmini Wickramasinghe<br>Freimut Bodendorf

## References

Brown, E. (2016, 13 September). *Who needs the Internet of Things?*. Linux.com. Retrieved 23 October 2016.

Brown, E. (2016, 20 September). *21 open source projects for IoT*. Linux.com. Retrieved 23 October 2016.

Hamid, U. Z. A., et al. (2019). Internet of Vehicle (IoV) applications in expediting the implementation of smart highway of autonomous vehicle: A survey. Performability in Internet of Things. Retrieved 28 August 2018.

Hendricks, D. (2015). *The trouble with the Internet of Things*. London Datastore. Greater London Authority. Retrieved 10 August 2015.

Hsu, C.-L., & Lin, J. C.-C. (2016). An empirical examination of consumer adoption of Internet of Things services: Network externalities and concern for information privacy perspectives. *Computers in Human Behavior, 62*, 516–527. https://doi.org/10.1016/j.chb.2016.04.023.

http://www.faz.net/aktuell/wirtschaft/diginomics/grosse-internationale-allianz-gegen-cyber-attacken-15451953-p2.html?printPagedArticle=true#pageIndex_1

Internet of Things (IoT). gatewaytechnolabs.com.

Internet of Things Global Standards Initiative. ITU. Retrieved 26 June 2015.

Lindner, T. (2015, 13 July). *The supply chain: Changing at the speed of technology*. Connected World. Retrieved 18 September 2015.

Magrassi, P., & Berg, T. (2002, 12 August). A world of smart objects. Gartner research report R-17-2243.

Mattern, F., & Floerkemeier, C. (2010). From the internet of computer to the Internet of Things (PDF). *Informatik-Spektrum, 33*(2), 107–121. https://doi.org/10.1007/s00287-010-0417-7. Retrieved 3 February 2014.

Mattern, F., & Floerkemeier, C. (2016). *From the internet of computers to the Internet of Things (PDF)*. ETH Zurich. Retrieved 23 October 2016.

Nordrum, A. (2016, 18 August). *Popular Internet of Things forecast of 50 billion devices by 2020 is outdated*. IEEE.

Santucci, G. (2016). *The Internet of Things: Between the revolution of the internet and the metamorphosis of objects(PDF)*. European Commission Community Research and Development Information Service. Retrieved 23 October 2016.

The "Only" Coke Machine on the Internet. Carnegie Mellon University. Retrieved 10 November 2014.

Vermesan, O., & Friess, P. (2013). *Internet of Things: Converging technologies for smart Environments and Integrated Ecosystems (PDF)*. Aalborg, Denmark: River Publishers. ISBN 978-87-92982-96-4.

Weiser, M. (1991). The computer for the 21st century (PDF). *Scientific American, 265*(3), 94–104. Bibcode:1991SciAm.265c..94W.
https://doi.org/10.1038/scientificamerican0991-94. Archived from the original (PDF) on 11 March 2015. Retrieved 5 November 2014.

Wickramasinghe, N., & Schaffer, J. (2010). *2010 realizing value driven patient centric healthcare through technology*. IBM Center for the Business of Government, DC.

Wickramasinghe, N., Johns, B., George, J., & Vogel, D. (2017). Achieving value-based care in chronic disease management: The DiaMonD (Diabetes monitoring) solution. *JMIR Diabetes forthcoming.*

Wigmore, I. (2014, June). Internet of Things (IoT). TechTarget.

# Acknowledgements

This work would not be possible without the support and contributions of many people, too numerous to name individually. First and foremost, we thank all the authors who have contributed such thought-provoking chapters that fill the following pages. We are indebted to our respective institutions for affording us with the time to work on this book project. In addition, we are most grateful to the wonderful assistance provided by the team at Springer, especially Janet Kim. Finally, we thank our families for their ongoing support.

Professors Nilmini Wickramasinghe and Freimut Bodendorf, December 2018

# Contents

# About the Editors

**Nilmini Wickramasinghe** is the inaugural Professor-Director, Health Informatics Management, at Epworth HealthCare and the Professor of Digital Health and Deputy Director of the Iverson Health Innovation Research Institute at Swinburne University as well as an Honorary Professor at the Peter MacCallum Cancer Centre. After studying at the University of Melbourne, she accepted a full scholarship to undertake PhD studies in the United States in Management of Information Systems as they relate to healthcare contexts. For over 20 years, she has been actively researching and teaching within the health informatics domain. Her particular focus is on designing and developing suitable models, strategies, and techniques grounded in various management principles to ensure that technology solutions effect superior, value-based, patient-centric care delivery. He collaborates with leading scholars at various premier healthcare organizations throughout Australasia, the United States, and Europe and has well published with more than 380 refereed, peer-reviewed scholarly articles, more than 12 books, numerous book chapters, an encyclopaedia, a patent, and a well-established funded research track record. He is the Editor-in-Chief of two scholarly journals published by Inderscience and has successfully taken four separate health technology solutions from idea to realization in the United States.

**Freimut Bodendorf** is Head of the Chair of Information Systems (Services, Processes, Intelligence) and Director of the Institute of Information Systems. He graduated from the University of Erlangen-Nuremberg (School of Engineering) with a degree in Computer Science. He obtained his doctor degree (PhD) in Information Systems. Subsequently, he was Head of an IS Department at the University Hospital and Medical School at the University of Freiburg, Germany; Professor at the Postgraduate School of Engineering in Nuremberg, Germany; and Head of the Department of Computer Science and Information Systems at the University of Fribourg in Switzerland. Actually, he is Director of the Institute of Information Systems at the University of Erlangen-Nuremberg and Faculty Member of the School Business and Economics as well as the School of Engineering. His research areas include service design and engineering, process management, and business intelligence. Current projects investigate and create solutions in the fields of healthcare, commerce/retail, industrial services (smart factory and supply), and education.

# About the Authors

**Mohammad Abdel-Rahman (Al-Ramahi)** is an Assistant Professor of CIS at the Department of Computing and Cyber Security, Texas A&M University, San Antonio. His research appears in the *ACM Transactions on Management Information Systems* journal. He also has published several refereed conference papers in the areas of text mining and IS design.

**Sasan Adibi** is a Lecturer in the School of Computer Science at Deakin University. He has interests in designing and developing technology solutions for various healthcare contexts.

**Monica Adya** is Professor of Management at Marquette University. Her research addresses knowledge-based systems with applications to business forecasting and IS workforce issues. She has published in *Communications of the Association for Information Systems*, *Decision Support Systems*, *Human Resource Management*, *Information Technology & People*, *International Journal of Forecasting*, and *Information Systems Research* among others.

**Karoly Bozan** is an Assistant Professor of Information Systems Management at Duquesne University. He spent over a decade as the Technical and Management Consultant before returning to academia. His research interests are technical empowerment of the elderly, health informatics, and digital innovations.

**David L. Branca** is the Director of Research and Development at IOMAXIS, LLC. He holds an MS degree from SUNY ESF and a BS degree from RIT. He is active in the research and application of securing IoT and containers, of which he holds five cyber security patents.

**Andrew Cooney** studied Information Systems at the University of Erlangen-Nuremberg. In 2017, he did his master thesis on readmission risk prediction at the Institute of Information Systems.

**Paul Cooper, PhD, FACHI, CHIA** specializes in health informatics management and teaches on this topic at Deakin University. He obtained a PhD at the Walter and Eliza Hall Institute of Medical Research (Melbourne University), BSc (Monash

University), and a Grad Dip Business (Organisational Development) at RMIT University.

**Judy Currey** is the Director, Active Learning, at Deakin University. Her research is focused on decisions that impact patient safety and quality of care. She is a qualified Team-Based Learning Consultant-Trainer and has been awarded numerous national and local teaching and innovation awards. She is a graduate of the AICD.

**S. A. Sjoerd de Vries** is Assistant Professor at the University of Twente, Enschede, Netherlands. His research is in the field of digital transformations in the public sector. His focus is on the development, adoption, and implementation of data-driven and currently AI-based services.

**Adam Dubrowski** obtained his PhD from the University of Waterloo in 2001. In 2002, he joined the University of Toronto and in 2013 the Memorial University of Newfoundland. He began his research career by studying factors that influence the acquisition, maintenance, and transfer of skills. More recently, his research has focused on the role of disruptive innovations, such as 3D printing and virtual reality in training healthcare professionals. He has worked with a wide scope of professionals including kinesiologists, physiotherapists, surgeons, paediatricians, emergency physicians, and nurses. The unifying theme of his work is the use of simulation for education and training, assessment, and patient safety.

**Cody Dunne** holds a Bachelor of Science degree from the Memorial University of Newfoundland and is currently a Doctor of Medicine (MD) candidate at this institution. His research interests focus on the use of simulation for the training of healthcare professionals and lay-rescuers.

**Isabella Eigner** studied International Information Systems at the University of Erlangen-Nuremberg. Since 2014, she is a Member of the research staff and PhD student at the Institute of Information Systems. In her research, she is focusing on predictive analytics in healthcare.

**Rima Gibbings** has worked as a System Designer and Developer in healthcare systems and has a PhD from the University of Illinois in Chicago. As a Health Informatics Specialist and Researcher, she offers a technical approach that links her background in application development with public health topics, addressing current technical requirements.

**Ranganath Gunawardane** received his bachelor's degree in Electronics from the University of Peradeniya, Sri Lanka. He has more than 14 years of professional experience in the field of information technology. He is also a Research Affiliate at Epworth HealthCare.

**Andreas Hamper** completed his PhD at the Friedrich-Alexander-University Erlangen-Nuremberg under the Chair of Information Systems. His focus is on mobile and sensors and their applications to health and wellness management.

**Timothy N. Harwood** is an Associate Professor of Anesthesiology and Departmental Director of Quality Improvement in the Department of Anesthesiology at Wake Forest University Health Sciences in Winston-Salem, NC, USA. He attended medical school at the University of Missouri-Columbia and residency at Indiana University School of Medicine in Indianapolis, IN.

**Harith Hassan** completed Master of Information Technology Advanced in 2017 from the University of Wollongong majoring in ICT Strategic Planning and Enterprise Networking. During his studies, he has a special interest in health informatics and conducted research on medication adherence.

**Kristopher Hoover** received a Bachelor of Engineering degree from the Memorial University of Newfoundland in 2018. His research interests are focused in the areas of holistic frameworks for medical technology development and simulation modalities for healthcare education. He intends to continue exploring these areas while studying Medicine at the Memorial University of Newfoundland.

**Feride Bahar Isin** is an Associate Professor of Management at the Faculty of Economics and Administrative Sciences, Baskent University, in Turkey. She has an MBA degree from Baskent University and a PhD degree from Gazi University, Turkey. Her research interests are consumer behaviour and services marketing.

**Jennifer Jewer** holds a MASc and a PhD in Management Sciences from the University of Waterloo. She is an Assistant Professor in Information Systems in the Faculty of Business Administration at the Memorial University of Newfoundland. Her research focuses on the management and use of IT in organizations, and her recent projects have been in healthcare settings. She has a diverse range of consulting experience in risk management and software development.

**Blooma John** is a Lecturer at Canberra University, Australia. She completed her PhD from Nanyang Technological University (NTU), Singapore. Her research interest is in the area of text mining, question answering services, and healthcare social media. Her research has been published in journals such as the *Online Information Review*, *IEEE Internet Computing* and *Computers & Education*. She has published in various international conferences. She also has international experience in teaching leading information systems and business analytics courses in the Middle East, Singapore, India, and Vietnam at both undergraduate and postgraduate level.

**Lars Kayser, MD, PhD** is a Board-Certified Specialist in Internal Medicine and has worked for the last 10 years in the field of innovation and informatics in healthcare. He has helped develop the Epital Care Model, as well as the eHealth literacy framework, and eHealth literacy questionnaire.

**Jianqiu Kou** is a PhD student of Deakin University, Australia, and has 20 years of working experience in the field of construction and management of information systems in healthcare institutions. Her main research interests lie in the application of information technology in healthcare field.

**K. Niki Kunene** is an Assistant Professor of Information Systems at Eastern Connecticut State University. Her research interests include the design, use of patient engagement tools for health outcomes, and information privacy. She's also published on decision support/data mining methods in healthcare and business applications.

**Edith Maier** has a background in social anthropology and information science and joined the University of Applied Sciences, St. Gallen, in 2003. Her research focuses on intercultural factors in human-computer interaction as well as behavioural change support systems, mainly in digital health.

**Michael J. McCoy** has had multiple executive roles in the healthcare information technology and interoperability world, consulting, and provider organizations such as CMIO and as the first Chief Health Information Officer for the Office of the National Coordinator for Health Information Technology (ONC). He is Co-Chair of IHE International, representing HIMSS.

**Hoda Moghimi, PhD, MSc, BEng** received her PhD from RMIT University, Australia. She operates and researches within the information systems domain including business analytics, data science, knowledge discovery, IT and governance, intelligent solutions, and smart systems. She's published over 30 peer-reviewed scholarly journal and conference papers as well as book chapters in information systems.

**David Mooney** is a graduate student at The Ohio State University and has an extensive industry experience in the health and engineering domain. He has been consulting internationally while performing academic research.

**Imran Muhammad, PhD, MIS, MBus (ERP Systems)** is a Lecturer at the School of Medicine, Deakin University, Melbourne. He has a broad range of research interests in the areas of management information systems, health information systems, and electronic health records and cross-cultural system implementations and adoptions.

**Lucas Neitzel** is a higher degree candidate at the Friedrich-Alexander-University Erlangen-Nuremberg under the Chair of Information Systems. His focus is on analytics in health and wellness.

**Lemai Nguyen** is an Associate Professor at the Department of Information Systems and Business Analytics, Deakin University (Melbourne Campus). She actively conducts research and publishes in information systems (IS) areas including health informatics, business requirement analysis, and business information systems. She has published widely in many international scholarly journals, book chapters, and fully refereed conferences of high-quality (ERA ranking A and B).

**Cherie Noteboom, PhD** is an Associate Professor of Information Systems in the College of Business and Information Systems at Dakota State University. She has extensive experience in information systems, healthcare, and leadership. She has refereed publications in variety of IS fields. She is an active Member of PMI, ACM, HIMSS, and AIS.

**Fatma Pakdil** is an Associate Professor of Management at Eastern Connecticut State University in CT, USA. She has a BS degree in Econometrics and MBA and PhD degrees in Management from Uludag University in Turkey. Her research interests are quality control, service quality, lean production, Six Sigma, and healthcare management.

**Michael Parsons, MD, CCFP (EM), FCFP** is an ER Physician who completed his EM Fellowship at the Memorial University of Newfoundland in 2007. He is Chair and Assistant Professor in the Discipline of Emergency Medicine, with a cross-appointment to the Discipline of Family Medicine. He currently serves as the Assistant EM Programme Director at the Memorial University of Newfoundland. His teaching interests focus on simulation, procedural skills training, toxicology, and point-of-care ultrasound. Ongoing research activities are focused on the delivery of simulation-based teaching to rural and remote areas through mobile tele-simulation.

**Napoleon C. Paxton** is the Director of Innovation at IOMAXIS, LLC. He holds a PhD from The University of North Carolina at Charlotte in Information Technology, concentrating in cyber security. He also has over 20 publications including 2 book chapters and holds 1 patent in cyber security.

**Bodil Rasmussen** research focuses to improve the quality of care and self-management in people with diabetes and complex conditions. Her research programme focuses on empowering people to self-manage their condition by using technologies and implementation of research evidence into practice to improve patient outcomes; numerous projects involve international teams.

**Ulrich Reimer** has a background in computer science and has been with the University of Applied Sciences, St. Gallen, since 2005. His research activities concern mobile health applications to support individuals in changing their behaviour to improve their health. This includes data mining approaches to make sense of sensor data.

**Zaid Sako** is a Master student who is focused on doing research around health informatics issues, in particular focusing on Big Data and healthcare.

**Jonathan L. Schaffer, MD, MBA** is Managing Director of the Distance Health team in the Office of Clinical Transformation and of eClevelandClinic in the Information Technology Division at Cleveland Clinic. He leads the MyConsult programme, an online second medical opinion programme that is used by consumers and corporations worldwide to provide access to care while removing the geographic barriers. He also leads the DrConnect series of portals which securely connect referring physicians, facilities, and external reviewers to their patient's care at Cleveland Clinic. An active Board-Certified Joint Replacement Surgeon, Dr. Schaffer is also a Staff Member in the Center for Joint Replacement in the Orthopaedic and Rheumatologic Institute where he is the Physician champion on the ICD-10 transition and leads the Advanced Operative Technology Group, developing novel technologies for the arthroplasty patient population including the physical and technology upgrading of the operating rooms.

**Luisa Schulz** (1990) is a Digital Healthcare Enthusiast with an academic background in innovation management and healthcare innovation. She has an international background and completed an MBA at the University of Twente and an MSc in Innovation Management and Entrepreneurship at the Technical University of Berlin. She lives and works in Berlin where she has gathered professional consulting experience in the pharma und healthcare industry. As she has specialized in healthcare, she dedicates her research to the area of digital health.

**Carlton H. Scott** is Professor of Operations and Decision Technologies at the University of California, Irvine. He specializes in optimization models for managerial decision-making. He has more than 100 articles in research journals that look at both theory and application of optimization models. Application papers cover diverse areas such as the harvesting of sugar cane, assignment of customers to a commercial satellite fleet, the sizing of blind fasteners for construction purposes, and the recycling of electronic components.

**Judy E. Scott, PhD** is an Associate Professor of Information Systems at the University of Colorado Denver. She received her PhD and MBA from the University of California at Irvine. Her research interests include information technology in healthcare, knowledge management, and analytics capability for decision-making. Her journal publications include the *Journal of Management Information Systems*, *Information and Management*, *Decision Support Systems*, *Information Systems Frontiers*, *MIS Quarterly Executive*, and *Communications of the ACM*. She serves on the Editorial Board of *MIS Quarterly Executive*, *Information Systems Management*, and the *International Journal of Strategic Information Technology and Applications*.

**Juergen Seitz** received his PhD from Viadrina European University, Frankfurt (Oder), Germany. He is Professor of Business Information Systems and Head of the Business Information Systems Department at Baden-Wuerttemberg Cooperative State University Heidenheim, Germany. He is Editor, Associate Editor, and Editorial Board Member of several international journals.

**Muhammad Nadeem Shaukat** achieved his BEng in 2001 from the University of Engineering and Technology, Lahore, Pakistan. He served as Research and Development Engineer at Faisal Spinning Mills and Blessed Textile Mills Ltd. from 2001 to 2004. He joined National Textile University, Faisalabad, as a Faculty Member in 2004 and meanwhile supervised a number of research and industrial projects. He completed his master's degree in Fibre and Polymer Engineering in 2009, from Hanyang University, Seoul, South Korea. He received his PhD in Nanotechnology from Deakin University, Australia. His research interests include nanotechnology, modelling, sensors, information technology, and healthcare.

**J. B. Silvers, PhD** is the John R. Mannix Medical Mutual of Ohio Professor of Health Care Finance and Professor of Banking and Finance at the Weatherhead

School of Management with a joint appointment in the Case Western Reserve University School of Medicine. His research in the areas of financial management and health services has been published in numerous journals including *The Journal of Finance*, *Financial Management*, *Journal of the American Medical Association*, *Medical Care*, *The Annals of Family Medicine*, the *Journal of Business Research*, *Journal of Consumer Marketing*, and *The New York Times*, MarketWatch, Crain's Cleveland Business, and others. Frequently, he contributes commentary to NPR's Marketplace and other national and local print and broadcast media.

**Elliot B. Sloane** has spent over 40 years in a dual career spanning IT and medical technologies that improve healthcare. He is also an Advisor and Consultant to several US federal agencies and the World Health Organization. Today, he is the Director of Drexel University's Health Systems Engineering Program in Philadelphia and is developing courses and research programmes in healthcare informatics, medical device interoperability, patient data and security, mobile healthcare, RFID in healthcare, and related fields. He also leads a non-profit Center for Healthcare Information Research and Policy (CHIRP), to assure the next generations of health IT products and services are truly patient-centric, safe, efficient, and effective.

**Andrew Smith** is a Family/ER Physician and Associate Professor at the Memorial University of Newfoundland with 10 years of clinical experience. Prior to Medicine, he completed a master's degree in Electrical Engineering. Presently, he is working to develop a formal biomedical engineering programme at the university and a medical technology sector for the province of Newfoundland and Labrador.

**A. A. M. Ton Spil** (1964) teaches in the area of business information systems for MBA and risk management and business and IT. He did his PhD thesis on the effectiveness of information strategies, and after that, he specialized in the area of healthcare application and professional organizations. He is e-health Track Chair in main conferences and published on ISI A level. He was (Guest) Editor of *JSIS* and Associate Editor of several health journals. In 2017, his main topics are adoption of IT, business modelling, serious gaming, and digital strategies applied on (tele) health, music, and banking. He creates digital strategies for (social) media and health. His new book is called *Digital Future* which appears in May 2018.

**Tom Ulmer** has been a Lecturer at the University of Applied Sciences, St. Gallen, since 2015. Before joining academia, he worked as a Product Manager and Software Architect with a focus on tracking and monitoring vital data for developing personalized health applications. This is now his main research focus.

**Elena Vlahu-Gjorgievska** received her PhD from the Faculty of Computer Science and Engineering at the "Ss. Cyril and Methodius" University in Skopje, in 2013. From 2007, she was working at the Faculty of Information and Communication Technologies, Sofia University "St.Kliment Ohridski" in Bitola, Macedonia. Currently, she is a Lecturer in the School of Computing and Information Technology, Faculty of Engineering and Information Sciences, at the University of Wollongong, Australia. Her research interests include information systems and e-health. She has published over ten papers in international journals and presented her work at several conferences.

**Khin Than Win** is Associate Professor at the School of Computing and Information Technology. As a medical doctor conducting health informatics research, she brings a wealth of clinical knowledge in information systems and health informatics research. Her research is in persuasive technology, IT-assisted health education, behavioural change support systems, health data analytics, information security in healthcare, and electronic health record systems. She is also a Director of the Centre for Persuasive Technology and Society at the University of Wollongong. She has published widely in health informatics and obtained competitive grants such as Australian Research Council Grant.

**John Zelcer** MB, BS, BMedSc (Hons), FANZCA, MAICD, CHIA is a Company Director, Corporate Strategy Advisor, Informatician, and Executive Coach whose passion and focus is helping organizations, executive leaders, and leadership teams generate outstanding results for all their stakeholders, particularly in the healthcare sector. His career has spanned clinical and academic medicine in Australia and the United States, including a Visiting Professor role at Stanford University in California and currently as an Adjunct Professor in the School of Medicine at Deakin University, Australia. He has also held strategy management consulting and senior executive management roles in both Australia and the United States. His business roles have included board chairmanship, board directorship, and senior executive roles in strategic planning and business development.

**Qianqian Zhang** has a master's degree major in Biomedical Engineering. She has 5 years of working experience in biobank information system construction and bioinformatics analysis. Her research interests and directions are the application of information technology, precision medicine, and molecular diagnosis in clinical practice.

**Vibeke Zoffmann** is a Professor at the University of Copenhagen and Rigshospitalet, JMC. She has developed theories and methods around PROM in translational empowerment. She has developed an analogue guided self-determination (GSD) in chronic conditions, digitized together with Sundhed.dk, the Danish national health platform, and being disseminated through GSD blended learning developed together with CHR in Copenhagen.

# Part I
# Mobile and Sensor-Based Solutions

This section consists of 10 chapters as follows:

Chapter 1: Towards a Medical Tricorder: A 3D-Map to Categorise Diseases for Self-Care with Mobile Technology by Hamper et al.
Chapter 2: Piloting a Mobile Tele-simulation Unit to Train Rural and Remote Emergency Health-Care Providers by Jewer et al.
Chapter 3: Drone Delivery Models for Medical emergencies by Scott, J and Scott C.
Chapter 4: Converting Disability into Ability Using IT/IS and Smart Textiles by Shaukat et al.
Chapter 5: A Mobile Nursing Solution by Kou et al.
Chapter 6: SmartCoping: A Mobile Solution for Recognizing Stress and Coping with IT by Reimer et al.
Chapter 7: Changing Behavior of Kids with Obesity with Gamified Wearables by Schultz et al.
Chapter 8: Precision Wellness: An Optimization Model by Cooper and Wickramasinghe
Chapter 9: The Development of a Wearable for an Automated Documentation and an Improved Staff Planning in Outpatient Care by Ma and Weissenbaeck
Chapter 10: Towards a Better Life for Diabetic Patients: Developing and Integrating a Non-invasive Self-Management Support Tool Within a Smart Digital Companion by Nguyen et al.

Taken together, these chapters enable us to begin to understand the potential and possibilities for mobile and sensor solutions to assist individuals and/or groups in the community to address specific health and wellness issues from chronic conditions to better medical adherence. As you read these chapters, it is useful to consider what it is about these solutions that make them successful and embraced by users. All these solutions are suitable because they exhibit key technology, process, and people requirements. Succinctly the technology components work and are at an appropriate level of accuracy and fidelity above standard solutions; the solutions simplify and streamline processes, and they are easy to use and patient/user centered. Hence, what we see is that successful mobile solutions need to have sound technical and clinical outcomes as well as support patient and clinical user needs. The potential and possibilities for incorporating mobile and sensor solutions into the health and wellness domain is only limited to our imagination, but getting these solutions right, adopted and ensuring that they do indeed enable superior patient-centered superior value-based care to result is more challenging.

# Towards a Medical Tricorder: A 3D Map to Categorise Diseases for Self-Care with Mobile Technology

**Andreas Hamper, Lucas Neitzel, Nilmini Wickramasinghe, and Freimut Bodendorf**

## 1 Introduction

The extensive spread of information and mobile technologies in the field of consumer electronics assists individuals in numerous areas. Health and fitness applications, fitness trackers and wearables show particularly high, sustained demand among private consumers. Highly efficient and cost-effective digital consumer electronic sensor technologies can also be used for medical purposes. Non-invasive technologies offer capabilities to detect, measure and analyse medical conditions by private consumers.

The increase in so-called civilization diseases has been recorded in industrialised countries such as Germany for many years (Robert Koch-Institut 2015b). Lifestyle-related cardiovascular diseases and metabolic disorders, such as diabetes, are problems which are growing rapidly (World Health Organization 2016b). The number of adults suffering from diabetes has risen worldwide from 108 million in 1980 to an estimated total of 422 million in 2014. This corresponds to an increase of 390% since 1980 (World Health Organization 2016c).

At the same time, self-care is increasing as well. The WHO defines self-care as "[ . . . ] the ability of individuals [ . . . ] to promote health, prevent disease, and maintain health and to cope with illness and disability with or without the support of a health-care provider" (World Health Organization 2009, p. 17). The raising awareness of self-care can be seen in the growing sales volume of fitness applica-

A. Hamper · L. Neitzel · F. Bodendorf
FAU Friedrich-Alexander-University of Erlangen-Nuremberg, Nuremberg, Germany

N. Wickramasinghe (✉)
Epworth HealthCare, Richmond, VIC, Australia

Swinburne University of Technology, Hawthorn, VIC, Australia



N. Wickramasinghe, F. Bodendorf (eds.), *Delivering Superior Health and Wellness Management with IoT and Analytics*, Healthcare Delivery in the Information Age, https://doi.org/10.1007/978-3-030-17347-0_1

tions for smartphones and fitness trackers in the field of consumer electronics. The core objective of these is to track and monitor vital parameters mainly for sports-related reasons (GfK SE 2016). Continuously increasing measuring accuracy and technical performance also makes non-invasive mobile technologies attractive to the medical sector.

With the evaluation of current technologies, a growing number of medical conditions, especially vital signs and lifestyle-related cardiovascular diseases, can be identified and monitored easily, precisely and non-invasively by consumers. This is of particular significance, given that consumer electronic technologies can no longer be seen only as a complementary element besides professional medical procedures but are increasingly able to provide medical diagnoses and monitor diseases without the help of medical experts. This implies that low-cost consumer electronic technologies empower consumers to better monitor their health and promote individual self-care in the near future.

Companies such as Google are already working towards products which could be used for medical purposes in the near future ("Digital Contact Lenses Can Transform Diabetes Care" 2016). Prizes for innovations in mobile health technologies and medical tricorders for the consumer electronics area are being offered. Particular commitment is shown by Qualcomm's XPRIZE. The prize was announced in 2013 and is endowed with USD 10,000,000. The aim of this prize is to develop a medical tricorder which will be able to detect a preselected set of vital signs and diseases non-invasively (XChallenge 2014; XPrize 2016).

Following the example of the XPRIZE, a preselected set of vital signs and health conditions will be examined in this chapter. The burden of disease will be determined using public health reports and medical journals as well as epidemiological measurement concepts. Furthermore, we analyse to what extent these diseases are lifestyle-related. Mainly lifestyle-related diseases have a higher self-care relevance since individuals could actively affect these medical conditions. Consumer Technologies are screened whether they are suitable for non-invasive or mainly non-invasive measurements of the predefined vital signs and diseases. Therefore, it should be possible for consumers to use these technologies independently without external help if possible, in order to promote self-care. The scope lies on accessible and easy-to-use technologies, which are open to the public and are already on the market or may come into the market in the near future.

The goal of the chapter is to evaluate suitable, non-invasive technologies for the measurement of medical conditions and to examine how these technologies may improve consumer self-care, especially in the field of lifestyle-related diseases.

The following methodical procedure is used to create a 3D self-care map to guide the development of mobile technologies for healthcare (Graph 1).

In the first step, essential terms and the theoretical medical background of selected conditions are described though a short literature review.

On this basis, the burden of disease is determined for each condition with the aid of quantitative epidemiological data such as incidence and prevalence rates. These serve as yardsticks for the classification of the burden of disease (Department of Health New York State 1999; Pai and Filion 2003).

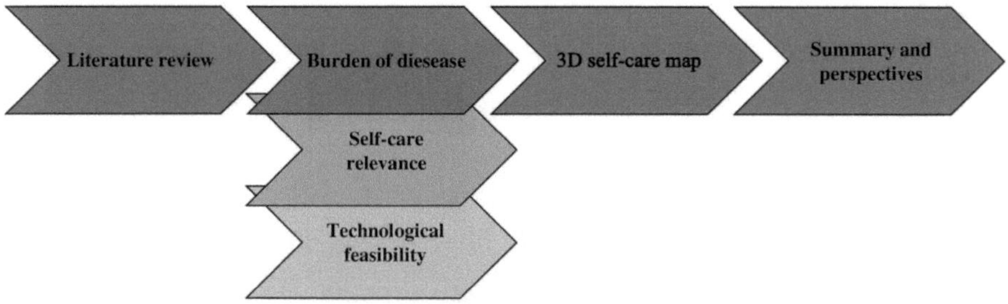


**Graph 1** Methodology of the section. (Source: authors' own graph)

We sort each condition into one out of three bubble sizes, according to their self-care relevance. The influence of lifestyle factors correlates directly with the relevance of self-care to consumers and thus the size of the bubble.

After that, technological feasibility and availability of sensor technology for these conditions are analysed. The obtained products and technical solutions are presented and their underlying technology is explained.

The 3D self-care map merges key findings of the previous sections in a three-dimensional bubble chart. The graphical model is described, major insights are discussed, and implications for distinct groups are given.

# 2 Theoretical Background

This section gives a description of fundamental terms and approaches which are necessary for the understanding of the work which follows.

## *2.1 Medical Perspective*

*Epidemiology* plays an important role in the evaluation of diseases. Last defines epidemiology as "[...] the study of the distribution and determinants of health-related states or events in specified populations, and the application of this study to the prevention and control of health problems." (Porta 1995, p. 55). This definition implies that epidemiology goes beyond pure statistical analysis of diseases by concerning itself with the prevention of health problems.

In the present work, epidemiological methods and key figures are used to determine the *burden of disease* for each of the selected diseases (Graph 2).

This work concentrates on the determination of basic epidemiological key figures for the selected diseases. Since not all diseases are measured with the same method, multiple epidemiological methods and units are applied and diseases are categorised afterwards to assure compatibility of varying diseases despite different

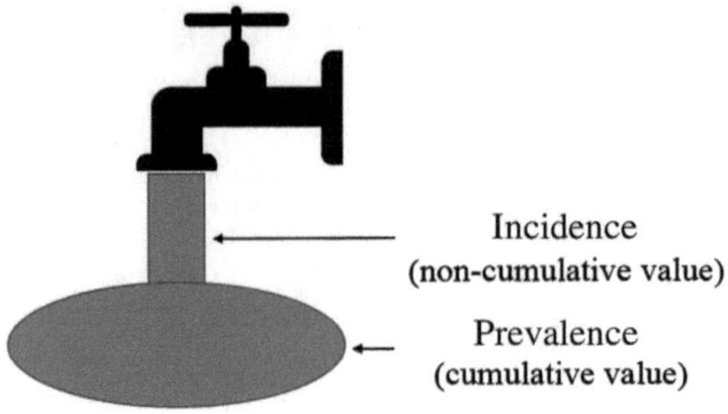


**Graph 2** Incidence and prevalence. (Source: authors' own graph)

epidemiological approaches. Unless specified differently, all key figures refer to Germany. The epidemiological key figures of prevalence and incidence are used to determine the burden of disease.

"The *incidence* of a disease is the rate at which new cases occur in a population during a specified period" (BMJ 2016). Incidence thereby measures the probability of occurrence of a disease within a specific time period, for example, 1 year (12-month incidence). Incidence is generally expressed as the number of cases per 10,000 or 100,000 people. Therefore, the most appropriate measure of disease frequency is incidence. A person newly diagnosed with a certain disease is an incidence case, whereas a person suffering from the same disease for a longer period of time is a prevalence case (BMJ 2016).

"The *prevalence* of a disease is the proportion of a population that are cases at a point in time" (BMJ 2016, p. 16). Prevalence represents the total number of individuals who are either suffering from the disease during a period of time (*period prevalence*), or who were suffering from the disease at a particular date (*point prevalence*). Prevalence is generally expressed as a percentage or as the number of cases per 10,000 or 100,000 people (National Institute of Mental Health n.d.). *Lifetime prevalence (LTP)* represents a particular form of period prevalence. It describes the relation between the number of people in the population of study, who at some point in their lives experienced the diseases, and the total population. Certain sources indicate a *total prevalence (TP)* for diseases without more detailed description. Within the scope of this work, it shall be assumed that total prevalence and lifetime prevalence describe the same and will therefore be equated for reasons of simplification.

## 2.2 Self-Care Perspective

The issue of health has gained high importance among OECD countries which is evidenced by the fact that public health expenditure is expected to increase significantly in those countries within the next 15 years (OECD 2015). In this context, the term *health literacy* was firstly discussed on a broad level in the

Anglo-Saxon area over the last 15 years but has received and continues to receive rising attention in Germany in the meantime. Depending on the context and focus, differing interpretations are associated with the term which makes a generally accepted definition difficult.

The US Department of Health and Human Services provides a widely used definition of the term by defining *health literacy* as "[...] the degree to which individuals have the capacity to obtain, process, and understand basic health information and services needed to make appropriate health decisions" (U.S. Department of Health and Human Services 2000).

In this context, health literacy concerns the ability of individuals to not only obtain and process health information but also understand them in order to make proportionate decisions for themselves (Wagner and Sparr 2012). The concepts of health literacy and self-care are closely linked to each other. Health literacy is a prerequisite for *self-care*, which was defined by WHO as "[...] the ability of individuals [...] to promote health, prevent disease, and maintain health and to cope with illness and disability with or without the support of a health-care provider" (World Health Organization 2009, p. 17).

On that score, *individual self-care* emphasises individuals and aims to prevent and inhibit diseases, preserve health and cope with diseases on an individual level. This means that decision-making competency and action competency is shifting away from traditional healthcare providers such as hospitals and doctors and towards individuals (Graph 3).

This development is very clearly seen in the field of lifestyle-related diseases. *Lifestyle-related diseases* or short *lifestyle diseases* besides social, economic, and environmental determinants refer to the occurrence of diseases which are primarily based on the habitual behaviour of individuals, including bad eating habits, incorrect posture and a lack of physical activity (Egger and Dixon 2014; Sharma and Majumdar 2009).

Consequently, the individual is at least partially responsible for their own health and the manifestation of lifestyle-related diseases. Derived from individual self-care, the term *consumer self-care* is introduced within this work to highlight the relationship between consumer habits, lifestyle and lifestyle-related diseases. In reverse, it means that a change in behaviour in the wake of an increase in health literacy and individual self-care may prevent lifestyle-related diseases and leads to a decline in those diseases (Willett et al. 2006).

It is necessary for consumers to know and understand the aetiology of these diseases. The word *aetiology* is a medical term and means the cause of a medical condition (Merriam-Webster n.d.-a). The analysis of aetiologies of diseases helps to assess which medical conditions are not, partially or mainly lifestyle-related.

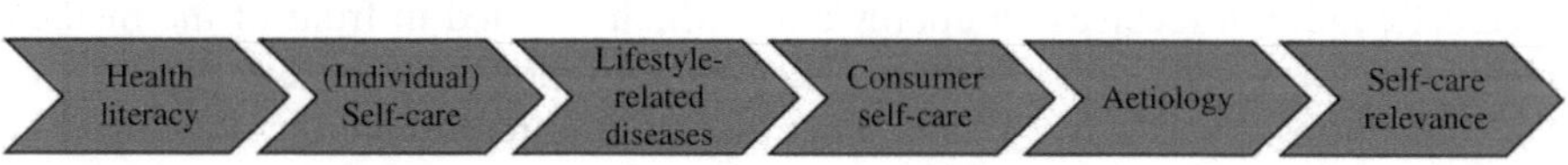


**Graph 3** Used self-care terms. (Source: authors' own graph)

For the purpose of this section, the term *self-care relevance* describes the extent to which a certain medical condition is influenced by lifestyle-related causes and factors and the influence consumers have on this condition.

### *2.3 Technological Perspective*

In order to empower consumers to monitor their medical condition without the help of healthcare professionals, it is necessary to provide precise, accessible and easy-to-operate technologies to enhance individual self-care. The Merriam-Webster dictionary defines *technology* as “a capability given by the practical application of knowledge” (Merriam-Webster 2016). These technologies need to be precise enough to recognise and monitor the predetermined medical conditions. To make operation easier and more convenient for consumers without medical knowledge or experience, only *non-invasive* and *minimally invasive technologies* will be presented and evaluated in this work. Non-invasive procedures and devices “[. . .] do not involve tools that break the skin or physically enter the body” (Vorvick 2015b). This includes, among other technologies, trackers, scanners and imaging techniques. Invasive procedures on the other hand enter the body with needles, scopes or other devices (Vorvick 2015a). Minimally invasive technologies require only small incisions or puncturing of the upper skin layers. Hence, those procedures empower a wide range of consumers to take advantage of technologies without the risk of injuries. In the context of this work, a technology is accessible to a wide range of consumers when this technology is offered on the market for consumer electronics. *Consumer electronics* comprises electronic products purchased for private needs (Cambridge Business English Dictionary n.d.). This includes software such as apps, or hardware in the form of devices, trackers, chips, fabrics and conductive products.

## 3 Burden of Disease

The first dimension of the 3D map focuses on medical conditions and the resulting burden of disease. This section gives a brief description of preselected medical conditions with the help of medical literature. These medical conditions consisting of vital signs and diseases are then categorised and the burden of disease is determined for each disease.

Diseases are substantially aligned with the medical conditions mentioned at Qualcomm’s XPRIZE (XPRIZE Foundation 2012). For a better overview, every medical condition receives an abbreviation which is listed in front of the medical name. Cardiovascular and metabolic diseases are represented by the abbreviation “C”, infectious diseases by “I” and other diseases are represented by the abbreviation “O”.

## 3.1 Criteria for Medical Conditions

During medical inspections, selected vital signs are measured to get a quick overview of the physical health of a patient. While they are necessary to evaluate health conditions, they are not included in this work, as they do not indicate a condition by themselves. The following diseases are selected as they either predominantly occur in industrialised nations and are expected to spread further, or because they have a strong impact on the affected person. In both cases, non-invasive detection and measurement of these diseases can help consumers to better monitor their medical conditions and help improve their self-care (Graph 4).

A total of 19 diseases are described and analysed in this section. These diseases are grouped into two major categories – cardiovascular/metabolic diseases and infectious diseases. Within this work, metabolic and cardiovascular diseases are merged into one category since they often appear jointly together and affect each other strongly. As malignant melanoma is neither a cardiovascular nor an infectious disease it will be presented below and falls under the category of other diseases (Graph 5).

Every disease is clearly defined and classified by the International Statistical Classification of Diseases and Related Health Problems (ICD-10). The ICD is published by WHO, and its current version ICD-10 is used for this work. It was chosen because this standard is recognised throughout the world and is the most important system for classification and medical diagnosis coding (World Health Organization 2010). For simplification purposes, only the three-digit code without

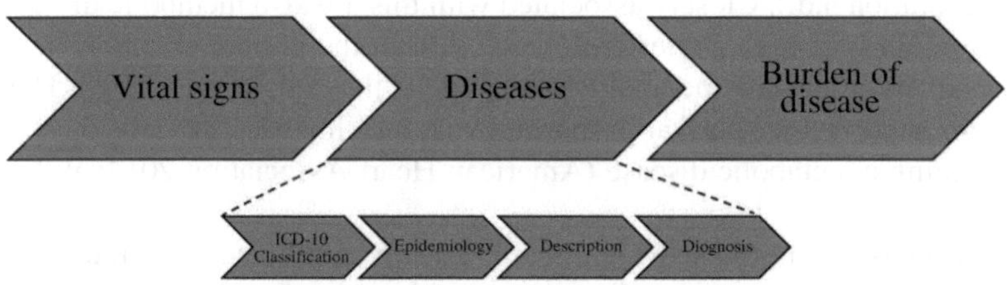


**Graph 4** Medical conditions methodology. (Source: authors' own graph)

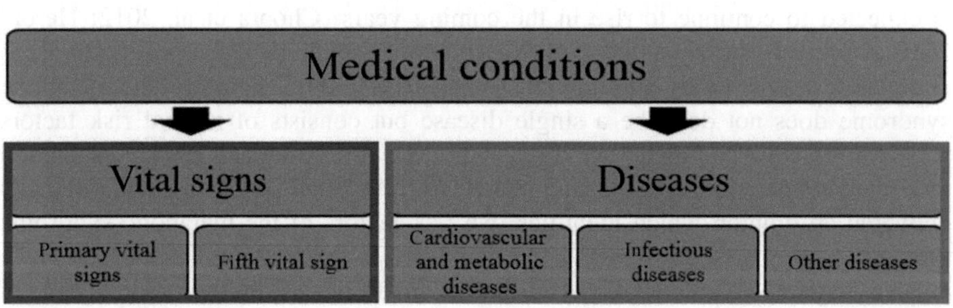


**Graph 5** Medical conditions composition. (Source: authors' own graph)

subcategories is applied for diseases in this work. If there is more than one disease linked to the three-digit code, only the most common disease will be covered and analysed. Following the classification, the epidemiology will be determined for each disease. This is necessary for the determination of the burden of disease. In order to detect the self-care relevance, it is important to be aware of the ethology, meaning the source of diseases. This is why every disease is shortly introduced and presented. Finally, for the technology selection and the technological feasibility, traditional measurement methods of the presented diseases are discussed.

On the following pages, eight preselected cardiovascular and metabolic diseases, ten infectious diseases and malignant melanoma are described and presented in detail.

## 3.2 *Categorisation of Medical Conditions*

### 3.2.1 Cardiovascular and Metabolic Diseases

According to the Robert Koch Institute, "Cardiovascular diseases are the leading cause of death in Germany, causing a total of approximately 40% of all deaths" (Rober Koch-Institut n.d., p. 1). Reasons for this can be found not only in genetic disposition but in preventable and changeable lifestyle-related factors which are responsible for a growing extent of cardiovascular diseases. An unhealthy diet, alcohol and tobacco consumption and physical inactivity contribute to this development (Rober Koch-Institut n.d.; World Health Organization 2016a; World Heart Federation n.d.). Closely associated with this are also metabolic diseases like diabetes mellitus or high cholesterol which disrupt the normal metabolism of food and negatively affect the metabolic processes of the cells (Enns 2016). Here too, lifestyle-related factors play an important role and increase the risk of acquiring or reinforcing a metabolic disease (American Heart Association 2014; Williamson 2009).

The metabolic syndrome, which is also known as affluenza, should be noticed because it is a widely common disease caused by an unfavourable lifestyle (Wirth et al. 2006). In 2010, prevalence of metabolic syndrome in Germany hit 20% and is expected to continue to rise in the coming years (Chopra et al. 2012; Herold 2010, p. 685). It is known that cardiovascular and metabolic diseases tend to occur and manifest together by building clusters (Huang 2009). Therefore, the metabolic syndrome does not describe a single disease but consists of several risk factors for cardiovascular and metabolic disease. In 2005, the American Heart Association (AHA) in cooperation with the National Heart, Lung, and Blood Institute (NHLBI) delivered an opinion which identifies five risk factors of the metabolic syndrome (Table 1).

The diagnosis of metabolic syndrome is made if at least three out of five of the factors mentioned above are present. This significantly increases the risk of comorbidities or secondary diseases in particular obesity, diabetes mellitus type II,

**Table 1** Risk factors for metabolic syndrome

| Risk factor | Defining level |
|---|---|
| Abdominal obesity, given as waist circumference | >102 cm in men (>88 cm in women) |
| Triglycerides | ≥150 mg/dl (Milligram per decilitre) |
| HDL cholesterol | <40 mg/dl in men (<50 mg/dl in women) |
| High blood pressure | ≥130/≥85 mm Hg (millimetre of mercury) |
| Fasting glucose or type II diabetes mellitus | ≥100 mg/dl |

Content: Grundy (2005). (Source: authors' own table)

obstructive sleep apnoea and other cardiovascular/metabolic diseases (Herold 2010, p. 686 f; National Heart, Lung, and Blood Institute 2016).

| | | |
|---|---|---|
| C1 cholesterol screen (high cholesterol) | ICD-10 classification: E78 | Epidemiology: total prevalence, 65% (men), 66% (women) Prevalence, 57% (men cholesterol >190 mg/dl), 61% (women cholesterol >190 mg/dl) (Robert Koch-Institut 2014b) |

Cholesterol appears in the form of high-density lipoprotein (HDL), low-density lipoprotein (LDL) and overall cholesterol. High cholesterol level is seen as a metabolic disease, which may harm blood vessels and is responsible for inadequate blood circulation, strokes and heart attacks (American Heart Association 2016b). Besides genetic predispositions and drug-drug interaction, lipid metabolism mainly develops on the basis of an unfavourable lifestyle, in particular, high calorie and sugar intake and excessive alcohol consumption. As a result, lipid metabolic disorder is frequently a comorbidity of diabetes mellitus, obesity and hypertension (Herold 2010, p. 668; Prinz and Ott 2012, p. 39). The most common and reliable method for examination of cholesterol is a blood sample.

| | | |
|---|---|---|
| C2 diabetes mellitus (type II) | ICD-10 classification: E10–E14 | Epidemiology: LTP, 9.3% |

Diabetes mellitus is a chronic metabolism disorder, which occurs as a result of reduced or missing insulin production of the pancreas or if the body cannot make effective use of insulin (World Health Organization 2016b). Diabetes substantially increases the risk for comorbidities and secondary diseases like heart attacks and sleep apnoea syndrome (Herold 2010, p. 686 f.; Robert Koch-Institut 2015a). While type I diabetes stems from a genetically predisposed autoimmune disease which presents around 5% of all diabetics, type II diabetes accounts for 95% of all cases and is predominantly lifestyle-related (Piper 2013, p. 466). Reason for type

II diabetes is an acquired autoimmune disease which stems from unhealthy diet, increased calorie intake and inactivity (Williamson 2009). As a result, 80% of type II diabetics develop obesity and other cardiovascular diseases like sleep apnoea syndrome (Herold 2010, p. 682). Glucose control takes place by invasive blood draw (American Diabetes Association 2016).

| | | |
|---|---|---|
| C3 hypertension | ICD-10 classification: I10 | Epidemiology: LTP, 25% (Herold 2010, p. 298) |

Hypertension, also known as high blood pressure, designates a steadily increased blood pressure as a result of diseases of the vascular system. High blood pressure cause permanent damages to major organs and may lead to life-threatening diseases like strokes and heart attacks (Deutsche Hochdruckliga e.V. DHL 2015). Hypertension is particularly benefiting from lifestyle-related factors, namely, an unhealthy diet, salt and sugar consumption, disproportionate caffeine intake and alcohol as well as tobacco consumption and stress. Being obese or having diabetes are significant risk factors for getting hypertension (Herold 2010, p. 290; Prinz and Ott 2012, p. 39). Blood pressure monitors and sphygmomanometers are most commonly used for non-invasive blood pressure detection but are less precise than invasive methods.

| | | |
|---|---|---|
| C4 chronic obstructive pulmonary disease (COPD) | ICD-10 classification: J44 | Epidemiology: prevalence 13% (at the age of 40+ and increasing) (Geldmacher et al. 2008) |

Chronic obstructive pulmonary disease (COPD) is a highly preventable bronchial obstruction with chronic cough, sputum and bacterial infestation (Herold 2010, p. 333 f.). COPD is the fourth most common cause of death worldwide and the most common respiratory disease. COPD is triggered by cigarette smoke in 60% of all cases (Piper 2013, p. 206; Prinz and Ott 2012, p. 65). Other lifestyle-related risk factors are diabetes mellitus and hypertension (Herold 2010, p. 281). In diagnostics, the lung function is tested with the help of a spirometer or by detecting the airway resistance (Johns et al. 2014).

| | | |
|---|---|---|
| C5 anaemia (iron deficiency) | ICD-10 classification: D50-D64 | Epidemiology: prevalence 10% (in Europe) (Herold 2010, p. 21) |

Anaemia describes a reduced concentration of haemoglobin content or erythrocytes under 8.06 mmol/l (13 g/dl) for men and 7.44 mmol/l (12 g/dl) for women in the blood (Herold 2010, p. 23). Low values of haemoglobin limit the oxygen uptake capacity which may result in fatigue, dizziness, shortness of breath up to tachycardia and heart attacks (National Heart, Lung, and Blood Institute 2014). 80% of anaemia are iron-deficiency anaemia caused by a lack of iron in the blood (Herold 2010, p. 21). Invasive blood counts provide insights about haemoglobin concentration, the number of erythrocytes and iron deficiencies (Herold 2010, p. 28).

| | | |
|---|---|---|
| C6 atrial fibrillation | ICD-10 classification: I48 | Epidemiology: prevalence 1% (increasing, 10% at age 75) (Schuchert et al. 2005) |

Atrial fibrillation (AF) is characterised by a quivering or irregular heartbeat which may lead to strokes, blood clots and heart-related complications (American Heart Association 2016a). The possible causes for the disease are diverse. Hypertension, alcoholism and thyroid disorders are other reasons which may lead to atrial fibrillation (Piper 2013, p. 106; Schnabel 2012). Atrial fibrillation is detected by long-term electrocardiograph (ECG) (Herold 2010, p. 274).

| | | |
|---|---|---|
| C7 hypothyroidism/hyperthyroidism | ICD-10 classification: E00-E07 | Epidemiology: prevalence (in general medical practice) 6% (iodine deficiency-related), 2% (hypothyroid), 2% (hyperthyroid) (Schulz et al. 2012) |

The thyroid plays an important role in the energetic metabolism, influences the formation of hormones and controls cell growth (Cheng et al. 2010; Pascual and Aranda 2013). Most common disorders are thyroid insufficiency (hypothyroidism) and overactive thyroid (hyperthyroidism). Lifestyle-related factors which affect thyroid diseases are tobacco use, alcohol and daily iodine intake (Prinz and Ott 2012, p. 269). Imaging techniques like sonography or laboratory diagnostics help to identify thyroid disorders. Thyroid hormones can be found in serum from patients or with the help of a thyroid-stimulating hormone (TSH) and free T4 (fT4) screening (Piper 2013, p. 537 f.).

| | | |
|---|---|---|
| C8 sleep apnoea (OSAS) | ICD-10 classification: G47 | Epidemiology: prevalence 4% of men/2% of women (middle aged) (Piper 2013, p. 198 f.). LTP 1,9% (OSAS) (Hein 2004). |

Sleep apnoea means the temporary interruption of breathing around mouth and nose for more than ten seconds. The complete interruption may last for up to two minutes (Piper 2013, p. 198 f.). In 90% of cases, an obstructive sleep apnoea (OSAS) is diagnosed, which means that parts of the respiratory tract are blocked by the thorax (Herold 2010, p. 325; Piper 2013, p. 199). It is notable that 80% of SAS patients are obese and every second smoker above the age of 40 suffers from obstructive ventilation disorders. Reasons for sleep apnoea are diabetes mellitus, obesity, smoking as well as alcohol and hypertension (Penzel et al. 2005; Piper 2013, p. 199). Diagnostic entails either ambulant screening unit which are able to measure oxygen saturation at night or polysomnography screening in the sleep laboratory (Herold 2010, p. 323).

### 3.2.2 Infectious Diseases

Worldwide, infectious diseases rank among the most frequent causes of death. In Germany, especially newly arising infectious agents and increasing resistance of bacteria to common antibiotic medicines present challenges (Robert Koch-Institut and Destatis 2015).

The Merriam-Webster Medical Dictionary defines infectious disease as "a disease caused by the entrance into the body of organisms (as bacteria, protozoans, fungi or viruses) which grow and multiply there" (Merriam-Webster n.d.-b, p. 1).

Not every infection necessarily manifest itself in the form of an infectious disease. Consumers' risk of developing an infectious disease depends on the type of infection, its pathogens and the physical condition of the consumer (Bundesministerium für Gesundheit 2016). High hygiene standards and monitoring as well as preventive vaccinations also reduce the risk of acquiring an infection (World Health Organization 2008).

The preselected infectious diseases are being described in the following section.

| | | |
|---|---|---|
| I1 food-borne illness | ICD-10 classification: A05 | Epidemiology: 12-month prevalence, 10% (per year in Germany) (Tschäpe 2000) |

After the consumption of contaminated food, a food-borne illness may occur. Typically, food poisoning lasts for 1 or 2 days. Depending on the cause and pathogens, first feelings of unease can be seen after only 1–16 hours after consumption. Besides viral and fungal infections and heavy metals impurities, it is primarily bacteria which cause food-borne illnesses. Medical anamnesis is sufficient to determine whether or not food-borne illness is present. Only laboratory samples can detect specific pathogens (Herold 2010, p. 825 f).

| | | |
|---|---|---|
| I2 human immunodeficiency virus (HIV) | ICD-10 classification: B20-B24 | Epidemiology: in total >83.400 infected people in Germany in 2014 (Koch-Institut 2015). Incidence: 4.5/100,000 (in Germany 2015) (Robert Koch-Institut 2016a) |

Human immunodeficiency virus (HIV)-infected patients suffer from a severe loss of CD4-T lymphocytes which are an important part of the immune system (Piper 2013, p. 770). This causes loss of the normal immune system function in the long term. This process is irreversible and entails other bacterial and viral infections such as herpes zoster, because of the weakening immune system. HIV is transmitted via blood, sperm, vaginal secretion and breast milk. HIV is mainly spread by unprotected sex. In 8% of the cases, HIV is transmitted by contaminated needles (needle exchange) between drug users (Herold 2010, p. 853). Positive detection of an HIV infection is detected within 6–12 weeks after the infection. Serological examination can prove HIV with the help of antigens (Robert Koch-Institut 2016b).

| | | |
|---|---|---|
| I3 infectious mononucleosis | ICD-10 classification: B27 | Epidemiology: prevalence of EBV 95% (up to age 30 in Western Europe) (Herold 2010, p. 816) |

Infectious mononucleosis or glandular fever is a widely prevalent infectious disease associated with strong swelling of the lymph nodes, high fever, fatigue and sickness. (Herold 2010, p. 816; Piper 2013, p. 849). Infectious Mononucleosis is triggered by the Epstein-Barr-virus (EBV), which belongs to the herpes virus family. At the age of 30, 95% of the population is infected with EBV. Even after full recovery, the virus stays in the body. The analysis of a blood sample detects activated T-lymphocytes and identify the virus. Antibody demonstration is performed with the Paul-Bunnel test (Herold 2010, p. 817; Piper 2013, p. 850).

| | | |
|---|---|---|
| I4 leucocytosis | ICD-10 classification: D72 | Epidemiology: incidence 8% (estimated) (Denzlinger 2014; Herold 2010). |

In case of excessive increase of white blood cells, which means more than $4 \times 10^9$/l to $11 \times 10^9$/l, then one talks of a leucocytosis. Aetiologies are manifold, especially for neutrophilia. Bacterial infections, resulting from inflammation and anti-inflammatory drugs are the most frequent reasons. Metabolic disorders such as diabetes mellitus and lifestyle-related factors, namely smoking, stress and physical strain are further reasons of neutrophilia (Herold 2010, p. 56). Blood count gives information about the amount of white blood cells and granulocytes.

| | | |
|---|---|---|
| I5 otitis media | ICD-10 classification: H65–H67 | Epidemiology: cumulative prevalence 61.4% (children, within the first 6 years) LTP up to 2.6% (in Great Britain) (Browning and Gatehouse 1992). |

Otitis media means a painful inflammation of the mucous membranes of the middle ear, commonly triggered by bacterial or viral infection of the respiratory passages (Thacker et al. 2006). The peak of cases occurs in early childhood, making it difficult to gain reliable epidemiological figures in adulthood (Deutsche Gesellschaft für Hals-Nasen-Ohrenheilkunde und Kopf und Hals-Chirurgie e.V. 2014). Otitis media is caused by viral or bacterial infections (pneumococci, haemophilus influenza) (Herold 2010, p. 361 ff.; Prinz and Ott 2012, p. 382). Key component of diagnosis is the physical microscopic examination of the ear by an otologist. In case of a positive result, a sputum analysis test can identify potential pathogens (Thomas et al. 2014).

| | | |
|---|---|---|
| I6 pertussis (whooping cough) | ICD-10 classification: A37 | Epidemiology: incidence 42/100,000 (in eastern Germany in 2011–2012) LTP: 1.9% (infants from 0 to 2 years in 2009–2012) (Neuhauser and Poethko-Müller 2014) |

Pertussis, colloquially called "whopping cough", is a severe or even deadly infectious disease which invades the airways (Herold 2010, p. 836). Pertussis is transmitted via droplet infection and is triggered by the bacterium *Bordetella pertussis* which settles at the respiratory tract (Robert Koch-Institut 2014a). Vaccinations at periodic intervals lower the risk of contracting pertussis significantly. Diagnosis of a pertussis infection is dependent on the stadium of the disease. At an early stage, bacterial pathogens are detected from nose swab or secretion. After 2 weeks, proof of antigens can be found in the serum after the use of an enzyme-linked immunosorbent assay (ELISA) (Riffelmann et al. 2008).

| | | |
|---|---|---|
| I7 pharyngitis (chronic) | ICD-10 classification: J02 | Epidemiology: incidence 30% (within 12 months) (Wächtler and Chenot 2011) |

Pharyngitis also referred to as strep throat means a throat infection accompanied by sore throat, fever and shivering. It is one of the most common diseases; 20% of people seeing a doctor show symptoms of a pharyngitis (Fink and Haidinger 2007). In 50–80%, viruses cause pharyngitis (Wächtler and Chenot 2011). The most frequent bacterial pathogen is *Streptococcus pyogenes* (Robert Koch-Institut 2009). Usually, endoscopic examination of the throat is sufficient for detection of a pharyngitis infection. Rapid antigen detection tests can be used for the detection of streptococcus (Robert Koch-Institut 2009).

| | | |
|---|---|---|
| I8 pneumonia | ICD-10 classification: J12–J18 | Epidemiology: LTP 20% (in Germany) (Prinz and Ott 2012, p. 66) |

Pneumonia is referred to as an inflammatory, infectious disease of the lung tissue often associated with cough, fever, sputum and chest trouble, which can cause death, if left untreated (Prinz and Ott 2012, p. 66). Pneumonia is the third leading cause of death worldwide (Herold 2010, p. 356). In 90% of all causes, a CAP is caused by a bacterial infection, namely, pneumococci, staphylococci or *Streptococcus pneumoniae*. Lifestyle-related risk factors, which contribute to pneumonia, are other lung diseases such as COPD, diabetes mellitus, alcoholism and a weakened immune system resulting from an HIV infection (Jacobi et al. 2009). Anamnesis and auscultation are first steps for the detection of pneumonia. In case of a suspected infection, radiographs and in some cases, a computer tomography is laid down. The exact pathogens can be identified in the blood or sputum (Piper 2013, p. 225 f).

| | | |
|---|---|---|
| I9 herpes zoster (shingles) | ICD-10 classification: B02 | Epidemiology: >95% of adults show antibodies against VZV (Herold 2010, p. 812) Prevalence 50% (for shingles in adults at the age of 85) (Robert Koch-Institut 2016c) |

Varicella zoster virus (VZV) causes two different clinical pictures: varicella (chickenpox) and herpes zoster (shingles). Primary infection which leads to varicella appears mostly during childhood. The virus stays inside the body and can be reactivated, which then leads to shingles (Piper 2013, p. 846; Robert Koch-Institut 2016c). Risk factors include melanoma, intense sun exposure and stress, which additionally weaken the immune system (Herold 2010, p. 812; Prinz and Ott 2012, p. 411). Diagnosis is based on anamnesis and visual diagnosis. In rare cases, the virus is detected via polymerase chain reaction (PCR) test.

| | | |
|---|---|---|
| I10 urinary tract infection | ICD-10 classification: N39 | Epidemiology: LTP >50% (for women in Germany) (Hummers-pradier and Schiemann n.d.; Wagenlehner 2015) LTP 12% (adult males in the USA) (Brumbaugh and Mobley 2012) |

Infectious pathogens of the urinary tract may lead to an urinary tract infection (Herold 2010, p. 592). In 80% of all cases, the bacterium *Escherichia coli* (*E. coli*) is responsible for the infection. Besides anatomic reasons, like an obstructive urinary flow disorder, immune deficiency, sexual activity and metabolic diseases favour infections of the urinary tract. Due to high blood and urine glucose levels, people with diabetes are at a higher risk of developing urinary tract infections (Saliba et al. 2015). Evidence of an infection is provided by urine samples, which prove leucocyturia and infectious bacteria.

### 3.2.3 Other Diseases

Malignant melanoma is neither a cardiovascular/metabolic disease nor an infectious disease and therefore listed separately as other disease.

| | | |
|---|---|---|
| O1 malignant melanoma | ICD-10 classification: C43 | Epidemiology: LTP 1%, incidence 15/100,000 (DKG, Deutsche Krebshilfe, and AWMF 2014; Sondak and Smalley 2009) |

Malignant melanoma is a malignant tumour of melanocytes, the pigment cells of the skin and the most malignant form of skin cancer. This tumour type grows rapidly and disseminates metastases via blood and lymphatic pathways from an early stage. Disease rates in Germany have steadily increased over the last decades (Kraywinkel et al. 2012). People with light skin and a high number of pigments are at particular risk. A lifestyle-related factor, which is responsible for malignant melanoma, is UV exposure (Zentrum für Krebsregisterdaten 2013). Malignant melanoma is generally detected by visual diagnosis or microscopic examination of a dermatologist.

Based on the epidemiological information of each disease given in the former section, the burden of disease is determined and each condition is then sorted into one of three categories, according to their prevalence and incidence rates.

For this work, the burden of disease indicates one person's probability for developing a disease within a certain period of time or of having a disease at a specific point in time. With the concept of prevalence and incidence described in Sect. 2, the burden of disease is determined. It should not be confused with the Global Burden of Disease, which evaluates mortality and disability caused by major diseases and injuries and which is used by several organisations, such as the WHO or Harvard School of Public Health since 1990 (World Health Organization n.d.). Due to difficult data collection and non-comprehensive basis of data for the determination of disability-adjusted life years (DALY), the concept of Global Burden of Disease was dispensed for this work (World Health Organization 2016d).

In order to be able to determine the corresponding burden of disease for each condition, three categories are being used. Each category is defined by its epidemiological thresholds for incidence, prevalence and lifetime prevalence/total

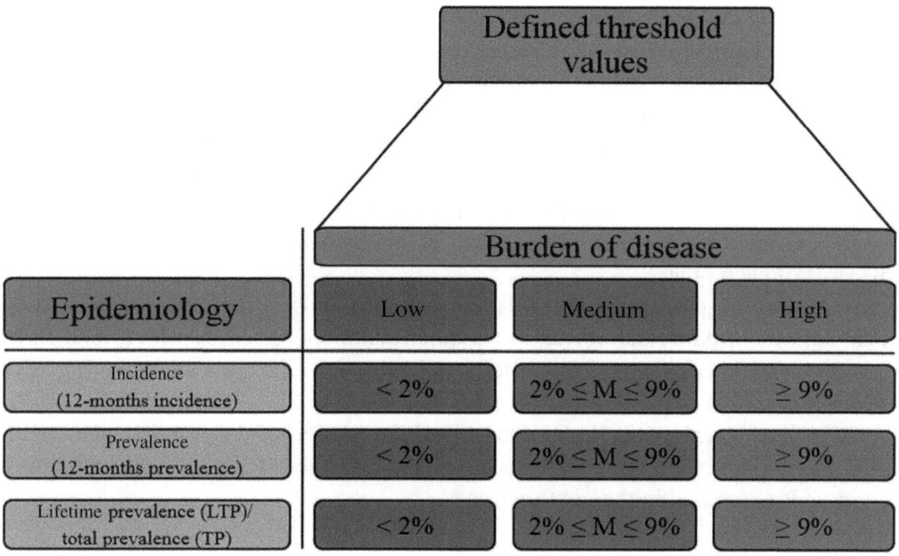


**Graph 6** Threshold values of the burden of disease. (Source: authors' own graph)

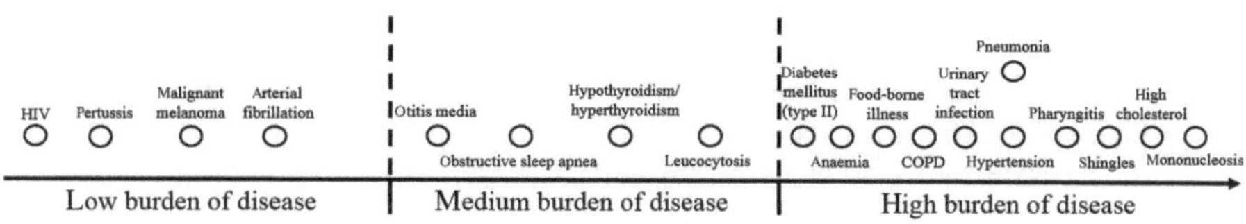


**Graph 7** Burden of disease map. (Source: authors' own graph)

prevalence. With this, each disease can be sorted into one category based on their incidence or prevalence rate, resulting in a low, medium and high burden of disease.

Thresholds and categories for individual epidemiological key figures are valued as follows (Graphs 6 and 7).

After categorising all diseases on the basis of their epidemiology and applying the criteria which were mentioned above, the burden of disease for each condition is presented in the graph:

On the X-axis, diseases are sorted according to their burden of disease in ascending order. On the left, there are diseases with a low burden of disease, while on the right, there are conditions which were categorised as conditions with a high burden of disease.

The low and medium burden of disease categories are each represented by four diseases. It must be observed that while the burden of disease might be low for some conditions, correct and prompt treatment for these diseases (HIV, pertussis, malignant melanoma) still is essential and a lack of treatment may even lead to death (AIDS.gov 2015; Sandru et al. 2014). Mortality and morbidity are not related to the burden of disease.

It turns out that more than half of the examined diseases, 11 out of 19% or 57.9%, show a high burden of disease. This includes both cardiovascular/metabolic diseases and infectious diseases, which tend to have high prevalence and incidence rates. This especially includes diabetes mellitus, hypertension and high cholesterol which are all components of metabolic syndrome. In addition, diabetes and hypertension are also main risk factors of COPD, which once again favours chronic pharyngitis. This confirms the statement claiming that cardiovascular and metabolic diseases often occur and manifest together and tend to be accompanying illnesses or secondary diseases of each other.

In a nutshell, more than half of the conditions show a high burden of disease. This applies in particular for cardiovascular/metabolic diseases which tend to be closely interrelated, whereas this is less applicable for infections with a high burden of disease.

After having categorised and determined the burden of disease for each condition, the impact of lifestyle-related causes and risk factors for the development of diseases is examined in the next section.

## 4 Self-Care Relevance

This section focuses on lifestyle-related causes and risks and their impact on diseases from the previous section.

With the help of medical literature and scientific studies, potential causes and risk factors of diseases are listed, and each condition is categorised in relationship to its lifestyle-related causes and risk factors. Finally, the self-care relevance for each disease is determined and graphically illustrated (Graph 8).

Self-care relevance describes the extent to which a certain medical condition is influenced by lifestyle-related causes and factors and the influence consumers have on this condition. Accordingly, a low self-care relevance indicates that a disease is not or only to a small extent induced by lifestyle and lifestyle-related factor of individual consumers. On the other hand, a high self-care relevance implies that a disease is highly or predominantly induced or even caused by lifestyle and lifestyle-related factors of individual consumers. The term has been chosen to point out which

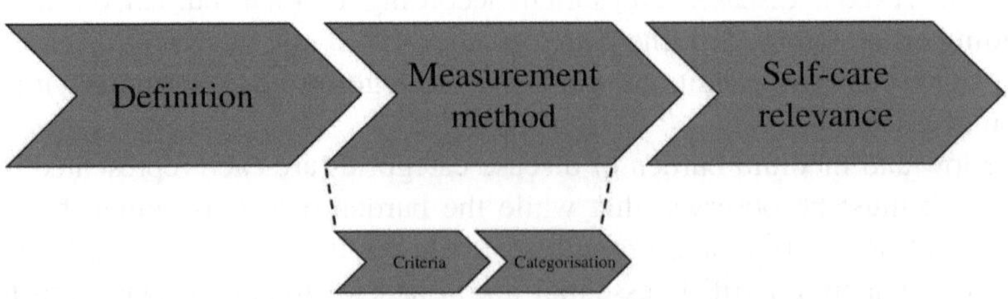


**Graph 8** Self-care relevance methodology. (Source: authors' own graph)

diseases are impacted by consumers. In other words, diseases with a high self-care relevance can be positively influenced by lifestyle changes made by consumers themselves without the need of medical institutions, while diseases with a lower self-care relevance may be the result of other non-lifestyle-related causes.

The aim of self-care relevance is twofold. On the one hand, the impact of lifestyle-related factors on diseases is examined. On the other hand, it is outlined which disease condition may improve or even heal due to lifestyle changes of the consumer. Two assumptions are made:

1. Consumers are in a position to make conscious changes in lifestyle.
2. High and medium self-care-relevant diseases are positively affected by specific changes of lifestyle.

If both assumptions apply, it can be said that the higher diseases are impacted by lifestyle-related factors, the higher is their self-care relevance and the more consumers can influence the progress of diseases on an individual level by making lifestyle changes.

## 4.1 *Criteria for Self-Care Relevance*

Self-care relevance for each disease is determined by evaluation of medical literature and studies. Based on the description of diseases in Sect. 3, additional literature is used to detect known causes and risk factors. A distinction is made between lifestyle-related and non-lifestyle-related causes and risk factors in order to determine the self-care relevance later on. Due to reasons of complexity, correlations and interdependencies of causes and risk factors are not taken into account, but causes and risk factors are listed in the same category.

Whereas the gaining of insights is the main purpose in this work, the level of details regarding methodical approaches and medical context has to be narrowed at some point. For this work, lifestyle-related causes and risk factors are subdivided into four areas which are defined as follows (Graph 9).

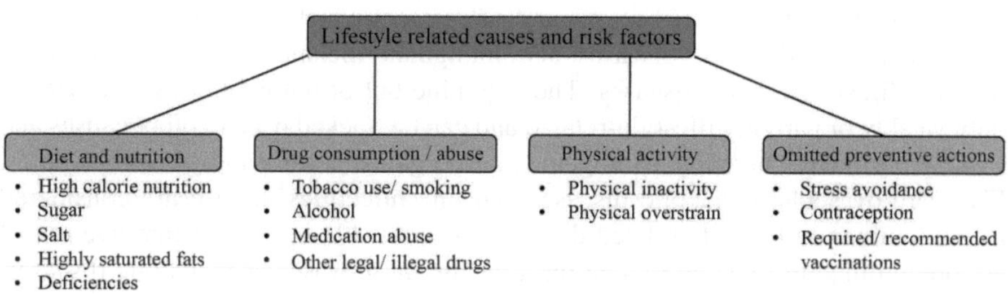


**Graph 9** Lifestyle-related causes and risk factors. (Source: authors' own graph)

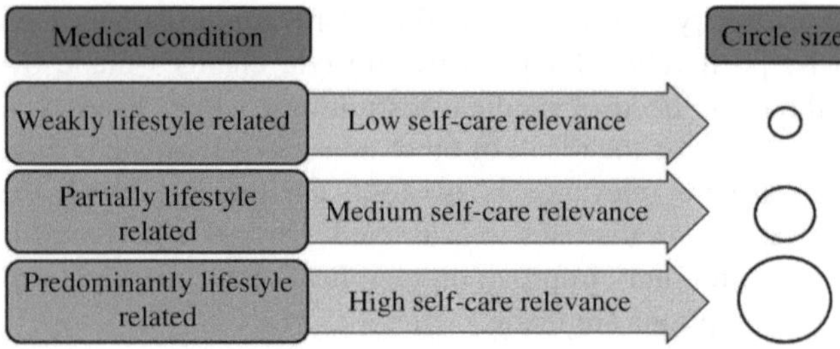


**Graph 10** Self-care relevance circle size mapping. (Source: authors' own graph)

Other factors which are not listed above are counted as non-invasive lifestyle-related causes and risk factors. In addition, it is recalled that occupational risks (increased risk of infections for doctors and staff working for healthcare institutions, sun exposure of construction workers) are not considered as lifestyle-related causes and risk factors. Also excluded are actions and risks which are not self-inflicted (accidents, passive smoking, childhood vaccination). All causes and risk factors apply to Germany unless otherwise noted.

As an analogy to Sect. 3 diseases are continued to be classified in cardiovascular/metabolic, infectious and other diseases.

Thereafter, every category is assigned to a degree of self-care relevance. Diseases which have been detected as being weakly lifestyle-related diseases receive a low self-care relevance. Diseases which have been detected as partially lifestyle-related diseases are marked with a medium self-care relevance, while diseases which have been detected as predominantly lifestyle-related diseases receive a high self-care relevance. Moreover, the circle size of each disease represents the self-care relevance of each condition (Graph 10).

## *4.2 Categorisation of Self-Care Relevance*

All diseases described in Sect. 3 are sorted into one of the three categorises on the basis of their lifestyle-related causes and risk factors which are described above.

Thus, three infectious diseases and malignant melanoma are considered as partially lifestyle-related diseases. Thereby, nine out of ten infectious diseases are only weakly or partially lifestyle-related and can be tracked back to other causes and risk factors such as heredity, pathogenic microorganisms and pathogenic molecules. Five cardiovascular/metabolic diseases and one infectious disease are considered as predominantly lifestyle-related diseases. Remarkable is the fact that five out of six predominantly lifestyle-related diseases are cardiovascular/metabolic diseases. Moreover, diabetes mellitus, hypertension, high cholesterol, COPD and OSAS share mainly the same causes and risk factors: This includes in particular an unfavourable diet (high in calories and sugar-rich nutrition), which often results in obesity, alcohol

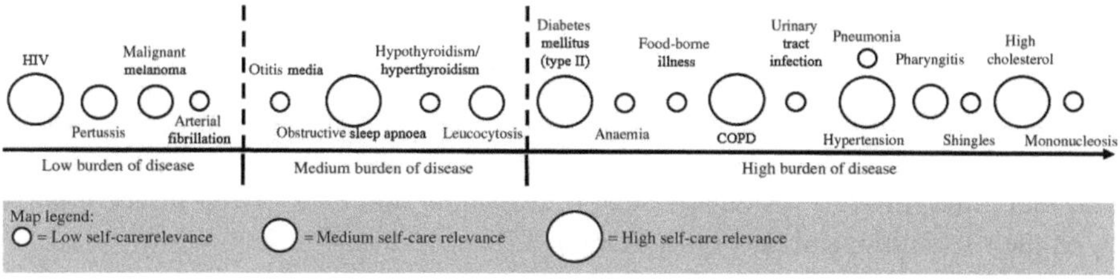


**Graph 11** Self-care relevance map. (Source: authors' own graph)

and tobacco use and physical inactivity. In conclusion, it must be said that especially diabetes mellitus, hypertension and high cholesterol are closely linked with each other, and these diseases are frequently accompanying, co-existing, or are secondary diseases.

The goal of the self-care relevance map is to visualise all diseases and to set them in relation to their lifestyle-related causes and risk factors in order to determine which diseases can be influenced by consumers. The self-care relevance for each disease is symbolised by the circle size of the condition. Furthermore, conditions are sorted into ascending order starting with a low burden of disease and growing rightwards (Graph 11).

The evaluation of results shows that non-lifestyle-related causes and risk factors for cardiovascular diseases are mainly advancing age and familial dispositions (heredity). It is also apparent that the main causes and risk factors for cardiovascular diseases are high caloric, poor diets with highly saturated fat and sugar intake; drug consumption, mainly represented by alcohol and tobacco use; as well as physical inactivity.

Regardless of the burden of diseases, the large majority of cardiovascular and metabolic diseases fall into the high self-care category (5/8 cardiovascular/metabolic diseases). Five out of six diseases which are located in the high self-care category are cardiovascular and metabolic diseases. Diabetes mellitus, hypertension, high cholesterol and COPD are predominantly lifestyle-related, and both show a high burden of disease and a high self-care relevance.

For this reason, the focus lies on those four cardiovascular and metabolic diseases and less on infectious diseases since these conditions are mainly caused by pathogenic microorganisms and pathogenic molecules and are less affected by lifestyle changes. Also the focus is on the measurement and monitoring of vital signs, which in many cases are necessary and suitable for the detection of diseases like high cholesterol or diabetes mellitus.

# 5 Technological Feasibility

This section focuses on technologies for detection and measurement of the medical conditions mentioned in Sect. 3. To achieve this, available and suitable technologies are identified through systematic search, evaluated and categorised in accordance with their technological feasibility (Graph 12).

For his work, "technology" refers to both software and hardware which are capable of detecting and measuring medical conditions in consumers. For the purpose of simplification, throughout this work, technology covers all approaches, concepts, prototypes, products, sensors and devices dealing with non-invasive technologies for the measurement of medical conditions. In addition, detected technologies need to be as precise as possible, easy to use and accessible for consumers in order to enhance individual self-care.

Sufficiently precise means that a technology is able to reliably identify and measure individual or multiple medical conditions. This can be ensured through governmental and independent institutions which certify devices. In particular, this includes the CE marking for medical devices within the European Union, ISO standards, and FDA-cleared or FDA-approved devices. In order to receive one of these certificates, devices must not only pass a harmlessness test but also need to fulfil capability tests, environmental compatibility and meet additional quality requirements.

Ease of use of a technology is ensured, when consumers are able to handle the technology without the help of medical staff or deeper technical understanding. To this end, only non-invasive and minimally invasive technologies are examined. Such technologies minimise the risk of injuries and are easier to handle than invasive methods.

"Availability" relates to whether or not the technology is accessible for the public and if it can be bought and used by consumers. This shall be the case if the technology is available and sold as consumer electronics. In Section 2.3 above consumer electronics were defined as electronic products purchased for private needs. Accordingly, these technologies are directed at private consumers.

The more these three criteria are fulfilled, the more advanced the technology is considered to be. Categorising these technologies serves as a qualitative assessment of the technology without evaluating technical details. The section derives the

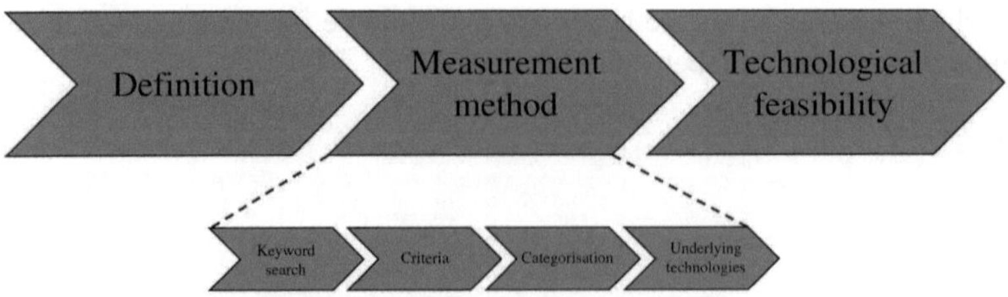


**Graph 12** Technology methodology. (Source: authors' own graph)

technological feasibility for each medical condition from the categorisation of technologies. The technological feasibility indicates how well a medical condition can be measured by private consumers with the use of non-invasive technologies. It is important to note that the technological feasibility does not directly evaluate specific technologies but indicates how advanced the measurement of a medical condition is.

## *5.1 Criteria for Categorisation*

The process of searching, identifying suitable technologies and determining of the technological feasibility of said technologies has four steps:

At first, search keywords help to search for technologies for respective medical conditions. In the second step, the results are aggregated and structured by sorting out duplicate reports, failed and outdated technology concepts and unproven claims of online sources. Following this, the detected technologies and products are filtered. Products which were available on the market before 2005 or which use invasive measurement methods are sorted out. If there is more than one product using the same technology, some manufacturers are selected representatively. The third step involves the explanation of the underlying principles and technologies of the products and devices being studied. Finally, the technological feasibility of each medical condition is determined. Section 5.2 graphically illustrates the technological feasibility of each medical condition and presents major findings (Graph 13).

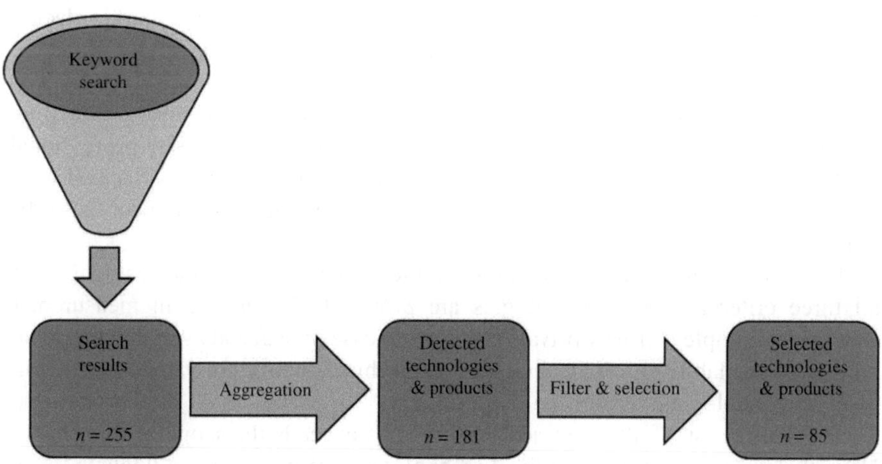


**Graph 13** Keyword search and selection process. (Source: authors' own graph)

Overall, 255 websites, conferences, societies, medical and technology blogs, online journals and commercial sites were screened using the following keywords in combination with medical condition:

**Keywords**: non-invasive, minimally invasive, consumer electronics, prototype, detection, diagnostic, measurement, monitoring, analysis, innovation, health device, technology, wearable, wearable technology, medical wearable, sensor, tracker, wristband, lenses, tech-tat, medical patch, biotech-tattoo

After aggregating and summarising 181 technology approaches, prototypes, scientific studies, products, sensors and devices were detected; 85 of them were selected after filtering. It has to be taken into account that different medical conditions can be measured by the same technology, implying that the number of technologies is less than the number of selected products.

## 5.2 *Categorisation of Technologies*

The technological feasibility is divided into the three categories: early stage, medical devices and consumer electronics. The classification is based on the criteria precision, ease of use and availability.

In the category early stage are medical conditions which either cannot be measured with non-invasive technologies or for which only prototypes are available. These technologies are characterised by the fact that they are not user-friendly (lack of ease of use) and are not available on the consumer electronics market (lack of availability). The early-stage category equates to a low technological feasibility.

In the medical device category, there are technologies for the precise, non-invasive measurement of medical conditions which are used in medical facilities (precision). Technologies in this section often need to be operated by professionals (lack of ease of use) and are not available as consumer electronics (lack of availability). The medical device category represents a medium technological feasibility (Graph 14).

Technologies for medical conditions in the consumer electronics category fulfil all three criteria. These technologies are proven to be precise in measurement (precision), simple to use by private consumers (ease of use) and available for private purposes (availability). Medical conditions in this category are valued with a high level of technological feasibility.

The full review of all examined technologies exceeds the scope of this chapter. Nevertheless, some of the examined technologies are presented as finalists for the XPRIZE challenge and are good examples for state-of-the-art technology. Each technology team works towards an own medical tricorder for detection and mea-

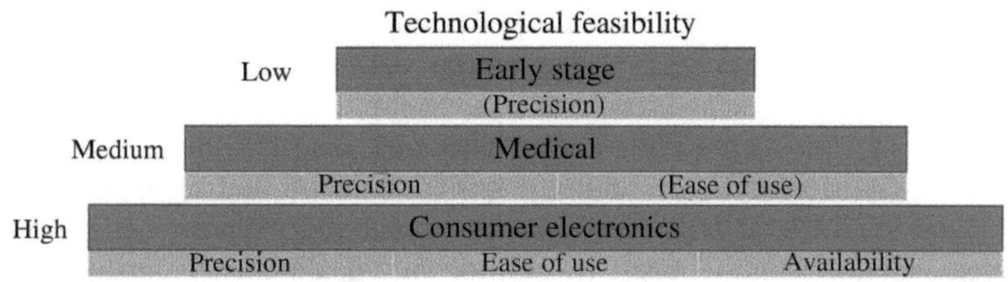


**Graph 14** Technological feasibility mapping. (Source: authors' own graph)

surement of the presented medical conditions ("Where No Healthcare Device Has Gone Before" 2015). To give an overview of the types of examined technology, the seven finalists are shown below as examples for innovative healthcare technology:

*Intelesense-Scanadu*: Intelesens, a medical technology company which specialises in intelligent wireless vital signs monitoring, and Scanadu Inc., a Silicon Valley-based medtech company, collaborate to win the XPRIZE challenge. The ultimate goal is to design a tricorder device for vital sign monitoring which is capable of changing the healthcare industry by putting medical devices back into the hands of individuals.

*Aezon*: After the XPRIZE challenge had been announced, Aezon was founded in September 2012. The team includes students from Johns Hopkins University in Baltimore, USA, and partner with Symcat, which recently developed a symptom checker for diseases. Other partners include SpiroSmart, Aegle, which provide big data analytics for healthcare and the developers of Biomeme, a smartphone-based DNA detection platform.

*Danvantri*: The team named Danvantri consists of seven professionals and is led by Sridharan Mani, an Indian entrepreneur and IT project manager. They are dedicated to developing affordable but high-quality healthcare solutions based on mobile technologies. The main focus lies on the prevention of diseases by early detection made possible with an all-in-one handheld device.

*DMI*: DMI comprises experienced scientists, engineers and designers in the area of Boston/Cambridge, USA, and has won the Nokia Sensing XCHALLENGE with their rHEALTH sensor technology. This device also received funding from NASA, the Gates Foundation and National Institutes of Health (NIH). A single drop of blood is taken by the new rHEALTH X1 prototype, which applies it to nanoscale test strips, analyses the sample and links it with possible diagnoses.

*Dynamical Biomarkers Group* (DBG): DBG is a research group of the Center for Dynamical Biomarkers and Translational Medicine (CDBTM) in Taiwan which includes clinicians, medical researches, scientists and engineers. The main objective is to design light-weight and portable diagnostic instruments with a high diagnostic accuracy and which offer a good user experience and a flexible exchange between medical data and the cloud.

*Final Frontier Medical Devices*: The team was formed by the brothers Basil and George Harris who founded Basil Leaf Technologies. For the XPRIZE challenge, the company is currently developing a portable consumer-level device called DxtER. This device collects and interprets data to diagnose selected medical conditions in real time and recommend appropriate actions.

*Cloud DX*: Dr. Sandeep Kohli, a physician and assistant clinical professor of medicine at McMaster University, Canada, founded and leads the Canadian XPRIZE team Cloud DX. The company is working towards a medical device which enables ordinary people to monitor their health status, diagnose selected diseases and get a sense of when they should seek professional medical treatment.

In summary, in the technology section of this chapter, promising concepts, devices and products were presented, underlying technologies were explained, and their technological feasibility was detected based on defined criteria; 85 technologies in the development state of prototypes, medical devices and consumer technology have been reviewed. Following an in-depth investigation and categorisation of technologies, diseases have been sorted into the technological feasibility map and presented graphically. The resulting findings will be explained shortly.

The graph shows each of the examined diseases by their technological feasibility. The vertical y-axis represents the technological feasibility for each disease and is divided into the three sectors: early stage, medical and consumer electronics. Each disease is categorised according to its technological feasibility in ascending order. The higher a disease is listed, the higher its level of technological feasibility. The graph only shows diseases and no vital signs (Graph 15).

It should be noted that medical conditions which are ranked in the same category may show minor differences in their technological feasibility due to the current development status of a technology. A technology, for example, which has undergone only basic tests is less advanced than a technology, which applied for FDA approval.

After assessing the study of the graphical technological feasibility, the following findings should be mentioned:

An even distribution of the examined technologies for each disease onto the early-stage sector (7 diseases), the medical sector (5 diseases) and the consumer electronics sector (7 diseases) can be seen. For almost any medical condition (including vital signs), at least early-stage approaches or research is conducted. However, there are big differences between detection and monitoring of vital signs, cardiovascular/metabolic diseases and infectious diseases. Devices for the measurement of vital signs are not only the most advanced and all located in the consumer electronics sector but are often able to monitor multiple vital signs simultaneously. In some cases, those technologies are also able to detect other medical conditions, such as atrial fibrillation. The majority (5/8) of the technologies for cardiovascular/metabolic disease detection and measurement have reached the consumer electronics sector. Early-stage technologies are almost exclusively occupied by infectious diseases. Apart from HIV detection, no other technology for infectious disease measurement is located in the consumer electronics sector. Reasons for this might be that infections are more difficult to

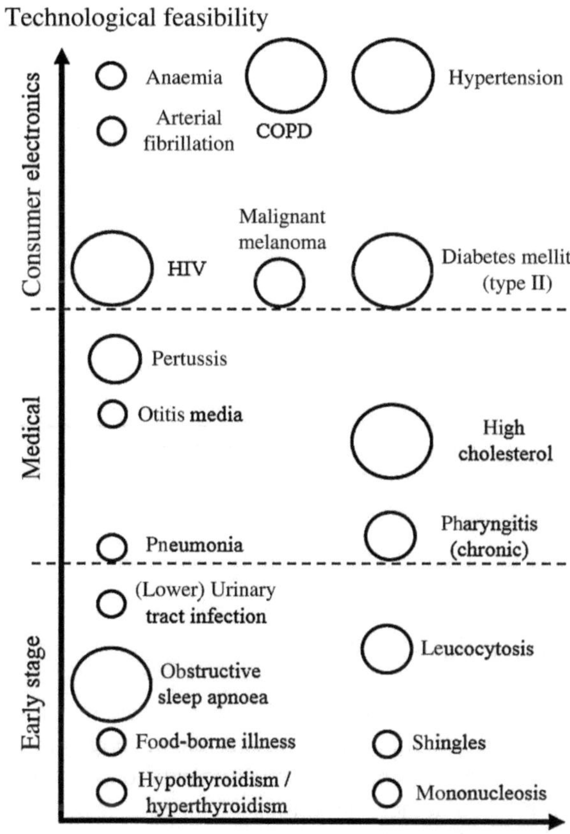


**Graph 15** Technological feasibility map. (Source: authors' own graph)

detect than cardiovascular/metabolic diseases due to lack of precision resulting from low sensitivity or specificity and possible impurities of samples.

Furthermore, it can be observed that the examined consumer electronics technologies are more often cleared or approved by the FDA, than technologies in their early stages. As expected, technologies for consumer electronics are usually more precise and better meet medical compliance than early-stage technologies. This is most likely traceable back to a higher degree of feasibility compared to technologies in the early-stage sector.

As a final point, it should be mentioned that some of the devices which are ranked as a consumer electronics device still might need known medical procedures or medical staff for evaluations. For example, the app for malignant melanoma detection or a professional second HIV test after a false positive result. The same applies to minimally invasive measurement methods like the FreeStyle Libre device which still needs a small needle in order to deliver results which are precise enough.

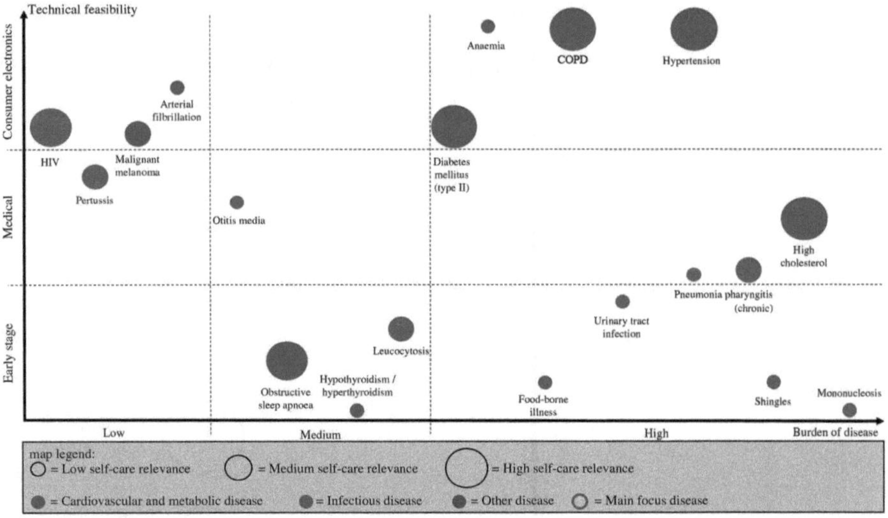


**Graph 16** 3D self-care map. (Source: authors' own graph)

### 5.2.1 3D Self-Care Map

Bringing together burden of disease, self-care relevance and technological feasibility, this section graphically presents the main results of previous sections, discusses them and derives practical recommendations for different groups. The 3D self-care map combines the burden of disease, self-care relevance and technological feasibility of each disease within a single graph (Graph 16).

The X-axis of the 3D self-care map is divided into three areas: low, medium and high, and shows the burden of disease ascending, from the left to the right. The burden of disease measures one person's probability for developing a disease within a certain period of time, or of having a disease at a specific point in time (see Sect. 3). The positioning of the disease on the X-axis correlated with the level of burden of disease. The rightmost position on the X-axis indicated the highest level of burden of disease.

The Y-axis indicates the technological feasibility, rising from bottom to top and is categorised into the three sectors: early stage, medical and consumer electronics. The technological feasibility indicates how well a medical condition can be measured by private consumers, through non-invasive technologies (see Sect. 5). The higher a disease is located on the Y-axis, the higher its technological feasibility.

Each disease is represented by a circle. The circle size provides insights into the self-care relevance of a disease. Self-care relevance describes the extent to which a certain medical condition is influenced by lifestyle-related causes and factors and the influence consumers have on this condition (see Sect. 4). The bigger the circle, the higher the self-care relevance of a certain disease. On the basis of this description, several statements can be made.

A high burden of disease was detected by 11 out of 19 diseases, equally distributed among cardiovascular/metabolic and infectious diseases. Major differences can be seen in terms of self-care relevance. Infectious diseases are hardly associated with lifestyle-related factors, mainly resulting in low self-care relevance (6/10 diseases) and medium self-care relevance (3/10 diseases). Lifestyle-related causes and risk factors of infectious diseases and malignant melanoma are mainly attributed to omitted preventive actions. This includes primarily unprotected intercourse (HIV), irregular vaccinations (pertussis) and intense sun exposure (malignant melanoma).

By contrast, cardiovascular/metabolic diseases are predominantly lifestyle-related (5/8 diseases); 5 out of 6 diseases with a high self-care relevance are cardiovascular/metabolic diseases. Lifestyle-related causes and risk factors of these diseases are diet/nutrition (high calories, sugar, and saturated fats), drug consumption (tobacco and alcohol) and physical inactivity. This applies in particular to comorbidities of the metabolic syndrome such as diabetes type II, hypertension and high cholesterol as well as COPD (Geldmacher et al. 2008).

Technologies for detection and measurement of vital signs are consistently high and are established in the consumer electronics sector. A growing number of specific technologies are able to detect and monitor multiple vital signs and diseases such as atrial fibrillation at once.

The technological feasibility of examined diseases is distributed evenly across the early-stage medical and consumer electronics sector (7/5/7 diseases). Aside from HIV, which has reached the consumer electronics sector, infectious diseases are mainly located in the early-stage sector or at the lower end of the medical sector (7/10 diseases). Cardiovascular/metabolic diseases, on the other hand, are represented mainly in the consumer electronics sector (5/8 diseases) and show a high technological feasibility.

Currently for most of the examined medical conditions, there are early-stage concepts and technologies which aim at the consumer electronics market and are planned for completion within the next 3 years.

Increasing precision in the field of detection and measurement of medical conditions, ease of use for consumers and availability of technologies in consumer electronics steadily raise the technological feasibility. Consequently, based on the made assumptions, increased technological feasibility entails growing self-care in private consumers.

## 6 Discussion

This chapter has demonstrated that cardiovascular and metabolic diseases tend to occur together and are predominantly the result of lifestyle-related factors. Infectious diseases, however, are far less affected by the same lifestyle-related factors. Furthermore, it was found that current research mainly focuses on the development of technologies for the non-invasive measurement of vital signs and cardiovascular/metabolic diseases. A growing number of these technologies are available in the field of consumer electronics and provide precise, non-invasive measurement and ease of use for private consumers. Some technologies already compare measured values with databases in order to provide diagnosis and recommendations for action. As a result, it is seen that individual self-care of consumers becomes increasingly independent of medical diagnostics and advice. Closely associated with this are consumers who gain growing influence on their personal health and decision-making. It is conceivable that these developments will lead to a decentralisation of the healthcare system and further transformation processes in the long term.

Technology solutions, in particular leveraging from the Internet of Things and incorporating sensors in combination with mobile, appear to offer a potential for more tailored and person-centric solutions to facilitate better and sustained health and wellness management. Perhaps while technology in many respects has led to the lifestyle-related factors that are leading to a more sedentary style and thus contributing to increases in the incidents of various chronic conditions, technology may also hold the answer for assisting in addressing the problem and finding appropriate, efficient and effective solutions such as the medical Tricorder.

## *6.1 Implications for Research and Practice*

On the basis of the results, the following implications for consumers, health insurance funds and technology manufacturers can be seen:

**Consumers:**

I. Consumers should initially focus on managing existing diseases which are categorised with a high self-care relevance such as diabetes type II, hypertension, high cholesterol and COPD. Since these diseases are closely lifestyle-related, consumers themselves could have a major impact by implementing simple lifestyle changes. The focus should lie on minimising lifestyle-related causes and risk factors by optimising diet and nutrition, discouraging drug consumption and enhancing physical activity.

II. Consumers should then aim to prevent diseases with high and medium self-care relevance. Cardiovascular/metabolic diseases may be prevented by proper diet and nutrition, reduced drug consumption and regular physical activity. Infectious diseases may be prevented by vaccinations (pertussis) or safe sex (HIV). The risk of malignant melanoma can be reduced by the avoidance of long periods of intense sun exposure.

**Health Insurance Funds:**

I. Health insurance funds should address diseases with a combined high self-care relevance and high burden of disease as those cause avoidable, high costs. Through further development of relevant technologies and promotion of health literacy among customers, expenditures can be reduced.

II. Health insurance funds should then deal with diseases with high and medium self-care relevance, due to high and long follow-up costs (HIV and malignant melanoma). Investing in technologies for the detection of these diseases and the preventive education of their customers could help to prevent diseases.
III. Health insurance funds are advised to deal with diseases with a high burden of disease since significant parts of their customers are affected. Technologies for quick and reliable detection of those diseases do not only save time and costs for expensive laboratory tests (infections), but also enable more efficient early treatments of patients.

**Technology Manufacturers:**

I. Technology manufacturers should implement a market penetration strategy for diseases which have reached a high technological feasibility because the related technologies are already available for consumers and cause only minor further development expenses.
II. Technology manufacturers are encouraged to invest in technologies which detect and monitor diseases with a high burden of disease. Consumer and medical demand is correspondingly large, and potential competitive and pioneer advantages can thereby be realised with early-stage and medical technologies.

## *6.2 Further Research*

This work mainly focused on the aspect of technological feasibility of non-invasive measurement of medical conditions and how medical diseases can be detected and measured, without the support of a healthcare provider. The assumption was made that an increased feasibility directly resonates with an enhanced understanding of self-care by consumers. However, technological enablement itself is only one part which defines self-care and should be complemented with actions to raise awareness and concrete recommendations for consumers. Measurement alone cannot increase self-care of individuals without further actions or lifestyle changes to promote and maintain health, and prevent disease.

Technology search and selection were based on a literature review with a fixed scheme, keywords and filters. The products and technologies represent only a limited selection of existing technical approaches. This is particularly due to the conflict between offering as many promising concepts as possible and in-depth examinations of technologies.

The early description of the medical conditions and epidemiological terms provided a good overview of reasons for their origin, their rate of disease and conventional detection and measurement. This established the basis for the detection of the burden of disease, self-care relevance and technological feasibility within the next sections. It would have been desirable to further limit the amount of technologies, conduct additional interviews with experts and focus more intensely on promising technologies instead.

The aforementioned issue could have been avoided by narrowing down the pool of diseases beforehand. In return, this would have made in-depth analysis of a smaller number of suitable technologies and identification of potential future key technologies possible.

Non-invasive technologies are not only capable of monitoring medical conditions but will also help to prevent diseases through early detection in the near future. It is to be expected that these technologies will no longer be restricted to the measurement of medical conditions but start searching for possible causes, making diagnoses and giving practical advice based on the measurement results. Particularly, patients with reduced mobility, people living in areas with little medical infrastructure and those suffering from diseases such as diabetics may benefit from those developments.

However, this chapter also raises a number of questions and challenges which need to be addressed by future research: For example, it should be examined how proper diagnosis and suitable recommendations can be made on the basis of measurement results and how accurate these diagnoses are. Furthermore, the potential benefits and fields of application of these technologies for hospitals and medical staff should be evaluated. Future research should also focus on subsequent development of technologies for the detection of infectious diseases. Developing countries with insufficient health services would benefit most from an extensive, fast and cost-efficient detection of infectious diseases.

Great challenges are expected particularly in the legal field. Regulations for security and privacy of patient data and exchange of sensible data content between devices and clouds need further discussion.

## References

AIDS.gov. (2015). *Stages of HIV infection*. Retrieved December 18, 2016, from https://www.aids.gov/hiv-aids-basics/just-diagnosed-with-hiv-aids/hiv-in-your-body/stages-of-hiv/

American Diabetes Association. (2016). *Checking your blood glucose*. Retrieved December 21, 2016, from http://www.diabetes.org/living-with-diabetes/treatment-and-care/blood-glucose-control/checking-your-blood-glucose.html

American Heart Association. (2014). *Lifestyle changes and cholesterol*. Retrieved December 21, 2016, from https://www.heart.org/HEARTORG/Conditions/Cholesterol/PreventionTreatmentofHighCholesterol/Lifestyle-Changes-and-Cholesterol_UCM_305627_Article.jsp

American Heart Association. (2016a). *What is atrial fibrillation (AFib or AF)?* Retrieved December 17, 2016, from http://www.heart.org/HEARTORG/Conditions/Arrhythmia/AboutArrhythmia/What-is-Atrial-Fibrillation-AFib-or-AF_UCM_423748_Article.jsp

American Heart Association. (2016b). *Why cholesterol matters*. Retrieved December 21, 2016, from http://www.heart.org/HEARTORG/Conditions/Cholesterol/WhyCholesterolMatters/Why-Cholesterol-Matters_UCM_001212_Article.jsp

BMJ. (2016). *Chapter 2. Quantifying disease in populations*. Retrieved December 17, 2016, from http://www.bmj.com/about-bmj/resources-readers/publications/epidemiology-uninitiated/2-quantifying-disease-populations

Browning, G. G., & Gatehouse, S. (1992). The prevalence of middle ear disease in the adult British population. *Clinical Otolaryngology & Allied Sciences, 17*(4), 317–321. Retrieved from https://doi.org/10.1111/j.1365-2273.1992.tb01004.x.

Brumbaugh, A. R., & Mobley, H. L. (2012). Preventing urinary tract infection: Progress toward an effective Escherichia coli vaccine. *Expert Review of Vaccines, 11*(6), 663–676. Retrieved from http://www.ncbi.nlm.nih.gov/pubmed/22873125%255Cn; http://www.pubmedcentral.nih.gov/articlerender.fcgi?artid=PMC3498450.

Bundesminiterium für Gesundheit. (2016). *Infektionskrankheiten*. Retrieved December 17, 2016, from http://www.bundesgesundheitsministerium.de/themen/praevention/gesundheitsgefahren/infektionskrankheiten.html

Cambridge Business English Dictionary. (n.d.). *Consumer electronics*. Retrieved December 17, 2016, from http://dictionary.cambridge.org/dictionary/english/consumer-electronics

Cheng, S.-Y., Leonard, J. L., & Davis, P. J. (2010). Molecular aspects of thyroid hormone actions. *Endocrine Reviews, 31*(2), 139–170. Retrieved from http://press.endocrine.org/doi/10.1210/er.2009-0007.

Chopra, R., Chander, A., & Jacob, J. J. (2012). Ocular associations of metabolic syndrome. *Indian Journal of Endocrinology and Metabolism, 16*(Suppl1), S6–S11. Retrieved from http://www.ncbi.nlm.nih.gov/pmc/articles/PMC3354923/.

Denzlinger, C. (2014). Leukozytose und Leukopenie rational abklären. *Der Allgemeinarzt*, (10). Retrieved from www.allgemeinarzt-online.de.

Department of Health New York State. (1999). *Basic statistics: About incidence, prevalence, morbidity, and mortality – Statistics teaching tools*. Retrieved October 5, 2016, from https://www.health.ny.gov/diseases/chronic/basicstat.htm

Deutsche Gesellschaft für Hals-Nasen-Ohrenheilkunde und Kopf und Hals-Chirurgie e.V. (2014). Chronisch-mesotympanale Otitis media. *AWMF-Leitlinien*, (017), 1–14.

Deutsche Hochdruckliga e.V. DHL. (2015). B*luthochdruck wirksam bekämpfen*. Retrieved December 17, 2016, from https://www.hochdruckliga.de/bluthochdruck.html

"Digital Contact Lenses Can Transform Diabetes Care." (2016). *Digital contact lenses can transform diabetes care*. Retrieved October 10, 2016, from http://medicalfuturist.com/googles-amazing-digital-contact-lens-can-transform-diabetes-care/

DKG, Deutsche Krebshilfe, & AWMF. (2014). S3-Leitlinie Prävention von Hautkrebs. *AWMF-Leitlinien*, (April), 1–231.

Egger, G., & Dixon, J. (2014). *Beyond obesity and lifestyle: A review of 21st century chronic disease determinants, 2014*. Retrieved from https://www-hindawi-com.pitt.idm.oclc.org/journals/bmri/2014/731685/

Enns, G. (2016). *Metabolic disease*. Retrieved December 17, 2016, from https://www.britannica.com/science/metabolic-disease

Fink, W., & Haidinger, G. (2007). Die Häufigkeit von Gesundheitsstörungen in 10 Jahren Allgemeinpraxis. *ZFA - Zeitschrift Für Allgemeinmedizin, 83*(3), 102–108. Retrieved from http://www.thieme-connect.de/DOI/DOI?10.1055/s-2007-968157.

Geldmacher, H., Biller, H., Herbst, A., Urbanski, K., Allison, M., Buist, A. S., et al. (2008). Die Prävalenz der chronisch obstruktiven Lungenerkrankung (COPD) in Deutschland TT – The prevalence of chronic obstructive pulmonary disease (COPD) in Germany. *Deusche medizinische Wochenschrift, 133*(50), 2609–2614. Retrieved from https://www.thieme-connect.de/DOI/DOI?10.1055/s-0028-1105858.

GfK SE. (2016). *Press release*. Retrieved October 5, 2016, from http://www.gfk.com/global-studies/global-studies-fitness-tracking/

Grundy, S. M. (2005). Diagnosis and management of the metabolic syndrome: An American Heart Association/National Heart, Lung, and Blood Institute Scientific Statement. *Circulation, 112*(17), 2735–2752. Retrieved from http://circ.ahajournals.org/cgi/doi/10.1161/CIRCULATIONAHA.105.169404.

Hein, H. (2004). Herz-kreislauferkrankungen und schlafbezogene obstruktive Atmungsstörungen. *Pneumologie*. Reinbek: Dr. Holger Hein. Retrieved from http://www.dr-holger-hein.de/index.php?id=12

Herold, G. (2010). *Innere Medizin 2010.* Köln: Gerd Herold.

Huang, P. L. (2009). A comprehensive definition for metabolic syndrome. *Disease Models & Mechanisms, 2*(5–6), 231–237. Retrieved from http://www.ncbi.nlm.nih.gov/pmc/articles/PMC2675814/.

Hummers-pradier, E., & Schiemann, G. (n.d.). *Diagnostik und Therapie der Harnwegsinfekte in der Allgemeinmedizin.* Universitätsmedizin Göttingen.

Jacobi, V., Lehnert, T., & Thalhammer, A. (2009). Pneumonien bei Immunsuppression. *Radiologie Up2date, 9*(04), 359–384. Retrieved from http://www.thieme-connect.de/DOI/DOI?10.1055/s-0029-1215330.

Johns, D. P., Walters, J. A. E., & Walters, E. H. (2014). Diagnosis and early detection of COPD using spirometry. *Journal of Thoracic Disease, 6*(11), 1557–1569. Retrieved from http://www.ncbi.nlm.nih.gov/pubmed/25478197.

Koch-Institut, R. (2015). Schätzung der Prävalenz und Inzidenz von HIV-Infektionen in Deutschland, Stand Ende 2014. *Epidemiologisches Bulletin,* (45).

Kraywinkel, K., Bertz, J., Laudi, A., & Wolf, U. (2012). Epidemiologie und Früherkennung häufiger Krebserkrankungen in Deutschland. *Gesundheitsberichterstattung Des Bundes (GBE),* (3).

Merriam-Webster. (n.d.-a). *Etiology.* Retrieved December 20, 2016, from https://www.merriam-webster.com/dictionary/etiology#medicalDictionary

Merriam-Webster. (n.d.-b). *Infectious disease.* Retrieved December 17, 2016, from https://www.merriam-webster.com/medical/infectious disease.

Merriam-Webster. (2016). *Technology.* Retrieved December 17, 2016, from https://www.merriam-webster.com/dictionary/technology

National Heart, Lung, and Blood Institute. (2014). *What are the signs and symptoms of hemolytic anemia?* Retrieved December 21, 2016, from https://www.nhlbi.nih.gov/health/health-topics/topics/ha/signs

National Heart, Lung, and Blood Institute. (2016). *What is metabolic syndrome?* Retrieved December 17, 2016, from https://www.nhlbi.nih.gov/health/health-topics/topics/ms

National Institute of Mental Health. (n.d.). *What is prevalence?* Retrieved December 17, 2016, from https://www.nimh.nih.gov/health/statistics/prevalence/index.shtml

Neuhauser, H., & Poethko-Müller, C. (2014). Chronische Erkrankungen und impfpräventable Infektionserkrankungen bei Kindern und Jugendlichen in Deutschland: Ergebnisse der KiGGS-Studie - Erste Folgebefragung (KiGGS Welle 1). *Bundesgesundheitsblatt - Gesundheitsforschung - Gesundheitsschutz, 57*(7), 779–788.

OECD. (2015, September 24). *Healthcare costs unsustainable in advanced economies without reform.* Retrieved December 20, 2016, from http://www.oecd.org/health/healthcarecostsunsustainableinadvancedeconomieswithoutreform.htm

Pai, M., & Filion, K. (2003). *AN OVERVIEW OF MEASUREMENTS IN EPIDEMIOLOGY EXPOSURES Disease Frequency exposure.* http://www.med.mcgill.ca/epidemiology/ebss/resources/Docs/An%20overview%20of%20measurements%20in%20epi_v3_2007.pdf

Pascual, A., & Aranda, A. (2013). Thyroid hormone receptors, cell growth and differentiation. *Biochimica et Biophysica Acta (BBA) - General Subjects, 1830*(7). Retrieved from http://linkinghub.elsevier.com/retrieve/pii/S0304416512000852.

Penzel, T., Peter, H., & Hermann, J. (2005). Schlafstörungen. *Gesundheitsberichterstattung Des Bundes (GBE), 27,* 49.

Piper, W. (2013). *Innere Medizin* (2nd ed.). Heidelberg: Springer Medizin.

Porta, M. (1995). *A dictionary of epidemiology* (3rd edit ed.). New York: Oxford University Press. Retrieved from file://www.oxfordreference.com/10.1093/acref/9780195314496.001.0001/acref-9780195314496.

Prinz, C., & Ott, I. (2012). Basiswissen Innere Medizin. In L. Dittgen, J. Gaa, U. Henke-Luedecke, I. Ott, F. Schneller, & A. Umgelter (Eds.), *Basislehrbuch Innere Medizin* (1st ed.). Heidelberg: Springer Medizin Verlag.

Riffelmann, M., Littmann, M., Hellenbrand, W., Hülße, C., & Wirsing von König, C. H. (2008). Pertussis - nicht nur eine Kinderkrankheit. *Dtsch. Ärztebl, 105*(37), 623–628. Retrieved from http://www.aerzteblatt.de/v4/archiv/artikel.asp?src=suche&id=61447%5CnC:%5CKarsten%5CPDFs%5CBakteriologie-PDFs%5CBakt-2008%5CRiffelmann et al.-Pertussis - nicht nur eine Kinderkrankheit.pdf.

Rober Koch-Institut. (n.d.). *Cardiovascular disease*. Retrieved December 17, 2016, from http://www.rki.de/EN/Content/Health_Monitoring/Main_Topics/Chronic_Disease/Cardiovascular_Disease/cardiovascular_disease_node.html

Robert Koch-Institut. (2009). *Streptococcus pyogenes-Infektionen*. Retrieved December 17, 2016, from https://www.rki.de/DE/Content/Infekt/EpidBull/Merkblaetter/Ratgeber_Streptococcus_pyogenes.html

Robert Koch-Institut. (2014a). *Keuchhusten (Pertussis)*. Retrieved December 17, 2016, from https://www.rki.de/DE/Content/Infekt/EpidBull/Merkblaetter/Ratgeber_Pertussis.html#doc2374534bodyText3

Robert Koch-Institut. (2014b). *Koronare Herzkrankheit. Faktenblatt zu GEDA 2012: Ergenisse der Studie >>Gesundheit in Deutschland aktuell 2012<<.*

Robert Koch-Institut. (2015a). *Faktenblatt zu DEGS1 : Studie zur Gesundheit Erwachsener in Deutschland (2008–2011) Prävalenz von Diabetes mellitus*. Berlin.

Robert Koch-Institut. (2015b). *Health in Germany. Federal Health Reporting. Joint service by RKI and Destatis. RKI, Berlin.*

Robert Koch-Institut. (2016a). Epidemiologisches Bulletin. *Epidemiologisches Bulletin, 38*, 177–186. Retrieved from http://www.rki.de/DE/Content/Infekt/EpidBull/Archiv/2015/Ausgaben/15_15.pdf?__blob=publicationFile.

Robert Koch-Institut. (2016b). *HIV-infektion/AIDS*. Retrieved December 17, 2016, from https://www.rki.de/DE/Content/Infekt/EpidBull/Merkblaetter/Ratgeber_HIV_AIDS.html#doc2374480bodyText9

Robert Koch-Institut. (2016c). *Windpocken, Herpes zoster (Gürtelrose)*. Retrieved December 18, 2016, from https://www.rki.de/DE/Content/Infekt/EpidBull/Merkblaetter/Ratgeber_Varizellen.html

Robert Koch-Institut, & Destatis. (2015). Gesundheit in Deutschland (2015). *Gesundheitsberichterstattung Des Bundes*, 13–76.

Saliba, W., Nitzan, O., Chazan, B., & Elias, M. (2015). Urinary tract infections in patients with type 2 diabetes mellitus: review of prevalence, diagnosis, and management. *Diabetes, Metabolic Syndrome and Obesity: Targets and Therapy, 8*, 129. Retrieved from http://www.dovepress.com/urinary-tract-infections-in-patients-with-type-2-diabetes-mellitus-rev-peer-reviewed-article-DMSO.

Sandru, A., Voinea, S., Panaitescu, E., & Blidaru, A. (2014). Survival rates of patients with malignant melanoma. *Journal of Medicine and Life, 7*(4), 572–576.

Schnabel, R. (2012). Vorhofflimmern und Schlaganfallrisiko in der Allgemeinbevölkerung. In *German Cardiac Society Web Presence*. Hamburg: Deutsche Gesellschaft für Kardiologie – Herz- und Kreislaufforschung e.V. (DGK).

Schuchert, A., Gerth, A., Näbauer, M., Steinbeck, G., & Meinertz, T. (2005). Vorhofflimmern Epidemiologie, Klinik und Prognose. *Med Welt*, 361–365. Retrieved from www.die-medizinische-welt.de.

Schulz, S., Freitag, M., & Gensichen, J. (2012). Schilddrüsenerkrankungen in der Hausarztpraxis. *Ärzteblatt Thüringen, 23*(9), 468–471.

Sharma, M., & Majumdar, P. K. (2009). Occupational lifestyle diseases: An emerging issue. *Indian Journal of Occupational and Environmental Medicine, 13*(3), 109–112. https://doi.org/10.4103/0019-5278.58912.

Sondak, V. K., & Smalley, K. (2009). Targeting mutant BRAF and KIT in metastatic melanoma: ASCO 2009 meeting report. *Pigment Cell & Melanoma Research, 22*(4), 386–387. Retrieved from http://doi.wiley.com/10.1111/j.1755-148X.2009.00593.x.

Thacker, S. B., Stroup, D. F., Carande-Kulis, V., Marks, J. S., Roy, K., & Gerberding, J. L. (2006). Measuring the public's health. *Public Health Reports, 121*(1), 14–22.

Thomas, J. P., Berner, R., Zahnert, T., & Dazert, S. (2014). Acute otitis media–a structured approach. *Deutsches Ärzteblatt International, 111*(9), 151–159; quiz 160. Retrieved from http://www.pubmedcentral.nih.gov/articlerender.fcgi?artid=3965963&tool=pmcentrez&rendertype=abstract%5Cn; http://www.ncbi.nlm.nih.gov/pubmed/24661591%5Cn; http://www.pubmedcentral.nih.gov/articlerender.fcgi?artid=PMC3965963.

Tschäpe, H. (2000). Lebensmittelbedingte Infektionskrankheiten durch Bakterien. *Bundesgesundheitsblatt - Gesundheitsforschung - Gesundheitsschutz, 43*(10), 758–769. Retrieved from http://link.springer.com/10.1007/s001030050353.

U.S. Department of Health and Human Services. (2000). Healthy People 2010: Understanding and Improving Health. Washington, DC: US Government Printing Office, November 2000. In *Book US Department of Health and Human Services. Healthy People 2010: Understanding and Improving Health* (Vol. 1, 2nd ed.). Washington, DC: U.S. Government Printing Office.

Vorvick, L. J. (2015a). *Invasive*. Retrieved December 17, 2016, from https://medlineplus.gov/ency/article/002384.htm

Vorvick, L. J. (2015b). *Noninvasive*. Retrieved December 17, 2016, from https://medlineplus.gov/ency/article/002269.htm

Wächtler, H., & Chenot, J.-F. (2011). Die Halsschmerz-Leitlinie der Deutschen Gesellschaft für Allgemeinmedizin und Familienmedizin. *HNO, 59*(5), 480–484. Retrieved from http://link.springer.com/10.1007/s00106-011-2263-6.

Wagenlehner, F. M. E. (2015). *Urinary tract infections epidemiology and management strategies*. Gießen: Clinic for Urology, Pediatric Urology and Andrology Universitätsklinikum Gießen.

Wagner, K., & Sparr, F. J. (2012). *Wie kann die Gesundheitskompetenz in der Bevölkerung verbessert werden?* Linz.

Where No Healthcare Device Has Gone Before. (2015). *The Pathologist*, (June).

Willett, W. C., Koplan, J. P., Nugent, R., Dusenbury, C., Puska, P., & Gaziano, T. A. (2006). Prevention of chronic disease by means of diet and lifestyle changes. In *Disease control priorities in developing countries* (pp. 833–850). Oxford University Press. Retrieved from http://www.ncbi.nlm.nih.gov/pubmed/21250366.

Williamson, R. T. (2009). Causes of diabetes. *The Practitioner, 253*(1718), 37.

Wirth, A., Pfeiffer, A., Steinmetz, A., Albus, C., Bjarnason-Wehrens, B., Cordes, C., et al. (2006). Das Metabolische Syndrom - Empfehlungen für die kardiologische Rehabilitation. *Herzmedizin, 23*(4), 187–196.

World Health Organization. (n.d.). *Global burden of disease*. Retrieved December 18, 2016, from http://www.who.int/topics/global_burden_of_disease/en/

World Health Organization. (2008). Vaccination greatly reduces disease, disability, death and inequity worldwide. *Bulletin of the World Health Organization, 86*(2), 81–160. Retrieved from http://www.who.int/bulletin/volumes/86/2/07-040089/en/.

World Health Organization. (2009). *Self-care in the context of primary health care*. Report of the World Health Organization Regional Consultation Bangkok, Thailand, 7–9 January 2009.

World Health Organization. (2010). *International statistical classification of diseases and related health problems* (10th ed.). World Health Organization.

World Health Organization. (2016a). *Cardiovascular diseases (CVDs)*. Retrieved December 17, 2016, from http://www.who.int/mediacentre/factsheets/fs317/en/

World Health Organization. (2016b). Global Report on Diabetes. In: G. Roglic. World Health Organization. 86.

World Health Organization. (2016c). *Global report on diabetes* (Vol. 978). Retrieved from http://www.who.int/about/licensing%5Cn; http://apps.who.int/iris/bitstream/10665/204871/1/9789241565257_eng.pdf

World Health Organization. (2016d). *Metrics: Disability-Adjusted Life Year (DALY)*. Retrieved October 5, 2016, from https://www.who.int/healthinfo/global_burden_disease/metrics_daly/en/

World Heart Federation. (n.d.). *What causes cardiovascular disease?* Retrieved December 17, 2016, from http://www.world-heart-federation.org/heart-facts/resources/videos/about-cvd/key-facts/what-causes-cardiovascular-disease/

XChallenge. (2014). *Nokia XChallenge Sensing*. Retrieved October 5, 2016, from http://sensing.xprize.org/

XPrize. (2016). *Qualcomm XPrize Tricorder*. Retrieved October 5, 2016, from http://tricorder.xprize.org/

XPRIZE Foundation. (2012). *Overview*. Retrieved December 17, 2016, from http://tricorder.xprize.org/about/overview

Zentrum für Krebsregisterdaten. (2013). *3.11 Malignes Melanom der Haut*. Berlin.

# Piloting a Mobile Tele-simulation Unit to Train Rural and Remote Emergency Healthcare Providers

**Jennifer Jewer, Adam Dubrowski, Cody Dunne, Kristopher Hoover, Andrew Smith, and Michael Parsons**

## 1 Introduction

Rural and remote practice of emergency medicine presents unique challenges, particularly when faced with infrequently encountered cases and procedures (Williams et al. 2001). These challenges are amplified by the fact that a large proportion of emergency care in rural areas must be provided by general practitioners, or by nurses and nurse practitioners (Williams et al. 2001; Casey et al. 2008). This poses a serious challenge to equitable health-care delivery when patients in rural areas do not have access to comparable levels of emergency care as those in urban centers (Rogers et al. 1999; Ireland et al. 2006). In emergency medicine, and rural and remote emergency medicine in particular, low-frequency occurrences of many clinical encounters limit the opportunity for skills to be developed and maintained through on the job experience alone. Therefore, a systematic approach to training personnel for these emergencies is required. Simulation-based medical education (SBME) has been identified as a valuable tool in the acquisition and maintenance of knowledge and skills (Rogers et al. 1999; Ireland et al. 2006; Issenberg et al. 2005; Cook et al. 2011; Scott and Dunnington 2008; Roy et al. 2011) because it facilitates deliberate practice without compromising patient safety (Ziv et al. 2003). However, simulators are often located in urban centers and are not easily accessible outside these centers due to geographic, cost, and time constraints (Ikeyama et al. 2012; McCoy et al. 2017; Rosen et al. 2012). Mobile tele-simulation has the potential to overcome these barriers by bringing the simulation training to the trainees, but challenges such as a comfortable learning environment, technical issues, and ability to teach desired content via tele-simulation must be addressed.

J. Jewer (✉) · A. Dubrowski · C. Dunne · K. Hoover · A. Smith · M. Parsons
Memorial University of Newfoundland, St. John's, NL, Canada
e-mail: jenniferj@mun.ca

 
N. Wickramasinghe, F. Bodendorf (eds.), *Delivering Superior Health and Wellness Management with IoT and Analytics*, Healthcare Delivery in the Information Age, https://doi.org/10.1007/978-3-030-17347-0_2

We have developed a mobile tele-simulation unit (MTU) prototype that enables mentor and trainee emergency health care workers, to connect and access SBME on procedural skills in rural and remote settings. This study focuses on a proof of concept regarding the acceptability, feasibility, and effectiveness of the proposed intervention. The goal is to determine whether using this unit, in areas where simulation training would otherwise not be available, is acceptable to all parties given the proposed advantages that an MTU can offer in terms of flexibility, convenience, and costs. The specific objectives of this project are:

1. Acceptability and feasibility: to gather feedback on the design and function of each iteration of the MTU prototype and incorporation into the finalized MTU
2. Effectiveness: to examine learning outcomes and assess if the outcomes in the MTU are comparable to face-to-face training

This study takes place in Newfoundland and Labrador (NL), Canada where 40 percent of the population lives in rural areas. NL has a population of around 525,000 that is geographically dispersed across a land mass of approximately 405,000 $km^2$, or almost one and three quarters the size of Great Britain. However, Great Britain has more than 100 times the number of people. NL has a relatively new simulation lab, located at the medical school in the capital city; however, the geographic dispersion of medical facilities across the province makes it expensive, time consuming, and often impractical for trainees to visit and access resources at the urban facility. In addition, the simulation lab operates at near capacity, with preference given to medical students and limited access to outside groups.

## 2 Background

Mobile tele-simulation is a combination of tele-simulation and mobile simulation. Tele-simulation involves using Internet protocol-based teleconference software to give trainees access to simulators and/or mentors in a different location. It couples the principles of simulation with remote Internet access to teach procedural skills (Mikrogianakis et al. 2011). Mobile simulation enables access to simulation training by bringing necessary equipment, and sometimes even the training environment, directly to the remote teaching site. Research in tele-simulation and mobile-simulation is limited but has been growing in recent years.

Tele-simulation is particularly useful when there are distance limitations, time constraints, or a lack of skilled mentors that constrain access to training at simulation centers (McCoy et al. 2017). Tele-simulation has been shown to be an effective means of teaching medical skills. It has been used to teach procedural and surgical skills such as intraosseous line insertion (Mikrogianakis et al. 2011), laparoscopic surgery (Okrainec et al. 2009, 2010; Henao et al. 2013), treatment of ventricular fibrillation or desaturation in an intubated patient (Ikeyama et al. 2012), pediatric resuscitation (Ohta et al. 2017), and performance of ultrasound-guided anesthetic techniques (Burckett-St Laurent et al. 2016). Some of these studies examined the use of tele-simulation to provide training to physicians in resource-restricted regions, for example, laparoscopic surgery to surgeons in Botswana, Africa (Okrainec et al. 2009, 2010), in Colombia (Henao et al. 2013), and in Puerto Rico (Small et al. 1999; Treloar et al. 2001).

Tele-simulation has also been found effective in the remote assessment of skills (Burckett-St Laurent et al. 2016; Okrainec et al. 2013; Choy et al. 2013). This is important as it has the potential to decrease costs without impacting assessment validity. For example, Burckett-St et al. (2016) found that the evaluations of ultrasound-guided anesthetic procedures conducted remotely were consistent with those conducted on-site. Also, Okrainec et al. (2013) found that the results of remote administration and scoring of the exam for laparoscopic surgery were consistent with standard on-site testing.

Emergency personnel have also been trained using tele-simulation. Treloar et al. (2001) used a high-fidelity human patient simulator (HPS) to provide an educational program for emergency personnel. The personnel had access to the HPS and they received both on-site and remote instruction. They found a significant overall improvement in both perceived preparedness and self-efficacy. Von Lubitz et al. (2003) also studied the use of HPS and took it one step further by examining remote access to the simulators to train physicians in three emergency scenarios. They found a statistically significant improvement in all testing measures and that trainees' confidence in performing the procedures also improved.

A challenge with tele-simulation is that trainees may not have access to simulation equipment or the training environment necessary for tele-simulation. These challenges are addressed with mobile simulation. Mobile simulation can make use of a specialized unit with portable simulation equipment that effectively represents a safe, immersive classroom environment for simulation training. For example, a patient simulator and an audiovisual system were set up at rural health centers in Australia to teach trauma teams (Ireland et al. 2006). They used multidisciplinary team training to combine scenario-based learning with mobile patient simulation to practice technical and behavioral skills in the actual work environments. Weinstock et al. (2009) recognized the high setup cost and the need for a dedicated space when establishing the simulation training at the health center and sought to create a low-cost method for delivering mobile simulation. They created a self-contained mobile simulation cart that contained a laptop to display vital signs and had audiovisual equipment to allow for video-based debriefing. A systematic review of the mobile simulation literature found only 29 papers that conducted a study of the use of mobile simulation to train physicians (Rosen et al. 2012). They found that the studies covered a limited range of clinical topics, with the majority focusing on surgical and obstetrical areas. Most studies focused on evaluating learner reactions and changes in attitudes and found positive results, and several studies found improvements in care (Steinemann et al. 2011; Riley et al. 2011; Hunt et al. 2008).

Another type of mobile simulation is one in which a self-contained unit containing the simulation and other materials is transported to the trainees, rather than setting up the simulation equipment in the hospital itself. In this model, the equipment is sent to the training site and the training is conducted in the self-contained unit. Some examples of a mobile simulation unit include: a modified van with simulation equipment to practice laparoscopic skills in Australia (Xafis et al. 2013); a modified van with simulation equipment and camera to record and train emergency clinicians in Italy (Ullman et al. 2016), and; a modified ambulance with camera and simulation equipment in the US to teach endotracheal intubations (Bischof et al. 2016). There is limited information available on the learning or patient outcomes associated with use of such mobile simulation tools to train physicians. However, the preliminary results provide some evidence of the potential power of mobile simulation. With Ullman et al.'s (2016) modified van, all participants in the study expressed interest in participating in future training sessions. Furthermore, Xafis et al. (2013) found that learning with their modified van was comparable to training at a fixed simulation center. In fact, there was a trend toward superior participant performance with the mobile unit. The authors speculated that this may be because of the convenience of having the unit deliver training at the hospital instead of trainees travelling to a skills center, or because of the novelty of skills training in a vehicle.

Proponents of mobile simulation suggest that enabling trainees to learn in their work environment with their own clinical team fosters individual, team, unit and organizational learning. Also, it saves staff time and money as staff does not have to travel to a physically separate training environment (Rosen et al. 2012). For rural areas, or those without access to a dedicated simulation center, mobile simulation is an especially valuable resource for the delivery of medical training (Ireland et al. 2006; Rosen et al. 2012; Xafis et al. 2013; Ullman et al. 2016; Bischof et al. 2016; Weinstock et al. 2009; Pena et al. 2015). However, bringing the mentor, experienced in the subject area and in effective SBME and debriefing, to the learner can often prove to be quite expensive and prohibitive due to the time needed to travel to the training site. Since accessibility to both an expert mentor, along with the appropriate training environment and equipment, can be obstacles to simulation training in rural and remote areas, merging the two concepts of tele-simulation and mobile simulation presents an innovative solution. To our knowledge, research on the concurrent application of tele-simulation and mobile simulation to deliver medical training has yet to be conducted.

## 3 Description of MTU Prototype

Using the MTU, procedural skills training sessions would be delivered remotely to emergency health-care providers in rural or remote locations using content developed by mentors experienced in the subject area and in SBME. The MTU would be transported to the location and is designed to require minimal technical support for setup and execution of the training session. Educational content of the modules delivered can be variable and tailored to the site-specific needs of the learners. The geographically separated mentor would deliver the training session remotely via a live broadcast with two-way video and audio. The importance of a mentor with experience in the clinical environment and with delivering simulation training remotely cannot be underestimated (McGaghie et al. 2010). All sessions would consist of a pre-briefing, simulation scenario, and deliberate practice with feedback. Relevant review materials would be sent out to learners prior to each session to allow pre-session familiarization with key information.

## 4 Methods

The iterative development and piloting of the MTU prototype was carried out through a mixed-methods approach and with input from a multi-disciplinary team with backgrounds in emergency medicine, clinical simulation, health informatics, engineering, computer science and research. We followed Haji et al.'s (2014) adapted Medical Research Council (MRC) framework to develop programs in simulation education for training of health professionals. The MRC framework was developed to help researchers of SBME develop programs of research, rather than a project-based strategy, with the goal of optimizing instructional design of SBME. Research on a "study-by-study" basis can result in "a body of evidence that is at times chaotic, contradictory, and limited in advancing ... understanding" (Haji et al. 2014, p. 250). Our primary goal of following the MRC framework is so that we are moving beyond studying *if* the MTU is effective and toward an understanding of *why* it is effective or not. Also, future research will be able to build upon this program of research to advance the understanding of SBME.

The MRC emphasizes a theory-based, iterative systematic approach to the design, refinement, evaluation, and implementation of SBME. The emphasis is on the design of programs of research in SBME that are theoretically based and methodologically transparent. The MRC framework (Fig. 1) was originally created for development of complex clinical interventions and has been successfully applied in that area (Campbell et al. 2000).

The MRC framework consists of four research process cycles:

1. Cycle A – Theory and Modeling: theory and/or evidence identification and modeling of the MTU.
2. Cycle B – Piloting: following a reflective approach, collect data to determine appropriateness of MTU, outcome measures, comparison groups and understanding of the context within which the MTU will operate.
3. Cycle C – Evaluation: conduct a summative evaluation of the MTU.
4. Cycle D – Implementation: implement MTU into the health-care setting.

We followed an iterative approach and have completed cycles A and B. The necessary institutional ethics review board approval was obtained before the project began and initial results of this study have been presented at academic conferences (Parsons et al. 2016a, b, 2017a, b; Jewer et al. 2018).

Fig. 1 MRC framework (Haji et al. 2014)

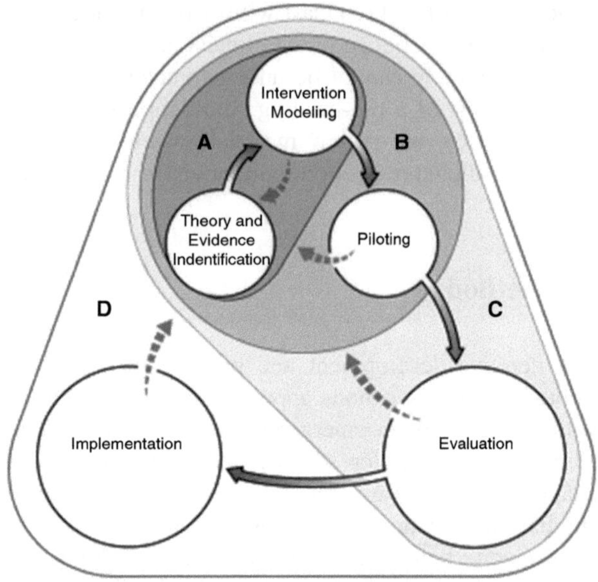


## 4.1 Cycle A – Theory and Modeling

We started by identifying the need for improved rural and remote emergency healthcare providers' access to training. We then set about determining how to address this need and deliver the training remotely. As previously discussed, a review of the literature revealed some research on tele-medicine, and mobile simulation; however, we did not find any research on mobile tele-simulation units. Using the Aim-FineTune-FollowThrough (AFT) process to guide the design of the MTU prototype, we moved through the iterative development process (Cristancho et al. 2011). The AFT process is grounded in learning theory and was developed to aid the development of simulation training programs. The AFT process has been used to successfully design a simulation-based program to train surgeons (Cristancho et al. 2012). In the "Aim" stage of the AFT process, we selected the procedural skill to be taught, broke the design into main components, and developed a concise, measurable definition of each component. We then used motor and cognitive modeling diagrams (MCMD) to determine processes, decisions, and logic required to complete the components of the MTU prototype on three main areas – comfort, technology, and human factors. Refer to Appendix A for an overview of the AFT process and a sample of the MCMD diagrams we constructed. In the "FineTune" stage, we used the Delphi method to collect input from experts with experience in emergency medicine, simulation training, and medical education on potential applications and key design components (Dunne et al. 2018). The prevailing opinion was that mobile tele-simulation would be useful for those in rural or remote locations. Key design components identified included: a reliable connection and

**Table 1** Features of the MTU prototype

| Feature | Description |
|---|---|
| Physical – size/layout of the MTU | Balance portability of unit with available work space |
| | Ability to adapt space to variety of simulation scenarios |
| Technical – telecommunications | Real-time communications – simple and easy to use |
| | Displays and quality of audiovisual communications |
| | Infrastructure – either cell or broadband network |
| | Low-cost software communications platform |
| Practical considerations | Efficient heating, ventilation, wiring, lighting, power supply |

competent technical support, a knowledgeable mentor, and content relevant to the trainee's location. We also revised the MCMDs and determined evaluation points and performance measures. In the "FollowThrough" stage, we finalized the MCMDs and developed and validated the MTU prototype.

### 4.1.1 Development of MTU Prototype

We designed the MTU prototype to ensure an efficient arrangement and operation of telecommunications and simulation equipment to allow ease of instruction, procedural performance, and assessment. Table 1 identifies the design and technical features that guided the design of the MTU prototype.

As the main focus of the study design was to assess educational effectiveness of a mobile tele-simulation unit, an inflatable rapid deployment tent was determined to be the most practical solution (Fig. 2). Vehicle- and trailer-based units were much more expensive and felt to be impractical at this point in time. The MTU tent was obtained locally in NL from Dynamic Air Shelters.[1] Its robust construction makes it suitable for transport and deployment in a variety of harsh environmental settings.

Table 2 and Fig. 3 show an overview of the equipment used in the most recent iteration of MTU prototype. Off-the-shelf and low-cost equipment was used to keep the design of the MTU accessible and practical.

### 4.1.2 Development of Training Program

We applied the best practices of SBME pedagogy outlined by McGaghie et al. (2010), including: feedback, deliberate practice, outcome measurement, simulation fidelity, and skill acquisition and maintenance. Educational content was provided through presession delivery of background information to the learner followed by hands-on teaching during instructional sessions. Prior to the teaching day, presession information consisted of an online New England Journal of Medicine

[1]https://www.dynamicairshelters.com/

**Fig. 2** Rapid deployment tent designed to function as the MTU

**Table 2** General equipment for setup of the mentor base station and the remote MTU station

| Mentor | Remote trainee |
|---|---|
| Technology<br>Computer with communications software<br>Web camera – dual function: (1) mounted on desk to show mentor and (2) worn as head cam to demonstrate skills; headset vs external speaker and microphone | Technology<br>Laptop with communications software (VSee)<br>Display monitor<br>External web camera, speaker, and microphone<br>Portable wireless Internet hub |
| Simulation materials<br>Medical instruments and supplies for procedure<br>Match the setup to the remote trainee station | Simulation materials<br>Medical instruments and supplies for procedure<br>Replication of materials used by mentor |

video demonstrating the procedure and important details about chest tube insertion including indications, contraindications, complications, and necessary equipment (Dev et al. 2007). On the teaching day, learners were given a brief clinical scenario on details necessitating insertion of the chest tube on their "patient." We designed the session to allow for deliberate practice, which has been found to be an important part of SBME (Cordray and Pion 2006). During the hands-on sessions, learners received guidance and real-time feedback on their performance and had the opportunity to ask questions. The real-time two-way communication between the mentor and trainees enabled this feedback.

Our session was geared toward teaching an important procedural skill, with joint reductions at Session A and tube thoracostomy (chest tube) at Sessions B and C. Joint reductions were taught with trainees doing hands-on practice on each other. In contrast, chest tube placement was taught using a low-fidelity setup: 3D-printed ribs secured to a plexiglass stand with low-cost simulated skin and subcutaneous tissue (later in the project the skin was also 3D printed) (Fig. 3). There is evidence of SBME as an effective method for teaching the chest tube procedure (Hutton et al. 2008).

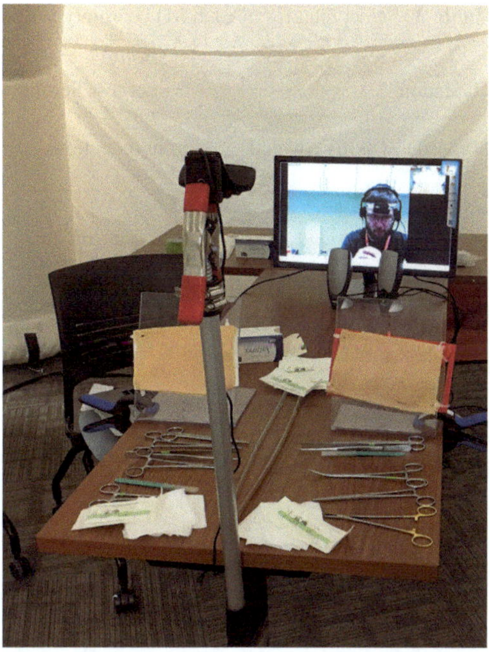

**Fig. 3** MTU with low-fidelity simulation setup at the remote site and mentor presence via telecommunication

## *4.2 Cycle B – Piloting*

Piloting is divided into four sub-phases: (1) establish feasibility and acceptability; (2) clarify uncertainties in the design of the intervention and outcome assessment; (3) identify and design the training protocol for a comparison group, and; (4) address methodological issues. These sub-phases are independent and not completed in any particular order. We held three prototype evaluation sessions to complete these four sub-phases and pilot the MTU prototype. This also involved iteratively applying the AFT process. The descriptions of the sessions are presented in Table 3, and Fig. 4 presents an overview of the piloting cycle.

### 4.2.1 Session A

The purpose of the first session, Session A, was to evaluate the feasibility and acceptability of the MTU and to clarify uncertainties in the design of the intervention. We considered possible barriers to the prototype implementation and addressed technical issues. We also evaluated and documented the setup and takedown of the MTU and all related components since the MTU would require setup by a technician at a remote site. The MTU prototype was deployed at a wilderness training course attended by 35 family medicine residents. Groups of approximately 9 residents were instructed on select joint reductions, skills relevant to the rural

**Table 3** Select features of each MTU prototype evaluation session

| | Session A | Session B | Session C |
|---|---|---|---|
| Location | Wilderness Setting (5 °C) | Wilderness Setting (−20 °C) | Indoors (23 °C) |
| Procedural skills covered | Joint reductions | Tube thoracostomy (chest tube insertion) | |
| Number of trainees | 35 | 6 | 18 |
| Trainee background | Family medicine residents | Family medicine residents, nurses and paramedics | Medical students |
| Number of trainees who performed procedure before | Did not ask | 3 (1–2 times) | 0 |
| Past exposure to telemedicine | 30 (86%) | 5 (83%) | 3 (17%) |
| Past exposure to low-fidelity SBME | 20 (57%) | 6 (100%) | 17 (94%) |

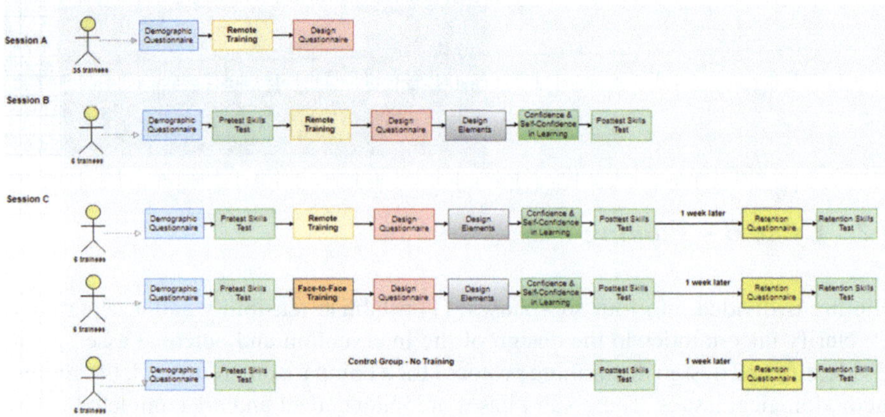


**Fig. 4** Overview of cycle B – piloting

practitioner. Following the format for the curriculum described in Sect. 4.1.2 of this paper, an experienced emergency medicine physician (the mentor) remotely instructed learners on elbow dislocation via a telecommunications link. Trainees had the opportunity to interact directly with the mentor during the session. The mentor's camera was displayed on the laptop screen in the MTU. The mentor observed trainees by two cameras stationed in the MTU. An experienced emergency medicine physician, located in the MTU, provided support and led training on finger and shoulder dislocations.

As shown in Fig. 4, students were asked to fill out a demographic questionnaire at the beginning of the session and a design questionnaire at the end of their session. The demographic questionnaire collected information on demographics and past experience with the procedure, SBME, and tele-medicine training. The design questionnaire focused on design and telecommunications features of the MTU, and perceptions of learning experiences. The features were rated on a five-point Likert scale from strongly disagree (1) to strongly agree (5).

### 4.2.2 Session B

Prototype B incorporated feedback from Session A, involving family medicine residents at the wilderness training course, and also took into consideration the comments of research team members with respect to improvements. The purpose of Session B was to continue examination of the feasibility and acceptability of the MTU and clarify uncertainties in the design of the intervention and outcome assessment.

Session B saw the MTU transported by airplane to Labrador, a more remote northern region of the province. It was necessary to address challenges of packaging and transport with this deployment. The extreme environment, with its very cold temperatures (−20 °C) presented additional challenges to the effective delivery of our educational content. Chest tube insertion was chosen as the procedure for this session as it was felt to be an important skill for learners and it was amenable to low-fidelity simulation setup and effective demonstration by the remote mentor. Learners were instructed remotely on the completion of a chest tube insertion procedure on 3D-printed low-fidelity models following the curriculum described in Sect. 4.1.2. No on-site mentor was present in this session, learners received only remote instruction. We reduced the number of trainees receiving training in the MTU in each session from nine to two, acting on feedback from Session A with respect to learner to instructor ratios. Additionally, due to the lag with two cameras on the trainees in Session A, we decided to use just one camera per site in Session B.

As with Session A, trainees completed the demographic questionnaire before the session and completed the design questionnaires after the session (see Fig. 4). Additional information was also collected. Design elements of the training session were evaluated using the adapted National League for Nursing (NLN) Simulation Design Scale (National League for Nursing 2005). Questions about the objectives and information provided, support, problem-solving and feedback were asked (see Appendix C). Learning outcomes were also evaluated in this session. Trainees were given a set of procedural skills questions, based on the presession materials (Dev et al. 2007), to answer before and after the session (see Appendix B). These questions were used to assess whether there were differences in the knowledge of the chest tube procedure within the group. They were also included to measure learning after the session. These materials were evaluated by an experienced physician to determine if differences existed pre and post-session. Measures of self-reported learning outcomes were adapted from the NLN Student Satisfaction and Self-Confidence in Learning scales (National League for Nursing 2005) to measure beliefs and attitudes about learning in simulation (see Appendix C). These NLN scales have been widely used and have been found to have sufficient reliability and validity to be used in education research (Franklin et al. 2014). We found that the extreme cold temperatures presented challenges. The space heater could not overcome the −20 °C temperatures, and some related discomfort was noted by participants. As well, low temperature resulted in compromised seals on the tent components and caused a slow air leak requiring reinflation during the session – a process requiring air blowers and potentially a generator, all leading to significant noise interference.

### 4.2.3 Session C

This session continued to build upon information gathered from the earlier prototype design cycle. We continued to evaluate the design and function of the MTU but also worked to complete the third and fourth sub-phases of Cycle B, the design of the training protocol for the comparison group and addressing any methodological issues. Because the overall purpose of this MTU prototype is to deliver training comparable to face-to-face training, we designed the training session for the comparison group to be given in this manner. The same procedure was taught (i.e., chest tube insertion), using the same medical instruments, supplies, and low-fidelity setup. The session was given the same amount of time for the face-to-face and tele-medicine groups. It also took place in the MTU tent to minimize any environmental influences as compared to Session B, although this round of testing was in a warm environment. Eighteen first and second year medical students were the subjects for this session. Three groups of equal sizes were created: the intervention group (tele-medicine), the comparison group (face-to-face), and the control group (no training session). Since this is a noninferiority study, a control group was needed to confirm that not only is the intervention group not inferior to the comparison group but that both treatments are actually effective (Greene et al. 2008). Trainees were randomized to each group based on the order of their reply to request for participation and we delivered the session to two trainees at a time. A third student per group was put in the control cohort and did not receive training (either remote or face-to-face). Instead they worked on solving a game puzzle for 20 minutes and then completed the post-tests and questionnaires.

Upon arrival at the session, the trainees completed the demographic and design questionnaires as in the previous sessions. Satisfaction and self-confidence in learning was evaluated using the instruments from Session B. To evaluate skill and knowledge maintenance over time, trainees were tested 1 week after the training session using the procedural skill questions (see Appendix B), and they were asked to rate how competent they perceived themselves with performing a chest tube insertion. We also asked if they had performed, witnessed, or received training in chest tube insertions in the week prior to doing the retention test.

## 5 Results

Through each successive session, the MTU was evaluated on physical design of the unit, function of the telecommunications equipment and overall impression on the utility of the MTU. All trainees completed these questions with the exception of the six control and six face-to-face trainees in Session C, because they did not receive remote training. The trainees' ratings on a scale of 1 (lowest) to 5 (highest) regarding design features, telecommunications, and overall satisfaction with the MTU are shown in Figs. 5, 6, and 7, respectively. Appendix C shows the means and standard deviations.

The design and telecommunications Figs. 5, 6, and 7, features and the overall satisfaction with the MTU were rated at around 4 or higher for all sessions, except for noise level and audio clarity. A Kruskal-Wallis test (non-parametric ANOVA) was used to determine if there were statistically significant differences between the sessions on these features. Table 4 shows the results of the analysis. Trainees' ratings were only statistically significant on the noise level and the clarity of the audio. Pairwise comparisons were performed using Dunn's (1964) procedure with a Bonferroni correction for multiple comparisons for these two features. This post hoc analysis revealed statistically significant differences in ratings on noise level between Session A (27.13) and Session B (10.17) ($p = 0.007$), and on clarity of the audio between Session A (27.26) and Session B (12.42) ($p = 0.019$).

In addition to examining the acceptability and feasibility of the MTU, in Sessions B and C, we also examined the design elements of the training session with questions on the objectives and information, support, problem-solving, and feedback. Appendix C shows the means and standard deviations. The results in

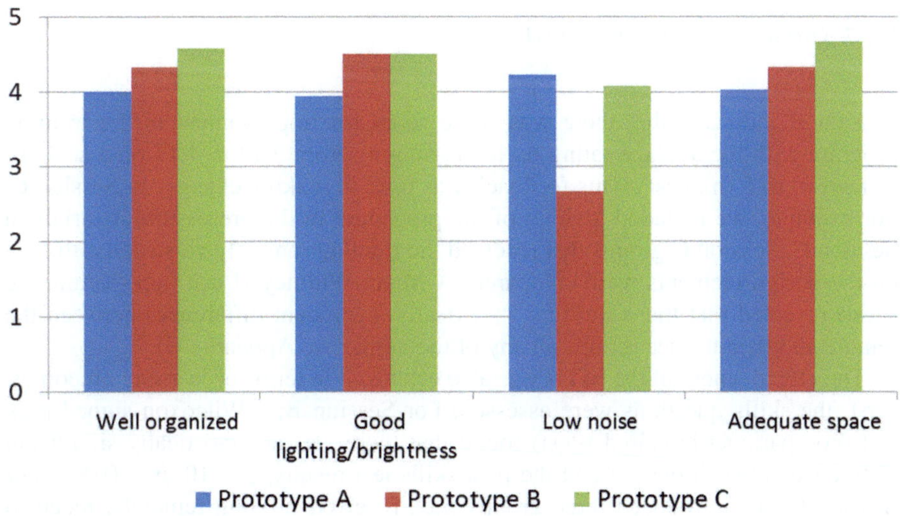


**Fig. 5** Feedback on physical MTU design features

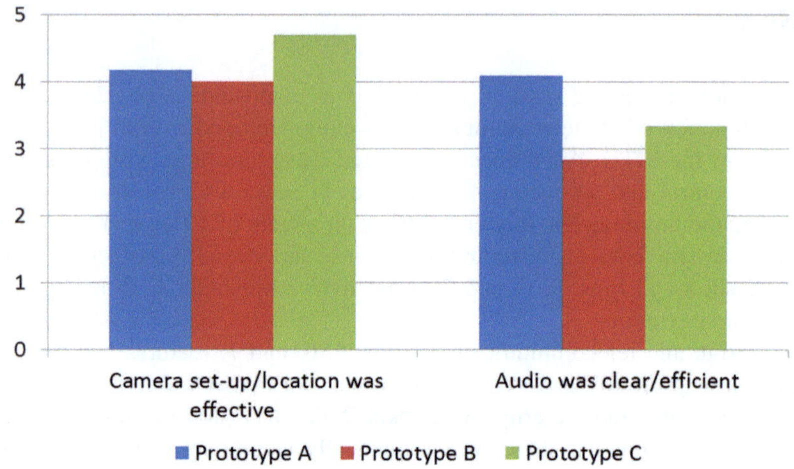


**Fig. 6** Function of telecommunications

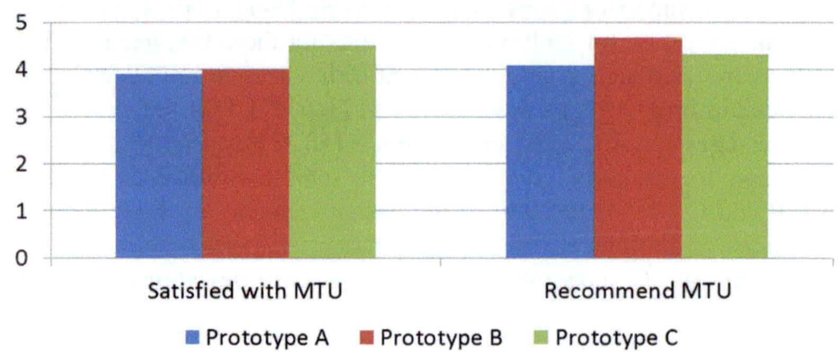


**Fig. 7** Overall satisfaction with the MTU

Session B indicated that there was some room for improvement in the training program with the average rating on items ranging from 3.17 to 4.17 on a scale of 1 (lowest) to 5 (highest). This feedback was used in making changes to Session C. For example, we included a video of the procedure in the presession materials in Session C. Session C groups that received the training remotely versus face-to-face on the design elements were compared. A Mann-Whitney U test (non-parametric t-test) revealed that there were no statistically significant differences between the remote and face-to-face groups on any of the items (see Appendix C).

The effectiveness of the MTU was also evaluated in terms of learning outcomes. First, the skills questions were assessed. For Session B, a Wilcoxon signed-rank test (non-parametric paired t-test) found that there was no statistically significant difference between the pre and the post-skills test results, $z = 10$, $p = 0.066$. For Session C, we introduced three groups (i.e., received training remotely, received training face-to-face, and the control group that did not receive any training) and a

**Table 4** Analysis of design, telecommunications, and overall satisfaction with the MTU

| | | | Mean rank | | | | |
|---|---|---|---|---|---|---|---|
| | $\chi^2(3)$ | $p$ | Session A ($n = 35$) | Session B ($n = 6$) | Session C ($n = 6$) | | Pairwise comparisons |
| *Design features* | | | | | | | |
| Noise level | 10.135 | 0.006 | 27.13 | 10.17 | 19.58 | | A–B $p = 0.007$, B–C $p = 0.586$, A–C $p = 0.526$ |
| Equipment well organized | 2.675 | 0.263 | 22.34 | 28.67 | 29.00 | | n/a |
| Good lighting | 4.508 | 0.105 | 21.77 | 30.50 | 30.50 | | n/a |
| Adequate space | 5.773 | 0.056 | 21.60 | 27.83 | 34.17 | | n/a |
| *Function of telecommunications* | | | | | | | |
| Camera setup | | 0.099 | 0.952 | 23.80 | 23.58 | 25.58 | n/a |
| Audio | | 9.958 | 0.007 | 27.26 | 12.42 | 16.58 | A–B $p = 0.019$, B–C $p = 1$, A–C $p = 0.149$ |
| *Overall satisfaction with MTU* | | | | | | | |
| Overall satisfaction with MTU | | 3.839 | 0.147 | 20.58 | 21.33 | 30.00 | n/a |
| Recommend MTU to colleagues | | 4.959 | 0.084 | 21.22 | 31.67 | 28.25 | n/a |

**Table 5** Mean and standard deviation for self-reported competency in Session C

| Group | Pre-test: self-reported competency | Retention test: self-reported competency | Increase in self-reported competency |
|---|---|---|---|
| Remote | 1.167 (0.408) | 2.167 (0.408) | 1 |
| Face-to-face | 1.000 (0) | 1.833 (0.408) | 0.833 |
| Control | 1.000 (0) | 1.5 (0.548) | 0.5 |

Self-reported competency was rated on a five-point Likert scale from "Not competent at all" (1) to "Expert" (5)

retention test. We created two new variables: one variable to calculate the difference between the pre- and post-skills tests and between the post and retention skills tests. A Kruskal-Wallis test on these new variables indicated that there were no statistically significant differences between the groups on the pre and post-skills test (($\chi^2(2) = 4.150$, $p = 0.126$) or between the post and retention skills tests ($\chi^2(2) = 2.485, p = 0.289$). Next, the self-reported learning measures (adapted from the NLN scales) that were asked in the post-test were analyzed (see Appendix C for the means and standard deviations). A Mann-Whitney U test revealed that there were no statistically significant differences between the remote and the face-to-face groups on any of the items (see Appendix C). Finally, self-reported competency in performing the chest tube procedure was examined for Session C using a two-way repeated measures ANOVA. The effect of training methodology (or no training in the case of the control group) on perceived competency shows no significant difference (*F(2, 10) = 2.059, p = 0.178*), nor is there a difference between the groups (*F(1,10) = 2.317, p = 0.149*). However, it should be noted that there is a larger increase in the mean self-rated competency level from the pre-test to retention tests for the remote group than for any of the other groups. In addition, the control group has the lowest mean competency level increase (see Table 5). There is a statistically significant increase in self-reported competency within the groups (*F(1,5) = 122.5, p < 0.005*). All but one of the trainees in the remote and face-to-face groups reported improved competency in the retention tests; whereas, only three of the six trainees in the control group reported an improvement.

## 6 Discussion and Future Research

To our knowledge, this is the first report of MTU development for remote training of emergency health-care providers. It was helpful to follow the four cycles of the adapted MRC framework to develop and pilot the MTU prototype. This framework enabled us to follow a theory-based approach, to identify challenges in the prototype, and to address these challenges iteratively in the piloting phase.

Overall, the trainees in each session were satisfied with their experience and would recommend the MTU to their colleagues for SBME. Additionally, the design and telecommunication features were rated highly in all sessions except for the noise level and the audio clarity of the telecommunications equipment. Specifically, issues were noted with the noise level and audio clarity during Session B. During this session, the extreme cold was associated with air leaks in the MTU structure and required pausing instruction to reinflate the unit, a process we feel contributed to the lower ratings on satisfaction with respect to noise and audio quality. The other two deployments required no reinflation. Built-in laptop speakers provide adequate audio in most circumstances, but external speakers of better quality proved advantageous in Sessions A and C.

One of the key challenges of the development of the prototype was minimizing costs and keeping the MTU easy to deploy with little technical experience, while maximizing the value for trainees. We used off-the-shelf communications software to keep costs low. The challenge with this was that it is developed for high bandwidth; however, the rural or remote locations may not have access to high bandwidth. Setting video quality at low-resolution helped avoid choppy audiovisual transmission but was associated with compromise of fine detail and made assessment of some components of the skill (e.g., suturing) more difficult. Using single camera setups at each of the mentor and remote stations in Session C helped to solve some of the delays seen in Session A when two cameras were used in the remote station.

Feedback on design elements of the training in Session B (i.e., objectives and information provided, support, problem-solving, and feedback) were used to modify the training in Session C. On average, the trainees consistently rated these design elements highly. Furthermore, a comparison of the ratings of these elements provided by trainees in Session C who completed the training remotely versus face-to-face revealed that there were no statistically significant differences on the ratings between these groups. This is important as these design elements can impact the instructional effectiveness and learning outcomes.

Learning outcomes were measured in three ways. All findings support the remote training as comparable to face-to-face training. First, examination of the written procedural skills tests revealed that there were no statistically significant differences between the groups (remote, face-to-face, or control). This indicates that the effects of training methodology (or no training) on skills necessary to perform the chest tube insertion procedure between the groups are not different, with respect to performance on a written test. The ability to physically and capably complete a procedural skill therefore relies on deliberate practice of that skill (Ericsson 2008). There are two distinct key areas of knowledge with respect to competent procedural skills performance; one relating to factual background information and the second being the ability to complete all necessary steps. The second learning outcome measure was self-reported confidence in learning and personal satisfaction with the learning experience. Again, no statistically significant differences were found between the groups in Session C. The third learning outcome measure was self-reported competency level in procedural performance. No statistically significant differences between the groups were found. However, there was a larger increase in the mean self-reported competency rating with the remote group than with the other groups, and the control group had the smallest mean increase in self-reported competency rating. This is encouraging and it would be interesting to investigate this with a larger sample. A statistically significant difference was found between subjective competency level before the hands-on training and after the retention test. These results are encouraging and indicate that self-perceived learning appears to have occurred during the training, and that it did not matter if the training was delivered remotely or face-to-face. These findings are consistent with other studies, which have compared SBME with other instruction, and with no intervention (Ilgen et al. 2013).

The main limitation of the study to this point has been the small sample sizes at each stage of prototype development. This is mitigated by the use of the Delphi method which enabled the inclusion of experts' opinions on potential applications and key design components of the MTU. Another limitation is the use of self-reported learning measures; however, the use of self-reported performance measures is common practice in educational research and such measures tend to be consistent with objective measures (Anaya 1999). Additionally, the use of the NLN scales that have been shown to be reliable and valid, help to alleviate some of these concerns. Further sessions are planned in the evaluation stage of this research to study the MTU with more subjects and objective measures of learning outcomes to enable more robust collection of data and analysis of results.

The next steps are to follow Cycle C of the MRC framework: evaluation. We will evaluate the educational effectiveness of the MTU's use with a larger group of subjects and the application of objective assessments to obtain quantitative data amenable to statistical analysis. These results will allow comparison of the pre, post, and retention tests with respect to learning outcomes. If we find that the learning outcomes from sessions delivered remotely are comparable to face-to-face, then we will proceed with Cycle D, implementing the MTU into broader practice settings. The ultimate goal is the delivery of the simulation-based training remotely through the use of a larger, self-contained vehicle outfitted with simulation equipment necessary for provision of a wider range of scenarios. This will present an opportunity to overcome geographic, cost, and time barriers to emergency medical education provision in rural and remote areas.

Future research will investigate the challenges faced in this study with audiovisual transmission and explore the use of a purpose-built efficient communications system designed for low bandwidth. Elements in the delivery of the training program, such as objectives and information provided during the session, could also be studied. There were no statistically significant differences found between the remote and face-to-face groups on any of these elements; however, several of the elements around the cues and information provided during the session (see Appendix C) were rated lower by the remote than by the face-to-face group. This increase in communication ambiguity as media naturalness decreases is a known issue in remote training (Kock 2005). Ways to reduce this ambiguity could be explored. Additionally, the impact of adding debriefing to the training program on learning outcomes could be examined, as debriefing has been found to be essential to SBME (Cheng et al. 2014). Debriefing could be added following an approach such as the four-step model presented by Rudolph et al. (2008). Future research will also examine possible collaboration between urban and rural clinicians using the MTU to learn from each other with the goal of improving the delivery of care in both regions. The potential delivery of mobile tele-simulation-based training to other medical disciplines could be examined. For example, use of the MTU could enable surgeons to train with their own clinical team, and thus foster the benefits of mobile simulation training, while saving time and money. Furthermore, as with tele-simulation, the MTU could also be beneficial for the remote assessment of skills, especially in domains that are poorly covered by traditional written and oral examinations. Concepts explored in the MTU project also have the potential to be useful for the provision of training in less developed regions of the world. Research on conditions specific to these regions that may impact learning outcomes, such as cultural differences or low bandwidth could be conducted.

## 7 Conclusion

Following a theory-based approach of the MRC framework and the AFT process has helped us to conduct the iterative development and piloting of an MTU prototype targeted to meet the learning needs of emergency health-care providers in rural and remote areas. Designing a complex intervention, such as the MTU, poses substantial challenges to investigators; however, the use of frameworks that harness qualitative and quantitative methods should improve the intervention, study design, and generalizability of results. The MTU prototype has been improved through ongoing evaluation, reflection, and redesign. Feedback to ensure a quality learning experience in the MTU has directed key features of physical design, technical performance and the training program that have been applied in deployment of the unit in each evaluation session. The MTU prototype appears to be an effective means to make quality simulation training on procedural skills more accessible to emergency health-care providers in rural and remote areas, while addressing the challenges of simulation, tele-simulation and mobile simulation. Further evaluation of design, telecommunications, and learning outcomes will help to determine the full potential of the MTU and help to address some of the challenges to equitable health-care delivery by transcending the barriers of distance, time, and costs.

Effective applications of tele-simulation, mobile simulation, and particularly mobile tele-simulation are in their infancy and the opportunities that these platforms provide for innovative training are limitless. Challenges such as inadequate exposure to infrequently encountered medical cases and procedures, and lack of access to SBME may be addressed through the use of these techniques. The MTU in particular may provide advantages to those with limited access to simulation training centers by providing them access to experienced mentors and enhanced quality SBME experiences.

**Acknowledgments** This project has been supported by an Ignite RDC grant awarded by the Research and Development Corporation of Newfoundland and Labrador. Thanks to the following organizations at Memorial University of Newfoundland: the Tuckamore Simulation Research Collaborative (TSRC) for research support and advice, the Clinical Learning and Simulation Center (CLSC) for equipment and operational support, and MUN Med 3D for the provision of simulation models. Thank you to the following people for their assistance during this research project: Research Assistants Megan Pollard, Samantha Noseworthy and Sarah Boyd; Tate Skinner (technical support), Joanne Doyle (Emergency Medicine discipline secretary), and Memorial University's Emergency Medicine Interest Group (EMIG).

## Appendix A – Aim-FineTune-FollowThrough Approach

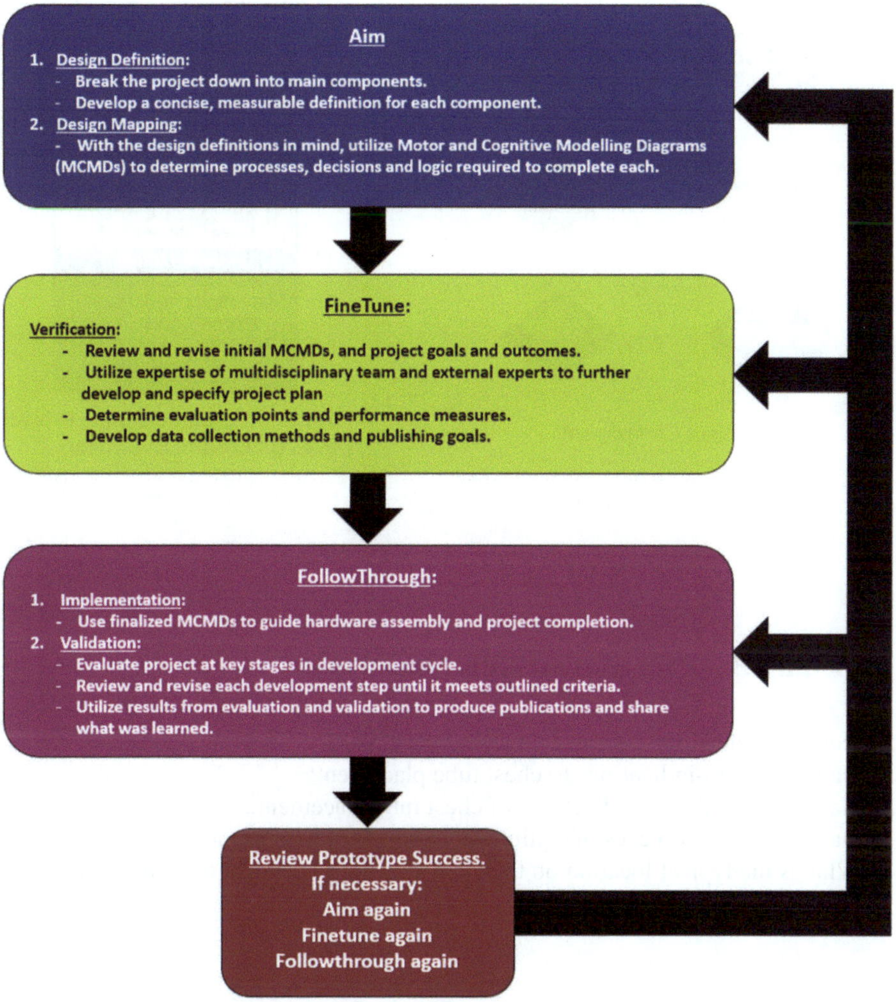

*Sample of MCMD to identify the design factors to consider during prototype development of the MTU*

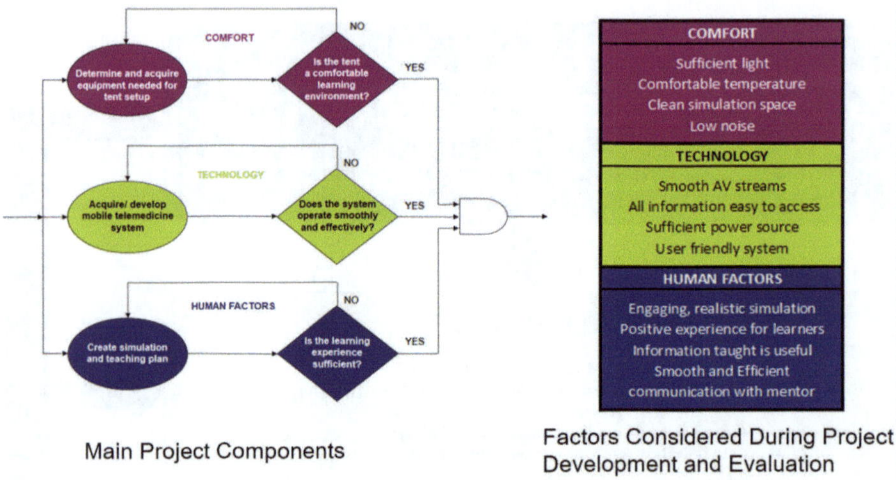


## Appendix B – Procedural Skills Questions

1. Name **3** indications for chest tube placement:
2. Name **2** contra-indications to chest tube placement:
3. Name **4** potential complications of chest tube placement:
4. Name **5** essential pieces of equipment for chest tube placement:
5. What is the typical location on the chest wall for placement of a chest tube?

## Appendix C – Mean and Standard Deviation of MTU Characteristics

| Characteristic | Session A $N = 35$ | Session B $N = 6$ | Session C $N = 6$ (Remote) | Session C $N = 6$ (Face-to-face) | Mann-Whitney U test Session C remote vs. Face-to-face |
|---|---|---|---|---|---|
| *Design features of MTU* | | | | | |
| Well organized | 4.00 (0.594) | 4.33 (0.516) | 4.33 (0.816) | 4.83 (0.408) | |
| Good lighting/brightness | 3.94 (0.873) | 4.50 (0.548) | 4.50 (0.548) | 4.50 (0.548) | |
| Low noise | 4.23 (0.646) | 2.67 (1.211) | 3.5 (1.378) | 4.67 (0.516) | |
| Adequate space | 3.89 (0.867) | 4.33 (0.516) | 4.67 (0.516) | 4.67 (0.516) | |
| *Function of telecommunications* | | | | | |
| Camera setup/location | 4.17 (1.465) | 4.00 (1.095) | 4.17 (0.408) | N/A | |
| Audio | 4.09 (0.853) | 2.83 (1.169) | 3.33 (0.816) | N/A | |
| *Satisfied with MTU* | 3.90 (0.746) | 4.00 (0.632) | 4.50 (0.548) | 4.5 (0.837) | |
| *Recommend MTU* | 4.09 (0.712) | 4.67 (0.516) | 4.50 (0.548) | 4.17 (1.169) | |
| **Design elements of the training session** | | | | | |
| Adapted from the NLN Simulation Design Scale (NLN 2005) | | | | | |
| *Objectives and information* | | | | | |
| There is enough information provided before the session to provide direction and encouragement. | | 3.83 (1.169) | 3.67 (0.816) | 4.17 (1.329) | $U = 70$, $z = 1.158$, $p = 0.310$ |
| I clearly understood the purpose and objectives of the session. | | 4.17 (0.408) | 4.33 (0.516) | 4.50 (0.837) | $U = 59$, $z = 0.371$, $p = 0.770$ |
| The session provided enough information in a clear matter for me to problem-solve the situation. | | 3.83 (0.983) | 3.67 (0.816) | 4.17 (0.408) | $U = 65.5$, $z = 1.01$, $p = 0.454$ |
| I learn from the comments made by the teacher before, during, or after the simulation. | | 3.5 (1.049) | 4.17 (0.408) | 4.17 (0.753) | $U = 66.5$, $z = 0.962$, $p = 0.415$ |
| The cues are appropriate and geared to promote my understanding. | | 3.67 (1.033) | 3.83 (0.408) | 4.5 (0.548) | $U = 80.5$, $z = 2.142$, $p = 0.077$ |

(continued)

| Characteristic | Session A $N = 35$ | Session B $N = 6$ | Session C $N = 6$ (Remote) | Session C $N = 6$ (Face-to-face) | Mann-Whitney U test Session C remote vs. Face-to-face |
|---|---|---|---|---|---|
| There is enough information provided to me during the session. | | 3.67 (1.033) | 3.83 (0.408) | 4.50 (0.548) | $U = 80.5$, $z = 2.142$, $p = 0.077$ |
| *Support* | | | | | |
| My need for help was recognized. | | 3.67 (0.816) | 4.33 (0.516) | 4.33 (0.516) | $U = 60$, $z = 0.468$, $p = 0.721$ |
| I felt supported by the teacher's assistance during the session. | | 3.83 (0.983) | 4.17 (0.753) | 4.17 (0.753) | $U = 65.5$, $z = 0.853$, $p = 0.454$ |
| *Problem-solving* | | | | | |
| Independent problem-solving was facilitated. | | 3.67 (1.033) | 3.67 (0.516) | 4.33 (0.816) | $U = 80$, $z = 1.922$, $p = 0.090$ |
| *Feedback* | | | | | |
| Feedback provided was constructive. | | 3.83 (0.753) | 3.83 (0.408) | 4.00 (0.894) | $U = 74$, $z = 1.459$, $p = 0.199$ |
| Feedback was provided in a timely manner. | | 3.83 (0.983) | 3.83 (0.408) | 3.83 (0.753) | $U = 74.5$, $z = 1.551$, $p = 0.177$ |
| The session allowed me to analyze my own behavior and actions. | | 3.5 (1.049) | 4.00 (0.632) | 4.33 (0.816) | $U = 66.5$, $z = 0.905$, $p = 0.415$ |
| There are enough opportunities in the session to find out if I clearly understand the material. | | 3.17 (1.169) | 3.50 (0.837) | 4.00 (0.632) | $U = 71$, $z = 1.245$, $p = 0.280$ |
| **Learning outcomes** | | | | | |
| Adapted from the NLN Student Satisfaction and Self-Confidence in Learning scales (NLN 2005) | | | | | |
| *Satisfaction with learning* | | | | | |
| The teaching methods used were helpful and effective. | | 4.00 (0.632) | 4.83 (0.408) | 4.50 (0.548) | $U = 42$, $z = -0.979$, $p = 0.454$ |
| I enjoyed how the teacher taught the session. | | 3.50 (1.225) | 4.67 (0.516) | 4.67 (0.516) | $U = 47$, $z = -0.539$, $p = 0.673$ |
| *Self-confidence in learning* | | | | | |
| I am confident that I am developing the skills and obtaining the knowledge needed to understand this procedure. | | 3.17 (1.169) | 4.17 (0.408) | 4.33 (0.516) | $U = 54.5$, $z = 0.044$, $p = 0.965$ |

## References

Anaya, G. (1999). College impact on student learning: Comparing the use of self-reported gains, standardized test scores, and college grades. *Research in Higher Education, 40*(5), 499–526. Retrieved from http://www.jstor.org/stable/40196370.

Bischof, J. J., Panchal, A. R., Finnegan, G. I., & Terndrup, T. E. (2016). Creation and validation of a novel mobile simulation laboratory for high fidelity, prehospital, difficult airway simulation. *Prehospital and Disaster Medicine, 31*, 465–470. 1–6.

Burckett-St Laurent, D. A., Cunningham, M. S., Abbas, S., Chan, V. W., Okrainec, A., & Niazi, A. U. (2016). Teaching ultrasound guided regional anesthesia remotely: A feasibility study. *Acta Anaesthesiologica Scandinavica, 60*, 995–1002.

Campbell, M., Fitzpatrick, R., Haines, A., et al. (2000). Framework for design and evaluation of complex interventions to improve health. *BMJ, 321*, 694Y696.

Casey, M. M., Wholey, D., & Moscovice, I. S. (2008). Rural emergency department staffing and participation in emergency certification and training programs. *The Journal of Rural Health, 24*(3), 253–262.

Cheng, A., Eppich, W., Grant, V., Sherbino, J., Zendejas, B., & Cook, D. A. (2014 Jul 1). Debriefing for technology-enhanced simulation: A systematic review and meta-analysis. *Medical Education, 48*(7), 657–666.

Choy, I., Fesco, A., Kwong, J., Jackson, T., & Okrainec, A. (2013). Remote evaluation of laparoscopic performance using the global operative assessment of laparoscopic skills. *Surgical Endoscopy, 27*, 378–383.

Cook, D. A., Hatala, R., Brydges, R., Zendejas, B., Szostek, J. H., Wang, A. T., Erwin, P. J., & Hamstra, S. J. (2011). Technology-enhanced simulation for health professions education: A systematic review and meta-analysis. *Journal of the American Medical Association, 306*(9), 978–988.

Cordray, D. S., & Pion, G. M. (2006). Treatment strength and integrity: Models and methods. In R. R. Bootzin & P. E. McKnight (Eds.), *Strengthening research methodology: Psychological measurement and evaluation* (pp. 103–124). Washington, DC: American Psychological Association. xix, 299 pp.

Cristancho, S. M., Moussa, F., & Dubrowski, A. (2011). A framework-based approach to designing simulation-augmented surgical education and training programs. *The American Journal of Surgery, 202*(3), 344–351.

Cristancho, S., Moussa, F., & Dubrowski, A. (2012). Simulation-augmented training program for off-pump coronary artery bypass surgery: Developing and validating performance assessments. *Surgery, 151*(6), 785–795.

Dev, S. P., Nascimiento, B., Jr., Simone, C., & Chien, V. (2007). Chest-tube insertion. *New England Journal of Medicine, 357*(15), e15.

Dunn, O. J. (1964). Multiple comparisons using rank sums. *Technometrics, 6*, 241–252.

Dunne, C., Jewer, J., & Parsons, M. (2018). *Application of the Delphi method to refine key components in the iterative development of a mobile tele-simulation unit (MTU)* (p. 039). Calgary, AB: Canadian Association of Emergency Physicians Conference. May 2018.

Ericsson, K. A. (2008). Deliberate practice and acquisition of expert performance: A general overview. *Academic Emergency Medicine, 15*, 988–994.

Franklin, A. E., Burns, P., & Lee, C. S. (2014). Psychometric testing on the NLN student satisfaction and self-confidence in learning, simulation design scale, and educational practices questionnaire using a sample of pre-licensure novice nurses. *Nurse Education Today, 34*(10), 1298–1304.

Greene, C. J., Morland, L. A., Durkalski, V. L., & Frueh, B. C. (2008). Noninferiority and equivalence designs: Issues and implications for mental health research. *Journal of Traumatic Stress, 21*(5), 433–439.

Haji, F. A., Da Silva, C., Daigle, D. T., & Dubrowski, A. (2014). From bricks to buildings: Adapting the medical research council framework to develop programs of research in simulation education and training for the health professions. *Simulation in Healthcare, 9*(4), 249–259.

Henao, O., Escallon, J., Green, J., et al. (2013). Fundamentals of laparoscopic surgery in Columbia using telesimulation: An effective educational tool for distance learning. *Biomédica, 33*, 107–114.

Hunt, E. A., Walker, A. R., Shaffner, D. H., Miller, M. R., & Pronovost, P. J. (2008). Simulation of in-hospital pediatric medical emergencies and cardiopulmonary arrests: Highlighting the importance of the first 5 minutes. *Pediatrics, 121*(1), e34–e43.

Hutton, I. A., Kenealy, H., & Wong, C. (2008). Using simulation models to teach junior doctors how to insert chest tubes: A brief and effective teaching module. *Internal Medicine Journal, 38*(12), 887–891.

Ikeyama, T., Shimizu, N., & Ohta, K. (2012). Low-cost and ready-to-go remote-facilitated simulation-based learning. *Simulation in Healthcare, 7*(1), 35–39.

Ilgen, J. S., Sherbino, J., & Cook, D. A. (2013). Technology-enhanced Simulation in Emergency Medicine: A Systematic Review and Meta-Analysis. *Academic Emergency Medicine., 20*(2), 117–127.

Ireland, S., Gray, T., Farrow, N., Danne, P. D., & Flanagan, B. (2006). Rural mobile simulation-based trauma team training-an innovative educational platform. *International Trauma Care, 16*, 6–12.

Issenberg, S. B., McGaghie, W. C., Petrusa, E. R., Lee Gordon, D., & Scalese, R. J. (2005). Features and uses of high-fidelity medical simulations that lead to effective learning: A BEME systematic review. *Medical Teacher, 27*(1), 10–28.

Jewer, J., Dubrowski, A., Hoover, K., Smith, A., & Parsons, M. (2018). Development of a mobile tele-simulation unit prototype for training of rural and remote emergency health care providers. In *Proceedings of the 51st Hawaii International Conference on System Sciences*.

Kock, N. (2005). Media richness or media naturalness? The evolution of our biological communication apparatus and its influence on our behavior toward e-communication tools. *IEEE Transactions on Professional Communication, 48*(2), 117–130.

McCoy, C. E., Sayegh, J., Alrabah, R., & Yarris, L. (2017). Telesimulation: An innovative tool for health professions education. *AEM Education, Training, 1*, 132–136.

McGaghie, W. C., Issenberg, S. B., Petrusa, E. R., & Scalese, R. J. (2010). A critical review of simulation-based medical education research: 2003–2009. *Medical Education, 44*(1), 50–63.

Mikrogianakis, A., Kam, A., Silver, S., Bakanisi, B., Henao, O., Okrainec, A., & Azzie, G. (2011). Telesimulation: An innovative and effective tool for teaching novel intraosseous insertion techniques in developing countries. *Academic Emergency Medicine, 18*(4), 420–427.

National League for Nursing. (2005). Retrieved from http://www.nln.org/professional-development-programs/research/tools-and-instruments/descriptions-of-available-instruments.

Ohta, K., Kurosawa, H., Shiima, Y., Ikeyama, T., Scott, J., Hayes, S., Gould, M., Buchanan, N., Nadkarni, V., & Nishisaki, A. (2017). The effectiveness of remote facilitation in simulation-based pediatric resuscitation training for medical students. *Pediatric Emergency Care, 33*, 564–569.

Okrainec, A., Smith, L., & Azzie, G. (2009). Surgical simulation in Africa: The feasibility and impact of a 3-day fundamentals of laparoscopic surgery course. *Surgical Endoscopy, 23*, 2493–2498.

Okrainec, A., Henao, O., & Azzie, G. (2010). Telesimulation: An effective method for teaching the fundamentals of laparoscopic surgery in resource-restricted countries. *Surgical Endoscopy, 24*, 417–422.

Okrainec, A., Vassiliou, M., Kapoor, A., Pitzul, K., Henao, O., Kaneva, P., Jackson, T., & Ritter, E. M. (2013). Feasibility of remote administration of the Fundamentals of Laparoscopic Surgery (FLS) skills test. *Surgery Endoscopy, 27*, 4033–4037.

Parsons, M., Wadden, K., Pollard, M., Dubrowski, A., & Smith, A. (2016a). P098: Development and evaluation of a mobile simulation lab with acute care telemedicine support. *CJEM, 18*(S1), S111–S111.

Parsons, M., Wadden, K., Pollard, M., Dubrowski, A., & Smith, A. (2016b). Mobile simulation lab with acute care tele-medicine support. In *Medical education scholarship forum proceedings* (Vol. 3). Available at: Cureus: Archives of Scholarship in Simulation and Educational Techniques-Annual Collections 2016. http://www.cureus.com/posters/1118.

Parsons, M., Smith, A., Hoover, K., Jewer, J., Noseworthy, S., Pollard, M., Dunne, C., & Dubrowski, A. (2017a). P100: Iterative prototype development of a mobile tele-simulation unit for remote training: An update. *CJEM, 19*(S1), S112–S112.

Parsons, M., Smith, A., Rogers, P., Hoover, K., Pollard, M., Dubrowski, A. (2017b). *Outcomes of prototype development cycle for a mobile simulation lab with acute care telemedicine support – work in progress*. 17th international meeting on simulation in healthcare, Orlando, FL.

Pena, G., Altree, M., Babidge, W., Field, J., Hewett, P., & Maddern, G. (2015). Mobile Simulation Unit: Taking simulation to the surgical trainee. *ANZ Journal of Surgery, 85*(5), 339–343.

Riley, W., Davis, S., Miller, K., Hansen, H., Sainfort, F., & Sweet, R. (2011). Didactic and simulation nontechnical skills team training to improve perinatal patient outcomes in a community hospital. *Joint Commission Journal on Quality and Patient Safety, 37*(8), 357–364.

Rogers, F. B., Shackford, S. R., Osler, T. M., et al. (1999). Rural trauma: The challenge for the next decade. *The Journal of Trauma, 47*, 802–821. https://doi.org/10.1097/00005373-199910000-00038.

Rosen, M. A., Hunt, E. A., Pronovost, P. J., Federowicz, M. A., & Weaver, S. J. (2012). In situ simulation in continuing education for the health care professions: A systematic review. *The Journal of Continuing Education in the Health Professions, 32*, 243–254. https://doi.org/10.1002/chp.21152.

Roy, K. M., Miller, M. P., Schmidt, K., & Sagy, M. (2011). Pediatric residents experience a significant decline in their response capabilities to simulated life-threatening events as their training frequency in cardiopulmonary resuscitation decreases. *Pediatric Critical Care Medicine, 12*(3), e141–e144.

Rudolph, J. W., Simon, R., Raemer, D. B., & Eppich, W. J. (2008). Debriefing as formative assessment: Closing performance gaps in medical education. *Academic Emergency Medicine, 15*(11), 1010–1016.

Scott, D. J., & Dunnington, G. L. (2008). The new ACS/APDS skills curriculum: Moving the learning curve out of the operating room. *Journal of Gastrointestinal Surgery, 12*, 213–221.

Small, S. D., et al. (1999). Demonstration of high-fidelity simulation team training for emergency medicine. *Academic Emergency Medicine, 6*, 312–323.

Steinemann, S., Berg, B., Skinner, A., et al. (2011). In situ, multidisciplinary, simulation-based teamwork training improves early trauma care. *Journal of Surgical Education, 68*(6), 472–477.

Treloar, D., Hawayek, J., Montgomery, J. R., & Russell, W. (2001). Onsite and distance education of emergency medicine personnel with a human patient simulator. *Military Medicine, 166*, 1003–1006.

Ullman, E., Kennedy, M., Di Delupis, F. D., Pisanelli, P., Burbui, A. G., Cussen, M., Galli, L., Pini, R., & Gensini, G. F. (2016). The Tuscan Mobile Simulation Program: A description of a program for the delivery of in situ simulation training. *Internal and Emergency Medicine, 11*(6), 837–841.

von Lubitz, D. K., Carrasco, B., Gabbrielli, F., et al. (2003). Transatlantic medical education: Preliminary data on distance based high-fidelity human patient simulation training. *Studies in Health Technology and Informatics, 94*, 379–385.

Weinstock, P. H., Kappus, L. J., Garden, A., & Burns, J. P. (2009). Simulation at the point of care: Reduced-cost, in situ training via a mobile cart. *Pediatric Critical Care Medicine, 10*(2), 176–181.

Williams, J. M., Ehrlich, P. F., & Prescott, J. E. (2001). Emergency medical care in rural America. *Annals of Emergency Medicine, 38*(3), 323–327.

Xafis, V., Babidge, W., Field, J., Altree, M., Marlow, N., & Maddern, G. (2013). The efficacy of laparoscopic skills training in a Mobile Simulation Unit compared with a fixed site: A comparative study. *Surgical Endoscopy, 27*(7), 2606–2612.

Ziv, A., Wolpe, P. R., Small, S. D., & Glick, S. (2003). Simulation based medical education: An ethical imperative. *Academic Medicine, 78*, 783–788.

# Drone Delivery Models for Medical Emergencies

**Judy E. Scott and Carlton H. Scott**

## 1 Introduction

Medical emergencies have typically involved transporting the patient to a hospital in an ambulance. However, in some circumstances, an ambulance cannot access the patient in a timely manner. For example, the roads are nonexistent, congested, or damaged by severe weather or natural disasters. When ground transport is compromised, transport by air could be an alternative. However, airplanes need an airport, and helicopters are usually prohibitively expensive and may not be available. Instead of flying a patient to a hospital, a drone could potentially deliver urgently needed lifesaving medications, medical equipment, blood, vaccines, or organs for a transplant, to the patient. The speed and cost of the drone delivery are likely to be favorable and more so as drone technology improves.

Drones are devices which are capable of sustained flight, which do not have a human on board, and are under sufficient control to perform useful functions (Clarke 2014; Villasenor 2012). Drones are commonly known as unmanned aerial vehicles (UAVs) (Choi-Fitzpatrick et al. 2016). Alternative terms include Unmanned Aircraft (UA), Remotely Operated Aircraft (ROA), Remotely Piloted Vehicle (RPV), and Remotely Piloted Aircraft (RPA) (Clarke 2014).

Although the definition of drones is complex because of the diversity of characteristics, the main types of drones are fixed-wing, multi-rotor, and hybrid (Wurbel 2017). Each of these types of drones has strengths and weaknesses. For

J. E. Scott (✉)
University of Colorado Denver, Denver, CO, USA
e-mail: judy.scott@ucdenver.edu

C. H. Scott
University of California Irvine, Irvine, CA, USA
e-mail: chscott@uci.edu



N. Wickramasinghe, F. Bodendorf (eds.), *Delivering Superior Health and Wellness Management with IoT and Analytics*, Healthcare Delivery in the Information Age, https://doi.org/10.1007/978-3-030-17347-0_3

example, fixed-wing drones look like small airplanes and travel faster and further than multi-rotor drones, which look more like small helicopters. On the other hand, multi-rotor drones are capable of vertical take-off and landing whereas fixed-wing drones need a landing strip and take-off using a catapult. Other advantages of the multi-rotor drones include generally lower cost, the ability to replace the battery on turn around and deliver items on the return journey. By contrast, the fixed-wing drones drop supplies by parachute and do not pick up supplies or replace the battery for the return trip. The hybrid drones combine the advantages of the other two types.

Recent innovations have taken place in drone-specific hardware, software, and networks. For example, light composite materials and global positioning systems (GPS) enable efficient flight. Furthermore, lithium batteries are rapidly improving so drones can fly further on a charge (Raffaello 2014). Drone software can use mobile phone or tablet apps for tracking and navigation (French 2015; Khazan 2016; Raptopoulos 2013). The drone operating system manages the network by monitoring weather data from all the ground stations and optimizing the routes of the drones (Raptopoulos 2013). The routes need to avoid both adverse weather conditions and other risk factors. Also, an on board webcam can enable communication with a control center (Prigg 2014).

Many drone applications involve surveillance using an on board camera. However, drones also are capable of carrying devices other than cameras and capable of delivering small loads. Drones have been used extensively by the military in combat and for humanitarian aid. Useful nonmilitary drone applications in different industries include agriculture surveillance and crop spraying, shark surveillance at beaches, monitoring wildlife for conservation, monitoring fires, scientific research and exploration, monitoring riots and international borders by police and governments, sports and entertainment event coverage, other media coverage, emergency services, and disaster response (Choi-Fitzpatrick et al. 2016; Clarke 2014; Esler 2015).

Arguably, humanitarian drone applications are the most useful since human lives are at stake. For example, in Nepal, after the 2015 earthquake, drones helped rescuers locate survivors (Choi-Fitzpatrick et al. 2016). Furthermore, a study used experiments to model automated response to a scenario of a San Francisco earthquake (Mosterman et al. 2014). The automated system included fixed-wing and rotorcraft drones integrated with autonomous ground vehicles.

In contrast to humanitarian drone applications, drones have been used for crimes, such as delivering contraband to prison inmates, firing weapons, terrorism, and hacking (Choi-Fitzpatrick et al. 2016). Also drones have been criticized for inadequate regulation, safety issues, and security and privacy abuse (Balasingam 2017; Clarke 2014; Welch 2015). Celebrities and others are concerned that drones will spy on them. Furthermore, irresponsible drone owners have been a nuisance in cases related to photographing accidents or fires and interfering with emergency responders (Choi-Fitzpatrick et al. 2016). Other concerns are that people would shoot down drones and steal drone packages (Welch 2015).

Regulatory bodies, such as the Federal Aviation Administration (FAA), usually ban commercial drones because they are wary of collisions in airline air space.

Nevertheless, there have been reports of hundreds of collision close calls, many due to noncommercial drones (Esler 2015). Since December 2015, drones weighing between 0.55 and 55 pounds need to be registered with the FAA and marked with the registration number, so owners can be identified if there is an accident or criminal use of the drone (Choi-Fitzpatrick et al. 2016). Owners need to follow FAA restrictions, which currently include only daylight hours, within line of sight, and not over people.

In the US, commercial drone use is illegal without a Section 333 exemption (French 2015). FAA exemptions have been granted for surveillance applications, for example, BP pipelines, rather than delivery. This has been frustrating to Amazon and other companies that are planning drone delivery (Welch 2015). Regulation is an important limitation to the acceptance of drone healthcare delivery. As discussed, Flirtey has had FAA approval for dropping medical supplies in rural Virginia with drones (Hackman and Nicas 2015; Pepitone 2015). However, the approval was only for a weekend. In contrast, countries with urgent healthcare needs, such as Rwanda, are more likely to quickly overcome regulatory hurdles (French 2015).

Worthwhile drone functions include delivery of small items that are urgently needed in locations with difficult access. Critical healthcare drone delivery applications include timely delivery of medications, blood, and vaccines. Despite the urgency, locations needing the delivery may have difficult access due to poor transportation infrastructure, or roads blocked by severe weather, disasters, or traffic congestion. Drones are not dependent on a well-developed road infrastructure because they fly to their destination. Since a drone can fly over an inaccessible road, innovative organizations have begun to use drones for healthcare delivery.

While modeling drone delivery has mainly focused on parcel delivery, this chapter looks at drone delivery in a healthcare context. The next section overviews drone applications for medical emergencies. In section three, we provide a review of documented examples of drone healthcare delivery for medical emergencies, followed in Models for Drone Delivery by a review of previous modeling work in drone delivery. In the final section, we discuss concluding remarks.

## 2 Drone Applications for Medical Emergencies

Medical emergencies are varied. In more developed countries, heart attacks and asthma attacks are common. In less developed countries, vaccines are desperately needed. Blood is urgently needed after an accident or injuries associated with disasters anywhere. When the medical emergency is in a remote place, such as a mountainous region or island, then delivery of healthcare items by drone could be faster than other modes of transportation and save lives. Examples of drone healthcare delivery include over water, such as rivers or seas to islands; bypassing poor or damaged roads from disasters or extreme weather; or avoiding traffic congestion.

**Table 1** Medical emergencies and drones

| | Medical emergency | Drone | Where | Why |
|---|---|---|---|---|
| Medicine | Natural disasters | Matternet | Switzerland | Save lives after earthquakes, hurricanes, etc. |
| Blood | Hemorrhaging after childbirth | Zipline | Rwanda | Save lives with blood transfusions |
| Vaccines | Rabies Malaria | Zipline | Rwanda, Madagascar and Tanzania | Help prevent up to 1.5 million deaths each year (Wurbel 2017) |
| Defibrillators | Heart attacks | "Ambulance drone"; Flirtey | Amsterdam, Stockholm, Toronto, Reno | Survival rate potentially improves from 8% to 80%. |
| Inhaler | Asthma attack | CyPhy Works | Children's Island, Massachusetts | Save lives by fast response to remote locations |

Drone applications in healthcare include delivery of medicine, defibrillators, blood samples, and vaccines (Agatz et al. 2015; Choi-Fitzpatrick et al. 2016; DHL 2016; Hackman and Nicas 2015; Haidari et al. 2016; Khazan 2016; Pepitone 2015; Preimesberger 2016; Prigg 2014; Raptopoulos 2013; Tilley 2016; Varnholt 2016) (Table 1).

Improvements in global vaccination coverage would help prevent up to 1.5 million deaths each year (Wurbel 2017). One of the challenges is providing access in the last few miles where the road infrastructure is limited or nonexistent. Vaccine deliveries in Rwanda, Madagascar, and Tanzania (Haidari et al. 2016; Hotz 2017; Wurbel 2017) often use drones to complement other forms of transport in the supply chain, such as motorcycles and trucks (Wurbel 2017).

## 3 Examples of Drones for Medical Emergencies

In this section, we discuss examples of drone applications for medical emergencies. Drone companies that are providing or testing service for medical emergencies include Matternet, DHL Parcel, CyPhy Works, Zipline, Flirty, and Deft University.

### *3.1 Matternet*

Autonomous drones, such as those employed by Matternet, use GPS and other sensors to navigate between automated ground stations in order to deliver medications in remote locations that lack adequate roads (Raptopoulos 2013). Matternet has delivered medications in Haiti following the 2010 earthquake and in the Dominican Republic (Choi-Fitzpatrick et al. 2016), as well as in Bhutan, New Guinea (French 2015; Peters 2017). The company works with UNICEF and Doctors without Borders.

These Matternet drones can carry 1–2 kg (2.2–4.4 pounds) and transport items about 10 km, traveling up to 40 km per hour, taking about 18 minutes including lift off and landing (French 2015; Peters 2017; Raffaello 2014; Raptopoulos 2013). A smartphone application enables senders to choose from a list of possible destinations. The drone then automatically generates a route based on the terrain, weather, airspace, and population density. The route will avoid airports, schools, and public squares, as well as hills and buildings. A parachute will deploy if the drone malfunctions.

Matternet also has begun the first urban network of drone base stations in Switzerland's hospitals and labs (Peters 2017). Drone delivery, which facilitates urgent delivery of blood and tissue samples, is more cost-effective than ground transportation, saving 20% to 50%. This form of delivery also potentially saves lives by taking 3 minutes instead of 25 minutes. The drones are quadcopters with duplicate sensors and autopilot systems on board. If all on-board electronics fail, the drones release a parachute and land (Kolodny 2017b). Matternet expects to lease each of its drones for $2000 a month and charge the same rate for the vehicle's automated base station (Hotz 2017). The drones fly from and land on the base station where batteries can be recharged between flights. The base station scans the package in a shoebox-sized storage container using a QR reader and loads it automatically into the drone. The base also serves as a docking port and transmits an infrared signal to ensure a pinpoint landing. The Matternet M2 drone can travel about 12 miles (20 km) at 22 miles per hour (36 km/hour) while carrying around 4 pounds (Stewart 2017). The drones use the same airspace as emergency helicopters and constantly broadcast their locations.

## *3.2 DHL Parcel*

In Germany, DHL Parcel has researched three generations of medical drone delivery called Parcelcopter (DHL 2016). The first generation traveled 1 km to deliver blood samples across the Rhine at Bonn. The second generation tested drone delivery of medications and other urgently needed material for three months in 2014 to Juist, one of Germany's remote North Sea Islands, traveling 12 km across open sea (Agatz et al. 2015; Varnholt 2016). From January to March 2016, DHL's third generation Parcelcopter tested delivery of over 130 parcels of urgently needed medicines or sporting goods between automated Skyports in two Bavarian Alpine villages. Drone delivery took 8 minutes compared to a 30 minute road trip in winter (DHL 2016). The time difference could be significant in a medical emergency.

### 3.3 *CyPhy Works*

UPS has partnered with CyPhy Works to do a drone test delivery of an asthma inhaler to a remote summer camp on Children's Island (24/7Staff 2016). The 8-pound 40-inch drone launched from the ground at CyPhy Works' Beverly, Massachusetts headquarters and landed on the ground at Children's Island three miles off the Atlantic coast. The inhaler which was inside a capsule was released by the drone after it landed (Quittner 2016). The trip took 8 minutes whereas a boat trip would take 30 minutes. The time saved could be critical if a child was having a severe asthma attack.

### 3.4 *Zipline*

UPS is also working with Zipline on a drone network to deliver vaccines and blood to 20 clinics in remote locations in Rwanda (Khazan 2016; Preimesberger 2016; Tilley 2016). Fifteen Zipline all-weather battery-powered drones logged 2000 medical flights in 2017 (Hotz 2017). One-third of the flights have been emergency deliveries of blood for women hemorrhaging after childbirth or for accident victims. Malaria, infant deaths, and mothers dying in childbirth are common in Rwanda. When rabies vaccine is needed urgently, drone delivery is not hindered by washed out roads during rainy season. Only a third of Africans live within two kilometers of a road that functions year round (Khazan 2016). Zipline is preparing to open a second drone base in Rwanda, bringing its fleet there to 60 drones (Hotz 2017). In 2018, Zipline will open four bases in Tanzania, where it expects to operate 120 autonomous aircraft to supply blood, vaccines, and other medical supplies to 1000 clinics that together serve about ten million people (Hotz 2017; Landhuis 2017).

Zipline drones are launched from a nest and make a delivery by dropping items with a paper parachute. After the drone returns to the nest, a sim card and new battery are inserted along with the blood or vaccines for the next delivery. Zipline 30-pound drones are the size of a large dog and can carry three pounds (Khazan 2016; Landhuis 2017). They can fly at 68 mph to health centers up to 50 miles away (Landhuis 2017). Their route is tracked and changed with a tablet app and costs about the same amount as delivery using traditional road vehicles.

### 3.5 *Flirtey*

The first FAA-approved drone delivery successfully dropped medical supplies to a health clinic in rural southwest Virginia (Pepitone 2015). The clinic services about 3000 patients one weekend each year (Hackman and Nicas 2015). Flirtey drones delivered prescription items from the Wise County Regional airport to the clinic in the remote fairgrounds in about 3 minutes. Delivery is usually 90 minutes along a winding bumpy road from the pharmacy in Oakwood 35 miles away. An experimental National Aeronautics and Space Administration (NASA) manned drone delivered supplies from Oakwood to the Wise County airport since Flirtey drone batteries at that time were limited to about 20 miles. Flirtey drones have also delivered healthcare items in Nevada, Australia, and New Zealand.

Another first for Flirtey was a ship-to-shore drone delivery demonstration (Knight 2016). Partnering with the John Hopkins University School of Medicine and the nonprofit Field Innovation Team (FIT), the drones carried medical samples for emergency testing, between an onshore medical relief camp at Cape May, N.J., and a test facility on a vessel off the New Jersey coast. The drones also transported medical supplies from the vessel to the onshore medical camp.

Flirtey drones are partnering with REMSA Health, a provider of ambulance and emergency health services in the state of Nevada, to dispatch a drone to deliver a portable automatic external defibrillator (AED) whenever a 911 caller in the area reports symptoms of cardiac arrest (Kolodny 2017a). The drones will launch from locations that are owned and operated by Flirtey's commercial clients and health partners. After the drone lowers a defibrillator on a line, a bystander will unbox it and apply the pads inside to a patient's chest. The pads contain sensors that read what's happening with a patient's heart, and the system determines whether the patient needs an electrical shock. The defibrillator can either prompt a user to press a button or automatically deliver a shock after a brief warning. REMSA and Flirtey are working with local regulators in Nevada and targeting a start date in early 2018 for defibrillator deliveries. They are starting in Northern Nevada because the state is an FAA-approved drone test site.

All this testing has been on the outskirts of Reno-Sparks (Hidalgo 2018). Partners have been added including City of Reno, City of Sparks, Reno Police Department, Reno Fire Department, Washoe County, AirMap, and tribal government which has less access to healthcare. AirMap will help Flirtey integrate into the national airspace. Future plans include delivery of EpiPens for relief of asthma attacks and Narcan for opioid overdose.

## *3.6 Delft University "Ambulance Drone"*

Alec Mormot began work on developing the first prototype with a defibrillator as a graduate industrial design student at TU Delft University in Holland (Knight 2016). The prototype ambulance drones, traveling at speeds of up to 60 miles per hour, can reach patients within a 4.6 square mile radius in a minute versus an average of 10 minutes for traditional emergency services, increasing the chance of survival to 80% from 8% (Balasingam 2017; Claesson et al. 2016; Momont 2014; Prigg 2014; Van de Voorde et al. 2017). The drones track emergency mobile calls and use GPS to navigate. A paramedic, from a control room connected to a livestream web camera on the drone, can instruct a lay person assisting the patient.

**Table 2** Comparison of drone delivery for medical emergencies

| Drone company | Healthcare items | Delivery location | Launching pad | Delivery method | Payload | Range | Speed |
|---|---|---|---|---|---|---|---|
| CyPhy Works | Asthma inhaler | Children's Island, Massachusetts | Ground | Ground landing | 4 lb | >3 miles | 30 mph |
| Delft University | Defibrillators | Netherlands | Hospital, clinic | Ground landing | 4 kg (8.8 lb) | 12 km (7.5 miles) | 60 mph |
| DHL Parcel | Blood, medications | Germany | Automated Skyport | Automated Skyport | 4.4 lb | 12 km (7.5 miles) | >40 mph |
| Flirtey | Medications, defibrillators | Virginia, Nevada | Airport, medical offices | Dropped by rope | 2 kg (4.4 lb) | 20 miles | ? |
| Matternet | Blood, medications | Haiti, Dominican Republic, Papua New Guinea, Switzerland | Automated ground station | Automated ground station | 2 kg (4.4 lb) | 10 km (6.25 miles) | 40 kmph 25 mph |
| Zipline | Blood, vaccines | Rwanda | Nest (catapult) | Paper parachute | 3 lb | 45 miles | 90 mph |

An emergency simulation study of drone defibrillator delivery shows the importance of device-specific customization of emergency operator instruction (Fleck 2016). It was expected to take five years to develop an operational emergency drone network and address legal issues and improve the steering on the ambulance drones which are expected to cost $19,000 each (Prigg 2014). Although regulation issues have slowed progress, researchers at the University of Toronto and Stockholm are developing models for drone delivery of defibrillators (Claesson et al. 2016; Van de Voorde et al. 2017). There is potential for "flying medical toolbox" drones to carry other healthcare devices, such as oxygen masks or insulin injections for diabetes patients (Prigg 2014).

Table 2 shows that the healthcare items transported by the drones are predominantly medications, blood, and vaccines. Potentially, defibrillators, oxygen, and insulin could be transported by the ambulance drone, which is only a prototype currently in the Netherlands. However, testing is underway for defibrillator delivery by Flirtey in Nevada.

A drone's launching pad is often automated and called a ground station, Skyport or nest. The delivery method can be the same, as in the case of Matternet and DHL Parcel or the load can be dropped as in the case of Zipline, by a paper parachute, and Flirtey dropped by rope.

Table 2 also compares the payload, range, and speed for the different drone delivery platforms. Matternet, DHL, and Flirtey all carry about the same payload. Zipline carries a smaller payload but is much faster than the others because of its streamlined fixed-wing plane design versus the helicopter-like design of the others. The range varies from Matternet's 6.5 miles to Zipline's 45 miles. Note that these numbers are based on reported information in articles and websites and may not reflect current capabilities. Also, as stated earlier, technology, such as the duration of the charge of lithium batteries, is rapidly improving.

## *3.7 Drone Delivery Costs*

According to some sources, drone delivery is usually much more expensive than delivery by conventional transportation. For example, delivery costs per ton mile have been estimated at $2000 for drone delivery, versus $40 for same-day ground truck delivery, $1.37 for air freight, and $0.20 for national truck (French 2016). Furthermore, drones perform relatively poorly economically on route density and drop size (Wang 2016). Nevertheless, analysis of "value density" versus urgency shows instances when the price premium makes sense (French 2016). For example, an organ for transplant would have high value density and would be needed urgently to save a life.

On the other hand, researchers have shown using simulation modeling of the Gaza, Mozambique, vaccine supply chain that a system of drones could increase the availability of vaccines and decrease costs in a wide range of situations (Haidari et al. 2016). Sensitivity analyses showed that the major drivers of costs savings are speed of traditional land vehicles, the number of people needing to be vaccinated, and the distance that needs to be traveled.

One cost comparison estimates orders of magnitude less money and less time for delivery of blood with a drone, versus costs of $10,000 to emergency lift someone in a helicopter to go to the hospital if they need blood (Knight 2016). Instead of sending several people in a helicopter, the blood is sent directly to the person who needs it.

Another source states that the Zipline drone service costs about the same amount as delivery using traditional road vehicles (Landhuis 2017). In summary, although the cost-effectiveness of drone delivery for medical emergencies needs more research, the likelihood is that substantial savings will be realized in some circumstances.

The speed of response to a medical emergency varies with different forms of transport. Ambulances and motorcycles are dependent on an adequate road infrastructure and can also be slowed down by traffic congestion and weather issues that affect accessibility such as natural disasters. Airplanes cannot access locations without adequate air infrastructure such as airports. Helicopters are less dependent than airplanes on air infrastructure, but cannot fly or land in inhospitable conditions. In addition, helicopters are often not available because of their high cost. Boats need to be able to dock and are slowed down by adverse weather. Some examples have shown that drones are faster than other forms of transport. For example, the CyPhy drone took 8 minutes to deliver the asthma inhaler compared to 30 minutes in a boat (Quittner 2016). The Flirtey drone delivered medications in 3 minutes compared to road delivery of usually 90 minutes (Hackman and Nicas 2015). DHL Parcel Drone delivery of urgently needed medicines between automated Skyports in two

Bavarian Alpine villages took 8 minutes compared to a 30 minute road trip in winter (DHL 2016). Matternet drones' urgent delivery of blood and tissue samples between Swiss hospitals and labs is potentially saving lives by taking 3 minutes instead of 25 minutes by ground transportation (Peters 2017).

In the next section, we review models that form the basis for managerial decisions with respect to the operation of the drone fleet. First, we discuss models for drone delivery then models specific to healthcare and medical emergencies.

## 4 Models for Drone Delivery

Inspired by Google, Amazon, UPS, DHL, and other logistic companies, several researchers have developed various models for drone use primarily for parcel delivery. Some models use a truck as the primary delivery vehicle with attendant drones. For example, a study developed two models for delivery of parcels by drones to be launched from a truck (Murray and Chu 2015). They model this as a mixed integer program of a traveling salesman type. Computational issues associated with this model have been discussed by others (Wang et al. 2016). Related models use various meta-heuristic approaches to solve the resulting integer programs (Agatz et al. 2015; Ferrandez et al. 2016; Ponza 2015). These papers essentially conclude that there are cost savings to be realized when both trucks and drones work jointly to deliver parcels.

Furthermore, researchers extend vehicle routing problems to drone delivery by adding time and budget constraints to vehicle routing scenarios via integer programming formulations (Dorling et al. 2017). They explicitly recognize drone energy consumption and finite payload.

However, the above-described models assume that a drone flight is not impeded by obstacles such as high mountains or buildings or airport exclusion zones. Another study developed a sole drone-based delivery system for urban areas by positioning drone recharging stations and routing delivery paths around obstacles (Hong et al. 2015). The researchers address this issue by embedding a network in space such that travel is on the shortest path but confined to the network (Hong et al. 2015). They give a location coverage model to both maximize demand coverage and minimize flight distance by maximizing their ratio. Additionally, nodes of the network serve as refueling stations for drones, thus increasing the range of drones.

In this section, we review computer models that have been developed for the delivery of healthcare products by drones. Our focus is on operational decisions associated with supply chain management for drone delivery. Supply chain management decisions include location of service facilities, sizing of service facilities, and detailed routing decisions.

A model-based emergency response system for a natural disaster involves a heterogeneous fleet of vehicles (Mosterman et al. 2014). Integration of the components of their response system is achieved with internet technology and a graphical interface.

We look at a location of a supply warehouse and multiple drone nests to provide medical supplies in an underdeveloped country (Scott and Scott 2017). In such countries, the road network is generally not well developed and some roads can become impassable at various times of the year. This suggests a tandem truck-drone delivery system where trucks carry multiple loads to a region on available roads and "the last mile delivery" is by drone. In addition, such a strategy could be used in a developed urban area due to congestion near the delivery site. The location problem is formulated as a nonlinear program which determines locations for both a warehouse where supplies are kept, and for drone nests that serve outlying villages that are potentially inaccessible in a timely manner. Various sensitivity analyses are performed to look at the effect of demand and drone range on fast delivery.

A logistics network is designed to provide timely delivery of healthcare items using a tandem strategy involving land-based transportation and final delivery by drone. Both models focus on locating a warehouse with supplies and drone nests to complete final delivery. In the first model, our objective is to minimize the total weighted time for delivery; the second model seeks to minimize the maximum weighted time to any delivery point, thereby ensuring more equitable service to outlying regions. As such, the new models given are a contribution to continuous location theory with particular application to tandem truck drone delivery of healthcare products. For a review of location decisions, see Drezner and Hamacher (2004).

A recent study examines the use of drones to deliver routine health services to a rural area where medical options are more limited than in a highly populated area (Kim et al. 2017). The researchers develop two models; one at the strategic level and one operationally oriented. Their first step was to identify a number of potential locations for drone centers from which drones could fly to deliver healthcare services. These could be locations of a current healthcare facility such as a hospital or clinic or perhaps some vacant land which could accommodate a center. In order to provide service to a rural area, they employ as a requirement the maximum possible distance that a drone may travel. The strategic decision then is to identify the locations of the minimal cost number of drone centers necessary to meet this requirement so that all rural demand is covered. This is formulated as a set covering problem. The output of this model being the locations of the drone centers becomes the input to the operational model that routes the drones to provide service to the rural area. This is modeled as a variant of a multi-depot vehicle routing problem where the objective is to minimize the operational cost by appropriate routing to both deliver medical supplies and to return to base with exam kits for further analysis. A detailed algorithm is given as well as an application to a rural area in Texas.

Using a method based on geographic information systems (GIS), a study looks at locations for drones to deliver defibrillators to cardiac arrest victims for the county of Stockholm, Sweden (Lennartsson 2015). Traditional delivery by ambulance takes an average of 8–10 minutes in the city and 25–30 minutes outside the city. However, a patient needs a defibrillator shock within 3 minutes for a survival rate of 74%. As each additional minute reduces the chance of survival by 10%, a faster drone-based delivery system would greatly reduce the death rate.

The paper uses a multi-criteria evaluation method and GIS software to come up with two locations to base delivery drones: one is the city center where most incidents occur, the other in an outlying area. Defibrillators are delivered as needed by drone (termed a Deficopter), which then drops the defibrillator by parachute and returns to base. Note that while the defibrillator is traveling via the drone and on the patient, it is unavailable for immediate use elsewhere. Below we look at how many defibrillators and drones should be stocked at each base to meet demand; an issue that is not addressed in the paper under discussion.

We model the stocking decision as a multi-server queue where each defibrillator is a server. Demand for defibrillators is random and the time when a defibrillator is in use (unavailable for another patient) is also random due to the different spatial locations that cardiac arrests occur. We assume that when an incident occurs, both a drone carrying defibrillator and an ambulance will be dispatched simultaneously and after use, the defibrillator will be returned to base with the ambulance that is transporting the patient to a hospital. According to queueing theory, the probability that all defibrillators will be in use at any time is

$$P = \sum_{n=s}^{\infty} \frac{\left(\frac{\lambda}{\mu}\right)^n}{s!s^{n-s}} \left( \sum_{n=0}^{s-1} \frac{\left(\frac{\lambda}{\mu}\right)^n}{n!} + \frac{\left(\frac{\lambda}{\mu}\right)^s}{s!} \frac{1}{\left(1 - \frac{\lambda}{s\mu}\right)} \right)^{-1}$$

where $\lambda$ is the arrival rate and $\mu$ is the service rate and $s$ is the number defibrillators to be determined to satisfy $P \leq 0.01$. Here, we have specified a level of safety of 0.99 which implies that there will be a chance of at most 1% of a defibrillator not being available when needed. The same model with a different service rate can be used to size the drone fleet. This is a standard model used in the public sector to size an emergency fleet such as police cars or ambulances (Kolesar et al. 1975).

We use data from the county of Stockholm given in the above-referenced paper (Lennartsson 2015) to address the number of defibrillators and drones needed. Based on 10,000 incidents per year, the arrival rate for calls is 1.14 per hour. Since most occur in the city area, we split this rate into call rates of 1 and 0.14 for the city and outside areas respectively. The service rate for a defibrillator based on out and back times plus time to attend to the patient is 2 per hour in the city and 0.67 per hour outside the city. The results are given in Table 3 below.

Note that the reason the service rate of defibrillators in the city is much greater than in outlying areas is that although the defibrillators are delivered fast by drones,

**Table 3** Optimal number of defibrillators

| Location | Arrival rate $\lambda$ per hour | Service rate $\mu$ per hour | Number of defibrillators |
|---|---|---|---|
| City | 1 | 2 | 4 |
| Outlying area | 0.14 | 0.67 | 3 |

**Table 4** Optimal number of drones

| Location | Arrival rate $\lambda$ per hour | Service rate $\mu$ per hour | Number of drones |
|---|---|---|---|
| City | 1 | 6 | 3 |
| Outlying area | 0.14 | 6 | 2 |

the relatively slow return journey by ambulance that transports the patient is further for the outlying areas.

A similar analysis may be done for the optimal number of drones. We use a 6-minute flight time for a drone and 4 minutes to change out a battery. With the same safety level, the results are given in Table 4.

Since we have assigned resources to address all 10,000 incidents per year, this is likely an overestimate since some incidents will be dealt with by nearby defibrillators kept in fire or police stations, for example. Also, it is unlikely that two drone bases would be sufficient to serve the Stockholm area, but the methodology is readily scaled upward after more detailed analysis.

Similarly, a study examines the integrated use of optimization and queueing models to deliver defibrillators to out of hospital cardiac arrests in urban and rural areas in and around Toronto, Canada (Boutilier et al. 2017). The set covering model minimizes the number of zone bases and the queueing model (as above) sizes the number of drones assigned to each base. Using their models, they claim a dramatic improvement of survival rates particularly in a private or rural setting.

In summary, models for delivery by drones address a variety of healthcare issues. The most important healthcare issue is saving lives. In a medical emergency, lives can be saved by fast delivery of healthcare products and services. The models include optimized location of drones to deliver AED, blood and vaccines faster than ambulances or other emergency forms of transportation. In addition, models include detailed routing and fleet-sizing decisions.

## 5 Concluding Remarks

Drones, along with mobile technology, are enabling developing countries to leapfrog ahead with healthcare delivery to remote locations despite an unreliable road infrastructure. As stated earlier, only a third of Africans live within two kilometers of a road that functions year round (Khazan 2016). Even in developed countries, disasters, such as earthquakes and fires, can render roads inaccessible. Furthermore, extreme weather and city congestion can be an obstacle to emergency medical delivery.

Drone technology and its components, such as GPS and lithium batteries, are available and improving at a rapid pace. Despite issues related to privacy, security, safety, and regulation, drones can provide beneficial and humanitarian applications, especially related to healthcare. Consequently, drone healthcare delivery to inaccessible locations is likely to become more ubiquitous in the near future.

This chapter discusses some innovative applications of drones in healthcare and models for drone delivery in medical emergencies. Models discussed include tandem strategies involving traditional land transportation followed by drone delivery, optimized location of drones to deliver emergency products and services, and detailed routing and fleet-sizing decisions. Models could explicitly model the actual road network. Future research can also investigate how many drones are needed at a particular drone nest and how well a network provides coverage for a region. Models should account for the effect of obstacles such as a high mountain on a drone path and integrate it into the location model. For drone delivery in a metropolitan area, paths need to be routed around tall buildings and airports, for example.

While healthcare costs are a major concern in both developed and developing countries, technology and innovation are considered at least part of the solution. Drone technology and models for delivery can contribute to lowering costs associated with medical emergencies. Drones are likely to facilitate more timely, efficient, and economical healthcare delivery. Some models use a budget constraint while providing location decisions for warehouses and drone nests that enable timely delivery. Since time is of the essence in an emergency, faster response would prevent medical trauma and potential deaths.

Drone companies, delivery companies, healthcare organizations, humanitarian organizations, or governments could use models as described in this chapter to expedite delivery of urgently needed healthcare items. Vaccine delivery could help prevent up to 1.5 million deaths each year (Wurbel 2017). Governments could partner with drone companies, such as Rwanda does with Zipline to more economically deliver vaccines and blood to potentially save lives.

Implications for research include the need for further model development. Although extensive research has focused on sophisticated delivery models for traditional transportation, research on drone delivery models for medical emergencies is in its infancy. Nevertheless, the potential to save lives and facilitate lower healthcare costs are compelling incentives for furthering this research.

## References

24/7 Staff, S. (2016). UPS and CyPhy works test drone for urgent commercial last mile delivery. *SupplyChain 24/7*. Retrieved from https://www.supplychain247.com/article/ups_and_cyphy_works_test_drone_for_urgent_commercial_delivery

Agatz, N., Bouman, P., & Schmidt, M. (2015). *Optimization approaches for the traveling salesman problem with drone*. Retrieved from Erasmus University Rotterdam: http://hdl.handle.net/1765/78472

Balasingam, M. (2017). Drones in medicine—The rise of the machines. *International Journal of Clinical Practice, 71*(9), 1–4.

Boutilier, J. J., Brooks, S. C., Janmohamed, A., Byers, A., Buick, J. E., Zhan, C., Schoellig, A. P., Cheskes, S., Morrison, L. J., & Chan, T. C. Y. (2017). Optimizing a drone network to deliver automated external defibrillators. *Circulation, 135*, 2454.

Choi-Fitzpatrick, A., Chavarria, D., Cychosz, E., Dingens, J. P., Duffey, M., Koebel, K., Siriphanh, S., Tulen, M. Y., Watanabe, H., Juskauskas, T., Holland, J., & Almquist, L. (2016). Up in the air: A global estimate of non-violent drone use 2009–2015. Retrieved from http://digital.sandiego.edu/gdl2016report/1

Claesson, A., Fredman, D., Svensson, L., & Ringh, M. (2016). Unmanned aerial vehicles (drones) in out-of-hospital-cardiac-arrest. *Scandinavian Journal of Trauma, Resuscitation and Emergency Medicine, 24*, 124.

Clarke, R. (2014). Understanding the drone epidemic. *Computer Law & Security Review, 30*(3), 230–246.

DHL. (2016). Successful trial integration of DHL parcelcopter into logistics chain [Press release]. Retrieved from http://www.dhl.com/en/press/releases/releases_2016/all/parcel_ecommerce/successful_trial_integration_dhl_parcelcopter_logistics_chain.html

Dorling, K., Heinrichs, J., Messier, G. G., & Magierowski, S. (2017). Vehicle routing problems for drone delivery. *IEEE Transactions on Systems, Man, and Cybernetics: Systems, 47*(1), 70–85.

Drezner, Z., & Hamacher, H. (Eds.). (2004). *Facility location: applications and theory*. Berlin: Springer.

Esler, D. (2015, Oct). *Drone revolution*. Business & Commercial Aviation, 39–47.

Ferrandez, S. M., Harbison, T., Weber, T., Sturges, R., & Rich, R. (2016). Optimization of a truck-drone in tandem delivery network using K-means and genetic algorithm. *Journal of Industrial Engineering and Management, 9*(2), 374–388.

Fleck, M. (2016, May 7–12). *Usability of lightweight defibrillators for UAV delivery.* Paper presented at the Proceedings of the 2016 CHI conference extended abstracts on human factors in computing systems San Jose, CA.

French, S. (2015). Drone delivery is already here — And it works. *Marketwatch*. Retrieved from http://www.marketwatch.com/story/drone-delivery-is-already-here-and-it-works-2015-11-30

French, S. (2016). Amazon's first Prime Air delivery is just more drone hype *Marketwatch*. Retrieved from http://www.marketwatch.com/story/the-heavy-on-hype-light-on-substance-world-of-drone-delivery-2016-12-01

Hackman, M., & Nicas, J. (2015, July 17). Drone delivers medicine to rural virginia clinic. *Wall Street Journal*. Retrieved from http://www.wsj.com/articles/drone-delivers-medicine-to-rural-virginia-clinic-1437155114

Haidari, L. A., Brown, S. T., Ferguson, M., Bancroft, E., Spiker, M., Wilcox, A., Ambikapathi, R., Sampath, V., Connor, D. L., & Lee, B. Y. (2016). The economic and operational value of using drones to transport vaccines. *Vaccine, 34*, 4062–4067.

Hidalgo, J. (2018). Flirtey sees EpiPens, Narcan in future as AED drone delivery expands. *Reno Gazette Journal*. Retrieved from https://www.rgj.com/story/money/business/2018/02/20/flirtey-defibrillator-epipens-narcan-future-medical-drone-delivery/353627002/

Hong, I., Kuby, M., & Murray, A. (2015, May 20–23). *Deviation flow refueling location model for continuous space: Commercial drone delivery system for urban area.* Paper presented at the 13th International Conference on GeoComputation, Richardson, Texas.

Hotz, R. L. (2017, December 1). In Rwanda, drones deliver medical supplies to remote areas. *Wall Street Journal*. Retrieved from https://www.wsj.com/articles/in-rwanda-drones-deliver-medical-suppliesto-remote-areas-1512124200.

Khazan, O. (2016, April 4). A drone to save the world. *The Atlantic*. Retrieved from http://www.theatlantic.com/technology/archive/2016/04/a-drone-to-save-the-world/476592/

Kim, S. J., Lim, G. J., Cho, J., & Murray, J. C. (2017). Drone-aided healthcare services for patients with chronic diseases in rural areas. *Journal of Intelligent and Robotic Systems, 88*, 163–180.

Knight, R. (2016). Drones deliver healthcare. *Inside unmanned systems: Engineering, policy, and practice.* Retrieved from http://insideunmannedsystems.com/drones-deliver-healthcare/

Kolesar, P. J., Rider, K. L., Crabill, T. B., & Walker, W. E. (1975). A queuing-linear programming approach to scheduling police patrol cars. *Operations Research, 23*(6), 1045–1062.

Kolodny, L. (2017a). Drones will deliver defibrillators to 911 callers to help treat cardiac arrest. *CNBC*. Retrieved from https://www.cnbc.com/2017/10/10/watch-flirtey-and-remsa-deliver-defibrillators-by-drone-to-911-callers.html

Kolodny, L. (2017b). Matternet cleared-to-fly-blood-samples-in-delivery-drones-over-swiss-cities. *Techcrunch*. Retrieved from https://techcrunch.com/2017/03/31/matternet-cleared-to-fly-blood-samples-in-delivery-drones-over-swiss-cities/

Landhuis, E. (2017). Tanzania gears up to become a nation of medical drones. Retrieved from https://www.npr.org/sections/goatsandsoda/2017/08/24/545589328/tanzania-gears-up-to-become-a-nation-of-medical-drones

Lennartsson, J. (2015). *Strategic placement of ambulance drones for delivering defibrillators to out of hospital cardiac arrest victims*. Stockholm: KTH Royal Institute of Technology School of Architecture and the Built Environment.

Momont, A. (2014). *Drones for good*. (Masters), TU Delft. Retrieved from http://www.alecmomont.com/projects/dronesforgood

Mosterman, P. J., Sanabria, D. E., Bilginc, E., Zhangd, K., & Zander, J. (2014). Automating humanitarian missions with a heterogeneous fleet of vehicles. *Annual Reviews in Control, 38*(2), 259–270.

Murray, C. C., & Chu, A. G. (2015). The flying sidekick traveling salesman problem: Optimization of drone-assisted parcel delivery. *Transportation Research Part C: Emerging Technologies, 54*, 86–109.

Pepitone, J. (2015). First FAA-approved drone delivery drops medicine in virginia. Retrieved from http://www.nbcnews.com/tech/tech-news/first-faa-approved-drone-delivery-drops-medicine-virginia-n393986

Peters, A. (2017). Switzerland is getting a network of medical delivery drones. *Fast Company*. Retrieved from https://www.fastcompany.com/40467761/switzerland-is-getting-a-network-of-medical-delivery-drones

Ponza, A. (2015). *Optimization of drone-assisted parcel delivery.* (Honors Thesis), University of Padova. Retrieved from http://tesi.cab.unipd.it/51947/1/Andrea_Ponza_-_Analisi_e_ottimizzazione_nell'uso_di_droni_per_la_consegna_di_prodotti.pdf

Preimesberger, C. (2016, May 9). Drones will soon be dropping medicines to save lives in Rwanda. *eWeek*, 15306283, Retrieved from https://www.eweek.com/mobile/drones-will-soon-be-dropping-medicines-to-save-lives-in-rwanda

Prigg, M. (2014, October 28). The ambulance drone that could save your life. *Daily Mail*. Retrieved from http://www.dailymail.co.uk/sciencetech/article-2811851/The-ambulance-drone-save-life-Flying-defibrillator-reach-speeds-60mph.html

Quittner, J. (2016). Meet the woman working to make delivery by drone a reality. *Fortune*. Retrieved from http://fortune.com/2016/11/10/meet-the-woman-working-to-make-delivery-by-drone-a-reality/

Raffaello, D. A. (2014). Guest editorial can drones deliver? *IEEE Transactions on Automation Science and Engineering, 11*(3), 647–648.

Raptopoulos, A. (2013). *No roads? There's a drone for that.* Paper presented at the TEDGlobal 2013.

Scott, J. E., & Scott, C. H. (2017). *Drone delivery models for healthcare.* Paper presented at the 50th Hawaii International Conference on System Sciences, Hawaii.

Stewart, J. (2017). Switzerland is getting a delivery network for blood toting drones. wired.com. Retrieved from https://www.wired.com/story/switzerlands-getting-a-delivery-network-for-blood-toting-dro

Tilley, A. (2016, May 9). UPS experiments with drone delivery in partnership with zipline. *Forbes*. Retrieved from http://www.forbes.com/sites/aarontilley/2016/05/09/ups-experiments-with-drone-delivery-in-partnership-with-zipline/#13c4870f75da

Van de Voorde, P., Gautama, S., Momont, A., Ionescu, C. M., De Paepe, P., & Fraeyman, N. (2017). The drone ambulance [A-UAS]: Golden bullet or just a blank? *Resuscitation, 116*, 46–48.

Varnholt, H. (2016, January 19). DHL's drone demonstration fails to deliver. *Wall Street Journal*. Retrieved from http://www.wsj.com/articles/dhls-drone-demonstration-fails-to-deliver-1453226792

Villasenor, J. (2012). What is a drone, anyway? *Scientific American*. Retrieved from http://blogs.scientificamerican.com/guest-blog/2012/04/12/what-is-a-drone-anyway/

Wang, D. (2016, January 5). The economics of drone delivery. Retrieved from http://spectrum.ieee.org/automaton/robotics/drones/the-economics-of-drone-delivery

Wang, X., Poikonen, S., & Golden, B. (2016). The vehicle routing problem with drones: Several worst-case results. *Optimization Letters*, 1–19.

Welch, A. (2015). *A cost-benefit analysis of Amazon Prime Air.* (Honors Thesis), University of Tennessee at Chattanooga.

Wurbel, H. (2017). *Framework for the evaluation of cost-effectiveness of drone use for the last-mile delivery of vaccines*. Barcelona: Master of Global Health), University of Barcelona.

# Converting Disability into Ability Using IT/IS and Smart Textiles

**Muhammad Nadeem Shaukat, Nilmini Wickramasinghe, and Freimut Bodendorf**

## 1 Introduction

This is an era of healthy and long life owing to improved hygiene, innovations in technology and better nutrition which contribute to making our lives comfortable and lavish (Australia's Health 2016). The advancements in nanotechnology, medical imaging, and innovative treatments have enabled us to access treatments and remedies for the diseases which were not available to our ancestors; as a result, we enjoy healthier lives (Australia's Health 2016). These innovations in medicine and equipment have also helped us to live and manage illness and disease (Australia's Health 2012), whereas technology has also changed our socio-economical behaviour leading to increased number of accidents due to high-speed automobiles and drink-driving (Moan et al. 2013). These traffic accidents mostly involve younger people (Roozenbeek et al. 2013), whereas fall incidents are more common in elderly and fragile population (An Ageing Australia 2013). It was noted that the fall incidents exceeded the traffic accidents (Roozenbeek et al. 2013), both resulting in higher number of brain injuries (Teasdale and Jennett 1976). Brain injuries can be fatal in the first instant or can gradually worsen inflicting multiple disabilities (Smith et al. 1999) and may hinder the victims living a normal happy healthy life (Stocchetti and Zanier 2016).

M. N. Shaukat
Epworth HealthCare, Melbourne, VIC, Australia

N. Wickramasinghe (✉)
Epworth HealthCare, Richmond, VIC, Australia

Swinburne University of Technology, Hawthorn, VIC, Australia

F. Bodendorf
Friedrich-Alexander-University Erlangen-Nuremberg, Erlangen, Germany

 
N. Wickramasinghe, F. Bodendorf (eds.), *Delivering Superior Health and Wellness Management with IoT and Analytics*, Healthcare Delivery in the Information Age, https://doi.org/10.1007/978-3-030-17347-0_4

In this work, we have focused on physical disability due to brain injury (either acquired brain injury or traumatic brain injury) or neurological disease, specially loss of control over organs owing to complete loss or weak disrupted neurological signals to the healthy muscle (Rodriguez-Paez et al. 2005). We aim at developing a combination of smart textile and electronic circuitry to transfer signals to disconnected or damaged neuron of upper limbs/hands. The majority of the such sufferers possess complete natural organs with healthy muscles with connected network of bones and ligaments, but absence of signal from motor neuron does not permit muscles to do any movement or action (Osier et al. 2015), rendering the victim from brain to physical inactivity.

Subsequently, the extended periods of dormancy or immobility translate to a new set of problems, e.g. periarticular calcification, bladder and sphincter control issues (Rodriguez-Paez et al. 2005). This chapter serves to conceptualize the opportunity to use smart textile gloves/garments integrated with sensors and actuators to control limbs by sending signals from the processing unit utilizing an application (mobile or tablet based) or external control (e.g. joystick) to perform daily life activities.

In comparison to physical injuries, brain injuries/diseases are hard to observe and remain unknown and invisible (Edge 2010) until it starts worsening or affecting daily life activities, e.g. effective thinking, emotional or behavioural change or effect on focus, or multiple processing (Lezak et al. 2004), depending on the magnitude and area of injury or disease (McGinn and Povlishock 2016); therefore, it is not easy to determine brain injury/disease just by looking or observing a person. The physical injury can be seen or felt through pain or state of the wound thus highlighting the extent of injury (Faul et al. 2010a). Similarly, the recovery or improvement of the physical injury can be ascertained, while with brain injuries/diseases which may affect rational thinking, emotional or behavioural issues, memory loss or lack of concentration, sluggishness, persistent headache, extreme mental or physical fatigue, seizures, paralysis, etc., depending on the extent of injury/disease, it cannot be easily recognized (Brain Damage: Symptoms, Causes, Treatments 2018); therefore, brain injury is also termed as a hidden disability (Edge 2010). The brain injury occurred through either a sudden impact (traumatic brain injury) or due to other reasons (acquired brain injury) both can be as small as having no visible impairment on motor control or daily life activities, the sufferer may live with it for whole life until some sophisticated diagnostic techniques are used to identify brain injury/disease (Brain Injury is a "Hidden Disability" 2018).

Over 700,000 Australians have been diagnosed with brain injuries limiting some of the individuals from simple household activities to actively involving in workforce (About Brain Injury 2016). In only 1 year (2013–2014), almost 500,000 injury-related cases were registered for hospitalization (Australia's Health 2016). People over 85 years old were found to be in the highest number in the age category for injury-related hospitalization. Therefore, it was not surprising to note that fall-related incident for 2013–2014 constituted 41% of total injury-related hospitalization and it was more than 3 times to automobile accidents (13%) (Australia's Health 2016). Around 75% of these brain injured were in their 60s (65 and over) (About Brain Injury 2016), while around 67% of the population was under

25 years, of which 75% were male, needing care for rest of their life (About Brain Injury 2016). The Australian population is suffering injury-related problems way in early age, in particular male population (Mathers et al. 1999), which costs around 8.6 billion dollars per year for brain injured people (Economics 2009).

## 2 Brain Injury

Brain injury can be of two types external (due to any blow or impact) called as traumatic brain injury (TBI) or internal factors without any impact or jolt termed as acquired brain injury (ABI) (Faul et al. 2010b; Bigler 2013; Stocchetti and Zanier 2016; Acquired Brain Injury (ABI) 2018). In either case, it can cause almost negligible to severe impact on body movement or coordination, vision, swallowing, cognition or mood swing (Australia's Health 2016; Causes of Acquired Brain Injury 2018). This can also result in deterioration to cognitive, physical, emotional or independent functioning. In fact, brain injury can cause sudden death of specific brain part and make impairment of certain functions originating from particular brain part (Australia's Health 2016), or it can lead to multiple disabilities arising from worsening conditions of disease/injury (Bigler 2013; National Community Services Data Dictionary 2012) (Fig. 1).

### *2.1 Traumatic Brain Injury (TBI)*

Traumatic brain injury can be referred as a disruption in the normal function of the brain that can be caused by a bump, blow, or jolt to the head or a penetrating head injury (Faul et al. 2010a, b). As depicted in Fig. 2, traumatic brain injury (TBI) can be caused by a sudden high force giving jolt or blow to head or neck area, such situation can compel brain to wobble tremendously inside the skull or mutilate the skull or brain (Gaetz 2004). Causes of Traumatic Brain Injury (TBI) include (Brain Damage: Symptoms, Causes, Treatments 2018; Langlois et al. 2006a; Pervez et al. 2018):

- Traffic accidents due to high-speed vehicle
- External sudden blows to the head/upper neck area
- Sporting/leisure injuries
- Falls for elderly/slips on wet surfaces
- Physical/domestic violence by spouse or partner
- Unsuccessful suicide attempt
- Blast or explosion in battlefield
- Hard shaking (physical abuse) to an infant

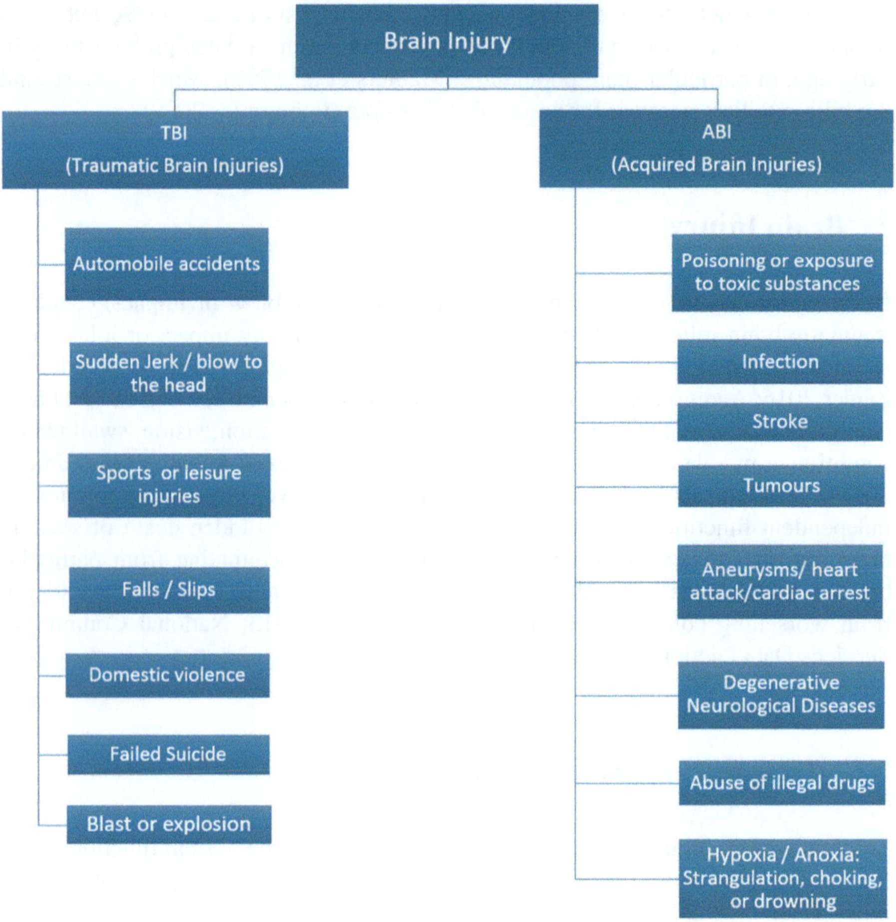


**Fig. 1** Types of brain injuries

As brain injury is not a visible damage unless there is a significant impact, there is not much perception and openness in the general public about the brain injury or problems arising from it, this leads to a fatal invisible epidemic (Faul et al. 2010b; Sample and Darragh 1998). The brain damage can lead to losing or forgetting best skills, or have difficulties in performing them at their level best owing to lack in rational thinking, mindfulness and capabilities in linguistic or sentiments (Faul et al. 2010a; Bigler 2013; Chew and Zafonte 2009).

**Fig. 2** Causes of traumatic brain injuries (TBI)

**Fig. 3** Different causes of acquired brain injury (ABI)

## *2.2 Acquired Brain Injury (ABI)*

The brain injury arising from internal reasons mostly happens at brain cell level without any damage or impact from outside elements as shown in Fig. 3 (Gaetz 2004), e.g. it can be as a result of stroke, brain tumour, infection, illegal drugs, poisoning, lack of oxygen due to drowning or choking/strangulation or degenerative neurological disease (Langlois et al. 2006b; Ragnarsson et al. 1999). The growth of tumour or neuron impairment or lack of oxygen supply can cause partial or segmental damage to brain (Brain Injury is a "Hidden Disability" 2018). The following diagram enlists different types of brain injuries in detail (Brain Injury is a "Hidden Disability" 2018; Acquired Brain Injury (ABI) 2018; Causes of Acquired Brain Injury 2018; How Brain Injury Occurs 2018) (Table 1).

**Table 1** Causes of acquired brain injuries (ABI) (Acquired Brain Injury (ABI) 2018; Causes of Acquired Brain Injury 2018; How Brain Injury Occurs 2018)

| Cause | Brief description |
|---|---|
| Poisoning/toxic exposure | Exposure to poison or toxic materials either in a domestic environment such as long-term misuse of alcohol and entertainment/illegal drugs or in industrial settings, e.g. toxic fumes or smoke |
| Infection | Inflammation to brain from infection can cause meningitis (inflammation to brain skull/cover) or encephalitis (inflammation to brain itself) |
| Hypoxia/anoxia | Limited or complete stoppage of oxygen to brain can cause damage to brain tissues, e.g. strangulation, choking, near-miss drowning or asthma |
| Stroke | An unexpected sudden disruption of blood supply to brain is termed as stroke. A haemorrhage or blocked blood vessel can be cause of stroke |
| Aneurysms, heart attacks/cardiac arrest | During any of heart problems, e.g. aneurysms and heart attacks/cardiac arrest, the blood supply from the heart to brain suffers blockage, depriving brain from oxygen and consequently damaging brain cells |
| Tumours | Presence or growth of tumour in brain inflicts mutilation (damage) to proximate neuron tissues and structures |
| Treatment or surgery | Some type of chemotherapy, radiotherapy and surgery can also result in injury to the brain |
| Neurological disability | It applies to impairments of the nervous system occurring after birth; it includes epilepsy and organic dementias |
| Disease related | Some diseases can cause brain damage depending upon severity, e.g. AIDS, Alzheimer, cancer, multiple sclerosis (MS) or Parkinson |

Tree table showing different causes of acquired and traumatic brain injuries.

Images of different types of traumatic brain injuries due to blow to head or neck area from external factors, e.g. automobile accident, slip, fall, domestic violence, sports injury, failed suicide, etc. (Honda City Hit by Motorcycle 2019; Sports Photo by Chris Chow 2019) Caution wet floor (2019).

Different images of factors involved in acquired brain injuries. There is no external impact on head or neck area; rather, it involves depriving of oxygen due to drowning, strangulation, exposure to poison, stroke, tumour, illegal drugs, etc. (Papillary Glioneuronal Tumor 2019; Drowning Photo by Carlos Koblischek 2019; Drug Addiction 2019).

**Table 2** Classification of brain injuries and their different subclasses (Pervez et al. 2018)

| Type | Symptoms |
|---|---|
| Cognitive | Difficulty in communication or having a conversation, e.g.<br>Information processing<br>Expressing emotions and feelings<br>Understanding others |
| | Limited ability to focus (problems with concentration or attention, slow-thinking, disorientation or confused state of mind) |
| | Difficulty in comprehension (abstract ideas/thoughts/dyspraxia) |
| | Poor decision-making (difficulty with planning or organization) |
| | Difficulty in remembering events |
| Perceptual | Change in sensory abilities (vision, hearing or touch) |
| | Spatial disorientation |
| | Difficulty in time sense |
| | Disorder of smell/taste |
| | Poor balance (paresis/plegia) |
| | Intensified pain sensitivity |
| Physical | Continuous headaches |
| | Extreme tiredness (mental/physical) |
| | Paralysis |
| | Tremors/seizures/fits |
| | Weakness, shaking, stiffness or poor balance |
| | Changes in sleep patterns |
| | Sensitivity for light |
| | Slurred speech |
| Behavioural | Impatient/easily irritable |
| | Inability to bear stress |
| | Mood swings (being irritable or feeling on edge) |
| | Changes in personality |
| | Periodically decreased or low level of consciousness |
| | Denial of disability or condition |
| | Increased aggressiveness |

## *2.3 Symptoms and Types of Brain Injuries*

Brain injury can have immediate effect, and it can be a gradual worsening of conditions (Polinder et al. 2015). These symptoms can be of different nature, e.g. cognitive, perception related, affecting body or causing change in personality (Van Bost et al. 2017). A number of symptoms in people suffering from brain injuries/damages have been described in Table 2 (Brain Damage: Symptoms, Causes, Treatments 2018). There can be multiple symptoms at the same time, some of the symptoms could be supressed, while others can be more prominent. Table 3 details the clinical severity and duration of symptoms (Brain Damage: Symptoms, Causes, Treatments 2018; Pervez et al. 2018; Chew and Zafonte 2009).

**Table 3** Categorization of Clinical severity and duration of brain injury symptoms (Pervez et al. 2018)

| Severity | Consciousness loss | Memory deficit | Post-traumatic amnesia | Motor deficit |
|---|---|---|---|---|
| Concussion | No loss of consciousness | No or mild | Minutes – Hours | No or mild |
| Mild | Loss of consciousness for 6–24 hours | Mild to moderate | Hours | Mild |
| Moderate | Loss of consciousness for more than 24 hours | Moderate | Days | Moderate |
| Severe | Loss of consciousness lasting for days | Severe | Weeks | Severe |

### *2.4 Categories of Brain Injuries According to Clinical Severity*

Brain injuries can be classified into four different types depending upon the concussion, loss of memory and defect/discrepancy of muscles or motor control (Pervez et al. 2018). Table 3 details the effects of concussion type on these parameters (memory, amnesia and motor control).

## 3 Types of Assistive Technologies

The brain injury can impact a person in a number of ways, e.g. headaches, fatigue, seizures, trouble in balance or coordination, affecting logical decision-making, hearing and visual impairment, severe pain, paralysis, epilepsy, difficulty in mode or emotion control, partial disability or difficulty to control or move organs, trouble in memory and concentration and inability to upgrade skills or use learnt skills (Rushworth 2008). In most of brain injuries, it was noted that a person shows a noticeable change in behaviour, i.e. he/she becomes excessively violent, loses impulse control, lacks social regard and underestimates risk, aggregating multiple chances of further injury (Rushworth 2008). Our focus in this chapter is to enable a person who is suffering from physical disability because of neuron/nerve damage and faces difficulty in controlling organ specially arms/hands to perform daily routine activities. Such person does have healthy muscles with bones and ligaments intact. A technology based on smart textiles and IoT (Internet of things) in combination of sensors and actuators can help regain control (Zheng et al. 2014; Liang-Jie Zhang 2017). There is an extensive work going on in wearable textiles for vital signs and health monitoring for fitness and patient care (Zheng et al. 2014; Jin et al. 2017; Fritz et al. 2014; Fang et al. 2016; Khan et al. 2016; Imani et al. 2016). The main idea is to utilize IoT and sensor-embedded textile-based arm long glove to transfer brain or movement signal to healthy muscles to enable them for movement or control (Fig. 4).

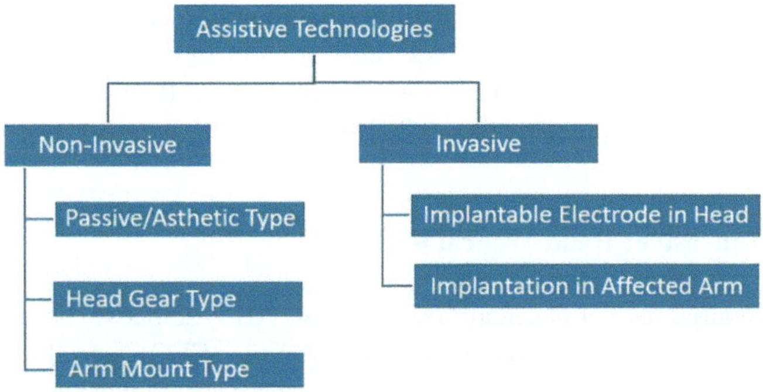


**Fig. 4** Types of assistive technologies for hand/arm

Diagram showing types of assistive technologies for arm/hand movement.

## *3.1 Non-invasive Technologies*

There are different types of technologies available for the control and movement of upper body limb. They include from a simple aesthetic purpose built passive limb (body powered harnessing system (From Below and Above Elbow Harness and Control System 2018; Bowker and Michael 1992)), to brain signal controlled systems. The brain signal powered system can be either non-invasive ones or invasive ones (requiring surgery to connect brain cells/neurons to the motion controlling system (Wickham 2012; Regalado 2014; Thought-controlled Prosthesis is Changing the Lives of Amputees 2012)). The brain signals powered devices can communicate with nerves/neurons and control the arm to complete required tasks.

### 3.1.1 Passive or Aesthetic Body Powered Limb

This is a slight modification of an old technique for amputees which was used to give aesthetic aspect to an amputee (From Below and Above Elbow Harness and Control System 2018). Previously, it was a piece of wood or plastic material shaped into full length of arm, but this new one has a set of straps, rings, wires and springs to utilize body power to lift up or move the arm (From Below and Above Elbow Harness and Control System 2018; Bowker and Michael 1992; Types of Upper Extremity Prostheses 2018). This system requires body power from adjacent muscles to create pull into spring which in turn moves the arm (Bowker and Michael 1992). Despite being an economical option, this system is not suitable for any person having pain or chances of increasing damage due to power required. For any brain injury sufferer, it needs to have enough power in shoulder and surrounding area to move this harness system (From Below and Above Elbow Harness and Control System 2018; Bowker and Michael 1992). Yet, with brain injury, it could be difficult to use such system (Types of Upper Extremity Prostheses 2018).

### 3.1.2 Non-invasive: Head Wearable Gear

Korea University and TU Germany (Kwak et al. 2015) scientists have successfully tested a system which uses electroencephalogram (EEG) signal capturing cap which serves as interface to communicate between a set of exoskeletons and brain (Kwak et al. 2015). By using this system, user can independently sit, stand, move forward and turn left or right by intentionally viewing a set of light emitting diodes (LEDs) (Kwak et al. 2015). This smart system comprises a powered exoskeleton, visual stimulus, wireless EEG transmitter and signal processing unit. There are 5 LEDs connected to each visual stimulus assigned to a different task. Each LED flashes at a distinctive frequency when a user gazes any one of these LEDs, EEG cap deciphers the EEG signal which is then converted into command to control exoskeleton (Kwak et al. 2015). This system utilizes a heavy wearable system to convert electrical signals into mechanical motion using exoskeleton through electromechanical devices. Nonetheless, the system itself is a wonderful technique for disabled person but this system is heavy and bulky, produces noise, which may hinder in smooth handling of EEG signals, consequently, might create delays in movements or inaccuracy in command processing (Kwak et al. 2015). In addition, this system requires assistance in putting on and off for system and may also not be comfortable (Kwak et al. 2015).

### 3.1.3 Non-invasive; Fixed to Limb

Myomo, medical robotics company based in Cambridge, Massachusetts has launched a robotic arm device which can be controlled by thoughts of a person (What is a MyoPro Orthosis 2018). MyoPro prosthetic developed by Myomo was initially developed by MIT and Harvard Medical School (Moving Your Hand and Arm with the MyoPro Motion G 2018). It has proved a blessing for hundreds of patients who suffered from weakness in their muscles or dead-arm due to stroke, spinal cord or nerve injury, e.g. brachial plexus injury (BPI) or similar neuromuscular diseases such as amyotrophic later sclerosis (ALS) or multiple sclerosis (MS) (What is a MyoPro Orthosis 2018). MyoPro detects myoelectric signals from hand without requiring any surgery. When a person thinks to move his hand, a myoelectric signal is generated in brain and transferred to relevant organ through nerves, MyoPro robotic prosthetic uses these nerve signals from limb, amplifies this signal and through motors controls the motion of weakened or dead-hand. This system has saved people from amputation, enabled them to live a high-quality life and reduced overall care cost.

The MyoPro is custom-made for each individual patient according to the weakness or state of the limb. MyoPro is a wearable motor powered brace, connected to a system capable of using nerve signal from weak or dead hand, amplifying these signal and then sending signal to motor to create motion/action (Moving Your Hand and Arm with the MyoPro Motion G 2018). MyoPro actually has helped patients in returning back to their work and living a high-quality life without disability, overall reduced cost of care on individual, family, government and economy. Although this is such a remarkable technique bringing a positive change to a person living a disabled life, it seems to be costly and requiring frequent battery charging. Furthermore, the device itself is huge and visible in such a way that it highlights a person's disability.

## 3.2 *Invasive: Implantable Technologies*

### 3.2.1 Invasive: Brain-Implanted Electrode

Researchers at the University of Pittsburgh Medical Center are working on an electrode which can help people who suffer disability and are unable to control or move their hands (Wickham 2012). They can automatically control prosthetic device using brain thoughts. The electrode was named as Utah Electrode; it consists of a pair of fine needle beds, which can be implanted into brain skull through surgery (Regalado 2014). This electrode connects with neurons and uses electric pulses when a person thinks and converts these electric signals into motion on patient's will to manipulate mechanical prosthetic. In this way, a person suffering from disability can successfully manoeuvre artificial prosthetic to perform different activities ranging from holding, lifting and stacking varying shapes (Wickham 2012; Regalado 2014). In a comparable research scientist from Brown University/Blackrock Microsystems, used brain signals to connect and operate a variety of internet of things (Brain-Computer Interface Lets You Control IoT Devices 2018). In the near future, the dream to control TVs, computers, wheel chairs and even independently driving automobiles with thoughts might become a reality (Brain-Computer Interface Lets You Control IoT Devices 2018).

### 3.2.2 Invasive: Implant into Arm

A scientist from Chalmers University Sweden has devised a system capable of bidirectional interface which uses natural and intuitive control system to communicate with healthy muscles and neurons of a disabled amputee person to control artificial limbs, as if controlling their own natural arm or hand just by thinking (Thought-controlled Prosthesis is Changing the Lives of Amputees 2012). In this technique, they developed and used Osseointegration electrode, which permanently connects muscles and neurons. A titanium implant (OPRA implant system) controls limb directly via nerves and muscles in arm (Thought-controlled Prosthesis is Changing the Lives of Amputees 2012). The electrodes are implanted into hand instead of brain but still amputated limbs can transfer signal to control the artificial hand (Thought-controlled Prosthesis is Changing the Lives of Amputees 2012).

These above-mentioned invasive systems (skull or arm implants) can help an amputee to control limbs autonomously through his/her own will (brain signals). A brain implant has its own benefits and demerits. The insertion of tiny electrode requires surgery to connect to brain cells (neurons). The wires are coupled to brain and require electrode to emerge out, posing a possible inconvenience and unacceptability by users. Moreover, the risk of losing signals partially or completely in the case of neuron damage/disconnection can lead to repeated surgery procedures to re-establish connection. The arm implant might have better acceptance as it might not hinder day-to-day activities, but it still requires surgery to insert electrode for brain connection.

## 4 Proposed Smart Textile Glove

The above-mentioned techniques no doubt have big impact on day-to-day life for a person suffering from brain injury/disease and living a disabled life, yet they either are not fully developed or are heavy and bulky (Kwak et al. 2015; Moving Your Hand and Arm with the MyoPro Motion G 2018) otherwise require surgery (Wickham 2012; Thought-controlled Prosthesis is Changing the Lives of Amputees 2012; Mind-controlled Artificial Limb Gives Patients Sense of Touch Again 2018) or are very expensive (What is a MyoPro Orthosis 2018; Moving Your Hand and Arm with the MyoPro Motion G 2018; My Own Motion 2018). We propose a full arm length glove embedded with sensors and actuators, which can sense myoelectrical signals from brain through neuron present in the healthy part of the arm and send command signal to corresponding muscle through actuators to perform a task. The anticipated glove will consist of smart textile material (capable of conducting electrical signals) with carefully integrated sensors (to collect neuro signals), actuators (to transmit electrical signal to healthy muscles) and relevant integrated circuits connected to a command processing unit. A number of scientists recorded electrical signals by adding sensors into textiles (Wickham 2012; Thought-controlled Prosthesis is Changing the Lives of Amputees 2012; Kwak et al. 2015; Moving Your Hand and Arm with the MyoPro Motion G 2018; Mind-controlled Artificial Limb Gives Patients Sense of Touch Again 2018; The Groundbreaking Mind-controlled BIONIC ARM that Plugs into the Body and has Given Wearers Back their Sense of Touch 2018).

There is an immense research going on into developing the textile-based gloves to sense the movement of the fingers and convert them into electrical signals (Low-cost Smart Glove Wirelessly Translates the American Sign Language Alphabet into Text 2018; Sensor Gloves 2018). Our proposed smart textile-based glove, instead of converting mechanical energy into electrical signal, will impart electrical/neuro

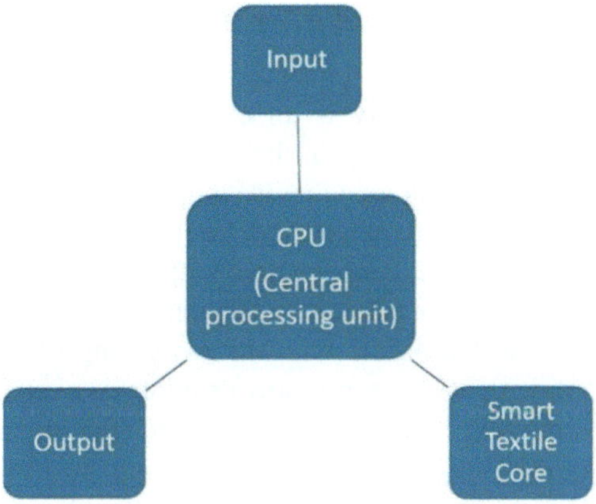


**Fig. 5** Basic elements of proposed textile smart glove

signals to the healthy muscles of a person for movement suffering from neuron disease or nerve damage and convert electrical signals into motion. We plan to control limbs non-invasively through obtaining brain signals either from healthy limb of affected person or another healthy person and convey them to the related limb muscle through an IT/IS solution. Such IT/IS solution should record, save and transmit brain signals using voice, touch or joystick type control system (depending upon the condition of the disabled person as depicted in Fig. 5). We hope, such system can enable a disabled person to gain organ control with the help of an app either on a mobile/tablet/computer or joystick by touch or using voice prompt.

The smart textile-based glove will have following basic elements:

1. Smart textile core
2. Processing unit
3. Input (sensing/command prompt) system
4. Output (motion/action) system

Various parts of signal input, processing and output system using smart textile glove and IT solution

The smart glove will be a better choice as it can be developed keeping in view individual's likes and dislikes for specific material, colour and most important body height and weight. Moreover, it is easier to replace or change textile-based material if the individual outgrows or need to change or replace for any reason.

Following are the possible steps involved in development of by smart textile-based IoT Glove.

## 4.1 Nerve and Muscle Actuation Mapping

The human body is an extremely complex structures of nerves, ligaments and muscles. First step in the process will determine the active and inactive nerves. It will also involve in precise mapping of positions and functions of nerves, ligaments and muscles to perform particular activity or group of activities. This mapping might vary from patient to patient suffering from different level and types of neurological disease or brain injury. Therefore, it might be complicated as the state of the patient deteriorate with the worsening of disease or state of brain injury or other related passive symptoms developing over the period of time but the flexibility of removing and attaching sensors/actuator on textile glove will help in fixing this issue. Similarly, change of sensing position might vary with growth of a child, the use of textile core will help in replenishing inexpensive new textile glove and reattaching the sensors/actuators.

## 4.2 Signals Conductions

The signals to the PU (Processing Unit) can be transmitted to the smart textile glove using three different ways. The input can be given using a joystick or GUI (Graphic User Interface) by the individual herself/himself or by the carer. Otherwise, the command signals can be sent through authorized voice in the system (especially for the persons lacking ability to use hands or having difficulty in using hands). This will need to register voice of physically challenged person or carer into the application and then use it as command input otherwise if person affected is not able to speak, eyeball tracking system might be another way out for command input.

## 4.3 Signal Communications

The signal to the processing unit can be connected either linking the input devices through wires or wirelessly (using different wireless technologies, e.g. Bluetooth, radio signal, etc.). There might be a possibility to convey commands over the Internet and connect physiotherapist or chiropractors to the physically challenged individual. This can help in achieving better outcomes without needing to travel and wait at chiropractor/physiotherapist's clinic. Another feature could be the recording/registering of exercises suggested by the physiotherapist and online analysis of correct posture, number of reps and missing or overdoing of such exercises.

### 4.4 *Signal Processing*

The information coming from input devices can be processed by signal processing unit which can be located with the joy stick or otherwise on or near body (wirelessly) processing signals and sending commands through wires or wirelessly. A group of multiple commands can be coded into a special code, e.g. bicep 3, palm clinching, arm lifting, etc., and reduce the effort to input the same command again and again for the purpose of exercise or treatment. This can also be done to perform a specific set of tasks to complete work-related tasks (grab, lift, place, loose and repeat), making it easy for a person getting back to work and integrating into society and gaining self-confidence and independence to live a better and happy life.

### 4.5 *Output/Actions*

The output from the processing unit can be in the form of motion or action. This output can be as simple as to open or close fist to grab something or drop or combination of different simple actions, e.g. move hand, grip something, lift hand to move over to another point, open fist and then drop the item. These actions or outputs can be a series of tasks interconnected and integrated to perform exercise or treatment; otherwise, it can also be done to perform a specific set of tasks to complete work-related or other entertainment activities (Fig. 6).

Flow diagram of the process highlighting different types of inputs and output tasks interconnected through processing unit

## 5 Benefits for Using Smart Textile Glove?

### 5.1 *Material Choices*

The benefits of smart textile glove will include customization in terms of choice of material, colours and design. With the ever-growing advancements in the textile materials by textile technologist, polymer scientist and genetic engineers, now there is a wide range of textile materials available to choose from. These materials can suit according to the sensitivity of a person to specific material say allergic to wool, cotton, etc. or someone's fashion appetite.

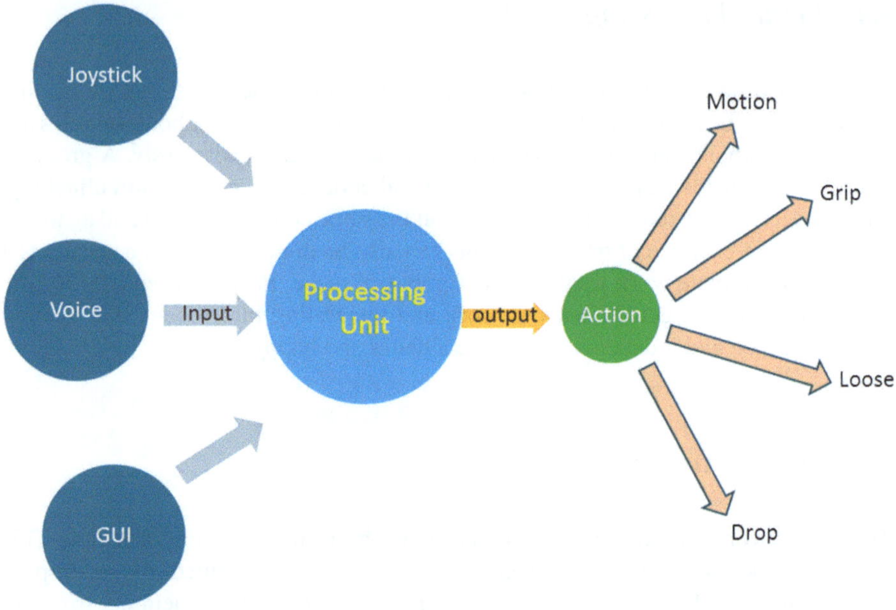


**Fig. 6** Projected types of inputs and outputs through smart textiles glove

## *5.2 Aesthetic Aspect*

Currently available prosthetic devices for the persons suffering from disease or brain injury are often bulky with mechanical parts or visual appearance of different parts or requires surgery to implant electrodes to connect to brain signal. Our proposed smart textile will be virtually invisible and will be amalgamated into a person's garments.

## *5.3 Cost*

It is expected that as the smart textile glove will be using easily available materials, the overall cost for manufacturing should be lower, and they can be manufactured using a number of textile manufacturing techniques, e.g. weaving, knitting, etc.

## *5.4 Integration and Acceptance in Society*

Utilization of the smart textile glove will help in disabled person's acceptance to use it and feel integrated into society. As this glove will be invisible and constitutes textile material, the overall reaction from the society will be normal.

### 5.5 *Reduction in Depression and Suicidal*

From a disabled person's point of view, the overall rejection or cutting off from the society comes majority due to pity or sympathy for disability; once they are viewed as a normal person, it is believed that the attitude of wearer and overlooker will be different. Consequently, it will give boost to the confidence for a disabled person and help in reduction of depression and suicidal behaviour.

### 5.6 *Upgradation with Age or Need*

As a person grows, especially kids and young children, there will be a need to replace the prosthetic limb which will involve a huge cost. By replacing or changing textile smart glove with growth or other reasons should be inexpensive. The smart textile glove if fits someone else can be used by them, or otherwise, the sensors and actuators can be taken out and attached to a new custom-made textile glove.

### 5.7 *Reduction in Overall Care Need*

With a person using smart textile glove, there are more chances for the individual to take care of himself or get back into workforce, reducing overall cost and care burden on families and governments.

### 5.8 *Comfort*

In general, textile materials have better flexibility and improved bending, which make them ideal material for use in gloves. Most of the artificial prosthetics are made of a variety of materials, e.g. wood, metal, nylon, polyester or composite material, none of them can be as comfortable as textile material.

## 6 Conclusion

Continuous and rapid progress in technologies, medicines and medical equipment is helping us in living longer and healthier life. This results in enlarge ratio of older and frail population. The traumatic brain injury is the result zof a blow or jolt to the brain either on skull or brain itself. Even though the high-speed automobile accidents have been by and large on rise, fall incidents have outnumbered the brain injuries due to accidents. Along with brain injuries, there are diseases or factors affecting brain at cellular level known as acquired brain injuries. The brain injury either caused by trauma or by other factors leads to multiple disabilities ranging from unnoticeable to impairment of limbs. In this chapter, we proposed a smart textile glove embedded with sensors and actuators which can be used through an IT/IS solution to control effected limb. The system consists of input devices, processing unit, output devices and textile core. It is expected that the smart textile glove will be virtually invisible, comfortable, light weight, cost-effective, trendy in fashion, easy to upgrade in case of growth or need and above all accepted by the disabled and society for better integration and improved confidence. This small change will yield greater confidence and ability for a disabled person to live independently, in some case returning to work, requiring less or no care, saving individuals and governments enormous budget. It will also make it easy for a disabled person to live a high-quality life by socializing and reducing depression and suicidal thoughts.

## References

*About Brain Injury*. (2016) [cited 2018 30 Jan 2018]; Available from: https://www.braininjuryaustralia.org.au/

*Acquired Brain Injury (ABI)*. [22/1/2018]; Available from: https://www.healthdirect.gov.au/acquired-brain-injury-abi

An Ageing Australia. (2013). *Preparing for the future*. Canberra.

*Australia's Health 2012: In Brief*. (2012). Canberra: AIHW.

*Australia's Health 2016*. (2016). *Australian Institute of Health and Welfare 2016*. Canberra: AIHW.

Bigler, E. (2013). Traumatic brain injury, neuroimaging, and neurodegeneration. *Frontiers in Human Neuroscience,7*(395), 395.

Bowker, J. H., & Michael, J. (1992). *Atlas of limb prosthetics*. Boston: Mosby Year Book.

*Brain Damage: Symptoms, Causes, Treatments*. (2018, November 02). Available from: https://www.webmd.com/brain/brain-damage-symptoms-causes-treatments#2

*Brain Injury is a "Hidden Disability"*. [28/1/2018]; Available from: https://www.braininjuryaustralia.org.au/brain-injury-2/

*Brain-Computer Interface Lets You Control IoT Devices*. [30/1/2018]; Available from: https://wtvox.com/cyborgs-and-implantables/brain-computer-interface-implantable/

*Causes of Acquired Brain Injury*. [22/1/2018]; Available from: http://raisingchildren.net.au/articles/acquired_brain_injury_d.html

Caution wet floor. (2019). [cited 2019 27/02/2019]; Available from: https://commons.wikimedia.org/wiki/File:Caution_wet_floor.jpg and it is from commons to be used without any permission. https://commons.wikimedia.org/wiki/File:Caution_wet_floor.jpg

Chew, E., & Zafonte, R. D. (2009). Pharmacological management of neurobehavioral disorders following traumatic brain injury-a state-of-the-art review. *Journal of Rehabilitation Research and Development, 46*(6), 851.

*Drowning Photo by Carlos Koblischek*. 2019 [cited 2019 27/02/2019]; Available from: https://www.freeimages.com/photo/help-1314958

*Drug Addiction*. 2019 [cited 2019 27/02/2019]; Available from: https://pixabay.com/images/id-2713526/

Economics, A. (2009). *The economic cost of spinal cord injury and traumatic brain injury in Australia*. In *Report by access economics for the Victorian Neurotrauma initiative* (p. 31). Canberra: Access Economics.

Edge, L. (2010). Traumatic brain injury: Time to end the silence. *The Lancet Neurology, 9*(4), 331.
Fang, B., et al. (2016). BodyScan: Enabling radio-based sensing on wearable devices for contactless activity and vital sign monitoring. In *Proceedings of the 14th annual international conference on mobile systems, applications, and services*. ACM.
Faul, M., et al. (2010a). *Traumatic brain injury in the United States*. Atlanta: Centers for Disease Control and Prevention, National Center for Injury Prevention and Control.
Faul, M., et al. (2010b). *Traumatic brain injury in the United States. Emergency department visits, hospitalizations, and deaths* (p. 2010). https://www.cdc.gov/traumaticbraininjury/pdf/blue_book.pdf.
Fritz, T., et al. (2014). Persuasive technology in the real world: a study of long-term use of activity sensing devices for fitness. In *Proceedings of the SIGCHI conference on human factors in computing systems*. ACM.
*From Below and Above Elbow Harness and Control System. Evanston, Ill.* [29/1/2018]; Available from: http://www.oandplibrary.org/popup.asp?frmItemId=0B931D83-40E1-4363-9BAE-223AD4C612EB&frmType=image&frmId=10
Gaetz, M. (2004). The neurophysiology of brain injury. *Clinical Neurophysiology, 115*(1), 4–18.
*Honda City Hit by Motorcycle*. (2019). [cited 2019 26/02/2019]; Available from: https://commons.wikimedia.org/wiki/File:Honda_City_hit_by_motorcycle_1.jpg
*How Brain Injury Occurs*. [2018 8/3/2018]; Available from: https://www.betterhealth.vic.gov.au/health/conditionsandtreatments/acquired-brain-injury
Imani, S., et al. (2016). A wearable chemical–electrophysiological hybrid biosensing system for real-time health and fitness monitoring. *Nature Communications, 7*, 11650.
Jin, H., Abu-Raya, Y. S., & Haick, H. (2017). Advanced materials for health monitoring with skin-based wearable devices. *Advanced Healthcare Materials, 6*, 1700024.
Khan, Y., et al. (2016). Monitoring of vital signs with flexible and wearable medical devices. *Advanced Materials, 28*(22), 4373–4395.
Kwak, N.-S., Müller, K.-R., & Lee, S.-W. (2015). A lower limb exoskeleton control system based on steady state visual evoked potentials. *Journal of Neural Engineering, 12*(5), 056009.
Langlois, J. A., Rutland-Brown, W., & Thomas, K. E. (2006a). *Traumatic brain injury in the United States; Emergency department visits, hospitalizations, and deaths.*
Langlois, J. A., Rutland-Brown, W., & Wald, M. M. (2006b). The epidemiology and impact of traumatic brain injury: A brief overview. *The Journal of Head Trauma Rehabilitation, 21*(5), 375–378.
Lezak, M. D., et al. (2004). *Neuropsychological assessment*. New York: Oxford University Press.
Liang-Jie Zhang, C. L. (2017). Internet of things solutions. *Services Transactions on Internet of Things (STIOT), 1*(1), 1–22.
*Low-cost Smart Glove Wirelessly Translates the American Sign Language Alphabet into Text.* [30/1/2018]; Available from: https://phys.org/news/2017-07-low-cost-smart-glove-wirelessly-american.html
Mathers, C., T. Vos, C. Stevenson. (1999). The burden of disease and injury in Australia. Australian Institute of Health and Welfare.
McGinn, M. J., & Povlishock, J. T. (2016). Pathophysiology of traumatic brain injury. *Neurosurgery Clinics, 27*(4), 397–407.
*Mind-controlled Artificial Limb Gives Patients Sense of Touch Again*. [30/1/2018]; Available from: http://www.wired.co.uk/article/darpa-touch-sensitive-prosthetic
Moan, I. S., Norström, T., & Storvoll, E. E. (2013). Alcohol use and drunk driving: The modifying effect of impulsivity. *Journal of Studies on Alcohol and Drugs, 74*(1), 114–119.
*Moving Your Hand and Arm with the MyoPro Motion G*. [9/03/2018]; 2]. Available from: http://myomo.com/wp-content/uploads/2016/12/MyoPro-BPI-Patient-Trifold-PN-2348-Rev-3-12-11-16-v4.pdf
*My Own Motion*. 2018 [cited 2018 24/1/2018]; Available from: http://myomo.com/
*National Community Services Data Dictionary*. (2012). Canberra: Australian Institute of Health and Welfare.

Osier, N. D., et al. (2015). Chronic histopathological and behavioral outcomes of experimental traumatic brain injury in adult male animals. *Journal of Neurotrauma, 32*(23), 1861–1882.

*Papillary Glioneuronal Tumor*. 2019 [cited 2019 27/02/2019]; Available from: https://en.m.wikipedia.org/wiki/File:Papillary_glioneuronal_tumor.jpg

Pervez, M., Kitagawa, R. S., & Chang, T. R. (2018). Definition of traumatic brain injury, neurosurgery, trauma orthopedics, neuroimaging, psychology, and psychiatry in mild traumatic brain injury. *Neuroimaging Clinics, 28*(1), 1–13.

Polinder, S., et al. (2015). Health-related quality of life after TBI: A systematic review of study design, instruments, measurement properties, and outcome. *Population Health Metrics, 13*(1), 4.

Ragnarsson, K., et al. (1999). Rehabilitation of persons with traumatic brain injury. *Journal of the American Medical Association, 282*(10), 974–983.

Regalado, A. (2014, Jan 17). *Rewriting life the thought experiment*. [cited 2018 29/1/2018]; Available from: https://www.technologyreview.com/s/528141/the-thought-experiment/

Rodriguez-Paez, A. C., Brunschwig, J., & Bramlett, H. M. (2005). Light and electron microscopic assessment of progressive atrophy following moderate traumatic brain injury in the rat. *Acta Neuropathologica, 109*(6), 603–616.

Roozenbeek, B., Maas, A. I., & Menon, D. K. (2013). Changing patterns in the epidemiology of traumatic brain injury. *Nature Reviews Neurology, 9*(4), 231.

Rushworth, N., (2008). *Brain Injury Australia: Submission to the Australian Government's National Mental Health and Disability Employment Strategy*. https://www.braininjuryaustralia.org.au/wp-content/uploads/BRAININJURYAUSTRALIAsubmissionEMPLOYMENTSTRATEGYfinal.pdf.

Sample, P. L., & Darragh, A. R. (1998). Perceptions of care access: The experience of rural and urban women following brain injury. *Brain Injury, 12*(10), 855–874.

*Sensor Gloves*. [30/1/2018]; Available from: http://stdl.se/?p=3851

Smith, D., et al. (1999). Accumulation of amyloid β and tau and the formation of neurofilament inclusions following diffuse brain injury in the pig. *Journal of Neuropathology and Experimental Neurology, 58*(9), 982–992.

*Sports Photo by Chris Chow*. (2019). [cited 2019 27/02/2019]; Available from: https://unsplash.com/photos/zVmuTzhmvXY

Stocchetti, N., & Zanier, E. R. (2016). Chronic impact of traumatic brain injury on outcome and quality of life: A narrative review. *Critical Care, 20*(1), 148.

Teasdale, G., & Jennett, B. (1976). Assessment and prognosis of coma after head injury. *Acta Neurochirurgica, 34*(1–4), 45–55.

*The Groundbreaking Mind-controlled BIONIC ARM that Plugs into the Body and has Given Wearers Back their Sense of Touch*. [30/1/2018]; Available from: http://www.dailymail.co.uk/sciencetech/article-2785769/The-mind-controlled-prosthetic-limb-given-wearer-sense-touch.html

*Thought-controlled Prosthesis is Changing the Lives of Amputees*. (2012, December 05). [cited 2018 29/1/2018]; Available from: http://www.chalmers.se/en/news/Pages/Thought-controlled-prosthesis-is-changing-the-lives-of-amputees.aspx

*Types of Upper Extremity Prostheses*. (2018). Available from: http://www.pandocare.com/upper-extremity-types/#1471966938595-7757743e-c7b3

Van Bost, G., Van Damme, S., & Crombez, G. (2017). The role of acceptance and values in quality of life in patients with an acquired brain injury: A questionnaire study. *PeerJ, 5*, e3545.

*What is a MyoPro Orthosis?* [2401/2018]; Available from: http://myomo.com/what-is-a-myopro-orthosis/

Wickham, C. Mind-controlled robotic arm has skill and speed of human limb. [2012, December 17 29/1/2018]; Available from: https://www.reuters.com/article/us-science-prosthetics-mindcontrol/mind-controlled-robotic-arm-has-skill-and-speed-of-human-limb-idUSBRE8BG01H20121217

Zheng, Y.-L., et al. (2014). Unobtrusive sensing and wearable devices for health informatics. *IEEE Transactions on Biomedical Engineering, 61*(5), 1538–1554.

# A Mobile Nursing Solution

**Jainqui Kou, Qianqian Zhang, and Nilmini Wickramasinghe**

## 1 Introduction

In recent years, the severe consequences of medical error are realized by both healthcare practitioners and the public because of the critical direct injury towards the involved patients and the high indirect costs subsequently all around the world (Van et al. 2011). Thus, a critical problem is to reduce the occurrence of medical errors and improve the efficiency of clinical treatment at the same time (Wang et al. 2013). The utilization of information and communication technology (ICT), in particular the tools and techniques of the Internet of Things in the healthcare field, is recognized to have the potential of improving the quality of clinical treatment and reducing medical errors (Govind and Bhatt 2007). Moreover, information and communication technology has already shown its significance in improving the effectiveness and outcome of healthcare delivery (Haluza and Jungwirth 2015).

To achieve safe and high-quality healthcare service, the whole process of medication administration needs to be examined (Bowman 2013). Since there are various steps from the physicians prescribing orders to the orders being performed and documented according to the type of the order, the concept of closed-loop medication administration is a strategy to deliver medication safely (Henderson et al. 2013).

A closed-loop medication administration system is a medication administration system which records and traces every step through the prescribing to implementing of every single order using some emerging technology (Hwang et al. 2016). As

J. Kou · Q. Zhang
Deakin University, Nanjing, China

N. Wickramasinghe (✉)
Epworth HealthCare, Richmond, VIC, Australia

Swinburne University of Technology, Hawthorn, VIC, Australia



N. Wickramasinghe, F. Bodendorf (eds.), *Delivering Superior Health and Wellness Management with IoT and Analytics*, Healthcare Delivery in the Information Age, https://doi.org/10.1007/978-3-030-17347-0_5

a result, some errors can be identified and avoided with an automated check and remind technique (Voshall et al. 2013). Even if medical errors occur, the records can be retrieved, and the cause of the error can be quickly revealed so that appropriate corrective action can be quickly taken. Therefore, closed-loop medication administration is an efficient solution since all the processes are monitored and recorded by its information system (Franklin et al. 2007).

Among all the clinical staff, nurses play an integral key role in the whole patient care process in hospitals including medication administration, and most of the medical errors could be recognized, handled and prevented by nursing staff (Cornell et al. 2010a, b). According to (Gaffney et al. 2016), although the medication order is prescribed by the physicians and dispensed by the pharmacist, medication errors are a constant issue of nursing practice because the rest of the processes such as preparing, checking, giving medications, monitoring the effectiveness or the adverse reaction are more likely to lead to medical errors; hence all the errors have the possibilities to be found and prevented by nurses (Sherwood and Zomorodi 2014).

Mobile nursing system is the information system used by nurses to record their daily work and make it possible to enquire all the related patient information anywhere anytime in the clinical setting institutions with the use of technology such as WLAN, handheld terminal of personal digital assistant (PDA), bar-code scanning and so on (Wang et al. 2015). Mobile nursing systems can smooth the working flow and patient flow through transforming the information of patients and treatment to everywhere at anytime (Su and Liu 2012). The mobile nursing systems can also record every step of the implementation of the prescribing treatment timely and accurately (Wang et al. 2015). As a result, mobile nursing systems may have the possibility to improve the efficiency and patient safety through the achievement of closed-loop medication administration (Su and Liu 2012).

The research aims to explore the following major question:

*How can a mobile nursing system be used to achieve closed-loop medication administration to deliver high-quality and safe nursing care?*

In order to answer this key question, the following questions are also investigated:

(i) What are the key issues in designing and using mobile nursing system in the implementation of the closed-loop medication administration?
(ii) What are the benefits and problems when utilizing mobile nursing system to achieve closed-loop medication administration?

## 2 Materials and Methods

The method of designing and implementing a mobile nursing system to achieve close-loop medication administration is comprehensive and innovative (Su and Liu 2012). Consequently, how to design and implement a mobile nursing system that brings benefit to both the nurses and the patients is very significant. This chapter will just present a meaningful attempt to accomplish this goal – to achieve close-loop medication administration using mobile nursing system successfully.

## 2.1 Case Study

In this research, a single exemplar case study is appropriate since the case represents, illustrates and examines existing theory and insights from this case are used as a prelude to future research (Yin 2017). A single case study of designing and implementing a mobile nursing system with the purpose of completing closed-loop medication administration in a large hospital located in China has been conducted to investigate the research question. The case site is a large hospital that contains around 3000 inpatients beds so the situation there is very complicated, and therefore the solution would be reliable to be spread since various kinds of problem would be encountered and solved in this single case site. In the data collection phase, some kinds of qualitative research techniques including semi-structured interviews, focus group interview and site observation were utilized to achieve primary data, and secondary data was acquired through reading related files and archives. In addition to this, a questionnaire had provided some quantitative data to evaluate the effect of such a mobile nursing system.

## 2.2 Questionnaire

Since filling out a questionnaire is less time-consuming than interviewing, more nurses can participate in the questionnaire. The questionnaire was developed specifically for the study, based on a literature review, experts' suggestions and the results of a pre-survey study. The pre-survey study was conducted to collect feedback from respondents and survey experts, and the researchers reached consensus regarding the final questionnaire by making necessary revisions to the original version.

The structured questionnaire was designed to include the basic information and opinion of the users from the site hospital. The questionnaires used Likert scale items where respondents were asked to agree/disagree with various statements such as "Portability of tablets". Most scales use a five-point range – strongly agree/agree/not sure/disagree/strongly disagree – and include a set of items designed to cover a range of constructs and which have been piloted to test for reliability.

### 2.3 Interviews

In addition to the questionnaire, interviews were designed in order to include the opinion of the users from the Chinese hospital into the study. All interviews were audio-recorded for verbatim transcription prepared by a professional transcriptionist, checked for accuracy against the sound files by the interviewers and corrected where necessary (Yatim et al. 2017). The data collected was then interpreted through the lens of Activity Theory (Er and Lawrence 2011). In the data analysis phase, thematic analysis and arithmetic analysis were adopted to acquire the research results

## 3 Results

### 3.1 Questionnaire Survey

The questionnaire consisted of 18 questions pertaining to demographic information, satisfaction with mobile nursing information system and the effectiveness of the system. Five questions were used to collect the following demographic information: gender, age, educational background, role as a nurse and system usage time. Seven questions were used to collect the following information regarding satisfaction degree: feedback on the hardware and software of the system. Six questions were used to collect the following information regarding effectiveness of mobile nursing information system: degree of safety, accuracy, timeliness of information, sufficiency of nursing and optimize processes.

Led by hospital's contact person, trained investigators distributed questionnaires to nurses and asked them to complete it, independently, voluntarily and autonomously. Questionnaires were collected immediately after completion. All valid questionnaire responses were recorded and entered into Excel. SPSS (version 22.0, IBM, Chicago, IL, USA) was used to perform statistical analysis of the data.

Participants' basic demographic information is shown in Table 1. Regarding respondents' gender, the number of female included in the study was higher relative to that of male, with a male-to-female ratio of 1:5. It was due to the fact that a large majority of nurses are female in general. With respect to age group, most participants were aged between 20 and 40 years, and the largest group included those aged between 20 and 30 years. Regarding educational background, most respondents held bachelor degree at least, and very few held degrees below bachelor's level. Regarding the respondents' job titles, most residents were senior nurse, followed by supervisor nurse and above; the nurse is the lowest, but the difference is marginal. Regarding the length of usage of the mobile nursing information system, most of the respondents used the system for more than 181 days, followed by 11 respondents who used the system for a period between 91 and 180 days; there was only one nurse

**Table 1** Basic information of the interviewees

| Item | Selected options | Number of persons |
|---|---|---|
| Gender | Male | 5 |
| | Female | 25 |
| Age | 20–30 years | 19 |
| | 30–40 years | 10 |
| | 40–50 years | 1 |
| | Older than 50 | 0 |
| Educational level | Secondary degree | 0 |
| | College degree | 6 |
| | Bachelor degree and above | 24 |
| Titles | Nurse | 7 |
| | Senior nurse | 13 |
| | Supervisor nurse and above | 10 |
| Length of usage | Less than 30 days | 0 |
| | 31–90 days | 1 |
| | 91–180 days | 11 |
| | More than 181 days | 18 |

who used the system for a period between 31 and 90 days, while no result recorded less than 30 days.

Results about satisfaction with mobile nursing system are shown in Table 2. In total, 30 questionnaires were distributed, and all were collected, providing a response and valid rate of 100%. 1 represents great satisfaction, 5 represents very dissatisfied; the smaller the score, the more satisfied. Regarding seven questions, the mean satisfaction is 3 which reveals most people are satisfied with the mobile nursing information system. Most satisfying (2.07) is the training how to use mobile nursing information system and tablets, and the information display of tablets (2.50) was the second, leaving response speed of the system (3.90) last. It is worth mentioning that stability of wireless networks needs to be strengthened as well (3.90). Another interesting pattern is the lower variance, the more consensuses on the feedback which are consistent with the variance results.

Results regarding effectiveness of mobile nursing information system are shown in Table 3. In total, 30 questionnaires were distributed, and all were collected, providing a response and valid rate of 100%. 1 represents strongly agree, 5 represents disagree; the smaller the score, the more they agree with the view of the investigation. The question that scored lowest among the six points is improve the accuracy of patient identification (1.50), and the next is agree with improves patient safety (1.60). One difference is that average score is lower than satisfaction survey, that is, respondents pretty much agree that mobile nursing information system improves accuracy and safety. Consistent with the previous section, the basic pattern is the lower variance, the more consensuses on the feedback which are consistent with the variance results.

**Table 2** Satisfaction survey results

| | Great satisfaction | Satisfaction | General | Dissatisfaction | Very dissatisfied | Max | Min | Mean | Var |
|---|---|---|---|---|---|---|---|---|---|
| Portability of tablets | 1 | 6 | 14 | 6 | 3 | 5 | 1 | 3.13 | 0.95 |
| Stability of wireless networks | 0 | 1 | 5 | 18 | 6 | 5 | 2 | 3.90 | 0.71 |
| System navigation design | 2 | 5 | 20 | 3 | 0 | 4 | 1 | 2.80 | 0.51 |
| Information display of tablets | 1 | 13 | 16 | 0 | 0 | 3 | 1 | 2.50 | 0.33 |
| Convenience of information entry | 2 | 9 | 15 | 4 | 0 | 4 | 1 | 2.70 | 0.63 |
| Response speed of the system | 0 | 3 | 5 | 14 | 8 | 5 | 2 | 3.90 | 0.85 |
| System training | 3 | 22 | 5 | 0 | 0 | 3 | 1 | 2.07 | 0.27 |

**Table 3** Effectiveness survey results

| | Strongly agree | Agree | General | Partly disagree | Disagree | Max | Min | Mean | Var |
|---|---|---|---|---|---|---|---|---|---|
| Greatly improves the accuracy of patients identification | 18 | 9 | 3 | 0 | 0 | 3 | 1 | 1.50 | 0.47 |
| Greatly improves patient safety | 14 | 14 | 2 | 0 | 0 | 3 | 1 | 1.60 | 0.39 |
| Greatly increases the work efficiency | 0 | 4 | 10 | 10 | 6 | 5 | 2 | 3.60 | 0.94 |
| Greatly optimized the workflow | 2 | 9 | 9 | 7 | 3 | 5 | 1 | 3.00 | 1.24 |
| Greatly improved patient satisfaction | 2 | 9 | 14 | 3 | 2 | 5 | 1 | 2.80 | 0.92 |
| Storage data is more accurate and safe | 6 | 14 | 7 | 3 | 0 | 4 | 1 | 2.23 | 0.81 |

**Table 4** Outline of semi-structured interviews

| | |
|---|---|
| Benefits | What are the benefits of using a mobile nursing system?<br>Who can benefit from mobile nursing system?<br>Why and how they can benefit? |
| Problems | What are your dissatisfactions in the usage of this information system?<br>What needs to be improved in the system design and implementation? |
| Improvement | What needs to be improved do you think regarding to the mobile nursing systems? |

## *3.2 Semi-structured Interviews*

For this study, semi-structured interviews were conducted with 15 nurses in the site hospital. One of the objectives of this study was to conduct interviews to unearth further information on the issues in the questionnaire survey study regarding nurses' feedback information about utilizing mobile nursing system to achieve closed-loop medication administration. In previous study we learned that the degree of information demand was high, while degree of information satisfaction was general for nurses, and they expected high on the inclusion of new information technologies (Wen et al. 2017). As shown in Table 4, the interviews included three aspects comprising the benefits of mobile care system, the current situations and problems of information systems and evaluation on the improvement of mobile care through the usage of information technology

The interviews were conducted and recorded upon obtaining the approval of the interviewees, and the interview recordings do not reveal sensitive personal identifiable information such as the names and contact methods of the interviewees. Upon completion of the interviews, the interviews were transcribed from audio into text, and the interview contents were analysed using the inductive conventional content analysis approach (Hsieh and Shannon 2005). Researchers read the transcribed texts several times to understand the information conveyed in the interviews. The key information of the interview contents was extracted, coded and subsequently categorized into three themes: benefits, problems and advices

### 3.2.1 Benefits from Using Mobile Nursing Information System

The nurses were generally satisfied with the current situation of mobile nursing information system. First, nearly all of the interviewees mentioned that there were no obvious problems in terms of patient identity verification and implementation of doctor's orders, and the main reasons were that system training in the early stage was quite clear and comprehensive. Notably, some interviewees indicated that mobile nursing information system can automatically record the actual implementation time and operator number. If the information was wrong, the system displays a mismatch, and then a warning information would appear on the screen according to the reasons to remind nurses. Therefore, mobile nursing information system improves the safety of clinical medication, effectively prevents and reduces the incidence of medical care errors and improves the level of hospital nursing management

On the other hand, patient wristbands were equipped with important information signs for patient identification and verification. By scanning the two-dimensional code, identification information of patients can be identified and managed, which can effectively improve the quality and efficiency of patient information management. Thus patient compliance was increased by means of mobile nursing information system because of trust and professionalism

### 3.2.2 Dissatisfaction with the Mobile Nursing Information System

With regard to the usage of current mobile nursing information system, some of the interviewees said that the system is not running smoothly in some corners, which was likely due to weak network coverage in that area. On another note, the interviewees mentioned that they have encountered occasional system errors such as network disruptions and software crashes that significantly affected their work efficiency and regular operations. These situations even can lead to a loss of information, and the problem was required to be fixed

### 3.2.3 Anticipation of the Mobile Nursing Information System

As a system supplier, the programmers lack both in-depth understanding of the hospital information system and clinical experience, and they can't fully understand the requirements of the nurses. Therefore, the development and design of the system should be completed together with the nurses, so as to ensure relatively consistency with the clinical work of real environment. For the maintenance period after implementation, the interviewees generally recommend that technical personnel should take regular on-site visits to check and refine the system

## 4 Discussion

On the date of the survey, the questionnaires were sent randomly to nurses in multiple departments. The demographic data in Table 1 show that respondents' gender was essentially imbalanced, and age, educational background, title and usage days showed to be non-homogeneous. Nurses were generally young and highly educated with college's degrees, whereby the numbers of supervisor nurse in hospitals decrease as their seniority increases. In hospitals in Mainland China, senior nurses with rich experience are employed mainly for nursing patients, while those at the highest professional level are responsible for caring quality and management. The results here could have been affected by the fact that the sample size included in the study was small. The sociotechnical approach which combined quantitative and qualitative research methods proved to be a strong research policy, in which outcomes of interviews supported and explained the results of the questionnaire. The results of the study were meaningful; there was some difficulty in conducting large sample, simple random sampling of all nurses in hospitals because their roles varied, and they are very busy and change shifts relatively rapidly. Under the current conditions, it is difficult to draw conclusions regarding the general characteristics of all nurses for comparison and analysis, which was one of the limitations of this survey.

Overall, our studies obtained the results of nurses' satisfaction survey on mobile nursing system; although there were important discoveries revealed by these studies, there are also limitations. As shown in Table 2, there is a strong need for improving stability and speed of mobile caring information system. The nurses were moderately satisfied with portability with tablets and system design as average. Half of respondents (50%) were only moderately satisfied with convenience, which indicates that the mobile tablets required improvement. According to (Hilliard et al. 2014), the importance of a user-friendly design needs to be taken into account when developing new m-Health services. The combination of user-specific requirements and limited relative small-screen interfaces as on smart phones creates design challenges in usability, which includes learning ability, efficiency and satisfaction as the general requirements. In a systematic approach to design principles, Su et al. showed how a mobile nursing information system can be developed when taking care of the usability aspects and the user's cognitive model, and not only the technology aspects (Su and Liu 2012). 73.3% of nurses were satisfied with system training, which means mobile nursing system was well prepared before implementation.

According to Table 3, the following six indicators of the effectiveness of mobile nursing system were rated as average in the current study: accuracy (60% of nurses reported that systems were very accurate), safety (46.7% of nurses reported that the system can assure safety of patients identification), optimization (only 10% of nurses reported that the workflow of system is optimized), time sufficiency (none of the nurses reported having sufficient time), data safety (only 20% nurses reflected that the system can guarantee data security) and satisfaction (only 6.7% of nurses reported that they were very satisfied with the system).

In the current study, the scale was divided into only questionnaire survey and semi-structured interview; therefore, these feedback and anticipation of the mobile nursing information system should be subdivided further in future surveys. Even though the survey results showed that respondents were moderately satisfied with mobile nursing information system, it did not create obstacles to clinical work.

## 5 Conclusions

The results indicate that there was specific demand for mobile nursing information system in China, and there is no difference between the requirements according to gender, age, title, usage period and educational backgrounds. However, nurses were only moderately satisfied with current mobile nursing information system, suggesting that they require improvements. Further requirements are recommended for the system in the following aspects: further communication with clinical staff, regular on-site support and technical support.

## References

Bowman, S. (2013). Impact of electronic health record systems on information integrity: Quality and safety implications. *Perspectives in Health Information Management, 10*, 1c.

Cornell, P., Herrin-Griffith, D., Keim, C., Petschonek, S., Sanders, A. M., D'Mello, S., Golden, T. W., & Shepherd, G. (2010a). Transforming nursing workflow, part 1: The chaotic nature of nurse activities. *Journal of Nursing Administration, 40*, 366–373.

Cornell, P., Riordan, M., & Herrin-Griffith, D. (2010b). Transforming nursing workflow, part 2: The impact of technology on nurse activities. *Journal of Nursing Administration, 40*, 432.

Er, M., & Lawrence, E. (2011). *Using activity theory to examine information systems for supporting mobile work*. Slovenia: University of Maribor.

Franklin, B. D., O'Grady, K., Donyai, P., Jacklin, A., & Barber, N. (2007). The impact of a closed-loop electronic prescribing and administration system on prescribing errors, administration errors and staff time: A before-and-after study. *Quality & Safety in Health Care, 16*, 279–284.

Gaffney, T. A., Hatcher, B. J., & Milligan, R. (2016). Nurses' role in medical error recovery: An integrative review. *Journal of Clinical Nursing, 25*, 906.

Govind, P. & Bhatt, J. (2007). Information technology to reduce medical errors: An overview. *Indian Journal of Hospital Pharmacy, 64*(5), 169–173.

Haluza, D., & Jungwirth, D. (2015). ICT and the future of health care: Aspects of health promotion. *International Journal of Medical Informatics, 84*, 48–57.

Henderson, D., Lunak, R., Markiewicz, E., & Tobin, C. C. (2013). *Closed loop medication use system and method*. Google Patents. https://patentscope.wipo.int/search/en/detail.jsf?docId=WO2004001539

Hilliard, M. E., Hahn, A., Ridge, A. K., Eakin, M. N., & Riekert, K. A. (2014). User preferences and design recommendations for an mHealth app to promote cystic fibrosis self-management. *JMIR mHealth and uHealth, 2*, e44.

Hsieh, H. F., & Shannon, S. E. (2005). Three approaches to qualitative content analysis. *Qualitative Health Research, 15*, 1277–1288.

Hwang, Y., Yoon, D., Ahn, E. K., Hwang, H., & Park†, R. W. (2016). Provider risk factors for medication administration error alerts: Analyses of a large-scale closed-loop medication administration system using RFID and barcode. *Pharmacoepidemiology & Drug Safety, 25*, 1387.

Sherwood, G., & Zomorodi, M. (2014). A new mindset for quality and safety: The QSEN competencies redefine nurses' roles in practice. *Nephrology Nursing Journal, 41*, 15.

Su, K.-W., & Liu, C.-L. (2012). A mobile Nursing Information System based on human-computer interaction design for improving quality of nursing. *Journal of Medical Systems, 36*, 1139–1153.

Van, D. B. J., Rustagi, K., Gray, T., Halford, M., Ziemkiewicz, E., & Shreve, J. (2011). The $17.1 billion problem: The annual cost of measurable medical errors. *Health Affairs, 30*, 596–603.

Voshall, B., Piscotty, R., Lawrence, J., & Targosz, M. (2013). Barcode medication administration work-arounds: A systematic review and implications for nurse executives. *Journal of Nursing Administration, 43*, 530–535.

Wang, L.-R., Wang, Y., Lou, Y., Li, Y., & Zhang, X.-G. (2013). The role of quality control circles in sustained improvement of medical quality. *Springerplus, 2*, 141.

Wang, X., Luo, Y., Shang, M., Li, Z., Zhang, M., & Sun, Z. (2015). *Application of mobile nursing information system in high-quality nursing*. International Conference on Education, Management, Commerce and society (EMCS-15).

Wen, D., Zhang, X., Wan, J., Fu, J., & Lei, J. (2017). The challenges of emerging HISs in bridging the communication gaps among physicians and nurses in China: An interview study. *Bmc Medical Informatics & Decision Making, 17*, 85.

Yatim, F., Cristofalo, P., Ferrua, M., Girault, A., Lacaze, M., di Palma, M., & Minvielle, E. (2017). Analysis of nurse navigators' activities for hospital discharge coordination: A mixed method study for the case of cancer patients. *Support Care Cancer, 25*, 863–868.

Yin, R. K. (2017). *Case study research and applications: Design and methods*. Oakwood, California: Sage Publications.

**Jianqiu Kou** Ph.D. student of Deakin University Australia, and she has 20 years of working experience in the field of construction and management of information systems in healthcare institutions. Her main research interests lie in the application of information technology in healthcare field.

**Qianqian Zhang** Masters' major in biomedical engineering. She has 5 years working experience in biobank information system construction and bioinformatics analysis. Her research interests and directions are the application of information technology, precision medicine and molecular diagnosis in clinical practice.

# SmartCoping: A Mobile Solution for Recognizing Stress and Coping with It

**Ulrich Reimer, Edith Maier, and Tom Ulmer**

## 1 Introduction

Stress is the body's normal response to a real or implied threat. In small doses, stress can help us perform under pressure and make us stay focused, energetic, and alert. However, if stress symptoms persist, it starts causing major damage to our health, productivity, relationships, and quality of life. Chronic stress can cause hypertension, suppress the immune system, increase the risk of heart attack and stroke, and make people more vulnerable to anxiety, addictive behavior, and depression (Legendre and Harris, 2006; Ornish et al., 1990). Excessive and prolonged stress may cause burnout, a state of emotional, mental, and physical exhaustion.

We cannot completely eliminate stress from our lives, but we can learn how to cope with it by controlling stress-inducing situations and physiological reactions. This, however, requires that we are aware of the fact that we are stressed at a particular moment, by certain events or by encounters with specific persons. With this in mind, the publicly funded research project *SmartCoping*[1] set out to develop a mobile app for stress warning and relaxation support. The project has pursued the following main goals:

- Help users *enhance their self-perception* so that they are able to recognize when they suffer from stress. This is achieved by continuous monitoring of stress levels

[1]The project was funded by the Swiss Confederation's innovation promotion agency under grant number 14049.2 PFES-ES.

U. Reimer (✉) · E. Maier · T. Ulmer
University of Applied Sciences, St. Gallen, Switzerland
e-mail: ulrich.reimer@fhsg.ch; edith.maier@fhsg.ch; tom.ulmer@fhsg.ch



N. Wickramasinghe, F. Bodendorf (eds.), *Delivering Superior Health and Wellness Management with IoT and Analytics*, Healthcare Delivery in the Information Age, https://doi.org/10.1007/978-3-030-17347-0_6

and warning users when their stress level is high. This function helps users improve their self-perception and after some time be able to recognize on their own when they are stressed.

- Give users hints in which *situations they experience significantly higher stress* so that they may develop avoidance or coping strategies.
- Help users relax with a *biofeedback* component.

The SmartCoping app addresses two application scenarios:

1. *The prevention of chronic stress:* The target group consists of individuals who are or feel threatened by stress or who are interested in measuring and documenting their vital as well as contextual data so as to increase their self-awareness and long-term health (Swan, 2012).
2. *Therapeutic and rehabilitation support for conditions caused by stress:* Here the target group are in- or outpatients who need support in avoiding stress, e.g., patients after alcohol detoxification, burnout patients, or patients suffering from depression. In this scenario the therapist or nurse may have access to the data if the patient agrees.

This paper is an updated and considerably extended version of an earlier publication (Reimer et al., 2017b). In the following, Sect. 2 discusses the physiological parameters through which stress can be detected. Section 3 gives an overview of the state of the art in stress detection and management. Section 4 is the core of the paper and presents the design rationales and approaches underlying SmartCoping. Section 5 deals with the evaluation of the system, while Sect. 6 concludes the paper.

## 2 Physiological Indicators of Stress

Stress results in the release of adrenaline and, with a certain time delay, cortisol. *Cortisol* can be measured in the saliva (van Eck et al., 1996; Koh et al., 2007), where the change of cortisol level – not the absolute level – is a good indicator for medium-term stress (Vedhara et al., 2003). Short-term stress can best be determined by the adrenaline level in the blood, which requires taking blood samples. Another good indicator of short-term stress is *alpha-amylase*, which has the advantage that it can be quite easily tested in saliva (Noto et al., 2005; Koh et al., 2007; Harmon et al., 2008).

*Heart rate variability* (HRV) is generally considered a reliable indicator for stress (Delaney and Brodie, 2000; Orsila et al., 2008; Tanev et al., 2014). It correlates well with salivary measurements of cortisol and alpha-amylase as has been shown in Liew et al. (2015). Stress reduces the fluctuation of beat-to-beat intervals, leading to lower HRV values, whereas a reduction of stress lets the fluctuation increase again and thus corresponds to a higher HRV.

Stress causes increased perspiration so that it can be measured via *skin conductance*, preferably on palms and foot soles (Huang et al., 2011; Villarejo et al., 2012). Two types of electrodermal activity can be distinguished: *phasic skin con-*

*ductance response* which means a rapid change in skin conductance and *tonic skin conductance level* which refers to a slowly changing (background) skin conductance (Boucsein, 2012). Phasic skin conductance measurements typically occur as a result of short-term stimuli and reflect arousal. Tonic skin conductance is associated with stressful stimuli and can therefore be used to detect stress (Weinert et al., 2015). While these two types of electrodermal activity can be used to distinguish arousal from stress, skin conductance still remains a difficult parameter for determining stress since it also increases as a result of regulating body temperature and thus requires a controlled environmental setting.

Stress can further be detected from electroencephalogram (EEG) signals. The characteristic, stress-indicating wave patterns in the EEG can be identified with machine learning techniques as is shown in Vanitha and Krishnan (2016) and Gaurav et al. (2017). The training data needed for learning such stress classifiers is obtained by inducing stressful situations in a controlled laboratory setting.

Facial *skin temperature* is another parameter for measuring stress (Yamakoshi et al., 2008), as is blood pressure. Both, however, are much less sensitive and selective than, e.g., HRV (Hjortskov et al., 2004).

Stress levels as well as other emotional states can be detected from *facial expressions* (Dinges et al., 2005; Gao et al., 2014; Giannakakis et al., 2017; Jáuregui et al., 2017), body movements (Jáuregui et al., 2017; Mozos et al., 2017), and *speech* via automatic speech analysis (El Ayadi et al., 2011; van den Broek et al., 2013; Simantiraki et al., 2016). These indicators do not lend themselves to the continuous measurement of stress levels unless they are combined with other indicators (Mozos et al., 2017).

## 3 Stress Detection and Management

There is a plethora of health-related apps on the market including apps for coping with stress, e.g., apps which trigger stress warnings, offer progressive muscle relaxation, or give guided exercises. Whereas some apps determine a user's stress level by asking a series of questions, the more innovative apps use the sensors integrated in many of today's smartphones as well as external sensors to recognize and display stress symptoms and monitor them over time. An early approach was the Mobile Heart Health Project driven by Intel researchers. They used a wireless ECG to detect changes in stress levels as captured by heart rate variability (HRV) and to trigger mobile interventions such as breathing techniques (Morris and Guilak, 2009). So-called mood maps were used for subjective assessment to correlate HRV measurements with self-perception. In the end, the HRV measurement was discontinued because of the challenges posed by the continuous capturing of sensor data in everyday life. Since then wearable sensors have much improved, opening up new opportunities for realizing mobile stress recognition.

HRV is a physiological parameter that is frequently used for stress recognition. Several studies use HRV to learn a classifier for distinguishing between relaxed states and stressful situations (Melillo et al., 2011; Karthikeyan et al., 2013; Kaur et al., 2014; Alexandratos et al., 2014). A simpler approach is described in Andreoli et al. (2010) where stress is assumed to be present when the measured HRV parameters exceed predefined thresholds.

Using HRV as a stress indicator, however, is doomed to failure unless the correlation between HRV and stress level is calibrated to the specific user because each person has quite a different range of possible HRV values. The approach described in Muaremi et al. (2013) implements such a user-specific calibration where a daily stress score is calculated from multiple inputs, among them HRV. A stress level classifier is trained against a self-assessment questionnaire which the study participants have to fill in several times a day. The correctness of stress level classification is determined by comparing it with the stress level derived from the questionnaire. Combining all features, an accuracy of 61% was achieved. Just like in SmartCoping, the authors included an initial learning phase to train their classifier before switching to classification mode. Not surprisingly, user-specific classification models proved superior to classification models learned across all participants.

As already mentioned, skin conductance is another parameter for stress recognition and has been used in a project aimed at recognizing stress by means of a wrist sensor which apart from skin conductance also captures movements and mobile phone usage (Sano and Picard, 2013). A self-assessment questionnaire served as the ground truth for deriving stress classifiers.

Some projects focus on developing *therapeutic solutions* to stress. Interstress, for example, employs biofeedback, meditation, and systematic training of coping skills in a virtual environment to reduce stress (Gaggioli et al., 2011; Villani et al., 2016). An intervention program described in McCraty et al. (2009a) aims at developing stress-coping strategies by providing participants with specific techniques to effectively manage stress as it occurs and to prepare for upcoming stressful events. The program consists of several training modules, among them a module for improving communication skills and a module which is based on a positive emotion refocusing technique designed to improve decision-making skills, especially in stressful situations. Individuals learn to more effectively cope with a stress reaction in the moment it occurs, thus reducing its negative effects. The intervention program also includes a biofeedback component based on HRV (cf. Sect. 4.6).

A *conversational agent* that assists patients with stress and chronic pain is described in Shamekhi et al. (2017). The agent – an animated character – interacts with users in a face-to-face dialog that includes nonverbal behavior such as hand gestures and facial expressions. The agent promotes behavior change by providing a learning environment based on the principles of mindfulness-based stress reduction. The agent guides a user through yoga exercises and meditation sessions and assists in setting and revising behavioral goals.

The work described in Sarker et al. (2017) focuses on identifying the right moment for triggering an intervention to reduce the stress level. According to this study, it may be more opportune to intervene after the stress level has peaked and is already down again. The reasoning behind this is that in high-stress situations, an intervention to reduce stress may have the opposite effect.

An important application area for stress detection is *driver assistance*. There are several approaches for detecting driver stress in real-time using various stress indicators. HRV is used in the work described in Munla et al. (2015). In some studies HRV has been supplemented with further parameters such as skin conductance and respiration (Healey and Picard, 2005; Rigas et al., 2011) as well as head movements (Rigas et al., 2011). Certain changes in skin conductance are used in Affanni et al. (2018) to detect a driver's stress level. Another study uses differential skin temperature, measured, e.g., at the periphery with sensors and at the face with an infrared camera (Yamakoshi et al., 2008). A multitude of sensor inputs are used in the approach described in Mühlbacher-Karrer et al. (2017): electrodermal activity (EDA), electrocardiography (ECG), and electroencephalography (EEG). A completely different approach has been pursued by Lee and Chung (2017) who avoid using physiological signals because of their high variability between individuals. Instead a driver's stress level is predicted by evaluating the movement pattern of the steering wheel which is detected by an inertial motion sensor placed on a glove worn by the driver.

Some approaches to stress detection use algorithms based on thresholds because they employ physiological measures such as HRV which have a clear correlation to stress, but need calibration to the individual user to work properly. Other approaches use physiological indicators where stress is reflected by certain patterns or they even use a multitude of sensors. They therefore need to find characteristic, stress-indicating patterns in the sensor signals by *learning a corresponding stress classifier* (Melillo et al., 2011; Karthikeyan et al., 2013; Muaremi et al., 2013; Kaur et al., 2014; Alexandratos et al., 2014; Hovsepian et al., 2015; Simantiraki et al., 2016; Cho et al., 2017; Mozos et al., 2017). The labeling of stressful events needed for learning such a classifier is either obtained via a questionnaire (e.g., the Perceived Stress Questionnaire (PSQ)) or by inducing several levels of stress in an experimental setting and then measuring physiological responses. Recent approaches tend to use a wide variety of sensor input (e.g., EDA, ECG, EEG, differential pulse transit time) for learning stress classifiers (Mühlbacher-Karrer et al., 2017; Kaur et al., 2017; Gaurav et al., 2017).

The drawback of first having to collect a sufficient amount of training data before a stress classifier can be learned is circumvented by the approach described in Huysmans et al. (2018) which adopts an *unsupervised learning approach* to generate a stress classifier. The authors measured skin conductance and ECG of test persons and then built a self-organizing map (SOM) from the measured data. Subsequently, the SOM was clustered by organizing regions with similar characteristics into one cluster. The measurements in the two resulting clusters corresponded well with the values for low and for high stress levels, i.e., high skin conductance and high heart rates for indicating stress and vice versa. However, the advantage of not requiring labeled training examples may be set off by a lower classification accuracy.

So-called technostress has emerged as another interesting and growing application area for stress detection (Connolly and Bhattacherjee, 2011). While until recently usability was largely considered a design issue from the perspective of human-computer interaction, making the interaction with an IT system as stress-free as possible has come to be seen as an important issue. Several research projects aim at measuring technostress in a lab setting, comparable to usability tests in usability labs. Others focus on real-time stress measurements during information system usage (Zhai et al., 2005) and investigate how the information system can adapt according to a user's stress level (Adam et al., 2014). This requires unobtrusive measuring devices, e.g., a wristband or a smartwatch. Riedl (2013) provides an elaborate survey on the area of technostress.

# 4 The SmartCoping Approach

In this section we present the design rationales and algorithms underlying the SmartCoping system. It is organized as follows: Sect. 4.1 introduces the main features of SmartCoping, Sect. 4.2 discusses the choice of physiological stress parameters for the system, and Sect. 4.3 introduces our approach to calculating stress levels and presents the two variants of calibrating HRV to the individual user. Section 4.4 explains our approach underlying the generation of stress warnings, while Sect. 4.5 presents our approach to identify recurrent high-stress situations of a user. Section 4.6 finally introduces the biofeedback component of SmartCoping for stress relaxation.

## *4.1 Main Features of SmartCoping*

SmartCoping has been developed to provide the following *innovative features*:

*Real-time stress warnings* Since SmartCoping monitors stress levels continuously, it can give real-time stress warnings (Sect. 4.4). These warnings help users to become aware of stressful situations and increase their self-perception by matching stress warnings with how they feel. The last aspect is particularly important for people such as burnout candidates who have been under high stress for a long time as well as for people suffering from depression or addiction. All these people typically have poor self-perception.

*Calibrating HRV values to the individual user* Using HRV as a physiological stress parameter – as SmartCoping does – poses the major challenge that HRV varies greatly between individuals, depending on age, health status, and other factors. Thus it is necessary to calibrate the HRV intervals associated with high, medium, and low stress to match the range of HRV values of the individual user (Sect. 4.3). This feature distinguishes SmartCoping from other apps that offer stress detection via HRV measurements.

*Relaxation support* An app for stress detection should not just give stress alerts but help the user relax. The SmartCoping app therefore comprises a biofeedback component which guides the user through breathing exercises while at the same time visualizing the actual stress level (see Sect. 4.6). This has a relaxing effect, as has amply been shown in the literature (Lehrer, 2013; Sakakibara et al., 2013).

*Detection of recurrent high-stress situations* From the recorded stress level history of a user, SmartCoping identifies the time periods when high-stress situations typically occur. For example, if it turns out that every Tuesday from 8 to 10 pm the stress level is high, the user can think about what happens at that time which causes him or her stress and then develop appropriate avoidance and coping strategies (see Sect. 4.5).

*Therapeutic effectiveness* The SmartCoping app has been developed for therapeutic purposes. This is why we have involved medical experts right from the start and stress measurement is based on sound medical principles. An evaluation with healthy volunteers has been made, but to collect evidence for the therapeutic efficacy of the app, a randomized controlled open-label study with people suffering from frequent high stress levels, e.g., patients after alcohol detoxification, burnout patients, or patients with depression, is planned (see Sect. 5).

To implement the abovementioned features, SmartCoping needs to measure stress levels continuously in everyday life settings. For this purpose, most of the stress-indicating parameters discussed in Sect. 2 are not suitable because they either do not allow continuous measurement (cortisol and alpha-amylase in saliva, speech analysis, blood pressure) or require special sensor arrangements that are not suitable for everyday use but only in specific circumstances (skin temperature, facial expressions, speech analysis). Measuring skin conductance alone would be too unspecific because it is not only influenced by stress but also by factors such as clothing and outside temperature. This leaves only HRV as the most reliable and practical indicator for determining stress levels.

## 4.2 Obtaining Physiological Stress Parameters

The SmartCoping system comprises the smartphone app and a backend system (see Fig. 1). A body sensor transmits heart rate and interbeat (or RR) intervals in ECG quality with a sampling rate of 1000 Hz via a Bluetooth interface (BLE) to the smartphone app. At present, only chest straps guarantee such quality. Wristbands or smartwatches would be much more convenient and less obtrusive for continuous measurement but do not achieve the necessary accuracy. However, new devices are under development that are more comfortable to wear than a chest strap and offer ECG quality. One such device is the upper-arm sensor by a Swiss start-up company which is very comfortable to wear and with which we have already conducted some first promising experiments.

The SmartCoping system calculates heart rate and several HRV parameters from the ECG signals and derives the user's stress level from these measurements (see Sect. 4.3). The accelerometer and the GPS receiver built into the smartphone

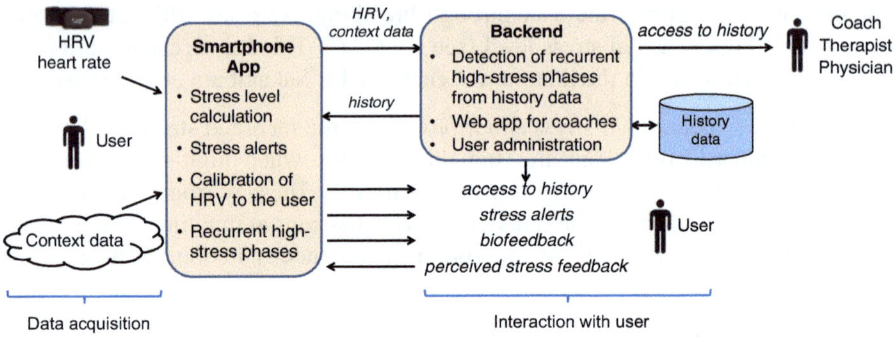


**Fig. 1** Data flow and architecture of the SmartCoping system

are used to collect *context data*. Since physical activity also decreases HRV, the accelerometer data is particularly important as it allows to suppress stress calculation when a user is physically active.

The HRV and heart rate data as well as the associated stress level and context data are stored on the backend and can be viewed by the user in the smartphone app or via a web browser. Transmission of data to and its storage on the backend are encrypted.

Every 30 s over a time window of 1 min, the SmartCoping app calculates the *average heart rate* (which increases with stress) and the following *HRV parameters*:

- SDNN: standard deviation of RR intervals (i.e., intervals between two heart beats);
- RMSSD: root-mean-square difference of successive RR intervals in the time frame;
- PNN50: percentage of pairs of adjacent RR intervals differing by more than 50 ms (Bilchick and Berger, 2006).

We have also experimented with a frequency-based HRV parameter, namely, the ratio between low- and high-frequency spectral powers (LF/HF) (Fagard et al., 1998). We decided to dismiss this parameter because it typically shows great fluctuations and is therefore not sufficiently reliable for short-term stress detection. Furthermore, breathing often falls into the LF band and can distort HRV measurement unless it is appropriately corrected (Choi and Gutierrez-Osuna, 2011).

Whenever making use of a body sensor, one is confronted with the problem of *artifacts*. Artifacts are either erroneous additional heart beats or beats that dropped due to poor connection of the chest strap to the body. Both kinds of artifacts result in HRV values that are too high and thus distort stress calculation. An effective *artifact elimination* is therefore quite critical.

SmartCoping eliminates artifacts in three steps: The first step is to reduce the probability of artifacts. Since movement is a major reason for artifacts and also affects HRV, we ignore segments with too much physical activity as detected by the accelerometers in the smartphone or chest strap. The second step is to detect artifacts and correct them automatically. A filter detects time intervals whose HRV values differ more than an amount $x$ from the surrounding intervals and overrides the measured value with the mean of the surrounding intervals. The value of $x$ ranges between 6% and 46%, depending on the HRV baseline of the user. In a third step, segments with 3% or more artifacts are excluded from stress calculation.

## *4.3 Stress Level Calculation*

In this section we will describe how we derive stress levels from measured HRV values. Similar to other projects, we started out by learning classifiers for low, medium, and high stress levels using supervised machine learning algorithms. To achieve sufficient accuracy, one needs many learning examples, i.e., HRV and heart rate measurements for each stress level. The learning examples are usually generated using controlled experiments where users are put in relaxing as well as in high-stress situations. In both kinds of situations, the corresponding physiological stress parameters, in our case HRV, are measured. Such an approach would be a high hurdle to the user compared to just having to obtain a wearable sensor device and downloading the SmartCoping app. Of course, users could conduct such experiments on their own by performing relaxation exercises to achieve a high degree of relaxation and then performing stress-inducing games on the smartphone to increase stress levels. While the latter would probably work quite well, the former case of lowering stress levels would need a human assistant to make sure that the user is indeed relaxed. This would be particularly important for target groups of highly stressed people for whom a relaxation exercise might not sufficiently lower their stress.

In principle, associating HRV values to stress levels is quite straightforward since the lower the measured HRV value, the higher is the stress level. However, the exact correspondence between HRV and stress is highly user-specific as people can have quite different ranges of HRV values, which has been shown both in lab and real-life settings (Morris and Guilak, 2009; Muaremi et al., 2013). Therefore we have implemented a *user-specific calibration for associating HRV values to stress levels*. We have done this in two variants – by prompting users to give feedback on their subjective stress levels (Sect. 4.3.1) and by determining the minimum and maximum HRV values for a user over some time period (Sect. 4.3.3). The former approach is more accurate while having the disadvantage that users must possess at least some rudimentary degree of self-perception to estimate their stress levels sufficiently well. Furthermore, it might be annoying for some users to have to provide feedback to the app before they can properly use it. The second variant avoids both disadvantages but at the risk of being less accurate in estimating stress levels.

Physical activity is considered as stress by the body, accordingly lowers the HRV, and would therefore trigger stress warnings. To prevent this, the SmartCoping app detects physical activity using the accelerometer built into the smartphone and suppresses warnings during and some time after physical activity. A similar approach is described in Sun et al. (2012).

### 4.3.1 Calibration Based on User Feedbacks on Stress Levels

One variant of calibration relies on user feedbacks of perceived stress levels. This approach imposes much less of a burden on the user as compared to filling in questionnaires or participating in stress-inducing sessions to calibrate physiological measurements against perceived stress levels.

Studies have shown that stress as experienced by a subject largely coincides with normalized physiological measurements (Mandryk and Atkins, 2007; Föhr et al., 2015) so that a user's stress feedbacks can be taken as valid (albeit not exact) stress assessments – at least as long as the user has not completely lost self-perception due to long-term stress exposure or other conditions such as depression and various kinds of addiction.

Knowing that typical users would not be willing to provide lots of feedback on perceived stress levels before starting to use the SmartCoping app, we exploited the fact that rising stress levels correlate with decreasing HRV and increasing heart rate. This allows the SmartCoping app to learn a user-specific stress recognition function with the necessary accuracy from a relatively small number of stress level feedbacks (between 25 and 30). For this purpose, we introduced an *initial learning phase* during which the app prompts the user for feedback on his or her perceived stress level (see Sect. 4.3.2 for more details).

The highest HRV values coincide with no stress and the lowest values with extreme stress. The simpler approach without user feedbacks splits the range between minimum and maximum HRV values into three equally large intervals to represent high, medium, and low stress levels, respectively (see Sect. 4.3.3). However, the correlation between perceived stress levels and HRV values is not linear so that by utilizing user feedbacks to adjust the intervals to the user's perception, we achieve more accurate results. Moreover, we make use of several HRV parameters as well as heart rate so that all measurements need to be combined to determine a final stress reading.

The *algorithm for determining a user-specific stress recognition function* comprises the following main steps (cf. Algorithm 1):

*Line 2* The feedbacks collected during the learning phase result in a set of tuples $\langle v_p, psl \rangle$ for each HRV parameter $p \in \{sdnn, rmssd, pnn50\}$ where $v_p$ is the HRV value measured when the perceived stress level $psl$ has been given by the user (cf. Formula (2)). Higher stress feedbacks correlate with lower HRV values, but this correlation is not perfect since the stress feedbacks are based on subjective perceptions. Therefore, dividing the range of measured HRV values into intervals that correspond to low, medium, and high stress typically leads to some misclassified stress feedbacks (cf. Fig. 2). The algorithm determines the thresholds between low and medium stress $tl$ and between medium and high stress $th$ in a way that the overall number of misclassifications is minimized. The misclassifications are measured by using the root-mean-square error function that for given thresholds $tl$ and $th$ adds up the distance of each misclassification to its proper threshold (cf. Formula (1)).

*Line 3* Based on the thresholds $tl$ and $th$, for each HRV parameter $p$, a function $slevel_p$ is defined that for a given HRV parameter yields the values 0, 1, and 2 for low, medium, and high stress, respectively.

*Line 6* We need to combine the resulting three stress level functions $slevel_p$ into a single function. This is done by computing the weighted mean of the three single $slevel_p$ values. The weights are determined according to the predictive accuracy of each HRV parameter in the set of user feedbacks. Our experiments showed that the accuracy of the parameters varied greatly between persons which is why the weights were determined for each user individually. We calculate the accuracy $w_p$ of an HRV parameter as 1 minus its relative error, i.e., its error divided by the sum of the errors of all parameters.

*Line 8* The final stress level is given as the weighted mean of the stress level functions for each HRV parameter. This results in a continuous value between 0 (low stress) and 2 (high stress).

In addition to HRV, heart rate is also used as a stress indicator and needs to be calibrated. This is done in the same manner as with Algorithm 1 except that higher stress levels correspond to higher heart rates whereas it is lower values for HRV. For the sake of simplicity, heart rate calibration is not included in Algorithm 1.

The formula for determining the error of misclassified stress feedbacks in a set of user feedbacks $fb$ and for thresholds $tl$ and $th$ is as follows (see also Fig. 2):

$$rmse(fb, tl, th) = \sqrt{\frac{\sum_{i=1}^{n}(v_i - ref(v_i, psl_i, tl, th))^2}{n}} \tag{1}$$

where $fb$ is the vector of measured values of a given HRV parameter and associated user feedbacks of low, medium, or high stress levels:

$$fb = (\langle v_1, psl_1\rangle, \ldots, \langle v_n, psl_n\rangle) \tag{2}$$

In case of a misclassified HRV value, the function $ref$ returns the threshold to the interval where it correctly belongs. If the HRV value $v$ is correctly classified, the function $ref$ returns that value itself so that it does not count as an error in Formula (1):

**Algorithm 1** Calibrate stress level calculation from HRV by utilizing user feedbacks

**Input:** tuples of HRV values with the associated perceived stress level feedback ($psl$) from the user, i.e. for $p \in hrvp = \{sdnn, rmssd, pnn50\}$:
$fb_p = (\langle v_{p,1}, psl_1 \rangle, \ldots, \langle v_{p,2}, psl_2 \rangle, \langle v_{p,n}, psl_n \rangle)$

**Output:** function $slevel$ for calculating the stress level from a given set of HRV values

**for all** $p \in hrvp$ **do**
    set thresholds $tl_p$ and $th_p$ on the interval of HRV values so that the error of misclassified stress feedbacks in $fb_p$ is minimal (see Formula (1)):
    $$\arg\min_{tl_p, th_p}(rmse(fb_p, tl_p, th_p))$$
    define a function $slevel_p(v_p)$ that gives for an HRV parameter value $v_p$ the values 0, 1, 2:
    $$slevel_p(v_p) = \begin{cases} 0: v_p > tl_p \\ 2: v_p \leq th_p \\ 1: else \end{cases}$$
**end for**
**for all** $p \in hrvp$ **do**
    calculate the weight for HRV parameter $p$ according to its accuracy:
    $$w_p \leftarrow 1 - \frac{rmse(fb_p, tl_p, th_p)}{\sum_{p' \in hrvp} rmse(fb_{p'}, tl_{p'}, th_{p'})}$$
**end for**
calculate the overall stress level by taking the weighted mean of all parameters:
$$slevel(v_{sdnn}, \ldots, v_{pnn50}) \leftarrow \frac{\sum_{p \in hrvp} w_p \cdot slevel_p(v_p)}{\sum_{p \in hrvp} w_p}$$

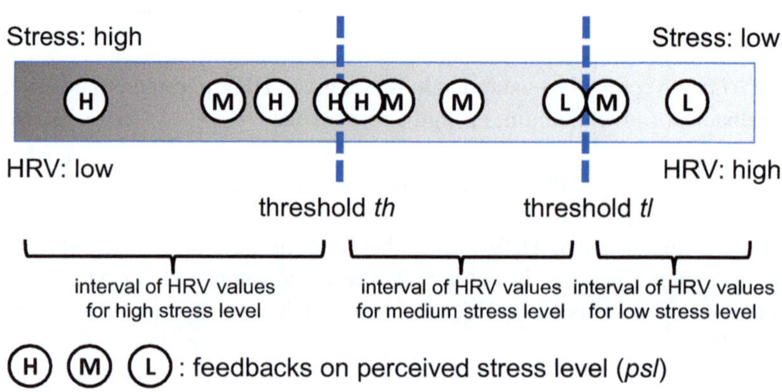


**Fig. 2** Separating stress level intervals

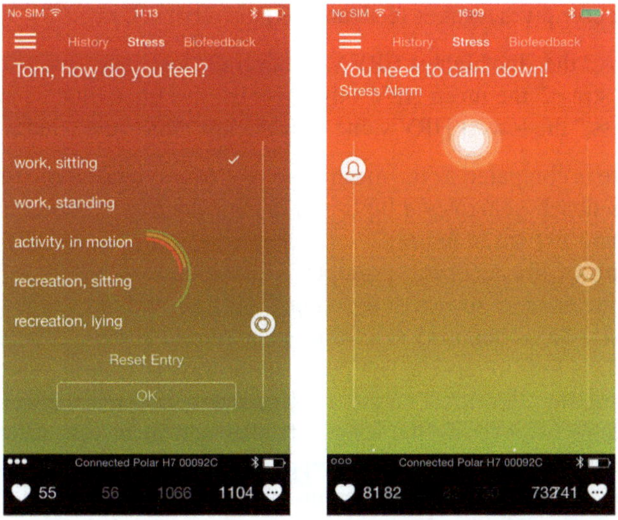


**Fig. 3** Feedback screen (calibration phase), left; stress warning screen (coaching phase), right

$$ref(v, psl, tl, th) = \begin{cases} th\colon psl = H \wedge v > th \\ th\colon psl = M \wedge v \leq th \\ tl\colon psl = M \wedge v > tl \\ tl\colon psl = L \wedge v \leq tl \\ v : else \end{cases}$$

### 4.3.2 Acquiring User Feedbacks in an Initial Learning Phase

When starting the SmartCoping app for the first time, the user is presented with a screen as shown in Fig. 3 (left). In the initial learning phase, feedback is given with the slider on the right side of the screen. The app maps the slider position to a continuous value between 0 (no stress) and 2 (extreme stress) and divides the interval [0, 2] into three equal subintervals which represent low, medium, and high stress.

Instead of using three subintervals, we have also experimented with five subintervals because we expected that a more fine-grained distinction would yield better stress recognition results. However, it turned out that this led to a higher overall error rate according to Formula (1) than with three intervals. This may be explained by the fact that users can estimate their stress levels only approximately, which is why trying to map their estimates to five subintervals is too fine-grained and causes more inconsistent feedbacks.

When the user gives a stress feedback and the feedback says

- "low stress," then the highest HRV value over the last three time windows is used
- "medium stress," the mean HRV value over the last three time windows is used
- "high stress," the lowest HRV value over the last three time windows is used

The heart rate is computed as the mean heart rate over the last three time windows.

A user feedback generates a tuple $\langle v_p, psl \rangle$ with $p \in \{sdnn, rmssd, pnn50, heartrate\}$ and $psl \in \{L, M, H\}$.

The timing of the feedback prompts is controlled by an algorithm which is inspired by the concept of *active learning* from the machine learning community (Settles, 2011). Active learning is a form of supervised learning where the learning algorithm asks a user to label a data point. In the case of SmartCoping, this means that the prompting algorithm keeps track of feedbacks and the associated HRV values and aims at covering the range of HRV and heart rate values as best as possible to obtain an informative and representative sample. In this way, learning is accelerated so that fewer learning inputs are needed.

The *generation of feedback prompts* is governed by the following principles:

1. Apart from the prompt generation algorithm, users are free to provide feedback any time, e.g., when they feel particularly stressed or relaxed.
2. During an initial measurement period, no prompts are generated so as to roughly determine the range of heart rate and HRV values first.
3. Users can adjust the number of prompts per day so they do not become a nuisance.
4. No prompts for feedbacks are generated during or some time after physical activity – as measured by the accelerometer in the smartphone or wearable sensor. This is to avoid contradictory information which would distort stress level calculation: Users might feel relaxed after physical activity while the HRV is still low as physical activity lowers the HRV.
5. Explore minimum and maximum HRV and heart rate values:
    - Extreme values, i.e., those that exceed previous minimum or maximum values and are not yet covered by a user feedback, generate a prompt.
    - If no new maximum or minimum values are encountered for some time, we allow prompts to be generated when only 90% of previous minimum or maximum values are reached. In other words, the app begins to generate prompts for values slightly below the maximum or slightly above the minimum. Thus the prompting algorithm starts from the extremes and increasingly approaches average values. If the relaxation to 90% does not lead to new prompts within a given time frame, a further relaxation by another 90% is performed and so on.
6. Explore maximum value changes:
    - A decrease or an increase in values within a certain period of measurements (typically three measurements over a time of 90 s) that is bigger than the amount of change so far observed prompts a feedback request.

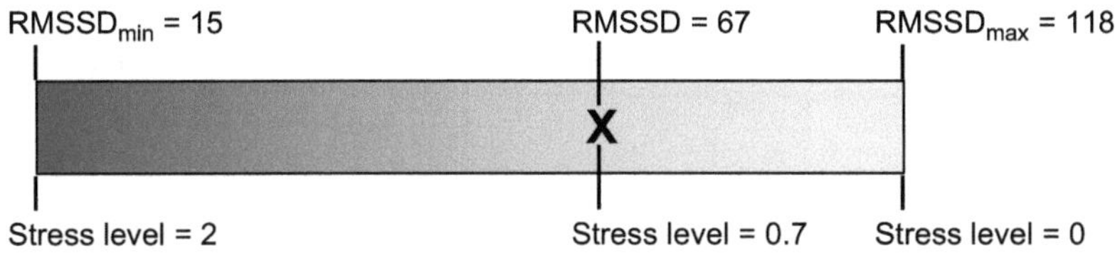


**Fig. 4** Associating HRV values with stress levels when there is no user feedback

- If no value changes are encountered for some time that are bigger than the ones so far observed, prompts are already generated when changes reach 90% of the maximal differences so far encountered. If this does not lead to new prompts, a further relaxation by another 90% is applied and so on.

The coaching phase (cf. Fig. 3 (right)) begins once a predefined number of low, medium, and high stress level feedbacks has been received so that the user gets stress level readings as soon as possible. The learning phase typically overlaps with the coaching phase and gradually comes to an end once the thresholds $tl$ and $th$ cease to show any major changes despite additional input (cf. Line 2 of Algorithm 1 and Fig. 2). Therefore it is possible that a user might get the occasional prompt even though he or she has already entered the coaching phase.

### 4.3.3 Calibration Without User Feedbacks

In some application scenarios, it is not possible or would be very inconvenient to acquire user feedbacks for calibrating stress level detection. This is the case when targeting users with reduced self-perception or when the target users are difficult to motivate to give feedback, e.g., because they are ill. In such application scenarios, calibration can be done without user feedbacks. The basic idea is to determine for each HRV parameter the highest and lowest values of a user over some initial time period. The resulting range of HRV values is then linearly associated with stress levels in the range of [0, 2], the highest HRV value corresponding to stress level 0 and the lowest to stress level 2 (see Fig. 4). This is done for all three HRV values (RMSSD, SDNN, PNN50). The final stress level is calculated as the mean of the three stress levels derived from the three HRV parameters.

The SmartCoping app continues to check for new minimal and maximal HRV values, and if one is detected, the intervals get readjusted accordingly.

When comparing this approach to the one utilizing user feedbacks, the difference is that the resulting HRV intervals for low, medium, and high stress are not equally divided when using user feedbacks as has already been explained above (cf. Fig. 2). Since the correspondence of HRV values to subjective stress perceptions is different for each person, the method of employing user feedbacks for calculating stress levels is more accurate and to be preferred.

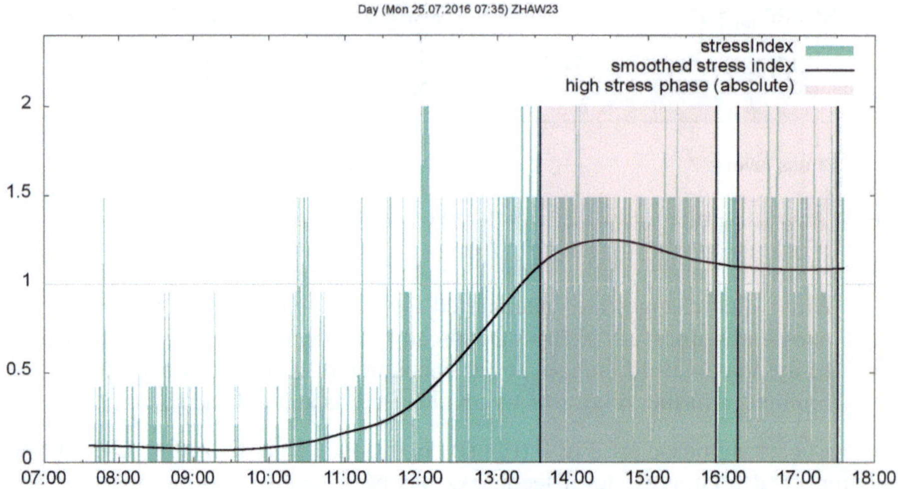


**Fig. 5** Smoothing stress level readings and identifying high stress phases

## *4.4 Generation of Stress Warnings*

It is not necessarily a good idea to generate a stress warning as soon as a high stress level is identified. On the one hand, this might not be helpful at that moment and even increase the user's stress; on the other hand, the detected high stress level might be transitory only in which case a stress warning would be rather pointless. Therefore we decided to delay stress warnings and trigger a warning only when the stress level stays above "medium" (=1) for more than 15 min. However, the problem with this criterion is that HRV and the derived stress level readings typically change a lot within a short time period (see Fig. 5) so that the trigger condition of a constant stress level above "medium" is rarely fulfilled. We thus apply a post-processing step on measured stress levels to smooth the plotted stress levels and generate a continuous graph from them. For smoothing we use the algorithm described in Cleveland (1979). The effect of this is illustrated in Fig. 5 where the smooth curve interpolates the plotted stress level bars. In this figure two high stress phases in the afternoon are indicated, which come with a 15 min delay as caused by the trigger condition.

Summarizing, the *condition for generating a stress warning* can now be specified as follows: A stress warning is triggered when the smoothed stress level curve stays above "medium" (=1) for more than $x$ minutes, where $x$ is set to 15 min by default but can be given a different value by the user.

## *4.5 Indication of Recurrent High-Stress Situations*

The primary goal of SmartCoping is not to tell users that they are stressed but improve their self-perception and point them to those situations where they have shown in the past to be stressed so that they may develop appropriate avoidance or coping strategies. These suggestions are generated from the history data stored on the backend. The history data consists of all stress level measurements combined with the following *context data*:

- *Time of day* and *day of week* are the most relevant attributes for describing recurrent stress situations.
- *Physical activity* is measured by an accelerometer in the smartphone or by the heart rate sensor.
- *Location* is a geographic cell that consists of a range of GPS coordinates. Whenever the user's GPS coordinates fall within that cell, he or she is considered to be at that place. Since a user would find it difficult to interpret GPS coordinates, they are labeled with names of locations relevant to a user, e.g., home, workplace, and favorite restaurant. For this purpose, the SmartCoping app asks the user to set a label for relevant locations.
- *Change of location* from one specific place to another one may be a further important indicator for characterizing stress situations.

Other contextual data that may help interpret high-stress situations are a user's communication patterns in terms of incoming and outgoing phone calls, e-mails, text messages, and even a semantic analysis of their content. However, this would not only raise data protection and privacy issues but is – depending on the smartphone being used – technically not possible because this kind of data may not be accessible as with iOS, for instance.

The first step in calculating recurrent high-stress phases is to overlay the measured stress levels of all weeks and derive the mean stress levels for each hour of each day of the week. This results in visualizations like the one shown in Fig. 6 from which a user can identify the recurrent high stress phases with one glance. For example, Fig. 6 shows that the stress level is usually high on Thursday afternoons, in particular between noon and 3pm.

A subsequent data analysis step can then determine if the identified high-stress phases are strongly correlated with some other context data, e.g., a particular place, thus giving a hint as to what the specific situation might be that causes the stress.

At the moment we are working on a user interface to visualize the typical high-stress situations in a clearer way than it is done with the visualization shown in Fig. 6. Furthermore, we are planning to implement a component that will work a bit like a human coach and actively point out recurring high-stress phases to a user thus giving some active support in developing avoidance and coping strategies.

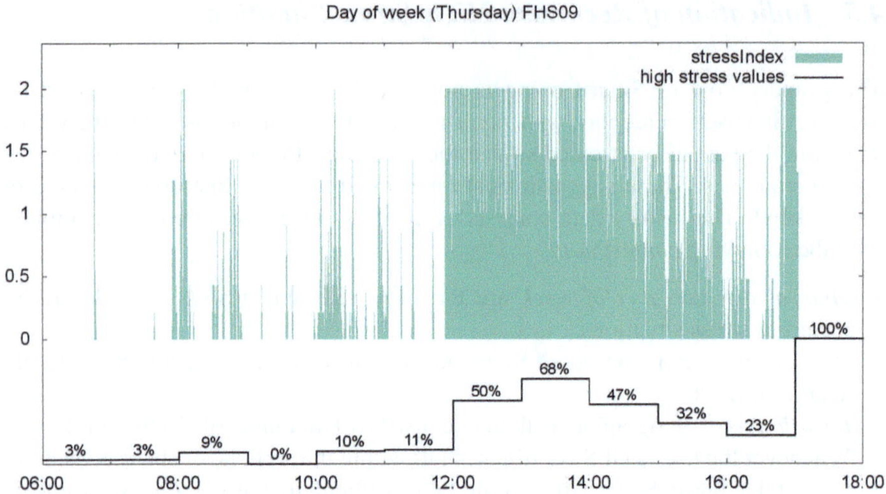


**Fig. 6** Visualization of recurrent high stress phases as extracted from all Thursday measurements

## *4.6 Biofeedback for Reducing Stress*

The SmartCoping solution comprises a biofeedback component to help users relax and reduce stress. The component guides the user through breathing exercises and at the same time visualizes the actual stress level by changing the background color from red to green to indicate increasing relaxation (cf. Fig. 7). The stress (or relaxation) level is computed every second within a sliding window of 30 s. This real-time computation and visualization establishes the biofeedback loop.

The relaxation effect is due to increasing coherence between breathing and heart rate (McCraty et al., 2009b). To achieve coherence, the biofeedback component offers a pulsating ball as an optional visual pacer for breathing in and out. The timing of the pacer can be set to a default value or be dynamic. In the latter case, based on the last couple of heartbeats, the four longest heartbeat intervals are used for timing breathing in, whereas for breathing out, the four shortest heartbeat intervals are used. The number of four heartbeats for the breathing cycle is inspired by yoga breathing (pranayama), where the heartbeat is also used as a pacer for breathing.

The relaxation effect of HRV biofeedback has been well demonstrated in various studies (Karavidas et al., 2007; Siepmann et al., 2008; Zucker et al., 2009; Lehrer, 2013; Sakakibara et al., 2013) and intervention programs aimed at improving stress-coping skills (Villani et al., 2016; McCraty et al., 2009a). For example, it has been shown that an HRV biofeedback approach is effective for treating posttraumatic stress disorder (Tan et al., 2011). The approach described in McCraty et al. (2009a) utilizes a biofeedback setup which is very similar to ours and aimed at enabling participants to experience how stress affects their heart rhythms and how they are able to increase HRV coherence by activating a positive emotional shift.

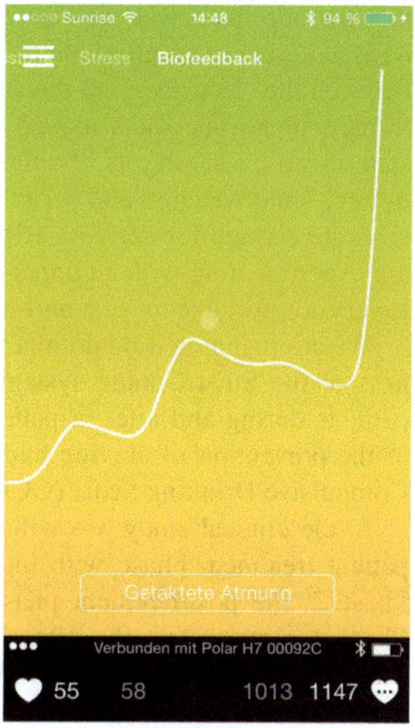


**Fig. 7** Screenshot of biofeedback component

## 5 Evaluation

The SmartCoping app was tested with 35 healthy individuals who had been recruited by the project partners from within their respective organizations. Apart from age, gender, and marital status, test persons were asked to specify their level of education and work status (self-employed vs. employed). After they had signed the informed consent form and received detailed information about the SmartCoping project, test persons were shown how to install and use the SmartCoping app. Each test person was given a unique identification number which was also used when they registered the app. This allowed us to link the data to the respective questionnaires.

The test phase lasted 4 weeks during which test persons had to fill in the German version of the Perceived Stress Questionnaire (PSQ) at the beginning and at the end of the pilot as well as after each week. PSQ has shown to be a valid and economical tool for stress research (Fliege et al., 2005). The individual assessment of perceived stress was done at three levels (low, medium, high stress). Test persons were also required to fill in the German stress-coping questionnaire SVF 120 at the start and at the end of the testing period. The SVF 120 comprises 120 items and is divided into 20 subtests. Each item offers the possibility to select the adequate answer from a five-level rating scale.

A major focus of the test phase was to evaluate the app with respect to usability (including the handling of the chest strap sensor) and overall acceptance, especially users' judgments of the everyday practicability and convenience of the system. From the interviews which we conducted at the end of the test phase, we can conclude that the app itself was well accepted but that many users had problems with the body sensor: not comfortable to wear, no sensor readings due to contact problems between the electrodes and the skin, and sometimes broken Bluetooth connections, in particular when users moved away from their smartphones. Therefore, we can state that a chest strap is ill-suited for long-term stress measurements and, as said before, smartwatches and wrist sensors are not an option because they are not accurate enough for deriving HRV from the measured heart beats. Meanwhile, we are experimenting with an upper-arm sensor from a Swiss start-up company which is very comfortable to wear and still accurate enough to deliver HRV measurements.

We are planning a randomized controlled open-label study to test the effectiveness of the SmartCoping system for supporting abstinence in alcohol-dependent patients during and after in-patient treatment. Perceived effectiveness with regard to the prevention of craving and thus relapse will be measured by the Obsessive Compulsive Drinking Scale (OCDS).

In the clinical study, we will distinguish between two phases: Phase 1, the in-patient treatment phase with instructor-mediated HRV biofeedback training, and Phase 2, the posttreatment phase supported by the SmartCoping system. While Phase 1 aims to test the effectiveness of instructor-mediated HRV biofeedback training to increase HRV and reduce craving in alcohol-dependent patients during in-patient treatment, Phase 2 will explore the therapeutic benefits and the usability of the SmartCoping system for aftercare support.

Another upcoming application scenario for SmartCoping is its use with COPD (chronic obstructive pulmonary disease) patients after they have suffered from an exacerbation of symptoms to measure the decreasing stress level over several weeks. We will then use the insights gained to make use of SmartCoping for detecting *increasing stress levels* as an early warning indicator of an imminent exacerbation.

## 6 Conclusions

We have presented the goals, design principles, and main algorithms underlying SmartCoping, a system for recognizing stress and helping users cope with it. One major goal of SmartCoping is to enhance people's self-perception by giving stress alerts so that they can learn how they feel when they are stressed. This is important for users with poor self-perception as is the case with people suffering from burnout, depression, or alcohol addiction. Moreover, the system identifies recurrent stress situations from the history data, i.e., which time of the day on which weekdays the stress level is high and how this may be associated with a particular place. Furthermore, a biofeedback component based on breathing exercises gives relaxation support. The SmartCoping app was tested with a group of healthy individuals for usability and overall acceptance.

As opposed to other stress apps, SmartCoping calibrates HRV measures to the user's individual stress levels. Calibration is crucial since HRV varies greatly among people. Calibration is implemented in two variants. One variant requires user feedbacks on perceived stress levels during an initial learning phase and one where no feedback is needed; instead a user's maximal and minimal HRV values are determined during an initial usage phase. The variant with user feedbacks results in stress warnings which better reflect a user's stress perception but is not applicable when target users have a poor self-perception.

The SmartCoping system primarily targets people who are in danger of chronic stress and high-risk groups such as patients after alcohol detoxification where stress can cause relapse. We are planning to conduct a clinical study to validate the app for these high-risk target groups.

The insights we have gained from SmartCoping are currently being transferred to other application areas where stress plays an important role, such as sleep. For example, we are investigating how one's behavior during the day (e.g., physical activity) as captured by body sensors influences sleep architecture and subjective sleep quality (Reimer et al., 2017a). As is the case with stress, the correlations between behavioral patterns and sleep vary greatly between individuals so that a user-specific approach is called for. The importance of taking into account individual preferences, attitudes, and behavior has been shown time and again in studies related to behavioral change support. In a recent paper Reimer and Maier (2016), we presented an application framework for a behavioral change support system that comprises various components to adapt system interventions to individual users. We are confident that with our approach, user acceptance as well as the effectiveness of interventions will be improved – be it for coping with stress, sleep problems, or other health issues.

**Acknowledgements** The work reported in this paper was funded by the Swiss Confederation's innovation promotion agency (CTI) under grant number 14049.2 PFES-ES.

## References

Adam, M., Gimpel, H., Maedche, A., & Riedl, R. (2014). Stress-sensitive adaptive enterprise systems: Theoretical foundations and design blueprint. In F. Davis, R. Riedl, J. vom Brocke, P. M. Leger, & A. Randolph (Eds.), *Gmunden Retreat on NeuroIS 2014 Proceedings*, Gmunden, Austria.

Affanni, A., Bernardini, R., Piras, A., Rinaldo, R., & Zontone, P. (2018). Driver's stress detection using skin potential response signals. *Measurement, 122*, 264–274.

Alexandratos, V., Bulut, M., & Jasinschi, R. (2014). Mobile real-time arousal detection. In *Proceedings of IEEE International Conference on Acoustics, Speech and Signal Processing (ICASSP)* (pp. 4394–4398).

Andreoli, A., Gravina, R., Giannantonio, R., Pierleoni, P., & Fortino, G. (2010). SPINE-HRV: A BSN-based toolkit for heart rate variability analysis in the time-domain. In A. Lay-Ekuakille & S. C. Mukhopadhyay (Eds.), *Wearable and autonomous biomedical devices and systems for smart environment* (pp. 369–389). Berlin: Springer.

Bilchick, K. C., & Berger, R. D. (2006). Heart rate variability. *Journal of Cardiovascular Electrophysiology, 17*(6), 691–694.

Boucsein, W. (2012). *Electrodermal activity*. Boston: Springer.

Cho, D., Ham, J., Oh, J., Park, J., Kim, S., Lee, N. K., & Lee, B. (2017). Detection of stress levels from biosignals measured in virtual reality environments using a kernel-based extreme learning machine. *Sensors, 17*(10), 2435.

Choi, J., & Gutierrez-Osuna, R. (2011). Removal of respiratory influences from heart rate variability in stress monitoring. *Sensors, 11*(11), 2649–2656.

Cleveland, W. S. (1979). Robust locally weighted regression and smoothing scatterplots. *Journal of the American Statistical Association, 74*(368), 829–836.

Connolly, A. J., & Bhattacherjee, A. (2011). Coping with the dynamic process of technostress, appraisal and adaptation. In *Proceedings of 17th Americas Conference on Information Systems, AMCIS*.

Delaney, J. P. A., & Brodie, D. A. (2000). Effects of short-term psychological stress on the time and frequency domains of heart-rate variability. *Perceptual and Motor Skills, 91*, 514–524.

Dinges, D. F., Rider, R. L., Dorrian, J., McGlinchey, E. L., Rogers, N. L., Cizman, Z., Goldenstein, S. K., Vogler, C., Venkataraman, S., & Metaxas, D. N. (2005). Optical computer recognition of facial expressions associated with stress induced by performance demands. *Aviation, Space, and Environmental Medicine, 76*, B172–B182.

El Ayadi, M., Kamel, M. S., & Karray, F. (2011). Survey on speech emotion recognition: Features, classification schemes, and databases. *Pattern Recognition, 44*(3), 572–587.

Fagard, R. H., Pardaens, K., Staessen, J. A., & Thijs, L. (1998). Power spectral analysis of heart rate variability by autoregressive modelling and fast fourier transform: A comparative study. *Acta Cardiologica, 53*(4), 211–218.

Fliege, H., Rose, M., Arck, P., Walter, O. B., Kocalevent, R. D., Weber, C., & Klapp, B. F. (2005). The perceived stress questionnaire (PSQ) reconsidered: Validation and reference values from different clinical and healthy adult samples. *Psychosomatic Medicine, 67*(1), 78–88.

Föhr, T., Tolvanen, A., Myllymäki, T., Järvelä-Reijonen, E., Rantala, S., Korpela, R., Peuhkuri, K., Kolehmainen, M., Puttonen, S., Lappalainen, R., et al. (2015). Subjective stress, objective heart rate variability-based stress, and recovery on workdays among overweight and psychologically distressed individuals: A cross-sectional study. *Journal of Occupational Medicine and Toxicology, 10*(1), 39.

Gaggioli, A., Raspelli, S., Grassi, A., Pallavicini, F., Cipresso, P., Wiederhold, B. K., & Riva, G. (2011). Ubiquitous health in practice: The interreality paradigm. *Studies in Health Technology and Informatics, 163*, 185–191.

Gao, H., Yuce, A., & Thiran, J. P. (2014). Detecting emotional stress from facial expressions for driving safety. In *Proceedings of IEEE International Conference on Image Processing* (pp. 5961–5965).

Gaurav, Anand, R. S., & Kumar, V. (2017). Non-invasive EEG-metric based stress detection. In *Proceedings 4th International Conference on Signal Processing, Computing and Control (ISPCC)* (pp. 516–520).

Giannakakis, G., Pediaditis, M., Manousos, D., Kazantzaki, E., Chiarugi, F., Simos, P. G., Marias, K., & Tsiknakis, M. (2017). Stress and anxiety detection using facial cues from videos. *Biomedical Signal Processing and Control, 31*, 89–101.

Harmon, A. G., Towe-Goodman, N. R., Fortunato, C. K., & Granger, D. A. (2008). Differences in saliva collection location and disparities in baseline and diurnal rhythms of alpha-amylase: A preliminary note of caution. *Hormones and Behavior, 54*(5), 592–596.

Healey, J. A., & Picard, R. W. (2005). Detecting stress during real-world driving tasks using physiological sensors. *IEEE Transactions on Intelligent Transportation Systems, 6*(2), 156–166.

Hjortskov, N., Rissén, D., Blangsted, A. K., Fallentin, N., Lundberg, U., & Søgaard, K. (2004). The effect of mental stress on heart rate variability and blood pressure during computer work. *European Journal of Applied Physiology, 92*(1–2), 84–89.

Hovsepian, K., al'Absi, M., Ertin, E., Kamarck, T., Nakajima, M., & Kumar, S. (2015). cstress: Towards a gold standard for continuous stress assessment in the mobile environment. In *Proceedings of the 2015 ACM International Joint Conference on Pervasive and Ubiquitous Computing, UbiComp '15* (pp. 493–504). New York: ACM.

Huang, T., Xiong, J., & Xuan, L. (2011) Design and implementation of a wireless healthcare system based on galvanic skin response. In D. Zeng (Ed.), *Applied informatics and communication* (pp. 337–343). Berlin/New York: Springer.

Huysmans, D., Smets, E., de Raedt, W., van Hoof, C., Bogaerts, K., van Diest, I., & Helic, D. (2018) Unsupervised learning for mental stress detection-exploration of self-organizing maps. In *Proceedings BIOSIGNALS*.

Jáuregui, D. A. G., Castanier, C., Chang, B., Val, M., Cottin, F., Le Scanff, C., & Martin, J. C. (2017). Toward automatic detection of acute stress: Relevant nonverbal behaviors and impact of personality traits. In *Proceedings Seventh International Conference on Affective Computing and Intelligent Interaction (ACII)* (pp. 354–361).

Karavidas, M. K., Lehrer, P. M., Vaschillo, E., Vaschillo, B., Marin, H., Buyske, S., Malinovsky, I., Radvanski, D., & Hassett, A. (2007). Preliminary results of an open label study of heart rate variability biofeedback for the treatment of major depression. *Applied Psychophysiology and Biofeedback, 32*(1), 19–30.

Karthikeyan, P., Murugappan, M., & Yaacob, S. (2013). Detection of human stress using short-term ECG and HRV signals. *Journal of Mechanics in Medicine and Biology, 13*(2), 1350038.

Kaur, B., Durek, J. J., O'Kane, B. L., Tran, N., Moses, S., Luthra, M., & Ikonomidou, V. N. (2014) Heart rate variability (HRV): An indicator of stress. In *Proceedings of Independent Component Analyses, Compressive Sampling, Wavelets, Neural Net, Biosystems, and Nanoengineering XII*, 91180V (2014)

Kaur, B., Hutchinson, J. A., & Ikonomidou, V. N. (2017). Visible spectrum-based non-contact HRV and dPTT for stress detection. In *Mobile Multimedia/Image Processing, Security, and Applications 2017* (Vol. 10221).

Koh, D. S. Q., & Koh, G. C. H. (2007). The use of salivary biomarkers in occupational and environmental medicine. *Occupational and Environmental Medicine, 64*(3), 202–210.

Lee, B. G., & Chung, W. Y. (2017). Wearable glove-type driver stress detection using a motion sensor. *IEEE Transactions on Intelligent Transportation Systems, 18*(7), 1835–1844.

Legendre, A., & Harris, R. (2006). Exaggerated response to mild stress in rats fed high-fat diet. *American Journal of Physiology – Regulatory, Integrative, and Comparative Physiology, 291*(5), R1288–1294.

Lehrer, P. (2013). How does heart rate variability biofeedback work? Resonance, the baroreflex, and other mechanisms. *Biofeedback, 41*(1), 26–31.

Liew, W. S., Seera, M., Loo, C. K., Lim, E., & Kubota, N. (2015). Classifying stress from heart rate variability using salivary biomarkers as reference. *IEEE Transactions on Neural Networks and Learning System*, Vol. 27, 2035–2046.

Mandryk, R., & Atkins, M. (2007). A fuzzy physiological approach for continuously modelling emotion during interaction with play environments. *International Journal of Human-Computer Studies, 6*(4), 329–347.

McCraty, R., Atkinson, M., Lipsenthal, L., & Arguelles, L. (2009a). New hope for correctional officers: An innovative program for reducing stress and health risks. *Applied Psychophysiology and Biofeedback, 34*(4), 251–272.

McCraty, R., Atkinson, M., Tomasino, D., & Bradley, R. T. (2009b). The coherent heart. Heart-brain interactions, psychophysiological coherence, and the emergence of system-wide order. *Integral Review: A Transdisciplinary & Transcultural Journal for New Thought, Research, & Praxis, 5*(2), 10–115.

Melillo, P., Bracale, M., & Pecchia, L. (2011). Nonlinear heart rate variability features for real-life stress detection. Case study: Students under stress due to university examination. *BioMedical Engineering OnLine, 10*(1), 1–13.

Morris, M., & Guilak, F. (2009). Mobile heart health: Project highlight. *IEEE Pervasive Computing, 8*(2), 57–61.

Mozos, O. M., Sandulescu, V., Andrews, S., Ellis, D., Bellotto, N., Dobrescu, R., & Ferrandez, J. M. (2017). Stress detection using wearable physiological and sociometric sensors. *International Journal of Neural Systems, 27*(02), 1650041.

Muaremi, A., Arnrich, B., & Tröster, G. (2013). Towards measuring stress with smartphones and wearable devices during workday and sleep. *BioNanoScience, 3*(2), 172–183.

Mühlbacher-Karrer, S., Mosa, A. H., Faller, L. M., Ali, M., Hamid, R., Zangl, H., & Kyamakya, K. (2017). A driver state detection system – combining a capacitive hand detection sensor with physiological sensors. *IEEE Transactions on Instrumentation and Measurement, 66*(4), 624–636.

Munla, N., Khalil, M., Shahin, A., & Mourad, A. (2015). Driver stress level detection using HRV analysis. In *Proceedings of International Conference on Advances in Biomedical Engineering (ICABME)* (pp. 61–64).

Noto, Y., Sato, T., Kudo, M., Kurata, K., & Hirota, K. (2005). The relationship between salivary biomarkers and state-trait anxiety inventory score under mental arithmetic stress: A pilot study. *Anesthesia & Analgesia, 101*(6), 1873–1876.

Ornish, D., Brown, S., Scherwitz, L., Billings, J., Armstrong, W., Ports, T., McLanahan, S., Kirkeeide, R., Brand, R., & Gould, K. (1990). Can lifestyle changes reverse coronary heart disease? *Lancet, 336*(8708), 129–133.

Orsila, R., Virtanen, M., Luukkaala, T., Tarvainen, M., Karjalainen, P., Viik, J., Savinainen, M., & Nygård, C. H. (2008). Perceived mental stress and reactions in heart rate variability – A pilot study among employees of an electronics company. *International Journal of Occupational Safety and Ergonomics, 14*(3), 275–283.

Reimer, U., & Maier, E. (2016). An application framework for personalised and adaptive behavioural change support system. In *Proceedings of 2nd International Conference on Information and Communication Technologies for Ageing Well and e-Health (ICT4AWE).*

Reimer, U., Emmenegger, S., Maier, E., Zhang, Z., & Khatami, R. (2017a). Recognizing sleep stages with wearable sensors in everyday settings. In *Proceedings of 3rd International Conference on Information and Communication Technologies for Ageing Well and e-Health (ICT4AWE).*

Reimer, U., Laurenzi, E., Maier, E., & Ulmer, T. (2017b). Mobile stress recognition and relaxation support with smartcoping: User-adaptive interpretation of physiological stress parameters. In *Proceedings of the 50th Hawaii International Conference on System Sciences.*

Riedl, R. (2013). On the biology of technostress: Literature review and research agenda. *SIGMIS Database, 44*(1), 18–55.

Rigas, G., Goletsis, Y., Bougia, P., & Fotiadis, D. I. (2011) Towards driver's state recognition on real driving conditions. *International Journal of Vehicular Technology, 2011*, 1–14.

Sakakibara, M., Oikawa, L. O., Katsamanis, M., & Lehrer, P. (2013). Heart rate variability biofeedback improves cardiorespiratory resting function during sleep. *Applied Psychophysiology and Biofeedback*, Vol. 38, 265–271.

Sano, A., & Picard, R. W. (2013). Stress recognition using wearable sensors and mobile phones. In *Proceedings of Humaine Association Conference on Affective Computing & Intelligent Interaction* (pp. 671–676).

Sarker, H., Hovsepian, K., Chatterjee, S., Nahum-Shani, I., Murphy, S. A., Spring, B., Ertin, E., al'Absi, M., Nakajima, M., & Kumar, S. (2017). From markers to interventions: The case of just-in-time stress intervention. In J. M. Rehg, S. A. Murphy, & S. Kumar (Eds.), *Mobile health* (pp. 411–433). Springer.

Settles, B. (2011). From theories to queries: Active learning in practice. In *Journal of Machine Learning Research: Workshop and Conference Proceedigns 16. Workshop on Active Learning and Experimental Design* (pp. 1–18).

Shamekhi, A., Bickmore, T., Lestoquoy, A., & Gardiner, P. (2017). Augmenting group medical visits with conversational agents for stress management behavior change. In *International Conference on Persuasive Technology* (pp. 55–67). Springer.

Siepmann, M., Aykac, V., Unterdörfer, J., Petrowski, K., & Mueck-Weymann, M. (2008). A pilot study on the effects of heart rate variability biofeedback in patients with depression and in healthy subjects. *Applied Psychophysiological and Biofeedback, 33*(4), 195–201.

Simantiraki, O., Giannakakis, G., Pampouchidou, A., & Tsiknakis, M. (2016). Stress detection from speech using spectral slope measurements. In N. Oliver, S. Serino, A. Matic, P. Cipresso,

N. Filipovic, & L. Gavrilovska (Eds.), *Pervasive Computing Paradigms for Mental Health* (pp. 41–50). Springer.
Sun, F. T., Kuo, C., Cheng, H. T., Buthpitiya, S., Collins, P., & Griss, M. (2012). Activity-aware mental stress detection using physiological sensors. In M. Gris & G. Yang (Eds.), *Mobile Computing, Applications, and Services* (pp. 211–230). Springer.
Swan, M. (2012). Health 2050: The realization of personalized medicine through crowdsourcing, the quantified self, and the participatory biocitizen. *Journal of Personalized Medicine, 2*(3), 93–118.
Tan, G., Dao, T. K., Farmer, L., Sutherland, R. J., & Gevirtz, R. (2011). Heart rate variability (HRV) and posttraumatic stress disorder (PTSD): A pilot study. *Applied Psychophysiological and Biofeedback, 36*(1), 27–35.
Tanev, G., Saadi, D. B., Hoppe, K., & Sorensen, H. B. D. (2014). Classification of acute stress using linear and non-linear heart rate variability analysis derived from sternal ECG. In *Proceedings of 36th Annual International Conference of the IEEE Engineering in Medicine and Biology Society* (pp. 3386–3389).
van den Broek, E. L., van der Sluis, F., & Dijkstra, T. (2013). Cross-validation of bimodal health-related stress assessment. *Personal and Ubiquitous Computing, 17*(2), 215–227.
van Eck, M., Berkhof, H., Nicolson, N., & Sulon, J. (1996). The effects of perceived stress, traits, mood states, and stressful daily events on salivary cortisol. *Psychosomatic Medicine, 58*(5), 447–458.
Vanitha, V., & Krishnan, P. (2016). Real time stress detection system based on EEG signals. *Biomedical Research, 2016*, S271–S275.
Vedhara, K., Miles, J., Bennett, P., Plummer, S., Tallon, D., Brooks, E., Gale, L., Munnoch, K., Schreiber-Kounine, C., Fowler, C., Lightman, S., Sammon, A., Rayter, Z., & Farndon, J. (2003). An investigation into the relationship between salivary cortisol, stress, anxiety and depression. *Biological Psychology, 62*(2), 89–96.
Villani, D., Cipresso, P., Gaggioli, A., & Riva, G. (2016). Positive technology for helping people cope with stress. In D. Villani, P. Cipresso, A. Gaggioli, & G. Riva (Eds.), *Integrating Technology in Positive Psychology Practice* (pp. 320–347). Hershey: IGI Global.
Villarejo, M. V., Zapirain, B. G., & Zorrilla, A. M. (2012). A stress sensor based on galvanic skin response (GSR) controlled by ZigBee. *Sensors, 12*(5), 6075–6101.
Weinert, C., Maier, C., & Laumer, S. (2015). What does the skin tell us about information systems usage? A literature-based analysis of the utilization of electrodermal measurement for is research. In F. D. Davis, R. Riedl, J. vom Brocke, P. M. Léger, & A. B. Randolph (Eds.), *Information Systems and Neuroscience* (pp. 65–75). Cham: Springer.
Yamakoshi, T., Yamakoshi, K., Tanaka, S., Nogawa, M., Park, S. B., Shibata, M., Sawada, Y., Rolfe, P., & Hirose, Y. (2008). Feasibility study on driver's stress detection from differential skin temperature measurement. In *Proceedings of 30th Annual International Conference of the IEEE Engineering in Medicine and Biology Society* (pp. 1076–1079).
Zhai, J., Barreto, A. B., Chin, C., & Li, C. (2005). Realization of stress detection using psychophysiological signals for improvement of human-computer interactions. In *Proceedings of IEEE SoutheastCon* (pp. 415–420).
Zucker, T. L., Samuelson, K. W., Muench, F., Greenberg, M. A., & Gevirtz, R. N. (2009). The effects of respiratory sinus arrhythmia biofeedback on heart rate variability and posttraumatic stress disorder symptoms: A pilot study. *Applied Psychophysiology and Biofeedback, 34*(2), 135–143.

# Changing Behavior of Kids with Obesity with Gamified Wearables

**Luisa Schulz, A. A. M. Ton Spil, and S. A. Sjoerd de Vries**

## 1 Introduction

In recent years, increase in body weight and obesity have become a widespread, prevalent societal problem. It is now recognized as an official chronic disease by the AMA and WHO.[1] Obesity leads to other chronic diseases such as diabetes or cardiovascular problems, the major cause of morbidity worldwide.[2] Obesity is a financial burden to the economy of the health-care system as well as the clients. It results in delayed skill acquisition in educational and cognitive abilities of children as young as 3 years old (Cawley 2010). The National Center for Health Statistics in the USA has published results indicating that the worldwide prevalence of obesity is rising steadily and has more than tripled since 1970, especially among children and adolescents (Ogden et al. 2006). Thus, childhood obesity can be considered as one of the most serious public health challenges in the twenty-first century. Thorpe et al. (2004) stated that this fact explains 27% of the increase in health-care expenditures between 1987 and 2001. WHO currently estimates the number of overweight or

[1] http://www.who.int/dietphysicalactivity/media/en/gsfs_obesity.pdf & American Medical Association, "Obesity as a disease," Policy Statement, vol. 420, no. 13, pp. 6–18, 2013, http://www.amaassn.org/assets/meeting/2013a/a13-addendum-refcomm-d.pdf

[2] http://www.who.int/mediacentre/factsheets/fs311/en/ & http://apps.who.int/iris/bitstream/10665/148114/1/9789241564854_eng.pdf

L. Schulz
University of Twente, Munich, Germany

A. A. M. T. Spil (✉) · S. A. S. de Vries
University of Twente, Enschede, The Netherlands
e-mail: a.a.m.spil@utwente.nl; s.a.devries@utwente.nl



N. Wickramasinghe, F. Bodendorf (eds.), *Delivering Superior Health and Wellness Management with IoT and Analytics*, Healthcare Delivery in the Information Age, https://doi.org/10.1007/978-3-030-17347-0_7

obese infants and young children globally to increase to 70 million by 2025.[3] As obesity in children is generally associated with a higher chance of premature death, disability in adulthood, and serious consequences regarding psychological health, there needs to be numerous effective intervention programs that aim at helping children, adolescents, and adults with obesity (Dietz 1998; Kelishadi and Azizi-Soleiman 2014). Despite these intervention programs, the number of obese children is increasing at a steady rate. The question that remains is, how can existing obesity intervention programs be successfully augmented?

One particular novel concept that helps to promote healthy lifestyles and can fight obesity is gamification, where the treatment of obese clients is supported through the use of video and online games or apps with game elements. Another promising concept is the use of wearables. Tracked health-related data can promote and enhance a healthy lifestyle, but currently, there is very little research available regarding the effectiveness of gamified wearables to reduce obesity in children, although several studies (Spil et al. 2017; Tong et al. 2015; Zhao et al. 2016a, b) suggest that the two concepts combined can lead to positive outcomes. A recent field aimed at helping to reduce obesity is the field of digital health. In principle, the concept of gamified wearables is a novel approach to motivate people to promote a change through the use of gaming elements which have been repurposed to engage the user playfully in a nongame context (Deterding et al. 2011).

The purpose of this study is to gain insight in the success of gamified wearables for in obesity therapy for youth by studying fundamental guidelines for such gamified wearables. The research question is: What are fundamental guidelines for gamified wearables to accomplish weight-loss and long-term behavior change in overweight and obese children and adolescents? First, based on a rigorous literature study, we present five fundamental guidelines. Second, we conduct interviews with clinicians and clients to ask their perception and experiences about the acceptance and effectiveness of these guidelines. Third, based on an in-depth analysis of the interviews, we conclude with four guidelines which are according to the literature study and interviews with the two main stakeholders in obesity treatments fundamental for the success of gamified wearables.

## 2 Literature Review

A rigorous literature review, a five-stage process based on the grounded theory approach as proposed by Wolfswinkel et al. (2013), is conducted to identify fundamental guidelines. Before the initial search progress through databases began, various possible search terms were determined as to ensure topicality and scientific relevance. The main search terms that were used were obesity/overweight in children, childhood obesity, weight loss, health apps, gam$_*$ which stands for serious

[3] ebd.

gaming and gamification, wearables, and smartphones, as these terms sufficiently reflect the scope of the chosen research area.

The initial search was carried out while navigating through the different databases, and duplicates were collected, which is why in stage 3 of Wolfswinkel et al. (2013) approach these doubles were filtered. Papers which were not applicable, identified by reading through abstracts and parts of the texts, were filtered and removed. During the selection process, the preliminary criteria regarding the digital health and obesity subfield were adjusted. The search process was repeated several times until no new relevant articles appeared.

In stage 4 of Wolfswinkel et al. (2013) approach, the selected texts were analyzed by first rereading the abstract and highlighting the most interest findings and insights. During this process, certain concepts began to emerge. These concepts were grouped into categories. During the process, the categories (1) obesity in children, (2) digital healthy obesity, (3) gamification, (4) wearables, and (5) gamified wearables were constructed. Lastly, while preparing the analysis of the literature review, relations between the categories were found, which lead to the last stage of Wolfswinkel et al. (2013) approach, organizing the structure of the literature review chapter. In the following, the processed and prepared data from the literature review will be presented.

## 2.1 *Obesity in Children*

First and foremost, it is necessary to define the term obesity. According to WHO, "Obesity is abnormal or excessive fat accumulation that may impair health." Obesity, regardless whether the patient is an adult or child, is the result of a chronically high-calorie intake where more calories are being consumed than expended daily (Pulgarón 2013).

Ebbeling et al. (2002), who reviewed many studies regarding obesity in children, name insufficient physical activity and excessive calorie consumption as cause for childhood obesity. They argue that the parent-child interactions and home environment have changed over the past decades, and this affects the lifestyle and behavior of children related to the risk of obesity.

Childhood obesity comes with considerable consequences for both the (1) individual and the (2) economy. From the perspective of the patient, childhood obesity can have substantial physical, psychological, and psychosocial consequences in the near and distant future (Ebbeling et al. 2002; Han et al. 2010). Bjørge et al. (2008) and Must et al. (1992) among many others have studied the relation between childhood and adolescent obesity and increased adult morbidity and mortality.

Singh et al. (2008) reviewed and examined publications, and found that the likelihood for obese children to grow into overweight and obese adults is moderate for overweight and obese youth. This indicates that obesity is a chronic disorder which is in need of persistent management and efficient treatment (Han et al. 2010).

It becomes clear from corresponding academic and scientific literature that smartphones and wearables pose as two burgeoning device categories that yield the power to promote healthy lifestyles and enhance treatment (Spil et al. 2017), which is the goal that clinicians would like to achieve after the initial intervention program.

## 2.2 Gamification

As mentioned above, gamification is a major field of research and interest for the health-care sector and has received worldwide attention in recent years due to pervasiveness of smartphones and computers (Zhao et al. 2016a; Zichermann and Cunningham 2011). In principle, the concept of gamification poses as novel approach to motivate people and promote a change in them through the usage of game elements which have been repurposed to engage the user playfully in a nongame context. The game elements trigger a sense of joy and thus enhance the users experience, ultimately training or teaching the user new knowledge trough the game elements (Deterding et al. 2011). Deterding et al. (2011) defined gamification as "the use of game design elements in nongame contexts."

With the increasing prevalence of smartphones, by 2021, 72% of the global population, meaning 5.6 billion people, are said to be mobile users.[4] Companies are developing gamified apps that are aimed at favoring certain behavioral traits through motivating the user with game elements, positive feedback, and rewards. McKeown et al. state in their article that the possible game elements which can be applied reach from overall game mechanics and dynamics over very specific components to aesthetics for such potential apps. In a broader sense, one can distinguish between two components which are part of every gamification app: (1) game mechanics and (2) game dynamics (Zichermann and Cunningham 2011). According to Zichermann and Cunningham (2011), game mechanics are functional components which one can also find in traditional video games, like pattern recognition (memory-game interaction), gifting (easy transferable virtual items), collecting, gaining status (trophies, badges), fame (leaderboards), or flirtation (e.g., poking, easy-to-ignore interactions), while game dynamics represent the reactions of engaged users as response to the used game mechanics like self-expression, altruism, or competition to name a few (Zichermann and Cunningham 2011). If game mechanics are implemented the right way, game dynamics can be exploited to inherently align the companies interest with the intrinsic motivation of the user (Zichermann and Cunningham 2011).

Cugelman (2013) elaborates that within behavior change, it is important to know which persuasive principles work for which application and that it needs to be assessed whether gamification itself is a suitable application for an intervention

[4]http://www.cisco.com/c/m/en_us/solutions/service-provider/vni-complete-forecast/infographic.html

and therefore assessed seven criteria that help identify the suitability. Cugelman's (2013) thought of process matches to some extent Cialdini who described six influential principles and distinct psychological means that can be used by designers of persuasive technologies to create an effective persuasive system. Cialdini outlined that the appropriateness of a principle varies with cultural and personal preferences due to underlying psychological processes. According to Cugelman (2013), the framework of gamification needs to have, among other things, the ability to foster motivation within the user and for them to overcome their challenges. This compares the persuasive architecture of gamification with the principle of coaching. Moreover, he calls upon the theory of "flow" by Csikszentmihalyi that is frequently used in video game design. It describes how people become bound up and engaged in certain activities given that their skill level perfectly matches the challenge level suggesting that a game should increase its degree of difficulty and excitement with the users' experience so the user becomes absorbed in the game. He states that gamification should comprise a game that is not only fun but also supports the user to set goals, provide increasing challenges and motivation in form of reinforcement, show progress and feedback, and give rewards (like status or badges for achievements) and should contain a fun factor with social connectivity. Zichermann and Cunningham (2011) defined the enjoyment and fun component as the "fun quotient."

It becomes evident that there are numerous scientific publications (Cugelman 2013; Hswen et al. 2013; Spil et al. 2017) that solely examine the usability of gamification in health care and specifically in relation to reducing childhood obesity. Several gamified apps which are available in the stores of iOS, Android, and Windows under the category of health and fitness are proof (Deterding et al. 2011).

It can be concluded from the literature review that gamification can be used in the health-care industry to motivate clients and stimulate their long-term behavior; effort and difficult challenges for the clients can be turned into more engaging and enjoyable activities provided the situation is assessed correctly; and adequate game mechanics and game dynamics, like rewards and skill-matching game approaches, are carefully combined and applied (Cugelman 2013; Deterding et al. 2011; Spil et al. 2017; Zichermann and Cunningham 2011).

## 2.3 *Wearables*

Wearables exist in various devices, such as smart watches, smart glasses, bracelets, gesture controllers, or belts, and are believed to change our lives (Zhao et al. 2016a). Prominent examples are the Apple Watch, the FitBit bracelet, the controller belt, and Google glasses (Spil et al. 2017).[5] The number of wearables is expected to grow from 9.7 million devices in 2013 to 135 million devices in 2018.[6] As society has recently started moving toward a healthier lifestyle, an increase in the adoption of such wearables has occurred (Nelson et al. 2016). "The terms 'wearable

[5] http://coronect.de

[6] http://www.ccsinsight.com/press/company-news/1944-smartwatches-and-smart-bands-dominate-fast-growing-wearablesmarket

**Table 1** Wearable technology, function/measurement, and data based on Spil et al. (2017)

| Wearable technology | Function/measurement | Data |
|---|---|---|
| Pedometer | Can accurately measure step count, also estimates distance and energy expenditure but less accurate | Distance and speed tracking, calorie counting |
| GPS | A Global Positioning System provides information on a person's location, mode of transportation, and speed | Distance and speed tracking, recording achievement |
| Accelerometer | Recording movement and speed. Thus, provides indication of frequency, duration, and intensity of physical activity | Speed tracking |
| Heart rate monitor (HR) | Lightweight and (small) monitoring devices for heart rate that indicate of energy expenditure | Heartbeat measuring and calorie counting |
| Bluetooth | Connects multiple devices with each other | Sharing capability and expandability |
| Wi-Fi | A wireless computer networking technology with which networking capabilities over various radio bandwidth is possible | Sharing capability and expandability |

technology,' 'wearable devices,' and 'wearables' all refer to electronic technologies or computers that are incorporated into items of clothing and accessories which can comfortably be worn on the body". Health-care and fitness-specific wearable devices can be identified as technological gadgets and activity trackers to monitor and record a person's physical fitness activity which are worn by the person. The most common functionalities of wearables are step counting, speed and distance tracking, heart rate tracking, and quality of sleep monitoring. The transfer of the data to other devices takes place through Bluetooth or Wi-FI. Tehrani and Michael pointed out in their publication that although wearables can perform many of the core tasks of mobile phones and computers, the focus of wearables lies in the provision of sensory, monitoring, and scanning features of health-related data which are typically not seen in mobile and computers, therefore outperforming them in this area.

Table 1 which is based on Spil et al. (2017) displays a summary of the identified wearable categories, their functions, and the data that is being collected.

Wearables can easily and comfortably be worn the entire day, and personal physical information can be collected throughout the day (Spil et al. 2017). Dontje et al. stated in their study that the self-monitoring of such physical activity on a daily basis potentially increases one's awareness to such an extent that it can pose as a prerequisite for behavioral change. This implies that wearable devices can serve as a client management tool and have enormous potential in the field of health and medicine. Nelson et al. (2016) conducted a survey that has shown that individuals wearing wristbands like FitBit, or Apple Watch, are positively influenced because an individual's feelings of empowerment are increased.

As can be seen in existing literature, wearable devices yield enormous potential for the health sector. The combination of gamification and wearables, gamified wearables, seems to yield enormous potential. Therefore, the following literature review subchapter 3.2.3 will investigate in detail gamified wearables.

## 2.4 *Gamified Wearables*

Zhao et al. (2016a) investigated to what extent the three general fields – (1) serious gaming and gamification, (2) wearable technologies, and (3) health and fitness – combined can help to enhance traditional obesity intervention. The investigated target field is thus the overlap of these three fields, named gamified wearables. Therefore, it is important to determine, similar to Zhao et al. (2016a), Wortley (2015), and Zhao et al. (2016b), to what extent wearable technologies can be utilized for interaction with gamification and whether that idea is feasible, motivating, and engaging. Spil et al. (2017) assessed in their publication that people have a positive attitude toward wearables and gamified health apps in general, but the adoption and usage rates of wearables are still relatively low especially in comparison with the adoption and usage rate of gamified health apps. Nonetheless, people show an increasing interest in wearables and their usefulness. Gamified apps are already a widely accepted concept, but the results of Spil et al. (2017) have shown that users show increasing doubts over the physical data quality. As has been already established by Spil et al. (2017) through the use of gamified wearables, these doubts could be eliminated; thus, the quality of the gathered data would increase. Tong et al. (2015) also suggested that game-based approaches can be utilized to encourage and promote physical activity and lifestyle improvement. Within the serious gaming and gamification, an incentive approach is being applied where virtual and physical rewards are set to decrease obesity and increase quality of life. Tong et al. (2015) stated that the effectiveness of research prototypes which use "gamified wearables" is rarely investigated. Zhao et al. (2016a) developed a pilot prototype based on their findings. The case study found that based on their results, it can be said that existing technologies do match the current user needs, indicating that the idea of deploying gamified wearables for exercise and fitness is feasible, motivating, and engaging.

Another case study regarding gamified wearables was carried out by Wortley (2015) over a time period of 2 years. He investigated the use of the wearable bracelet, Jawbone UP, which measured and collected data regarding physical activity, ones' sleep cycle, and calorie consumption (manual entering of nutrition data) and displayed the data within a free mobile application that gave feedback and visual support. The main conclusion Wortley (2015) drew from his own study was that disruptive technologies like gamified wearables, which provide accurate data measurement, feedback, and smart coaching, e.g., in the form of games, can create an engaging experience that yields the power to influence behavioral change. It has to be noted that the study was applied to adults. Therefore, it would be interesting to see whether same results could be achieved with children. Since the original draft of his publication, Wortley (2015) was involved in several gamified wearable projects. Among others, the PEGASO project[7] investigated the use of wearables and gamified apps to encourage and enable healthy lifestyles among teenagers.

When reviewing existing literature, it becomes evident that there have been several attempts to develop and implement gamified wearables to achieve weight loss, to raise realistic awareness of physical activity and calorie consumption (Ahtinen et al. 2010; Kniestedt and Maureira 2016; Tong et al. 2015; Xu et al. 2012; Zhao et al. 2016a).

Tong et al. (2015) have published a paper in which they state that utilizing a gamification strategy for an app together with a wearable can promote physical activity and substantially improve quality of life. In their paper, they describe a research prototype named "FitPet," which introduced the design of such a potential mobile game. A follow-up study could expose important issues such as how such games and systems can be effectively implemented into the everyday life and cast some light onto the user's reaction toward interactive games. Tong et al. (2015) have applied the mentioned game mechanics and dynamics from Zichermann and Cunningham (2011). Through its accelerometer, the smartphone records steps one takes in real time. The step data can be converted into coins, so with each step, coins can be accumulated which can help to maintain a healthy pet. Similarly, Kniestedt and Maureira (2016) proposed a concrete game, the Little Fitness Dragon (LFD), which is a game for smartphones and smart watches. It can work as a self-standing app or in combination with a smart watch to track and motivate the players to be physically active. Kniestedt and Maureira (2016) were not yet able to test the validity of the LFD but could evaluate the interest in the game design from their target audience. Byrne et al. (2012) showed within their randomized field experiment with 39 adolescents in the USA that virtual pets can influence changes in behavior. Xu et al. (2012) project AHPC is a competition game to increase physical activity for adolescents and consists of a wearable pedometer, web-based game, and personal feedback. The game challenge was intended to change physical activity behavior. Consequently, the results of the study focused only on increasing adolescent physical activity, not direct weight loss. The challenge was a multi-month school-based competition in the form of a virtual race against other participating schools within the USA. The pupils had wearables which counted their steps and the earned points into an online platform. Each student contributed to their school's rank and was notified via a virtual achievement reward about their progress and points. Xu et al. (2012) results indicated that the AHPC in fact did increase physical activity of students but that the effectiveness reduced over time especially in students who participated for a long period. The authors believe that the justification lies within the design scope of the gamified wearable, as interest cannot be sustained over a long time period. As has become evident, gamified wearables are a major field of research and discussion within the realms of the health-care system. Current research has

[7] http://www.pegaso4f4.eu/

already revealed several issues, especially to maintain the user's interest over a longer period.

Based on this literature review, we present five fundamental guidelines for gamified wearables for obesity therapy for youth.

1. Obesity management tool:

   Visually edited data helps to support self-improvement due to a learning curve.

2. Game approach:

   Gamified apps encourage physical activity and trigger lifestyle improvement.

3. Rewards:

   Virtual and physical rewards decrease obesity and increase life quality.

4. Emotional support:

   Virtual motivational reminders create engaging experiences and provide support that eventually trigger behavioral change.

5. Education:

   Nutrition and exercise and well-being education have a positive effect on body weight.

## 3 Empirical Analysis

### *3.1 Objectives*

Next in our study, we interviewed the most involved stakeholders in obesity therapy for youth, clients, and clinicians, about their perception of the acceptance and effectiveness of these guidelines. The objectives of this empirical study are to gain insight in the relevance of the guidelines.

### *3.2 Methodology*

We conducted qualitative interviews with 18 overweight and obese children and adolescents and 3 clinicians that work within the realms of interventions to achieve weight loss and long-term behavior change for the clients.

The qualitative analysis method follows the grounded theory approach as first proposed by Glaser and then Strauss in the twentieth century and the framework Matrix of Miles and Huberman (1994). In grounded theory, one starts with a very concrete and contextual vision, and then, based on interviews and emerging answers, underlying explanations can be revealed. The underlying explanations are revealed by obtaining and analyzing in-depth data. Eventually, an evolving theory will develop from the corpus of used data. The advantage of the applied qualitative method is that detailed backgrounds of the interview subjects can be disclosed to enrich the information and the results which is needed because the theoretical guidelines are not established yet. In the future, also quantified studies are needed for empirical validation.

The results have been obtained by means of a computer-based qualitative data analysis software, MAXQDA, by a two-step set of coding processes: substantive and theoretical coding. In substantive coding, open axial and selective coding is applied. The goal is to produce categories and their properties. Theoretical coding weaves the substantive codes together into a hypothesis and theory (Glasser and Strauss 1967; Strauss 1987; Walker and Myrick 2006).

## 3.3 Design and Data Collection

The client questionnaires were distributed in Germany and the Netherlands. The personal conversation lasted between 10 and 15 minutes. For clinicians, questionnaires were distributed in Germany and the Netherlands. These personal conversations lasted between 45 and 60 minutes. Questionnaires were also answered by mail. The questionnaires for the clinicians were kept general, as the clinicians were asked to elaborate on their own experience and difficulties within obesity therapy. The questionnaires were built on the identified critical guidelines of the literature analysis.

In the first part, the clinicians were asked to describe the intervention program they work with. This introduction directly led to the central and crucial discussion of the interview and the main question of the paper. What are the issues that arise within therapy? What are the best tactics and tools to handle these issues? And which parts of the therapy need improvement?

Here, clinicians gave in-depth opinions and experiences that revealed various problems. The interview guide for the overweight and obese children and adolescents covered a sports, nutrition, and technology section which was intended to detect the attitude of clients toward the use of gamification and wearables within obesity treatment. This included input from both the interviews with the clinicians and extended literature review.

To generate valid results, 18 overweight and obese children between the age of 11 and 16 with a BMI between 25 and 52 were interviewed. The characteristics can be found in the Matrix below (Table 2).

**Table 2** Interview matrix

| | | | | Gender | |
|---|---|---|---|---|---|
| BMI | | | | Boys | Girls |
| Age | 11–13 | Overweight | 25–30 | 3 | 4 |
| | | Obesity grade 1 (slight obesity) | 30–35 | 1 | 2 |
| | | Obesity grade 2 (obesity) | 35–40 | 2 | |
| | | Obesity grade 3 (strong obesity) | >40 | | |
| | 14–16 | Overweight | 25–30 | 1 | |
| | | Obesity grade 1 (slight obesity) | 30–35 | 1 | 1 |
| | | Obesity grade 2 (obesity) | 35–40 | 2 | |
| | | Obesity grade 3 (strong obesity) | >40 | 1 | |
| | | | Total | 11 | 7 |
| | | | | 18 | |

## 3.4 Data Analysis

The transcribed interviews were analyzed by using the coding software MaxQDA. In reference to the elaborated questioning, the essential statements were coded openly and through an in-vivo method, the direct takeover of a quotation. All in all, the 18 client interviews yielded 500 codes which were assigned to categories. In the beginning, there were six main categories. During the coding process, new subcategories were created and quotations assigned. Although this method of qualitative data analysis constrains generalizability, it allows developing theoretical constructs to confirm or reject existing theories. The same procedure was used for the medical professional interviews. Three interviews yielded 74 codes and were originally assigned to 5 main categories and 2 subcategories. During the coding process with the in vivo method, new categories evolved, and issues within the obesity therapy were revealed. Based on Miles and Huberman (1994), a content-analytic summary table was created which brings together all data from the interviews into a single form where all four target groups (based on the independent variable age and gender) are examined for their attitude toward the five fundamental guidelines.

## 3.5 Results

The empirical analysis revealed that all interviewed programs seem to experience the same issues when trying to change the behavior of overweight and obese children toward a balanced and well-sized diet due to a lack of intrinsic motivation. This often rules out any chance for a long-term behavior change, resulting in weight gain after the therapy ends, because parents often are not the role model they need to be. Clinicians indicated that lack of physical activity can be an issue.

The clinicians stressed that a potential smartphone app should feature tools that motivate children to eat healthy, workout, and give them the possibility to control themselves better, and should also give them the possibility to control the everyday progress of their clients. Based on these statements, children were presented several questions regarding their habits and preferences. They were also given several gamified wearable ideas for which they had to give their opinion. The table below shows a summary of the requested features in a potential gamified wearable based on the question asked in the interview (Fig. 1).

Both female and male clients were open to the idea of using a wearable device if it could help them to lose weight and data also showed that clients were also in favor of the idea to use a gamified app which helps them to be physically more active and commit to a healthy diet. Clinicians have stated that children seem to have an issue controlling themselves when it comes to healthy food. The children were asked if they needed emotional support. Both female and male respondents answered that they could use some emotional support reminders in various moments throughout the day to keep up a healthy diet. The proposed monster app was not the right tool they are looking for.

When asked whether clients can estimate the caloric intake of the food they consume, all but two respondents answered negatively and had a positive attitude toward the proposal of an electronic food diary and a picture feature to count calories on a plate. The analysis on children's attitude toward rewards for keeping up a healthy diet and increased physical activity resulted in the fact that personal

| | Female 11- 13 | Male 11 - 13 | Female 14 - 16 | Male 14 - 16 | Clinicains |
|---|---|---|---|---|---|
| Food Diary | requested | requested | requested | requested | requested |
| Picture Diary | requested | requested | requested | requested | n/a |
| Nutrition Game App | not requested | requested | requested | requested | requested |
| Sports Game App | not requested | requested | requested | requested | requested |
| Competition and Team Games | requested | requested | requested | requested | requested |
| Horse / Fighting Game | requested | requested | requested | requested | n/a |
| Nutrition Rewards | not requested | requested | requested | not requested | requested |
| Sport Rewards | Neutral | requested | not requested | Neutral | requested |
| App rewards | not requested | requested | requested | requested | requested |
| Step Game | Neutral | requested | requested | not requested | n / a |
| Support Reminders | requested | requested | requested | not requested | requested |
| Monster App | not requested | Not needed | not requested | not requested | n / a |
| Snack Estimation / Calories | requested | requested | requested | requested | requested |
| Nutrition game App | not requested | requested | requested | requested | requested |
| Hero App | Neutral | Neutral | requested | requested | n / a |

**Fig. 1** Overview of data in table

rewards and a combination of personal and virtual rewards are a good extrinsic motivator. However, it is debatable whether rewards are critical guidelines for gamified wearables. A discussion on this can be found in the following chapter. Four specific games were proposed. Regarding the monster companion which is a nutrition reminder, it can generally be stated that clients disliked the proposed monster companion game app. Three other proposed gamified wearables have received mainly positive feedback.

The content-analytical summary table summarizes all results and shows clear indication of which fundamental guidelines need to become implemented in future app development. The results for the variable BMI have been omitted due to too small sample size.

It becomes evident from the empirical analysis that clinicians believe in the potential of gamified wearables and are positive that the concept can help them to tackle the challenge of a healthy diet and long-term behavior change. The empirical analysis has demonstrated that some guidelines would be adapted enthusiastically into the daily routine by the children; e.g., children have shown enormous interest in the idea of using *game approach* apps to lose weight and become healthy. The thought of getting displayed nutritional-related data and wearing a wearable device to track health-related data, as some form of *obesity management tool*, meets approval with most clients. The concept of *rewards* did not meet the approval as assumed by clinicians or myself through existing scientific literature, which comes as a surprise. One clinician explicitly stated that they successfully use the concept of rewards within their program. Three clinicians have stated that in their experience, children respond well to rewards (Table 3).

The clinicians and literature have indicated the importance of nutritional impact education on the degree of obesity. The data shows that *education* is wanted and needed by the clients. One theoretical guideline was not supported by the data of the empirical analysis: *emotional support*.

The analysis has shown that children and adolescents do not think a gamified wearable solution that is intended to provide emotional support is a helpful tool for them, thus indicating it is not an effective guideline.

It became clear that a potential app must be well structured and tailored to gender and age. Clinicians have confirmed within the questionnaire that the traditional therapies are also adjusted for age variable. Within the realms of this study, there have been clear differences between preference within the variables age and gender, but often only a very slight or uninterpretable difference for the variable BMI, which could be due to the small number of clients within these groups.

## 4 Discussion

The literature analysis has shown that overweight and obese children have an illogical and abnormal relation to food and meal sizes. Further, unhealthy food and more calories are being consumed in less time (Sarah E. Barlow 2007; Cawley 2010). This phenomenon has also been observed by clinicians, which is why they

**Table 3** Fundamental app guideline content-analytical summary table

| | Critical guidelines working for gamified wearables? Target groups | | | | |
|---|---|---|---|---|---|
| Critical guidelines from the literature analysis | Female 11–13 | Male 11–13 | Female 14–16 | Male 14–16 | Clinicians |
| 1. *Realization of obesity management tool* | *Implement the guideline* | *Implement the guideline* | *Implement the guideline* | *Implement the guideline* | *Implement the guideline* |
| Food diary | Needed | Needed | Needed | Needed | Expressed deep interest in an accurate food diary |
| Picture diary | Needed observation: have expressed doubts of accuracy and functionality | Needed | Needed | Needed | n/a |
| 2. *Realization of game approach* | *Implement the guideline* | *Implement the guideline* | *Implement the guideline* | *Implement the guideline* | *Implement the guideline* |
| Nutrition game app | Not needed | Needed | Needed | Needed | Elaborated on importance of nutrition education within therapy |
| Sports game app | Needed observation: have expressed doubts of accuracy and functionality | Needed | Needed | Needed | Elaborated on sport being a crucial component of therapy, a successful treatment method, believe in motivation of |
| Competitive and team games | Needed observation: smaller focus on competition and more on team work and friendship games | Needed observation: a few clients have expressed wish to play alone | Needed | Needed | One clinician has explained that team competition elements are already integrated into the therapy program. The others have expressed the strong belief in |

| | | | | | |
|---|---|---|---|---|---|
| Horse/fighting game | Needed observation: some complaints, have expressed regarding the setting of the game and competitiveness | Needed | Needed | Needed observation: loved the competitive focus of the game | n/a |
| 3. *Realization of rewards* | *Don't implement the guideline* | *Implement the guideline* | *Implement the guideline* | *Guideline indecisive* | *Implement the guideline* |
| Nutrition rewards | Not needed observation: clients looking for approval of their parents | Needed observation: in particular virtual rewards | Needed | Not needed | Believe in the power and potential of rewards, one has already integrated such feature |
| Sport rewards | Neutral observation: seems to dependent on the personality and attitude of the client | Needed observation: some have expressed a particular interest in personal rewards | Not needed | Neutral | Believe in the power and potential |
| App rewards | Not needed | Needed observation: some have expressed a particular interest in personal | Needed observation: only personal rewards | Needed | Believe in the power of rewards, one has already integrated such feature |
| Step game | Neutral observation: seems to be dependent on the personality and attitude of the client | Needed | Needed | Not needed | n/a |
| 4. *Realization of emotional support* | *Don't implement the guideline* | *Don't implement the guideline* | *Don't implement the guideline* | *Don't implement the guideline* | *Implement the guideline* |
| Reminders and support in moments of doubt | Needed | Needed | Needed | Not needed | Think that clients need a stable and safe environment with guardians reminding them to lead a healthy |

(continued)

**Table 3** (continued)

| | Critical guidelines working for gamified wearables? Target groups | | | | |
|---|---|---|---|---|---|
| Critical guidelines from the literature analysis | Female 11–13 | Male 11–13 | Female 14–16 | Male 14–16 | Clinicians |
| Monster app | Not needed | Not needed | Not needed | Not needed | n/a |
| 5. *Realization of education* | *Guideline indecisive* | *Implement the guideline* | *Implement the guideline* | *Implement the guideline* | *Implement the guideline* |
| Snack estimation/calories | Needed observation: expressed to see the benefit of extending their knowledge in this field | Needed observation: expressed to see the benefit of extending their knowledge in this field | Needed observation: expressed to see the benefit of extending their knowledge in this field | Needed observation: expressed to see the benefit of extending their knowledge in this field | Elaborated that it is of importance that children have an idea on nutritional value of food they consume |
| Nutrition game app | Not needed | Needed | Needed | Needed | Elaborated on importance of nutrition education within therapy |
| Hero App | Neutral observation: seems to be dependent on the personality and attitude of the client, but is not conceived as to childish | Neutral observation: game should not be to childish | Needed | Needed | n/a |
| *Results: which guidelines are critical to implement into the app* | *1,2, and maybe 5* | *1,2,3, and 5* | *1,2, 3, and 5* | *1,2,5, and maybe 3* | *Have expressed the necessity of a gamified wearable to include all the above named guidelines to resolve the current issues 1,2,3,4, and 5* |

have expressed the need for an obesity management tool with a control function to access and watch the child's daily routine. The clinician's suggestion is to implement a food diary into a potential app, as the ease of food logging and exercise, will help ensure a more precise and frequent documentation of the caloric consumption and energy expenditure. Clients have expressed interest in such a tool. Children have also reacted positively toward the idea of a food photo recognition feature which would provide an easy way for children to estimate calories in a meal using the smartphone camera instead of manually entering the nutritional value of their food. There are companies currently working on such a service.[8] Clinicians believe that parents should have access to such a tool. Literature showed that wearable devices can give access to these data points (Nelson et al. 2016; Spil et al. 2017; Zhao et al. 2016a). The data analysis has revealed that children would be willing to wear such a device. Provided clinicians can have a third-party access to an app used by clients on a regular basis, this could pose as an obesity management tool with a control function. The data implies that an obesity management, in the form of a food and picture diary tool, serves as a basic and effective guideline that is both requested and needed by clinicians and clients.

Clinicians and literature agree that game approaches yield the potential to not only complement already established game elements within current therapy formats but also motivate and encourage clients outside of therapy to behave accordingly. Gamification does indeed have the potential to engage users in normal effort tasks. Previous projects have shown that the level of physical activity can increase (Deterding et al. 2011; Xu et al. 2012). The problem is that the effectiveness can decrease over time, emphasizing that gamification needs to pique the users interest over a longer period. This can be achieved through matching skill and entertainment level. The difficulty of the game has to and new incentives have to be offered to maintain the suspension and interest in achieving new goals and, e.g., new rewards (Hswen et al. 2013; King et al. 2013). Based on the literature findings and the wish to integrate nutrition and physical activity games from the clinician's side, clients were offered several gamified wearables. Children were offered (1) a nutrition reminder app, (2) a food education game, (3) a Tamagotchi-based lifestyle game for female and male patients, and (4) a step-based workout game with rewards. As Cugelman (2013) has reported, it is important to assess beforehand whether gamification provides a potential tool for obesity interventions and which persuasive principles need to be applied in detail. Clinicians have confirmed that potential gamified wearables need to consider gender and age differences. Therefore, all interviews were analyzed considering age, gender, and BMI, the third variable. The purpose of this study was to gain insight in the success of gamified wearables in obesity therapy for youth by studying fundamental guidelines for such gamified wearables. The research question was: What are fundamental guidelines

[8]https://www.sri.com/engage/products-solutions/food-recognition-technology & http://foodai.org/ & http://www.caloriemama.ai/ & http://www.popsci.com/google-using-ai-count-calories-food-photos

for gamified wearables to accomplish weight-loss and long-term behavior change in overweight and obese children and adolescents? In conclusion, it can be said that the interviews have revealed that when the four effective guidelines (game approach, obesity management tool, rewards, and education) are being applied and potential gamified wearable are adjusted for gender and age, they have the potential to be good support tools for traditional obesity interventions.

## *4.1 Potential Gamification App: Body Balance*

A prototype app which was created with the help of Yulia Byron-Moiseenko will be presented. The empirical findings were first translated into sketches and then into a functioning, well-designed prototype.

The prototype app "Body Balance" was first created on paper and then transformed into a clickable dummy app using the platform Invision to demonstrate how a prototype obesity management and gamified app with a wearable would appear.[9] The link for the online app will be available at the end of this chapter. When following the link, the static screens will transform into a clickable interactive prototype. The link can also be opened on a smartphone where it will simulate a real-life mobile application.

The idea behind the app is to incorporate it into current obesity therapy programs. The clients, their family, and the clinicians will download a version of the app to support and enhance treatment. Results have indicated that the content of a potential app should vary with gender and age. A prototype for the specific target group male aged 11–13 was created. This was a random choice for reasons of time and space. Only one clickable prototype could be created.

After having downloaded the app from the corresponding mobile app store, the clients will enter their age and gender into the phone. The design and content of the app will adapt based on these two data points. Once the data points are entered, the screen will change and children are given the ability to fill out their personal profile. Clients will need to provide the following: name, birthdate, patient number, doctor, weight, and height. The BMI is automatically calculated (Figs. 2 and 3).

At the bottom of the screen, four tabs are visible which gives the user the ability to go to their profile, the achievement hall, the game room, or their food diary. The app is an obesity management tool, utilizing the achievement hall, food diary, and games in the game room. All sections of the app contain gamified elements which synchronize with a wearable, a gamification app.

Within the achievement hall, clients are able to see their progress and their physical health status. It was decided to integrate this tool because the interviews made it obvious that clients interpret their own skills and health status inaccurately. Clients can also display the average value of the past week/month and see the value of their caloric intake and the physical steps taken. This section also monitors weight loss progress. All three information screens are visually implemented using a color

[9]https://www.invisionapp.com/

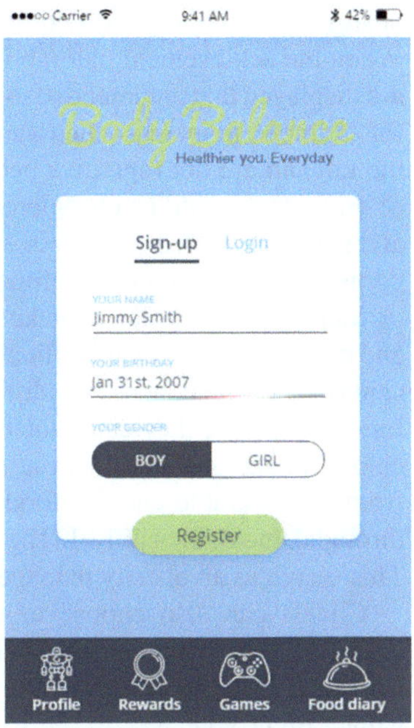


**Fig. 2** Landing page

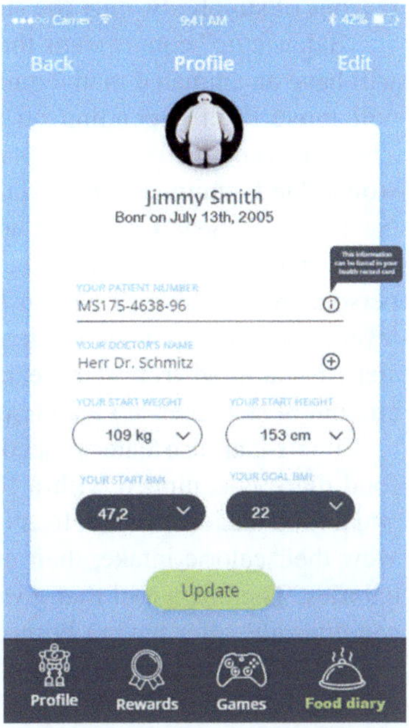


**Fig. 3** Profile page

scheme to display the positive or negative current state of health. All data points have the potential to earn badges and also provide the opportunity for self-improvement. The gamified element badges are being used to verify the reality of the client's health status and to provide them with a sense of amusement and accomplishment. When the achievement hall tab is opened, brief feedback and motivating messages are displayed that are intended to grab the attention of the client and encourage them, for example, to increase their step count. These messages are only displayed within the tab and are not repetitive because clients have stated that an excessive number of reminders would be considered annoying. Male clients aged 11–13 prefer virtual and personal rewards, which is why the tool was developed. In the lower right hand corner, a *Help button* can be found which can be used if a client needs support.

Step goals and caloric intake are discussed and predetermined by the clinicians and parents. These can be difficult goals to achieve leading to dissatisfaction if goals cannot be attained, and the color scheme indicates this to the client over a period of time. Therefore, they can contact their clinicians or parents to adjust the step and caloric intake goal through the *Help button*. The data is being captured primarily from the wearable and the food diary. The data will be transferred automatically through Bluetooth or Wi-FI. The weight loss progress needs to be entered manually. This serves as an obesity management tool.

Clients can also choose to enter the game room. The game room contains a section that will increase their physical activity. In the prototype app "Body Balance," three games are displayed: the steps game named "Balance up," the fight game "Fight your way to the top," and the hero game "From Zero to Hero." For reasons of simplicity, only the step game will be illustrated.

"Balance up" counts steps that can be changed into rewards and games.[10] Users will have an animated mannequin, a so-called virtual player, at their disposal which will move itself depending on the steps the client takes on a virtual landscape. The landscape of the virtual player adjusts depending on the season (e.g., winter wonderland, summer camp, Halloween haunted castle, etc.). Every time a client has reached a new destination which correlates with a predetermined step goal that has been set by the parents and clinicians, children will not only be rewarded with a personal reward by their parents but will also receive virtual rewards such as badges, level rank, small animated videos, or accessible mini games. Again, the number of steps is captured from the wearable device. It is important to mention that a low amount of step does not lead to penalties (Figs. 4 and 5).

Clients can also enter a second obesity management tool, the food diary. The food diary is equipped with a food photo recognition feature, a barcode scanner, and a food database that delivers nutritional values for any type of food. Clients can view their caloric intake, their energy consumption, and the remaining number of calories. There is a tool that gives them information on what additional foods they can consume after having burned a certain number of calories. The proposition of the app will be based on the foods the user eats regularly as this will be classified as favorite food. Further, they can see the macro level category of their nutritional intake, referencing fats, carbohydrates, or proteins. They can log food into the corresponding section, e.g., breakfast, snack, and so forth. Additionally, clients are

[10]https://itunes.apple.com/us/app/ibitz-by-geopalz-kids/id588227932?mt=8

**Fig. 4** Game room

**Fig. 5** Balance up – Halloween edition

being provided a button to express their emotions toward a meal with an emoji and can add notes into the additional free space. These emojis can be viewed by

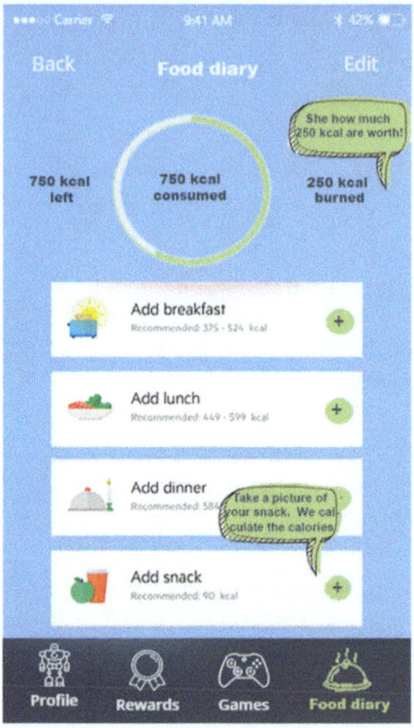


**Fig. 6** Food diary

clinicians and parents to offer the client emotional support when experiencing a bad day. Also, patterns regarding their food habits can be revealed (Fig. 6).

It should be mentioned that both the clinicians and parents will be able to download a corresponding app that will provide them access to the client's data by entering the name, client number, and a PIN. Both parties then have access to all health-related data from the wearable. This includes step counting, speed/distance tracking, heart rate, and quality of sleep monitoring. Through this app, clinicians and parents can see whether the child has activated the help button in the achievement hall to request emotional support.

Follow the link:

https://projects.invisionapp.com/share/UJCGGGP5A#/screens in order test the clickable prototype online. Or insert the link into your internet browser on a smartphone.

# 5 Conclusion

The results have shown that gamified wearables are expected to help to manage obesity by helping clients to reduce weight.

The qualitative analysis has also revealed that triggering long-term behavior change will, despite utilizing a gamified wearable, still pose a major challenge. Rewards, according to the clinicians and literature, pose as an effective tool to entice user to remain committed to a gamified wearable. Surprisingly, it has become clear that children are not sure what they think of rewards as incentive to commit to healthy behaviors like a healthy diet and regular physical activity.

In conclusion, clinicians and patients need a multifaceted technological solution to achieving weight loss and enduring behavior change which will complement obesity therapy. Results indicate that the theoretical effectiveness of the guidelines will differ with age and gender of the clients. For the variable, BMI results were inconclusive due to a too small number of clients within these groups. Based on this, it was established that a potential gamified wearable will have to be developed for specific target groups. Further empirical quantitative analyses, ideally with a mix-method research and a bigger target sample size, are needed to confirm or disconfirm the fundamental guidelines in the future. Future study can also explore other chronic diseases, like diabetes, to use the same kind of gamification and wearables.

**Acknowledgements** We would like to thank Yulia Byron-Moiseenko for her drawings in the prototype.

# References

Ahtinen, A., Huuskonen, P., et al. (2010). *Let's all get up and walk to the North Pole: Design and evaluation of a mobile wellness application*. Paper presented at the proceedings of the 6th Nordic conference on human-computer interaction: Extending boundaries.

Barlow, S. E. (2007). Expert committee recommendations regarding the prevention, assessment, and treatment of child and adolescent overweight and obesity: Summary report. *Pediatrics, 120*(Supplement 4), S164–S192. https://doi.org/10.1542/peds.2007-2329C.

Bjørge, T., Engeland, A., et al. (2008). Body mass index in adolescence in relation to cause-specific mortality: A follow-up of 230,000 Norwegian adolescents. *American Journal of Epidemiology, 168*(1), 30–37.

Byrne, S., Gay, G., et al. (2012). Caring for mobile phone-based virtual pets can influence youth eating behaviors. *Journal of Children and Media, 6*(1), 83–99.

Cawley, J. (2010). The economics of childhood obesity. *Health Affairs, 29*(3), 364–371.

Cugelman, B. (2013). Gamification: What it is and why it matters to digital health behavior change developers. *JMIR Serious Games, 1*(1), e3.

Deterding, S., Dixon, D., et al. (2011). *From game design elements to gamefulness: Defining gamification*. Paper presented at the proceedings of the 15th international academic MindTrek conference: Envisioning future media environments.

Dietz, W. H. (1998). Health consequences of obesity in youth: Childhood predictors of adult disease. *Pediatrics, 101*(Supplement 2), 518–525.

Ebbeling, C. B., Pawlak, D. B., et al. (2002). Childhood obesity: Public-health crisis, common sense cure. *The Lancet, 360*(9331), 473–482.

Glasser, B., & Strauss, A. (1967). *The discovery of grounded theory: Strategies for qualitative research*. New York: Adline De Gruyter.

Han, J. C., Lawlor, D. A., et al. (2010). Childhood obesity. *The Lancet, 375*(9727), 1737–1748.

Hswen, Y., Murti, V., et al. (2013). Virtual avatars, gaming, and social media: Designing a mobile health app to help children choose healthier food options. *Journal of Mobile Technology in Medicine, 2*(2), 8–14. https://doi.org/10.7309/jmtm.2.2.3.

Kelishadi, R., & Azizi-Soleiman, F. (2014). Controlling childhood obesity: A systematic review on strategies and challenges. *Journal of Research in Medical Sciences, 19*(10), 993.

King, D., Greaves, F., et al. (2013). Gamification': Influencing health behaviours with games. *Journal of the Royal Society of Medicine, 106*(3), 76–78. https://doi.org/10.1177/0141076813480996.

Kniestedt, I., & Maureira, M. A. G. (2016). *Little fitness dragon*: *A gamified activity tracke*. Paper presented at the international conference on entertainment computing.

Miles, M. B., & Huberman, A. M. (1994). *Qualitative data anulysis: An expanded sourcebook*. Thousand Oaks: Sage.

Must, A., Jacques, P. F., et al. (1992). Long-term morbidity and mortality of overweight adolescents: A follow-up of the Harvard Growth Study of 1922 to 1935. *New England Journal of Medicine, 327*(19), 1350–1355.

Nelson, E. C., Verhagen, T., et al. (2016). Health empowerment through activity trackers: An empirical smart wristband study. *Computers in Human Behavior, 62*, 364–374.

Ogden, C. L., Carroll, M. D., et al. (2006). Prevalence of overweight and obesity in the United States, 1999–2004. *JAMA: Journal of the American Medical Association, 295*(13), 1549–1555.

Pulgarón, E. R. (2013). Childhood obesity: A review of increased risk for physical and psychological comorbidities. *Clinical Therapeutics, 35*(1), A18–A32.

Singh, A. S., Mulder, C., et al. (2008). Tracking of childhood overweight into adulthood: A systematic review of the literature. *Obesity Reviews, 9*(5), 474–488.

Spil, T., Sunyaev, A., et al. (2017). *The adoption of wearables for a healthy lifestyle: Can gamification help*?. Paper presented at the Proceedings of the 50th Hawaii international conference on system sciences.

Strauss, A. L. (1987). *Qualitative analysis for social scientists*. Cambridge: Cambridge University Press.

Thorpe, K. E., Florence, C. S., et al. (2004). The impact of obesity on rising medical spending. *Health Affairs, 23*, W4.

Tong, X., Gromala, D., et al. (2015). *Encouraging physical activity with a game-based mobile application*: *FitPet*. Paper presented at the Games Entertainment Media Conference (GEM), 2015 IEEE.

Walker, D., & Myrick, F. (2006). Grounded theory: An exploration of process and procedure. *Qualitative Health Research, 16*(4), 547–559.

Wolfswinkel, J. F., Furtmueller, E., et al. (2013). Using grounded theory as a method for rigorously reviewing literature. *European Journal of Information Systems, 22*(1), 45–55.

Wortley, D. (2015). *Gamification and lifestyle technologies for personal health management*. Paper presented at the European conference on games based learning.

Xu, Y., Poole, E. S., et al. (2012). *Designing pervasive health games for sustainability, adaptability and sociability*. Paper presented at the proceedings of the international conference on the foundations of digital games.

Zhao, Z., Etemad, S. A., et al. (2016a). *Gamification of exercise and fitness using wearable activity trackers*. Paper presented at the proceedings of the 10th international symposium on computer science in sports (ISCSS).

Zhao, Z., Etemad, S. A., et al. (2016b). *Motivational impacts and sustainability analysis of a wearable-based gamified exercise and fitness system*. Paper presented at the proceedings of the 2016 annual symposium on computer-human interaction in play companion extended abstracts.

Zichermann, G., & Cunningham, C. (2011). *Gamification by design: Implementing game mechanics in web and mobile apps*. Sebastopol: O'Reilly Media, Inc.

# Precision Wellness: An Optimization Model

**Paul Cooper and Nilmini Wickramasinghe**

## 1 Introduction

Over the past century, global success in scientific research has facilitated the realization of a new era of precision healthcare delivery and cure (Ashley 2015). The healthcare industry has experienced a rise of inventions directed toward enhancing life expectancy and diagnostic and cure options and improvement of the quality of life along with the efficacy and cost-effectiveness of the entire healthcare systems (Baird 1990; Ashley 2015). The emergence of precision health, wellness and medicine practices is playing a crucial role in the innovation and development of better healthcare systems (Ashley 2015; Jameson and Longo 2015; Juengst and McGowan 2018). This focus on precision practices in healthcare has been enabled through the leveraging of various technology developments especially those connected with the IoT (Katsnelson 2013) and advances in genomics and high speed data analysis (Helwande 2018). Now, we are witnessing the subsequent patient-driven demand for precision wellness, medicine and health which is serving to push the technology developments further and also impact many current as well as future aspects of healthcare delivery (Katsnelson 2013; Kaur et al. 2017).

P. Cooper
Epworth HealthCare, Melbourne, VIC, Australia

N. Wickramasinghe (✉)
Epworth HealthCare, Richmond, VIC, Australia

Swinburne University of Technology, Hawthorn, VIC, Australia

 
N. Wickramasinghe, F. Bodendorf (eds.), *Delivering Superior Health and Wellness Management with IoT and Analytics*, Healthcare Delivery in the Information Age, https://doi.org/10.1007/978-3-030-17347-0_8

## 2 Background

The terms precision health, precision medicine, and precision wellness are now peppering the literature. In order to develop a suitable model, it is first important to understand these respective terms, their distinctions and similarities.

### *2.1 Precision Health*

According to Severi et al. (2014), precision health is defined as a new practice motivated by digital high-tech developments that allow more specific explanations and examines persons and populace sets, with an intention to improve the general wellness of people. Precision health involves robust surveillance information, prompt use of sophisticated evaluations to track the geographic spread of illness, and the ability to perform analysis on such data (Dowell et al. 2016). This practice is therefore effective since it not only uses more exact methods to measure illness, disease-causing organisms, exposures, behaviors, and risks at an individual level but also allows better evaluation of public health along with the improvement of policies and directed prevention plans (Jamerson and Longo 2015). Khoury et al. (2016) state that this practice centers on increasing precision and resolute agendas in outlining population cohorts and providing objective involvements of numerous forms.

### *2.2 Precision Medicine*

Precision medicine can be regarded as a new and emerging field that brings together medical studies, social science, along with computer science (Juengst and McGowan 2018). According to Leff and Yang (2015), precision medicine purposes to combine an all-inclusive data collected over time about a person's genetics, environment, as well as lifestyle in order to improve disease understanding and interception, assist in drug discovery, and guarantee delivery of suitable therapies. Its primary aim is to constantly improve the medications suitable for exact mutations and illnesses (ibid). The goal of precision medicine is a techno-utopia established on evidence-based practices of how the evaluation of data can transform healthcare delivery and the healthcare system (Juengst et al. 2016). Emphasis is thus placed on health as determined not only by biology but also on a complex relationship of genetics and social and economic aspects (ibid). It is important to note that the causes of illnesses are complex and heavily dependent on multiple biological, social, and economic factors (Kaur et al. 2017; Khoury 2018). Therefore, precision medicine interventions need to have sufficient capacity to address these complexities (ibid). This medical practice uses the biomedical research where the findings help to shape the future directions of precision medicine which will use understandings from multiple forms of data (Khoury et al. 2016; Lyles et al. 2018; Khoury 2018). Precision medicine can thus be described as the effort to collect, combine, and evaluate multiple sources of data with an aim to develop individualized insights in regard to health and disease (ibid). This practice helps to capture what used to be traditionally termed as medical data including lab results and integrates the data with other forms of nonmedical data like air and lifestyle quality measures (ibid).

Ginsburg et al. (2016) state that precision medicine has been implemented to improve patients' safety and maintains a great promise to create a safer regime across the globe. Thus, the argument is that costs connected with adverse events are currently being reduced following the increasingly strategic use of tools that strengthen precision medicine (Minor 2018). Through the effective standard cure offered by precision medicine there is the possibility to offer hope and better cures to many individuals but there are also many challenges (Meagher et al. 2017). It is clear, however, that precision medicine is transforming the future of healthcare sector driving it toward better and at times more expeditious care (ibid).

Precision medicine has value in supporting and enabling superior medical decisions (Lyles et al. 2018) although its cost effectiveness has been found to be somewhat difficult to demonstrate in practice due to deficiencies in the clinical and economic evidence base (Gavan et al. 2018). Its increasing popularity is attributed to its exciting influence on medical research and patient care (Leff and Yang 2015). As precision medicine continues to grow, the healthcare industry is likely to be tasked with the responsibility of offering numerous treatment options that are increasingly customized to a person's needs (University of Michigan 2018). Advances in precision medicine have already generated results with some of the recent challenging illnesses such as genetically related diseases (Adams and Petersen 2016).

## 2.3 Precision Wellness

Precision wellness is an evolution of precision medicine (Juengst and McGowan 2018). Fernandez (2017) argues that precision wellness focuses on providing highly significant and targeted data-driven plans to reduce costs and improve end results. The practice is relatively nascent in the healthcare system and thus it is at a very embryonic stage. However, based on the current trends, it is clear that precision wellness will be fully adopted into the healthcare system in the next few years to come (Juengst and McGowan 2018). Precision wellness is, indeed, an important practice in healthcare as it equips people with data and tools to modify behaviors, whether those of diet and physical exercises or their genes, to maximize personal health (Juengst et al. 2016). It aims at improving an individual's well-being by creating medicines that are preventive and personalized and reversing the increasing cost trends in healthcare (ibid). Fernandez (2017) maintains that this practice offers practitioners' offices with techniques and services that improve patient results and offers additional revenue for the practice with no open investment or long-term contracts needed. Further, some of the most impactful technologies utilized in precision wellness focusing on precision oncology, molecular diagnostics, and predictive biomarkers focus on illness prevention, pain alleviation, and affordability (Fernandez 2017). Therefore, it is important to embrace the added value and personalization that precision wellness brings to population health management programs and practices. Many medical institutions are already establishing roadmaps to integrate pharmacogenomics, real-time data and diagnostics, and exogenous influencers of healthcare into what is referred to as precision wellness platforms (ibid).

## 2.4 Current Trends

According to Kaur et al. (2017), the notion of making healthcare systems more precise is central to many healthcare agendas around the world. The healthcare standards for the treatment of various key illnesses such as cancer are shifting slowly over time (Ashley 2015). Currently, health practitioners are establishing a broad perspective to cure illnesses, often leaving behind the traditional conventional approach commonly known as "one size fits all" as they are moving toward treatment-based methods which focus on the uniqueness of each individual (Baird 1990; Adams and Petersen 2016).

The focus of "precision" in medicine/health/wellness is to guarantee delivery of directed cures to the right people and at the right time (Ashley 2015). This, however, depends on the development of precise biomarkers (Ginsburg and Phillips 2018). In the present times, all scientists are working toward this direction in order to subcategorize different illnesses in accordance with these biomarkers. Similarly, the development of appropriate medicine in precision medicine is also largely influenced by these biomarkers (Kaur et al. 2017). While biomarkers-based prognostic examinations serve as the foundation for molecularly directed therapies, the appropriate regulatory restriction is needed over these investigations to make certain their validity, dependability, precision, and application in medical society (Ginsburg and Phillips 2018). However, substantial advancement has been realized in the establishment of biomarkers in different illnesses such as cancer and rare genetic, metabolic, and cardiovascular conditions (Ginsburg et al. 2016).

Kaur et al. (2017) argue that all aspects of precision healthcare are currently working toward integration which calls for multiple collaboration with other players in order to attain full potential. For instance, the incorporation of precision medicine with other forms of technologies such as RNA involvement and nanomedicine toward establishing therapeutic methods has immense potential to generate new concepts in this discipline (ibid). This incorporation of precision medicine and nanobiotechnology has proved to yield great success following its ability to detect diseases early such as cancer as well as establish a more efficient, secure, and directed cure with higher chances of treatment (ibid). Without a doubt, the era of precision practices in healthcare largely depends on us. However, the constant growth of these practices heavily relies on increasing integration between research and healthcare industry (Juengst et al. 2016). Therefore, the success of a precision medicine vision and the development of a sustainable and beneficial system relies on collaborative determinations of all parties involved in the healthcare industry as well as research (ibid).

Precision practices are being transformed from the money-oriented approach to a value-based plan that seeks to identify cost-effective solutions (Minor 2018). Precision practices, therefore, focus on reducing the redundant and needless examining and obligatory hospital admissions (ibid). As a result, the cost of a value-based set is increased (ibid). With the novel reimbursement approaches, precision practices will have the infrastructure to achieve its aim of offering exact care that a particular patient requires or might not necessarily need based on his or her individual genetic makeup and wellness history (Kaur et al. 2017). Through more cost-effective treatments and monitoring of patients, savings are realized hence leading to an increase in medication adherence (Kaur et al. 2017).

Blockchain technology is another trend in precision medicine that handles big data on the populace's genome structure (Helwande 2018). According to Helwande (2018), this technology works on spread networks along with shared ledgers that can be applied to ensure the security of data and ethical use of information and is among the latest technologies projected to reach the optimum of an inflated prospect in 2018. However, for precision medicine, the technology is proffered to have real potential (Meagher et al. 2017). Since machine learning tools are more influential with big learning sets and more enormous data, precision practices require methods to gather more data (Leff and Yang 2015). The contribution of this trend impacts the larger sector of health and genomic data required for precision, wellness, health, and medicine (ibid). It guarantees security to genetic data as the information needs to be disseminated across varying platforms. Additionally, blockchain operates as a good auditing technology and could have a role in treatment coordination (ibid).

Artificial intelligence has also been implemented in the current precision practices with an intention to assess patterns in the data and offer insights to clinical professionals in relation to the condition of a person (Helwande 2018). Furthermore, genome sequencing is also advanced using artificial intelligence (Meagher et al. 2017; University of Michigan 2018). This technological trend can also be applied in clustering and segregation of populaces into sets that can assist in the generation of tailored medicine, wellness, and health (Helwande 2018).

With the great advances in research and technology that inspires patients to expect and demand for better, precision practices will continue to enable a new period of medicine and strategies in which health providers, scientists, and patients work in collaboration to improve personalized care (Meagher et al. 2017). With the millennial determination to attain wellness, varying technologies have been established to educate and track their wellness strategies (Minor 2018). For instance, millennials are currently using applications to trace training information as well as search for healthier living data from the Internet (Lyles et al. 2018). This trend has already begun to influence numerous non-millennial people who now examine healthcare products through a varying set of values and criterion than in the previous years.

## 2.5 Technologies Applied

Precision health, precision medicine, and precision wellness are all emerging practices leveraging ground-breaking tools and data science, i.e., the technologies of the IoT, to tailor disease prevention, detection, as well as treatment. These practices seek to gain insights on health threats and tailor cure for a particular and homogenous subpopulaces mainly through the use of new data, digital health tools, and approaches. Big data is, therefore, one aspect that has constantly assisted to attain these objectives, over its capacity to deliver to physicians a large and multiplicity of both structured and unstructured information that was in the past impossible (Adams and Petersen 2016). In addition, the large data has enhanced more extensive and explicit study of stratifying and segmenting populaces at risks for a range of health issues (ibid). The constant development in capacity and range of existing data, the cost-effectiveness of data collection, and developing computational tactics mean that the achievement of big data will probably be an essential support in precision health, wellness, and precision medicine in the days to come (Baird 1990).

According to Leff and Yang (2015), big data has a very great impact on precision health practices, in precision medicine and in precision wellness. Moreover, a report by the Institute for Health Technology Transformation in 2013 notes that America's healthcare institutions have established 150 Exabyte of healthcare data, and its continued growth is expected in the upcoming times (National Academy of Medicine 2018). This unique volume of data, once implemented in a meaningful way, can offer significant understanding to numerous illnesses and their respective progressions. This insight helps precision practitioners to avoid unnecessary treatments, minimize drug adverse effects, and optimize general safety which ultimately leads to a more effective and proficient health care system. As a result, a path that realizes the objectives of personalized medicine, health, and wellness needs to be developed.

Advances in precision health, medicine, and wellness have already led to new discoveries and numerous new treatments that are personalized to specific physiognomies of individuals, such as an individual's genetic makeup (Baird 1990; Fernandez 2017; University of Michigan 2018). These innovations are greatly influencing transformation in some treatments such as cancer. These efforts are allowing precision physicians to select treatments that can improve survival and reduces the probability of having adverse effects (Khoury et al. 2016).

## 2.6 Barriers to Implementation

In spite of precision medicine, wellness and health's potential to influence the health delivery, the adoption of these precision practices has been slow (Dowell et al. 2016). The slow implementation of these practices has been caused by lack of supporting IT infrastructure, the absence of evidence of medical utility offered by investigations, the absence of data standards along with interoperability, inadequate decision support system, and inadequate funding for translational health investigations (Horgan et al. 2014). Other consistent factors hindering precision health, medicine and wellness implementation includes, insufficient evidence of cost-effectiveness, inadequate insurance coverage, and absence of a suitable regulatory model (ibid).

Currently, in the USA, there are no incentives for payers to fund precision medicine improvement (ibid). There lacks a sense of earnestness among payers and research developers to gather enough data to prove that extensive investigation rather than a narrow research of person's analysis will assure usefulness (ibid). In addition, inadequate regulatory structure hinders the implementation of precision practices (Juengst et al. 2016). Excessive regulation, however, stifles innovation, increases development expenditures, and thus causes patients to receive less efficient results. On the other hand, there is inadequate regulation due to lack of uniformity in data availability, IP security, trouble with reproducibility, wrong research administration, and fraud. This inadequacy impacts the absence of practitioners' confidence (Juengst et al. 2016; Horgan et al. 2014). Moreover, barriers do not only occur as a result of the insufficient application of existing regulations but also due to lack of consensus in rules on elucidation and application of precision tests in healthcare (Horgan et al. 2014).

Clinical utility is regarded as the effectiveness of an assessment to provide data relating to diagnosis, cure, management, results, and prevention of illnesses (Adams and Petersen 2016; Dolley 2016). This is, however, a challenge facing precision tools inventors and venture capitalists of current molecular experiments (Horgan et al. 2014). Lack of long and costly medical research influences the decisions that have to be made mainly based on the evaluation of the peer-reviewed study (ibid). Payers will, therefore, refrain from reimbursing precision in the absence of value-based costs and outcomes (Horgan et al. 2014). While many investors believe that a precision medicine, wellness, and health can contribute to health care system, practitioners and payers who are in most cases reluctant to transform policies and strategies without having a well convincing evidence of medical value tend to have mixed views on the diffusion of these technologies and techniques (ibid). However, it is unclear how evidence ought to be best established and distributed for optimum impact (ibid). Similarly, economic barriers such as insufficient insurance covers and lack of resources have hindered the success of precision practice implementation (Ginsburg and Phillips 2018). These barriers are intertwined in the entire process of precision research to the implementation of the practice in healthcare. Developing and implementing the precision practice, therefore, poses a great hindrance, since the healthcare sector has limited resources set apart for implementation, research, maintenance as well as sustainability (Ginsburg and Phillips 2018).

Lack of education and awareness among patients and throughout the healthcare delivery system is another one of the greatest barriers to integrated precision practices implementation in healthcare (Horgan et al. 2014). Though community strategies are straightforward and clear, it is important to note that building awareness and skills is not an easy thing especially when it comes to health practitioners (Ginsburg and Phillips 2018). Lack of education among physicians limits their ability and sense of know-how in interpreting precision medicine tests and in communicating the results to the victims affected (ibid). The current lack of skills and awareness increases uncertainty on how to interpret a biomarker examination, and this may lead to underappreciation of therapeutic benefit (Horgan et al. 2014).

## 2.7 Facilitators

For precision practices in healthcare to realize success in attaining results and developing capabilities at various levels, a number of facilitators need to be considered (Pritchard et al. 2017). Critical facilitators include collaborative decision-making factor; agreement of goals and aims; local design and action; the creation of trust among payers, investors, and health providers; sufficient resource availability; and presence of an expert and well-trained staff (ibid).

Ramsey et al. (2017) claim that the presence of an expert and well-trained staff within the healthcare system helps to facilitate precision medicine, wellness, and health. Precision health practitioners help to promote the implementation of precision practices by generating and applying novel skills (ibid). The National Academy of Medicine (2018) in the US has thus promoted learning within the health care system where clinical care, as well as research, are incorporated.

Resource availability facilitating factor harnesses the implementation of precision practices in healthcare systems thus enhancing achievement of objectives. In addition, action-based implementation of precision practices in healthcare system depended heavily on stakeholder inclusivity (Adams and Petersen 2016). Therefore, it is important to engage all stakeholders as this will help in promoting a considerable planning and action toward applying precision medicine, health, and wellness in the healthcare sector. We contend that all these initiatives will be more successful if a precision wellness model is developed.

## 3 The Current Need to Develop an Operationalized Model for Precision Wellness

In considering the broader aspects of healthy living, there is evidence of the interconnectedness between mind and body and what have been called "Biobehavioural Factors" (Board on Neuroscience and Behavioral Health 2001). For example, there have been several investigations into the factors resulting in patients being readmitted to hospital in the period immediately after treatment (i.e., within 30 days of discharge). Key factors of importance in this context include not just the medical condition but also the patient's functional status, coping abilities, and level of social and community support for their caregiving needs (Razali et al. 2017). Without support, the patient is more likely to be readmitted, and while this may be for a whole range of reasons, a key aspect is that in considering a patient's health, it is increasingly evident that one should consider the circumstances of the person's life (ibid).

Given this, we believe that there is an opportunity for a holistic precision wellness model to be developed that serves to operationalize a person's life circumstances including factors such as age, medical status, and psychosocial factors including available social support, social engagement, and personality in order to design an individualized co-designed intervention program. Evidence shows that interventions that are co-designed with the individual enable those involved to feel more engaged and comfortable (Green et al. 2018), and thus we propose that such a co-design approach be an integral part of the presented precision wellness model (Fig. 1).

This model focuses on presenting a dashboard of wellness parameters to an individual to enable a co-design process of activities to be developed in support of providing an optimization pathway for an improved level of wellness. In addition to the dashboard, input factors, and sources, the use of an optimization engine and key feedback processes and outputs are also noted. The goal of our precision wellness model is to provide a system where the individual concerned can avail themselves of appropriate tools and capabilities to optimize their health and wellness over an extended period with the aim of reducing the number and frequency of medical interventions and at the same time provide them with the maximum opportunity for health and wellness (physical and mental) and thereby a high quality of life.

The first step in the presented model is for the system to receive key inputs relating to the individual concerned (such as demographics, relationship status, physical and medical status, personality factors and lifestyle preferences), which can be derived from user-provided inputs, instrumentation and even potentially sourced from personal health records. In Australia, the evolving national MyHealthRecord might eventually be a convenient possible source, or private health record systems such as the Apple Healthcare Record.

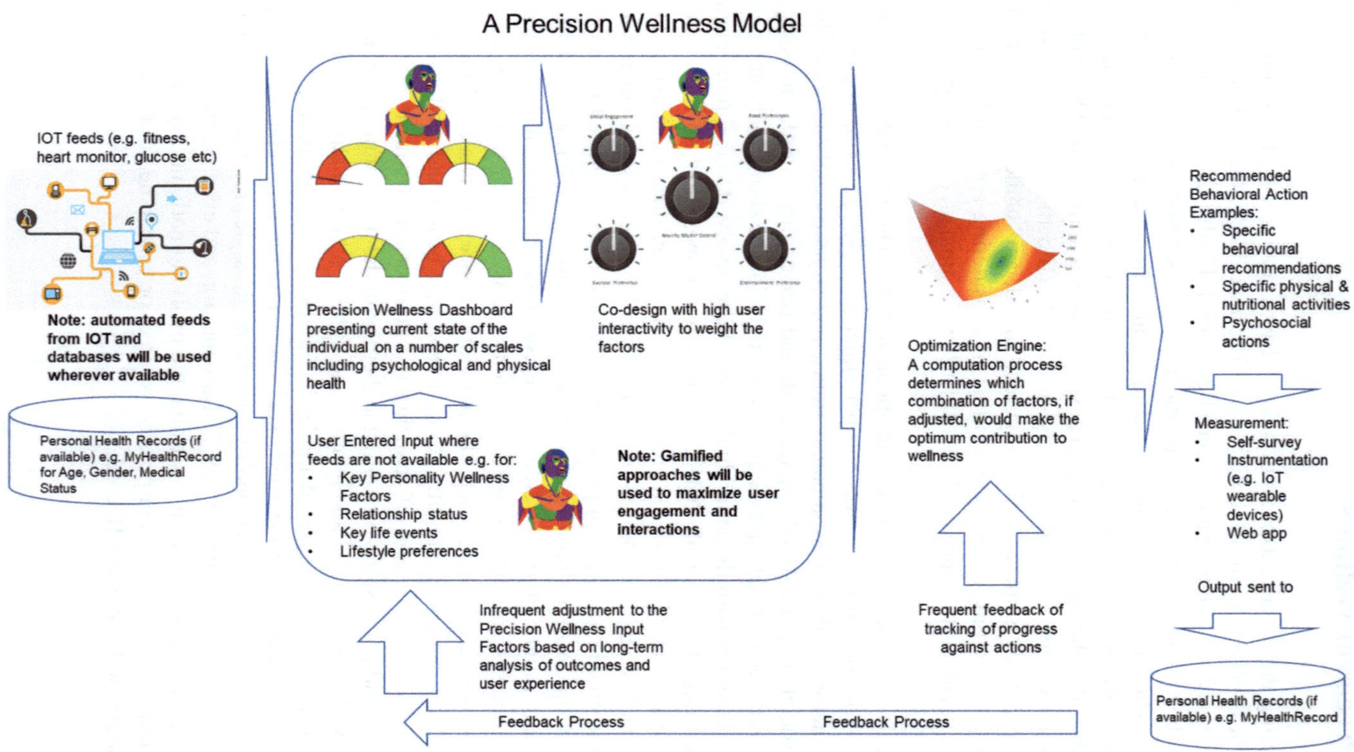


**Fig. 1** Precision wellness model

## 4 Identifying Parameters for Input to, and Presentation of a Precision Wellness Dashboard

We might consider a spectrum of parameters ranging from those focused on physical health to those considering mental health. In considering the physical health parameters, we could consider, for example, behaviors that have been identified as contributing to poor health outcomes (such as cardiovascular heart disease) such as the seven behavioral factors called by the American Heart Association "Life's Simple 7" (Lloyd-Jones et al. 2010):

- Manage blood pressure.
- Control cholesterol.
- Reduce blood sugar.
- Get active.
- Eat better.
- Lose weight.
- Stop smoking.

These behavioral factors have also recently been positively validated in the Australian general population (Peng and Wang 2017) and have been correlated with dementia (Samieri et al. 2018). We could hence envisage an input questionnaire or input feeds from a personal health record that would enable a readout for an individual against some or all of these aspects.

However, for a precision wellness model to be comprehensive and useful in optimizing an individual's wellness in the long term, we anticipate that mental wellness including social aspects such as isolation (loneliness) will be important to consider in addition to such physical health factors. For example, there has been recognition of the impact to health of having limited social engagement to the extent that the UK Government has recently appointed a Minister for Loneliness to directly address the health, economic, and societal impacts of social isolation. This launch included a quote on the health impact of loneliness: "It's proven to be worse for health than smoking 15 cigarettes a day, but it can be overcome and need not be a factor in older people's lives".[1]

Precision wellness parameters relating to psychological wellness might well include those factors identified by Seligman (2011) as leading to "flourishing": the PERMA factors (positive emotion, engagement, relationships, meaning, accomplishments). Alternatively the measures of Diener et al. (2009) might prove useful.

In our presented model, we do not currently specify the input factors for the model, but rather we anticipate that further research and analysis of the extant literature will identify all key factors that will fall into the following two primary categories:

[1]UK Government 2018: retrieved from: https://www.gov.uk/government/news/pm-commits-to-government-wide-drive-to-tackle-loneliness

1. Highly conserved or slowly changing factors - in effect constants.
2. Dynamic factors.

*Example typical highly conserved or slowly changing factors (input constants):*
- Age.
- Gender.
- Personality factors.

*Example dynamic factors:*
- Social support networks/social engagement/relationship status.
- Health, fitness, and medical status including relevant personal history. Fazio (2009) identifies practical methods of facilitating growth through loss and adversity and Seery et al. (2010) have revealed that for some people who have experienced adversity (for example, several negative events or life-changing moments) they may end up ultimately happier (and less distressed, traumatized, stressed, or impaired) than those who have experienced no adversity at all.
- Income and work status. Depending on which types of measures are used and which comparisons are made, income is a complex association with our well-being and some literature has reported that it is the comparative nature of income that causes concern for individuals rather than the quantum itself (Senik 2009). Unemployment negatively affects well-being (Nezlek and Gable 2001). Thus, income and work status would be expected to be important input parameters.

Whatever parameters are ultimately decided upon, our presented precision wellness model is flexible in that a variety of input measures including social, demographic, health, and psychological (including personality) could be entered as inputs sourced from user inputs, medical records, health records, and IoT devices such as fitness tracking or glucose monitoring devices. We propose that a strongly user-centric model for any user input should be taken, and the ideal platform would be via smartphone or similar given the engagement interfaces available on those platforms and their current ubiquity.

## 5 Precision Wellness Dashboard

Following the input of the key factors, a current state precision wellness dashboard is then presented to the individual based on a number of scales that are expected to include psychosocial and physical measures related to wellness in its broadest use. Such an overall precision wellness dashboard and the most appropriate input factors would need to be established based on appropriate research and literature.

### *5.1 Co-design and Gamification*

In our model, the individual is presented with a series of controls enabling them to dial up and down, the weight applied to the parameters (e.g., potentially relating to willingness to undertake behavioral activity suggestions for physical and psychological wellness). In addition, there is potential to use a gamified approach to obtain inputs (e.g., personality parameters) as well as in the co-design so that different levels of engagement and experience with the platform might unlock more advanced features or potentially could involve supportive engagement with family or friends (much like fitness trackers permit you to jog with friends and to compare performance in a positive manner).

## 6 Optimization Engine

Following presentation of the precision wellness dashboard to the individual, an optimization engine then takes these measurements together with the weightings applied by the individual and computes an optimum wellness activity program for the individual concerned (one might anticipate weight loss, exercise, and social engagement factors as prime examples, but the actual factors would need to be derived from appropriate research and field testing).

The idea is that the co-designed optimized set of actions would enable the individual to engage in a variety of tasks, the goal being to help them progress on a pathway to the best possible wellness state that might be achievable for them based on their life circumstances, personality, and health state – thus operationalizing precision wellness.

## 7 Tracking and Measurement

Following the optimization computation, an expected future state dashboard would be presented to the individual so they could see the likely benefit of the recommended behavioral actions. Activities would be suggested and ideally measured through automation to the largest extent possible while preserving some degree of engagement with the individual. Some simple interfaces would be required for monitoring how the person is feeling at several times a day – e.g., smiley face button or frowny face. That could then be correlated with the activities the person is engaged in and a gamified approach to improved wellness can thus achieved.

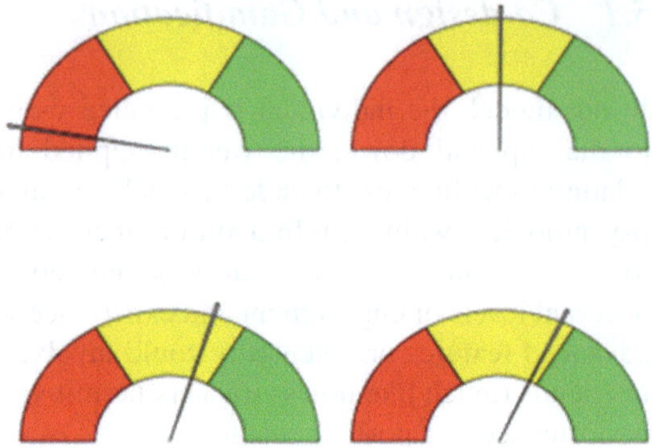

Fig. 2 Precision wellness dashboard

# 8 Feedback Points

This model has two key feedback points – the first being to ascertain progress on the specific variables of wellness that have been co-designed on a very regular basis (e.g., exercise or nutrition or social engagement). The second feedback pathway considers over the longer term the efficacy of the input factors and the precision wellness dashboard parameters to enable the overall optimization program to be modified as required based on what has been working and what has not been working for the individual concerned (ideally based on feedback from a large cohort of users). This second feedback loop might be activated every 6 months (say) as part of a longitudinal study of population health and wellness.

An example is shown in Fig. 2.

This figure shows a mock-up of a presentation to the individual of parameters pertaining to individual wellness based on data inputs (via feeds and/or questionnaires). Such parameters might include physical and psychosocial metrics.

One potential benefit of a model with high levels of user engagement and feedback might be expected to be in avoiding or perhaps slowing down "hedonic adaptation," a frequent phenomenon where individuals become attuned to the current state (e.g., of wellness), whatever the current level. Lyubomirsky (2011) has extensively reviewed the literature on hedonic adaptation to positive and negative experiences.

# 9 Challenges

Challenges of this model are expected to be in creating and validating the input factors and creating an engaging display of parameters for a precision wellness dashboard, validating the optimization engine, and validating proposed interventions that would be safe and appropriate for the individual. Given the variety of possible input factors into the precision wellness model, we may overwhelm the user with the requirement to enter in too many factors, and it may turn out that many are correlated and thus redundant to the task of optimization. Thus, we suggest that a next step in the development and testing of the model is to establish through research the minimum set of input variables and constants and to develop an appropriate set of input factors so that a practical precision wellness model can operate for a specific individual, thereby once again highlighting the precision and tailored nature of the model.

## 10 Discussion and Conclusion

As the tools and technologies of the IoT develop and mature, we are able to more accurately capture data, monitor activities, and communicate these results between key stakeholders so that better health and wellness can ensue. The methods for co-design and user engagement via gamification also continue to evolve and we anticipate excellent engagement via smartphones and similar devices to be achievable.

The preceding has served to outline an operationalized model for precision wellness that we contend will assist to harness the developments with the tools and technologies of the IoT to focus them to enable targeted and appropriate wellness management solutions to be designed and developed for each individual so that they may enjoy a high quality of life and at the same time manage their health and wellness effectively and efficiently.

In operationalizing the presented model, however, there are still areas that need to be considered which we will expand upon in our future work. In particular, we note that it is necessary to assess the health literacy and general literacy level of an individual so that their wellness monitoring is appropriately designed to develop a suitable business model to ensure sustainability of the solution and ensure that the developed solution adheres to relevant policies and legal aspects around data privacy and security as well as compliance to current health and wellness standards and protocols.

Given the rapid growth of chronic and noncommunicable disease coupled with an aging population, wellness management and empowering individuals to take responsibility for their health and wellness will become more important as the twenty-first century progresses. To do this successfully, appropriate tools and technologies will need to be embraced. We believe that our operationalized precision wellness model will serve to fill an important current void and encourage further research in this domain.

## References


Adams, S., & Petersen, C. (2016). Precision medicine: Opportunities, possibilities, and challenges for patients and providers. *Journal of the American Medical Informatics Association, 23*(4), 787–790.

Ashley, W. (2015). The precision medicine imitative a new national effort. *JAMA, 313*(21), 2119–2120. https://doi.org/10.1001/jama.2015.3595.

Baird, P. (1990). Genetics and health care: A paradigm shift. *Perspectives in Biology and Medicine, 33*(2), 203–213.

Board on Neuroscience and Behavioral Health; Institute of Medicine (US) Committee on Health and Behavior. (2001). *Health and Behavior: The interplay of biological, behavioral, and societal influences. Committee on Health and Behavior: Research, Practice and Policy*. Washington, D.C.: National Academies Press (US).

Diener, E., Wirtz, D., Tov, W., Kim-Prieto, C., Choi, D., Oishi, S., & Biswas-Diener, R. (2009). New measures of well-being: Flourishing and positive and negative feelings. *Social Indicators Research, 39*, 247–266.

Dolley, S. (2018). Big Data's role in precision public health. *Frontiers in Public Health, 6*, 68.

Dowell, S. F., Blazes, D., & Desmond-Hellmann, S. (2016). Four steps to precision public health. *Nature News, 540*(7632), 189.

Fazio, R. J. (2009). Growth consulting: Practical methods of facilitating growth through loss and adversity. *Journal of Clinical Psychology: In Session, 65*, 532–543.

Fernandez, M. (2017). Top innovative health and wellness technologies to accelerate global adoption of precision medicine. Retrieved from: https://ww2.frost.com/news/press-releases/top-innovative-health-and-wellness-technologies-accelerate-global-adoption-precision-medicine/. Accessed Sept 2018.

Gavan, S. P., Thompson, A. J., & Payne, K. (2018). The economic case for precision medicine development. *Expert Review of Precision Medicine and Drug Development, 3*(1), 1–9.

Ginsburg, G. S., & Phillips, K. A. (2018). Precision medicine: From science to value. *Health Affairs, 37*(5), 694–701.

Ginsburg, G. S., Finkelman, E., Balatbat, C., Flott, K., Prestt, J., & Dzau, V. (2016). *Precision medicine: A global action plan for impact*. Doha: World Innovation Summit for Health.

Green, S. A., Carnegie, W., Barber, S., Matthew, D., & Matthew, R. (2018). Co-designing within quality improvement initiatives: Notes from the field. *Journal of Health Design, 3*(1), 49–56.

Helwande, S. (2018). Precision medicine: The future of healthcare. Retrieved from: https://www.technologynetworks.com/diagnostics/articles/precision-medicine-the-future-of-healthcare-305639

Horgan, D., Jansen, M., Leyens, L., Lal, J. A., Sudbrak, R., Hackenitz, E., & Brand, A. (2014). An index of barriers for the implementation of personalized medicine and pharmacogenomics in Europe. *Public Health Genomics, 17*(5–6), 287–298.

Jamerson, L., & Longo, D. (2015). Precision medicine – personalized, problematic and promising obstetrical & gynecological survey., *70*(10), 612–614. https://doi.org/10.1097/01.ogx.0000472121.21647.38.

Jameson, J. L., & Longo, D. L. (2015). Precision medicine–personalized, problematic, and promising. *N Engl J Med, 372*(23), 2229–2234. https://doi.org/10.1056/NEJMsb1503104. Epub 2015 May 27.

Juengst, E., & McGowan, M. (2018). Why does the shift from "personalized medicine" to "precision health" and "wellness genomics" matter? *AMA Journal of Ethics, 20*(9), 881–890.

Juengst, E., McGowan, M. L., Fishman, J. R., & Settersten, R. A., Jr. (2016). From "personalized" to "precision" medicine: The ethical and social implications of rhetorical reform in genomic medicine. *The Hastings Center Report, 46*(5), 21–33. https://doi.org/10.1001/amajethics.2018.881.

Katsnelson, A. (2013). Momentum grows to make "personalized" medicine more "precise". *Nature Medicine, 19*(3), 249.

Kaur, J., Rahat, B., Thakur, S., & Kaur, J. (2017). Trends in precision medicine. In *Progress and challenges in precision medicine* (pp. 269–299). London: Academic Press.

Khoury, M. J., Feero, W. G., Chambers, D. A., Brody, L. C., Aziz, N., Green, R. C., Janssens, A. C. J. W., Murray, M. F., Rodriguez, L. L., Rutter, J. L., Schully, S. D., Winn, D. M., & Mensah, G. A. (2018). A collaborative translational research framework for evaluating and implementing the appropriate use of human genome sequencing to improve health. *PLoS Med., 15*(8), e1002631. https://doi.org/10.1371/journal.pmed.1002631. eCollection 2018 Aug.

Khoury, M. J., Iademarco, M. F., & Riley, W. T. (2016). Precision public health for the era of precision medicine. *American Journal of Preventive Medicine, 50*(3), 398.

Leff, D. R., & Yang, G. Z. (2015). Big data for precision medicine. *Engineering, 1*(3), 277–279.

Lloyd-Jones, D. M., Hong, Y., Labarthe, D., et al. (2010). Defining and setting national goals for cardiovascular health promotion and disease reduction: The American Heart Association's strategic Impact Goal through 2020 and beyond. *Circulation, 121*, 586–613.

Lyles, C., Lunn, M., Obedin-Maliver, J., & Bibbins-Domingo, K. (2018). The new era of precision population health: Insights for the all of Us research program and beyond. *Journal of Translational Medicine, 16*, 211.

Lyubomirsky, S. (2011). Hedonic adaptation to positive and negative experiences. In S. Folkman (Ed.), *The Oxford handbook of stress, health, and coping* (pp. 200–224). New York: Oxford University Press.

Meagher, K. M., McGowan, M. L., Settersten, R. A., Jr., Fishman, J. R., & Juengst, E. T. (2017). Precisely where are we going? Charting the new terrain of precision prevention. *Annual Review of Genomics and Human Genetics, 18*(1), 369–387.

Minor, L. (2018). We don't just need precision medicine, we need precision health. Stanford Medicine. Retrieved from: http://med.stanford.edu/school/leadership/dean/dean-minor-on-precision-health.html. Accessed 19 Mar 2018.

National Academy of Medicine. (2018). Retrieved from: https://nam.edu/. Accessed Sept 2018.

Nezlek, J. B., & Gable, S. L. (2001). Unemployment alters the set point for life satisfaction. *Psychological Science, 15*, 8–13.

Peng, Y., & Wang, Z. (2017). Association of Life's Simple 7 and presence of cardiovascular disease in general Australians. *Open Heart, 4*, e000622.

Pritchard, D., Moeckel, F., Villa, M. S., Housman, L. T., McCarty, C. A., & McLeod, H. L. (2017). Strategies for integrating personalized medicine into healthcare practice. *Personalized Medicine, 14*(2), 141–152.

Ramsey, L. B., Mizuno, T., Vinks, A. A., & Margolis, P. A. (2017). Learning health systems as facilitators of precision medicine. *Clinical Pharmacology & Therapeutics, 101*(3), 359–367.

Razali, M., Chong, Y., Mustapha, N., Xu, Y., Ayoob, S., Lim, M., & Aloweni, F. (2017). Identifying patients with high risk of readmission from the patient navigators' perspectives: A descriptive qualitative study. *Proceedings of Singapore Healthcare, 26*(3), 166–171.

Samieri, C., Perier, M.-C., Gaye, B., et al. (2018). Association of cardiovascular health level in older age with cognitive decline and incident dementia. *Journal American Medical Association, 320*(7), 657–664.

Seery, M. D., Holman, E. A., & Silveer, R. C. (2010). Whatever does not kill us: Cumulative lifetime adversity, vulnerability and resilience. *Journal of Personality and Social Psychology, 99*, 1025–1041.

Seligman, M. E. P. (2011). *Flourish*. New York: Simon & Schuster.

Senik, C. (2009). Direct evidence on income comparisons and their welfare effects. *Journal of Economic Behavior & Organization, 72*, 408–424.

Severi, G., Southerly, D., English, D., et al. (2014). Epigenome-wide methylation in DNA from peripheral blood as a marker of risk for breast cancer. *Breast Cancer Research and Treatment, 148*(3), 665–673. https://doi.org/10.1007/s10549-014-3209-y. Epub 2014 Nov 19.

University of Michigan. (2018). Retrieved from: Precision health. http://precisionhealth.umich.edu. Accessed 9 Mar 2018.

# The Development of a Wearable for Automated Documentation and Improved Staff Planning in Outpatient Care

Tuan Huy Ma and Markus Weissenbäck

## 1 Introduction

Outpatient nurses increasingly suffer from physical and psychological complaints as a result of their work. Physical complaints are the result of high physical stress and usually manifest themselves in the form of musculoskeletal diseases. Psychological complaints are manifested, for example, in low job satisfaction, chronic exhaustion, or a depressive illness as a result of latent deadline pressure, a lack of scope for action, or conflict-laden care situations. The burdens and problems do not go unnoticed by the person in need of care and their relatives, as they can be reflected in a lack of motivation, unpunctuality, stress, and a lack of care of the nursing staff.

In a research project funded by the German Federal Ministry of Education and Research ("Dynasens," Dynamic sensor-based personnel deployment and route planning in outpatient care), solutions for the physical and psychological relief of outpatient nursing staff were investigated between 2012 and 2015. In the following, the problem and the state of the art will be discussed in detail. Subsequently, the technical solutions and the project results will be presented. After the discussion, implications will be derived and a conclusion drawn.

T. H. Ma (✉) · M. Weissenbäck
Fraunhofer Center for Applied Research on Supply Chain Services, Nuremberg, Germany
e-mail: tuan-huy.ma@scs.fraunhofer.de



N. Wickramasinghe, F. Bodendorf (eds.), *Delivering Superior Health and Wellness Management with IoT and Analytics*, Healthcare Delivery in the Information Age, https://doi.org/10.1007/978-3-030-17347-0_9

# 2 Theoretical Background

## *2.1 Problem Definition*

In Germany, the number of people in need of long-term care is rising. According to the Federal Statistical Office, 2.86 million people were in need of care in 2015 as defined by the Nursing Care Insurance Act. More than two thirds (2.08 million) of all people in need of long-term care were cared at home (Destatis 2018). At the same time, the demand for outpatient nursing staff is growing. On the other hand, there is a shortage of specialists in the field of health care and a high level of sickness among professional nursing staff. The current absenteeism report of the AOK (Allgemeine Ortskrankenkasse, a German health insurer) (Badura et al. 2018) shows that employees in nursing professions have an above-average sickness rate of 6.8% compared to the average of insured persons (5.3%). In addition to mental illnesses, this is a result from diseases of the musculoskeletal system. The problems are the result of heavy physical strains placed on the nursing staff when lifting, turning, or supporting their patients. In a survey conducted by Grabbe et al. (2006) of 728 outpatient nurses surveyed, 46% stated that it was difficult to lift more than six times a day. In the opinion of Grabbe et al. (2006), this can be remedied by expanding workplace health promotion and striving for a burden-oriented distribution of tasks.

The incidence of mental and psychosomatic illnesses among outpatient nursing staff has increased in recent years. About 11% of all illnesses are mental illnesses such as burnout, depression and anxiety (Grabbe et al. 2006). A comprehensive literature analysis by Geiger-Brown and Lipscomb (2010) attributes the psychological stress to long working hours, shift work, lack of communication and support in the team, as well as lack of work-life balance. Grabbe et al. (2006) identify organizational problems as the main cause. For example, 66.8% of respondents are often or very often under time pressure, 54.3% are unable to take breaks, and 42% feel that they are under pressure to perform. Another aspect of psychological stress is the close contact with the person in need of care and his or her social environment. The suffering or death of the person to be cared for can be a burden as well as aggression and violence emanating from the person in need of care. Understandably, for the purpose of relief, the desire for more a flexible staff and tour planning systems and a greater influence in staff planning is expressed (Krämer et al. 2005; Simon et al. 2005). It is, however, a great challenge to reconcile the requirement profile of the clients with the qualification profile of the caregivers to adjust the staff schedule accordingly and to ensure adequate care even in the event of short-term rearrangements (e.g., staff absences or wishes of those in need of care). In addition to the personal contact and the professional care, there is also the desire for reliability and personnel continuity on the part of the persons in need of care. Language skills and cultural aspects should also not be neglected (Hasseler and Görres 2005). In addition, the increase in treatment care requires a higher degree of specialization on the part of the nursing staff (Osl and Benz 2010).

A trend negatively rated by outpatient nurses, and a possible cause of the time pressure, is the increase in documentation requirements. Complete documentation is necessary to enable the exchange of information between all parties involved, to provide proof of performance and to create a basis for invoicing (Kamm 2004). Careful documentation is therefore essential. For example, 56.9% of outpatient nursing staff spends 30–60 minutes a day on documentation and 18.2% spend more than 60 minutes. 47.5% of those surveyed agree that this time expenditure leaves less time for the care and nursing of patients (Grabbe et al. 2006). Both caregivers and those in need of care would therefore benefit from a reduction of documentation procedures.

It becomes clear that solutions are necessary that increase the health and job satisfaction of outpatient nursing staff. They are essential factors for the quality of care and therefore have a direct influence on the satisfaction of those in need of care (Grabbe et al. 2006; Geiger-Brown and Lipscomb 2010).

## 2.2 *State-of-the-Art Solutions*

Work-related diseases of the musculoskeletal system represent a significant burden on the health care system and industry. At present, there are hardly any validated measurement methods that can objectively record postures and movements in everyday working life. This makes it very difficult to investigate how these factors affect physical complaints (Straker et al. 2010; Li and Buckle 1999).

Developments in the field of microelectromechanical systems (MEMS) open up new possibilities by integrating small sensors for motion and posture detection directly into clothing. In contrast to the camera-based systems for motion detection frequently used in laboratory environments, the user's range of motion is no longer limited to the area observed by cameras, but expands at will.

Some solutions already exist based on MEMS-based inertial sensors, e.g., xsens mvn (www.xsens.com) or animazoo IGS-190 (www.animazoo.com). These are primarily designed to meet the requirements of the film industry and science. By combining different sensors (acceleration sensors, gyroscopes, magnetic field sensors), their accuracy is almost comparable to that of optical and electromagnetic systems (Thies et al. 2007; Saber-Sheikh et al. 2010). In terms of price, however, these systems are in a segment that makes them unattractive for a widespread use in nursing care. Another disadvantage is the lack of integration into work clothing, a basic requirement for an easy-to-use solution.

In addition to the evaluation of ergonomically relevant postures and movements, systems for motion detection also offer the potential to relieve nursing staff of documentation tasks. Activity monitors based on inertial sensors can already detect activities of daily life such as walking, running, and climbing stairs (Rulsch et al. 2009). Compared to these activities, care procedures are much more complex. The recording and classification of elementary care steps with the aid of sensors in the clothing of the nursing staff could, however, lead to an automated documentation of care and thus to significant time savings. At that time, however, this question had not yet been considered scientifically.

The creation of a duty roster is a highly complex task. From a mathematical point of view, it can be formulated as an optimization problem and can therefore be assigned to the field of Operations Research. A number of algorithms and software support tools have been developed in recent years to support this planning activity (Van den Berg and Hassink 2008). However, current systems and algorithms only support rigid, inflexible planning that only rudimentarily takes into account the individual needs of nursing staff and persons in need of care, but also unexpected events such as delays or staff absences.

Dynamic resource planning, taking into account individual needs (e.g., physical health, desired appointments), is an extension of the fundamental problem of duty planning in the outpatient sector. It is based on the concept of dynamic scheduling. This concept is already used in other service areas, such as breakdown service (Rambau and Schwarz 2007) or the service industry (Kiechle and Stadler 2009), but not in the nursing sector. Due to the specific requirements prevailing in the nursing sector, the development of an optimization algorithm for the dynamic deployment planning of nursing staff represents a new and demanding application. In comparison to other industries, a number of additional restrictions have to be taken into account which complicates the problem of resource planning. For example, there has to be a high degree of agreement between the patient's requirement profile and the nurse's qualification profile. If the client requires treatment care, the deployment of a certified nurse must be guaranteed. If the client only wants to be cared for by a male nurse, it makes no sense for a female nurse to step in as a substitute. In addition to fulfilling these specific requirements, general restrictions have to be considered as well, e.g., working time regulations or compliance with statutory breaks. The inclusion of the stress profiles collected in the project in the personnel deployment planning can lead to an additional physical relief of the nursing staff but at the same time increases the demands of the task.

## 3 Description of the Planned Solutions

The aim of the project was to reduce mental and physical stress of outpatient nurses with the help of technical assistance systems. The aim was to promote health, increase job satisfaction and improve the quality of care. The developments of the project started at three points:

- Physical relief through individual measures in the field of work ergonomics
- Reduction of documentation procedures
- Mental relief by reducing time pressure and increasing room for maneuver

For the physical relief, sensors were developed which were integrated into the work clothing in order to record postures and movements of outpatient nurses and to derive stress profiles from the data. It was intended to use the data for developing individualized training and education programs to promote workplace health. Until then, stress profiles could only be created by using questionnaires or video recordings. However, these methods are time-consuming, too dependent on the subjective perception of the respondent and inaccurate.

In the project, the sensor data were tested for a solution to automate nursing documentation. For this purpose, possible nursing activities were derived from the movement patterns and compared with the nursing services defined for the client. From the comparison, the probability of the actual performance of the service could be determined. These data could be viewed directly via mobile terminals, confirmed by the nurse in a timely manner and transferred to the nursing software of the nursing service.

The advantages of such a system lie in the prompt recording of the services provided (time of service provision = time of documentation). Especially in comparison to paper-based recordings, an automated recording enables a more accurate documentation of the services, whereby the accounting and negotiation basis is significantly improved compared to the cost bearers. The targeted solution also increases the quality of the documentation and the transparency toward relatives and the German medical service of the health insurance funds.

The third component of the project was the development of a dynamic software solution for staff deployment and route planning for outpatient nurses. Dynamic means that the route plans can be adapted to short-term, unplannable events such as staff absences or delays. In addition, further components were to be integrated into duty roster and tour planning. These include the stress and qualification profiles of the employees and the requirement profiles of the clients. This was intended to ensure that nursing patients are also cared by suitable nursing staff in the event of short-term changes. The advantages of the targeted solution are manifold: not only does it relieve the mental and physical strain on caregivers, but it also enables clients and their relatives to make new appointment requests at a short notice. This strengthens the autonomy of the person in need of care and supports the compatibility between the care situation, work, and private life of the relatives. In order to ensure user-friendly solutions, potential users and target groups were actively involved in the development and tests.

## 4 Project Developments and Results

### *4.1 Sensors and Wearable*

The sensor network consists of ten sensor nodes (see Fig. 1). Each sensor node contains an acceleration sensor, a gyroscope and a magnetic field sensor (see Fig. 2). In addition, two sensors have a pressure sensor. The number of sensor nodes was based on the assumption that four sensors were required for both arms. The remaining six sensor nodes should be used to cover the back. The results of

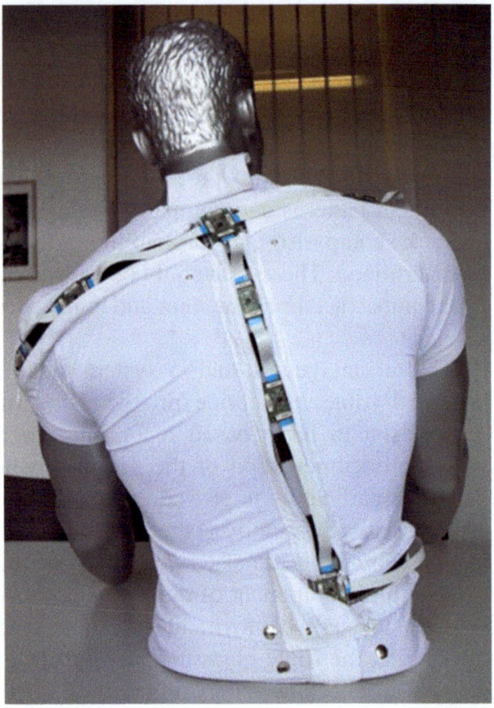

**Fig. 1** The entire sensor network integrated into the shirt

a literature search suggested that this number of sensor nodes could be used to draw conclusions about posture. However, practice has shown that four sensors are sufficient for the reconstruction of the back posture. For this purpose, own approaches and the state of the art were evaluated against each other, and a new concept for posture reconstruction was developed and implemented.

For the determination of the sensor positions, the five selected care measures were:

- Transfer of the patient from the bed to the (wheel)chair
- Transfer of the patient from (wheel)chair to bed
- Tightening support stockings
- Positioning – optimum lying position for the patient
- Changing diapers including intimate care

The selected nursing procedures for which exposure profiles are to be created during execution are identical to the activities for automated documentation. Furthermore, practical experience with clothing was gathered within the framework of reference measurements and during the development of algorithms for an automated documentation. Then, a t-shirt for accommodating all sensors could be developed. For the reference measurement, the sensor system was integrated into a developed PC application for recording sensor data. Subsequently, an Android parser was implemented for the sensor network in order to be able to use it in connection with

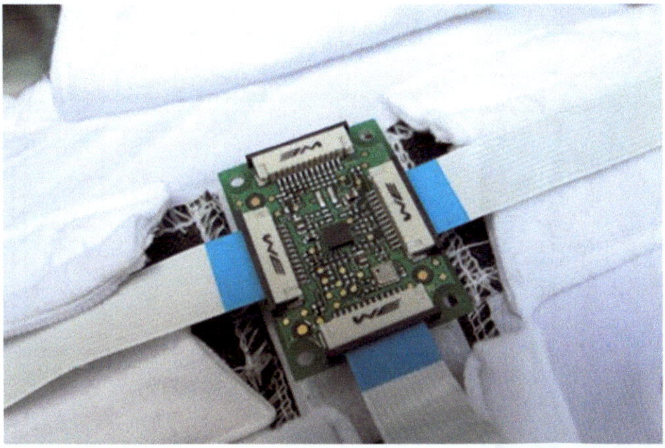

**Fig. 2** A sensor node of the network for recording the movement sequences during nursing activities

Android smartphones. Measurement data of the sensor network can be received at the PC and on Android smartphones.

The five selected nursing activities were carried out by eight nurses by wearing the sensors which where integrated into specially developed nursing clothing. At the same time, their movements were recorded by hand and additionally recorded by a Vicon system for motion capturing. Finally, the posture in the back area was reconstructed from the sensor data. A new concept for posture reconstruction was developed for this purpose. The collected data was stored via an interface in the Dynasens data warehouse. On this basis, a software library for the determination of the back posture and an Android-based demonstrator for the recognition and visualization of these postures could be realized (see Fig. 3). The reconstruction of the posture will use the position of the individual sensors in space. Vertical position changes are calculated from a combination of the acceleration sensor data with the gyroscope data. A plug-in has been developed for a PC application to record sensor network data and store it in an XML file. For further processing of the data in Matlab, Python scripts were written to import the data.

## 4.2 Automated Documentation

The nursing services selected for the automated documentation where already presented in Sect. 4.1. When making the selection, special attention was paid to characteristic movement sequences that facilitate an automatic recognition. Interface specifications, communication flow charts and integration concepts for the groups of people involved were developed. The interface between automated

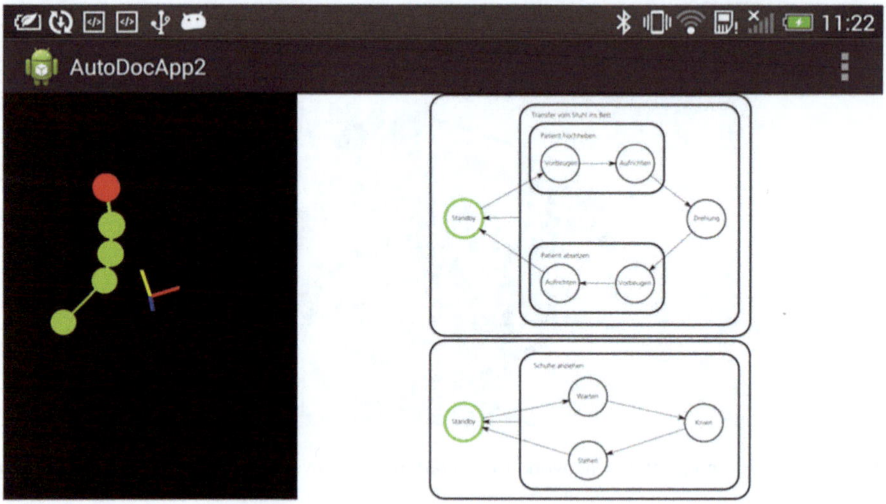


**Fig. 3** Screenshot of the Android app for visualizing the back posture (left) and the state machine (right)

documentation and care software was defined. The identified subprocesses of care measures were passed on to the documentation software.

The classification was validated by a comparison with a reference system. Based on the findings, a software library for real-time classification of the two selected nursing measures – transfer of the patient from the bed to the wheelchair and vice versa – was implemented. In order to ensure reusability in more complex processes, the measures were divided into substeps (e.g., kneeling when putting on shoes, preventing, erecting, or turning during the transfer). Using a state machine, it was possible to visualize the progress of the individual substeps (see Fig. 4).

With the aid of a concept for transferring the measurement data into the middleware, a hardware-related connection to the measurement system, a module for filtering and enriching the data, a module for transparent access to filtered and enriched data and a functioning communication and data platform were implemented.

An Android service was developed which transmits the measurement data of the t-shirt during a visit via middleware to the data warehouse. The developed Android service is able to be addressed directly by the automatic documentation service of the documentation application. It consists of a data logger and a data transmitter. In the data logger, the inventory data of the visit are entered as well as the measurement data during the visit. Subsequently, this measurement data is compressed by the data transmitter and divided into smaller packets. These packets are then transmitted to the middleware depending on the available radio connection. In the middleware, a consistency check is carried out after the transmission of a packet and, depending on the result, packets are requested again. If all packets of a visit have been successfully transferred, they are unpacked and stored as measurement data together with the

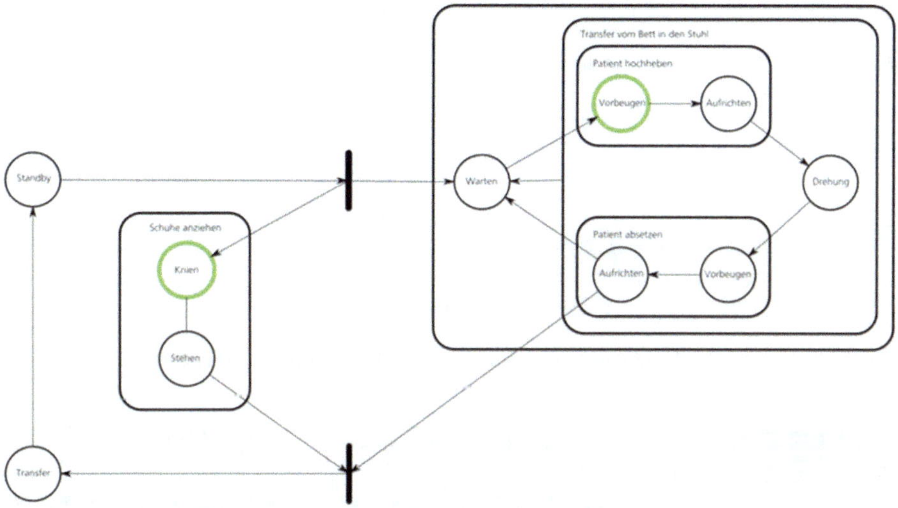


**Fig. 4** Status machine for real-time visualization of the steps passed through for the care measure – transfer from bed to chair

inventory data of the visit in the data warehouse. From there, they can be called up for further processing.

In cooperation with the nursing service, five practice-relevant nursing measures were selected for an automated documentation. From the five selected nursing services, the best to be detected by the sensors in real-time where transfer actions from the bed to the wheelchair and from the wheelchair back to the bed. Algorithms were developed in Matlab for these two activities. The partial steps (kneeling when putting on shoes, preventing, straightening up and turning during transfer) of the two transfer care measures can also be detected in Matlab (for an example, see Fig. 5). The automated documentation of the classified care activities was integrated into the documentation software of the project partner as an Android service (see Fig. 6). The mock-up and the interface provided by the service were implemented and the integration of the Android service in the nursing documentation software was completed. Finally, a software-based demonstrator including an integrated automated documentation was created.

## *4.3 Staff Planning*

The main question of personnel deployment planning is which person does which task at which time. The planning is usually done by a single person. Many nurses believe that their schedules are unfair so that the staff planner gets complaints. The goal in the project was to show how to help planners to find objectively fair and

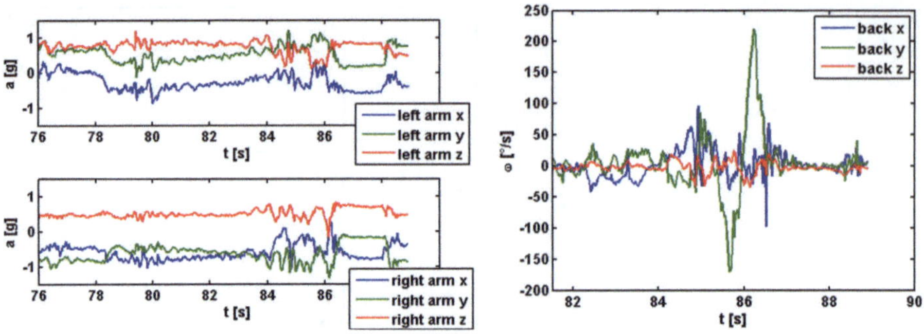


**Fig. 5** Sensor data of the left and right arm and of the back during the transfer of a patient from the bed to the chair

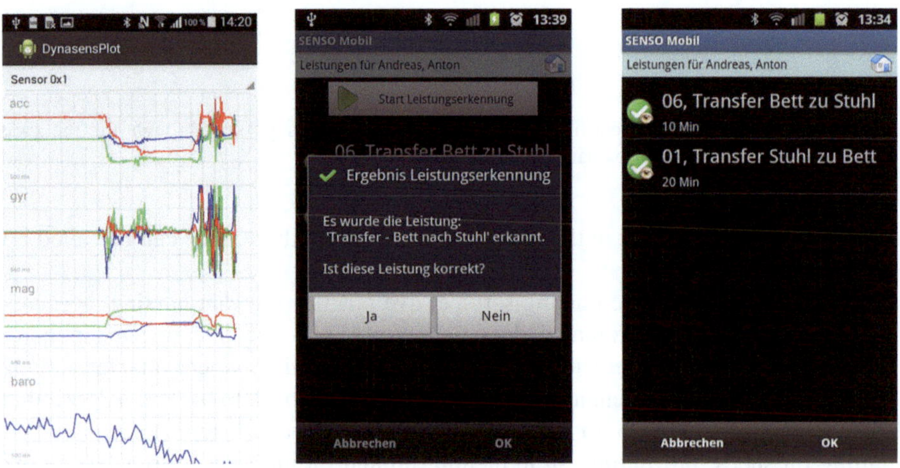


**Fig. 6** Mock-ups of the Android app: sensor data, the recognition of the nursing service "transfer from bed to chair" and the automated documentation including the time required

balanced schedules for professional caregivers by using mathematical optimization and sensor data.

The essential requirements for a fair personnel deployment planning task could by clarified by interviewing patients, relatives, nursing staff and staff planners. Finally, the derivation and verbal description of the concrete mathematical planning task were recorded. It turned out that the focus was on an automated deployment planning which considers fairness as well as personal requests and preferences of the staff. The dynamic replanning of tour schedules, on the other hand, did not play the central role. The result of this project phase was an extensive catalogue of requirements with detailed specifications clustered into *must*, *should*, and *optional*. All requirements that were only classified as *nice-to-have* were not implemented within the research project, but could be considered and implemented in the future.

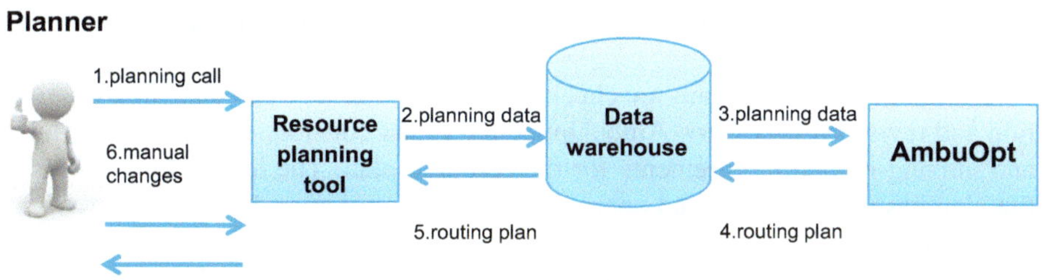


**Fig. 7** Intended software usage

A balanced allocation of physical strains among caregivers played the central role. Nursing activities were weighted with stress points. With this weighting, a fair deployment plan could be created. It was planned that the sensor system should automatically detect selected activities and helps determine the amount of stress points. For each participant, individual load profiles were created. These load profiles contained a maximum value for physical and mental stress points per period that should not be exceeded, initially covering 1 week. In the future, larger periods such as 1 or 6 months can be considered as well.

On this basis, a formal mathematical model, specifically a mixed integer program (MIP) was developed, which was subsequently solved algorithmically. A software tool for the flexibility and solution of personnel resource planning (PRP) and route planning (RP) was developed. The tool was connected to the personnel planning software for outpatient care. The typical usage procedure is outlined in Fig. 7. The call to the optimization kernel can be done manually by a registered user at any time. The optimization kernel (called AmbuOpt) can be used by any number of users; however, only one plotting can be made at the same time. The optimization kernel is executed on a specially equipped computer. Distribution over multiple servers is not envisaged. No LAN or Internet connection is required to use the optimization core. The maximum calculation time for a calculation run is 48 hours. The data generated by the optimization kernel was made accessible via the database interface FIMAS.

The optimization core runs on conventional computers (Windows operating systems with 32 or 64 bit). At least 3 GB of RAM and 50 GB of free hard disk space are required. AmbuOpt must have the licensed software GAMS and a valid usage license for CPLEX or XPRESS installed. The database used is preferably MS SQL Server 2005. The central functionality of the optimization core is the determination of a route plan for the nursing staff of an outpatient nursing service. For this, AmbuOpt reads in the data provided by the resource planning tool via the database interface specified above. From the underlying mathematical model, the optimization kernel creates an instance and solves it with the implemented algorithm. The calculation result is then entered into the above-mentioned database interface and thus made available to the resource planning tool. Users of the optimization core are exclusively resource planners. Knowledge about the underlying mathematical model and algorithm is not required for using AmbuOpt.

The created resource plans are checked by an experienced person before publication and revised if necessary. AmbuOpt contains and considers a collection of legal and nursing-related requirements for the resource plan. The legal admissibility of the plan has to be verified and approved by the planner after the completion of the calculation.

The requirements are also referred to below as "restrictions." In addition to the prioritization, the requirements for "hard restrictions," "soft restrictions" and target criteria are differentiated. "Hard restrictions" describe requirements that must be met during planning so that the tour plan is functional and feasible. When requesting that the travel times between the individual patients are included in the tour planning, for example, it is a hard restriction, because without them no functioning plan will be created. With the help of the "soft restrictions," the tour plan is made more accommodating for all participants. An example of a soft constraint is that language skills should also be considered when assigning a caregiver to a patient.

A comprehensive description of all important restrictions is given in Table 1. This makes it possible to mathematically model the problem. Almost all criteria that were marked with "can" could not be implemented. The main reason was the running time behavior of the optimization core.

The planning problem was classified as Vehicle Routing Problem (VRP). The extension of the existing procedures with an objective fairness was a real innovation in this context. The problem was mathematically modeled (Shao 2011; Cheng and Rich 1998; Rasmussen et al. 2012) and compared with regard to efficiency and solvability. An exact optimization method was selected. It was modeled with the

**Table 1** Restrictions for the operative tour planning

| | Hard restrictions | Soft restrictions |
|---|---|---|
| Must | Planning period<br>Complete service delivery<br>Duration of the service provision<br>Travel time between patients<br>Maximum daily working time<br>Compensation for overtime<br>Daily rest with 6–9 hours working time<br>Daily rest with more than 9 hours of work<br>Distribution and duration of daily rest breaks<br>Duration of rest between working shifts | Time restrictions for patient<br>Performance time window of the nurses<br>Average daily working hours |
| Should | Type of vehicle<br>Fixed dates<br>Compliance with the service requirement | Calculation time<br>Priority of the services<br>Gender of caregiver<br>Load changes for caregiver |
| Can | Double care | Language skills of the caregiver<br>Cultural/religious background of the caregivers |

modeling language GAMS and solved with the commercial solver Cplex. This approach was obvious here, since it is possible to describe soft criteria very well such as the preference of the patients concerning the caregivers' gender. However, since there are countless mathematically equivalent ways to model the problem, the challenge was to find a model that was as efficient as possible in terms of running time and quality.

In addition, a multicriterial objective function was considered, which also enhanced the use of a commercial solver. The solution method was implemented as a generic, encapsulated software component in the GAMS modeling language.

Concerning the modeling, three different model approaches were implemented in the modeling and optimization system GAMS and analyzed for running time and solution behavior. The most promising approach according to these criteria was chosen and gradually refined. The defined design principles were implemented in the form of a functional calculation kernel that complies with all interface and communication regulations. The technical implementation of the mathematical model is not trivial, since the type of implementation can also have a significant influence on the running time behavior of the calculation. All components of the multicriterial objective function needed to be weighted and normalized. Therefore, the criteria were compared to decide which one is more important or less important as shown in Table 2. If the line is more important than the column, the "1" flag is "0" for equally important, "-1" for less important. Using this table, it was possible to rank the criteria in order to weight the individual summands of the objective function.

As mentioned earlier, there are tons of equivalent modeling approaches for the same mathematical problem. The key to efficient modeling essentially consists of three areas: knowledge about underlying methods of the solution algorithm, modeling experience and testing. Constraints had to be formulated in such a way that the time-varying variables were handled as efficiently as possible. Another complexity driver was the multicriterial objective function. All soft factors such as overtime and load balancing are normalized and weighted as part of objective function. The emphasis was therefore not purely on the costs, but rather on the criteria that employees and patients of the outpatient nursing service communicated in advance as important. From a mathematical point of view, this causes several problems:

- There are conflicting target criteria
- Normalization can lead to numerical difficulties. The handling of very large values (e.g., the overtime hours of a caregiver in minutes) on one side, very small values with many decimal figures on the other side, can cause computer problems.
- Finding a suitable normalization at all can be very difficult, since the linearity of the objective function in a MIP always needs to be ensured.

Therefore, the optimization system was regularly tested and fine-tuned until the end of the project. The model with the best results was used for the prototype.

**Table 2** Prioritization of criteria

| | Cost of nursing staff | Compensation of physical stress | Compensation of mental stress | Compensation working time | Service too early | Service too late | Starting work too early | Starting work too late | Overtime nurse |
|---|---|---|---|---|---|---|---|---|---|
| Cost of nursing staff | 0 | 0 | 0 | −1 | −1 | 0 | −1 | −1 | −1 |
| Compensation of physical stress | | 0 | 0 | 0 | −1 | 1 | −1 | −1 | 0 |
| Compensation of mental stress | | | 0 | 0 | 0 | 0 | 1 | 1 | 0 |
| Compensation working time | | | | 0 | −1 | −1 | 1 | 1 | 0 |
| Service too early | | | | | 0 | 1 | −1 | −1 | −1 |
| Service too late | | | | | | 0 | 0 | 0 | −1 |
| Starting work too early | | | | | | | 0 | 1 | 0 |
| Starting work too late | | | | | | | | 0 | −1 |
| Overtime nurse | | | | | | | | | 0 |

For the evaluation, 20 benchmark test records were created, which were subsequently used to evaluate changes in the implementation. The maximum number of nurses was seven, with the maximum number of patient services being 102. A visualized result is shown in the figure below. Since no integration with the actual planning interface was implemented at that time of the project, the results were visualized as seen Fig. 8.

The patient services are marked v1–v34, while the time is displayed in minutes (between 470 and 1270). The green bars represent the periods when the services need to be provided, while the other colors represent the individual nurses working for the outpatient service. The length of these fields corresponds to the duration of the service. However, we soon realized that instances with a number of nurses larger than four and a number of nursing services greater than 30 could not be resolved in an acceptable time. An acceptable time for the use case may not exceed the duration of a manual daily schedule, so it should be less than 1 hour. The developed software demonstrator was used in an outpatient nursing service to evaluate its suitability. For this purpose, actual planning data of the nursing service were connected to the demonstrator. However, with the calculation kernel, no daily planning could be carried out within an acceptable running time. Therefore, the

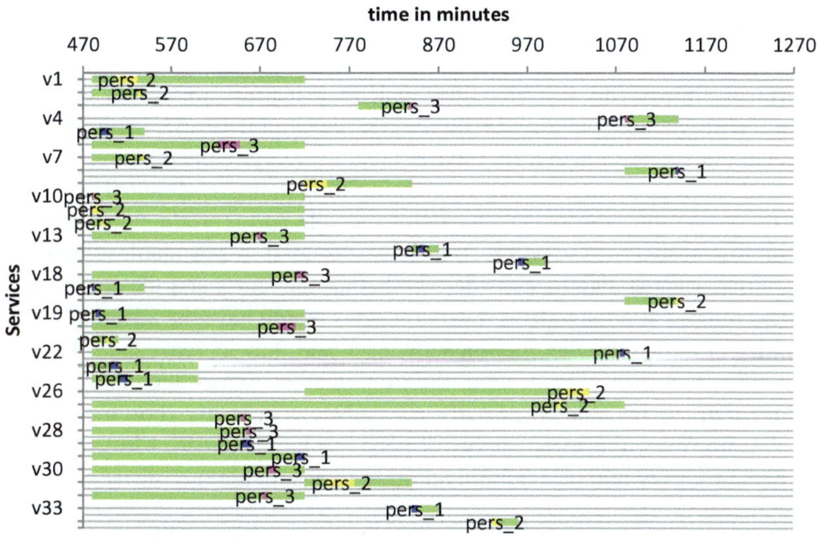


**Fig. 8** Visualized result of the technical evaluation of the optimization

demonstrator was not tested in real operation. In order to be able to carry out a useful evaluation, ten already existing plans were rescheduled in an anonymous form with the optimization kernel and then evaluated by the personal planner of the nursing service in a joint interview.

The evaluation showed that all hard criteria were well implemented. Above all, working hours and delivery times of the nursing services were consistently well respected. The time was also well distributed to the existing nursing staff. Soft criteria, in part despite formal optimality, can be better achieved through manual planning by the personnel planner with years of experience. Above all, the open part of the questionnaire was used to constructively criticize the PRP planner. For example, in some created plans, patient services were more likely to be spread throughout the day for the same patient, although from the point of view of the planner, it makes more sense to bundle them. It was also noted that old patients, in particular, are keen to be treated at the same time every day. In addition, the order of service provision was criticized. For example, it does not make sense for the patient to first wear support stockings and then wash them. It would be better to wash it first and then put it on. In the current form, it is generally possible to implement such patient requests. The biggest criticism, however, was the time it took to generate such a plan and that it was not possible to plan the entire shift plan but only a subset of the nursing staff and the patient services to be provided. In general, the staff planner was very impressed by the idea that an automated day-to-day planning would generate suggestions that would have to be fine-tuned by hand. She herself was a nurse in addition to her work as a planner. She could invest more time in nursing by saving time at the planning task.

## 5 Discussion, Limitations, and Conclusion

The project showed for the first time that, in principle, wearables can be used to identify movement patterns of nursing services and to use them both for an automated documentation and for a fairer personnel deployment planning.

Several findings were obtained concerning the nursing clothing. The shirt must fit tightly so that the sensors can also deliver usable and reproducible measurement data. In this respect, a very precise fit is essential for a stable classification. Furthermore, it must also be considered whether the shirts and the sensor network should be offered as a long-sleeved version. This would have the advantage of being able to detect more complex motion sequences and thus increase the robustness of the classification results. In general, the implemented sensor technology provided sufficient data to be able to classify the substeps robustly. For more complex processes, however, the sensor network needs to be expanded. Movements of the arms, hips or legs would be a helpful addition to ensure a more robust classification of care measures.

The wireless connection via Bluetooth to send the sensor data to a smartphone proved to be unstable. In the future, an evaluation of alternatives for Bluetooth in order to establish a stable, permanent connection is necessary. Other variants for the plug connections and cables used should also be considered. The wearable showed signs of wear. Especially the connectors and cables were subject to increased abrasion. These connections and cables became fragile or could not ensure a clean line contact. Furthermore, the handling should be made more user-friendly, e.g., by integrating an on/off button.

The modular implementation of the software library for determining the back posture makes it easy to integrate and use it for further work in the field of ergonomics and motion reconstruction. The extension of the sensor network represents an important issue in order to be able to carry out a more comprehensive evaluation with regard to ergonomic aspects and corresponding load profiles.

For the automated documentation, the movements of the upper body could be well recorded by the four sensors along the spine. Movements of the arms, however, can only be insufficiently observed by the four sensors in the upper arms and shoulders area. This means that even complex motion sequences cannot be detected with sufficient accuracy. As a result, only two nursing services could be detected. However, the classification of the other three care measures requires a great deal of additional effort – for the (further) development of the algorithms as well as for the extension of the sensor network through further nodes or sensors. On the other hand, the classification of the partial steps (prevention, erection, and rotation of the patients' body) delivered very good results and forms a solid basis for the further development of algorithms for more comprehensive motion sequences.

Subsequently, the development and evaluation of algorithms should focus more on conditions that are as real as possible. Due to the measurement results recorded under laboratory conditions, the state machine delivered high classification rates for test persons with "trained" care measures, while deviations in the execution led to poorer results.

In the requirements analysis for a dynamic personnel deployment planning, it turned out that almost all stakeholders, in particular the nursing staff, see potential for improvement. It is hardly possible for the staff planner to fulfill all requests by hand. In the project, automated plans were generated which could consider some of these requests. With the optimization methods used, it was not possible to achieve an acceptable running time that makes such a tool practical, but it was shown that it is generally possible and that there is a demand for such a solution. The evaluation showed that a large part of the soft restrictions were considered in the plans. In the detailed examination of the plans, however, there are sequences of patient services that should be planned differently in practice. These include, in particular, the sequence and bundling of different services for the same patients. These points could be realized in further research. An exclusive mixed-integer-program (MIP) formulation, as it was used in this context, does not seem to be the means of choice for a planning tool. With the help of the catalogue of requirements and the exact problem specification, it is possible to develop advanced methods that require less computing time. For example, based on the developed MIP, a hybrid procedure could be applied, i.e., a heuristic that only solves partial problems with a commercial MIP solver and then skillfully combines them. It might also be possible to adapt common heuristics for the vehicle routing problem with time windows (VRPTW) so that some of the desired constraints can be considered. If it is possible to focus on the running time component while maintaining the quality of the solution, this planning tool will significantly relieve the burden on outpatient nursing staff in Germany.

Empirical results on the acceptance of outpatient nursing staff for the project solutions are not available. However, the resonance with the test persons shows that the potential users are open-minded about the project solutions, regardless of gender and age. Particular emphasis was placed on the aspect that research and development work was carried out with the aim of relieving the burden on outpatient nursing staff. This showed appreciation for the nursing staff. The test persons in particular found it positive to be involved in research and development work "up close" in the truest sense of the word. The benefit of the intended solution was not questioned. Concerns exist, however, regarding the possible total surveillance of outpatient nurses. Here, it must be ensured that there is no possibility of abuse of the sensors.

In the economic evaluation, exemplary model calculations were carried out which showed that a financial benefit can be expected both for the nursing service and for society if sick days and staff turnover can be reduced. In our estimation, the social benefit will be even greater if the consequences of a reduction in the earning capacity of nursing staff are taken into account. Based on our experiences in the project, the price of the potential solutions, and thus the return-on-investment for the care service, appears to be a greater factor than acceptance problems of the end users. A viable product would lower manufacturing costs, which would ultimately reduce the return-on-investment period.

## References

Badura, B., Ducki, A., Schröder, H., Klose, J., & Meyer, M. (Eds.). (2018). *Fehlzeiten-Report 2018, Schwerpunkt: Sinn erleben – Arbeit und Gesundheit*. Berlin: Springer.

Cheng, E., & Rich, J. L. (1998). *A home health care routing and scheduling problem*. Technical report CAAM TR-04. Rice University.

Destatis. (2018). *Statistisches Bundesamt*. Wiesbaden. https://www.destatis.de/DE/ZahlenFakten/GesellschaftStaat/Gesundheit/Pflege/Tabellen/PflegebeduerftigePflegestufe.html. Accessed 13 Dec 2018.

Geiger-Brown, J., & Lipscomb, J. (2010). The health care work environment and adverse health and safety consequences for nurses. *Annual Review of Nursing Research, 28*, 191–231.

Grabbe, Y., Nolting, H. D., Loos, S., & Krämer, K. (2006). *AK-BGW Gesundheitsreport 2006 Ambulante Pflege, Arbeitsbedingungen und Gesundheit in ambulanten Pflegediensten*. Berlin: IGES.

Hasseler, M., & Görres, S. (2005). *Was Pflegebedürftige wirklich brauchen, Zukünftige Herausforderungen an eine bedarfsgerechte ambulante und stationäre pflegerische Versorgung*. Hannover: Schlütersche Verlagsgesellschaft.

Kamm, J. (2004). *Rechtliche Bewertung der Pflegedokumentation. Teilbericht des Projekts ‚Entbürokratisierung der Pflegedokumentation"*. München: Bayerisches Staatsministerium für Arbeit und Sozialordnung, Familie und Frauen.

Kiechle, G., & Stadler, H. (2009). Konzeption einer Servicearchitektur für die Nutzung von Echtzeit-Verkehrsinformationen zur dynamischen Flottendisposition. In *Geoinformatik 2009* (pp. 51–57). Osnabrück: Akademische Verlagsgesellschaft AKA.

Krämer, K., Petzold, C., & Sulmann, D. (2005). Empfehlungen und Forderungen zur Verbesserung der Qualität der Betreuung und Pflege: Zu den Ergebnissen der Arbeitsgruppen am Runden Tisch Pflege. *Informationsdienst Altersfragen, 32*(6), 5–11.

Li, G., & Buckle, P. (1999). Current techniques for assessing physical exposure to work-related musculoskeletal risks, with emphasis on posture-based methods. *Ergonomics, 42*(5), 674–695.

Osl, P., & Benz, A. (2010). *Evangelische Hauskrankenpflege der Diakoniestation Frankfurt am Main GmbH*. St. Gallen: Universität St. Gallen.

Rambau, J., & Schwarz, C. (2007). *Zwei auf einen Streich: Optimierte dynamische Einsatzplanung für Gelbe Engel und Lastenaufzüge*. Working paper. Bayreuth.

Rasmussen, M. S., Justesen, T., Dohn, A., & Larsen, J. (2012). The home care crew scheduling problem: Preference-based visit clustering and temporal dependencies. *European Journal of Operational Research, 219*(3), 598–610.

Rulsch, M., Benz, M., Arzt, C., Podolak, C., Zhong, J., & Couronné, R. (2009). A lightweight approach for activity classification on microcontrollers. In *World congress on medical physics and biomedical engineering* (pp. 190–193). Munich: Springer.

Saber-Sheikh, K., Bryant, E. C., Glazzard, C., Hamel, A., & Lee, R. Y. W. (2010). Feasibility of using inertial sensors to assess human movement. *Manual Therapy, 15*(1), 122–125.

Shao, Y. (2011). *Home therapist network modeling*. Dissertation, University of Texas at Austin.

Simon, M., Tackenberg, P., Büscher, A., Hasselhorn, H. M., & Müller, B. H. (2005). Zukunft der Pflege gestalten. *Krankenpflege Journal, 43*(1–3), 5–7.

Straker, L., Maslen, B., Burgress-Limerick, R., Johnson, P., & Dennerlein, J. (2010). Evidence-based guideline for wise use of computers by children: Physical development guidelines. *Ergonomics, 53*(4), 458–477.

Thies, S. B., Tresadern, P., Kenney, L., Howard, D., Goulermas, J. Y., Smith, C., & Rigby, J. (2007). Comparison of linear accelerations from three measurement systems during "reach & grasp". *Medical Engineering & Physics, 29*(9), 967–972.

Van den Berg, B., & Hassink, W. H. (2008). Cash benefits in long-term home care. *Health Policy, 88*(2–3), 209–221.

# Towards a Better Life for Diabetic Patients: Developing and Integrating a Non-invasive Self-Management Support Tool Within a Smart Digital Companion

Lemai Nguyen, Sasan Adibi, and Nilmini Wickramasinghe

## 1 Introduction

Diabetes is a chronic disease that occurs when there is too much glucose in the blood because the body is not producing insulin or not using insulin properly (Diabetes Australia 2007). As noted by the WHO (World Health Organization 2016), diabetes is at epidemic proportions globally and needs to be addressed. Diabetes management involves a combination of both medical and non-medical approaches with the overall goal for the patient to enjoy a life which is as normal as possible (Australian Institute of Health and Welfare 2007, 2008). As there is no cure for diabetes, diabetes must be regularly managed and monitored. Critical to this management regimen is the systematic monitoring of blood glucose levels. However, achieving this goal can be challenging because it requires effective lifestyle management as well as careful, meticulous attention and monitoring by the patient and health professionals (Britt et al. 2007). There is a need for identifying a simple and convenient non-invasive approach to monitoring blood glucose (So et al. 2012). This forms the focus of this research.

L. Nguyen
Department of Information Systems and Business Analytics, Deakin University, East Kew, VIC, Australia

S. Adibi
School of Information Technology, Deakin University, East Kew, VIC, Australia

N. Wickramasinghe (✉)
Epworth HealthCare, Richmond, VIC, Australia

Swinburne University of Technology, Hawthorn, VIC, Australia



N. Wickramasinghe, F. Bodendorf (eds.), *Delivering Superior Health and Wellness Management with IoT and Analytics*, Healthcare Delivery in the Information Age, https://doi.org/10.1007/978-3-030-17347-0_10

# 2 Background

## *2.1 Invasive, Semi-invasive and Non-invasive Solutions*

A key factor in the management of diabetes has been found to be the patient's self-blood glucose monitoring (SBGM) (Karter et al. 2001; Haller et al. 2004; Guerci et al. 2003). As a result of recent research (Malanda et al. 2012; Farmer et al. 2009), General Practice Management of type 2 diabetes (RACGP and Diabetes Australia 2014-2015) recommends SMBG for patients with type 2 diabetes who are on insulin. Currently, the dominant method to test blood glucose level is invasive. It requires a blood glucose meter, a lancet device with lancets and test strips. Further, the patient must prick their finger with the lancet sometimes more than four times a day. Finger pricking in SBGM has been found to have several clinical and psychological disadvantages. These are described below.

Clinically, there is a risk of skin infection and tissue damage. Repeated finger pricking associated with the depth and possible vibrations of the needle tip while penetrating the skin were found to cause soreness (Burge 2001) and damage the skin and in severe cases could lead to ulcer on their patient fingers (Dahiya et al. 2012; Giannini and Mayr 2004). Therefore, this SBGM practice can result in damage to the patient body site.

Further, SBGM using a finger prick glucometer is not practical for continuous monitoring of blood glucose (So et al. 2012). As blood glucose levels of a patient change overtime, possible occurrences of hyperglycaemia or hypoglycaemia between measurements may not be recorded. Thus, the measurements may not truly reflect the patient blood glucose pattern (Kannampilly 2013).

Psychologically, inconvenience and anxiety of constant performing finger pricking and extracting a drop of blood in patients' daily lives, and associated physical and emotional pain have always been troublesome in SBGM (Pacaud et al. 1999; Burge 2001; Karges et al. 2008; Wainstein et al. 2013). In a cross-sectional questionnaire survey with 315 patients with diabetes in the UK, about one-third of general diabetes patients were found to have anxiety to finger pricking for SBGM (Shlomowitz and Feher 2014). Positive correlations were found between anxiety due to finger pricking and avoidance of testing as well as between anxiety due to finger pricking and general anxiety. In previous studies (Koschinsky 2007; Cradock and Hawthorn 2002; Burge 2001), pain and discomfort were consistently found to cause a natural resistance to SBGM, and subsequently result in a lack of adherence to this procedure. Anxiety due to the finger prick method and avoidance of testing were found across different ethnic groups and female patients were found to have greater anxiety to finger pricking (Shlomowitz and Feher 2014).

The aforementioned disadvantages served to motivate the need for new approaches to SBGM. They can be categorised into two groups. The first group is to support for measuring blood glucose levels in a less painful manner. Wainstein et al. (2013) used a CoolSense device to reduce local pain sensation due to finger pricking. They conducted an experiment with 177 adult patients with type 2 diabetes and concluded that the CoolSense device significantly reduced subjective pain felt by the patients while maintaining the same level of clinical accuracy. Other studies suggest that instead of pricking fingers, patients can prick other areas such as the forearm, knee, earlobe, thigh and abdomen skin (Castilla-Peón et al. 2015; Nakayama et al. 2008; Heinemann 2008). While pricking alternative body sites were commonly found to reduce pricking fingers to some extent, it did not eliminate the pain completely. Disadvantages of pricking other body sites include lack of accuracy, inconvenience and difficulty of pricking in public, and technology switching costs to purchase new equipment for pricking and measurement (Heinemann 2008; Castilla-Peón et al. 2015; Cradock and Hawthorn 2002).

The second group of approaches to SBGM is to developing semi-invasive and non-invasive technologies for blood measuring without needle pricks (So et al. 2012; Makaram et al. 2014). Du et al. (2016) propose a biosensor that can detect low-level glucose in saliva. They conducted a study of ten healthy human subjects and conducted that the proposed biosensor can be seen as a potential alterative to SBGM using finger pricking. The protocol of use is still rather complex, consisting of nine steps requiring the patient to chew a sponge in his/her mouth to collect saliva, and later squeeze the collected saliva into the device with a sensor, thus is not easy to use. More studies are required to investigate the accuracy of sensors in detecting low salivary glucose levels, efficiency and practicality of the proposed approach. Zhang et al. (2011) review current developments of non-invasive continuous SBGM methods using ocular glucose. These authors review studies in ocular glucose monitoring: (1) using contact lens-based sensors and (2) using nanostructured lens-based sensors. They concluded that lens sensors have the potential to monitor a wide range of glucose levels quickly and accurately; however, there is a safety concern because boronic acid and concanavalin A may be released from the lens into the patient body. Nanostructured lens-based sensors have several advantages (e.g. better accuracy and sensitivity, less interference with patient vision); further studies are required to improve resolution and sensitivity of the lens and to determine physiologically relevance and baseline tear glucose concentration (Zhang et al. 2011). Another review of current nanomaterial-based solutions using saliva, sweat, breath and tears as a medium for SBGM suggests that they are far from optimal; further nanotechnology sensing devices need to be manufactured at a low cost to compete with established blood glucose meters (Makaram et al. 2014).

One of the most recent advances in the area of semi-invasive glucose monitoring is Abbott's Freestyle Libre System (Hindustan Times, HT Correspondent 2015) which is based on a body attached sensor and a smartphone loaded with the application (Fig. 1, adapted from HT Correspondence 2015). This disposable body attached device (sensor) is equipped with a thin and flexible fibre needle, which is the only invasive part; however, the fibre is inserted only once and under the skin of the back of the arm. The sensor can be used continually for 14 days without the need to be replaced. The sensor captures glucose concentration information and once the smartphone that runs the required app scans the sensor, the current and up to 8 hours of glucose levels are read and uploaded to the smartphone (Timothy et al. 2015).

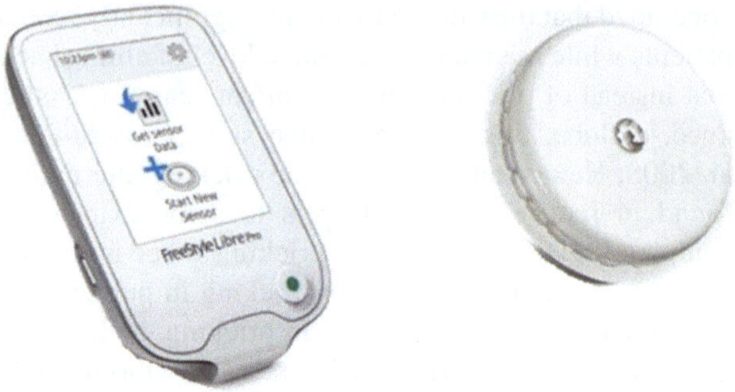

**Fig. 1** Freestyle Libre System. (Adapted from Hindustan Times, HT Correspondent (2015)

According to Timothy et al. (2015), the accuracy of the FreeStyle Libre system as dependent function of a number patient-related factors (e.g. diabetes type, gender, insertion site/administration, body mass index, haemoglobin A1c "HbA1c", age and rate of change) has been found to be above 85.2% up to 14 days of testing.

Overall, studies in issues associated with SBGM using finger pricking devices and learnings from in current developments of new SBGM methods suggest the following factors: technology (technical soundness, sensitivity of device and sufficient quantity and reproduction of medium), clinical accuracy (accuracy of measures, continuity of monitoring), clinical and safety interference with vision patient vision (infection, damage to patient body sites, release of chemicals to patient body, interference with patient vision), ease of use (level of invasiveness, practicality of sample collection and protocols of use), psychological effects (pain and discomfort, anxiety, fear, inconvenience and difficulty of performing tests in public) and costs (manufacturing costs and patient technology switching costs). We develop Table 1 to present the key aspects of problems and concerns identified in the extant literature relating to SBGM.

We conceptualise the problem in diabetes care management with the above factors. Therefore, we propose an approach to solution development consisting of patient analytics to focus on the targeted patient cohort, design science engaging patients in testing when it's safe, not early design, to finally propose a new solution to address their concerns and ensuring appropriate monitoring of the care management plan.

## 2.2 *Non-invasive Terahertz Technology Solution*

The ultimate approach in managing diabetes is based on non-invasive solutions, one of which is based Terahertz technology that is the focus of this section. Terahertz refers to the electromagnetic waves with the frequency range between millimetre-wave and infrared, approximately from 100 GHz up to 10 THz (see Fig. 2, adapted from Adibi 2013). The THz spectrum, also known as the "terahertz gap", is the last portion of the electromagnetic spectrum which has not been fully explored

**Table 1** Problems and concerns in current SBGM methods

| Key aspects | Factors |
|---|---|
| Technical feasibility | Sufficient quantity and reproduction of medium (Makaram et al. 2014).<br>Sensitivity of device (Zhang et al. 2011)<br>Technical soundness (Makaram et al. 2014; So et al. 2012) |
| Clinical accuracy | Accuracy of result (Timothy et al. 2015; Nakayama et al. 2008; Castilla-Peón et al. 2015)<br>Continuity of monitoring (Kannampilly 2013; So et al. 2012) |
| Side effects and safety | Infection (Dahiya et al. 2012; Giannini and Mayr 2004)<br>Physical damage to the body site (Dahiya et al. 2012; Giannini and Mayr 2004; Burge 2001)<br>Release of chemical to body (Zhang et al. 2011)<br>Interference with vision (Zhang et al. 2011) |
| Ease of use | Level of invasiveness (Castilla-Peón et al. 2015; Wainstein et al. 2013; Heinemann 2008; So et al. 2012).<br>Practicality of device, sample collection and protocols of use (Du et al. 2016; Heinemann 2008) |
| Psychological effects | Pain and discomfort (Wainstein et al. 2013; Burge 2001; Heinemann 2008; Karges et al. 2008; Koschinsky 2007; Pacaud et al. 1999)<br>Anxiety and distress (Shlomowitz and Feher 2014; Cradock and Hawthorn 2002)<br>Fear of needles and blood (Shlomowitz and Feher 2014; Burge 2001)<br>Inconvenience and difficulty of performing tests in public (Heinemann 2008; Castilla-Peón et al. 2015) |
| Costs | Manufacturing costs (Makaram et al. 2014)<br>Patient technology switching costs (Heinemann 2008) |

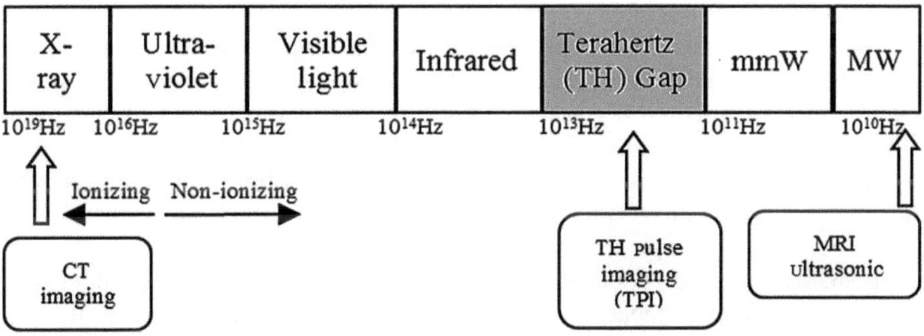


**Fig. 2** Frequency spectrum of EMR imaging technologies. (Adapted from Adibi (2013))

and exploited (Tonouchi 2007). Terahertz technology is a fast-growing field with applications in biology and medicine, medical imaging, material spectroscopy and sensing, security, monitoring and spectroscopy in pharmaceutical industry and high data-rate communications.

In biomedicine, Terahertz technology has so far been used in a variety of medical applications, including skin/breast cancer detection, wound inspection and dental imaging (Yang et al. 2016; Panwar et al. 2013).

The THz studies have uniquely revealed that medical image diagnoses are possible over wide range of tissues; however, much further detailed analyses are required to identify the degree of precision achieved in monitoring blood glucose concentration using THz technology.

## 3 Proposed Solution

The technology methods behind the operation of this solution's proof of concept are based on the following approaches (Jackson et al. 2011):

- Terahertz time-domain spectroscopy (THz-TDS)
- Terahertz frequency-domain spectroscopy (THz-FDS)
- Terahertz imaging using non-destructive evaluation (NDE)

These methods are considered to pinpoint the best option for blood glucose level monitoring from the transmitter/receiver perspectives. The THz transmitter, the operational power, energy consumption level and safety factors are of significant importance since the solution is ultimately deployed in a smartphone application.

From the mentioned approaches, the THz-TDS approach has shown promising behaviour since it was used to measure the full dielectric-based function representing as the absorption coefficient and the refraction index of glucose and galactose between 0.2 THz and 3.0 THz (Zhang 2008). A few distinct absorption features are identified as the signatures of intra- and intermolecular modes of the hydrogen bonded crystalline structure.

The design of the related app that runs on the smartphone platform requires a number of features, including: a fast Digital Signal Processing (DSP) system based on the microcontroller system used in the Arduino platform. The hardware side of the system consists of an open-source 32-bit Atmel ARM processor, which is capable of running fast concurrent processes. The software system features a fast and optimised image processing algorithm aided with Kalman filtering for higher accuracy. The high accuracy results are needed due to highly variable testing environment (handheld application). The system also features remote monitoring and cloud-computing capability. The proof of concept involves the identification of the optimal THz sensor, power, frequency spectrum and reflection analysis for the most optimal application of the Terahertz technology in monitoring under-skin blood glucose levels.

The framework is the continuation of the work of reference (Shen et al. 2002), which is based on the Fourier Transform Infrared (FTIR) Spectroscopy. This reference shows the successful deployment of infrared in detecting glucose level of the blood. The lessons learned from this work can directly be used in this project.

Once the specific THz approach is selected, the methodological approach is to model the existing solutions and study the physical behaviour of Terahertz technology when radiated onto the human skin and study the depth of penetration and the variations in the reflections. This requires sophisticated THz lab equipment to run experiments on a test dummy, which mimics the human skin and underlying soft tissues. The results between the traditional needle-based sensing are then compared and the precision figures for the Terahertz-based method are evaluated. Then the approach needs to be fine-tuned, and also other health issues (e.g. technology implications, training, health hazards, etc.) are then considered.

## 4 Methodology

A Design Science Research Methodology (DSRM) will be used to operationalise this research. This approach is particularly appropriate where improving an existing solution is desired and/or there is a need for a new solution to address specific unsolved or unique aspects (Hevner et al. 2004; Hevner and Chatterjee 2010). DSRM as a process model to carry out research is widely used in the information systems research to create new solutions or to improve existing ones. DSRM process model consists of six process elements (Peffers et al. 2007), starting from identifying the problem and the motivation to conduct research, and concluding with communicating the results and outcomes of the research. Table 2 maps the proposed project to DSRM process elements (Peffers et al. 2007).

## 5 Problem Exploration and Solution Visualisation: Work in Progress

We have conducted a systematic literature review and a series of workshops to explore the problem space and develop a vision for our technology enabled solution. Our systematic review of the literature to date has identified problems and concerns with the current methods for blood glucose monitoring as well as the need for a simple, accurate non-invasive solution to it. To address this void we have proffered a technology solution using sensor technology combined with a mobile phone application. The mobile application will act as a user-friendly interface to (1) enable the diabetic patient to interact with the sensor and (2) as a smart digital companion to provide personalised assistance in their monitoring and management of their conditions. This section describes our work in progress, specifically our visualisation of the problem space and the proposed solution.

### *5.1 Requirements Workshops*

Through a series of three requirements workshops facilitated by Navy Design, an independent company specialised in designing health technologies, the research

**Table 2** Mapping the proposed research to DSRM

| DSRM process elements | DSRM description | Application on this study |
|---|---|---|
| Problem identification and motivation | Defining the specific research problem and justifying the value of a solution based on knowledge of the state of the problem | With the increased diabetes population, and the disadvantages of conventional blood glucose tests, the lack of a reliable and easy- to-use non-invasive technology to monitor blood glucose motivates this research. Further, how such a solution will interact with the diabetic patient and support them through their journey living with and managing their diabetic conditions |
| Definition of objectives of the solution | The objectives can be qualitative or quantitative i.e. create or improve an artefact respectively based on knowledge of the state of the problem and current solutions, if any, and their efficacy | The objective is to create and refine an artefact (i.e. the sensor) and how it interacts with the diabetic patient (for example via a digital companion mobile application) |
| Design and development | Creating the artefact, including the desired functionality and its architecture based on knowledge of theory that can be used to bear in a solution. This is usually an iterative process | Through several iterations, the exact range for the frequency for the needed terahertz wave beam will be identified. Once the narrow range for the frequency for the terahertz wave is identified, the CAD/CAM programming will occur |
| Demonstration | Demonstrate the use of the artefact to solve the problem | Simulation-based: This will be done in a lab using their simulation set up and mannequins<br>Clinical: Demonstration of the use of a new device to a sample of targeted patient population |
| Evaluation | Iterate back to better design the artefact if needed | Simulation-based: As needed, iterations will take place, to fine tune the needed range for the terahertz wave projections<br>Clinical: Iterative evaluations to ensure that the prototype is truly tailored to meet clinical requirements for the targeted population |
| Communication | Publish and let the value of the solution talk about itself | This will include conference publications, journal publications and other presentation activities |

team examine the problem space including a range of problems faced by the diabetic patients and self-care instructions, care programmes and health information sources provided by health organisations, primarily Diabetes Australia and the National Diabetes Services Scheme (NDSS). We also applied design thinking in examining

**Another day in the life of a diabetic**

Jane Brown is a diabetic in her early-70s. She lives by herself. Here is a typical day in her diabetes dairy.

Jenny gets up at 7am. She measures her blood glucose at 7.45am. She has one cup of cereal, a cup of skim milk, and one small serve of orange or apple for breakfast at 8am and has a short walk in a local park near her house. She comes back and has a wholemeal biscuit with a glass of apple juice for morning tea at 10.30am. She prepares her own lunch and has lunch at 12.30pm. Her lunch usually includes a whole-wheat bread slice, cheesy scrambled eggs (one or two eggs) or ham, salad, one small cup of skim milk, a cup of apple juice.

She has rest at home until 3pm to measure her blood glucose 2 hours after having lunch. In the afternoon, she often goes to local shops to have a chat with some ladies and watch people walking by. Sometimes they enjoy some street walking together. Occasionally, they drink tea or have fruits (banana, apple, etc.). She gets back home around 4pm, has some rest and starts making her dinner. She measures her blood glucose before dinner. She often has a slice of bread or a cup of brown rice, some vegies such as carrots, broccoli, or spinach with chicken or fish, a small cup of banana yogurt.

Before going to bed, she measures her blood glucose again.

Jane's food intakes and activities vary from day to day. Sometimes, Lara (Jane's daughter) and her family visit her. They cook dinner for Jane or take her out for dinner. On 23rd February 2017, Jane went out with Lara's family to a local restaurant for dinner. Jane's daughter does weekly shopping and replenishes her fridge for her on Sunday.

| Day | When | Readings | Food diary | Activity diary | Digital Companion response |
|---|---|---|---|---|---|
| 21/2/2017 | Get up | 5.6 | | | |
| 21/2/2017 | 2h After lunch | 7.4 | Breakfast: White tea and cake; Lunch: Scrambled eggs, Bread, Water melon | | Jane, this is a little high ☹ |
| 21/2/2017 | Before dinner | 6.2 | | Afternoon: walked 20min | Jane, this is better ☺ |
| 21/2/2017 | Bed time | 7.5 | Dinner: While bread, chicken, Banana, and cake. | | Jane, try to cut down on cakes and walk a little longer. |
| 22/2/2017 | Get up | 6.1 | | | |

**Fig. 3** Extract of research team's work in exploring the problem space

and mapping our non-invasive approach with possible diabetic patient journeys. Next, we further explored and refined the problem space through visualising a day in the life of a patient (Fig. 3).

\looseness-1Based on such examples, we specified 2 themes, 14 epics and numerous user stories based on scenarios of patient day journals. At this stage, 33 cards were developed to depict 33 touch points and activities in patient journeys (see Appendix A).

## 5.2 *Mili: A Smart Digital Companion for the Diabetic Patient*

Mili (My Insulin Level Information) was developed by Navy Design as a prototype to visualise our proposed approach to support non-invasive monitoring of glucose level and to provide personalised assistance and encouragement to the diabetic patient in their everyday management of their diabetic conditions.

There are two themes describing patient view and clinician view of the overall solution. This paper focuses on the patient view. It starts with the first stage of Diagnosis, Hearing about Mili, Ordering Mili and Receiving Mili. Next, there are two uses of Mili –setting up Mili including the digital companion and the sensor, and using Mili including interacting with the digital companion and the sensor.

During the setting up of Mili, the patient should enter personal information, diabetic condition and a care plan. During this process, the patient will also activate the sensor (see Fig. 4).

**Fig. 4** Screen shots of sample Mili set-up interactions

After all the setting up steps, Mili will become a smart digital companion for the patient through the management of their conditions in their daily life. High-level user stories (adapted from the Agile development approach) were developed to visualise the self-care activities undertaken by patients and how they can be supported by Mili. The user stories were expressed in the format of

***As a Social Actor*** *(such as diabetes patient or clinician),*
***I want to*** *(do something)*
***so that I can*** *(achieve or get some goal).*

The user stories were organised into epics each consisting of related activities. Key epics include Patient Profile management, Blood glucose measure, Dynamic food diary, Dynamic activity diary, Dynamic insulin diary, Ask my Digital Companion, Share and Care. Figure 5 lists examples of user stories to illustrate our vision.

The user stories were organised into epics each consisting of related activities. Key epics include Patient Profile management, Blood glucose measure and Dynamic food diary. Figure 6 provides screenshots of Mili interactions to support non-invasive glucose measurement and monitoring.

There are user stories in the clinician view to enable the clinician to monitor their patients, view and edit care plans, or communicate with and arrange appointments with them as needed. The clinician view will have an analytics capability and a dashboard to monitor all their patients as well as an individual patient. Further details about the work described in this section can be found in Appendix A. A series of Mili simulated interactions can be found in Appendix B.

**Epic 2: Blood glucose measure**

1. As a diabetes patient, I want to be able to measure blood glucose level so that I am informed of my diabetes conditions
2. As a diabetes patient, I want to be able to store and view the blood glucose readings and date/time (e.g. pre-meal, 2 hours post-meal, and before bedtime) taken so that I can monitor glucose levels
3. As a diabetes patient, I want to know what the last and previous blood glucose readings mean – Normal range, Pre-diabetes range (Impaired glucose tolerance (IGT), Impaired fasting glucose (IFG)), At risk range, High, Very High.
4. As a diabetes patient, I want to receive reminders to measure blood glucose at set times (pre-meal, 2 hours post-meal, and before bedtime) to adhere to the care plan (structured monitoring)

**Epic 3: Dynamic food diary**

1. As a diabetes patient, I want to view, edit and store my food diary so that I can monitor glucose impact
2. As a diabetes patient, I want to view glucose impact of food intakes to adjust my diets

**Epic 4: Dynamic activity diary**

1. As a diabetes patient, I want to view, edit and store activity diary (e.g. 20mins or #km of walking, running, biking) so that I can monitor glucose impact
2. As a diabetes patient, I want to view glucose impact of activities to adjust my activities (e.g. 20mins or #km of walking, running, biking)

**Epic 5: Dynamic insulin diary**

1. As a diabetes patient, I want to view, edit and store insulin dose and intake date/time so that I can monitor glucose impact

**Fig. 5** Sample Mili user stories

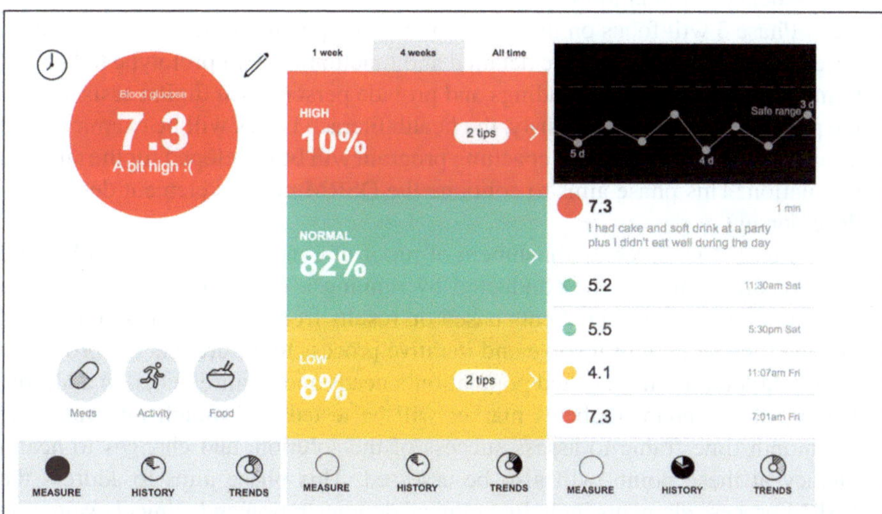


**Fig. 6** Screen shots of sample Mili glucose monitoring interactions

## 6 Next Steps

The next steps now include the design and development of the solutions coupled with establishment of proof of concept, usability and fidelity. This involves four key steps as follows:

Phase1: This phase will involve the conducting of multi-dimensional analysis of provided data sets from a diabetic population (for example in China). The results will provide a clear picture of the current state, reveal critical trends and important patterns regarding this population and assist to identify the patient sample. It is anticipated that target patient cohorts and their demographics as well as geographic and clinical characteristics will be identified for the project to identify who would benefit most from the non-invasive smart solutions. This phase aims to address the first DSRM process elements including problem identification and motivation, and definition of objectives of the solution (see Table 2).

Phase 2: Phase 2 will include designing and developing the appropriate sensor technology. The patented solution we have developed uses THz to identify blood sugar readings but this requires critical analysis to isolate the specific THz band. Inputs for this are derived from aspects of the data analytics performed in phase 1 above. Once this is done the required specifications for designing the sensors must be generated. Thus, the results at the end of this phase include the design specification for the sensors to be used in the specific context, so they are truly tailored to that context. This phase aims to address the subsequent DSRM process elements including design and development, simulation-based demonstration and associated simulation-based evaluation.

Phase 3: Phase 3 will focus on the design and development of the software solution necessary to develop the more detailed and functioning Mili prototype to be used to measure blood glucose readings and provide personalised digital assistance to the patients. Contemporaneously, the health literacy issues will be examined and an appropriate education and coaching program will be developed for the targeted population. This phase aims to continue the DSRM process element design and development.

Phase 4: Phase 4 involves establishment of proof of concept, usability, fidelity and functionality. This will be conducted by running a field study on the selected patient population of 50 patients based on results from phase 1. It is anticipated that the filed study will involve and iterative process to ensure that the prototype is truly tailored to the selected population's needs and requirements. In addition, HbA1C the standard diabetes marker will be tested at 3-month intervals over a 6-month time frame to assess success of the solution, and changes to health literacy at these points will also be assessed. This phase aims to address the DSRM process elements including clinical demonstration and clinical evaluation using the key aspects (technical feasibility, clinical accuracy, clinical side effects and safety, ease of use, psychological effects and costs) and relevant factors presented in Table 1. As a result, the list of factors will be refined to inform future implementations and evaluations.

The DSRM process element communication will take place through the whole project when findings from each phase become available.

## 7 Discussion and Conclusions

Concurrent and independent from the exponential rise of diabetes has been the rise of mobile and sensor technology. The maturing and sophistication of these technologies has enabled them to be sued in many aspects of healthcare and wellness management. The preceding has served to outline another potential area for the adoption of mobile and sensors; namely, to assist with a non-invasive approach for the monitoring and management of diabetes. Specifically, we have identified an opportunity to use Terahertz frequencies to detect blood glucose levels in individuals. Further, we envisage designing and developing this solution by combining sensors with a mobile phone so that detection of blood glucose can not only be non-invasive but truly pervasive.

The implications for this for theory and practice are wide and far reaching. From a theoretical perspective, we combine two technology genres mobile and sensors to address a healthcare issue – detection of blood glucose levels using a design science research methodology. From the perspective of practice, diabetes as noted by WHO (World Health Organization 2016) is global and at epidemic proportions with an estimated of 422 million adults living with diabetes in 2014 and 1.5 million deaths caused by diabetes in 2012. Complications from diabetes can lead to other serious conditions such as heart attack, stroke, blindness, kidney failure and lower limb amputation. Further, the number of pre-diabetic individuals is also considerable. Monitoring and management is the only recognised strategy to maintaining appropriate blood glucose levels and thereby managing diabetes and/or preventing a pre-diabetic becoming a diabetic. Given the problems and criticisms of finger pricking and other invasive approaches to SBGM, the most prevalent approach to testing blood glucose using a non-invasive Terahertz technology solution which we propose is very attractive to individuals. When such a solution is truly pervasive, it becomes even more attractive. Thus, we believe that the proffered solution will enable diabetic and pre-diabetic individuals to enjoy a better quality approach to monitoring and managing their blood glucose levels.

Alongside with developing a non-invasive Terahertz technology solution to measure blood glucose, further work has been conducted to design an integration of the digital companion component within a patient journey with AI-enabled capabilities. Our future work will focus on establishing usability, fidelity and acceptability of the proffered non-invasive pervasive solution.

## Appendix A

Mili – A Co-design Product of Navy and Deakin University https://app.milanote.com/1DdHpg11lPgFag/navy-design-deakin-mili

# Appendix B

Simulated Interactions with Mili https://e6qa6p.axshare.com/home.html

# References

Adibi, S. (2013). *Mobile health (mHealth) biomedical imaging paradigm*. Paper presented at the 35th annual international conference of the IEEE engineering in medicine and biology society (EMBC'13), Osaka, Japan.

Australian Institute of Health and Welfare. (2007). National indicators for monitoring diabetes: Report of the diabetes indicators review subcommittee of the national diabetes data working. *Diabetes series no. 6. Cat. no. CVD 38. Canberra: AIHW.* Canberra: Australian Institute of Health and Welfare.

Australian Institute of Health and Welfare. (2008). *Diabetes: Australian facts Canberra*. Australian Institute of Health and Welfare.

Britt, H., Miller, G. C., Charles, J., Pan, Y., Valenti, L., Henderson, J., Bayram, C., O'Halloran, J., & Knox, S. (2007). *General practice activity in Australia 2005–06*. Cat. no. GEP 16. Australian Institute of Health and Welfare.

Burge, M. R. (2001). Lack of compliance with home blood glucose monitoring predicts hospitalization in diabetes. *Diabetes Care, 24*(8), 1502–1503.

Castilla-Peón, M. F., Ponce-de-León-Rosales, S., & Calzada-León, R. (2015). Knee skin pricking to monitor capillary blood glucose is less painful than finger pricking in children with type 1 diabetes. *International Journal of Diabetes in Developing Countries, 35*(4), 620–623. https://doi.org/10.1007/s13410-015-0342-z.

Cradock, S., & Hawthorn, J. (2002). Pain, distress and blood glucose monitoring. *Journal of Diabetes Nursing, 6*(6), 188–191. 184p.

Dahiya, S., Voisine, M., & Samat, A. (2012). Gangrene from finger pricking. *Endocrine, 42*(3), 767–767. https://doi.org/10.1007/s12020-012-9787-8.

Diabetes Australia. (2007). *Diabetes facts*. New South Wales: Diabetes Australia.

Du, Y., Zhang, W., & Wang, M. L. (2016). Sensing of salivary glucose using nano-structured biosensors. *Biosensors, 6*, 10.

Farmer, A. J., Wade, A. N., French, D. P., Simon, J., Yudkin, P., Gray, A., Craven, A., Goyder, L., Holman, R. R., Mant, D., Kinmonth, A. L., & Neil, H. A. (2009). Blood glucose self-monitoring in type 2 diabetes: A randomised controlled trial. *Health Technology Assessment, 13*(5), iii–iiv, ix-xi, 1-50. https://doi.org/10.3310/hta13150.

Giannini, O., & Mayr, M. (2004). Finger pricking. *The Lancet, 364*(9438), 980. https://doi.org/10.1016/S0140-6736(04)17022-0.

Guerci, B., Drouin, P., Grangé, V., Bougnères, P., Fontaine, P., Kerlan, V., Passa, P., Thivolet, C., Vialettes, B., & Charbonnel, B. (2003). Self-monitoring of blood glucose significantly improves metabolic control in patients with type 2 diabetes mellitus: The Auto-Surveillance Intervention Active (ASIA) study. *Diabetes & Metabolism, 29*(6), 587–594.

Haller, M. J., Stalvey, M. S., & Silverstein, J. H. (2004). Predictors of control of diabetes: Monitoring may be the key. *The Journal of Pediatrics, 144*(5), 660–661. https://doi.org/10.1016/j.jpeds.2003.12.042.

Heinemann, L. (2008). Finger pricking and pain: A never ending story. *Journal of Diabetes Science and Technology, 2*(5), 919–921.

Hevner, A., & Chatterjee, S. (2010). *Design research in information systems: Theory and practice* (1st ed.). New York: Springer.

Hevner, A., March, S., Park, J., & Ram, S. (2004). Design science in information systems research. *MIS Quarterly, 28*(1), 75–105.

HT Correspondent. (2015, 02 April). Now, keep a check on your diabetes without a prick. *Hindustan Times*. Retrieved from URL: http://www.hindustantimes.com/mumbai/now-keep-a-check-on-your-diabetes-without-a-prick/story-sB15fHUDHc0FzsmsTqczFP.html.

Jackson, J. B., Bowen, J., Walker, G., Labaune, J., Mourou, G., Menu, M., & Fukunaga, K. (2011). A survey of terahertz applications in cultural heritage conservation science. *IEEE Transactions on Terahertz Science and Technology, 1*(1), 220.

Kannampilly, J. J. (2013). Continuous glucose monitoring system. In D. A. Muruganathan (Ed.), *Medicine update 2013 – contents* (Vol. 23). Mumbai: The Association of Physicians of India (API).

Karges, B., Muche, R., Moritz, M., Riegger, I., Debatin, K. M., Heinze, E., Wabitsch, M., & Karges, W. (2008). Low discomfort and pain associated with intensified insulin therapy in children and adolescents. *Diabetes Research and Clinical Practice, 80*(1), 96–101.

Karter, A. J., Ackerson, L. M., Darbinian, J. A., D'Agostino, R. B., Jr., Ferrara, A., Liu, J., & Selby, J. V. (2001). Self-monitoring of blood glucose levels and glycemic control: The Northern California Kaiser Permanente Diabetes registry*. *The American Journal of Medicine, 111*(1), 1–9. https://doi.org/10.1016/S0002-9343(01)00742-2.

Koschinsky, T. (2007). Blood glucose self-monitoring report 2006 reveals deficits in knowledge and action. *Diabetes, Stoffwechsel und Herz, 16*, 185–192.

Makaram, P., Owens, D., & Aceros, J. (2014). Trends in nanomaterial-based non-invasive diabetes sensing technologies. *Diagnostics, 4*(2), 27–46. https://doi.org/10.3390/diagnostics4020027.

Malanda, U. L., Welschen, L. M., Riphagen, I. I., Dekker, J. M., Nijpels, G., & Bot, S. D. (2012). SMBG in patients with type 2 diabetes mellitus who are not using insulin. *Cochrane Database of Systematic Reviews, 18*(1), CD005060. https://doi.org/10.1002/14651858.CD005060.pub3.

Nakayama, T., Kudo, H., Sakamoto, S., Tanaka, A., & Mano, Y. (2008). Painless self-monitoring of blood glucose at finger sites. *Experimental and Clinical Endocrinology & Diabetes, 116*(4), 193–197.

Pacaud, D., Lemay, J. F., Buithieu, M., & Yale, J. F. (1999). Blood volumes and pain following capillary punctures in children and adolescents with diabetes. *Diabetes Care, 22*(9), 1592–1594.

Panwar, A. K., Singh, A., Kumar, A., & Kim, H. (2013). Terahertz imaging system for biomedical applications: Current status. *International Journal of Engineering & Technology, 13*(2), 33–39.

Peffers, K., Tuunanen, T., Rothenberger, M., & Chatterjee, S. (2007). A design science research methodology for information systems research. *Journal of Management Information Systems, 24*(3), 45–77. https://doi.org/10.2753/mis0742-1222240302.

RACGP and Diabetes Australia. (2014-2015). *General practice management of type 2 diabetes*. The Royal Australian College of General Practitioners.

Shen, Y. C., Davies, A. G., Linfield, E. H., Taday, P. F., Arnone, D. D., & Elsey, T. S. (2002). *Determination of glucose concentration in whole blood using FTIR spectroscopy*. Paper presented at the THz-bridge conference, Capri, Italy.

Shlomowitz, A., & Feher, M. D. (2014). Anxiety associated with self monitoring of capillary blood glucose. *British Journal of Diabetes and Vascular Disease, 14*(2), 60–63.

So, C.-F., Choi, K.-S., Wong, T. K. S., & Chung, J. W. Y. (2012). Recent advances in noninvasive glucose monitoring. *Medical Devices (Auckland, N.Z.), 5*, 45–52. https://doi.org/10.2147/MDER.S28134.

Timothy, B., Bode, B. W., Christiansen, M. P., Klaff, L. J., & Alva, S. (2015). The performance and usability of a factory-calibrated flash glucose monitoring system. *Diabetes Technology & Therapeutics., 17*(11), 787–794. https://doi.org/10.1089/dia.2014.0378.

Tonouchi, M. (2007). Cutting-edge terahertz technology. *Nature Photonics, 1*(2), 97–105.

Wainstein, J., Chimin, G., Landau, Z., Boaz, M., Jakubowicz, D., Goddard, G., & Bar-Dayan, Y. (2013). The use of a CoolSense device to lower pain sensation during finger pricking while measuring blood glucose in diabetes patients—A randomized placebo. *Diabetes Technology & Therapeutics, 15*(8), 688–694.

World Health Organization. (2016). Global report on diabetes. *WHO Library Cataloguing-in-Publication Data.* © World Health Organization 2016.

Yang, X., Zhao, X., Yang, K., Liu, Y., Liu, Y., Fu, W., & Luo, Y. (2016). Biomedical applications of terahertz spectroscopy and imaging. *Trends in Biotechnology, 34*, 810. https://doi.org/10.1016/j.tibtech.2016.04.008.

Zhang, J., Hodge, W., Hutnick, C., & Wang, X. (2011). Noninvasive diagnostic devices for diabetes through measuring tear glucose. *Journal of Diabetes Science and Technology, 5*(1), 166–172.

Zhang, T. (2008). *Terahertz Time-Domain Spectroscopy of Crystalline Glucose and Galactose*. Paper presented at the The 2nd International Conference on Bioinformatics and Biomedical Engineering, ICBBE 2008.

# Part II
# Opportunities to Incorporate Critical Aspects of Analytics to Provide Superior Insights and Thus Support Better Decision-Making

Arguably, analytics and big data have become the key buzzwords and panacea in healthcare delivery in this second decade of the twenty-first century. In this section, seven chapters are presented within this broad topic as follows:

Chapter 11: Intelligent Risk Detection in Healthcare: Integrating Social and Technical Factors to Manage Health Outcomes by Moghimi et al.
Chapter 12: A Literature Review on Predicting Unplanned Patient Readmissions by Eigner and Cooney
Chapter 13: Using KM to Develop Superior Online Health Decision Support Solutions: The Case of Allergy Care by Wickramasinghe
Chapter 14: Opportunities for Using Blockchain Technology in e-Health: e-Prescribing in Germany by Seitz and Wickramasinghe
Chapter 15: Knowledge Acquisition of Consumer Medication Adherence by Vlahu-Gjorgievska et al
Chapter 16: Addressing Data Accuracy and Information Integrity in mHealth Solutions Using Machine Learning Algorithms by Sako et al
Chapter 17: Enabling Value-based Healthcare with Business Analytics and Intelligence by Wickramasinghe

One of the key issues that has developed around analytics and big data in healthcare delivery is can the promises be delivered or are we being given snake oil? What we try to achieve in this section is presenting a miscellany of chapters around analytics and big data for healthcare delivery that provide realistic expectations of what can and cannot be achieved and how to go about ensuring a successful outcome. Data, information, and knowledge hold great potential for supporting better decision-making and thereby better care delivery, but designing and developing appreciate algorithms and solutions requires careful planning and judicious application of key tools and techniques. It is our hope that after reading the chapters in this section, you will have a better and more balanced perspective of what can and cannot be achieved from analytics and big data in healthcare, and while it is not a panacea, it certainly affords many possibilities.

# Intelligent Risk Detection in Health Care: Integrating Social and Technical Factors to Manage Health Outcomes

**Hoda Moghimi, Nilmini Wickramasinghe, and Monica Adya**

## 1 Introduction

Healthcare organizations have increasingly been leveraging information technologies (IT) to enhance patient care through improved real-time decision-making (Trucco and Cavallin 2006; Tseng et al. 2003), customized care delivery (Tiwari et al. 2013), and managed healthcare costs (Thomas et al. 1983). Although the quality of healthcare in the USA is documented to be improving, new regulations and standards relating to systematic and organizational performance are, however, casting pressure on healthcare providers to improve real-time decisions and minimize adverse incidents while optimizing patient care costs. Essentially, care providers are under mandate to identify more optimal models that deliver care commensurate with the patient's risk level (Wickramasinghe et al. 2012). As patient care evolves, patient risk assessment is shifting from managing individual causes of incidents, e.g., physician error, to managing organizational factors where an adverse incident is viewed as stemming from a chain of events (Vincent et al. 2000). Each component of this chain has the ability to generate a 360° view of the patient and provide unique insights into delivery of patient care.

The confluence of decision sciences and business intelligence (BI) technologies offer numerous opportunities for a deeper, organizational-level understanding of

H. Moghimi
Health Informatics Management, Epworth HealthCare, Richmond, VIC, Australia

N. Wickramasinghe (✉)
Epworth HealthCare, Richmond, VIC, Australia

Swinburne University of Technology, Hawthorn, VIC, Australia

M. Adya
Management, Marquette University, Milwaukee, WI, USA



N. Wickramasinghe, F. Bodendorf (eds.), *Delivering Superior Health and Wellness Management with IoT and Analytics*, Healthcare Delivery in the Information Age, https://doi.org/10.1007/978-3-030-17347-0_11

patient risk management (Wickramasinghe and Schaffer 2010). Increasing maturity and demonstrated successes in the BI space are driving healthcare organizations to examine intelligent, analytical technologies to improved insights into operations, with an eye toward risk management (Wickramasinghe et al. 2008). Specifically, the application of data mining and knowledge discovery (KD) techniques are increasingly being considered for assessing patient health risks and assisting medical professionals in making optimal care decisions in real time (ibid). Although successfully used for risk management in other industries such as banking and finance, the utilization of these technologies in healthcare context has lagged for reasons discussed next.

Health risk management is informed by many patient-level dimensions, such as pathological process, physiological variables, general health perceptions, and quality-of-life considerations (Rizzo and Kintner 2013). Furthermore, risk assessment has, most often, been applied at the disease level, for instance, in the management of cardiovascular disease (Grundy et al. 1999) or at the patient/care provider level (Barrett et al. 2001). Simultaneously, healthcare providers are becoming increasingly aware of the relationship between organizational factors and riskiness of healthcare provision (Yagi et al. 2011; Yen and Bakken 2012). Gaining insights into patient risk requires leveraging large amounts of disparate data (Chen et al. 2012; Elbaum et al. 2005) and developing a comprehensive BI and data science-driven risk management framework that synergizes various levels of risk engagement. However, there exists no comprehensive framework that yields insights into how such an integrative solution could be architected. To address this need, this study is designed with the following goal: to conceptualize and develop, for the healthcare context, a risk detection framework that integrates knowledge- and intelligence-based technologies with patient, clinical, and organizational factors to provide real-time and continuous risk management capabilities.

Specifically, this article addresses the following questions:

(a) What design principles should underlie an intelligent risk detection (IRD) framework?
(b) What decision models may be most central and suitable to the healthcare domain?
(c) What technical infrastructure will be necessary to support such real-time decision-making, and, more critically, how are BI techniques integrated within this framework?
(d) What is the role of key stakeholders such as physicians, patients, and other care providers in facilitating and managing this real-time risk environment?

Accomplishing these objectives is complex for two reasons. First, considering the dearth of literature in this domain, conceptualizing an architecture that integrates both human and technological factors is challenging. Second, the manifestation of risk can vary significantly from one disease state to another, as might be for a hip surgery as opposed to cardiovascular disease. As such, the application of intelligent detection requires some adaptation to specific domains. In this article, the research questions are addressed within the context of congenital heart disease

(CHD) surgeries in children. CHD provides a useful domain to initiate and test this research as decision-making in this context is multifaceted and complex. Surgery is not always considered a final cure in CHD because it can result in a considerably high rate of complications, side effects, and comorbidities, such as cancer or bowel disease. There is also the direct adverse impact on patients and their families. Using the CHD context, this article makes several contributions:

(a) It elicits an intelligent risk detection framework, called the HOUSE (Health Outcomes around Uncertainty, Stakeholders, and Efficacy) model that elaborates on dual benefits of data/decision science and IT for developing health risk management capability. HOUSE integrates knowledge assets and capabilities with process capabilities of hospitals and links these to performance (Wu and Hu 2012). Although it builds on analysis from a single healthcare context, the CHD, the article demonstrates it to be applicable to other contexts.
(b) It emphasizes the role of BI and dynamic analytic techniques that support management of massive and disparate data as well as risk discovery and management from such data. This includes techniques for identifying key clinical and surgical risks, decision milestones, and participating actors.
(c) It uses design principles that emphasize collaboration between information receivers (patients) and generators (healthcare providers) as well as knowledge sharing them to facilitate clinical decision-making and risk management. Although presented in the healthcare context, these principles can extend to real-time risk detection in other domains.
(d) Finally, it provides a mechanism for routinizing organizational workflow while creating a framework that can support personalized medicine and care. Specifically, it describes initial architectures and analytics needed to store and process large-scale data necessary for personalized care as called for in recent studies (Fichman et al. 2011).

The next section provides background literature on clinical decision-making and clinical decision support systems (CDSS) that currently define the field. The section thereafter provides an overview of the research methodology, data collection, analysis, and key findings related to CHD that go into developing the HOUSE model which is discussed thereafter with particular emphasis on key organizational requirements, components, barriers, and issues related to the design and development such a risk detection model. The article concludes with implications for future research and practice.

## 2 Theory and Background

Patient care is replete with difficulties associated with identifying risk factors related to healthcare provision (Wetters et al. 2013), predicting outcomes prior to procedures (Kim et al. 2010), estimating quality of life (Quintana et al. 2009), and planning relevant wellness services (Kersten et al. 2012) to reduce the mortality and morbidity rates in healthcare. These challenges stem from myriad sources, both social and clinical, information about which is fragmented and inadequately leveraged to improve patient care. A robust IRD solution must comprehensively integrate decisions and information from clinical decision support systems (CDSS) with knowledge discovery from disparate and massive data sources, as well as social factors surrounding care in order to deliver real-time risk management.

## *2.1 Clinical Decision Support Systems (CDSS)*

CDSS support physician, nurse, and patient decisions by enabling medical knowledge to be applied to patient-specific information (De Backere et al. 2012). Such systems assist clinical staff with a variety of decisions such as drug dosing, health maintenance, and clinical diagnosis among others (Laine et al. 2007). Studies have confirmed that CDSS enhance quality, safety, and effectiveness of medical decisions, improve patient care, and lead to more effective clinical service through improved support of medical staff (Fichman et al. 2011; Restuccia et al. 2012). Additionally, in recent years, CDSS research in chronic disease care, particularly diabetes, has received greater attention with demonstrable gains in improved care provision (Jeffery et al. 2013). However, numerous gaps exist in design of most CDSS today. First, CDSS are often not fully integrated with clinical workflows. To support real-time risk management, CDSS must be integrated with clinical systems that are already present in hospitals and medical centers (Sanchez et al. 2013). Second, CDSS are challenging to maintain and extend (Peleg and Tu 2006). They often require significant investment, both financial and human capital. Once developed, updating these systems to reflect current knowledge and practices becomes a deterrent as organizations consider additional costs of enhancement. Hence, low-cost solutions are needed to maintain the underlying systems to fully support intelligent and real-time clinical decisions and risk management (Sanchez et al. 2013).

Third, to enhance value to clinicians, real-time medical and risk assessment should be embedded within CDSS (Peleg and Tu 2006). Currently, very few CDSS consider risk assessment, and those that do do not do so in real time. Fourth, most CDSS lack a robust mechanism for quantitative and qualitative evaluation of system performance, as well as for evaluation of decision outcomes supported (Liu et al. 2006; Peleg and Tu 2006; Sim et al. 2001). Finally, CDSS should be presented as complete solutions that assist clinicians during a variety of different tasks and daily responsibilities of physicians, and not merely during specific activities (Sanchez et al. 2013). This is not currently so. These limitations, the criticality of real-time outcomes, and the multispectral nature of care teams necessitate the following features for an intelligent CDSS to support risk assessment and management (Wickramasinghe et al. 2012):

- Multidimensional views of data – An intelligent CDSS must be able to cross-refer data sets and tables to support complex decision-making. This requires integration of data from various and disparate data sources.
- Calculation-intensive capabilities – CDSS must utilize mathematical, statistical, and machine-learning techniques such as cluster analysis and artificial neural networks to provide in-depth analysis of massive and disparate data.
- Intelligent timing – Finally, a CDSS must time the selection and application of the above techniques and data to needs of the decision task. Such systems must also be able to determine which data sets are most suited for the decision and tools at hand.

## *2.2 Health Risk Management*

Medical decisions always require a trade-off between benefit and risk (Cerrito 2011). To help patients understand potential risks and benefits of a procedure and identify options that best accommodate their unique needs, risk information must be coupled with high-quality decision counselling (Kuhn et al. 2006). Unfortunately, many decisions are based on incorrect knowledge (Cerrito 2011) and differing viewpoints of risk, both of which can lead to divergent treatment recommendations (Horn et al. 1985; Kuntz and Goldie 2002), even as such recommendations become increasingly technology supported. Identifying and characterizing basic risk types can not only lead to greater agreement between patient and clinical providers (Cerrito 2011; Lacour-Gayet 2002) but also provide a consistent and robust framework for risk management. Table 1 provides a summary and characterization of basic clinical risk types.

In treatment risk detection, multiple dimensions are important to consider (Yoshio et al. 2012). These include pathological process, physiological variables, general health perceptions, social paradigms, and quality of life (Rizzo and Kintner 2013). To address this, numerous risk-adjuster systems such as ACGs (Ambulatory Care Group) (Weiner et al. 1996), chronic disease, and disability payment system (CDPS) (Kronick et al. 2002) have been developed since the 1980s. These systems, which have been implemented by the Medicare Choice program, several U.S. states, and employer coalitions, are based on a variety of factors, including medical diagnosis, prior utilization, demographics, persistent diseases, and self-assessments of health and/or functional status. However, consistent with our earlier discussion on CDSS limitations, these systems lack a dynamic health risk assessment system (Greenland 2012; Ryan et al. 2012) and multidimensional risk detection models (Anderson et al. 2012; Staal et al. 2013). As such, applying BI-based techniques such as knowledge discovery and data mining could greatly enhance the performance of current risk assessment and adjustment methods (Guikema and Quiring 2012; Karaolis et al. 2010).

**Table 1** Basic types of clinical risk and their characteristics

| Types of clinical risks | Characteristics | Example |
|---|---|---|
| Common risks | There are common risks generally when patients have comorbidity that need to be considered | A patient with congestive heart failure is asked to avoid NSAID, while those with rheumatoid arthritis are often prescribed these. Case studies indicate that patients with both CHF and arthritis are at increased risk of mortality |
| Rare risks | A rare occurrence could mean that an event happens in out of every 100, 100,000, or 1,000,000 patients, and they are often not known at the time of treatment | Colonoscopies are recommended by physician to detect colon cancer at an early stage. However, colonoscopies have a rare complication of resistant infection that is 2–3 times the average. Unfortunately, patients are rarely informed of this increased risk because physicians and health agencies are convinced that the benefit of the colonoscopy outweighs risks |
| Unknown risks | Usually some risks are dismissed as unimportant. However, these risks may be important to the individual patient and should be considered | Mental impairment can be a problem after open heart surgery. However, only risk of stroke is considered because impairment has not been well studied |
| Uniform risks | Public health decisions are generally based on "societal risk" or what choice is of benefit to society generally. This public health perspective almost always assumes that risk is uniform for every individual and that treatment decisions should be made based on this uniform perspective | To prevent cervical cancer in all girls aged 9–18, the new HPV vaccine is recommended. In the VAERS (Vaccine Adverse Event Reporting System), some deaths have been attributed to the HPV vaccine. From a public health perspective, such deaths do not change recommendations because the risk is very small |
| Assumed risks | It is possible that treatments are given based on assumed risks or misperception of the actual risks | In 2010, it was discovered that the risk of allergy to medicine was much lower than previously assumed. The assumed risk was 30% of the adult population when just 5% actually had allergies |
| Biased risks | It is very difficult to consider overall or disease-free survival in a clinical trial, yet clinical trials are considered the gold standard of medical research. Also the research results should be validated by an acceptable number of patients' data in reasonable conditions | There are 1000 patients enrolled in a trial of drug A. Suppose that the study closes after 2 years and 500 patients are given an average survival of 2 years. The remaining patients will have an average that is less than 2 years, giving an overall average less than 2 years |

Adapted from Cerrito and Cerrito (2006)

## 2.3 Knowledge Discovery

Knowledge discovery (KD) is another crucial component of any intelligent risk management model. Data mining, as an approach to knowledge discovery, has been utilized in various healthcare contexts, including text mining (Safran et al. 2007), predicting health insurance fraud, controlling infection, managing physician order entry (Wright and Sittig 2006; Batal and Hauskrecht 2010), and identifying high-risk patients (Marschollek et al. 2012) (see table in Appendix A for detailed review). The KD process involves using a data source along with any required selection, preprocessing, subsampling, and transformations of the data source, applying data mining methods to elicit patterns from it, and evaluating the products of discovery to identify elements that can be deemed as knowledge (Han et al. 2006; Cios et al. 2007).

Recent studies have also recognized BI and business analytics (BA) as a data-centric approaches to the long-standing KD process (Turban et al. 2008). The term intelligence has been used by researchers in artificial intelligence since the 1950s. However, in the late 2000s, business analytics was introduced to represent the key analytical component in BI (Davenport 2006). More recently, big data analytics and predictive analytics have been used to describe the data sets and analytical techniques in applications that are so large (from terabytes to exabytes) and complex (from sensor to social media data) that they require advanced and unique data storage, management, analysis, and visualization technologies (Siegel 2013). Hence, in this study, we use the unified term BI to encompass BI/BA and data mining and treat predictive analytics as a related field that offers a new direction for developing such a solution in the healthcare context.

As with trends in other domains, the healthcare community has been inundated with opportunities and challenges stemming from big data and BI/BA. A complex set of diverse patient data is generated by the minute from medical devices, multiple points of patient contact, prescriptions, organizational operations, and web-based communities among others (Chen et al. 2012). Along with associated challenges, the ability to extract deep knowledge about patient care using such data is the next big opportunity for healthcare. With the adoption of electronic health records, organizations can now examine longitudinal patterns in patient care risk management. These systems, though, are still lacking an effective mechanism by which to integrate patient data with organizational data, and social factors and care providers, often constrained by patient protections such as HIPAA and technological investments, have inadequately leveraged the opportunity to do so.

## 2.4 Social Factors in Patient-Centric Risk Detection

The review above has largely addressed the technological and infrastructural limitations of clinical decision-making that pose challenges to the development of effective patient-centric risk management. However, for a holistic treatment of the patient, an equally important emphasis is needed on the integration of individual and organizational factors given the nature of healthcare decision-making (Carter et al. 2011; Banerjee and Ingate 2012). Consideration of social factors can lead to better management of factors that lead to pre- and postprocedure risk management, recidivism, and delivery of long-term care. Our examination of the literature has led to three key social factors that facilitate patient and patient-risk management. The first of these, presented in Table 2, centers on communication between key stakeholders in the healthcare setting. The remaining two relate to the need for, and utilization of, information systems to support the procedures.

A final theme in our review of social issues in healthcare relates to acceptance of health systems. Numerous studies have pointed to the relatively low adoption of healthcare systems (Garwood et al. 2018; Wickramasinghe and Gururajan 2016; Simon et al. 2015). Barriers often include concerns with diminished quality of patient encounter and relationships due to distractions from computer use (Sulaiman and Wickramasinghe 2014) and perceived inefficiencies and limitations of the systems (Simon et al. 2015). As such, physicians find workarounds to aid their decisions (Kent et al. 2015), leading to challenges with information integration and risk management (Wickramasinghe and Gururajan 2016).

Patient-centric design requires a sociotechnical approach to systems development and, hence, must inherently consider human, social, and organizational factors as

**Table 2** Social factors in patient-centric risk management

| Social factors | Description | Sample studies |
|---|---|---|
| Real-time communication/relationship and knowledge transition between care parties | Lack of communication facilitators to share information, knowledge, and decision between clinicians and between clinicians to patients of their families in order to support clinical decisions by understanding the treatment risk factors and anticipated outcomes in real time | Virshup et al. (1999), Kripalani et al. (2007), Elwyn et al. (1999), and Krones et al. (2008) |
| Sociotechnical issues | Lack of specific sociotechnical components in healthcare to make a successful implementation of a new technology to address data importance and timing issues | Baxter and Sommerville (2011), Schuele et al. (2015), Durst et al. (2013), and LeRouge and Wickramasinghe (2013) |
| Technology acceptance issues | Resistance between healthcare professionals to accept new compute-based technologies due to the importance of accuracy and privacy | Lawler et al. (2011), Wickramasinghe and Gururajan (2016), Kent et al. (2015), Garwood et al. (2018), Simon et al. (2015), and Sulaiman and Wickramasinghe (2014) |

well as technical factors in design of system artifacts (Baxter and Sommerville 2011; LeRouge and Wickramasinghe 2013). Thus, an integrated risk management solution should rely on the joint optimization of the technical and social subsystems, the lack of which can lead to suboptimal system performance and utilization (Schuele et al. 2015).

## 3 Eliciting a BI Framework: The Context of Congenital Heart Disease (CHD)

To design an appropriate BI framework, it is initially necessary to focus on a specific clinical domain because risk management is context dependent. We identified the case of congenital heart disease (CHD) in children as an appreciate domain. CHD is a common health problem affecting many children around the world (Marino et al. 2012). It provides a rich and complex context with risks that require a series of decisions to be taken by both clinicians and parents of patients at various stages during the treatment. CHD involves a multifaceted set of considerations that typically include immediate medical intervention, ongoing and increased risk of sudden death, exercise intolerance, neurological developmental, psychological problems, and long-term stresses on the family unit (Long et al. 2012). These considerations are important because of the far-reaching consequences of postsurgery outcomes and situations of mortality or morbidity. The decision process in CHD surgery can be divided into three broad phases: preoperative, operative, and postoperative.

In the preoperative phase, the surgeon, having received information about the patient and his/her medical condition, needs to make decisions relating the need for surgical procedures. Once this decision is made, but before surgery, the parents must consider predicted outcomes and decide whether to accept the surgeon's decision. Typically, by this time, parents have met numerous medical staff and specialists before they meet the cardiac surgeon. Thus, already in the first phase, preoperative, two key decisions must be made. Once parents and surgeons agree to proceed, in the operative phase, critical decisions pertaining to the unique situations that may arise during the surgery must be addressed. Finally, in the postoperative phase, decisions are made primarily at two levels: (a) strategies to ensure a sustained successful result for the patient during aftercare and beyond and (b) a record of lessons learnt for use by clinicians in future similar cases.

To further elaborate on the complexity of this decision framework across CHD surgery, we elaborate on the phases and the associated decision-making with reference to one of the common CHD classifications: hypoplastic left heart syndrome (HLHS). HLHS patients usually have three types of surgery during their treatment – Norwood, BCPS, and Fontan – at different ages and clinical condition groups. However, the Norwood surgery is much more complex and risky with a high rate of mortality and morbidity. As shown in Fig. 1, the current state typically

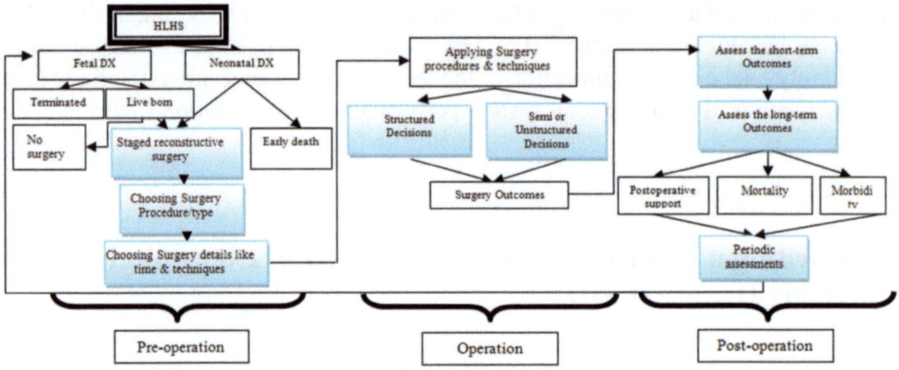


**Fig. 1** Flow diagram of key steps with CHD surgery in the case of HLHS to demonstrate the importance of IRD

involves parental decisions, and while these decisions are significant, they are often not recognized as key elements of the healthcare outcome. The surgical decision-making process, on the other hand, is clear in all steps. Thus, an IRD solution could be effectively employed in the highlighted steps in Fig. 1 for all three surgery phases and types to predict surgery outcomes.

## *3.1 Research Methodology and Key Outcomes*

Prior to developing the IRD framework discussed later in this article, we used the CHD context to identify initial parameters that would go into its design. Specifically, using CHD as an exemplar context enabled us to identify key clinical risks, critical decision milestones, typical surgical decisions and outcomes, and feasibility, benefits, and concerns with an integrative and intelligent risk solution. The research design is presented in Fig. 2. The methodology, discussed next, consists of three main phases – study design, qualitative data collection, and data analysis – and uses an approach that incorporates well-established qualitative data collection and analysis techniques (Boyatzis 1998; Glesne and Peshkin 1992; Yin 2003).

Qualitative data were collected from focus groups of surgeons, cardiologists, clinicians, and patient's parents; from hospital databases, reports, and documents; and from observations of routine clinical meetings at the study site. The research study was strictly confined to ethical research processes and was approved by the Human Research Ethics Committee (HREC) as a low-risk study. Content analysis was used to analyze semistructured discussions with expert focus group members as it provided a structured way of eliciting themes (Boyatzis 1998). Themes were extracted from transcripts, surveys, databases, reports, and clinical documents and were mapped to a priori themes identified in the literature. Key findings, organized around clinical and nonclinical perspectives, are summarized later in this section.

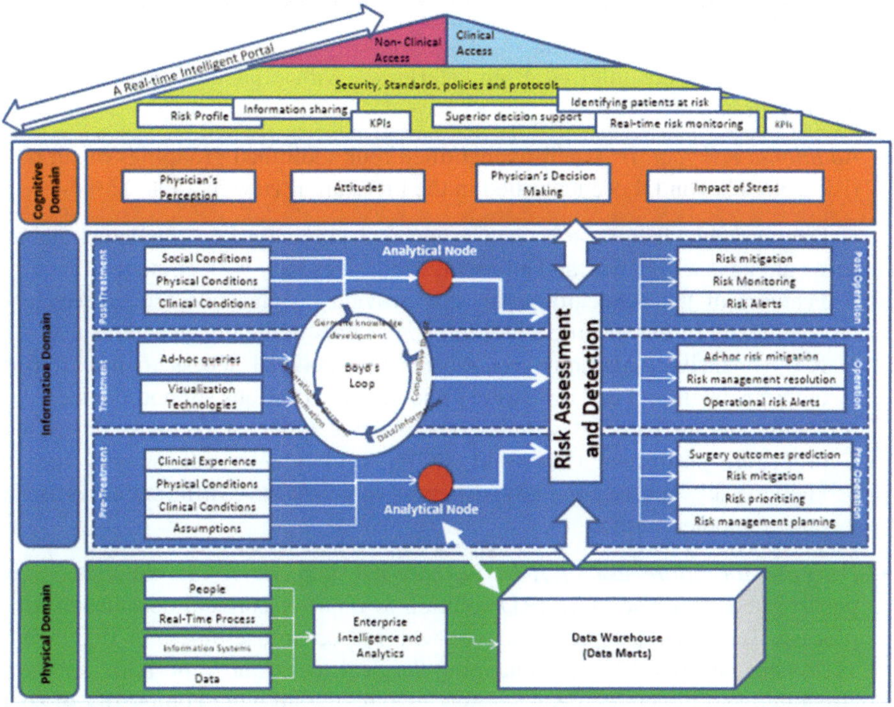


**Fig. 2** Intelligent risk detection: the HOUSE model

The CHD data set was constructed based on the patient procedures in all major surgeries and included patient diagnoses and operation, echo cardiology, and cardiac catheter reports. These databases, reports, and documents identified main components of the proposed IRD solution (such as risks from the clinical perspective, e.g., comorbidities and parent's perspectives such as quality of life and impact on siblings), which is discussed later. The variety of risk factors and their risk level in each surgical technique and phase were identified using these sources. Content from routine, weekly clinical meetings about complex clinical cases in the cardiology department were analyzed to identify issues faced by experts during treatment and to discuss decisions related to time and type of surgery. The meetings were useful in capturing critical milestones across three phases of surgery (preoperation, operation, and postoperation), confirming the potential benefits of applying risk detection to the CHD treatment process, and, most importantly, identifying barriers to successful application of intelligent risk detection to CHD treatment.

### 3.1.1 Clinical Perspective

Two key themes were inferred from interviews with clinicians and analysis of hospital records:

- *Additional risk factors*: Analysis identified four additional groupings of surgical risk factors beyond those identified in the literature review in Table 1.
  - Demographic factors, e.g., age group and ethnicity, i.e., premature babies would have a high risk and artery and vein sizes vary with ethnicity.
  - Features of the echocardiogram that provide insights into the severity and complexity of the specific condition, e.g., errors in the image and results.
  - ICU (intensive care unit) components that delayed access to life support including staff and bed availability and access to various support devices.
  - Nature of the surgery with reference to the types of recognized surgical procedure performed (Norwood, BCPS, or Fontan). Each surgery has its known risks, but, additionally, different surgeons have different expertise in one or another method.
- *Potential for future use of IRD*: Seventy-five percent of the clinicians were positive about the use of an IRD solution to improve surgical decision-making process. Ten percent expressed doubt with respect to usefulness of IRD, while 15% were negative. Rapidly changing surgical techniques, the need for more structured and integrated decision-making, and the growing complexity of such decisions were key drivers for clinical support. Resistance to IRD largely stemmed from concerns with automated risk analysis and decision-making as well as greater confidence in one's own experience and capabilities.

### 3.1.2 Nonclinical Perspective

The nonclinical perspective was gleaned from completed questionnaire responses of 35 parents (either mother or father or both) who signed appropriate consent form. The questionnaire sought parent's attitude toward surgical risk factors and their intention to be involved in the surgical decision-making process. Questions also pertained to their children's quality of life after surgery, the possible role of knowledge sharing, and the potential of IRD in improving their home care management and children's quality of life. The following key themes emerged from this analysis:

- *Surgical decisions, outcomes, and risks*: Most parents were satisfied that they had adequate understanding of surgical risks and outcomes, whether obtained directly from surgeon or from sources such as the internet. Parents who inadequately understood these factors often attributed it to lack of knowledge about clinical terms. Interestingly, all parents responded that they accepted undertaking surgery without consideration for risks and outcomes.

- *Perceived benefits of IRD*: Seventy percent of parents had a positive response to an IRD solution; 20% expressed a lack of understanding about such an approach, while 10% had a negative response, choosing to rely on physician expertise. Positive responses primarily related to being able to better predict outcomes and manage postoperative care.
- *Social impact on family life*: Sixty-five percent of parents anticipated negative impacts on their social lives after surgery in the form of depression, unpredictable health conditions, school/work absences, and increased hospital visits. About 26% expected positive outcomes, primarily because of improved patient health and family bonds. The remaining 9% prepared for mixed outcomes.

Tables 3 and 4 summarize key findings. In Table 4, expected findings (EF) were those already supported by the literature review. In developing the IRD framework, factors shown as EF are necessary for conceptualizing the initial framework. The significant findings (SF) were identified specifically through the data analysis described earlier based on interview and questionnaire data as well as observation and document reviewing. SFs substantially eliminated difficulties and barriers to developing and implementing the IRD solution in a hospital setting. Finally, emerging themes (ET) are categories that require further discussion in future studies.

**Table 3** Summary of key findings in each step of CHD data collection

| Data collection | Data sources | Summary of key findings |
|---|---|---|
| Interviews | Expert group of:<br>Cardiac surgeons<br>Cardiologists<br>ICU clinicians | Benefits of IRD for facilitating CHD decision-making<br>Barriers to applying IRD to CHD treatment<br>Facilitators of IRD adoption<br>Importance of IRD to the CHD decision-making process<br>Importance of analytic techniques such as data mining in the study context<br>Practical issues with design and development of IRD |
| Questionnaire | Patients' parents | Benefits of IRD to facilitate parental decision-making during the care and to assist with assessment of child's quality of life |
| Observation | Clinical meetings | Benefits and barriers to applying IRD to the CHD treatment process |
| Database/documents | Relevant clinical databases and reports | Importance of analytic techniques such as data mining in the study context |

**Table 4** Key findings from CHD analysis

| *Themes – clinical perspective* | *Category of findings*[a] |
|---|---|
| Clinicians' intention to use IRD | SF |
| Clinicians' recommendations | ET |
| Enables toward risk factor screening | EF |
| Contributing factors to assess patients' quality of life | ET |
| Clinicians' barriers to use IRD | ET |
| Clinicians' confident to address surgical risk factors | EF |
| *Themes – nonclinical perspective* | *Category of findings* |
| Social impact on the family life | SF |
| Patients' quality of life after the surgery | EF |
| Parents' understanding about surgery outcomes and risks | ET |
| Parents' intention to use IRD | SF |
| Parents' attitude toward surgery decision-making process | EF |

[a]*EF* Expected findings (already anticipated based on literature review), *SF* Significant findings (from data analysis in this article), *ET* Emerging themes (recommended for future studies)

## 4 A Framework for Intelligent Risk Detection: The HOUSE Model

Findings from the CHD domain were used to develop the IRD framework, called the HOUSE (Health Outcomes around Uncertainty, Stakeholders, and Efficacy) model, introduced in this section. A visual representation of the model and its underlying architecture is presented in Fig. 2. Along with the technical elements of this architecture, social factors discussed in the literature review are central to the effectiveness and relevancy of this model. As such, the model is built on three key principles that incorporate these social elements. The first is user-centered design (UCD), an approach that emphasizes design of a system based on understanding and supporting the interests, needs, and work practice of the intended users (Avison and Fitzgerald 2008). The second element is network-centric healthcare operations (NHCO) as an unhindered networking operation within and among all domains that govern all activities conducted in healthcare space (Von Lubitz and Wickramasinghe 2006a). NHCOs are based on free, multidirectional flow and exchange of information and utilizing all available means of (Information Computer/Communication Technology) IC2T to facilitate such operations (Von Lubitz and Wickramasinghe 2006a, b). The third element is the intelligence continuum (IC), defined as a collection of key tools, techniques, and processes used in the knowledge economy of today (Wickramasinghe and Schaffer 2010). These include but not limited to data mining, BI/BA and KM, and taken together, they represent a powerful mechanism for refining raw data materials stored in data marts and/or data warehouses by maximizing the value and utility of these data assets for any organization (Wickramasinghe and Schaffer 2006).

## 4.1 Underlying Principles: The Foundation of the HOUSE Model

**User-Centered Design (UCD)** UCD is an approach based on an explicit understanding of users, tasks, and environments (Avison and Fitzgerald 2008). UCD is relevant to the HOUSE model as it captures explicit interactions between users' tasks, environments, and multidisciplinary skills and perspectives. For instance, inclusion of clinician and patient's/parent's perspectives can shape their collective attitudes toward the potential of IRD as an enabler in healthcare. Similarly, clinician participation in design can enhance solution acceptance. As such, UCD is particularly important for addressing concerns about clinical knowledge sharing, critical thinking, and complex decision-making.

**Network-Centric Healthcare Operations (NHCO)** NHCO, based on the work of Boyd (1987), views the whole treatment process to identify critical data and germane knowledge and information to aid systematic analysis in complex and dynamic decision situations (Von Lubitz and Wickramasinghe 2006a). NHCO enables analysis of decision-making within, interaction with, and control of a dynamic and unpredictable environment. Successful application of NHCO enhances competitive advantage through information superiority and enables characterization of the operational environment such as on diversity of infrastructure, security, social structure, policies, and economy (Von Lubitz and Wickramasinghe 2006b). NHCO supports full, unhindered knowledge transfer and information sharing among clinicians and with patients/parents. As such, it can be effective in addressing risk detection in complex real-time decisions. The goal of NHCO is to quickly analyze complex scenarios and identify the best decision and strategy to effect decisions in real time.

**The Intelligence Continuum (IC)** The IC, developed by Wickramasinghe and Schaffer (2006), assists with understanding suboptimal healthcare processes and identifying, through the analysis of relevant data and information, prudent strategies to ensure optimal operations. The IC applies sociotechnical components as well as techniques to aid complex decisions. In complex clinical decisions, IC can maximize the value of data for process improvement through application of tools such as data mining and knowledge discovery (Wickramasinghe and Schaffer 2006). As suggested by CHD, real-time processing is a significant driver in developing intelligent solutions in the healthcare context. IC can support this by assimilating data, people, technology, process, and environment because in clinical contexts, particularly during surgical phases, receiving and updating all available information is important.

### *4.2 The OODA Loop: Clinical Decision-Making and Risk Management*

To capture the principles of UCD, NHCO, and IC effectively within the IRD framework, Boyd's decision-making loop, OODA (observe–orient–decide–act), is a central element of the IRD solution. OODA has been effective in promoting organizational creativity and initiative in competitive environments and harmonizing actions to achieve the organization's goals. The OODA loop provides a useful foundation for NHCO as it governs the process of information extraction, development of competitive thrust, and generation of germane knowledge (Von Lubitz and Wickramasinghe 2006a). The OODA loop incorporates both people- and technology-centric concepts given that it integrates three domains of network-centric analysis – information, physical, and cognitive (Boyd 1987). For instance, in the context of hip and knee arthroplasty, combining critical considerations about the patient with technical specifications of the prosthesis as an OODA loop analysis facilitates better pre-, peri-, and postoperative results. The dynamic aspects of OODA have led to its rapid adoption for managing unpredictable changes within operation theatres. For CHD, adopting OODA within the IRD framework provides a useful technique for integrating risk detection as it involves a rapid cycle of risk assessment, risk detection, decision-making, analytical reporting, and evaluation.

Risk detection involves observing data and information to determine situational surgical risk factors and forward them to the risk assessment process. In the OODA loop, such detection would typically be evident during the observe phase to identify hidden risk factors. The orient phase entails predicting surgical outcomes based on risk detection and present outcome predictions to patients' parents, cardiologists, surgeons, and intensive care clinicians. Based on these predictions, physicians and parents can make decisions regarding surgery. The final element entails taking action on the decision and developing analytical reports to evaluate actual and anticipated results. Considering this, the OODA loop is a central component to any intelligent tool designed to facilitate rapid decision-making.

As evidenced through interviews in the CHD case, often surgical decisions are made based on doctors' intuition and experience rather than on the knowledge-rich data hidden in the database. However, the interviews also point to numerous risk factors, cardiac conditions, and diagnoses across CHD spectrum of surgery, which make surgical decisions and the knowledge sharing process complex for all involved stakeholders. These factors emphasize the importance of discovering hidden patterns in data using data mining techniques such as decision trees and ANNs to inform the OODA loop regarding surgical risk detection and outcome prediction. The involvement of domain experts, however, is important to support extraction of specific domain-based knowledge and user requirements. Examples of data mining applications include:

- Identifying surgical risk factors through integrated examination of patient's medical profile, including data from medical reports, prior history, and environmental factors
- Identifying the significant relationships between surgical risk factors and their combined effects on outcomes related to mortality, morbidity, and success
- Given a patients' medical profile, predicting surgical outcomes related to mortality, morbidity, and success

The CHD case clearly suggests that there is no decision sharing between physicians and patients' parents as there is no alternative for parents other than surgery. As such, the OODA loop does not apply to parental engagement. Rather, a knowledge sharing, as opposed to a decision-sharing process, is recommended particularly postsurgery.

## *4.3 The Physical Domain*

To accommodate the underlying principles of IC, NHCO, and UCD, the HOUSE model is founded on three domains – physical, information, and cognitive. The physical domain is the collection of all physical assets (platforms), such as data management facilities, hospitals or clinics facilities and activities, administrative entities, as well as all other physical subcomponents (Von Lubitz and Wickramasinghe 2006c). People assets include all clinical and nonclinical actors such as patients' parents, cardiologists, surgeons, intensive care clinicians, nurses, and administrators that have clinical or administrative inputs to the current information systems available in the hospital. Real-time processes, identified by experts in the CHD analysis, encompass the operational and functional processes that support successful conduct of healthcare work. Information systems include IT infrastructure – hardware, software, procedures, and policies, all of which can define the HOUSE model to specific contexts.

Data, commonly viewed by physicians in our study as the most important element of an IRD, includes both clinical and administrative data. The rapid shift from paper-based to electronic health records since 2001 presents profound opportunities for integrated, validated, and holistic presentation of health data. Recently, there has been a noteworthy upswing in activities related to the adoption of personal health record (PHR) systems for patients. This trend continues through the prevalence of mobile applications, RFID chips, and medical sensors. These personal records, though external to the hospital environment, are becoming a crucial element of any care provision. As such, any IRD framework must consider the inclusion of both external and internal sources of data.

The true benefit of any risk detection environment lies in the integration of disparate data. External reports, risk sources, government regulations and guidelines, and social media can be transformative for risk detection. These diverse data necessitate data warehouses or integrated data marts for holistic risk management through discovery of knowledge and patterns in multidimensional relationships. As such, enterprise intelligence and analytical tools are essential for deriving value from these massive data. Such tools range from statistical and predictive models to data mining and artificial intelligence that can identify individual and organizational risks for particular patients, procedures, or treatments to improvement in organizational operations (Koh and Tan 2011). The true benefit of such tools lies in being able to inform decisions through integration of risks identified within these extremes. For this, integration with the other two domain layers, information and cognitive, is essential.

## *4.4 The Information Domain*

The information domain hosts all elements required for generation, storage, manipulation, and transformation and sharing as knowledge in all its forms (Von Lubitz and Wickramasinghe 2006c). All communications about the state of healthcare take place through interactions within this domain. As such, this layer is critical for successful interpretation and utilization of data and information extracted from the physical domain. To this end, the OODA loop is central to this domain to support execution of real-time solutions that produce and share knowledge and inform the complex and integrated decision-making spanning pre- to posttreatment phases.

In the pretreatment phase, analytical nodes are central to risk detection and assessment. These nodes should be comprised of techniques that support classification and categorization such as ANNs, decision trees, and case-based reasoning. In any IRD, analytical nodes should have risk monitoring, risk re-assessment, and risk alerts as the main outputs. The CHD case analysis suggests three critical inputs to these nodes: clinical experience, physical conditions of the patient, and the clinical environment. The OODA loop is useful in case-based risk ranking so that certain risks should be flagged as more critical to a specific case than for others. As such, OODA plays a critical role in risk prioritization and planning.

In the treatment phase, the emphasis shifts to observation, document review, and ad-hoc querying and decision-making. Interviews with CHD physicians suggested that the use of any computer-based solutions during the surgical and treatment phases would fall to disuse due to interference with the surgical procedure. As such, during the treatment phase, the HOUSE model shifts control to the physician who can drive decision-making and actions through queries initiated by the OODA loop as opposed to an analytical node. In this phase, the physicians and clinical staff will, most effectively, be supported with technologies that enable visualization of various aspects of the treatment environment. Care, however, is necessary for avoiding information overload as, under conditions of stress, physicians can experience overload and may disuse or misapply relevant information. The output of these actionable strategies should, however, still be subjected to risk assessment to identify additional risk factors stemming from real-time and ad-hoc physician decisions.

The CHD analysis indicates that, in the posttreatment phase, three functional factors come into play: social conditions such as quality of life and impact on other family members, physical components including likelihood of life-long physical limitations to the patient after surgery, and clinical conditions, such as unpleasant comorbidities, that might result. These factors must be applied to the risk detection process via an analytical node. The analytical node in this phase should function similarly to those in the pretreatment phase, i.e., this node should integrate the three factors to support classification and categorization of posttreatment risks.

## 4.5 The Cognitive Domain

The cognitive domain contains all human factors that affect the surgical procedures, including experience, training, personal engagement, or even intuition of individuals involved in relevant activities (Von Lubitz and Wickramasinghe 2006b). The cognitive domain also integrates social attributes such as attitudes, behaviors, and interactions. Physician's perceptions of treatment situation are a critical factor in informing judgment and mitigating risks of care provisioning. For instance, physician perceptions influence incidence reporting (Schectman and Plews-Ogan 2006) and quality of life (Rodriguez et al. 2003). Similarly, physician's attitudes towards certain disease states, patients, and treatments can shape risk response. For instance, care provider's attitudes toward childhood obesity influence recommendations for training and behavioral management strategies (Story et al. 2002). As revealed in CHD interviews, physicians' experience and environment are critical factors that shape such attitudes and perceptions and have the potential of influencing riskiness of care provision.

## 4.6 A Real-Time Intelligent Portal

Analysis of CHD interviews and observations indicated that the IRD solutions should be accessible in real time for both clinicians and patients with different level of access and security and a strong emphasis on stakeholder communications. It is important to note the organizational factors at play here as opposed to data-, people-, and technology-driven focus for the other layers. Many of these factors essentially constrain or enable, most often the latter, any IRD solution. For instance, government regulations around patient privacy will tend to limit and pull clinical and nonclinical access to patient risk and treatment data. In contrast, a desire to improve decision-making and deliver effectively on KPIs such as cost cutting, reducing patient recidivism, and increasing patient satisfaction through effective communication will push greater transparency. For any healthcare organization, these top down drivers require careful elaboration and critical consideration prior to design of any IRD solution.

# 5 Discussion and Implications

This article has described an exploratory research with the intent of developing a framework that combines a real-time IRD solution with decision support in a healthcare context. The proposed HOUSE model has an advantage in its continuous nature. By comparing anticipated results and actual outcomes and conducting risk audits, the important task of amending risk factors to improve future predictions is attainable. Another important feature of the proposed HOUSE model is the integration of the three IT solutions – risk detection, knowledge discovery, and CDSS – to mitigate the complex issue of defining and assessing patient outcomes. This framework emphasizes the importance of:

- Dynamic techniques that support management of massive and disparate data as well as risk discovery and management using such data. This includes techniques for identifying key clinical and surgical risks, decision milestones, and participating actors.
- Embracing knowledge sharing among clinicians, staff, and patients to facilitate decision-making and risk management process.
- Clinicians' involvement in systems development.
- High demand of outcome predictions to improve decision efficiency.
- Mindfulness to principles of UCD, NHCO, and IC in developing real-time solutions that emphasize the collaboration necessary between information receivers and generators.

## *5.1 Implications for Research*

This article cannot claim to have determined all risks versus benefit of developing an IRD solution. This is because it is currently challenging to identify long-term risks and rare occurrences using randomized controlled trials. These risks and occurrences can be identified using methods specifically developed for large, observational, and longitudinal data sets. This study, therefore, only presents components of the HOUSE model to as an initial IRD framework. This, in itself, presents opportunities for future research on assessing how IRD solutions could impact the bottom line for patient care and healthcare provisioning. Essentially, numerous opportunities exist for assessment of the value of IRD and contrast with expected returns. To this end, we suggest the examination of the following propositions to justify IRD investments

### 5.1.1 Outcome, Opportunities for IRD Systems in Healthcare

In this section, we develop propositions related to the assessment of IRD in its ability to deliver on health and patient care outcomes. Specifically, we focus on real-time and shared decision-making, improved knowledge sharing and communications, mitigation of posttreatment risks, standardization through routinization of workflow, and personalized care.

*Improved Patient Care and Decision-Making* Health providers are routinely being tasked with active decision-making that requires them to access, evaluate, and incorporate data and evidence into their clinical decision-making (Thompson et al. 2004). This demands sources of information that reflect holistic patient status in real time. Physician reliance on information systems in such real-time situations is often limited due to concerns with errors introduced while entering information and those from retrieving and communication of this information (Ash et al. 2003). The information architecture proposed for the IRD overcomes these limitations because data capture and processing as well as information dissemination will primarily be automated. In facts, risk factors that may be suppressed in the complexity of the healthcare space can come to the forefront through a viable integration of sociotechnical platforms. As such, the IRD framework enhances opportunities for real-time decision-making to improve health outcomes. To this end, we propose:

**Proposition 1** An integrated IRD solution can improve patient care and outcomes through real-time decision-making.

*Shared Decision-Making* User-centric design, a crucial principle underlying IRD, supports shared decision-making between patient and healthcare providers such that decisions are responsive to patient needs while being medically sound. Such shared decisions improve patient outcomes (DeMeester et al. 2016). However, few healthcare providers consistently involve patients in shared decision-making and even fewer accommodate clinical decisions to patient preferences (Couet et al. 2015). To a large extent, it has been hindered by the lack of integrated health information systems and decentralized healthcare systems (Ahlfeldt et al. 2016) and because information production and dissemination has largely been focused on meeting educational needs of physicians (Agoritsas et al. 2015). These issues can be overcome with an architecture that supports integrated processes and technologies, factors that are addressed in the design of the proposed HOUSE model. To this end, we expect that:

**Proposition 2** IRD solutions can improve patient care thorough shared decision-making.

*Mitigation of Posttreatment Risks* Early detection of risks, as aided by the HOUSE model, can enable clinicians to provide patients with appropriate strategies to mitigate these risks. For example, in the context of CHD patients when the patient returns home, it is necessary to know the home environment including flora and fauna as these provide airborne allergens that in turn can cause major respiratory complications for CHD patients postsurgery. Armed with this knowledge, the care team can determine appropriate strategies for home monitoring and care. On surface, this might appear to be a simple issue but its oversight can lead to life-threatening complications. Missed criteria are a highly cited cause for unplanned readmissions (Porter and Teisberg 2009).

**Proposition 3** IRD solution can mitigate posttreatment outcomes through improved management of posttreatment risks.

*Multidisciplinary Knowledge Sharing* The adoption of centralized health information systems can enhance multidisciplinary care for patients. A wide variety of healthcare providers adopt and adapt the use of such systems to their own practices while orienting use to those of other care providers in the team (Oborn et al. 2011). In healthcare today, an area where the role of knowledge sharing is vital is in cancer care. Especially, as cancer is now considered a chronic disease and more people survive cancer today than ever before, on-going management of cancer patient survival depends heavily on coordination and collaboration of a multidisciplinary team including oncologists, PCP, surgeon, radiation, and nurses to name just a few key members of typical care teams (Sullivan et al. 2011). Identifying, managing, and mitigating risks in this sphere is a key part of successful cancer survivorship, and risks include toxicity levels of chemo-therapy drugs, suppressed immune systems, and thus a higher propensity for infection and likely falls risks (Clemer and Spuhler 1998).

**Proposition 4** IRD can promote knowledge sharing between stakeholders to lead to improved patient risk management and decision-making.

*Stakeholder Communications* Research has consistently continued to support the indisputable link between patient–physician communication and improved patient care and self-management (Heisler et al. 2002). Furthermore, physicians use active listening techniques and respond to patients' emotional needs, lower risks, and thereby losses through malpractice suits (Virshup et al. 1999). Similarly, communications between physicians and staff, e.g., nurses, directly impacts patient care. For instance, found that while nurses and obstetricians agree on common goal of healthy child and mother, they disagreed on how such care was to be accomplished and communicated minimally when both patients were faring well. In such cases, it may be common to overlook transference of critical patient knowledge. The information and cognitive layers of the HOUSE model, we suggest, can equip an IRD solution to facilitate the sharing of knowledge and enhance communication between patient, physician, and staff. To this end, the following are proposed:

**Proposition 5** IRD solutions can improve communications among clinicians and healthcare providers to improve treatment decision-making process.

**Proposition 6** IRD solutions can improve communications among patients and clinical staff and thereby improve treatment decision-making process.

*Routinizing Workflow While Personalizing Care* Routinizing physician workflow is a way to attain "meaningful use" of HIS (Agarwal et al. 2010). Such routinization can enhance physician efficiency and minimize risks. Routines allow hospitals to deliver quality in healthcare through adoption of consistent standard operating procedures. Routinized workflow management improves patient care and organizational performance. Uniform use of HIS, however, runs the risk of applying standardized care and overlooking the need to customize care to patient needs. Personalized care that relies on knowledge about the patient's unique medical history and physiological makeup to tailor care can provide cheaper and more effective treatment (Fichman et al. 2011). The combination of multiple data sources and intelligent analytics techniques as embedded within the HOUSE model can deliver the ability to balance the two needs, i.e., routinization of workflow while providing customized care through the use of intelligent routines embedded within the workflow. As such, we propose that:

**Proposition 7a** IRD can result in improved organizational workflow to support physician and organizational performance.

**Proposition 7b** The intelligent routines in IRD can support personalized patient care while gaining from organizational workflow efficiencies.

### 5.1.2 Examining Critical Factors in the Success of IRD

Healthcare IT adoption and success has been rife with issues such as resistance to adoption by physicians (Simon et al. 2015), healthcare providers (Porter and Teisberg 2009; Kent et al. 2015), and patients (LeRouge and Wickramasinghe 2013) among other stakeholders. Concerns have also been raised about challenges associated with meaningful integration of isolated health systems (Kaplan and Porter 2011) and agility in IT investments (Sulaiman and Wickramasinghe 2014). Considering these concerns, there is a need to understand and enable factors that will be critical to successful use of IRD systems in healthcare. In light of the newness of this domain, we extrapolate factors from existing HIS literature and elaborate on how these might apply to IRD implementations in the future.

*Integration of Diverse and Disparate Systems* For decades, research has attributed challenges related to integration of isolated systems to the prevalence of small providers in healthcare who have been effective in meeting individualized needs of health areas as opposed to the entire organization or workflow (Kim and Michelman 1990). With the emergence of health enterprise systems, that barrier is mitigated, but new barriers related to integration of disparate data sources, not always internal to the organization, are emerging. Integration of internal systems with each other and with external data sources may also be especially relevant as new opportunities for personal health record and national health information networks emerge (Ozdemir et al. 2011). As such, integrated systems are central to successful implementation of IRD systems, particularly in response to the growing need to understand challenges and needs of individual patients within the space of personalized medicine. To this end, we propose:

**Proposition 8** Successful implementation and use of IRD depend on meaningful integration of disparate health information systems and underlying processes.

*Minimizing IRD Avoidance and Promoting Use* User-centric design of IRD systems presupposes the engagement of key stakeholders that deliver relevant data for such systems as well as those that adopt related information to support patient care and organizational performance. Yet, literature is replete with narratives of avoidance behaviors, despite the need and opportunity to use HIS. These avoidance behaviors are directly linked to negative impact on patient care (Kane and Labianca 2011) and increased risks related to patient readmission (Bardhan et al. 2014). In contrast, physician's use of HIS contributes to patient satisfaction with care quality (Venkatesh et al. 2011). To this end, several mechanisms have been proposed for overcoming these limitations of adoption challenges. Health organizations are encouraged to consider conscious adoption of brokering strategies that allow HIS avoiders to have someone else interact with the system or clustering users who have similar usage patterns (Kane and Labianca 2011). However, in those cases, significant communication processes must be in place to share both patient and process information. The central principle of NHCO assumes a free, multidirectional flow and exchange of information. Accordingly, social and technological mechanisms for facilitating communications to overcome avoidance will be central to IRD success. To this end, we propose:

**Proposition 9a** Successful implementation of IRD depends on minimizing avoidance among health professionals and developing mechanisms that promote use.

**Proposition 9b** Successful implementation of IRD depends on developing sociotechnical communication mechanisms that promote user or overcome limitations of avoidance behaviors.

*Stakeholder Readiness and Information Sharing* Individuals protect their electronic health records and apply different rules to determine whether or not to disclose their information depending on the risks relevant to their healthcare context (Anderson and Agarwal 2011). Furthermore, emotion plays an important role in disclosure as individuals with negative emotions about their health are more willing to disclose their private health records (ibid). Studies have suggested that financial or social incentives might encourage individuals to willingly participate in providing information necessary for a holistic IRD implementation (ibid). We take this idea further and suggest that the opportunity for improved knowledge sharing with health professionals and improved communications (as discussed in Propositions 4, 5, and 6) could be effective in increasing individual participation in sharing healthcare data.

**Proposition 10** Successful implementation of IRD requires improved information sharing among stakeholders.

*Adaptation for Disparate Domains* Although the HOUSE model describes elements of an IRD solution for healthcare, many issues regarding its implementation into specific healthcare contexts such as stroke treatment, brain surgery, and less severe conditions must be determined through domain-specific research. Our experience with both the CHD and an alternate domain of total hip arthroplasty

(THA) and total knee arthroplasty (TKA) clearly suggests that while the HOUSE model holds generally true, some modifications will be necessary to fully support each unique domain. Therefore, the next phase for this research is prototyping and simulation of the solution to trial the model in a selected clinical environment. To this end, we suggest the following:

**Proposition 11** The IRD will require modification for less severe treatments to better support the treatment decision process.

# 6 Conclusion

This study has contributed to theory and the body of knowledge by proving the importance of IC, NHCO, and UCD in developing a feasible IRD solution in the healthcare context. The study also contributes to practice at many levels. First, it provides insights for healthcare delivery wherein there as always an informational asymmetry that provides a barrier to clinicians in ever having a complete picture and understanding of all the critical underlying issues leading to the malady from which the patient is suffering. Better data that is integrated with tacit and explicit knowledge serves to significantly reduce this information asymmetry so that the clinician can make the best possible decision in the dynamic and complex context with which he/she is confronted. Further, given the general trend in healthcare delivery in the USA and globally to move from a focus on volume to value, data now becomes key. Specifically, data on predicted risks and risk factors such as BMI, HA1C, and/or smoking are critical in the assessments of value (ibid); simply stated better data directly leads to better outcomes, and the contributions of this paper address how to ensure systematic strategies so that better data is in fact generated rather than generating voluminous noise.

This research demonstrates that the selection of the risk detection, prediction by data mining tools as one of the data science techniques, and then decision support are important for decision-making in the complex treatments. IRD, in practice, can also be used as a training tool for nurses and medical students to detect treatment risk factors and their impact on outcomes. Moreover, it can also provide decision support to assist doctors in making better clinical decisions or, at least, provide a second opinion. Furthermore, IRD can be used as a knowledge sharing and information transferring mechanism between clinicians, clinicians and patients or their families, and between patients. The study makes a contribution to practice and theory, namely, confirming a role for data mining models and data science technologies in healthcare contexts. Given the move in healthcare currently in the USA to move to bundled payments and values-based care risk assessment, management and mitigation are now more critical than ever before, and without IRD, it will be most challenging to develop a comprehensive risk profile of patients and then decide on appropriate management and mitigation strategies, hence making this research of central importance. In closing, the study contends that real-time IRD is critical for many areas in healthcare where complex and high-risk decisions must be made and, in doing so, calls for further research.

# Appendix A

| Name of study | Disease state | Application of KD |
|---|---|---|
| Kaur and Wasan (2006) | Diabetes | Rule-based, decision tree, and artificial neural network were used to classify diabetic patients |
| Meyer et al. (2014) | Type 2 diabetes mellitus | Data mining classification techniques were applied to improve treatment of patients with type 2 diabetes through better prediction and elimination of treatment failures |
| Chandola et al. (2013) | Healthcare fraud | Three case studies using big data and data mining are developed for identifying fraudulent healthcare providers using insurance claims data. Hadoop/Hive data platforms were used for text mining, network analysis, and analysis of claims sequences |
| Marschollek et al. (2012) | Identification of patients with risk of falling | Data mining techniques were used to detect patients with high risk of falling. Techniques classified patients as fallers or nonfallers. Prediction accuracies ranged from 55% to 68% |
| Holzinger et al. (2012) | Strokes | Use of data visualization techniques for mapping brain activity patterns and improved sense-making of brain functions |
| Patrick et al. (2011) | Extraction of clinical data | Machine-learning and rule-based system is used to extract clinical data from text and natural language notes. A pipeline system was developed to extract clinical data |
| Batal and Hauskrecht (2010) | Antibody test orders | Using data mining, authors elicit minimally predictive rules (MPRs) that are applied to predict heparin–platelet factor 4 (HPF4) antibody test orders from electronic health records. MPR-based classification models were comparable with those obtained from using a complete set of association rules |
| Srinivas et al. (2010) | Cardiovascular disease | Data mining techniques were used to predict the likelihood of patients experiencing heart disease. Medical profiles such as age, gender, blood pressure, and blood sugar were used |
| Peng et al. (2006) | Breast cancer detection | Use of genetic algorithms to search for bright spots in mammograms improved detection |

(continued)

(continued)

| Name of study | Disease state | Application of KD |
|---|---|---|
| Wright and Sittig (2006) | Physician order entry | Data mining techniques were used to learn from past ordering behaviors. Compared to manual ordering, data mining increased order efficiency, better accounted for local preferences, and enabled improved integration into existing clinical systems |
| Gago et al. (2005) | Organ failure and mortality assessment | KD and agent-based technologies were used to predict organ failure in hospital ICU |
| Ceglowski et al. (2005) | Emergency department data | Mining of emergency data resulted in discovery of definitive treatment pathways that fully modelled patient treatment. Analysis provided insights into emergency department workload and types of procedures carried out |
| Kraft et al. (2003) | Spinal cord injuries | Nursing diagnosis and neural networks were used to predict length of stay for patients with spinal cord injuries. Algorithms correctly predicted 77% of the stays |
| | Cardiology | Predictive analytics model was effective in predicting readmission of patients diagnosed with congestive heart failure |
| Abidi and Goh (1998) | Infectious diseases | Neural networks were used to forecast bacterial infections using past data |

# References


Abidi, S. S. R., & Goh, A. (1998). Applying knowledge discovery to predict infectious disease epidemics. In *PRICAI'98: Topics in artificial intelligence* (pp. 170–181). Berlin/Heidelberg: Springer.

Agarwal, R., Gao, G., DesRoches, C., & Jha, A. K. (2010). Research commentary—The digital transformation of healthcare: Current status and the road ahead. *Information Systems Research, 21*(4), 796–809.

Agoritsas, T., Heen, A. F., Brandt, L., Alonso-Coello, P., Kristiansen, A., Akl, E. A., Neumann, I., Tikkinen, K. A., Van Der Weijden, T., Elwyn, G., & Montori, V. M. (2015). Decision aids that really promote shared decision making: The pace quickens. *British Medical Journal, 350*, 7624.

Åhlfeldt, R. M., Persson, A., Rexhepi, H., & Wåhlander, K. (2016). Supporting active patient and health care collaboration: A prototype for future health care information systems. *Health Informatics Journal, 22*(4), 839–853.

Anderson, C. L., & Agarwal, R. (2011). The digitization of healthcare: Boundary risks, emotion, and consumer willingness to disclose personal health information. *Information Systems Research, 22*(3), 469–490.

Anderson, O., Brodie, A., Vincent, C. A., & Hanna, G. B. (2012). A systematic proactive risk assessment of hazards in surgical wards: A quantitative study. *Annals of Surgery, 255*(6), 1086–1092.

Ash, J. S., Stavri, P. Z., & Kuperman, G. J. (2003). A consensus statement on considerations for a successful CPOE implementation. *Journal of the American Medical Informatics Association, 10*(3), 229–234.

Avison, D. E., & Fitzgerald, G. (2008). *Information systems development: Methodologies, techniques and tools* (4th ed.). Berkshire: McGraw-Hill.

Banerjee, A. K., & Ingate, S. (2012). Web-based patient-reported outcomes in drug safety and risk management. *Drug Safety, 35*(6), 437–446.

Bardhan, I., Oh, J. H., Zheng, Z., & Kirksey, K. (2014). Predictive analytics for readmission of patients with congestive heart failure. *Information Systems Research, 26*(1), 19–39.

Barrett, J., Gifford, C., Morey, J., Risser, D., & Salisbury, M. (2001). Enhancing patient safety through teamwork training. *Journal of Healthcare Risk Management, 21*(4), 61–69.

Batal, I., & Hauskrecht, M. (2010). Mining clinical data using minimal predictive rules. In *American Medical Informatics Association annual symposium*, pp. 31–35.

Baxter, G., & Sommerville, I. (2011). Socio-technical systems: From design methods to systems engineering. *Interacting with Computers, 23*(1), 4–17.

Boyatzis, R. E. (1998). *Transforming qualitative information: Thematic analysis and code development*. Thousand Oaks: Sage.

Boyd, 1987. available at https://danford.net/boyd/

Carter, E. L., Nunlee-Bland, G., & Callender, C. (2011). A patient-centric, provider-assisted diabetes telehealth self-management intervention for urban minorities. *Perspectives in Health Information Management/AHIMA, American Health Information Management Association, 8*(Winter), 1b.

Ceglowski, A., Churilov, L., & Wassertheil, J. (2005, January). Knowledge discovery through mining emergency department data. In *System sciences, 2005. HICSS'05. Proceedings of the 38th annual Hawaii international conference* (pp. 142c–142c).

Cerrito, P. B. (2011). *Data mining to determine risk in medical decisions*. Amsterdam: IOS Press.

Cerrito, P., & Cerrito, J. C. (2006). Data and text mining the electronic medical record to improve care and to lower costs. In *Proceedings of SUGI* (Vol. 31, pp. 26–29). Retrieved from https://pdfs.semanticscholar.org/a4e0/0a006becd0df35163c1d8a4b612dcc7cea07.pdf

Chandola, V., Sukumar, S. R., & Schryver, J. C. (2013, August). Knowledge discovery from massive healthcare claims data. In *Proceedings of the 19th ACM SIGKDD international conference on Knowledge discovery and data mining* (pp. 1312–1320).

Chen, H., Chiang, R. H., & Storey, V. C. (2012). Business intelligence and analytics: From big data to big impact. *MIS Quarterly, 36*(4), 1165–1188.

Cios, K. J., Pedrycz, W., Swiniarski, R. W., & Kurgan, R. A. (2007). *Date mining and knowledge discovery approach*. New York: Springer.

Clemer, T., & Spuhler, V. (1998). Developing and gaining acceptance for patient care protocols. *New Horizon, 6*(1), 12–19.

Cleveringa, F. G., Gorter, K. J., Van Den Donk, M., & Rutten, G. E. (2008). Combined task delegation, computerized decision support, and feedback improve cardiovascular risk for type 2 diabetic patients a cluster randomized trial in primary care. *Diabetes Care, 31*(12), 2273–2275.

Couët, N., Desroches, S., Robitaille, H., Vaillancourt, H., Leblanc, A., Turcotte, S., et al. (2015). Assessments of the extent to which health-care providers involve patients in decision making: A systematic review of studies using the OPTION instrument. *Health Expectations, 18*(4), 542–561.

Davenport, T. H. (2006). Competing on analytics. *Harvard Business Review, 84*(1), 98.

De Backere, F., De Turck, F., Colpaert, K., & Decruyenaere, J. (2012). Advanced pervasive clinical decision support for the intensive care unit. In *6th International conference on pervasive computing technologies for healthcare 2012.*

DeMeester, R. H., Lopez, F. Y., Moore, J. E., Cook, S. C., & Chin, M. H. (2016). A model of organizational context and shared decision making: Application to LGBT racial and ethnic minority patients. *Journal of General Internal Medicine, 31*(6), 651–662.

Durst, C., Viol, J., & Wickramasinghe, N. (2013). Online social networks, social capital and health related behaviours: a state-of-the-art analysis CAIS(Communications of the AIS), *32*, Retrieved from http://aisel.aisnet.org/cais/vol32/iss1/5

Eichner, J. S., & Das, M. (2010). *Challenges and barriers to clinical decision support (CDS) design and implementation experienced in the Agency for Healthcare Research and Quality CDS demonstrations*. Rockville: Agency for Healthcare Research and Quality.

Elbaum, S., Rothermel, G., Karre, S., & Fisher, M. (2005). Leveraging user-session data to support web application testing. *IEEE Transactions on Software Engineering, 31*(3), 187–202.

Elwyn, G., Edwards, A., & Kinnersley, P. (1999). Shared decision-making in primary care: The neglected second half of the consultation. *The British Journal of General Practice, 49*(443), 477–482.

Fichman, R. G., Kohli, R., & Krishnan, R. (2011). The role of information systems in healthcare: Current trends and future research. *Information Systems Research, 22*(3), 419–428.

Gago, P., et al. (2005). INTCare: A knowledge discovery based intelligent decision support system for intensive care medicine. *Journal of Decision Systems, 14*(3), 241–259.

Garwood, V., Claydon-Platt, D., Wickramasinghe, N., Mackay, J. R., & Smart, P. J. (2018). Smart phones and surgeons: Privacy and legal issues in the Australian healthcare. *International Journal of Networks and Virtual Organisations., 19*(1), 21.

Glesne, C., & Peshkin, A. (1992). *Becoming qualitative researchers: An introduction*. New York: Longman White Plains.

Greenland, P. (2012). Should the resting electrocardiogram be ordered as a routine risk assessment test in healthy asymptomatic adults? *JAMA, 307*(14), 1530–1531.

Grundy, S., Pasternak, R., Greenland, P., Smith, S., Jr., & Fuster, V. (1999). Assessment of cardiovascular risk by use of multiple-risk-factor assessment equations: A statement for healthcare professionals from the American Heart Association and the American College of Cardiology. *Journal of the American College of Cardiology, 34*(4), 1348–1359.

Guikema, S. D., & Quiring, S. M. (2012). Hybrid data mining-regression for infrastructure risk assessment based on zero-inflated data. *Reliability Engineering & System Safety, 99*, 178–182.

Han, J., Kamber, M., & Pei, J. (2006). *Data mining: Concepts and techniques*. San Francisco: Morgan Kaufmann.

Heisler, M., Bouknight, R. R., Hayward, R. A., Smith, D. M., & Kerr, E. A. (2002). The relative importance of physical communication, participatory decision making, and patient understanding in diabetes self-management. *Journal of General Internal Medicine, 17*(4), 1525–1497.

Holzinger, A., et al. (2012). Computational sensemaking on examples of knowledge discovery from neuroscience data: Towards enhancing stroke rehabilitation. In *Information technology in bio-and medical informatics* (pp. 166–168). Berlin/Heidelberg: Springer.

Jeffery, R., Iserman, E., & Haynes, R. (2013). Can computerized clinical decision support systems improve diabetes management? A systematic review and meta-analysis. *Diabetic Medicine, 30*(6), 739–745.

Kane, G. C., & Labianca, G. (2011). Is avoidance in health-care groups: A multilevel investigation. *Information Systems Research, 22*(3), 504–522.

Kaplan, R., & Porter, E. (2011). How to solve the cost crisis in health care. In *Harvard business review*. Boston: Harvard Business School Publishing.

Karaolis, M. A., Moutiris, J. A., Hadjipanayi, D., & Pattichis, C. S. (2010). Assessment of the risk factors of coronary heart events based on data mining with decision trees. *IEEE Transactions on Information Technology in Biomedicine, 14*(3), 559–566.

Kaur, H., & Wasan, S. K. (2006). Empirical study on applications of data mining techniques in healthcare. *Journal of Computer Science, 2*(2), 194–200.

Kelley, H., Chiasson, M., Downey, A., & Pacaud, D. (2011). The clinical impact of eHealth on the self-management of diabetes: A double adoption perspective. *Journal of the Association of Information Systems, 12*(Special issue), 208–234.

Kent, B., Redley, B., Wickramasinghe, N., Nguyen, L., Taylor, N. J., Moghimi, H., & Botti, M. (2015). Exploring nurses' reactions to a novel technology to support acute healthcare delivery. *Journal of Clinical Nursing*. https://doi.org/10.1111/jocn.12881.

Kersten, R. F., Stevens, M., van Raay, J. J., Bulstra, S. K., & van den Akker-Scheek, I. (2012). Habitual physical activity after total knee replacement. *Physical Therapy, 92*(9), 1109–1116.

Kim, K. K., & Michelman, J. E. (1990). An examination of factors for the strategic use of information systems in the healthcare industry. *MIS Quarterly, 14*, 201–215.

Kim, G., El Rouby, S., Thompson, J., Gupta, A., Williams, J., & Jobes, D. (2010). Monitoring unfractionated heparin in pediatric patients with congenital heart disease having cardiac catheterization or cardiac surgery. *Journal of Thrombosis and Thrombolysis, 29*(4), 429–436.

Koh, H. C., & Tan, G. (2011). Data mining applications in healthcare. *Journal of Healthcare Information Management, 19*(2), 65.

Kraft, M. R., Desouza, K. C., & Androwich, I. (2003, January). Data mining in healthcare information systems: Case study of a veterans' administration spinal cord injury population. In *System sciences, 2003. Proceedings of the 36th annual Hawaii international conference* (pp. 9–13).

Kripalani, S., LeFevre, F., Phillips, C. O., Williams, M. V., Basaviah, P., & Baker, D. W. (2007). Deficits in communication and information transfer between hospital-based and primary care physicians. Implications for patient safety and continuity of care. *JAMA., 297*(8), 831–841. https://doi.org/10.1001/jama.297.8.831.

Krones, T., Keller, H., Sönnichsen, A., Sadowski, E. M., Baum, E., Wegscheider, K., et al. (2008). Absolute cardiovascular disease risk and shared decision making in primary care: A randomized controlled trial. *The Annals of Family Medicine, 6*(3), 218–227.

Kronick, R., Gilmer, T., Dreyfus, T., & Ganiats, T. (2002). CDPS-Medicare: The chronic illness and disability payment system modified to predict expenditures for Medicare beneficiaries. *Final Report to CMS.*

Kuhn, K., Wurst, S., Bott, O., & Giuse, D. (2006). Expanding the scope of health information systems. *IMIA Yearbook of Medical Informatics*, 43–52.

Kuntz, K. M., & Goldie, S. J. (2002). Assessing the sensitivity of decision-analytic results to unobserved markers of risk: Defining the effects of heterogeneity bias. *Medical Decision Making, 22*(3), 218–227.

Lacour-Gayet, F. (2002). Risk stratification theme for congenital heart surgery. Pediatric cardiac surgery annual of the seminars in thoracic and cardiovascular surgery. *American Society for Thoracic Surgery, 5*(1), 148–152.

Laine, C., Horton, R., DeAngelis, C. D., Drazen, J. M., Frizelle, F. A., Godlee, F., Haug, C., Hébert, P. C., Kotzin, S., Marusic, A., Sahni, P., & Schroeder, T. V. (2007). Clinical trial registration — looking back and moving ahead. *New England Journal of Medicine, 356*, 2734–2736. https://doi.org/10.1056/NEJMe078110.

Lawler, E. K., Hedge, A., & Pavlovic-Veselinovic, S. (2011). Cognitive ergonomics, socio-technical systems, and the impact of healthcare information technologies. *International Journal of Industrial Ergonomics, 41*(4), 336–344.

LeRouge, C., & Wickramasinghe, N. (2013). A review of user-centered design for diabetes-related consumer health informatics technologies. *Journal of Diabetes science and Technology, 7*(4), 1–18.

Liu, J., Wyatt, J. C., & Altman, D. G. (2006). Decision tools in health care: focus on the problem, not the solution. *BMC Medical Informatics and Decision Making, 6*(1), 4.

Long, S. H., Galea, M. P., Eldridge, B. J., & Harris, S. R. (2012). Performance of 2-year-old children after early surgery for congenital heart disease on the Bayley Scales of Infant and Toddler Development. *Early Human Development, 88*(8), 603–607.

Marino, B. S., Lipkin, P. H., Newburger, J. W., Peacock, G., Gerdes, M., Gaynor, J. W., Mussatto, K. A., Uzark, K., Goldberg, C. S., & Johnson, W. H. (2012). Neurodevelopmental outcomes in children with congenital heart disease: Evaluation and management a scientific statement from the American Heart Association. *Circulation, 126*(9), 1143–1172.

Marschollek, M., Gövercin, M., Rust, S., Gietzelt, M., Schulze, M., Wolf, K.-H., & Steinhagen-Thiessen, E. (2012). Mining geriatric assessment data for in-patient fall prediction models and high-risk subgroups. *BMC Medical Informatics and Decision Making, 12*(1), 19.

Meyer, G., Adomavicius, G., Johnson, P. E., Elidrisi, M., Rush, W. A., Sperl-Hillen, J. M., & O'Connor, P. J. (2014). A machine learning approach to improving dynamic decision making. *Information Systems Research, 25*(2), 239–263.

Oborn, E., Barrett, M., & Davidson, E. (2011). Unity in diversity: Electronic patient record use in multidisciplinary practice. *Information Systems Research, 22*(3), 547–564.

Ozdemir, Z., Barron, Z., & Bandyopadhyay, S. (2011). An analysis of the adoption of digital health records under switching costs. *Information Systems Research, 22*(3), 491–503.

Patrick, J. D., Nguyen, D. H., Wang, Y., & Li, M. (2011). A knowledge discovery and reuse pipeline for information extraction in clinical notes. *Journal of the American Medical Informatics Association, 18*(5), 574–579.

Peleg, M., & Tu, S. (2006). Decision support, knowledge representation and management in medicine. *Yearbook of Medical Informatics, 45*, 72–80.

Peng, Y., Bin, Y., & Jiang, J. (2006). Knowledge-discovery incorporated evolutionary search for microcalcification detection in breast cancer diagnosis. *Artificial Intelligence in Medicine, 37*(1), 43–53.

Porter, M., & Teisberg, E. (2009). Redefining health care. In *Harvard business review*. Boston: Harvard Business School Publishing.

Quintana, J. M., Escobar, A., Aguirre, U., Lafuente, I., & Arenaza, J. C. (2009). Predictors of health-related quality-of-life change after total hip arthroplasty. *Clinical Orthopaedics and Related Research®, 467*(11), 2886–2894.

Restuccia, J. D., Cohen, A. B., Horwitt, J. N., & Shwartz, M. (2012). Hospital implementation of health information technology and quality of care: are they related? *BMC Medical Informatics and Decision Making., 12*, 109. https://doi.org/10.1186/1472-6947-12-109.

Rizzo, V. M., & Kintner, E. (2013). The utility of the behavioral risk factor surveillance system (BRFSS) in testing quality of life theory: An evaluation using structural equation modeling. *Quality of Life Research, 22*(5), 987–995.

Rodríguez, L. V., et al. (2003). Discrepancy in patient and physician perception of patient's quality of life related to urinary symptoms. *Urology, 62*(1), 49–53.

Ryan, P. B., Madigan, D., Stang, P. E., Marc Overhage, J., Racoosin, J. A., & Hartzema, A. G. (2012). Empirical assessment of methods for risk identification in healthcare data: Results from the experiments of the observational medical outcomes partnership. *Statistics in Medicine, 31*(30), 4401–4415.

Safran, C., Bloomrosen, M., Hammond, W. E., Labkoff, S., Markel-Fox, S., Tang, P. C., & Detmer, D. E. (2007). Toward a national framework for the secondary use of health data: An American Medical Informatics Association White Paper. *Journal of the American Medical Informatics Association, 14*(1), 1–9.

Sanchez, E., Toro, C., Artetxe, A., Graña, M., Sanin, C., Szczerbicki, E., et al. (2013). Bridging challenges of Clinical Decision Support Systems with a semantic approach. A case study on breast cancer. *Pattern Recognition Letters., 34*(14), 1758–1768.

Schectman, J. M., & Plews-Ogan, M. L. (2006). Physician perception of hospital safety and barriers to incident reporting. *Joint Commission Journal on Quality and Patient Safety, 32*(6), 337–343.

Schuele, M., Widmer, T., Premm, M., Criegee-Riech, M., & Wickramasinghe, N. (2015). Using a multiagent organisational approach to improve knowledge provision for shared decision making in patient-physician relationships: An example from Germany. *Health and Technology, 5*(1), 13–23. https://doi.org/10.1007/s12553-015-0102-6.

Siegel, E. (2013). *Predictive analytics: The power to predict who will click, buy, lie, or die*. Hoboken: Wiley.

Sim, I., Gorman, P., Greenes, R. A., Haynes, R. B., Kaplan, B., Lehmann, H., & Tang, P. C. (2001). Clinical decision support systems for the practice of evidence-based medicine. *Journal of the American Medical Informatics Association, 8*(6), 527–534.

Simon, A., Ebinger, M., Flaiz, B., Heeskens, K., & Wickramasinghe, N. (2015). A multi-centred empirical study to measure and validate user satisfaction with hospital information services in Australia and Germany. In *iHEA congress* (pp. 12–15).

Srinivas, K., Kavihta Rani, B., & Govrdhan, A. (2010). Applications of data mining techniques in healthcare and prediction of heart attacks. *International Journal on Computer Science and Engineering (IJCSE), 2*(2), 250–255.

Staal, I. I., Hermanns, J., Schrijvers, A. J., & van Stel, H. F. (2013). Risk assessment of parents' concerns at 18 months in preventive child health care predicted child abuse and neglect. *Child Abuse & Neglect, 37*(7), 475–484.

Story, M. T., et al. (2002). Management of child and adolescent obesity: Attitudes, barriers, skills, and training needs among health care professionals. *Pediatrics, 110*(1), 210–214.

Sulaiman, H., & Wickramasinghe, N. (2014). Assimilating healthcare information systems in a Malaysian hospital. *Communications of the Association of Information Systems (CAIS), 34*, 1291–1318.

Sullivan, R., Peppercorn, J., Zalcberg, J., Meropol, N., Amir, E., & Khayat, D. (2011). Delivering affordable cancer care in high-income countries. *The Lancet Oncology, 12*(10), 933–980.

Thomas, J. C., Moore, A., & Qualls, P. E. (1983). The effect on cost of medical care for patients treated with an automated clinical audit system. *Journal of Medical Systems, 7*(3), 307–313.

Thompson, C., Cullum, N., McCaughan, D., Sheldon, T., & Raynor, P. (2004). Nurses, information use, and clinical decision making—The real world potential for evidence-based decisions in nursing. *Evidence Based Nursing, 7*(3), 68–72.

Tiwari, R., Tsapepas, D. S., Powell, J. T., & Martin, S. T. (2013). Enhancements in healthcare information technology systems: Customizing vendor-supplied clinical decision support for a high-risk patient population. *Journal of the American Medical Informatics Association: JAMIA, 20*(2), 377–380.

Trucco, P., & Cavallin, M. (2006). A quantitative approach to clinical risk assessment: The CREA method. *Safety Science, 44*, 491–513.

Tseng, C.-W., Carter, G., Kapur, K., Keeler, E., & Rastegar, A. (2003). Medicare calibration of the clinically detailed risk information system for cost: University College Dublin.

Turban, E., Sharda, R., Aronson, J. E., & King, D. (2008) *Business intelligence: A managerial approach*. Pearson/Prentice Hall. ISBN: 978-0-13-234761-7.

Venkatesh, V., Zhang, X., & Sykes, T. A. (2011). Doctors do too little technology: A longitudinal field study of an electronic healthcare system implementation. *Information Systems Research, 22*(3), 523–546.

Vincent, C., Taylor-Adams, S., & Chapman, E. J. (2000). How to investigate and analyse clinical incidents: Clinical risk unit and association of litigation and risk management protocol. *British Medical Journal, 320*(7237), 777–781.

Virshup, B. B., Oppenberg, A. A., & Coleman, M. M. (1999). Strategic risk management: Reducing malpractice claims through more effective patient-doctor communication. *American Journal of Medical Quality, 14*, 153–159.

Von Lubitz, D., & Wickramasinghe, N. (2006a). Healthcare and technology: The doctrine of network centric healthcare. *International Journal of Electronic Healthcare, 2*(4), 322–344.

Von Lubitz, D., & Wickramasinghe, N. (2006b). Network centric healthcare: Applying the tools, techniques and strategies of knowledge management to create superior healthcare operations. *International Journal of Electronic Healthcare, 2*(4), 415–429.

Von Lubitz, D., & Wickramasinghe, N. (2006c). Creating germane knowledge in dynamic environments. *International Journal of Innovation and Learning, 3*(3), 326–347.

Wetters, N. G., Murray, T. G., Moric, M., Sporer, S. M., Paprosky, W. G., & Della Valle, C. J. (2013). Risk factors for dislocation after revision total hip arthroplasty. *Clinical Orthopaedics and Related Research, 471*(2), 410–416.

Weiner, J. P., Dobson, A., Maxwell, S. L., Coleman, K., Starfield, B., & Anderson, G. (1996). Risk-adjusted Medicare capitation rates using ambulatory and inpatient diagnoses. *Health care financing review, 17*(3), 77.

Wickramasinghe, N., & Gururajan, R. (2016). Innovation practice using pervasive mobile technology solutions to improve population health management: A pilot study of gestational diabetes patient care in Australia. *Journal for Healthcare Quality, 38*(2), 93–105.

Wickramasinghe, N., & Schaffer, J. (2010). *Realizing value driven patient centric healthcare through technology*. Washington, DC: IBM Center for the Business of Government.

Wickramasinghe, N., & Schaffer, J. L. (2006). Creating knowledge-driven healthcare processes with the intelligence continuum. *International Journal of Electronic Healthcare, 2*(2), 164–174.

Wickramasinghe, N., Bali, R., Gibbons, M., Choi, J., & Schaffer, J. (2008). A systematic approach: Optimization of healthcare operations with knowledge management. *Journal of healthcare information management: JHIM, 23*(3), 44–50.

Wickramasinghe, N., Bali, R., Kirn, S., & Sumoi, R. (2012). *Creating sustainable e-health solutions*. New York: Springer.

Wright, A., & Sittig, D. F. (2006). Automated development of order sets and corollary orders by data mining in an ambulatory computerized physician order entry system. In *Proceedings of the AMIA symposium* (pp. 819–823).

Wu, I., & Hu, Y. (2012). Examining knowledge management enabled performance for hospital professionals: A dynamic capability view and the mediating role of process capability. *Journal of the Association of Information Systems, 13*(2), 976, Article 3.

Yagi, M., Akilah, K. B., & Boachie-Adjei, O. (2011). Incidence, risk factors and classification of proximal junctional kyphosis: Surgical outcomes review of adult idiopathic scoliosis. *Spine, 36*(1), E60–E68.

Yoshio H., Koji I., Hitoshi T., Masakazu I., Yasuo W., (2012), Value of general surgical risk models for predicting postoperative liver failure and mortality following liver surgery, *Journal of Surgical Oncology*, Early View (Online Version of Record published before inclusion in an issue), Online ISSN: 1096-9098.

Yen, P.-Y., & Bakken, S. (2012). Review of health information technology usability study methodologies. *Journal of the American Medical Informatics Association, 19*(3), 413–422.

Yin, R. K. (2003). *Case study research: Design and methods*. Thousand Oaks: Sage.

# A Literature Review on Predicting Unplanned Patient Readmissions

**Isabella Eigner and Andrew Cooney**

## 1 Introduction

The changing demographics of developed countries are forcing medical practices to alter treatment behaviour. Whereas in the past, in which communicable diseases were a concern, it is now non-communicable diseases that strain both medical and economic resources. Diseases such as diabetes and dementia, which do not substantially reduce life expectancy but increase the cost of care per patient, are exacerbated by increasing life expectancy and decreasing population growth. As a population's growth rate falls, the injuries and diseases with the highest incidence and prevalence become those suffered by older populations (Murray and Lopez 1996). Moreover, decreasing population growth inhibits the working population to contribute economically to the welfare of older and frailer population groups. This leaves both publicly and privately funded healthcare systems vulnerable to operational inefficiencies.

In light of this, hospitals are forced and encouraged to improve treatment standards to become more efficient. One such factor to measure operational efficiency is a hospital's readmission rate, specifically, the rate at which patients, having been discharged, are readmitted to a hospital against planned intentions for the same or related admission reason. High readmission rates can result in a reduction of state or federal funding (Shams et al. 2015). Therefore, it is in the interest of avoiding monetary penalties that hospitals identify patients at higher risk of having unplanned readmissions.

Predicting patient readmission, however, often requires the joint analysis of multiple sources of data. Therein lies a problem of obtaining partial, fragmented and

I. Eigner (✉) · A. Cooney
University of Erlangen-Nuremberg, Institute of Information Systems, Nuremberg, Germany
e-mail: isabella.eigner@fau.de



N. Wickramasinghe, F. Bodendorf (eds.), *Delivering Superior Health and Wellness Management with IoT and Analytics*, Healthcare Delivery in the Information Age, https://doi.org/10.1007/978-3-030-17347-0_12

misaligned information. This limitation in data quality in turn historically hindered epidemiologists to extrapolate demographic information to within plausible limits (Murray and Lopez 1996). Additionally, fragmented data spread across multiple sources make it difficult for policy-makers to compare the relative cost-effectiveness of different interventions. Thus measuring, gauging and creating benchmarks for unplanned readmission have been hitherto difficult.

Rapid developments in the fields of data warehousing and data science have enabled researchers from multidisciplinary backgrounds to contribute to a growing body of knowledge in predictive analytics, namely, the building, training and application of predictive models to stratify patients into various risk groups based on information from administrative, insurance, clinical and government registry sources. Such contributions to this academic field are aimed at first aligning complex and sensitive information across multiple sources. This information is thereafter used to identify patients in need of additional healthcare resources by means of various intervention methods (Billings et al. 2013).

The emergence of predictive analytics in forecasting patient admission has a mixed approach in the current literature. Researchers of multidisciplinary backgrounds have approached this problem by analysing various sources of data, investigating the strength of association between socio-demographic and clinical variables with patient readmission. Depending on the strength of association and the objectives of the researchers, these variables are incorporated into a predictive model to correctly classify patients' risk of readmission. The goal of this literature review is thus twofold and structured according to the following tasks:

- Identify patient risk factors included in existing readmission prediction models.
- Collect, group and analyse the methods used to build the predictive models and their predictive performance.

## 2 Approach

In gathering the literature for this review, articles are obtained systematically from July 2017 to June 2017. Given the multidisciplinary nature of the subject matter, consistent search terms are applied to three different databases focusing on business, economics and health-related topics. The conducted search is not limited to a geographic region; however, it is assessed for eligibility through a set of criteria.

### *2.1 Eligibility Criteria*

Studies focusing on the building, training and validation of prediction models with a specific readmission-related outcome are considered for the literature review. Such readmission-related outcomes pertain to time-specific readmissions and disregard the specific place of the study samples' index admission and subsequent readmission. Articles are further subject to eligibility based on the type of prediction model used, that being an already recognised or adapted form of a prediction model. Finally, the use of discriminatory statistics, namely, the c-statistic or AUC or the rate at which a model correctly predicts in the derivation (or training) and validation (or testing) data set, is the basis of model performance comparison.

The type of split (e.g. 75:25 or 80:20) chosen by a researcher to divide a data set is not included as an exclusion criterion, as no exact science exists proving the impact on a model's performance (Kotu and Deshpande 2015). Additionally, studies included in the literature review have to be published after 2005 and in peer-reviewed journals only. Given that the existence of a prediction model is based upon the variables used to develop and train it, included studies already consist of risk factors and justifications for their use. Studies examining patients under the age of 18 or inflicted with psychological disorders are not included. This is based on the heterogeneity of risk factors among paediatric and mental illness patients (Kansagara et al. 2011). Further information on the exclusion criteria is given in Table 1.

**Table 1** Eligibility criteria for selected studies

| | Included studies | Excluded studies |
|---|---|---|
| Types of studies | Peer-reviewed journal articles | |
| Language | English | |
| Participants | Adult populations[a] | Missing or insufficient setting, population, data collection methods |
| Outcomes | Readmission-related outcome with a specified time between admission and readmission | Studies focusing solely on other predictable outcomes[b] |
| Statistics | Studies using the C-statistic[c] to assess the performance of the predictive model | |
| Derivation and validation | Contain, or at least mention, using internally or externally validated data[d] | |
| Other exclusion criteria | | Inclusion of patients dying in the index admission or post-discharge within the specified readmission time; model not sufficiently described |

[a]If not stated, then adult population is assumed
[b]For example, patient's expected length of stay, cost to the hospital or procedures needed
[c]C-statistic is a measure of model accuracy taking into account the ratio of correct vs. incorrect predictions made in the derivation and validation samples by a prediction model
[d]For example, using bootstrapping on the data as an internal validation and using new data as external validation

## 2.2 *Definition Criteria*

In the context of this review, a readmission is defined as an unexpected visit and subsequent admittance to a hospital for reasons corresponding to an earlier and related hospital admission. The specificity of timing (e.g. 30-day, 180-day readmission, etc.) in which this readmission occurs is not included in the eligibility criteria due to differing international standards. However, it is of importance that authors include a time-based parameter (henceforth referred to as an outcome) in which their patient samples are examined. Readmissions occurring outside of this specified time frame are hence not considered readmissions, but rather are new index admissions for related or different comorbidities.

A predictive model consists of associations between measurable variables to predict a certain outcome (Shmueli and Koppius 2011). Divergent sources of patient information are used in analysing patients' risk of readmission, and in some cases, this information is aligned across multiple sources (e.g. administrative and clinical data sources). Predictive models in the context of this chapter are thus directed towards the classification of patients' probability of being readmitted to the hospital.

Referred to frequently in this chapter is the c-statistic, otherwise known as the concordance statistic. The c-statistic often displayed alongside the receiver operating characteristic (ROC) plot is a reflection of both model accuracy and generalisability. It provides a good summary measure of model precision by providing a range of accuracy between 0 and 1. Within this range, a c-statistic of 0.5 is deemed as no better than chance, while a c-statistic between 0.7 and 0.8 indicates a modest or acceptable discriminative ability (Pencina and D'Agostino 2015). Moreover, the c-statistic provides for an easy-to-understand point of comparison, thus being suitable for the purpose of this section.

## 2.3 *Data Source and Search Method*

Literature searches are comprehensively conducted through the electronic databases EBSCO, AIS Electronic Library and PubMed. Publications matching the search terms "(readmission OR rehospitali$_*$) AND (predict$_*$ OR model)" in their title are included if they were published in a peer-reviewed journal after 2005. No restrictions regarding publication status and rank are used; however, the SJR and H Index are looked up and recorded in a research log. Studies using patient risk factors to build predictive models are found using a combination of search terms. These search terms are consistently applied across all electronic databases.

## 2.4 Data Collection

A total of 417 articles are collected in the initial search. The first stage of refining the papers is conducted by reading each of the titles and abstracts. Papers are assessed for eligibility based on the aforementioned exclusion criteria (cf. Table 1). 367 papers are excluded, and the remaining 50 papers are read in detail during which the exclusion criteria are once more applied. The resulting sample consists of 44 papers which met the eligibility criteria (cf. Table 2).

**Table 2** Identified studies on readmission prediction

| Study | Data source[a] | Disease under study[b] | Baseline test |
|---|---|---|---|
| Amalakuhan et al. (2012) | MR | COPD | N |
| Amarasingham et al. (2010) | EHR | HF | Y |
| Bardhan et al. (2015) | EHR | CHF | N |
| Betihavas et al. (2015) | Registry | CHF | N |
| Billings et al. (2013) | Registry | AC | N |
| Brown et al. (2013) | Mixed | AMI | N |
| Brzan et al. (2017) | MR | Obesity | Y |
| Choudhry et al. (2013) | EHR | AC | N |
| Demir et al. (2009) | Admin | COPD | N |
| Demir (2014) | Registry | COPD, Asthma | N |
| Donzé et al. (2013) | Mixed | AC | N |
| Dorajoo et al. (2017) | EHR | AC | N |
| Dugger et al. (2014) | EHR | AC | N |
| Fleming et al. (2014) | Mixed | HF | N |
| Golmohammadi and Radnia (2016) | Mixed | AC | N |
| Hasan et al. (2010) | Mixed | AC | N |
| Hilbert et al. (2014) | MR | AMI, HF, PN | N |
| Hummel et al. (2014) | Mixed | AC | N |
| Huynh et al. (2015) | Mixed | HF | N |
| Jamei et al. (2017) | EHR | AC | N |
| Fehnel et al. (2015) | MR | AC | N |
| Lee (2012) | MR | AC | N |
| Leeds et al. (2017) | EHR | Surgery | N |
| Lin et al. (2017) | Claims | AC | N |
| Lindenauer et al. (2011) | MR | PN | N |
| Mortazavi et al. (2016) | Mixed | HF | N |
| McLaren et al. (2016) | MR | HF | N |
| McManus et al. (2016) | MR | AMI | Y |
| Nguyen et al. (2016) | EHR | AC | Y |
| Pack et al. (2016) | Admin | AVR | N |
| Picker et al. (2015) | MR | AC | N |

(continued)

**Table 2** (continued)

| Study | Data source[a] | Disease under study[b] | Baseline test |
|---|---|---|---|
| Rana et al. (2014) | EHR | IHD | Y |
| Sawhney et al. (2017) | Mixed | AKI | N |
| Shulan et al. (2013) | Admin | AC | N |
| Shadmi et al. (2015) | HER | AC | N |
| Shams et al. (2015) | EHR | AC | N |
| Shams et al. (2015) | Admin | AMI, HF, COPD, PN | Y |
| Tabak et al. (2017) | EHR | AC | N |
| Taber et al. (2015) | Mixed | KTX | N |
| Tong et al. (2016) | EHR | AC | Y |
| Walsh and Hripcsak (2014) | MR | Specific + AC | N |
| Whitlock et al. (2011) | Mixed | AP | N |
| Yeo et al. (2016) | MR | Surgery | N |
| Yu et al. (2015) | Admin | AMI, HF, PN | Y |
| Zhu et al. (2015) | Admin | HF | N |

[a]*MR* Medical records, *Admin* administrative
[b]*AC* All-cause, *AKI* acute kidney injury, *AMI* acute myocardial infarction, *AP* acute pancreatitis, *AVR* heart valve surgery, *CHF* (chronic) heart failure, *COPD* chronic obstructive pulmonary disease, *IHD* ischaemic heart disease, *KTX* kidney transplant, *PN* pneumonia

During the latter two stages of the literature selection, data is collected and organised into concept matrices. A concept matrix, as defined by Webster and Watson (2002), is a method that effectively organises the concept-centric nature of a literature review. Moreover, the concept matrix provides a framework in which concepts can be aligned and compared between authors. The studies are separated into two groups for observation. These are segregated by the type of predictive model used and the risk factors used to train and validate them. In the literature, authors first assess the risk factors' correlation with readmissions within a data set. Thereafter, authors use these risk factors to build, train and test their predictive models. This chapter follows this layout by first presenting risk factors identified in the literature. The ensuing section will present the predictive models implemented by the various authors.

The first group concerning the assessment of patient risk factors are synthesised into eight subgroups. Variables included in the literature' final models are generally heterogenic and, when necessary, are qualitatively collapsed. Predictive models are initially grouped according to which family they belong (e.g. regression trees to the regression family) and synthesised hierarchically according to their relative performance (cf. Table 4). This performance evaluation is displayed for both the derivation (where applicable) and the validation cohorts.

# 3 Sample Statistics

## *3.1 Interest in the Subject Area*

To show the growing interest in the subject matter, the articles' publish date is displayed in Fig. 1. Interest in the creation of readmission models has been steadily increasing in recent years. Starting in 2005, the beginning of the search terms, and ending in 2017, a total of 44 papers matched the criteria specified in Table 1. The increased interest first started in 2010 and continued to develop until a peak is reached in 2015 and 2016. After this peak, interest in the topic subsided to four articles. The drop in interest, however, is more likely to be related to the end of the search period (June 2017) as opposed to a decline of interest in the study area.

The transfer towards electronic health records (EHRs) is still a recent trend with many hospitals moving away from traditional paper-based records. The digitisation of patient records increases researchers' abilities to apply predictive models focusing on patient readmission. Therefore, it is of interest to observe how the general number of published articles is impacted by the recent uptake of EHRs. It is evident from Fig. 1 that interest in the subject matter is being impacted by a sharp uptake in authors using EHRs as their data sources (peak of $n = 7$ in 2016). Moreover, the use of mixed and registry data sources remains stable over the search period, thus pontificating EHRs' positive impact on the academic field.

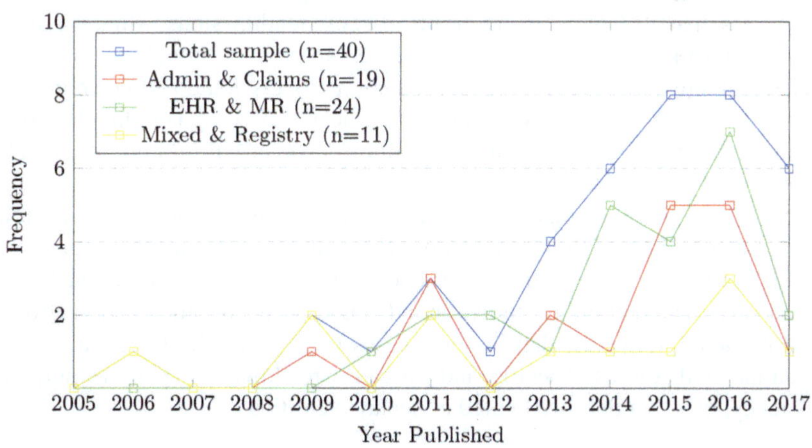


**Fig. 1** Development of published studies

## 4 Characteristics of Included Studies

The results present a sample containing much heterogeneity in research approaches. Data sources in this sample consist of administrative data (30%), EHR (28%), medical records (23%), registries (20%), claims data (10%) and interviews (10%). Authors additionally make use of multiple data sources (e.g. Donzé et al. 2013; Hasan et al. 2010; Hummel et al. 2014, etc.), linking, for example, medical health records with administrative information. In this sample, despite the recent growth of EHR-based studies, administrative data were most prevalent. Administrative records contain details of a patient's interaction or interactions with a hospital. The addition of clinical information to the demographic, behaviour and socio-economic information obtained from administrative data is possibly a reason for increased interest in the subject area. Evidence of this is observed in Fig. 1 whereby the cumulative interest in the topic is driven by studies based on EHRs published in 2016 and 2017. Interestingly, data coming from registry sources are least used in this sample. This may be concerned with the difficulty in obtaining these data, as they are typically intended for purposes related to clinical trials or intervention testing.

### *4.1 Disease Specificity*

Authors in this sample use mixed approaches in targeting specific and broad patient groups. Less than 50% of papers focus on all-cause readmissions, while the majority of papers (63%) focus on a specific disease category. Although not specifically correlated, interest in disease-specific prediction models could be in response to the 2012 Centers for Medicare and Medicaid Services (CMS) update directed towards lowering the readmission rates of heart failure, myocardial infarction, COPD and pneumonia (CMS 2016).

Papers with disease-focused predictive models consist mostly of heart-related conditions ($n = 12$). Within this group, heart failure makes up the largest cohort ($n = 8$). Following on from this, four studies focus on COPD, three on PNE and three on AMI. The remaining studies are all single counts of other disease groups. It should be noted, however, that in some cases (e.g. Yu et al. 2015) authors chose to examine both disease-specific and all-cause readmission. For the purpose of consistency, these models are treated as separate entities.

### *4.2 Study Design*

Much of the time spent in developing prediction models is ensuring the quality of the data (Kotu and Deshpande 2015). Raw data taken directly from information sources are likely to contain anomalies and noise that would affect the results produced by the predictive models. Possibly for the reason of long processing time that the majority of studies (70%) have retrospective study designs.

## 4.3 *Homogeneity of Papers' Samples*

To improve the generalizability of results, some authors use data from health networks and alliances. As the name suggests, health networks and alliances denote a group of connected hospitals and healthcare centres in a certain region. Leaving aside the administrative and financial advantages to link a group of hospitals in a vicinity, data obtained across multiple centres nullifies peculiarities specific to one treatment facility. In turn, depending on the incidence and prevalence of this peculiarity across the alliance, results can be normalized to a value that reflects real-world instances.

In this sample, just over half of articles use data originating from more than one treatment facility (60%). Studies using data originating from health alliances are solely located in the USA, using either the Advocate Health Care Hospitals located in the Chicago metro and suburban region or a series of veteran hospitals. The latter is not specifically an alliance when considering the underlying bureaucratic aspect of the hospital network. However, they are considered so in this thesis, as veteran hospitals provide an accumulation of data from multiple facilities and regions.

## 4.4 *Country*

Authors publishing studies were predominately doing so from data sources originating in the USA (75%). While Australia and the United Kingdom account for 15% of the sample collectively, interest in the topic is clearly highest in the USA. Given the existence of the aforementioned health networks, obtaining heterogeneous data from established collections may yield more reliable results than using data from a single source. This may, in turn, attract authors to seek out data being made available by these alliances.

## 4.5 *Type of Comparison*

Papers within this sample utilize a variety of different methods in building and implementing predictive models. While the majority of papers include no comparison of models (the purpose of their paper being solely to develop and test a predictive model; refer to, for example, Hummel et al. 2014; Dugger et al. 2014), just over a third of the sample incorporate some sort of comparison (37%). Such an aggregate level observation gives insight into the motivations behind the various authors. Moreover, the scattered approach is possibly a testament to the immaturity of the academic research field insofar as a consistency of approach is concerned.

Incorporating a baseline test into a study involves using known industry benchmarks such as HOSPITAL score, LACE index, and CMS. Within this sample, many papers refer to these industry benchmarks by providing comparisons in model performance. This comparison is typically constructed on the grounds of the resulting c-statistic. Only six studies within this literature review (15%) include an industry baseline test as a benchmark from which to develop, validate and compare their own predictive model (Amarasingham et al. 2010; Rana et al. 2014; Shams et al. 2015; Tong et al. 2016; Nguyen et al. 2016; McManus et al. 2016).

Some authors in this literature review direct their focus towards comparing the performance of various predictive models. Leaving issues such as heterogeneity of sample groups aside, these studies provide empirical evidence of how various modelling approaches perform on population groups. Within this sample, four authors develop, validate, and compare various predictive models (Shams et al. 2015; Tong et al. 2016; Mortazavi et al. 2016; Demir et al. 2012). In Shams et al. (2015), four prediction models are implemented before developing their own and comparing it to their own generated baseline model (derived from a tree-based classification method). In other cases, such as Tong et al. (2016) and Mortazavi et al. (2016), boosting algorithms (such as LASSO and AdaBoost) are tested on pre-existing predictive models.

## 4.6 Outcome

As mentioned previously in the exclusion criteria (cf. Table 1), inclusion into the literature review is necessitated by a time-based outcome. Within this sample, the overwhelming majority of papers use 30-day unplanned readmissions (83%). Given that the majority of papers published in this literature review originate from US-based data sources, this statistic may be indicative of CMS's influence on the academic field and mandate of hospital quality measures (Fehnel et al. 2015).

Despite this established industry standard, however, some studies dedicate some, if not all, of their research to the comparison of other unplanned readmission outcomes. Eleven such papers included the following outcomes: 15-day (Dorajoo et al. 2017), 28-day (Betihavas et al. 2015), 38-day (Demir et al. 2009), 45-day (Demir et al. 2012), 90-day (Fehnel et al. 2015; Pack et al. 2016; Sawhney et al. 2017) and 365-day (Fehnel et al. 2015; Amalakuhan et al. 2012; Billings et al. 2013). In Demir et al. (2009), a process-oriented viewpoint was adopted, treating patient risk as moving from high to low based on time from discharge to time spent in the community. In this particular paper, a 38-day outcome is found to be optimal based on minimizing the classification error between high- and low-risk patients and using the intersection of time-to-readmission curves. In other cases, the outcome is determined by disease-specific guidelines (e.g. Sawhney et al. (2017) focus on acute kidney injury and hence follow the recommended 90-day risk assessment).

# 5 Patient Risk Factors

## *5.1 Characteristics of Patient Risk Factors*

The studies included in the literature review present a vast collection of heterogeneous risk factors. In total, 175 different risk factors are observed, recorded and thereafter condensed into eight groups (cf. Fig. 2). Risk factors are qualitatively assessed for similarity and grouped accordingly.

From an aggregated perspective, authors frequently incorporate patients' history of comorbidities and concomitant diseases into their final models (81%). Possibly due to the ease of accessibility, demographics (which entail age, sex, etc.) are frequently included in predictive models (70%). Risk factors relating to hospital visitation are used by roughly two-thirds of the studies and consist largely of a patient's length of stay and number of admissions and discharges. More specific social determinants of health, which may possibly be derived from the aforementioned social demographics, are used less (50%). Similarly, health behaviour involving patient habits (e.g. smoking status or inactive lifestyle) are used less often (39%). Lab and clinical results are less prevalent in this data set and are present in only 44% of the final models. Again, this may stem from the difficulty of obtaining such information.

## *5.2 Socio-demographics and Social Determinants of Health*

Papers in this sample use a range of socio-demographic and social determinants of health data. Among these, age and gender are consistent predictors of readmissions (65% and 55% respectively). However, it should be noted that six studies focus on populations older than 60 years of age (Hummel et al. 2014; Lindenauer et al. 2011; Fehnel et al. 2015; Zhu et al. 2015; Lin et al. 2016; McManus et al. 2016; Yeo et al. 2016). Ethnicity and race are less considered as predictors of readmission and are included in less than a third (28%) of the studies. According to Shams et al. (2015), veteran hospitals do not capture this specific metric.

Social determinants of health are even less considered in the sample. This may in part be due to the difficulty in obtaining such metrics (e.g. calculating the distance from healthcare facility (Fleming et al. 2014; Huynh et al. 2015) or measuring education and health literacy (Demir et al. 2009; Lin et al. 2016; McManus et al. 2016). Of particular interest in this group, however, is a patient's regular access to care. This typically consists of a patient's access to a doctor by means of the type of residence (e.g. hospice). This risk factor is present in 25% of studies. Only 20% of the articles included insurance as a predictor of readmission.

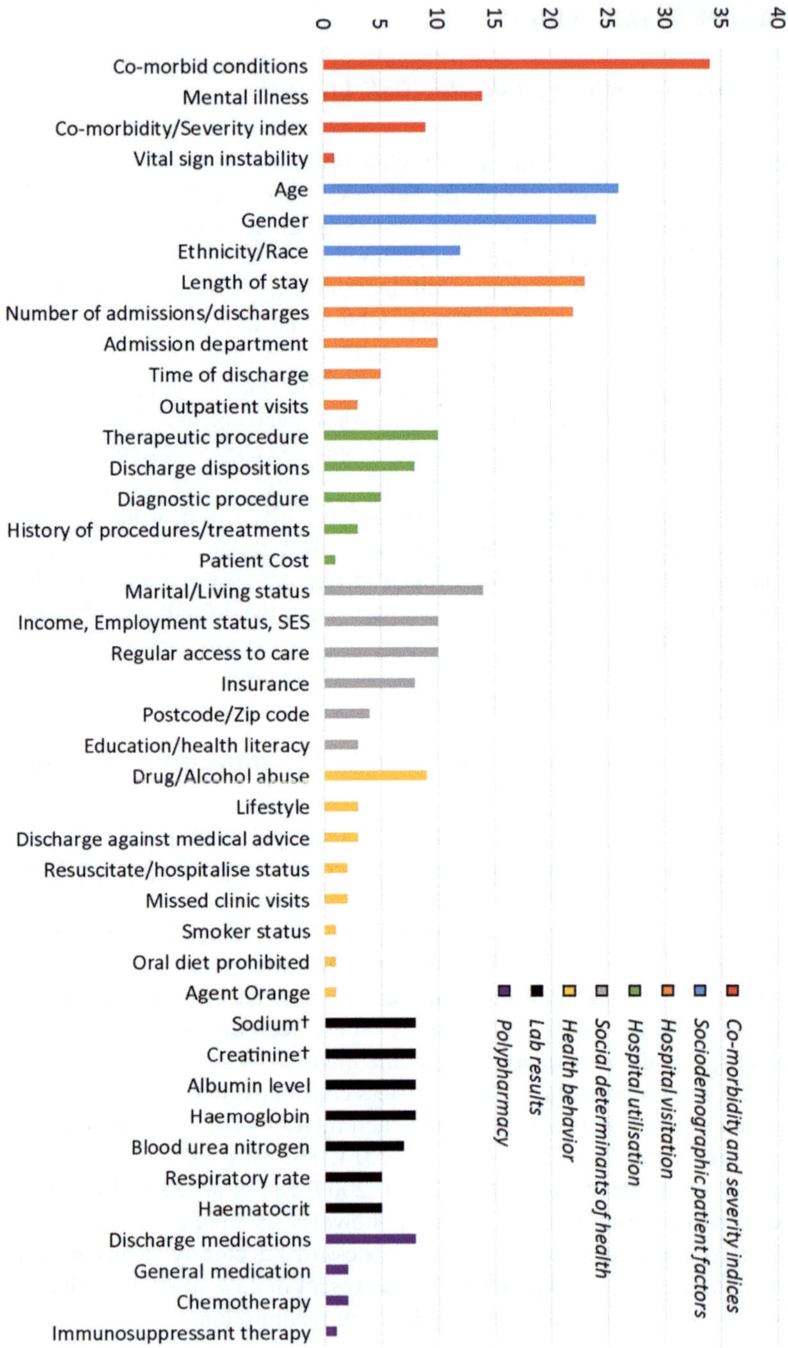


**Fig. 2** Frequency of risk factors

## 5.3 Comorbid Conditions and Clinical Results

The presence of patient comorbidities is frequently included as risk factors. Of the included studies, 88% of the sample use patients' comorbid conditions in their predictive models. 20% of studies use patients' comorbidity index or severity index as a measure of readmission. Examples of comorbidity and severity indices include CMI, CHESS and the Elixhauser score; the specific score itself is derived from previous and or coexisting diagnoses. Fourteen papers (31%) in the sample include a presence of mental illness as a risk factor in their predictive models; however, none report finding a significant relationship.

Although broad in available options, using clinical data obtained from laboratory tests results in varying effects. This information is mostly taken from EHRs, whereby use of broad and extensive clinical information is now available. Given the variation in published results, it is clear that authors are still investigating how this information can be best analysed or integrated with other sources of data. Within this category, however, some clear indicators of readmission are observed. For example, sodium ($\pm$ <130 mmol/L) and creatinine (<2.5 mg/dL) are used in 20% of studies as indicators for heart failure patients. Another strong indicator of readmissions is a patient's albumin level (recorded either at admission or discharge), which is used by 20% of the included studies.

## 5.4 Polypharmacy Risk Factors

Factors relating to a patient's use of polypharmacy during admission and at discharge are seldom used in this sample ($n = 9$). In Picker et al. (2015), Donze et al. (2013) and Dorajoo et al. (2017), a significant influence of polypharmacy on possible readmissions is observed. However, these associations, despite their statistical significance, are generally weak, and hence other factors in these studies are examined. Only Picker et al. (2015) investigate the association between polypharmacy risk factors and patient readmissions.

Research in Picker et al. (2015) is dedicated to observing the influence of discharge medication on 30-day readmission. Based on a population of 5507 patients, a statistically significant correlation between the number of discharge medications and a 30-day readmission is observed. Within their sample, 30% of their readmission cohort consists of patients who have prescribed 12 or more discharge medications ($p < 0.001$). However, when examined independently with 30-day readmission, the association was found to be non-significant.

Despite the overall weak performance of polypharmacy in the observed sample, authors nonetheless examine their effect on readmission. Yeo et al. (2016) investigate chemotherapy in both age- and non-age-stratified cohorts. In the age cohorts 65–74 and 75–84, patients are at higher risk of readmission. Both risk factors are statistically significant ($p < 0.001$). Huynh et al. (2015) examine eight polypharmacy risk factors, of which two, ACEI/ARB5 and diuretic use, are statistically significant ($p < 0.001$). In Leeds et al. (2017), although they found a significant association ($p < 0.01$), the magnitude of this is minuscule.

Calculating the association between polypharmacy and unplanned patient readmission may be difficult due to the multiplicity of medications that patients receive. Moreover, examining polypharmacy covers a wide range of pharmaceutical compounds with some compounds having more severe side effects than others (e.g. statins in endocrinology vs. biologics in rheumatology). This possibly affects a researcher's ability to examine the influence of a generalized polypharmacy definition on patient health.

## 5.5 *Hospital Visitation and Utilization*

Use of patient factors relating to hospital visitation and utilization is prevalent in this sample. These figures are generally obtained from administrative sources and prove to be relevant risk factors. Within hospital visitation, over half of the included studies use the length of patient stay (LOS) and the number of (previous) emergency admissions or discharges (58% and 55%, respectively). Within healthcare utilization, therapeutic procedures (e.g. surgery, dialysis, physical therapy, etc.) are observed to be strong determinants of unplanned patient readmissions. Despite this, just over half of the studies (56%) use these risk factors. The examination of patients' LOS in this sample returned homogenous results; however, there are varying methods of approaching this particular risk factor. Within the studies that include length of stay (58%), only ten have odds ratios (OR) to investigate the association with patient readmission, six of which report a significant association (Donzé et al. 2013; Hasan et al. 2010; Nguyen et al. 2016; Huynh et al. 2015; Bardhan et al. 2015). The risk of readmission, as reported by Demir et al. (2009), increases with a greater number of emergency days spent in the hospital.

In some studies, length of stay is altered from continuous data into dichotomous sets of days. Donze et al. (2013), who investigate potentially avoidable readmission (PAR), dichotomize length of stay into two bins, 1–4 days and >4 days. Within this particular sample, patients in the PAR cohort have hospital stays longer than in the non-readmission cohort (41% vs 56% for >4 days in hospital respectively). Similarly, patient cohort characteristics, with exception of the patients identified with PARs, are also observed in Tong et al. (2016), Nguyen et al. (2016) and Pack et al. (2016). This consistent finding in papers suggests that patients identified as high risk stay in care longer and that LOS may be a stronger predictor of patient readmission.

The number of discharges and admissions within a certain time frame are included in approximately half of the studies (55%). In this category, authors observe patients' interactions with hospitals over a specified time frame (e.g. 1, 3 or 5 years). As expected, patients belonging to unplanned readmission cohorts belong to the group with frequent hospital visits in this time frame. For example, in Donze et al. (2013), approximately 70% of the study sample with 1–5 and >5 hospital admissions in the previous year belong to the readmission cohort. This is compared to 51% for the non-readmission cohort. Sawhney et al. (2017) report a similar statistic, whereby 38% of the readmission cohort had one or more admissions in the past year compared to 22% in the non-readmission cohort.

Within hospital utilization, therapeutic and diagnostic procedures are separated based on the sequence in which they are conducted. Thus, if therapeutic procedures are administered after a diagnostic procedure, there is something to treat or, in some cases, operate on. Less than one-third (25%) of the included papers investigate the association between therapeutic procedures during admission and readmission. Of this proportion, three papers report a significant relationship (Donzé et al. 2013; Lindenauer et al. 2011; Bardhan et al. 2015). In Donze et al. (2013), the PAR group have a slightly larger amount of procedures compared to the non-readmission cohort during the index admission (70% vs 30% respectively). However, Bardhan et al. (2015) report a weak association with the number of patient procedures and readmission ($p < 0.013$). Furthermore, Pack et al. (2016) observe that more readmitted patients require physical therapy (73%) than the non-readmitted cohort.

## 5.6 *Health Behaviour*

Within risk factors in health behaviour, papers examine the influence of patients' activities outside of the hospital on readmission. Alcohol and substance abuse, lifestyle habits and discharge behaviour are among the included risk factors. Although the majority of papers obtain this information from administrative or EHRs, there are some papers that utilized interview methods either during or post-discharge (e.g. McManus et al. 2016). Implementation of health behaviour-related risk factors is used in approximately one-third of the study sample (39%). Within this scope, papers predominately focus on alcohol or substance abuse (23%) and, to a lesser extent, other lifestyle risk factors (7%). Reporting behaviour relating to alcohol or substance result in weak correlations with patient readmission, where only Pack et al. (2016) report a significant relationship for their 30-day readmission model. Reporting lifestyle as a risk factor involves observing patients' BMI and sedentary status. Only three authors include this in their predictive models (Betihavas et al. 2015; Leeds et al. 2017; Shadmi et al. 2015). Taking into consideration the small sample size ($n = 280$, 28-day re-admittance group = 32), Betihavas et al. (2015) report a significant relationship ($p < 0.001$). In contrast, Shadmi et al. (2015) note that a 1-unit increase in BMI is associated with lower odds for readmission (0.99), the significance of this association is, however, not reported.

Results on health behaviour risk factors are mixed among papers in the sample. Data of this nature originate typically from administrative sources. In only one case, primary data is obtained from patient interviews, however, with no significant improvement in predicting 30-day readmissions. Despite the few instances of significant associations, results observed within this group of risk factors are not consistent enough to generalize the health behaviour risk factors' influence on patient readmissions.

# 6 Risk Prediction Models

## *6.1 Characteristics of Prediction Models*

The approaches to model implementation vary widely in this sample, resulting in a total of 57 models that fit the criteria specified in Table 1. While there is some testing into more recently conceptualized models, the majority of authors fit regression-related models. The following section provides an in-depth overview of the types of prediction models in this sample and their respective discriminatory statistics. This section concludes with an in-depth look at the set of best-performing models that exist in this sample and what outcomes can be derived from them. Papers in this review most frequently fit models belonging to the regression family (71%). While there exists a variety in both the non-regression and regression-related models, logistic regression is most commonly used to investigate patient readmissions (44%). Following this, multiple logistic regression and random forests/trees are most commonly used (12% and 11% respectively). Decision trees are implemented solely by Hilbert et al. (2014) mostly for demonstrative purposes in the disease areas acute myocardial infarction (AMI), heart failure (HF) and pneumonia (PNE) to produce visually appealing and easily understandable risk hierarchies. Shams et al. (2015) and Jamei et al. (2017) attempt to fit a neural network in HF, AMI, PNE, chronic obstructive pulmonary disease (COPD) and all-cause sources of readmission. Similarly, support vector machines are fit with data in all-cause and HF, AMI, PNE and COPD sources of readmission in Shams et al. (2015).

## *6.2 Cohort Model Evaluation*

Model performance can be bracketed into three categories as denoted by various authors (Kansagara et al. 2011; Dugger et al. 2014; Demir et al. 2009; Sawhney et al. 2017). In the context of predicting patient readmission, fair and poor performance have limited reliability in their application. Achieving a c-statistic of greater than or equal to 0.8, however, suggests that a prediction model has the reliability to be applied in the real world. In this sample, only 14% of the included prediction models achieve such a score. The majority of risk prediction models (47%) perform poorly (lower than 7%), while the remainder results in fair c-statistic scores (39%). This result supports the finding made by Kansagara et al. (2011, 2016) stating that the majority of predictive models perform with either fair or poor performance.

**Table 3** Frequency of prediction methods

| Model type | Frequency | Good performance (%) | Fair performance (%) | Poor performance (%) |
|---|---|---|---|---|
| Logistic regression | 25 | 12 | 36 | 52 |
| Multivariate logistic regression | 7 | | 57 | 43 |
| Random forest/regression trees | 6 | 33 | 17 | 50 |
| Boosting | 3 | | 33 | 67 |
| Decision trees | 3 | | | 100 |
| Stepwise logistic regression | 2 | | 100 | |
| Artificial neural networks | 2 | | 100 | |
| Support vector machines | 2 | 50 | 50 | |
| LASSO | 1 | | 100 | |
| Conditional logistic regression | 1 | | | 100 |
| Cox hazards proportions | 1 | 100 | | |
| Other | 4 | 25 | 25 | 50 |

**Table 4** Prediction models with good predictive performance

| Study | Model | Disease area | Sample size | C-statistic | Data source |
|---|---|---|---|---|---|
| Demir (2014) | RT | Specific | 963 | 0.93 | Registry |
| Demir (2014) | LR | Specific | 963 | 0.92 | Registry |
| Walsh and Hripcsak (2014) | SVM | Specific | 118, 491 | 0.92 | MR |
| Shams et al. (2015) | Other | Specific | 2985 | 0.84 | Admin |
| Whitlock et al. (2011) | LR | Specific | 593 | 0.83 | Mixed |
| Shams et al. (2015) | RF | Specific | 2985 | 0.80 | Admin |
| Betihavas et al. (2015) | CX | Specific | 280 | 0.80 | Registry |
| Leeds et al. (2017) | LR | All-cause | 20, 970 | 0.80 | EHR |

Given the indubitable inclination towards regression-related prediction models (cf. Table 3), an overview of this collection is necessary to assess differences in approaches in both the regression family and the larger sample. Just under one half (44%) of papers use logistic regression, with performance predominately distributed between fair and poor performance. Whitlock et al. (2011), Demir et al. (2009) and Leeds et al. (2017) showcase high-performing predictive models using logistic regression (c-statistic of 0.83, 0.92 and 0.92 respectively). Multivariate regression is the second most used modelling technique both within the regression prediction models (18%) and the entire sample (12%). This sample, however, presents with lower c-statistic scores (56% and 43% in fair and poor performance categories respectively). C-statistics for the two groups reflect that of the total sample with the majority of the subsample resulting in either fair or poor model performance (39% and 47% respectively). As stated by Kotu and Deshpande (2015), regression models are used due to their relative ease of implementation. However, as is evident in Table 4, ease of implementation is not generally synonymous with good model performance.

Within this sample, eight prediction models in six studies have a c-statistic score greater than or equal to 0.80. Approaches vary with the majority using regression-related prediction models (50%). Sample diversity, or single or multiple-hospital studies, vary with slightly over half of this subsample using a data set originating from one hospital. Sample size differs greatly from $n = 593$ (Whitlock et al. 2011) in the smallest sample to $n = 118{,}491$ (Walsh and Hripcsak 2014) in the largest sample. No consistency in data sources is observed when examining the top-performing prediction models.

The disease focus, however, is noticed as a consistent observation in Table 4. Of the prediction models achieving a c-statistic equal to or over 0.8, only one study focuses on all-cause disease specificity. The remainder chooses specific disease groups to study. Moreover, authors' examination of readmissions typically includes the observation of chronic and, to an extent, terminal disease categories. The majority of which pertain to COPD, heart-related diseases, AMI and PNE. The one exemption to this generalization is Leeds et al. (2017) who focus on patient readmissions as a result of surgical complications.

The collation of prediction models in this sample and their characteristics present a plethora of findings. Prediction models relating to regression are widely used in the sample with logistic regression as the most prevalent. The performance of the said predictive models, however, present mostly with either fair or poor performance. This performance is not specific to regression-based models, but rather reflects the performance of the entire sample in this chapter. Additionally, it appears that data source, diversity, sample size or prediction model do not play an overly pivotal role in determining readmissions across the sample. Rather, and as evident, within the high-performing predictive models (cf. Table 4), disease specificity may be more decisive in developing better-performing prediction models.

# 7 Discussion

This review constructs an in-depth picture of the current state of models developed and tested to predict patient readmissions. Study characteristics including sample description used risk factors, and utilized prediction models are gathered to provide the reader with a current overview of how authors in this academic field have been active. The discovered findings of this literature analysis may serve as recommendations for future researchers.

Five categories of comparison are observed: baseline, model, risk factor, disease and outcome. Only a low prevalence of studies (38%) incorporates comparison against already established work. While providing such a comparison may not be the goal of a given study, incorporating such scores and indexes is necessary to reinforce benchmarks and building upon the hitherto acquired knowledge.

In the observation and collation of patient risk factors, the inclusion of patients' previous and current coexisting morbidities is consistently observed to be predictors of patient readmissions (cf. Fig. 2). This supports the findings by Kansagara et al.

(2011) and Dugger et al. (2014). Socio-demographic information, such as age and gender, is included in the majority of studies. This is most likely due to their ease of accessibility. Within social determinants of health, approximately half of the studies include a patient's regular access to care as an indicator of readmission. In studies examining older population groups, this proves to be easy and crucial information to obtain and include in prediction models.

Another major finding is the impact of a model's disease specificity on predictive performance. Although only a small portion of prediction models perform with good discrimination, this group uses mostly specific, as opposed to all-cause, prediction models. It may be the case that the combination of risk factors is confounding the predictive capability of all-cause prediction models. This finding is yet to be further tested; however, it serves as a legitimate finding not discussed or observed in other literature reviews or papers. Although more research is needed to find further points, this literature review identifies generalizations exhibited through researchers' work, that is, the inclusion of patient comorbidities, keeping a prediction model disease-specific and using a variety of statistics to provide the reader with an overall perspective of model performance.

## 8 Conclusion

The scope of this literature review covered 44 unique studies published between 2005 and 2017 with the shared goal of determining patient readmission. Each study developed, trained and tested at least one prediction model, and their results were assessed and compared using the c-statistic. In doing so, authors additionally published descriptive statistics about their sample and the origins of the data set, the risk factors or features used and why and specifics on how these prediction models were fine-tuned. Using methods proposed by Webster and Watson (2002), this information was sorted into respective concept matrices and was presented in the body of this review.

Interest in the subject matter has been increasing in recent years and is expected to continue as the use of EHRs becomes more widespread (cf. Fig. 1). With more patient information being digitized, the depth and wealth of research in this academic field will indubitably increase. This may, in turn, improve the cooperation between hospitals and traditionally non-medical researchers. As this academic field is still relatively young and dependent on data from hospitals, the transition towards EHR would indeed propagate the growth of this academic field.

The majority of studies collected in this review were conducted in health facilities in the USA. Data heterogeneity was determined by the source of data. Possibly due to increased digitization of health records in hospitals, the use of EHRs as a data source was high in this sample. Some authors used mixed approaches, combining and matching either health records and administrative records or health records with public registry information and so on. More than half of the papers trained their models on disease-specific areas with results favouring disease-specific prediction models as opposed to all-cause predictive models.

In assessing various patient populations, authors collated a plethora of information relating to socio-demographics, patient comorbidities, polypharmacy, hospital visitation and utilization and health behaviour. Within this sample, it is observed that assessing patient comorbid conditions is the strongest indicators of patient readmission. This applies specifically where the population under study belonged to an older age group and hence was more likely to have pre-existing conditions. Patient lab results were found to be useful in finding an association between blood urea nitrogen levels, creatinine levels and patient readmission in heart disease-related subgroups. Hospital visitation and utilization was also thoroughly investigated with just under half of the sample including some risk factors belonging to this group.

The majority of risk predictive models included in this study made use of logistic regression to examine the probability of patient readmission. This may be attributed to the relative ease and bivariate outcome of calculating patient readmission. Despite this, only a small portion of this sample's logistic regression predictive models performed at a level that would be acceptable in a hospital. This proportionality, however, was reflected in the performance of the entire sample as a whole with only 14% of the predictive models achieving a c-statistic score of 0.8 or higher. Interestingly, it was found that model diversity and size have less of an impact on model performance. Rather, when assessed on disease specificity, it was found that models on individual disease groups performed in general much better than those with an all-cause focus. This possibly indicates that models focusing on all-cause readmission are being confounded by a heterogeneous mix of patient risk factors.

Given that only 14% of the predictive models in this sample resulted in a good performance, much work is still needed to improve the predictive capability of readmission. This literature review covers a 12-year time span and is a testament to how this academic field is changing and adapting to advances made at the patient and hospital levels. However, further research is needed to validate the claims made in this review.

## References

Amalakuhan, B., Kiljanek, L., Parvathaneni, A., Hester, M., Cheriyath, P., & Fischman, D. (2012). A prediction model for COPD readmissions: catching up, catching our breath, and improving a national problem. *Journal of Community Hospital Internal Medicine Perspectives, 2*(1), 9915.

Amarasingham, R., Moore, B. J., Tabak, Y. P., Drazner, M. H., Clark, C. A., Zhang, S., Reed, W. G., Swanson, T. S., Ma, Y., & Halm, E. A. (2010). An automated model to identify heart failure patients at risk for 30-day readmission or death using electronic medical record data. *Medical Care, 48*(11), 981–988.

Bardhan, I., Oh, J.-h., Zheng, Z., & Kirksey, K. (2015). Predictive analytics for readmission of patients with congestive heart failure. *Information Systems Research, 26*(1), 19–39.

Betihavas, V., Frost, S. A., Newton, P. J., Macdonald, P., Stewart, S., Carrington, M. J., Chan, Y. K., & Davidson, P. M. (2015). An absolute risk prediction model to determine unplanned cardiovascular readmissions for adults with chronic heart failure. *Heart, Lung & Circulation, 24*(11), 1068–1073.

Billings, J., Georghiou, T., Blunt, I., & Bardsley, M. (2013). Choosing a model to predict hospital admission: an observational study of new variants of predictive models for case finding. *BMJ Open, 3*(8), e003352.

Brown, J. R., Conley, S. M., & Niles, N. W. (2013). Predicting readmission or death after acute ST-elevation myocardial infarction. *Clinical Cardiology, 36*(10), 570–575.

Brzan, P. P., Obradovic, Z., & Stiglic, G. (2017). Contribution of temporal data to predictive performance in 30-day readmission of morbidly obese patients. *PeerJ, 5*, e3230.

Choudhry, S. A., Li, J., Davis, D., Erdmann, C., Sikka, R., & Sutariya, B. (2013). A public-private partnership develops and externally validates a 30-day hospital readmission risk prediction model. *Online Journal of Public Health Informatics, 5*(2), 219.

CMS. (2016). *Readmissions Reduction Program (HRRP)*. Available at https://www.cms.gov/medicare/medicare-fee-for-service-payment/acuteinpatientpps/readmissions-reduction-program.html

Demir, E. (2014). A decision support tool for predicting patients at risk of readmission. A comparison of classification trees, logistic regression, generalized additive models, and multivariate adaptive regression splines. *Decision Sciences, 45*(5), 849–880.

Demir, E., Chahed, S., Chaussalet, T., Toffa, S., & Fouladinajed, F. (2012). A decision support tool for health service re-design. *Journal of Medical Systems, 36*(2), 621–630.

Demir, E., Chaussalet, T., Xie, H., & Millard, P. H. (2009). Modelling risk of readmission with phase-type distribution and transition models. *IMA Journal of Management Mathematics, 20*(4), 357–367.

Donzé, J., Aujesky, D., Williams, D., & Schnipper, J. L. (2013). Potentially avoidable 30-day hospital readmissions in medical patients: derivation and validation of a prediction model. *JAMA Internal Medicine, 173*(8), 632–638.

Dorajoo, S. R., See, V., Chan, C. T., Tan, J. Z., Tan, D. S. Y., Abdul Razak, S. M. B., Ong, T. T., Koomanan, N., Yap, C. W., & Chan, A. (2017). Identifying potentially avoidable readmissions: a medication-based 15-day readmission risk stratification algorithm. *Pharmacotherapy, 37*(3), 268–277.

Dugger, A., McBride, S., & Song, H. (2014). Can nurses tell the future? Creation of a model predictive of 30-day readmissions. *ANS. Advances in Nursing Science, 37*(4), 315–326.

Fehnel, C. R., Lee, Y., Wendell, L. C., Thompson, B. B., Potter, N. S., & Mor, V. (2015). Post-acute care data for predicting readmission after ischemic stroke. A nationwide cohort analysis using the minimum data set. *Journal of the American Heart Association, 4*(9), e002145.

Fleming, L. M., Gavin, M., Piatkowski, G., Chang, J. D., & Mukamal, K. J. (2014). Derivation and validation of a 30-day heart failure readmission model. *The American Journal of Cardiology, 114*(9), 1379–1382.

Golmohammadi, D., & Radnia, N. (2016). Prediction modeling and pattern recognition for patient readmission. *International Journal of Production Economics, 171*, 151–161.

Hasan, O., Meltzer, D. O., Shaykevich, S. A., Bell, C. M., Kaboli, P. J., Auerbach, A. D., Wetterneck, T. B., Arora, V. M., Zhang, J., & Schnipper, J. L. (2010). Hospital readmission in general medicine patients: a prediction model. *Journal of General Internal Medicine, 25*(3), 211–219.

Hilbert, J. P., Zasadil, S., Keyser, D. J., & Peele, P. B. (2014). Using decision trees to manage hospital readmission risk for acute myocardial infarction, heart failure, and pneumonia. *Applied Health Economics and Health Policy, 12*(6), 573–585.

Hummel, S. L., Katrapati, P., Gillespie, B. W., Defranco, A. C., & Koelling, T. M. (2014). Impact of prior admissions on 30-day readmissions in medicare heart failure inpatients. *Mayo Clinic Proceedings, 89*(5), 623–630.

Huynh, Q. L., Saito, M., Blizzard, C. L., Eskandari, M., Johnson, B., Adabi, G., Hawson, J., Negishi, K., & Marwick, T. H. (2015). Roles of nonclinical and clinical data in prediction of 30-day rehospitalization or death among heart failure patients. *Journal of Cardiac Failure, 21*(5), 374–381.

Jamei, M., Nisnevich, A., Wetchler, E., Sudat, S., & Liu, E. (2017). Predicting all-cause risk of 30-day hospital readmission using artificial neural networks. *PLoS One, 12*(7), e0181173.

Kansagara, D., Chiovaro, J. C., Kagen, D., Jencks, S., Rhyne, K., O'Neil, M., Kondo, K., Relevo, R., Motu'apuaka, M., Freeman, M., & Englander, H. (2016). So many options, where do we start? An overview of the care transitions literature. *Journal of Hospital Medicine, 11*(3), 221–230.

Kansagara, D., Englander, H., Salanitro, A., Kagen, D., Theobald, C., Freeman, M., & Kripalani, S. (2011). Risk prediction models for hospital readmission: a systematic review. *JAMA, 306*(15), 1688–1698.

Kotu, V., & Deshpande, B. (2015). *Predictive analytics and data mining: concepts and practice with RapidMiner*. Amsterdam: Elsevier Ltd.

Lee, E. W. (2012). Selecting the best prediction model for readmission. *Journal of Preventive Medicine and Public Health = Yebang Uihakhoe chi, 45*(4), 259–266.

Leeds, I. L., Sadiraj, V., Cox, J. C., Gao, X. S., Pawlik, T. M., Schnier, K. E., & Sweeney, J. F. (2017). Discharge decision-making after complex surgery. Surgeon behaviors compared to predictive modeling to reduce surgical readmissions. *American Journal of Surgery, 213*(1), 112–119.

Lin, K.-P., Chen, P.-C., Huang, L.-Y., Mao, H.-C., & Chan, D.-C. D. (2016). Predicting inpatient readmission and outpatient admission in elderly. A population-based cohort study. *Medicine, 95*(16), e3484.

Lin, Y. K., Chen, H., Brown, R. A., Li, S. H., & Yang, H. J. (2017). Healthcare predictive analytics for risk profiling in chronic care: a Bayesian multitask learning approach. *MIS Quarterly: Management Information Systems, 41*(2), 473–495.

Lindenauer, P. K., Normand, S.-L. T., Drye, E. E., Lin, Z., Goodrich, K., Desai, M. M., Bratzler, D. W., O'Donnell, W. J., Metersky, M. L., & Krumholz, H. M. (2011). Development, validation, and results of a measure of 30-day readmission following hospitalization for pneumonia. *Journal of Hospital Medicine, 6*(3), 142–150.

McLaren, D. P., Jones, R., Plotnik, R., Zareba, W., McIntosh, S., Alexis, J., Chen, L., Block, R., Lowenstein, C. J., & Kutyifa, V. (2016). Prior hospital admission predicts thirty-day hospital readmission for heart failure patients. *Cardiology Journal, 23*(2), 155–162.

McManus, D. D., Saczynski, J. S., Lessard, D., Waring, M. E., Allison, J., Parish, D. C., Goldberg, R. J., Ash, A., & Kiefe, C. I. (2016). Reliability of predicting early hospital readmission after discharge for an acute coronary syndrome using claims-based data. *The American Journal of Cardiology, 117*(4), 501–507.

Mortazavi, B. J., Downing, N. S., Bucholz, E. M., Dharmarajan, K., Manhapra, A., Li, S.-X., Negahban, S. N., & Krumholz, H. M. (2016). Analysis of machine learning techniques for heart failure readmissions. *Circulation. Cardiovascular Quality and Outcomes, 9*(6), 629–640.

Murray, C. J. L., & Lopez, A. D. (1996). *The global burden of disease: A comprehensive assessment of mortality and disability from diseases, injuries, and risk factors in 1990 and projected to 2020; summary, Global burden of disease and injury series* (Vol. 1., Published by the Harvard School of Public Health on behalf of the World Health Organization and the World Bank;). Cambridge, MA: Distributed by Harvard University Press.

Nguyen, O. K., Makam, A. N., Clark, C., Zhang, S., Xie, B., Velasco, F., Amarasingham, R., & Halm, E. A. (2016). Predicting all-cause readmissions using electronic health record data from the entire hospitalization. Model development and comparison. *Journal of Hospital Medicine, 11*(7), 473–480.

Pack, Q. R., Priya, A., Lagu, T., Pekow, P. S., Engelman, R., Kent, D. M., & Lindenauer, P. K. (2016). Development and validation of a predictive model for short- and medium-term hospital readmission following heart valve surgery. *Journal of the American Heart Association, 5*(9), e003544.

Pencina, M. J., & D'Agostino, R. B. (2015). Evaluating discrimination of risk prediction models. The C statistic. *JAMA, 314*(10), 1063–1064.

Picker, D., Heard, K., Bailey, T. C., Martin, N. R., LaRossa, G. N., & Kollef, M. H. (2015). The number of discharge medications predicts thirty-day hospital readmission. A cohort study. *BMC Health Services Research, 15*, 282.

Rana, S., Tran, T., Luo, W., Phung, D., Kennedy, R. L., & Venkatesh, S. (2014). Predicting unplanned readmission after myocardial infarction from routinely collected administrative hospital data. *Australian Health Review, 38*(4), 377–382.

Sawhney, S., Marks, A., Fluck, N., McLernon, D. J., Prescott, G. J., & Black, C. (2017). Acute kidney injury as an independent risk factor for unplanned 90-day hospital readmissions. *BMC Nephrology, 18*(1), 9.

Shadmi, E., Flaks-Manov, N., Hoshen, M., Goldman, O., Bitterman, H., & Balicer, R. D. (2015). Predicting 30-day readmissions with preadmission electronic health record data. *Medical Care, 53*(3), 283–289.

Shams, I., Ajorlou, S., & Yang, K. (2015). A predictive analytics approach to reducing 30-day avoidable readmissions among patients with heart failure, acute myocardial infarction, pneumonia, or COPD. *Health Care Management Science, 18*(1), 19–34.

Shmueli, G., & Koppius, O. (2011). Predictive analytics in information systems research. *MIS Quarterly, 35*(3), 553–572.

Shulan, M., Gao, K., & Moore, C. D. (2013). Predicting 30-day all-cause hospital readmissions. *Health Care Management Science, 16*(2), 167–175.

Tabak, Y. P., Sun, X., Nunez, C. M., Gupta, V., & Johannes, R. S. (2017). Predicting readmission at early hospitalization using electronic clinical data: an early readmission risk score. *Medical Care, 55*(3), 267–275.

Taber, D. J., Palanisamy, A. P., Srinivas, T. R., Gebregziabher, M., Odeghe, J., Chavin, K. D., Egede, L. E., & Baliga, P. K. (2015). Inclusion of dynamic clinical data improves the predictive performance of a 30-day readmission risk model in kidney transplantation. *Transplantation, 99*(2), 324–330.

Tong, L., Erdmann, C., Daldalian, M., Li, J., & Esposito, T. (2016). Comparison of predictive modeling approaches for 30-day all-cause non-elective readmission risk. *BMC Medical Research Methodology, 16*, 26.

Walsh, C., & Hripcsak, G. (2014). The effects of data sources, cohort selection, and outcome definition on a predictive model of risk of thirty-day hospital readmissions. *Journal of Biomedical Informatics, 52*, 418–426.

Webster, J., & Watson, R. T. (2002). Analyzing the past to prepare for the future: writing a literature review. *MIS Quarterly, 26*(2), xiii–xxiii.

Whitlock, T. L., Tignor, A., Webster, E. M., Repas, K., Conwell, D., Banks, P. A., & Wu, B. U. (2011). A scoring system to predict readmission of patients with acute pancreatitis to the hospital within thirty days of discharge. *Clinical Gastroenterology and Hepatology, 9*(2), 175–180. quiz e18.

Yeo, H., Mao, J., Abelson, J. S., Lachs, M., Finlayson, E., Milsom, J., & Sedrakyan, A. (2016). Development of a nonparametric predictive model for readmission risk in elderly adults after colon and rectal cancer surgery. *Journal of the American Geriatrics Society, 64*(11), e125–e130.

Yu, S., Farooq, F., van Esbroeck, A., Fung, G., Anand, V., & Krishnapuram, B. (2015). Predicting readmission risk with institution-specific prediction models. *Artificial Intelligence in Medicine, 65*(2), 89–96.

Zhu, K., Lou, Z., Zhou, J., Ballester, N., Kong, N., & Parikh, P. (2015). Predicting 30-day hospital readmission with publicly available administrative database. A conditional logistic regression modeling approach. *Methods of Information in Medicine, 54*(6), 560–567.

**Isabella Eigner** studied International Information Systems at the University of Erlangen-Nuremberg. Since 2014, she is a member of the research staff and PhD student at the Institute of Information Systems. In her research, she is focusing on predictive analytics in healthcare.

**Andrew Cooney** studied Information Systems at the University of Erlangen-Nuremberg. In 2017, he did his master thesis on readmission risk prediction at the Institute of Information Systems.

# Using Knowledge Management to Develop Superior Online Health Decision Support Solutions: The Case of Allergy Care

**Nilmini Wickramasinghe**

## 1 Introduction

The numbers of patients suffering from allergy is globally increasing (AAAAI 2016). Recent data generated from the local HealthNuts study has identified an increase of 20% sensitisation and a 10% challenge confirmed food allergy rate in 12-month-old Melbourne infants (Osborne et al. 2011, 2014; Osborne et al. 2011), and the picture of allergy in Australia in general looks even more worrying (Abels and Cogdill 2004). According to a report made for the Australasian Society of Clinical Immunology and Allergy (ASCIA) in 2007 (Access Economics Pty Limited 2007), there were 4.1 million Australians (19.6%) having at least 1 allergy and the average Australian allergic person with 1.74 allergies, forecasting of 70% in the number of Australians with allergies affected from 4.1 million now to 7.7 million by 2050 and an increased proportion affected from 19.6% to 26.1% if the trends back then continued (Access Economics Pty Limited 2007; HealthDirect 2018).

The financial impact of allergy treatment is significant and growing; 7.8 billion Australian dollars was the calculated cost of allergies in Australia in 2007, and this number is projected to increase exponentially in the next 20 years (HealthDirect 2018). This is due to different interrelated factors like lower productivity ("presenteeism" $4.2 billion), direct medical costs ($1.2 billion), lower employment rates ($1.1 billion), absenteeism and lost household productivity ($0.2 billion) and premature death ($83 million) (Access Economics Pty Limited 2007). Hence, focussing on identifying superior cost-effective solutions to support allergy care, as this study sets out to do, is a key need in the Australian healthcare environment.

N. Wickramasinghe (✉)
Epworth HealthCare, Richmond, VIC, Australia

Swinburne University of Technology, Hawthorn, VIC, Australia



N. Wickramasinghe, F. Bodendorf (eds.), *Delivering Superior Health and Wellness Management with IoT and Analytics*, Healthcare Delivery in the Information Age, https://doi.org/10.1007/978-3-030-17347-0_13

## 2 Background

Food allergy diagnosis is confirmed by a positive allergy test (skin prick test [SPT] or serum-specific IgE [ssIgE] blood test) in conjunction with a history of an immediate allergic reaction to that food (Osborn et al. 2014; Osborne et al. 2011; Peters et al. 2013). However, screening allergy testing is frequently performed in the community prior to a patient having known exposure to a food allergen, particularly in at-risk patient groups, e.g. patients' with atopic dermatitis (AD) [ibid]. The wheal size of the SPT and level of ssIgE are used as guides as to the likelihood or otherwise of clinical food allergy (Hill et al. 2004).

The gold standard for food allergy diagnosis in this setting is by a physician-supervised food challenge (Osborn et al. 2014; Osborne et al. 2011; Peters et al. 2013). Most food allergies in childhood resolve over time (e.g. egg, milk, wheat, soy); however, allergies to peanut, tree nut, fish and shellfish tend to be ongoing into adult life in up to 80–90% of patients (Osborn et al. 2014; Osborne et al. 2011; Peters et al. 2013). To follow up for the development of potential clinical tolerance, patients are re-evaluated every few years with repeat allergy testing (either SPT or ssIgE) (Peters et al. 2013). The wheal size of the SPT and level of ssIgE are used as guides as to the likelihood or otherwise of the development of clinical tolerance (Osborn et al. 2014; Osborne et al. 2011). Serial allergy testing over several years is common before the allergy test (SPT or ssIgE) is at a level where it may be considered appropriate to proceed with an oral food challenge (Fig. 1) (Sampson 2002).

It is not an unusual clinical practice for clinicians to provide patients with a copy of their SPT results each time the test is performed as well as keep a record of the test in either written or electronic form in their patient database (Osborn et al. 2014; Osborne et al. 2011; Peters et al. 2013). As allergy services are currently stretched, there are frequently prolonged waiting times for food allergy review, particularly in the public sector (HealthDirect 2018). Further, it is often the case for patients to be seen by different allergy practitioners for follow-up testing in the longer term (HealthDirect 2018). In the interim patients are frequently encouraged to visit their general practitioner for food allergy follow-up that may include intermittent evaluation of yearly or second yearly ssIgE to the food allergen(s) in question (HealthDirect 2018). In the event that these levels are low or approaching negative, re-referral for follow-up and consideration of formal inpatient challenge may be appropriate (HealthDirect 2018). In this situation, it is common that previous test results are not readily available for comparison (Duncavage and Hagaman 2013). Especially given that there is some evidence to suggest that the rate of change in SPT wheal size or ssIgE levels over time may help in predicting the development of tolerance, this is clearly a suboptimal scenario (Osborn et al. 2014; Osborne et al. 2011; Peters et al. 2013).

Providing allergy care thus seems to lack accurate and reliable data within and between different allergy care providers. According to Ho-Chang, Chang and Prybutok (2014), more than 57% of 89 countries around the world had no food

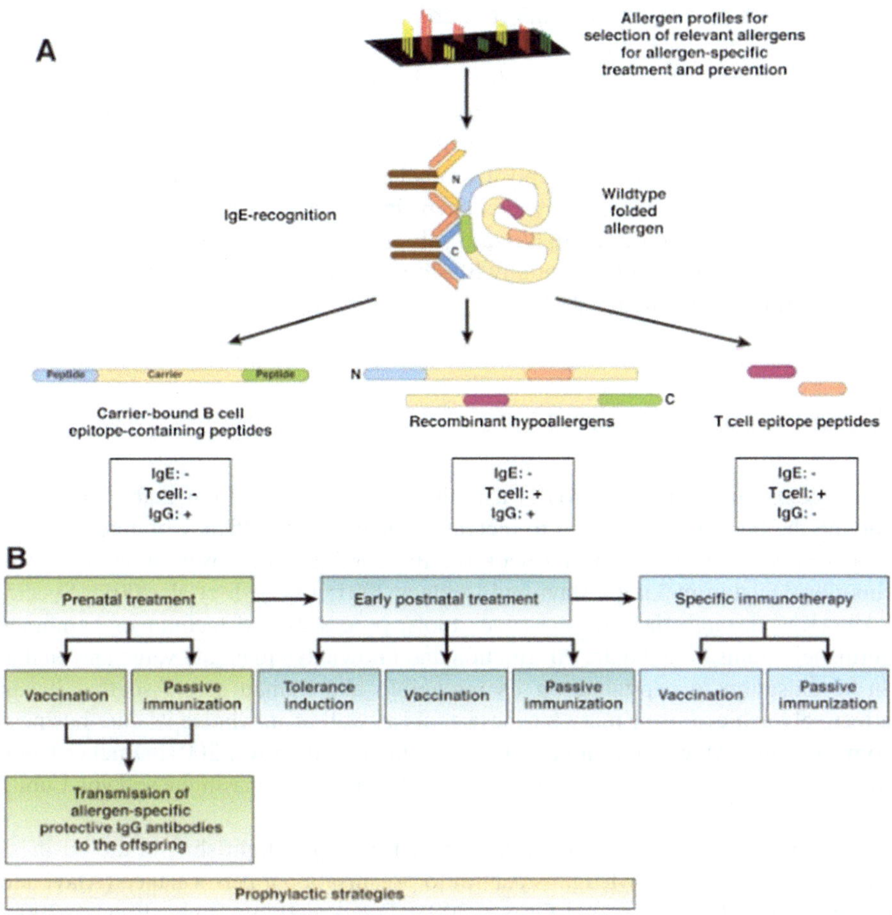


**Fig. 1** Key aspects in allergy management (Osborne et al. 2011)

allergy prevalence data of any kind, and just 10% of these countries had this sort of data. Hence, analysing the current approaches to treat allergy (Picker Institute 2000) and then designing and developing a technology solution to support and enable high value superior care in this domain appears to be a prudent approach. It would appear that the main deficiencies relate to the lack of documenting serial allergy test results (both SPT size and ssIgE levels) over time in a readily available and consistent dataset by both allergists and their patients. So offering this possibility would be an invaluable tool in the long-term management of these patients. Thus, this study aims to develop an online solution for use by both allergists and their patients to ensure prospective accurate and secure collection of SPT and ssIgE data for long-term follow-up of patients with food allergy.

This study was conducted in collaboration with an Australian private healthcare provider. Within this hospital, an allergy clinic was established in 2010 to provide a comprehensive service for children, adolescents and adults with allergic and immune disorders including the long-term follow-up and management of patients with food allergy. This clinic is the first such service to provide safe inpatient food challenges in the private sector in Australia. To date, this clinic has performed over 2000 food challenges to confirm or refute the development of tolerance to previously identified food allergens with a positive challenge rate of 20%. These challenges are not without risk with anaphylaxis occurring in 2% of all challenges currently performed at the study site, highlighting the importance of a service with expert staff equipped to deal with potentially life-threatening anaphylaxis (Ho-Chang et al. 2014; Institute of Medicine 2015).

## 3 Knowledge Management

Knowledge management (KM) is a multidisciplinary paradigm that is aimed at solving the business challenges to increase efficiency and efficacy of core business processes while simultaneously incorporating continuous innovation and enabling superior decision-making (Alavi and Leidner 2001; Dwivedi et al. 2001). Specifically, KM through the use of various tools, processes and techniques combines germane organisational data, information and knowledge to create value and enable an organisation to capitalise on its intangible and human assets so that it can effectively achieve its primary business goals as well as maximise its core business competencies (Alavi and Leidner 2001; Davenport and Grover 2001; Drucker 1999; Muhammad et al. 2013; Wickramasinghe 2006a, b; Wickramasinghe and von Lubitz 2006).

The premise for the need for KM is based on a paradigm shift in the business environment where knowledge is central to organisational performance (Alavi and Leidner 2001; Davenport and Grover 2001; Drucker 1999; Dwivedi et al. 2001; Muhammad et al. 2013; Wickramasinghe 2006a, b; Wickramasinghe and von Lubitz 2006).

Knowledge management deals with distinct, but related, concepts, starting from data (raw knowledge) to information (data arranged into a meaningful pattern), to knowledge (contextualised information) and to wisdom; upon its existence beyond knowledge, there is a wide agreement (Wickramasinghe 2006a, b). Central to KM is naturally the knowledge construct. Knowledge is not static; rather it changes and evolves during the life of an organisation (Alavi and Leidner 2001; Dwivedi et al. 2001). Moreover, it is possible to change the form of knowledge; i.e. turn existing tacit knowledge (or experiential "know-how") into new explicit knowledge (or factual "know what") and existing explicit knowledge into new tacit knowledge (Nonaka 1994; Nonaka and Nishiguchi 2001; Nonaka and Takeuchi 1995; Wickramasinghe 2006a). This process of changing the form of knowledge is known as the knowledge spiral (Wickramasinghe 2006a, b; Wickramasinghe and von Lubitz 2006), in which, according to Nonaka (1994; Nonaka and Takeuchi 1995, 2001): (1) Tacit to tacit knowledge transfer usually occurs through apprenticeship type relations where the teacher or master passes on the skill to the apprentice. (2) Explicit to explicit knowledge transfer usually occurs via formal learning of facts. (3) Tacit to explicit knowledge transfer usually occurs when there is an articulation of nuances; for example, if a famous surgeon is questioned as to why he does a particular procedure in a certain manner, by his articulation of the steps, the tacit knowledge becomes explicit. (4) Explicit to tacit knowledge transfer usually occurs as new explicit knowledge is internalised; it can then be used to broaden, reframe and extend one's tacit knowledge.

These transformations are often referred to as the modes of socialisation, externalisation, internalisation and combination, respectively (Nonaka 1994; Nonaka and Takeuchi 1995, 2001). Integral to this changing of knowledge through the knowledge spiral is that new knowledge is created (Wickramasinghe 2006a, b; Wickramasinghe and von Lubitz 2006) and this can bring many benefits to organisations. In the case of transferring tacit knowledge to explicit knowledge, for example, an organisation is able to capture the expertise of particular individuals; hence, this adds not only to the organisational memory but also enables single loop and double loop organisational learning to take place (Wickramasinghe 2006b). In healthcare, this may translate, for example, to the developing of better protocols and codes or even further refinement to existing DRGs (Diagnosis Related Groupings). While the transformations of the knowledge spiral provide a theoretical framework (that facilitates the understanding of the complex knowledge), in order to be useful in practice (i.e. for our purposes), these transformations must occur and/or be invoked to assist in the superior diagnosis and consequent prognosis and treatment of paediatric allergies.

## *3.1 Knowledge Management for Healthcare*

With the ever-increasing volume of data being produced daily in the electronic medical records (EMR) and clinical databases, knowledge management approaches provide a tool-rich platform to perform pattern identification tasks, such as detecting associations between certain risk factors and outcomes, ascertaining trends in healthcare utilisation or discovering new models of disease in populations; the tools and techniques of KM are very relevant to healthcare and this particular context, allergy detection, prediction and on-going treatment management (Wickramasinghe and von Lubitz 2006).

In their comprehensive assessment of applying knowledge management in the healthcare industry, Wickramasinghe et al. (2009) noted that the gap between data collection and data comprehension and analysis is becoming more problematic, given the increased volume and complexity of clinical data, which, in one way or other, reflects the complexity of the healthcare itself, and thus the tools of KM are applied to this research.

### 3.2 *The Application of Knowledge Management on This Study*

The tools, tactics and techniques of KM are relevant as follows: first, the designed database to store results and tests is expected to produce high volumes of data on different types of allergy in different age groups. Not only is the volume demanding, but also the complexity of the produced data is an issue. Those two factors combined, make the use of know management prudent to maximise the benefit of using the designed knowledge base. Second, KM techniques will help bridge the gap between data collection as a routine procedure and data comprehension and analysis as an innovative and iterative process. This is important based on the aforementioned explanation. Third, it will help clinicians to better understand their patient data with less effort and time; i.e. they can focus on pertinent information and germane knowledge, which, in turn, increases the efficiency and efficacy of their daily business. Fourth, the aim of this study is to create reliable and exchangeable knowledge among different allergy treatment providers, rather than merely creating the knowledge base. This is a central aspect of any knowledge management approach, by moving from raw knowledge (data), which is much context-dependent, to knowledge and then ideally wisdom, which is much more context-independent (Nonaka 1994).

## 4 The Proffered Solution

An essential aspect of the proffered solution is an electronic knowledge base which will ensure prospective accurate collection of allergy test results over time. In doing so, this will enable ready access to these results by both patients and clinicians at each point of contact with health service providers to aid in the long-term follow-up and management of these patients with food allergies. The knowledge base will be designed to alert patients to appropriate timing for re-evaluation and potential consideration for food challenge; thereby, allowing timelier introduction of food to which they had previously reacted or had allergy test results predictive of likely true clinical allergy.

The longer-term plan is to develop the knowledge base into a usable application for smartphone/handheld devices increasing portability/usability and autonomy of patient groups with allergic disease so that access to critical and potentially life-saving information is readily available at their fingertips in real time.

## 5 Study Design Methodology

This study adopts a single case study (Yin 2003) and Design Science Research Methodology (DSRM) (Peffers et al. 2002). Case study research is appropriate when a study is exploratory in nature and it is desired to capture the knowledge of practitioners and developing theories from it (Benbasat et al. 1987). Further, as a main focus of this study is to design an online solution that facilitates knowledge transfer and sharing among different allergy care providers, a DSRM approach is also adopted (Fig. 2a, b) (Peffers et al. 2002). DSRM is a process model that helps carry out research focussed on designing artefacts (Hevner 2007). This methodology is widely used quantitatively to improve existing solutions or qualitatively to create new solutions to unsolved problems (Gregor and Hevner 2013). In this research, this methodology is used qualitatively given the nature of this study as it aims at creating a novel solution for an existing problem (von Alan et al. 2004), namely, the fragmented chain of knowledge sharing among allergy care providers. Hevner (Peffers et al. 2002) conceptualises DSRM in three cycles, including relevance cycle, rigour cycle and design cycle. According to this framing, DSRM starts with the relevance cycle which role is centred on providing the requirements for the research by identifying the opportunity/problem to be addressed (Peffers et al. 2002). The rigour cycle is centred on providing past knowledge to the research project to ensure its innovation, while the design cycle concentrates on building the artefacts and processes and evaluating them (Peffers et al. 2002). As this research is in its initial stages, the relevance and design cycles are relevant to this study at this stage. Upon the building of the solution, the rigour cycle will then be addressed. Table 1 summarises how the relevance and design cycles are mapped to this study.

This study has two distinct phases:

1. Designing the technology solution, the knowledge base.
2. Testing the fidelity and usability of the designed knowledge base. Key activities include:

    - Storyboarding with head of clinic to clearly understand needs and requirements.
    - Investigating the possibility to securely store photos of results of skin prick tests with data.
    - Creating an app that the clinicians can use to enter skin prick test data.
    - Creating a secure encrypted knowledge base.
    - Ensuring data is backed up automatically.
    - Calculating the "mean" for each allergen.
    - Generating a trend analysis for each patient.
    - Creating a printout results to replicate the existing skin prick test.
    - Printing out an informative results sheet for parents/patients.
    - Testing and validation of the product.
    - Testing the fidelity and validity of the knowledge base.

The testing will be done over a 3-month period through a two-arm non-blinded study, during which patients who agree to participate will be randomly selected into the technology arm (knowledge base plus standard care protocols) or standard care arm (current practice method), respectively. Clinicians will then perform the required allergy testing and consult using, respectively, the technology solution or standard care approach. We note that for the technology arm, the research assistant

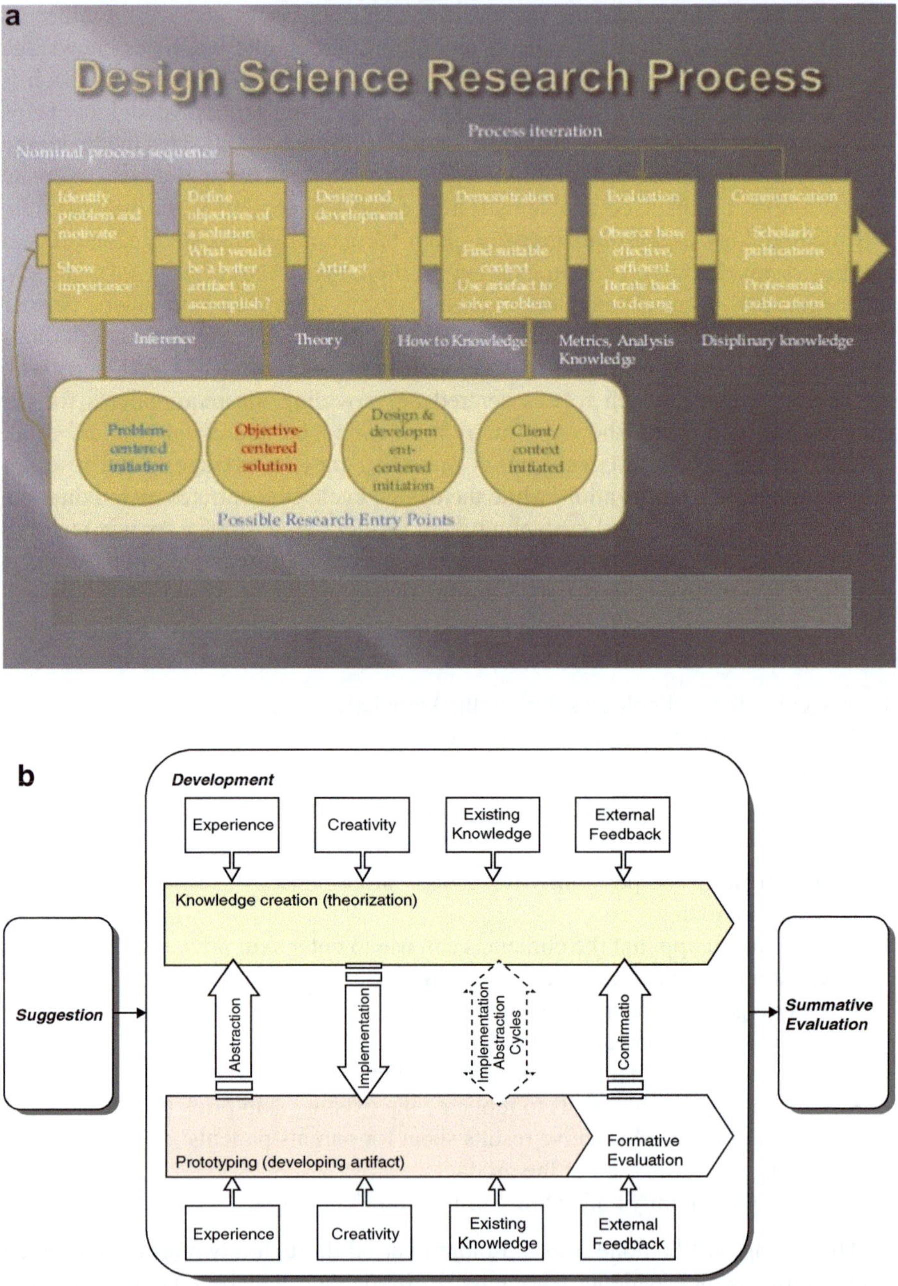


**Fig. 2** (**a**) DSRM, (**b**) DSRM for prototyping. (Adapted from Peffers et al. 2002)

**Table 1** Mapping DSRM (Peffers et al. 2002) to the proposed system

| DSRM activity | Activity description | Application on this study |
|---|---|---|
| Problem identification and motivation | Define the specific research problem and justify the value of a solution. | Increasing rates of food allergy diagnosis, the lack of a computerised information and communication system to support knowledge flow among allergy care and the need to address this gap. |
| Definition of objectives of the solution | Create or improve an artefact respectively based on knowledge of the problem and current solutions, if any, and their efficacy. | The objective is to create an artefact to address the fragmented nature of allergy care. |
| Design and development | Create the artefact, including the desired functionality and its architecture based of knowledge of theory that can be used to bear in a solution. | The artefact is to be designed to facilitate knowledge sharing and flow among different allergy care providers To do so, the design phase is conducted collaboratively between the researchers and the clinicians at the selected case. |
| Demonstration | Demonstrate the use of the artefact to solve the problem. | The artefact is to be tested and tried, and then in-depth analysis will be performed to measure the extent to which the proposed system helps solve the problem. |
| Evaluation | Iterate back to better design the artefact if needed. | As needed, iteration back will take place, based on in-depth analyses to identify rooms of improvement |
| Communication | Publish and let the value of the solution talk about itself. | Conference and journal publications and other presentation activities to develop the project further and share the findings with interested stakeholders. |

will still complete the paper work as directed by the clinician, so at all times all patients will at a minimum receive the same standard care.

On the completion of the trial, a focus group with the clinicians will be conducted to capture their insights on the usability of the proposed system, as well as their comments, recommendations and other feedback. Fidelity of the system will also be assessed by comparing standard care results with the technology arm results. To ensure the validity of the data collected through the focus group, an online survey will be administered at the conclusion of the focus group to ensure data triangulation.

### 5.1 Setting and Participants

As stated by Hung and Cheng (2013), individual's positive attitude, such as optimism, toward the Technology Readiness Index, positively affects acceptance of technology; an individual's negative perception of the Technology Readiness Index, such as discomfort, has a negative effect on his or her sense of the perceived ease of technology and compatibility in regard to prior experience and technology. Thus, during the research design phase, addressing the technological concerns of the allergy care providers represented the cornerstone in this study.

Both qualitative and quantitative data will be collected from three different sources:

1. Initial discussions with the head of a Melbournian allergy clinic to specify the initial design requirements based on the standard allergy care protocol.
2. Randomly selected de-identified existing patients' records from this clinic to help establish the main structure of the proposed system.
3. A focus group with up to five clinicians from this clinic to have their insights on the use of the proposed system, recommendations and comments on and about the prototype once it is developed.

For data triangulation, an online survey of clinicians will be designed and administrated using SurveyGizmo.

### 5.2 Data Collection Plan and Techniques

Initial discussions have been conducted with the head of the allergy clinic. Through these discussions we have established that the need for an electronic knowledge base with intelligent analysis tools and techniques plus the mobile App is urgently required. Patients' records will be randomly selected with full compliance with ethical requirements. This data is then used to populate the developed knowledge base. The focus group will target up to five clinicians who work for the selected clinic. They are recruited based on their daily interaction with patients and their records. Those who participate in the focus group will be requested to complete a follow-up survey for data triangulation. Patients who visit the clinic during the 3-month trial period will be asked if they wish to participate in the study (again subscribing to all ethical requirements) and, if so, will then be randomly selected into the respective arms of the two arm trial (as described above). All patients will receive equal care and attention irrespective of their participation or not in the study.

Data will be collected using the methods of (1) de-identified patient records, (2) focus groups and (3) an online survey (using SurveyGizmo). All patient data will be handled using standard Australian Privacy Principles; including using double de-identified data to ensure confidentiality and anonymity. In addition, all patient data collected will be disclosed only with participants' permission, except as required

**Table 2** A Summary of data collection plan and techniques

| Key participants/key milestones | Participants/records | Participant numbers |
|---|---|---|
| Initial discussions | Specialist doctor (allergist) | 1 |
| Patients' records | De-identified patients' records (historical records) randomly selected | Patients' records: $\leq$ 50 records |
| Focus group | Clinicians work in allergy care.<br>This will be conducted after the solution is trialled | An expert group: ~=5 |
| On-line survey | Clinicians work in allergy care.<br>This will be conducted after the completion of the focus group | An expert group: ~=5 |

**Fig. 3** Initial prototype

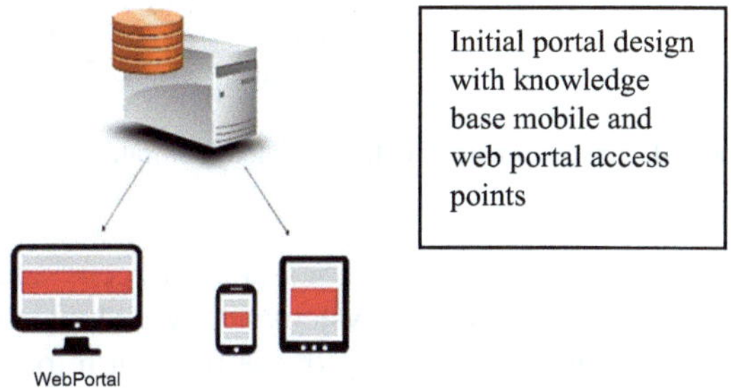


by law. All information will be stored securely at the study site in a locked office in a filing cabinet and/or password-protected computer. Table 2 summarises data collection plan and techniques.

## 6 Results to Date and Discussion

Currently, the initial prototype has been designed (Fig. 3). This includes design of the backbone knowledge base and the web portal and testing them both by using a dummy dataset. The initial results show that the knowledge base is robust and capable to be the back-bone supporting system. The next step is to use real de-identified data to test its scalability in real time. The majority of design requirements were met during the design, plus subscribing to the DSRM approach. This will be further confirmed during the second phase of the study.

This research has a number of implications for both theory and practice. From a theoretical perspective, this research addresses an urgent need to address the inter-organisational knowledge flow and the potential role of an online system in this regard, specifically in a healthcare context. In addition, it extends the application of the tools and techniques of KM in the healthcare context. Specifically (Fig. 4), the knowledge transformation in the knowledge spiral have been addressed. In particular, the knowledge in the knowledge communities (who have the knowl-

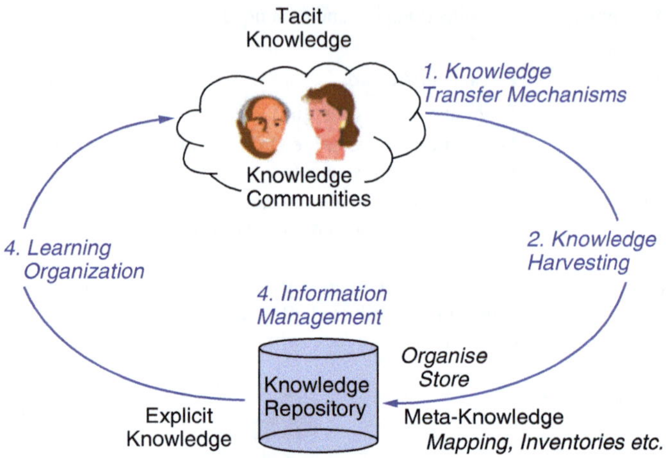


**Fig. 4** Transformations in the knowledge spiral. (Adapted from Wickramasinghe 2006b)

edge of the various allergies and allergens and how they, respectively, must be treated) is now stored systematically and succinctly in the electronic knowledge base, while the harvesting of the relevant information and pertinent knowledge on a case-by-case basis to assist specific individuals will be done via the mobile app and/or web portal; in addition, the knowledge will be added to the existing extant knowledge base to support the continuous learning cycles.

It also aims at highlighting the technical challenges that may affect the use of such systems and limit their usability. In addition, a further theoretical aspect is the application of DSRM in a healthcare context.

From the perspective of practice, we highlight several key aspects as follows: first, the proposed solution is an attempt to address the three problems facing the current model of allergy care, namely, the lack of awareness of allergy care requirements (by enabling educational contents for allergy patients), difficulty accessing specialised allergy care (by increasing the efficiency of care delivery through the use of information technology) as well as the nature of allergy care and the need for higher coordination among different care providers. Secondly, this solution has a potential to enhance the long-term follow-up of patients attending for comprehensive food allergy management at healthcare contexts. Thirdly, this project has great potential to be patented and commercialised again bringing kudos and financial benefits to healthcare contexts. Fourthly, the outputs of this study are far reaching to patients and their communities in developed countries, particularly in Australia, where allergy rates are among the highest in the world. It is expected that there will be instant and long-term clinical benefits by facilitating proactive and protective allergy care and management practices. The next step in this research is to examine the usability and fidelity of the proposed system at the selected case. The outputs of this testing will be used to further enhance different design and functionality aspects of the proposed system and complete the rigour cycle of the DSRM approach.

# 7 Conclusion

This chapter has served to outline a research in progress study that serves to address a healthcare need drawing upon the tools, techniques and tactics of knowledge management. Specifically, a knowledge-based online solution is designed and developed (Fig. 3) using a DSRM to address current key issues in the delivery of care for patients suffering from allergies. The developed prototype now is to be tested to establish usability, fidelity and patient-centeredness. In this way, the impact of the proffered system as a decision support tool powered by integrating key aspects of the tools, techniques and technologies of knowledge management to enable superior value-based care delivery will be identified and realised. Further, the study highlights the enabling role of technology solutions in the context of allergy care and supporting high value high quality on going care that is patient-centred. In this way it becomes possible to realise the critical role to be played by collaborative technologies in healthcare context. Finally, one that is to be addressed relates to the financial aspect. As noted at the start of this paper, the current approaches for treating, managing and monitoring allergy care especially paediatric allergy care is expensive and increasing exponentially. The developed prototype has the potential to minimise waste and provide a high value high quality solution. Currently, it is only possible to outline area where waste minimisation and cost containment can take place. Once the final prototype is operational, it will also be possible to perform an econometric analysis to identify specific areas of cost savings and thus the true financial implication for allergy care.

# 8 Acknowledgement

The insights and advice from Dr. Joann Smart are acknowledged with thanks.

# References

AAAAI. Allergy statistics. Retrieved May 10, 2016., from http://www.aaaai.org/about-aaaai/newsroom/allergy-statistics

Abels, E. G., Cogdill, K. W., & Zach, L. (2004). Identifying and communicating the contributions of library and information services in hospitals and academic health sciences centers. *Journal of the Medical Library Association, 90*(3), 46–55.

Access Economics Pty Limited. (2007). *The economic impact of allergic disease in Australia: Not to be sneezed at*. Balgowlah: Australasian Society of Clinical Immunology and Allergy (ASCIA).

Alavi, M., & Leidner, D. (2001). Review: Knowledge management and knowledge management systems: Conceptual foundations and research issues. *MIS Quarterly, 25*(1), 107–136.

Benbasat, I., Goldstein, D. K., & Mead, M. (1987). The case research strategy in studies of information systems. *MIS Quarterly, 11*(3), 369–386.

Davenport, T., & Grover, V. (2001). Knowledge management. *Journal of Management Information Systems, 18*(1), 3–4.
Drucker, P. (1999 Oct). Beyond the information revolution. *The Atlantic Monthly*, 47–57.
Duncavage, J., & Hagaman, D. D. (2013). Practical aspects of integrating allergy and pulmonology management into a rhinology practice: The Vanderbilt ASAP experience. *Current Opinion in Otolaryngology & Head and Neck Surgery, 21*(1), 74–78.
Dwivedi, A., Bali, R. K., James, A. E., & Naguib, R. N. G. (2001). Telehealth systems: considering knowledge management and ICT issues. In *Proc. IEEE EMBC-2001 23rd annual international conference of the IEEE engineering in medicine and biology society (EMBS)* (pp. 3614–3617). Istanbul.
Gregor, S., & Hevner, A. R. (2013). Positioning and presenting design science research for maximum impact. *MIS Quarterly, 37*(2), 337–355.
HealthDirect. (2018). https://www.healthdirect.gov.au/allergy-testing. Accessed 28 Apr 2018.
Hevner, A. R. (2007). A three cycle view of design science research. *Scandinavian Journal of Information Systems, 19*(2), 4.
Hill, D. J., Heine, R. G., & Hosking, C. S. (2004). The diagnostic value of skin prick testing in children with food allergy. *Pediatric Allergy and Immunology, 15*(5), 435–441. https://doi.org/10.1111/j.1399-3038.2004.00188.x.
Ho-Chang, C., Chang, E. K., & Prybutok, V. R. (2014). Information technology capability and firm performance: Contradictory findings and their possible causes. *MIS Quarterly, 38*(1), 305–A314.
Hung, S. W., & Cheng, M. J. (2013). Are you ready for knowledge sharing? An empirical study of virtual communities. *Computers & Education, 62*, 8–17. https://doi.org/10.1016/j.compedu.2012.09.017.
Institute of Medicine. (2015). *Building a better delivery system: A new engineering/health care partnership*. Washington, D.C.: Institute of Medicine.
Muhammad, I., Teoh, S., & Wickramasinghe, N. (2013). The need for a socio-technical analysis in E-health: The case of the PCEHR. *International Journal of E-Health and Medical Communications (IJEHMC), 4*(2), 65–79. https://doi.org/10.4018/jehmc.2013040105.
Nonaka, I. (1994). A dynamic theory of organizational knowledge creation. *Organizational Science, 5*, 14–37.
Nonaka, I., & Nishiguchi, T. (2001). *Knowledge emergence*. Oxford: Oxford University Press.
Nonaka, I., & Takeuchi, N. (1995). *The knowledge creating company*. Oxford: Oxford University Press.
Osborn, R., Moulds, D., Squires, D., Doty, M., & Anderson, C. (2014). International survey of older adults finds shortcomings in access, coordination, and patient-centered care. *Health Affairs (Project Hope), 33*(12), 2247–2255. https://doi.org/10.1377/hlthaff.2014.0947.
Osborne, N. J., Koplin, J. J., Martin, P. E., Gurrin, L. C., Lowe, A. J., Matheson, M. C., & Allen, K. J. (2011). Prevalence of challenge-proven IgE-mediated food allergy using population-based sampling and predetermined challenge criteria in infants. *The Journal of Allergy and Clinical Immunology, 127*(3), 668–676.e662. https://doi.org/10.1016/j.jaci.2011.01.039.
Peffers, K., Tuunanen, T., Rothenberger, M. A., & Chatterjee, S. (2002). A design science research methodology for information systems research. *Journal of Management Information Systems, 24*(3), 45–77.
Peters, R. L., Gurrin, L. C., Dharmage, S. C., Koplin, J. J., & Allen, K. J. (2013). The natural history of IgE-mediated food allergy: Can skin prick tests and serum-specific IgE predict the resolution of food allergy? *International Journal of Environmental Research and Public Health, 10*(10), 5039–5061. https://doi.org/10.3390/ijerph10105039.
Picker Institute. (2000). *Eye on patients: A report by the picker institute for the American Hospital Association*. Boston: Picker Institute.
Sampson, H. A. (2002). Improving in-vitro tests for the diagnosis of food hypersensitivity. *Current Opinion in Allergy and Clinical Immunology, 2*(3), 257–261.
von Alan, R. H., March, S. T., Park, J., & Ram, S. (2004). Design science in information systems research. *MIS Quarterly, 28*(1), 75–105.

Wickramasinghe, N. (2006a). Knowledge creation: a meta-framework. *International Journal of Innovation and Learning, 3*(5), 558–573.

Wickramasinghe, N. (2006b). Fostering knowledge assets in healthcare with the KMI model. *International Journal of Management and Enterprise Development, 4*(1), 52–65.

Wickramasinghe, N., & von Lubitz, D. (2006). *Knowledge-based enterprises: Fundamentals and theories*. Hershey: Idea Group Hershey.

Wickramasinghe, N., Bali, R., Lehany, B., Gibbons, C., & Schaffer, J. (2009). *Healthcare knowledge management primer (Routledge series in information systems)*. New York: Routledge. ISBN-13: 9780415994446.

Yin, R. K. (2003). *Case study research: Design and methods* (3rd ed.). Thousand Oaks: Sage Publications.

# Opportunities for Using Blockchain Technology in e-Health: e-Prescribing in Germany

**Juergen Seitz and Nilmini Wickramasinghe**

## 1 Introduction

Blockchain is a revolutionary technology which has taken the world by storm and changed the way data is transferred, managed, and shared with different parties (Gartner 2016). Its impressive success in various sectors has left the healthcare industry and its stakeholders with many questions regarding whether this advancement can be effectively applied with success to enable key activities in this sector. The answers to these questions are not simple, but we believe that it is important to investigate given the challenges facing healthcare delivery globally today. In particular, one area we assert that may benefit from blockchain technology is e-prescriptions as will be discussed in the following.

Gartner proposed in July 2016 that blockchain would reach the peak of inflated expectations very soon (Gartner 2016) (Fig. 1).

Bslockchain as a concept of a distributed database was described for the first time by Nakamoto in 2008. He developed the concept of Bitcoin, a virtual currency based on encryption (Nakamoto 2008). The concept behind Bitcoin is one of the most advanced blockchain applications and the most discussed concept of blockchain technology in academia (Gartner 2016; Nakamoto 2008). But there exist more opportunities and possibilities for blockchain in other areas and in other

J. Seitz
Baden-Wuerttemberg Cooperative State University Heidenheim, Heidenheim an der Brenz, Germany
e-mail: seitz@dhbw-heidenheim.de

N. Wickramasinghe (✉)
Epworth HealthCare, Richmond, VIC, Australia

Swinburne University of Technology, Hawthorn, VIC, Australia



N. Wickramasinghe, F. Bodendorf (eds.), *Delivering Superior Health and Wellness Management with IoT and Analytics*, Healthcare Delivery in the Information Age, https://doi.org/10.1007/978-3-030-17347-0_14

**Fig. 1** Hype cycle for emerging technologies (Gartner 2016)

industries. For example, there are also concepts developed in the area of e-health. The most sophisticated approach seems to be MedRec prototype for electronic health records and medical research data by Ekblaw et al. (2016), which describes blockchain technology for medical data access and permission management (Azaria et al. 2016). Electronic health records (EHR) are a highly sensitive area. Although there exist rules like HIPAA Privacy Rules and EU General Data Protection Regulation, in general, people have a low level of trust of EHR systems, especially in Germany (Haas 2016; Morton 2008). Therefore, the application of blockchain for e-prescriptions, i.e., to begin with an application where a lot of structured data are automatically processed by several interest groups: The process of electronic prescription could be very attractive. Before we can understand the potential for blockchain in e-prescriptions, let us first understand basic aspects around blockchain technology and its potential in the healthcare context.

## 2 Background

As more efforts are geared toward this data-based technology in healthcare, it is prudent to first unpack blockchain and focus on a few key questions in turn in order to fully comprehend its applications in healthcare sector.

## 2.1 What Is Blockchain?

Blockchain is a public ledger where recording of digital events takes place in a chronological order (Lemieux 2016). This ledger is a write-once-read-only type of record which is always shared in a peer-to-peer network (Lemieux 2016). All transactions taking place via this ledger are recorded in a database that can only be shared with approved users who can add onto it whenever necessary (Lemieux 2016). However, no person even the authorized users will be able to delete or amend any record without validation by all the users (Lemieux 2016). This technology was developed back in 2008, and it has been the backbone of digital currencies such as Bitcoin and Ethereum (Lemieux 2016).

Thus a blockchain is a sequence of data blocks where each data block consists of a data set and a hash value of the previous block (see Figs. 2a, b). These blocks

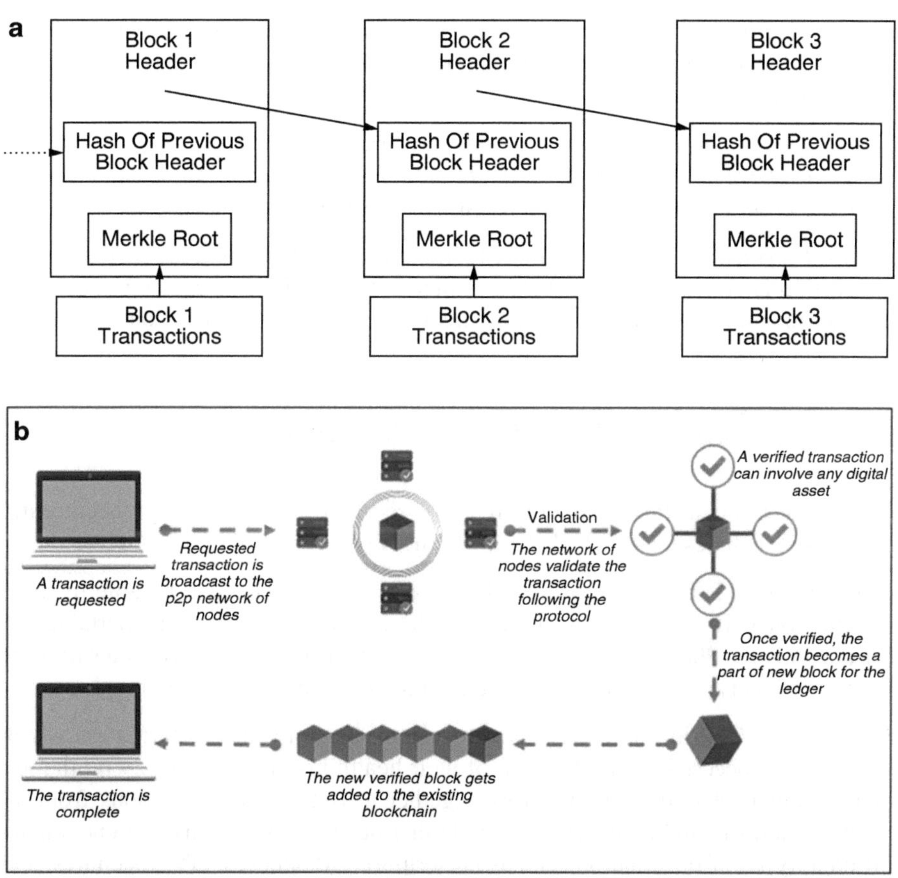


**Fig. 2** (**a**) Blockchain (Mettler 2016), (**b**) Summary of how blockchain works. (Adapted from How Blockchain Technology can Revolutionize 2018)

are stored on each node of the network (Swan 2015). Data blocks are encrypted so that it is not possible to make changes later without changing all following nodes. Therefore, a blockchain can serve as an open, distributed ledger that can record transactions between two parties efficiently and in a verifiable and permanent way. The ledger itself can also be programmed to trigger transactions automatically (Iansiti and Lakhani 2017).

## 2.2 How Does the Blockchain Function?

In order to function effectively, blockchain is designed to have three major components which complement each other. These are (How Blockchain Technology can Revolutionize 2018):

*Distributed network* This is a decentralized peer-to-peer architecture which has got various network participants. In this network, each participant stores a matching copy of blockchain and is therefore authorized to validate any transaction taking place in that network.

*Shared ledger* Participants of a given network record transactions in a shared ledger. However, for a transaction to be added in the ledger, algorithms have to be run in order to offer verification for any proposed transaction

*Digital transaction* This is any information which is stored in a blockchain. Each transaction is organized into a "block" which is then assigned a cryptographic hash in order to assist in linear and chronological arrangement of transactions.

## 2.3 Applications of Blockchain in Healthcare Sector

In the healthcare sector, the major areas where blockchain technology is currently being used includes (How Blockchain Technology can Revolutionize 2018):

*Clinical trials* Through the use of blockchain, various firms can be able to share information gotten from clinical trials such as adverse reactions as pertained to certain patients. Sharing of patient data profiles either from virtual trial sites or other multiple trials is possible through the utilization of blockchain technology.

*Patient records* With the current health system, the amount of information which a patient can access is just a fraction of their health history. However, with the use of blockchain, it is possible to bring together a lifetime of transactions from all health systems including pharmacies and other health centers. This will encompass different types of information such as prescriptions, diagnoses, lab procedures, and many other information.

*Insurance logistics* The ability to provide accurate insurance claims and prevent fraudulent activities in the healthcare insurance is one of the most important aspect within the healthcare sector. It ensures there is maintenance of trust among the various coordinated sectors such as the government, banks, and insurance providers. Through the use of blockchain, it is possible to achieve this trust considering tracking of all activities within the healthcare plan without any alteration is possible with this technology.

*Supply chains* blockchain is very crucial when it comes to regulation of various parameters in the line of supply chain. In most cases, biomedical products will require standardized environmental conditions during their transits to retain their integrity. Blockchain technology can be able to effectively record all these conditions during shipment and store that information in a secure manner. This technology can also be used to guard against counterfeits as well as record interactions with regulators.

### 2.3.1 Blockchain in Different Healthcare Regions

Although blockchain is still relatively new in healthcare, we are witnessing uses of blockchain around the world in various healthcare contexts.

United States

In the US, healthcare services mostly tend to come from a patchwork of private companies (Wickramasinghe and Schaffer 2010). This simply means that healthcare records and data are fragmented. John Halamka at Beth Israel Deaconess Medical Center in Boston, Massachusetts together with colleagues set about to find solutions to the challenges of the blockchain technology proposing a solution project MedRec (Chowdhury 2012). This system would use the Ethereum software (ibid). This software is unique to the one used in bitcoin in the sense that it can integrate and execute smart contracts (ibid). This allows for the building of a private yet secure blockchain that links the healthcare providers together (ibid). This allows one to adequately and efficiently share data on a secure platform (ibid). It is worth noting that on this blockchain, each set of instructions from a patient tend to create a specific smart contract on the blockchain (ibid). Only the patient can cryptographically sign the contract (ibid). The system is secure as there is no single place of failure hence no vulnerabilities (ibid). It ensures that medical providers are able to run the program module on their computer so as to access the database which is in turn instructed by smart contracts that have been initiated by the patient (ibid). Thus, the module has three major responsibilities: (1) allowing the healthcare provider to access data from the blockchain when it is instructed, (2) executing patients' instructions when needed, and (3) allocating computing resources to the framework so as to maintain the blockchain (Brodersen et al. 2016).

### South America

In South America, a cryptocurrency mobile platform (Timicoin) that uses blockchain for storing and accessing medical records has in early 2018 been able to get Miguel Esparza to lead its Mexican and Latin American expansion (https://blocktribune.com/blockchains-timicoin-tabs-miguel-esparza-lead-mexico-latin-america/). Esparza is currently the company's chief technology officer and will mainly serve as Spanish liaison for the innovative mobile platform (ibid). Timicoin states that Miguel will work with a team that will in turn help create awareness on the company's activities in South America (ibid). The company seeks to transform and safeguard healthcare in both Latin America and Mexico (ibid). In this, both patients and medical professionals in South America are to be made aware that they can safely store and access medical records from literally anywhere on the globe (ibid). The company makes use of crypto token and blockchain decentralize information and shows that blockchain holds a great potential for the healthcare sector in South America and the world at large (ibid).

### Europe

Currently, the European Union is rapidly changing its approach toward e-health. The new goal for the union is to have a Europe-wide ecosystem for data-driven healthcare (Mettler 2016). As such, a component of the European Health Record Recommendation is the blockchain technology (ibid). Although nothing is decided yet, discussions within the eHealth Network are ongoing. However, early research in European countries such as Estonia has seen them link up EHR access files with blockchain (ibid).

It is worth noting that the attraction of the European Union in blockchain with respect to EHR records stems from a project dubbed My Health My Data (ibid). This is a project that is aimed at creating a health network and systems that are secure and private at the same time (ibid). A look at specifically the United Kingdom shows that pilot programs in the adoption of the blockchain technology in healthcare are already in place (https://www.telegraph.co.uk/business/business-reporter/blockchain-trial-in-healthcare/; Stagnaro 2018; Pisa 2018) such as MedicalChain (a medical blockchain platform that enables the transparent exchange of medical records between doctors and patients) that has been able to partner with the London-based Groves Medical Group to pilot a blockchain program in four medical institutions and the MediLedger Project – a project that was backed up by giants in the healthcare industry such as Pfizer and uses blockchain tools to secure the supply chain through tracking of medicine.

Asia

In the Asian continent, healthcare companies, as well as other stakeholders, are also considering the great possibilities of blockchain and the numerous problems this technology could help solve (Pilkington 2017). As such, countries including the United Arab Emirates, China, and South Korea have taken huge steps in implementing the technology in healthcare.

A look at China shows that a Fintech company known as Radarwin has been at the center of many explorations in China focused on the creation of a standardized medical record system with the use of the blockchain technology (Peters et al. 2017).

MediBloc, the South Korean-based start-up, has set to create a new patient-centered health database platform with the use of the blockchain technology focusing on the medical record-keeping sector (Kho 2018). As such, MediBloc plans to roll out this new platform to the Korean health market by the end of 2018 and later to the rest of the world. In order to encourage users to join the platform, the company plans on rolling out a reward system that will be based on its cryptocurrency, "Medi Token" (ibid).

It is also encouraging to find out that various stakeholders in the blockchain technology are working together across the globe. This has been perfectly illustrated by Guardtime, a company that offers health records that are based on blockchain in Estonia (Hughes et al. 2018). The company has been able to collaborate with Du, a correspondent in the United Arab Emirates, to bring the new technology into the region (ibid). Du through a senior executive stated that they were proud to be the ones bringing this kind of innovation to the UAE as it has the potential to transform the healthcare and specifically on health records. This project is among the many that have been highlighted by the Dubai Future Foundation.

Clearly it is early days for the use of blockchain in healthcare; however, the preceding observations demonstrate that numerous regions of the world are looking forward to embrace the new technology in healthcare.

## *2.4 Strengths of the Blockchain Technology in the Healthcare Sector*

According to Mettler (2016), the healthcare sector loses more than $80 billion dollars in fraud activities involving the health insurance companies and other stakeholders in this sector. With the use of blockchain, it is possible to mitigate such losses considering all payments made via the use of this technology is highly secure and is not subject to any kind of revision without involving all the parties (Mettler 2016). Therefore, when effectively applied, blockchain technology can be very crucial in reducing the wastage of resources within healthcare organizations. This in return will promote efficiency leading to quality healthcare.

Another strength of blockchain is promotion of patients' privacy. In this case, privacy means confidentiality of medical records and information of the patients. According to Parino et al. (2018), healthcare professionals should ensure that personal information of the patients is always kept confidential and it should never be shared with a third party without the consent of the patient. This helps in maintaining trust between the caregiver and the patient as well as ensuring that the patients have got control on how their information is flowing (Parino et al. 2018). Through the use of cryptography, it is possible to achieve the highest level of privacy within a healthcare organization (How Blockchain Technology can Revolutionize 2018; Parino et al. 2018). Cryptography is the backbone of blockchain technology and it is applied in almost all blockchain-related activities (Parino et al. 2018). Through the use of cryptography, it is possible to mathematically encrypt information so that only the few selected individuals, who have got decryption keys, can be able to read the information (Parino et al. 2018). However, this can only work on those organizations which have implemented electronic health recording systems (Parino et al. 2018).

## *2.5 Weaknesses of Blockchain in Healthcare*

Blockchain use in healthcare is expected to gain more interest over the next few years as the implementation of the technology becomes more of a reality (How Blockchain Technology can Revolutionize 2018). Most healthcare organizations are inclined toward the new technology as it promises to transform service delivery which translates to quality healthcare for the masses (How Blockchain Technology can Revolutionize 2018). However, it is important to have a comprehensive understanding of the technology and the aspects that are critical to a successful deployment of the technology (Mettler 2016). In this, there are numerous weaknesses of blockchain in healthcare which ought to be understood. These include (Linn and Koo 2016; Tasca and Widmann 2017; Brodersen et al. 2016):

- *Difficulties in data modification:* A recent NIST (National Institute of Standards and Technology) publication indicated that there are numerous infrastructural challenges that organizations need to be aware of while intending to implement blockchain technology in their organizations. These include the difficulty in modifying the data that have been stored. It is worth noting that, in blockchain, it is much more difficult to update the database software as well as change the data. As such, changes implemented on the blockchain may end up causing forking effects which are, of course, distractive to the databases. Additionally, there is the need for organizations in the healthcare sector to consider the protocol in which participants converge and agree that a specific transaction is valid. Moreover, recent developments in the blockchain technology allow for permissionless implementation. This shows that anyone can read and write on it which is not an ideal move in the healthcare sector.

- *Requires collaboration:* Since the blockchain reputation was initially associated with Bitcoin, professionals in the healthcare sector struggle to find the correlation and as a result, tend to equate the technology with a privacy risk. As such, majority have stated that the blockchain technology in the healthcare sector can only be managed in a semi-private environment with a high level of collaboration between various stakeholders. In this, the organizations supporting patient care should work on improving how data is managed. Moreover, these organizations should be willing to collaborate in the blockchain-enabled patient data management systems which are transparent in which the exchange of information is fast and efficient. However, most organizations are yet to embrace these dynamic changes especially when it comes to collaboration with other parties.
- *Storage capabilities:* It is obvious that the adoption of blockchain technology in healthcare is and will mainly consist of documents, medical records, lab reports, and even images. These require a large amount of cyber space. This is justified by the fact that each patient in the chain will have a complete profile that has their records. In this, the volume of the data could potentially exceed the storage capacity of the presently available blockchain technology.
- *Data ownership and costs*: Another weakness with the blockchain technology in healthcare is that it is not yet clear on who will own the data or who will grant the permission to share the data between the various stakeholders. Additionally, the costs of establishing efficient healthcare blockchain are yet to be known. However, the technology is bound to attract huge costs of implementation and maintenance which not many organizations will be willing to risk.
- *Rules and regulations:* Currently there lacks a rigid legal framework that addresses the use of blockchain in the healthcare sector. This presents a dilemma in that the technology is bound to be used globally and hence the creation of rules and regulations with regards to blockchain will tend to attract concerns globally. Moreover, it is not certain on how new policies will tend to conform to the already existing ones, such as the HIPAA act.

## 2.6 Barriers Hindering Adoption of Blockchain in Healthcare

There are numerous barriers that hinder the adoption of blockchain in healthcare by the various organizations. These include (Tasca and Widmann 2017; Brodersen et al. 2016; Deshpande et al. 2017; Nugent et al. 2016):

- *Immature technology and insufficient skills:* Although digitization in the healthcare sector is taking root, it is worth noting that numerous healthcare organizations and institutions still depend on paperwork in the implementation of several processes. Due to this, asking the stakeholders to randomly shift to blockchain technology which involves electronic health records would be an uphill task. This is because the available technology in the majority of the organizations and institutions is still immature and cannot efficiently handle the shift. The process of changing people's behaviors as well as their routine activities is time-consuming and challenging to any industry, including the healthcare sector. In this, there are bound to be harsh resistance from employees who tend to enjoy the current comfort zone. Moreover, most of the employees lack adequate skills to deal with the blockchain technology. This prompts healthcare organizations to invest heavily in more mature and blockchain friendly infrastructure and technology. Consequently, this calls for adequate training to the employees before the implementation of the technology. Through this, employees will embrace it and the transition will be a smooth one.

- *Uncertainties and lack of proven credibility:* The blockchain technology in healthcare may sound to be exciting to the stakeholders. However, it is a matter of fact that it is yet to be implemented successfully anywhere. As such, the healthcare sector is a delicate yet distributed sector. Due to this, changes cannot be implemented overnight. Stakeholders in the sector have successfully pilot run the technology and determined the viability of adopting it. Additionally, it is worth noting that the blockchain technology has been found to mainly address the issues that are concerned with data access and sharing (Brodersen et al. 2016). However, uncertainties such as low data quality, lack of technical and semantic interoperability, and reliability tend to indicate that this technology might fail to improve quality healthcare after all. Such issues prevent the adoption in the healthcare sector until some concepts and uncertainties are well defined and dealt with efficiently.
- *Profit wars among stakeholders and legal constraints:* In some instances, health institutions and organizations might not be comfortable to share their real implementation costs data with insurance firms. The reason behind this is that the insurance companies work in a framework that sees different rates charged to different patients. In this kind of scenario where institutions are at profit wars, it becomes almost impossible to implement complex technologies such as blockchain. This is simply due to the unwillingness to disclose data. Moreover, the blockchain technology is completely decentralized. This means that the permissions to read and write data onto the chain are shared equally by all the stakeholders that come into a consensus before the data are stored in the chain. However, the permissionless blockchain technology in the healthcare sector would tend to be disruptive and almost unable to fit the existing legal frameworks. This is because, current healthcare systems are highly regulated and therefore could end up blocking the adoption and implementation of the technology. An alternative to this would be to introduce permissioned blockchains. This way, institutions would be at liberty to provide only the information needed by their counterparts and profit wars would be minimized.
- *Government policy:* This remains to be a key barrier to the adoption of the blockchain technology. This is because the healthcare sector tends to be owned by respective governments as it is broad. Due to this, it is worth noting that only through meaningful policies toward the implementation of blockchain will this technology gain root in the sector. However, not all governments of specific

nations have the same level of capabilities. This way, the technology will tend to be fast implemented in some countries while in some it may take time. Additionally, some governments will be coupled with issues such as corruption which may lag the adoption and implementation of blockchain technology.

- *Transparency and accountability Issues:* This is related to the healthcare professionals. Most are nervous about the transparency of the technology. This is because medical decisions, both good and bad, are recorded and locked in the patient's blockchain portal. This way, the decisions of the healthcare professionals can be scrutinized at a later date with the specific details available to all. Although this is a good thing on the part of the patients, the healthcare professionals may feel that their decisions have been put under criticism and could even attract lawsuits if there were errors in treatment. Therefore, professionals will tend to be repulsive to this kind of technology.

## 2.7 Factors Facilitating the Adoption of Blockchain in Healthcare

Technology in the healthcare sector has been transformed in recent years. The advent of the Internet and currently the blockchain technology has demonstrated the potential of problem-solving innovations especially in the healthcare sector which in turn will work toward changing livelihoods. It is worth noting that the success of the blockchain technology depends almost entirely on the extent of its adoption (Morabito 2017). In this, some of the factors that facilitate its adoption in healthcare include (Lemieux 2016; Nugent et al. 2016; Morabito 2017):

- *Social and cultural factors:* These mainly narrow down to the familiarity with technology among the masses. The best way to determine the level of technological familiarity in a society is the extent to which the technology impacts the day to day activities. It is worth noting that the blockchain technology is complex and more so if institutions are not well acquainted with the basic concepts on which it is built on. In this, societies that have already assimilated technological innovations and have experienced a change in value in their lives will tend to be more receptive. The world is in the twenty-first century, the digital era. Almost everybody has access to a smartphone. This shows that the world at large is familiar with modern technology and therefore could be ready for blockchain in the health sector. Currently, there are already technological tools in the healthcare sector such as mHealth apps and electronic health record systems which have improved to a large extent the delivery of quality healthcare. It is therefore evident that the majority of the people are familiar with new technology and continue to yearn for more such as blockchain.
- *Willingness to experiment and propensity of risk:* A combination of technological comfort and expectations is bound to bring forth a technologically curious society. This perfectly defines the current age and era. The current society

is able to appreciate prototypes and accept failure/delays that are expected in the experimental stage of technology. Moreover, the current world is full of individuals that possess the entrepreneurial spirit who are ready to take risks and try again if they fail. This is the mindset that the blockchain pioneer, Satoshi Nakamoto (Nakamoto 2008) had when he conceptualized the idea which came to be a huge success in some sectors. The current society that has a similar mindset provides a perfect environment in which blockchain can gain ground in the healthcare sector. It has already shown that it has the potential to transform the healthcare sector, with only a few challenges that can be dealt with sufficiently. Moreover, sponsorships by governments and enabling policies have proven to fuel the mainstream adoption of the blockchain technology in certain regions of the world. This is supported strongly by the huge success of the blockchain technology in the financial sector. This shows that its potential in sectors such as healthcare is huge and yet to be tapped fully.

- *Information flow and privacy concerns:* An important aspect of blockchain and what additionally brings value to the healthcare sector is its ability to be transparent and accountable. This may, however, make some to view technology with some level of suspicion. In this, people will tend to be more cautious about the information they share. This, on the other hand, is tied to their interests, beliefs, and culture. While transparency that comes with blockchain may be viewed as a gesture of openness and hence trust, some people may prefer to have privacy over their data. However, more work is being carried out on the issue of confidentiality and identity management in blockchain.
- *Cooperation among stakeholders:* Blockchain is a technology that delivers best when multiple parties participate and collaborate. This is a similar concept to care coordination in healthcare that most stakeholders are calling for. As such, when stakeholders have similar goals and mindsets, they will tend to embrace new technologies and methods in unity and this will add value in sectors like healthcare. Moreover, dealing with the challenges of technologies such as blockchain becomes easy.

Many of the risks and concerns associated with EHR management can be mitigated with blockchain. A blockchain is a distributed database system that keeps track of records (Molteni 2017; Tapscott and Tapscott 2016). As records are added to the blockchain they are ordered in blocks and each block contains timestamp links to the related blocks. Blockchain records are secure and easily verifiable. As events or transactions that are captured as records occur, decentralized verification of their authenticity is carried out by majority consensus in networks (Ekblaw et al. 2016).

Blockchains allow designing smart contracts: contracts between several parties. These contracts trigger defined actions (Bitcoin 2017; Szabo 1994). Thus, a record capturing an event becomes part of the blockchain if and only if significant effort is made by players in the network validating its genuineness and authenticity. Additionally, because network consensus is always required, alteration of records

**Table 1** Application areas for blockchain in healthcare

| Area/issue | Key points |
|---|---|
| Fragmented data | Computer networks aid in accurate patient data through decentralized storage<br>Data can be shared across networks and nodes<br>Decentralized source of Internet |
| Timely access to patient data | Distributed ledger system facilitates distributed, secure access to patient health data<br>Updates to patient's shared data done in real time |
| System interoperability | Decentralized Internet and computer networks across boundaries<br>High-level authenticity |
| Data security | Digital transaction security ensured; this protects patient identity |
| Patient-generated data | Holistic patient data collected through data from wearable IoT |
| Access and data inconsistency | Select healthcare companies enjoy consistent and rule-based method to access and analyze patient data through smart contracts |
| Cost-effectiveness | System turns highly efficient through real-time processing and reduced transaction costs<br>No third-party applications, so no time-lag in accessing data |

Adapted from https://hackernoon.com/how-blockchain-technology-can-revolutionize-the-healthcare-sector-31fe9301575

becomes very difficult and expensive. Hence, blockchain technology ensures that the effort required to alter a record (e.g. for the purpose of committing fraud, etc.) always exceeds the benefits or gains that result from attempts to alter the record. This reduces incentives of individuals or groups to change a blockchain record which indicates that what is in the blockchain is accurate and authentic (Molteni 2017).

As the blockchain is managed autonomously in peer-to-peer network information, it is not stored in a single location and is always available to use and verify and not susceptible to loss (e.g. because part of the network fails). Additionally, because verification and recording of information is carried out by the network, the need for intermediating role of trusted authority or central server is significantly reduced or, depending on application, even eliminated (Ekblaw et al. 2016).

While in its purest form information in the blockchain is available to anyone, it is possible to create a blockchain where permissions concerning the right of individuals to add to, record in, and read information on a blockchain can be easily controlled. Private blockchains, as opposed to public blockchains, will be suitable tools for EHR management (Buterin 2015).

We contend that the benefits of blockchain technology should not be limited to EHR and have as much or even more potential in many healthcare areas (Table 1) and especially in the area of electronic prescriptions.

## 3 Electronic Prescriptions

Many countries have considered electronic prescriptions as a key component of their e-health solution. The electronic prescription, for example, was initially in step 2 of the implementation of the German electronic health card a compulsory administrative solution (see Fig. 3).

This timeline was stopped in 2009 (Buterin 2015; Krüger-Brand 2005). There were a lot of open questions with the telematics infrastructure and the concept. It was planned to store prescriptions in the storage on the card. But the size of the storage was small. The number of prescriptions was limited. It was also not possible to write electronic prescriptions during a visit at the patient's home. For writing prescriptions card readers and access to the telematics infrastructure was necessary. The idea of a total replacement of paper-based prescriptions by electronic prescriptions was not possible. It was expected that round about 700 million paper-based prescription could be replaced every year (Schweim 2007). The process of medication management (see Fig. 4) is only a part of electronic prescription. The whole administration and clearance with insurance companies and other third-party regulators are missing here. The process of electronic prescriptions concerns all parties of the web of healthcare players (see Fig. 5).

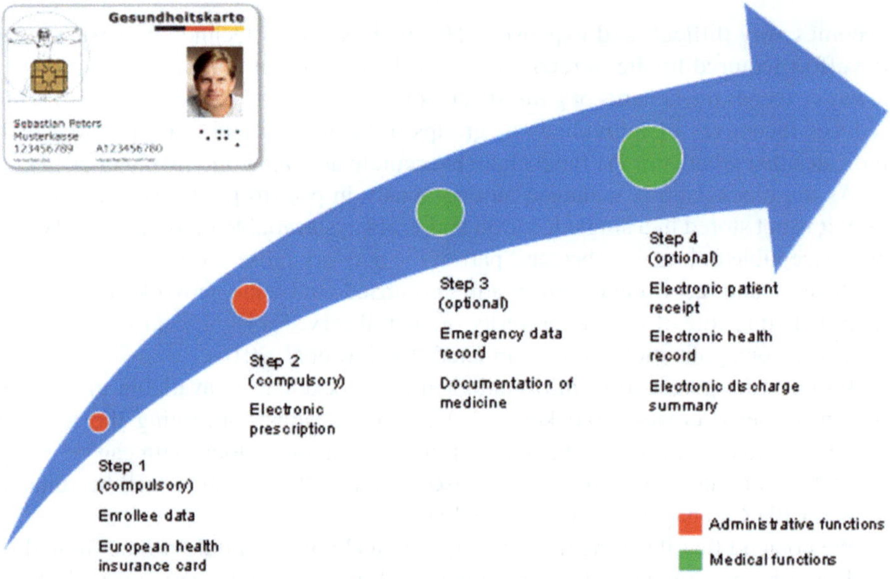


**Fig. 3** Initial implementation steps of the German electronic health card. (Adapted from Krüger-Brand 2005)

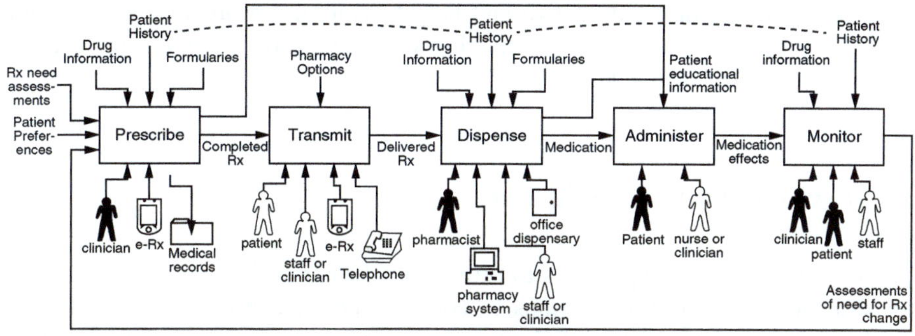


Fig. 4 Process of medication management (Bell et al. 2004)

Fig. 5 Web of healthcare players [30]

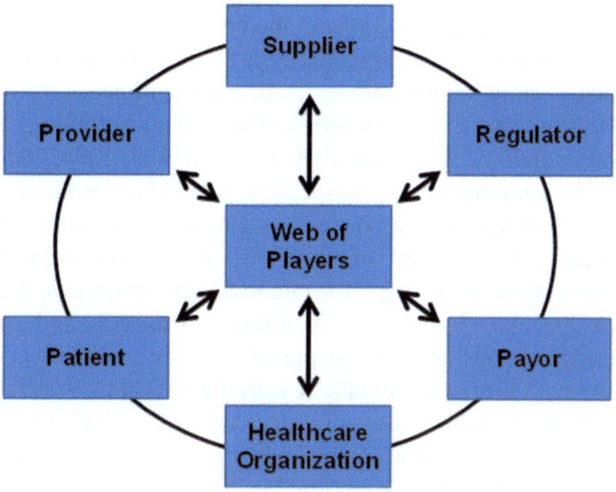


## 4 Electronic Prescription Based on Blockchain Technology

In the process of electronic prescription blockchain technology allows to design medication management as a smart contract. A medical doctor does not only prescribe medication he/she can also control and in some sense supervise the medication process remotely. A medication dispenser as an intelligent Internet of Things device can either remind the patient to take medicine to block overdoses and/or to report to the doctor patient's behavior. Smart contracts not only allow taking care of regular delivery but also interruption of dispensation if the patient or the insurance company does not pay. Also the whole clearing process can be automatized.

As long as the patient is able to take medicine physically by himself or herself, blockchain technology and smart contracts can not only support the medication process, but it can also automatize and help to control the process. For medium heavy diseases, this also means less nursing staff, especially as long as patients are able to live on their own at home.

## 5 Discussion and Conclusion

The objective of this chapter was to proffer blockchain technology as suitable for assisting in electronic prescriptions to address current concerns and challenges. To date, as with the example of Germany most initiatives to implement electronic prescriptions have been met with significant opposition from either doctors and/or pharmacists for many valid concerns. We contend though that these legitimate concerns should not forestall the benefits of electronic prescriptions and thus a new approach and superior technology solution needs to be developed rather than totally abandoning the idea of electronic prescriptions as too difficult and too hard. Hence, we suggest the incorporation of blockchain technology for e-prescribing and various other areas of healthcare delivery (Bell et al. 2004; Deloitte Briefing Paper 2018).

From a theoretical perspective, blockchain technology and smart contracts offer a large number of opportunities. However, from the perspective of practice, clearly there are still a lot of questions and unsolved problems regarding the use of blockchain technology for electronic prescriptions. This is only the technological perspective. There are also additional aspects and unsolved problems including critical issues around acceptance, barriers, and ethics. To answer all these issues, we need further systematic research in this area. We are confident that such research will both enable successful electronic prescription to be possible as well as many other e-health opportunities so that patients can benefit from a higher quality of care delivery and at the same time healthcare can be delivered efficiently and effectively to all. Conversely, we believe that by ignoring the potential of blockchain technology for healthcare, we run the risk of never realizing the true potential of the myriad of e-health solutions currently being developed and implemented.

## References

Azaria, A., Ekblaw, A., Vieira, T., & Lippman, A. (2016). MedRec: Using blockchain for medical data access and permission management. In I. Awan & M. Younas (Eds.), *Proceedings of the 2nd international conference on open and big data (OBD) August 22–24, 2016, Vienna, Austria* (pp. 25–30).

Bell, D.-S., Cretin, S., Marken, S., & Landman, A.-B. (2004). A conceptual framework for evaluating outpatient electronic prescribing systems based on their functional capabilities. *Journal of the American Medical Informatics Association (JAMIA), 11*(1), 60–70.

Bitcoin. (2017). Bitcoin developer guide. Retrieved July 5, 2017 from https://bitcoin.org/en/developer-guide#block-chain

Brodersen, C., Kalis, B., Leong, C., Mitchell, E., Pupo, E., Truscott, A., & Accenture, L. L. P. (2016). *Blockchain: securing a new health interoperability experience*. Accenture LLP. https://www.healthit.gov/sites/default/files/2-49-accenture_onc_blockchain_challenge_response_august8_final.pdf

Buterin, V. (2015). On public and private blockchains. Ethereum Blog: Ethereum Foundation. Retrieved July 9, 2017 from https://blog.ethereum.org/2015/08/07/on-public-and-private-blockchains/

Chowdhury, J. (2012). Hacking health: Bottom-up innovation for healthcare. *Technology Innovation Management Review, 2*(7), 31–35.

Deloitte Briefing Paper. https://www2.deloitte.com/us/en/pages/public-sector/articles/blockchain-opportunities-for-health-care.html. Accessed 2 July 2018.

Deshpande, A., Stewart, K., Lepetit, L., & Gunashekar, S. (2017). Distributed ledger technologies/blockchain: Challenges, opportunities and the prospects for standards. Overview report The British Standards Institution (BSI). https://www.bsigroup.com/PageFiles/508003/BSI_Blockchain_DLT_Web.pdf

Ekblaw, A., Azaria, A., Halamka, J., & Lippman, A. (2016). *A case study for blockchain in healthcare: 'MedRec' prototype for electronic health records and medical research data*. White paper. MIT Media Lab and Beth Israel Deaconess Medical Center. Retrieved June 26, 2017 from https://www.healthit.gov/sites/default/files/5-56-onc_blockchainchallenge_mitwhitepaper.pdf

Gartner. (2016). *Hype cycle for emerging technologies*. Stamford: The Gartner Group. Retrieved June 26, 2017 from. Retrieved June 26, 2017 from http://www.gartner.com/newsroom/id/3412017

Haas, P. (2016). *Elektronische Patientenakten: Einrichtungsübergreifende Elektronische Patientenaktenals Basis für integrierte patientenzentrierte Behandlungs management-Plattformen*. Bertelsmann-Stiftung. https://www.bertelsmann-stiftung.de/fileadmin/files/BSt/Publikationen/GrauePublikationen/VV_eEPA_Expertise_final.pdf

How blockchain technology can revolutionize the healthcare sector available at https://hackernoon.com/how-blockchain-technology-can-revolutionize-the-healthcare-sector-31fe9301575. Accessed 30 June 2018.

Hughes, E., Graham, L., Rowley, L., & Lowe, R. (2018). Unlocking blockchain: Embracing new technologies to drive efficiency and empower the citizen. *The JBBA, 12382018*(1), 3741.

Iansiti, M., & Lakhani, K-R. (2017). The truth about blockchain. In *Harvard business review*. Harvard University. Retrieved July 5, 2017 from https://hbr.org/2017/01/the-truth-about-blockchain

Kho, W. (2018). Blockchain revolution in healthcare: The era of patient-centered dental information system. *International Journal of Oral Biology, 43*(1), 1–3.

Krüger-Brand, E. (2005). Elektronische Gesundheitskarte. *Deutsches Ärzteblatt, 102*(37), 2432.

Lemieux, V. L. (2016). Trusting records: Is blockchain technology the answer? *Records Management Journal, 26*(2), 110–139.

Linn, L. A., & Koo, M. B. (2016). Blockchain for health data and its potential use in health it and health care related research. In *ONC/NIST use of Blockchain for healthcare and research workshop*. Gaithersburg: ONC/NIST.

Mettler, M. (2016). Blockchain technology in healthcare: The revolution starts here. In *e-Health networking, applications and services (Healthcom), 2016 IEEE 18th international conference on* (pp. 1–3). IEEE. https://www.researchgate.net/publication/310622367_Blockchain_technology_in_healthcare_The_revolution_starts_here

Molteni, M. (2017). Moving patient data is messy, but blockchain is here to help. Retrieved July 9, 2017 from https://www.wired.com/2017/02/moving-patient-data-messy-blockchain-help/

Morabito, V. (2017). Blockchain practices. In *Business innovation through blockchain* (pp. 145–166). Cham: Springer.

Morton, M-E. (2008). Use and acceptance of an electronic health record: Factors affecting physician attitudes. Ph.D. Dissertation. Philadelphia: Drexel University.

Nakamoto, S. (2008). *Bitcoin: A peer-to-peer electronic cash system*. Bitcoin Foundation. Retrieved June 26, 2017 from https://www.bitcoin.org/bitcoin.pdf

Nugent, T., Upton, D., & Cimpoesu, M. (2016). Improving data transparency in clinical trials using blockchain smart contracts. *F1000Research, 5*, 2541.
Parino, F., Gauvin, L., & Beiro, M. G. (2018). Analysis of the bitcoin blockchain: Socio-economic factors behind the adoption. arXiv preprint arXiv:1804.07657.
Peters, A. W., Till, B. M., & Meara, J. G. (2017). Blockchain technology in health care: A primer for surgeons. *Bulletin of the American College of Surgeons, 12*, 1–5.
Pilkington, M. (2017). Can blockchain technology help promote new tourism destinations? The example of medical tourism in Moldova.
Pisa, M. (2018). Reassessing expectations for blockchain and development. *Innovations: Technology, Governance, Globalization, 12*(1–2), 80–88.
Schweim, H.-G. (2007). Die unerträgliche Geschichte der Gesundheitskarte in Deutschland. *GMS Medizinische Informatik, Biometrie und Epidemiologie, 3*(1), Doc04.
Stagnaro, C. (2018). White paper: Innovative blockchain uses in health care.
Swan, M. (2015). *Blockchain. Blueprint for a new economy*. New York\Sebastopol: O'Reilly.
Szabo, N. (1994). Smart contracts. Retrieved July 5, 2017 from http://www.fon.hum.uva.nl/rob/Courses/InformationInSpeech/CDROM/Literature/LOTwinterschool2006/szabo.best.vwh.net/smart.contracts.html
Tapscott, D., & Tapscott, A. (2016). *Blockchain revolution: How the technology behind bitcoin is changing money, business, and the world*. New York: Penguin Random House LLC.
Tasca, P., & Widmann, S. (2017). The challenges faced by blockchain technologies–Part 1. *Journal of Digital Banking, 2*(2), 132–147.
Wickramasinghe, N., Schaffer, J.-L. (2010). Realizing value driven e-health solutions. In *IBM center for the business of government*. Washington DC.

# Knowledge Acquisition of Consumer Medication Adherence

**Elena Vlahu-Gjorgievska, Harith Hassan, and Khin Than Win**

## 1 Introduction

Evidence-based medicine and healthcare have been a focus of the health domain in recent years. Many knowledge-based systems have been developed in healthcare for assisting healthcare decision-making and health policy development. The concept of learning health system with understanding the data, analyzing the data and feedback to healthcare, would assist in evidence-based healthcare and decision making.

Understanding of how medication adherence has been considered in the health domain is an important aspect in data acquisition for knowledge discovery of medication management. Studies have shown that nonadherence rates are noteworthy across different categories of disease and medications, especially among patients with chronic diseases (like heart failure and diabetes). The biggest nonadherence challenge is the gap between the perception and reality of individual adherence levels. The evolution of big data, data analysis, and predictive methods is becoming an essential factor in adherence management in order to promote patient medication adherence, improve patient health, prevent prescription drug abuse and drug waste, and reduce the medical pathway costs. Healthcare managers and their organizations hope to mitigate nonadherence to medication behavior within the patients by encouraging patient's self-management (Win et al. 2016).

Even though nonadherence to medication has been studied widely, there is a lack of comprehensive study on different medication adherence measures. The gap in the research is on how medication adherence should be measured and what methodologies are more efficient to identify, predict, and improve adherence rates, relying on the available medical, pharmaceutical, and other administrative data. In

E. Vlahu-Gjorgievska · H. Hassan · K. T. Win (✉)
University of Wollongong, Wollongong, NSW, Australia
e-mail: elenavg@uow.edu.au; hsh020@uowmail.edu.au; win@uow.edu.au



N. Wickramasinghe, F. Bodendorf (eds.), *Delivering Superior Health and Wellness Management with IoT and Analytics*, Healthcare Delivery in the Information Age, https://doi.org/10.1007/978-3-030-17347-0_15

knowledge engineering, knowledge acquisition can be through data sources, data preprocessing, and data analysis using data mining techniques for the prediction of a specified condition (Shukla et al. 2018). Thus, prediction studies of medication adherence involve identifying data sources through knowledge elicitation, variable selection for data preprocessing and data analysis through different data mining techniques.

The aim of this study is to identify the medication adherence measures for knowledge acquisition and analytics that can improve medication adherence. This study will identify the knowledge sources, variables essential for knowledge preprocessing and data analytics techniques used in medication adherence studies. Metrics used to measure a patient's medication adherence and approaches used to predict adherence behavior patterns will also be reviewed.

## 2 Methodology

The papers were selected by searching relevant publications in two electronic databases: Scopus and Web of Science. The search on the databases was conducted in March 2017 for journal articles published in the English language in the period from 2007 to 2017. The keywords used in the search string were "medication adherence," "analytics," "prediction," and "improve adherence," combined with logical connection keywords "and" and "or."

Conference papers, letters, notes, and editorials were excluded from the search. Papers written in another language than English were excluded as well.

In order to be included in this review, the articles needed to explore adherence metrics (electronic and traditional), medication adherence improvement, and prediction methodologies related to the use of information technology and analytical methods.

From both databases, 491 articles were identified, 368 in Scopus and 123 in Web of Science. Three articles were duplicates, and an additional 391 articles were excluded based on the title and abstract assessment. From 97 full-text articles assessed for eligibility, 59 articles were further excluded. Thirty-eight articles met the selection criteria and were included in this study. The flow diagram for the process of papers selection is shown in Fig. 1.

## 3 Results

Adherence to medication is an important aspect in consumers' self-care management, and several causes could be identified.

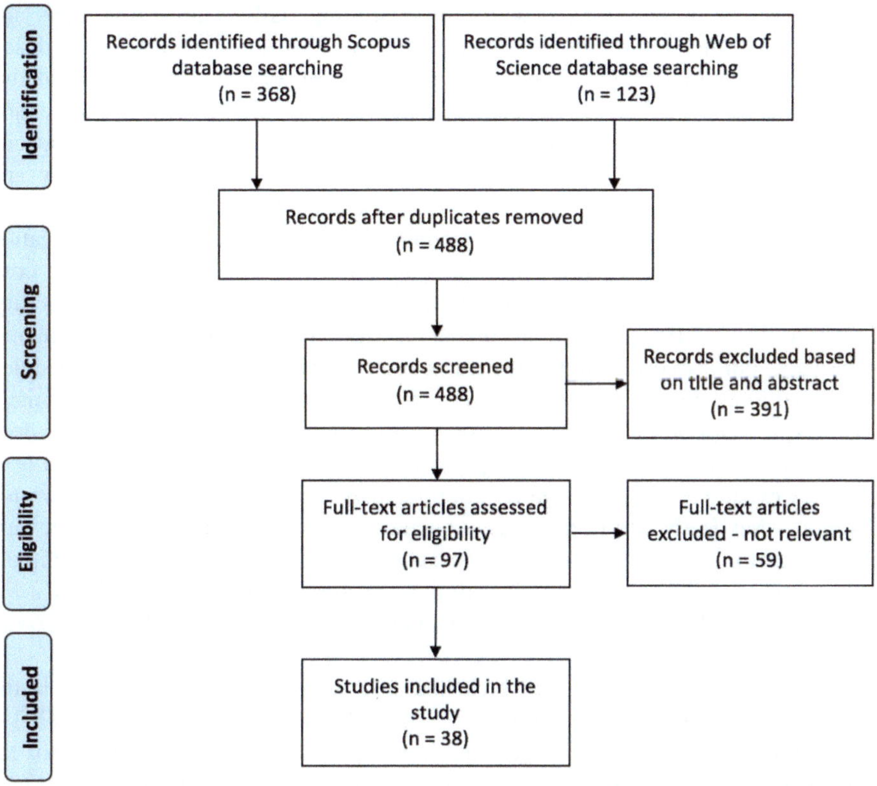


**Fig. 1** Flow diagram for the process of papers selection

## *3.1 Causes of Medication Nonadherence*

Several factors can be identified as the causes of patients' nonadherence to medications. They are patient forgetfulness and communication (Maulucci and Somerville 2011; Serdaroglu et al. 2015), forgetfulness and beliefs about efficacy and health condition (Piette et al. 2015; Wu et al. 2008), medication knowledge (Son et al. 2010), and under- and over-reporting of adherence (Sayner et al. 2015).

Chang et al. (2013) noted that younger age, black race, worse general health status, shorter duration of therapy, lower self-reported adherence, and not following doctors' orders are associated with nonadherence, while Ritchey et al. (2016) found out in their study that ethnicity, perceived social support, and financial status are the most important factors for nonadherence, and Wu et al. (2008) identified the importance of socioeconomic factors for the medication adherence. Davis and Kendrick (2014) identified that age is associated positively with better adherence ratios. However, the study of Coletti et al. (2015) identified that the overall patient's health plays an important role in the medication adherence. In addition, adherence timing (Sayner et al. 2015), dose-timing\mistiming (Gill et al. 2015), gaps and overlaps in medication regimen (Bjarnadóttir et al. 2016), prescription (fixed dose), and index dose (highest mean dose reduce adherence) (Eby et al. 2015) should also be considered in medication adherence. Drug class (Farr et al. 2014), medication regimen complexity (high number of medications, complicated schedules, special instructions) (McDonald et al. 2013), and medication regimen complexity and communication (Dixon et al. 2014) are also to be considered.

Ritchey et al. (2016) identify adherence factors like forgetting medication time, medication not in patient's possession when it is needed, medication cost, and patient's knowledge about dosage and beliefs about medication feasibility. Franklin et al. (2016) noted that initial prescription length in days can influence adherence, but also dose timing, the importance of taking medication as scheduled and mistiming dosage intake can affect adherence (Gill et al. 2015). Payment source (prescription paid per patient of health provider), medication classes, dose combination, average days' supply per fill, the average number of pills and how complex or simple the treatment regimen is, does the patient have another chronic disease, what is the drug class used by the patient previously, were there any hospitalization (Curtis et al. 2009), geographic region, urban residence, and clinical characteristics (Cheng et al. 2015) – all these factors can help to identify what are the areas of weakness in the treatment plan and how they can be addressed in order to improve adherence rates.

Nonadherence is related to different factors (younger age, female, living in the southern region of the USA) (Farr et al. 2014); the result in Lafeuille et al. (2014) concludes that based on their study and literature, older age is not a predictor for nonadherence, while Davis and Kendrick (2014) concluded that mean adherence increases with age and there are different adherence rates across different drug classes prescribed for the same medication. That has been due to health literacy, poor physical condition, and subjective well-being from the patient perspective (Coletti et al. 2015). Adherence pattern varies also based on the medication plan and the frequency of the dose. Cheng et al. (2015) noted that less frequent dosing plan for osteoporosis patients has better compliance.

### *3.2 Data Sources for Adherence Measures*

There are many measures to evaluate patient adherence to a medication, the literature includes several categories of these measures, including self-reporting methods, automated measures, and calculation measures based on pharmacy or administrative data.

### 3.2.1 Administrative Claims Data

One of the most common sources of data used in the literature is administrative claims data. This kind of data is frequently collected by large health systems like government or private health providers and insurance companies which offer the advantage of evaluating adherence patterns in large population samples that cover different population regions and demographics. In some cases, administrative claims data have a limited ability to identify a person with high compliance, as there is the inability to confirm the actual usage of prescribed medication. Administrative claim data were used by Lafeuille et al. (2014), Franklin et al. (2014), Curtis et al. (2009), Bjarnadóttir et al. (2016), Eby et al. (2015), Lo-Ciganic et al. (2015), Molfenter et al. (2012), Yu et al. (2010).

### 3.2.2 Electronic Medical Record Data

A great number of patients with poor adherence can be identified through the analysis of electronic prescription using electronic medical records (EMR) or electronic health records (EHR) as a data source (Mabotuwana et al. 2009). Such analysis requires consideration of the relationship between prescription classifications and evaluation period (e.g., 1 month, under 3 months, over 1 year) (Coletti et al. 2015; McDonald et al. 2013).

### 3.2.3 Pharmacy Refill Data

Pharmacy refill data is another common indicator of patient's medication dispensing. Similar to administrative and insurance data records, pharmacy refill data may not reflect the actual compliance of the patient to the prescribed medication regimen. In this context having a record of adherence to a specific chronic disease medication can supplement other baseline data (Davis and Kendrick 2014; Franklin et al. 2013, 2014, 2016; Tucker et al. 2015; Kozma et al. 2014).

### 3.2.4 Electronic Adherence Monitoring

Electronic medication monitoring is generally more accurate than self-reported adherence and pills count (pharmacy refill data) since the accuracy came from the real-time logging of taking medication events. Electronic adherence monitoring device (EAMD) or Electronic Medication Packaging (EMP) is a device or add-in device that collects medication-taking behavior data. There are few variations of electronic adherence monitoring tools, but most of them share the same concept - the prescribed medication is enclosed within a "smart enclose" that sense and log every bottle/container opening that can represent a medication intake. It must be mentioned that not all EMPs are in a medication bottle and cap; some products are in different "smart" forms like medication sleeve or variation of smart cards. Even though opening a medication bottle is not absolute evidence of a patient taking a medication, the outcomes are substantially more accurate than other measures.

Many studies evaluate adherence data to improve medication adherence and plan for better interventions. Researchers seek to measure patients' adherence to a certain medication as early as possible, so they can investigate the obstacles and find solutions. There are few objectives and many subjective methods to measure adherence: direct (objective) method as observation of patient taking a medication or automatic pills count and indirect (subjective) method including patient self-reporting scales, pills count (manually), review of dispensing or prescribing records, or claims data. Depending on if its objective or subjective method only, the adherence measurement could be inaccurate. The indirect method always has the limitation that the patient who has been prescribed and dispensed medication is actually not taking it. Sometimes the behavior of the patient is unchanged, but the difference in the measure will show different patterns of "frame error."

## *3.3 Medication Adherence Metrics*

### 3.3.1 ASK-12

Studies have found that patients tend to be more adherent when using self-reporting method. But that can be taken into consideration only if the adherence is also measured by the additional objective method, so under/over self-reporting will be reduced and more consistent results will be produced.

Self-reporting adherence indicators are mostly questionnaires designed to be answered by patients in order to discover the barriers to adherence. Self-reported methods are usually more cost-effective than other methods, but the accuracy of the results is not as effective. ASK-12 (Adherence Starts with Knowledge-12) is a survey that includes 12-main item scale to test the adherence treatment (adherence behavior, health beliefs, inconvenience/forgetfulness) (Bjarnadóttir et al. 2016; Coletti et al. 2015). Ask-12 investigates the demographic indicators like age, sex, and level of education as well.

### 3.3.2 Visual Analog Scale

Visual Analog Scale (VAS) questionnaire is another self-reporting adherence measurement method. Using advanced analytics to compare the accuracy of VAS with the adherence data from MEMS (Medication Event Monitoring System), Sayner et al. (2015) found out that patient's self-measured adherence doses (VAS results) are not associated accurately with MEMS (percentage or timing adherence) over their 60 days of observation. In addition, most of the patients with chronic disease or complicated medication regimen will more likely overestimate the medication schedule. Sayner et al. (2015) suggest that it is important to assess the adherence earlier and simplify the patient-prescribed medication if possible, so patients' over-reporting of their medication can be reduced.

### 3.3.3 The Morisky Scale

The Morisky Medication Adherence Scale (MMAS-8) is another version of self-reporting adherence measurement method that can accurately predict the patient's compliance with medication. This tool is very simple, has 8-item questions, and is feasible when requesting for real-time feedback (Curtis et al. 2009; Stewart et al. 2014).

### 3.3.4 Tool for Advance Behaviour Screening

Tool for Adherence Behaviour Screening (TABS) as an adherence self-reporting method has adherent and nonadherent subclasses and measures the intentional and unintentional deviations from treatment recommendations. TABS can have a better validity than other alternative measures and distinguishes intentional nonadherence focusing on screening adherence beliefs, experiences, and behavior (George et al. 2006; Stewart et al. 2014).

### 3.3.5 MedsIndex

MedsIndex measures compliance to treatment plan by calculating the prescription-refill history. The score is based on how many dosages the physician has prescribed the patient to take, compared to the actual interval between the medication dispensing date for that patient (Stewart et al. 2014).

### 3.3.6 Medication Adherence Report Scale-5

Medication Adherence Report Scale-5 (MARS- 5) is another form of measures used to estimate patient adherence to medication through self-reporting. This instrument developed by Horne and Weinman (2002) has been used in quantitative measures of self-reported adherence for hypertension, diabetes, COPD, and asthma (Sandy and Connor 2015). The advantage of MARS-5 is that it can be used to conduct self-reporting adherence on a wider population.

### 3.3.7 Medication Possession Ratio (MPR)

This adherence measure is defined as the number of days of medication supplied within the specific period. MPR can be calculated within the refill interval or a fixed interval. MPR ratios can be easily calculated from EMR (electronic medical record) data which provides insights into a significant number of cases. Most of the studies identified a patient as adherent to medication if the MPR score was 80% or higher. This measure has been used to calculate the adherence\compliance of patients based on their data (e.g., administrative claim data) and discover medication adherence patterns (Steinberg et al. 2014). Some of the studies use the same ratio with another name Medication Coverage Ratio (MCR) (Cheng et al. 2015).

Using the EMR data Mabotuwana et al. (2009) observed blood pressure medications (ACE-inhibitor and/or ARB) in patients with hypertension or diabetes conditions as a part of the computational model to identify patients with poor adherence. They found that nonadherent patients (MPR < 80) are three times more likely to have poor blood pressure. Bisphosphonates medication adherence has been measured by Curtis et al. (2009) using MPR in patients with osteoporosis condition. The data was extracted from an administrative claims database for a period of 12 months.

Administrative healthcare claims data of 6 months examined using MPR to calculate the abiraterone acetate and concomitant prednisone adherence rate for patients with prostate cancer is analyzed by Lafeuille et al. (2014). In another study, Steinberg et al. (2014) examined the data collected from insurance record, medical claims, pharmacy claims, laboratory results, and health risk assessment responses over 1 year cycle to calculate the adherence of antidiabetics using MPR. Investigating prescription fill patterns (fills and gaps) employing a visualization tool, Bjarnadóttir et al. (2016) perform their MPR calculations based on the administrative claims data of 790,609 patients with hypertension using diuretics medication. MPR was also used to measure diuretics adherence in patients with hypertension in the study of Malik et al. (2016).

MPR can fail in some cases to assess the actual patient adherence over a specific period of time; it may indicate that a patient is not adherent if there is a gap in the prescription history where the possibility of early prescription refill is not expected.

The medication gap in MPR does not necessarily reflect the facts. For example, the patients can intentionally refill their medication before going for a long holiday and keep taking their medication as prescribed even though this will leave a gap in the prescription records. On the other hand, the patient can stop taking medication or switch to another one just after recent refill, so his/her medication history will incorrectly indicate that the patient regularly took his/her prescription.

### 3.3.8 The Proportion of Days Covered (PDC)

PDC is the most common measure, which calculates the proportion of days within a fixed interval when the patient possesses a recommended supply of medication dosages. This adherence measure calculates the period covered by a prescription claim for the same or similar medication in the same category. When this percentage achieves 80%, the adherence is considered reasonable and can acquire the expected medical benefits (Franklin et al. 2016; Mabotuwana et al. 2009).

It is important as well to consider medication pack sizes and the number of refills in the prescription. PDC uses values between 0 and 1, which provide a more conservative estimate of medication adherence when multiple medications are prescribed and avoids double counting days of medication coverage. PDC is more accurate because it will adjust any overlapping days covered (Franklin et al. 2014). PDC has been used by Franklin et al. (2016) to measure the adherence of multiple statin initiators over a year of medical and pharmacy claims. Similarly Yen et al. (2012) used PDC method to measure ulcerative colitis patients to 5-aminosalicylic acid over 12 months. In the same way, PDC was used to calculate DPP-4, SU, and TZD initiators adherence in patients with diabetes mellitus type 2 (T2DM) (Farr et al. 2014). On the other hand, Molfenter et al. (2012) integrated self-reported questionnaire to pharmacy refill data to measure the adherence of statin medication for patients with heart disease over a period of 6 months. They considered patients as adherent when PDC score is greater than or equal to 85%, while all other abovementioned studies were based on 80% (or 0.8) PDC score to count a patient as adherent. Disease-modifying drugs (DMD) adherence in patients with multiple sclerosis was calculated using PDC measure on the health insurance database (HIPAA) (Kozma et al. 2014).

The retrospective analysis study by Eby et al. (2015) examined the connection between dosing and health outcomes; they examined the administrative data for patients with diabetes test insulin dosage and the connection to other characteristics.

Specific Medication Proportion of Days Covered (spPDC), a new measure for medication adherence based on the previous existing method (PDC), focuses on the adherence characteristics of patients' behavior. The method includes performing initial adherence calculations and running the parallel algorithm for each data partition to calculate the medication adherence (Davis and Kendrick 2014). Lo-Ciganic et al. (2015) have used the PDC to observe the medication level of their patients with diabetes while developing a prediction algorithm (machine learning framework) to examine the medication adherence threshold to reach optimal discrimination of hospitalization risks.

### 3.3.9 Medication Regimen Complexity Index

Medication regimen complexity index (MRCI) is a tool for quantifying multiple features of drug regimen complexity (McDonald et al. 2013). With the support of electronic health records, this tool automates and evaluates the medication treatment complexity and improves the medication management. MRCI helps healthcare providers with complexity assessment of their treatment plans, and by using health decision support system, they can focus on personalization of prescription regimen to better improve adherence.

### 3.3.10 MEMS (Medication Event Monitoring System)

One of the most important objective methods used to monitor the medication adherence is electronic medication adherence monitoring. The system consists of a regular medication bottle (container) in addition to a "smart" bottle cap or closure mechanism that enables electronically to record each time and date the container is opened or closed. Electronic drug monitor is a direct method for measuring patient adherence to medication off the clinical site, and it is a comprehensive tool to discover medication adherence patterns over time (Gill et al. 2015). These systems come in a few variations depending on the manufacturer and the medication form. The most widely used one is MEMS (the medication event monitoring system) (AARDEX Ltd, Union City, CA, USA), which is a medication bottle cap with a microprocessor unit that calculates the occurrence and time of each bottle opening to dispense medication.

MEMS system provides transfer of the dosing data (occurrence and times log) to MEMS reader and may also contain a simple digital screen to display the number of dosages taken in the past hours indicating how many hours from the last bottle opening (dose taken). Assuming that bottle openings represent medication intake, MEMS provides a detailed profile of the patient's adherence behavior based on the log between two dates. By examining this information, clinicians and/or researchers can track daily adherence to dosages, inter-dose intervals, and drug gaps, clarifying patient adherence behavior and apply interventions to improve adherence. Thus, incorporating the EDM data in medication adherence analysis can significantly improve medication adherence pattern discovery and adherence rates (Gill et al. 2015).

Wu et al. (2008) designed the Multidimensional Adherence Model (MAM) to discover the factors that contribute to medication adherence/nonadherence in patients. MAM model uses MEMS data as a foundation to test which variable measures medication adherence more accurately; the indicators are dose count, days covered, and time. The results showed that the major factor in nonadherence was perceived barriers and combination of minorities, and low income is another strong factor of nonadherence.

Petersen et al. (2015) demonstrated in their study that MEMS combined with clinical data analyzed via a super learner algorithm can successfully classify virological failure in HIV patients.

Sayner et al. (2015) conducted a comparison study to compare VAS (subjective) against MEMS (objective) and conclude that patients tend to over-report their actual adherence over VAS, while (MEMS) shows less adherence for the same reported period of time.

MEMS has been used for different medication adherence studies. Petersen et al. (2015) employed MEMS medication measure in their study of antiretroviral therapy (ART) medication in HIV-positive patients and "virological failure patients," discovering nonadherence patterns which indicate the need for an intervention to improve the outcomes.

MEMS was used to measure three types of heart failure medications (diuretic, beta blocker, and digoxin) over a 3-month study period conducted by Wu et al. (2008) to compare which drug class has more adherent patients.

A simulation system to improve medication adherence through mobile health suggests that "reinforcement learning" system can use the daily adherence patterns through bottle opening data provided by MEMS (Piette et al. 2015). MEMS data combined with data from EHR were used in Chang et al.'s (2013) study to improve adherence.

The TDA Travalert (Alcon, Fort Worth, TX) is another adherence electronic monitor device (for eye drops) used in the study of Chang et al. (2013). A bottle of travoprost was placed in the device and a lever was used to administer a drop. A built-in memory chip recorded the time and date each time the lever was depressed. The TDA can only provide data on the use of travoprost because no other medication bottles fit in the device (Chang et al. 2013).

"ECaps" is another electronic drug monitor device that enables specific and well-timed interventions while examining early nonadherence. ECaps (Information Mediary Corp, Ottawa, Canada) is designed as electronic pill bottles with microchip and sensor activated by the pressure when opening the bottle. Similar to MEMS, it records the time and date of every opening. ECaps offers visualization capabilities via its software and transfers the information to another data format for further analysis as well. Electronic drug monitors are considered as one of the most precise tools for measuring adherence that provides comprehensive data with regard to timing or mistiming of doses that were taken, so nonadherence can be identified when mistiming repeatedly occurs (Gill et al. 2015).

Medication adherence was measured indirectly through disease condition in Parkinson's disease patients. Human gait data (gesture) with the assistance of the motion sensor was used to measure motor dysfunction in neurological diseases like Parkinson's disease. This method monitors a patient's motion and concludes if the patient has consumed his/her prescribed medication or not. Monitoring patients through their gait dysfunctions can help health provider to monitor adherence rate and apply the best intervention that can improve patients' adherence (Tucker et al. 2015).

### 3.3.11 Cost-Related Measure

Few studies considered cost factor on adherence, using big data to explore cost-related medication nonadherence. Zhang and Meltzer (2016) used survey, electronic health records, and social security records data (over a period of 2 months) to propose an approach that can measure the medication adherence and has predictive powers as well. The model can identify patients with cost-related nonadherence and discover the ways that cost can change, delay, or prevent patients from adhering to their medication.

## *3.4 Adherence Predictors and Patient Characteristics*

When measuring or trying to predict adherence, research has identified major variables to base their analysis on, as there are so many factors that can affect medication adherence. The following section is to identify the frequent and more effective variables, which can be recognized in further researches to have more considerate assessment of their population samples.

Many factors have been identified for causes of nonadherence to medication in a population sample, which includes demographic factors such as age, sex, and race; socioeconomic factors (patients have different clinical, educational, economic, and situational factors); location; medication cost\sharing cost; behavioral factors; and additional factors such as mental state, e.g., depression, multiple therapies, intake timing regime, persistence time, initiation time, and gaps.

Studies have found that patients with chronic diseases discontinue their medications within 1–2 refills after initial treatment (Curtis et al. 2009); medication switch from drug subclass to another may also negatively affect adherence (Yen et al. 2012).

What has been identified as factors that affect medication use patterns are sociodemographic indicators like age, gender, region, mediation payment type, medication class, and other factors (Franklin et al. 2016; Molfenter et al. 2012). Medication formulation and complexity were important factors related to nonadherence, and other clinical factors were also significant (Coletti et al. 2015). Different classes of drugs returned different outcomes of medication adherence as well (Farr et al. 2014).

Knowing the factors and variables associated with adherence and persistence to medication can help providers and physicians to make informed decisions to target risk factors and variables of adherence in a patient in order to design better interventions and improve adherence. Some of the studies focus on examining the patient's characteristics and evaluating medication use patterns to assess risk factors that can lead to nonadherence. Initial adherence observation and initial filling patterns (Franklin et al. 2014; Tucker et al. 2015) and early adherence (Tucker et al. 2015) were factors that indicated future adherence.

### 3.4.1 Prediction and Improvement of Adherence Methods

Data analytics methods are important to compare pathways of different patient groups discovering exceptional paths for medical services, analyzing if the patients follow healthcare provider's guidelines. Some medication adherence improvement approaches are behavioral (like a mobile reminder, social assistance, dosing changes, interview or call with a healthcare professional), and the others are educational, social, and economic (written or verbal treatment instructions based) or a combination of both (Sayner et al. 2015). Data analysis and predictive modeling are methods to assist in decision support regarding prevention of drug waste, prescription drug abuse, and promoting patient medication adherence.

Advancements in medical, pharmaceutical, and mobile health sequentially created enormous and diverse data records as a result of different systems like electronic health records, personal health records, prescription data, pharmacy dispensing data, pharmacy refill data, and health and pharmaceutical claims data. These data created an opportunity to extract deeper insights, discovering trends, and predicting future outcomes to enhance the medical process and improve patient health.

The advanced analytics in healthcare context consist of data processing approaches that feasibly create different insights compared to other methods; advanced analytic methods process all kinds of data using sophisticated quantitative methods including statistic, descriptive, simulation, visualization, predictive, and optimization techniques. These advanced tools and techniques discover new insights that benefit the medical process by showing what happened in the past according to the records, why that happened (what are the factors, variables, characteristics), and whether we can predict the future. Analytics is not only used by the medical institutions, but also others like the pharmaceutical industry use health decision science to enable providers to make a better decision regarding their clients (patients). For example, big data analytics can utilize patients' medication refill pattern data to identify the individual patient with risks of low or nonadherence for a specific medication and increase treatment outcomes.

Researchers have used different strategies and analytical models to group the patients based on their characteristics and other relevant data collected from clinics, pharmacies, and devices in order to find trends and patterns. These patterns discover new insights into factors that cause nonadherence and facilitate prediction of future trends.

Advanced analytics can possibly help healthcare providers to consider and reduce risk factors that can be problematic. This will improve adherence rates among patients by allowing the providers to better manage personalized focused treatment plan and direct the efforts to intervene and communicate with specific patients before the problem arises.

When evaluating the patient compliance to the treatment plan, the focus should not be directed to demographic factors, comorbidities, and pattern of health service, but other compliance pattern data with medication data for other conditions (diseases) should be also considered and integrated.

There are different strategies for interventions, including self-reporting, counseling (patient–clinician communication), and reminders. The literature also includes measurements used in medication assessment, and it differs between subjective behavior reports (like self-assessment) and objective reports (like electronic drug monitor) (Curtis et al. 2009).

The incoherence between the medical information system, physicians, and environmental contexts may result in inefficiencies in treatment delivery and affect patient health quality. To provide better treatment for the patients, which promise improving adherence significantly, the system should integrate most of the variables that affect functionality. The study of Stewart et al. (2014) suggested a method to collect and display all the information that can help to identify the medication adherence barriers and reduce misunderstanding to improve adherence. Using a unified dashboard, the system uses data from EHR, PHR, PDC, and pharmacy claim to create a tailored assessment of patient medication adherence; thus, the outcomes will enhance health provider-patient communication to increase adherence levels (Stewart et al. 2014).

The most common medication adherence prediction and improvement methods were focused on the initial pattern of medication filling (Stewart et al. 2014; Mabotuwana et al. 2009; Franklin et al. 2013, 2014); MPR/time lapse reports (Franklin et al. 2016), the correlation between patient discrepancies (predication to spate nonadherent patients) (Yen et al. 2012); and improvement to the traditional proportion of days covered (PDC) measure made it more predictive (Farr et al. 2014). Selection of adherence variables using visual analytic tools to define medication possession through investigating the implications can improve the prediction (Molfenter et al. 2012) and explore the early adherence for different drug class (Kozma et al. 2014).

There is an association between a patient's highest daily mean dose and adherence. For example, Dixon et al. (2014) utilized analyses controlled for patient characteristics and index dose of insulin. Results were consistent with the past literature. Adherence to specific medication decreases accordingly with medication complexity, and poor adherence is related to higher dosage.

Analysis results by Davis and Kendrick (2014) showed that there is a need to optimize patient adherence by reducing the gap between provider recommendations and the actual extent to which the patient will follow. A more specialized approach is needed to help patients to understand, maintain, and then stay adherent to the medication regimens. A health belief matrix (Piette et al. 2015) was used to test beliefs. This matrix is a traditional questioner that assesses the medication adherence and is based on health belief model and social cognitive theory.

Patient's prescription-refill statistics can predict the likelihood of future adherence; patient's past behavior like medication refill is a solid predictor for the patient's long-term medication regimen which can establish long-term, stable, healthy behavior. Past behavior information for patients is an important indicator on patient's future adherence (Chang et al. 2013). Researchers used Receiver Operating Characteristic (ROC) analysis method to define how the model fits between the patient's health beliefs and previous adherence behavior with the current actual medication use. Patients' long-term medication regimen demonstrates long-term stable behavior, and hence, past behavior is a strong predictor of future behavior. Similarly, predicting adherence using initial medication filling can be performed by estimating the future trend of initial medication filling. Some researchers investigate the same concept: how the initial filling of the prescribed medication can be used to predict the patient's future commitment and how long this initial state should remain. Petersen et al. (2015) conducted a study to evaluate the initial medication dispensing ability to predict longer terms of medication adherence patterns. The research investigates the adherence pattern based on the initial observation of medication behavior to find if the predictions based on this observation are more accurate than the investigator-specified clinical characteristics of grouping participants in different groups (trajectories).

The models that included early adherence data are better predictors of future adherence than models that only included baseline characteristics (Wu et al. 2008). The accuracy of the initial medication filling method and the exact period of time that predicts accurately future adherence were also investigated by Piette et al. (2015). An initial medication filling directly indicates the patient's ability to access the prescribed medication and suggests future commitment and adherence. Prediction of patient's adherence/nonadherence according to the initial filling can be employed by health providers to better target interventions to patients who are likely to be nonadherent (based on the initial 30-day filling information), providing more efficient use of resources. A study conducted by Steinberg et al. (2014) also identified the importance of relying on initial prescription data as a solid base to predict further adherence behavior.

Long-term adherence for the same population varies between patients with different drug classes and subclasses (Mabotuwana et al. 2009). There is also variation in adherence when the observed period varied (Curtis et al. 2009). Variation in timing/mistiming of dose may also affect adherence patterns, and hence, it is possible to take corrective interventions to improve such behavior (Bjarnadóttir et al. 2016).

Variation in medication adherence behavior across countries, regions, and populations can cluster patient behavior patterns and has predictive power; by clustering patients, we can discover the difference in adherence, and it can be used to prioritize future interventions (Tucker et al. 2015).

To identify patients with higher risks and improve medication adherence and patients' outcomes, medication regimen complexity index has been adopted and automated based on electronic health records. Identifying these patients and supporting health providers to improve their medication management and calculate the complexity of treatment plans, can be conducted through the capabilities of decision support systems that tailor a personalized and focused prescription treatment plan (Sayner et al. 2015).

It is hard to assess patient adherence outside the clinical context; advanced technologies of the internet of things, connected devices, and sensors (like visual and motion sensors or wearable devices and sensors) are alternatives to clinical objective adherent measures.

A system for tracking patients with Parkinson disease has utilized wearable sensors that monitor patient's adherence to medication outside healthcare organization via observing gait data (human motion data) (Franklin et al. 2016) to differentiate between patients who are adherent to their medications and patients who are not. Yen et al. (2012) proposed a framework that monitors patient medication intake in real time to ensure adherence. This framework was composed of motion detection data collected from a watch worn by the patient.

Systems design has been still evolving to create more effective interventions in medication adherence and treatment pathway. Farr et al. (2014) focused on designing an effective personalized diabetes management regimen (computerized decision support system CDSS) that uses machine learning to personalize patients' treatment, which can lead to more effective estimation of the medication requirements and improve adherence and treatment goals. Another system targeted for diabetes monitoring and management also used data mining for personalized treatment and potentially calculated patient medication adherence to inform patient's/physician's interventions (George et al. 2009).

Machine learning method with a logistic regression model is used by Molfenter et al. (2012) for data classification and categorizing medication variables. The findings from the study indicated that patient's medication refill predicts the likelihood of medication adherence in the future. Kozma et al. (2014) used multiple linear regression analysis to predict medication adherence in patients with multiple sclerosis. Similarly, Dixon et al. (2014) used informatics tool, a clinical decision support system, to measure medication adherence for heart patients.

Interventions to improve medication adherence can be further improved by personalized treatments and target reinforcement learning (RL). RL is a type of machine learning which enables software agents to automatically determine the ideal behavior within a specific context and maximize the outcomes. The RL system was able to learn and characterize SMS messages to different patients based on their true underlining grounds for nonadherence, adapt to the changes that nonadherence causes over time, and in addition, adapt the messaging frequency for each individual (Eby et al. 2015).

Davis and Kendrick (2014) concluded that machine learning provides an opportunity for optimizing treatment outcomes and personalization of interventions for improved medication adherence. The prediction algorithm tests patient adherence (measured in PDC) and identifies predictors of hospitalization as a result of nonadherence. Machine learning approach is valuable in the identification of patient-specific adherence threshold (measuring the quality of care) and targets nonadherent patients for interventions.

Analytical models can be used to gain insights and knowledge from different data sources like demographics, diagnoses, claim data, lab results, prescription, management, and other related patient data. In order to discover knowledge and new insight, Chang et al. (2013) suggested a reverse engineering and forward simulation (REFS) model that learns risk factors (for diabetes) and medication different timings (early and late adherence) to create knowledgeable insight about adherence faster.

### 3.4.2 Mobile Monitoring and Prediction of Adherence Level

Mobile medication system has the ability to monitor patient's intake of their prescribed medication and to connect with providers and physicians. Such "smart" system can monitor and analyze patient's medication patterns using a combination of sensors and wireless devices to report adherence to their physicians/health provider. Mabotuwana et al. (2009) and Franklin et al. (2013, 2014) have designed advanced warning systems based on the patient's patterns of medication use. The systems can notify health professional when unhealthy habits of adherence have been reported.

Curtis et al. (2009) aimed to find the correlation between adherence and medication discrepancies and test if nonadherence is associated with discrepancies. The model has the ability to separate adherent patients with 60% success rate. The first step in improving medication adherence should be accurately assessing the barriers to adherence, then suggesting effective interventions for improvements.

Discrepancies in medication treatment themselves do not indicate nonadherence but the need for a better provider–patient communication. Patient–provider agreement on the optimal medication regimen is necessary for the pursuit to improve adherence. This can be done through increasing patient knowledge about medication, simplifying the regimen, and addressing patient emotional function (Stewart et al. 2014).

Discrepancies between self-reported adherence and objectively measured medication adherence are investigated by Stewart et al. (2014). The researchers found that the correlation between the measured adherence ratios is not strong between the methods, but there is necessity of simplifying medication regimen to avoid patient's mistakes and confusion. Such discrepancies are a possible reason for low adherence, so optimal adherence can be achieved by improving the communication between the provider and the patient.

## 4 Discussion

Review of medication adherence measurment (Fig. 2) and analytical models for predicting medication adherence is an insightful way to present the realized and recommended clinical, pharmaceutical, and research finding in the area of medication adherence. The findings identified few of the common measurement

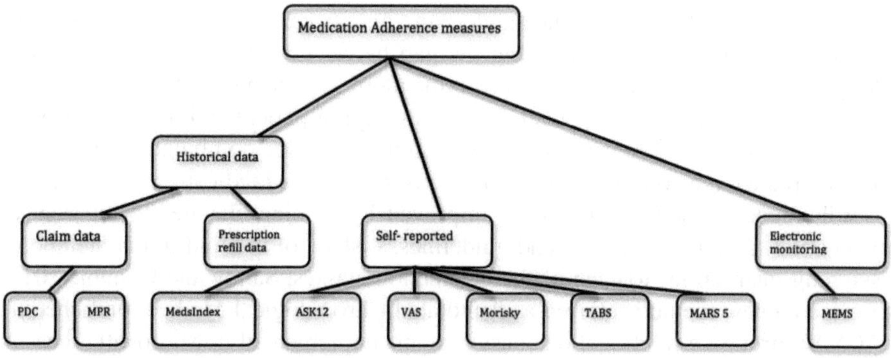


**Fig. 2** Medication adherence measures

tools that assess patient medication adherence ratio, varying between self-reported, refill data, and electronic adherence monitoring. When comparing self-reported methods to automatic measures, discrepancies in adherence were found, concluding that self-reported methods are not much accurate as standalone because patients tend to overestimate their intake over reality, either for social or personal reasons. Refill adherence (administrative) data is also a good measure to assess patients adherence accurately. Some refill adherence measure can overcome medications switching and other gaps by replacing PDC instead of MPR. Electronic adherence measures are so far the most accurate because of its ability to independently assess adherence for a variety of medications, providing convenient, easy-to-use, and efficient way of measurement. A triangulation of adherence data or at least supporting one data source with another is essential for the improvement of assessment efficiency.

It is important to integrate the data of electronic drug monitoring or MEMS to optimize reminding and alerting. A framework developed by Franklin et al. (2013) promotes the importance of patient context and preferences for optimizing alerts and reminders in order to increase the medication treatment adherence. Mobile health applications, such as mobile apps, and wearable, non-wearable, and sensors-enabled devices advanced the medication adherence measurement. Decision support system uses analytical methods and provides relevant information to physicians, as well as to patients in order to better understand their medication management. Decision support system connects the observations and trends with the knowledge to guide and assist the clinician to make better-informed decisions, and machine learning is an integral part of decision support systems used to predict outcomes, patterns recognition, and activity detection\estimation (Farr et al. 2014). Machine learning and predictive modeling can be used to study adherence patterns of a specific medication in a specific population and identify nonadherence\non-persistence factors. These systems assist in personalized interventions by classifying patients into smaller groups based on their identified barriers and provide effective medication measurements.

Nonadherence factors are not only patient- and socioeconomic-related (cost alone identified as a nonadherence measure and indicator), but it can also be due to the health condition, treatment, and healthcare system as a factor (Lafeuille et al. 2014; Gill et al. 2015). Adherence underlying beliefs can link population segments into specific interventions. Reducing the gap between provider recommendations along with medication complexity (prescribe more generic medication), medication instructions, and patient of pocket expenses (helping patients with the medication cost) can improve adherence. Prescribing simpler and fixed combination of medication can reduce the complexity and number of pills in the medication regimen.

Adherence to medication can be improved by making changes to the way of medication-taking instructions and guidelines. Instead of the traditional method of assigning instructions to each of the medication classes, Sandy and Connor (2015) proposed a new method referred as Annotated Chronological Medication Schedule (ACMS) through a relational database system (implemented in Microsoft Access). This system can be used by health providers to create a chronological medication schedule with all necessary usage direction only (e.g., upon rising, at breakfast time). This method can be helpful to include within mobile monitoring and alerting system and it is more useful in multiple medications context. Fewer medication adherence instructions can be simply delivered using the same electronic medication devices.

Chang et al. compared between the self-reported adherence and the adherence measure results from the electronic medical record (EMR) system. While participants' self-assessment results suggested high rates of adherence, the study showed discrepancies between patients' self-reported medication adherence and EMR results.

Most of the earlier tools used were a questionnaire style tool performed by the physician who interviews the patient and assesses the responses for further analysis. Examples of this kind of self-reported tools are ASK12, VAS, TABS, MARS-5, and health belief matrix. Pharmacy-focused and self-reported data like Medsindex score refill data; and administrative refill claim-focused tools like PDC and MPR. Then, there are electronic adherence measures such as MEMS and less common motion sensors system. Electronic tools to measure adherence are becoming more popular. These tools easily log medication-taking behavior (opening and closing the medication bottle cap), and recently, these electronic measures are becoming more available, easy to use, and cost-effective. Table 1 lists the identified medication adherence measures with their frequency of use in the reviewed studies.

Sandy and Connor (2015) identified some of the effective interventions like self-management and monitoring, simplified dosing, and medication review (pharmacist-delivered intervention) and concluded that not all interventions are equally effective. Exploring intervention effectiveness is critical for making more informed decisions on how to improve interventions, to meet patient needs and improve their treatment outcomes.

Process analytics can be integrated into healthcare processes. EHR data now can track more details about the patients in addition to healthcare processes. By including data proration, inspection, abstraction, clustering, mining, and validation, process analytics (PA) techniques can help to produce new insights into the healthcare environment (Davis and Kendrick 2014) and can create insights in applications for discovering the most frequent and exertional treatment path, comparing pathways of different patient groups or analyzing whether patients follow healthcare guidelines (like adherence instructions). By obtaining deeper insights using PA, health organization can improve their prediction for patient's outcomes. It can visualize which class of drugs are frequently consumed and if they are overlapping with different classes and successfully taken with other drugs.

Most of the literature explores adherence issues focused on nonadherence and more specifically on taking less dosage than the prescribed in the time frame. Using a different point of view, Chang et al. (2013) designed a model to discover the overall likelihood of medication addiction, where medication abuse was identified as taking more dosage or/and in more frequent times than recommended and has harmful implications. For the research, it was designed to monitor medication abuse based on the probability of multi-dosing, accidental dosing, current abuse, and the likelihood of addiction.

**Table 1** Measures/method for measuring medication adherence

| Source | Measure/methods | Definition | Representative study |
|---|---|---|---|
| Pharmacy claims/ prescription refill | The proportion of days covered (PDC) | PDC measure divides days covered by total days for a given evaluation period and indicates the maximum amount of medication a patient could have taken in a given time period | Eby et al. (2015), Farr et al. (2014), Franklin et al. (2013, 2014, 2016), Kozma et al. (2014), Lo-Ciganic et al. (2015), Molfenter et al. (2012), and Yen et al. (2012) |
| | Medication Possession Ratio (MPR) | MPR is obtained by dividing the sum of days of supply of medication by the total number of days on therapy | Cheng et al. (2015), Curtis et al. (2009), Lafeuille et al. (2014), Mabotuwana et al. (2009), Stewart et al. (2014), and Yen et al. (2012) |
| | Specific medication PDC (smPDC) | smPDC is calculated by dividing the number of days covered (by at least one drug in the class) with the number of days between the last and first fill, multiplied by 100 | Davis and Kendrick (2014) |
| | Medication regimen complexity index (MRCI) | MRCI is a tool for quantifying multiple features of drug regimen complexity. The index includes weighted components of the dosage form, dosing frequency, and additional administration instructions | McDonald et al. (2013) |
| | MedsIndex | MedsIndex scores patients for each of their long-term medications calculated from their prescription-refill history | Stewart et al. (2014) |

| | | | |
|---|---|---|---|
| Self-reported measure | Adherence Starts with Knowledge 12 (ASK-12) | ASK-12 is a condensed survey tool that offers quick identification of the most prevalent factors that influence medication adherence | Coletti et al. (2015) |
| | Visual Analog Scale (VAS) | VAS (psychometric response scale which can be used in questionnaires) is used as a self-report measure of adherence | Sayner et al. (2015) |
| | Morisky scale | The Morisky scale comprises four items and assesses both intentional and unintentional nonadherence. It is a validated scale designed to estimate the risk of medication nonadherence | Stewart et al. (2014) |
| | Tools for advanced behavior screening (TABS) | The TABS measures both intentional and unintentional deviations from treatment recommendations and has been shown to have greater incremental validity than other self-reported adherence measures | Stewart et al. (2014) |
| | Medication adherence report scale | MARS is an adherence instrument that has been used as a quantitative measure of self-reported adherence for many conditions | Sandy and Connor (2015) |
| | Five self-report questions | These items ask how well the patients knew the names, purposes, recommended doses, frequencies, and side effects of their medications. The five items with a 5-point Likert-type scale are used for assessing patients' medication knowledge | Son et al. (2010) |
| Electronic monitoring | Medication electronic monitoring system (MEMS) | MEMS is a computerized monitoring system that is used to indirectly measure medication adherence | Petersen et al. (2015), Sayner et al. (2015), and Wu et al. (2008) |
| | Electronic drug monitors (EDM) | EDM provides detailed data regarding the timing (or mistiming) of medication doses that were taken | Gill et al. (2015) |

Coletti et al. (2015) investigated the use of CoMac and Event Flow visualization tool to support the exploration of open-ended questions and experiment different scenarios of treatment. This methodology can provide useful insights into different treatment plans that have multiple drug classes or different diseases, for example, by testing multiple hypotheses and meaningfully visualizing the intermediate (expected) outcomes. Thus, the healthcare provider can consider patients medication knowledge and forgetfulness to simplify the prescribed regimen, integrate medication reminders functionality in devices or service, improve adherence offline or in real-time monitoring, and leverage the importance of health provider communication and engagement with the patient.

Discovering patterns of nonadherence when assessing patients data may not be entirely accurate. There could be changes in the medication regime by the doctor and the patient has been adhering to doctor instruction or verbal advice by delay/switch/discontinue a medication, so the prescription is not refilled. For that reason, it is recommended to include the physician's decision regarding each instruction given and integrate it with other types of data. Choosing the right or suitable medication adherence measure includes measure that is accurate, user-friendly, affordable, and flexible. In this context, subjective self-reported methods are still very useful in a population with low literacy or patients already identified as low/nonadherent, and physician–patient interaction during interviews can help health providers to identify nonadherence barriers further and try to customize the interventions and direct measures.

A combination of objective and subjective adherence assessment is needed and collected adherence data preferably should be linked or stored in patient health records. In that case, health providers will be able to quickly display past adherence behavior patterns, analyze the data, and alter or design personalized interventions in order to improve patient adherence. Reducing medication complexity or increasing the generic use of a medication dose can also be considered in assisting patients with the payment gaps, especially if the cost was identified as a barrier.

Decision support systems can depend on analytical techniques that can improve personalized intervention programs fitting the patient population and considering other medication adherence enablers. The personalized treatment regimen may include more suitable drug class for the condition and suggest personalized intake timing, personalized instructions, and follow-up preferences.

Patient segmentation is helpful in sorting patients by their level of adherence and prioritizing interventions based on this categorization. It enables patient–physician communications, so the patients who are nonadherent or most likely to be nonadherent can be contacted sooner in order to speed up interventions, increase their accuracy, and reduce the risks of any possible complications caused by nonadherence.

Poor or insufficient adherence to prescribed medication can have a major effect on the efficiency of the healthcare treatment. Medication nonadherence can also lead to patient hospitalization and mortality (Pavel et al. 2010).

Nonadherence to prescribed medications can occur when the patients believe that it is unnecessary to comply with medication instructions; the patient cannot physically or cognitively adhere to treatment regardless of the positive intention to do so. It could also be that patients may be willing and are able to adhere, but they may face some obstacles which did not allow them to take medication on schedule. In addition to these causes, there are few different reasons for nonadherence pointed out in the literature, and it varies between patient literacy (knowledge and education), coverage of the medication itself by the provider, side effects, medication cost, disease type, patient social support, affordability, transportation, medication class confusion, duplication of the therapy, difficulty in following treatment instructions, forgetfulness, and situational/environmental reasons.

Non- or poor adherence reduces treatment efficiency and time to treatment and increases healthcare cost in addition to affecting patient well-being. Knowing the factors that affect medication compliance, persistence and discontinuation may facilitate identifying prospective patients who are less likely to adhere to specific therapy and may allow to design specific medication treatments, interventions, and follow-up to specific patients where their adherence is prospectively low.

Many studies evaluate adherence data to improve medication adherence and plan for better interventions. Researchers seek to measure patient adherence to a certain medication as early as possible, so they can investigate the obstacles and find solutions. Multiple analytical, simulation, and visualization tools have been used to perform analysis, predict a pattern, simulate scenarios about potential future trends, and visualize the outcomes to discover hidden patterns and insights for valuable knowledge. Although several data analytics have been conducted in the reviewed articles, many of the articles (Chang et al. 2013; Curtis et al. 2009; Zhang and Meltzer 2016) use statistical packages including regression model and multivariate analysis (Coletti et al. 2015; Farr et al. 2014; Molfenter et al. 2012), and only a few of them use decision tree (Tucker et al. 2015), Random Forest (Lo-Ciganic et al. 2015), and a neural network and support vector machine (Son et al. 2010). With the advances in artificial intelligence and machine learning, medication adherence studies can be based on the rich dataset and provide the future adherence prediction.

## 5 Conclusion

Study results found a variation in the methods used for measuring, predicting, and improving patient medication adherence. The advancement in information technology has led to increased knowledge acquisition of health data and provide knowledge discovery. Healthcare systems and related administrative systems are combined to produce and store a massive amount of data about patients; systems like electronic health records store patient demographic and socioeconomic data, while administrative systems store medication, diagnoses, and hospitalization data, among others. Adherence measurements varied between relying on the administrative data such as refill data or health insurance data that can be calculated using PDC or MPR methods, self-reported measures that include questionnaires (VAS, MARS5, Ask12) and electronic medication adherence data (calculated using MEMS, Ecaps);

cost-related approach is also considered as a measure of nonadherence. Recently, more recommendations were stated to adopt electronic medication data mainly supported by the other two sources to assess adherence levels accurately. Analytical models include machine learning, simulation, data mining, visualizations, and some gesture detection. These models identify adherence patterns through the population and recognize nonadherence factors and their effect on treatment plans. Significant benefits are achieved from data mining and machine learning models and techniques especially to classify patient's adherence behavior and their segments. Identifying patient's segments and their level of adherence enables physician–patient communications and discussion regarding patient preferences, the underlying reasons for nonadherence, and the way to overcome them. Beyond the differences in the models and their implementation techniques, it is recommended to simplify medication regimen, promote the importance of patient self-management and participation through mobile monitoring, and also to use electronic adherence device system like MEMS to track adherence patterns.

This chapter outlined the methods and techniques identified in the literature for medication adherence improvement that were implemented in different studies and across the diverse population and disease context. Methods were grouped with similar concept studies to provide clear outlines of the current trends in medication adherence research. First, the methods and frameworks that aim to improve adherence depending on medication refill and dosing patterns were examined. Next medication discrepancies and how they affect adherence aspects and methods to overcome this issue were inspected. Finally emerging trends of using mobile technologies, data mining, machine learning, and internet of things concepts to monitor, predict, and assist the patient to become more adherent were presented.

The outcome of this research outlines medication adherence measure techniques and guides future research and health providers to adopt a specific combination of measures and prediction models. Adoptions of models already proven to be efficient can help health providers to direct their efforts into certain approaches enabling them to identify nonadherence factors and shed light on ways to improve adherence.

Identification of medication adherence patterns can assist healthcare providers to make better decisions about interventions. Being able to prospectively identify patients who are less likely to adhere to therapies would have important health implications. It might allow healthcare provider to tailor certain medications, treatment and follow-up strategies, or interventions to particular individuals who are at greater risk of non-compliance. It is more efficient and timely to be proactive than reactive when discussing medication adherence issues, as patients who adhere to their current medication will avoid transferring to more complex and expensive ones and for the health providers, early adherence can be more beneficial when they are planning into the future.

## References


Bjarnadóttir, M. V., Malik, S., Onukwugha, E., Gooden, T., & Plaisant, C. (2016). Understanding adherence and prescription patterns using large-scale claims data. *PharmacoEconomics, 34*, 169–179.

Chang, D. S., Friedman, D. S., Frazier, T., Plyler, R., & Boland, M. V. (2013). Development and validation of a predictive model for nonadherence with once-daily glaucoma medications. *Ophthalmology, 120*, 1396–1402.

Cheng, L. I., Durden, E., Limone, B., Radbill, L., Juneau, P. L., Spangler, L., Mirza, F. M., & Stolshek, B. S. (2015). Persistance and compliance with osteroporosis therapies among women in a commercially insured population in the United States. *Journal of Managed Care Pharmacy, 21*, 824–833.

Coletti, D. J., Stephanou, H., Mazzola, N., Conigliaro, J., Gottridge, J., & Kane, J. M. (2015). Patterns and predictors of medication discrepancies in primary care. *Journal of Evaluation in Clinical Practice, 21*, 831–839.

Curtis, J. R., XI, J., Westfall, A. O., Cheng, H., Lyles, K., Saag, K. G., & Delzell, E. (2009). Improving the prediction of medication compliance: The example of bisphosphonates for osteoporosis. *Medical Care, 47*, 334.

Davis, N. A., & Kendrick, D. C. (2014). An analysis of medication adherence of sooner health access network SoonerCare choice patients. In *AMIA annual symposium proceedings* (p. 457). American Medical Informatics Association.

Dixon, B. E., Jabour, A. M., Phillips, E. O. K., & Marrero, D. G. (2014). An informatics approach to medication adherence assessment and improvement using clinical, billing, and patient-entered data. *Journal of the American Medical Informatics Association, 21*, 517–521.

Eby, E. L., Van Brunt, K., Brusko, C., Curtis, B., & Lage, M. J. (2015). Dosing of U-100 insulin and associated outcomes among Medicare enrollees with type 1 or type 2 diabetes. *Clinical Interventions in Aging, 10*, 991.

Farr, A. M., Sheehan, J. J., Curkendall, S. M., Smith, D. M., Johnston, S. S., & Kalsekar, I. (2014). Retrospective analysis of long-term adherence to and persistence with DPP-4 inhibitors in US adults with type 2 diabetes mellitus. *Advances in Therapy, 31*, 1287–1305.

Franklin, J. M., Shrank, W. H., Pakes, J., Sanfélix-Gimeno, G., Matlin, O. S., Brennan, T. A., & Choudhry, N. K. (2013). Group-based trajectory models: A new approach to classifying and predicting long-term medication adherence. *Medical Care, 51*, 789–796.

Franklin, J. M., Krumme, A. A., Shrank, W. H., Matlin, O. S., Brennan, T. A., & Choudhry, N. K. (2014). Predicting adherence trajectory using initial patterns of medication filling. *The American Journal of Managed Care, 21*, e537–e544.

Franklin, J. M., Shrank, W. H., LII, J., Krumme, A. K., Matlin, O. S., Brennan, T. A., & Choudhry, N. K. (2016). Observing versus predicting: Initial patterns of filling predict long-term adherence more accurately than high-dimensional modeling techniques. *Health Services Research, 51*, 220–239.

Georga, E., Protopappas, V., Guillen, A., Fico, G., Ardigo, D., Arredondo, M. T., et al. (2009, September). Data mining for blood glucose prediction and knowledge discovery in diabetic patients: The METABO diabetes modeling and management system. In *2009 annual international conference of the IEEE engineering in medicine and biology society* (pp. 5633–5636). IEEE.

George, J., Mackinnon, A., Kong, D. C., & Stewart, K. (2006). Development and validation of the Beliefs and Behaviour Questionnaire (BBQ). *Patient Education and Counseling, 64*, 50–60.

Gill, C. J., Desilva, M. B., Hamer, D. H., Keyi, X., Wilson, I. B., & Sabin, L. (2015). Novel approaches for visualizing and analyzing dose-timing data from electronic drug monitors, or "how the 'broken Window'Theory pertains to ART adherence". *AIDS and Behavior, 19*, 2057–2068.

Horne, R., & Weinman, J. (2002). Self-regulation and self-management in asthma: Exploring the role of illness perceptions and treatment beliefs in explaining non-adherence to preventer medication. *Psychology and Health, 17*, 17–32.

Kozma, C. M., Phillips, A. L., & Meletiche, D. M. (2014). Use of an early disease-modifying drug adherence measure to predict future adherence in patients with multiple sclerosis. *Journal of Managed Care Pharmacy, 20*, 800–807.

Lafeuille, M.-H., Grittner, A. M., Lefebvre, P., Ellis, L., Mckenzie, R. S., Slaton, T., & Kozma, C. (2014). Adherence patterns for abiraterone acetate and concomitant prednisone use in patients with prostate cancer. *Journal of Managed Care Pharmacy, 20*, 477–484.

Lo-Ciganic, W.-H., Donohue, J. M., Thorpe, J. M., Perera, S., Thorpe, C. T., Marcum, Z. A., & Gellad, W. F. (2015). Using machine learning to examine medication adherence thresholds and risk of hospitalization. *Medical Care, 53*, 720.

Mabotuwana, T., Warren, J., & Kennelly, J. (2009). A computational framework to identify patients with poor adherence to blood pressure lowering medication. *International Journal of Medical Informatics, 78*, 745–756.

Malik, S., Shneiderman, B., Du, F., Plaisant, C., & Bjarnadottir, M. (2016). High-volume hypothesis testing: Systematic exploration of event sequence comparisons. *ACM Transactions on Interactive Intelligent Systems (TIIS), 6*, 9.

Maulucci, R. A., & Somerville, D. (2011). An automated medication adherence tool. Engineering in Medicine and Biology Society, EMBC. In *2011 annual international conference of the IEEE* (pp. 1165–1168). IEEE.

Mcdonald, M. V., Peng, T. R., Sridharan, S., Foust, J. B., Kogan, P., Pezzin, L. E., & Feldman, P. H. (2013). Automating the medication regimen complexity index. *Journal of the American Medical Informatics Association, 20*, 499–505.

Molfenter, T. D., Bhattacharya, A., & Gustafson, D. H. (2012). The roles of past behavior and health beliefs in predicting medication adherence to a statin regimen. *Patient Preference and Adherence, 6*, 643–651.

Pavel, M., Jimison, H., Hayes, T., Larimer, N., Hagler, S., Vimegnon, Y., et al. (2010). Optimizing medication reminders using a decision-theoretic framework. *Studies in Health Technology and Informatics, 160*(Pt 2), 791–795.

Petersen, M. L., Ledell, E., Schwab, J., Sarovar, V., Gross, R., Reynolds, N., Haberer, J. E., Goggin, K., Golin, C., & Arnsten, J. (2015). Super learner analysis of electronic adherence data improves viral prediction and may provide strategies for selective HIV RNA monitoring. *Journal of Acquired Immune Deficiency Syndromes (1999), 69*, 109.

Piette, J. D., Farris, K. B., Newman, S., An, L., Sussman, J., & Singh, S. (2015). The potential impact of intelligent systems for mobile health self-management support: Monte Carlo simulations of text message support for medication adherence. *Annals of Behavioral Medicine, 49*, 84–94.

Ritchey, M., Tsipas, S., Loustalot, F., & Wozniak, G. (2016). Use of pharmacy sales data to assess changes in prescription-and payment-related factors that promote adherence to medications commonly used to treat hypertension, 2009 and 2014. *PLoS One, 11*(7), e0159366.

Sandy, R., & Connor, U. (2015). Variation in medication adherence across patient behavioral segments: A multi-country study in hypertension. *Patient Preference and Adherence, 9*, 1539.

Sayner, R., Carpenter, D. M., Blalock, S. J., Robin, A. L., Muir, K. W., Hartnett, M. E., Giangiacomo, A. L., Tudor, G., & Sleath, B. (2015). The accuracy of patient-reported adherence to glaucoma medications on a visual analog scale compared with electronic monitors. *Clinical Therapeutics, 37*, 1975–1985.

Serdaroglu, K., Uslu, G., & Baydere, S. (2015). Medication intake adherence with real time activity recognition on IoT. In *2015 IEEE 11th international conference on wireless and mobile computing, networking and communications (WiMob)* (pp. 230–237). IEEE.

Shukla, N., Hagenbuchner, M., Win, K. T., & Yang, J. (2018). Breast cancer data analysis for survivability studies and prediction. *Computer Methods and Programs in Biomedicine, 155*, 199–208.

Son, Y.-J., Kim, H.-G., Kim, E.-H., Choi, S., & Lee, S.-K. (2010). Application of support vector machine for prediction of medication adherence in heart failure patients. *Healthcare Informatics Research, 16*, 253–259.

Steinberg, G. B., Church, B. W., Mccall, C. J., & Scott, A. B. (2014). Novel predictive models for metabolic syndrome risk: A "big data" analytic approach. *The American Journal of Managed Care, 20*, e221–e228.

Stewart, K., Mc Namara, K. P., & George, J. (2014). Challenges in measuring medication adherence: Experiences from a controlled trial. *International Journal of Clinical Pharmacy, 36*, 15–19.

Tucker, C. S., Behoora, I., Nembhard, H. B., Lewis, M., Sterling, N. W., & Huang, X. (2015). Machine learning classification of medication adherence in patients with movement disorders using non-wearable sensors. *Computers in Biology and Medicine, 66*, 120–134.

Win, K. T., Hassan, N. M., Oinas-Kukkonen, H., & Probst, Y. (2016). Online patient education for chronic disease management: Consumer perspectives. *Journal of Medical Systems, 40*(4), 88.

Wu, J. R., Moser, D. K., Chung, M. L., & Lennie, T. A. (2008). Predictors of medication adherence using a multidimensional adherence model in patients with heart failure. *Journal of Cardiac Failure, 14*, 603–614.

Yen, L., Wu, J., Hodgkins, P., Cohen, R. D., & Nichol, M. B. (2012). Medication use patterns and predictors of nonpersistence and nonadherence with oral 5-aminosalicylic acid therapy. *Journal of Managed Care Pharmacy, 18*, 701–712.

Yu, W., Liu, T., Valdez, R., Gwinn, M., & Khoury, M. J. (2010). Application of support vector machine modeling for prediction of common diseases: The case of diabetes and pre-diabetes. *BMC Medical Informatics and Decision Making, 10*, 16.

Zhang, J. X., & Meltzer, D. O. (2016). Identifying patients with cost-related medication non-adherence: A big-data approach. *Journal of Medical Economics, 19*, 806–811.

Son, Y.-J., Kim, H.-G., Kim, E.-H., Choi, S., & Lee, S.-K. (2010). Application of support vector machine for prediction of medication adherence in heart failure patients. *Healthcare Informatics Research, 16*, 253–259.

Steinberg, G. B., Church, B. W., McCall, C. J., & Scott, A. B. (2014). Novel predictive models for metabolic syndrome risk: A "big data" analytic approach. *American Journal of Managed Care, 20*, [illegible]

[illegible] adherence. [illegible]

Tucker, C. S., Behoora, I., Nembhard, H. B., Lewis, M., Sterling, N. W., & Huang, X. (2015). Machine learning classification of medication adherence in patients with movement disorders using non-wearable sensors. *Computers in Biology and Medicine, 66*, 120–134.

Wu, [illegible] disease management [illegible]

[illegible]

[illegible]

Zhang, [illegible] Identifying patients with [illegible]

# Addressing Data Accuracy and Information Integrity in mHealth Solutions Using Machine Learning Algorithms

**Zaid Sako, Sasan Adibi, and Nilmini Wickramasinghe**

## 1 Introduction

Noncommunicable diseases (NCD), also known for being chronic diseases, are 'conditions which are known to affect individuals over an extensive period of time, and for which there are no known causative agents that are transmitted from one individual to another' (Bovell-Benjamin 2016. p. 5). The duration of chronic diseases is very long, and the causes of chronic diseases are a combination of genetic, physiological, environmental and behavioural factors (World Health Organization 2018a, b, c, d). According to World Health Organization (WHO), chronic diseases affect individuals, and it is estimated that the number of people dying from chronic diseases will rise to 52 million by 2030, compared to 38 million in 2012 (WHO 2014). WHO classifies the top four diseases as cardiovascular diseases, cancers, respiratory diseases and diabetes. NCDs are now posing as the greatest threats to health and development, especially in the developing world (WHO 2018a, b, c, d). This threat must be dealt with as failure to implement proven interventions is proving to be increasing health care costs, and continued lack of investment in actions against NCDs will also have a big effect on enormous health, economic, and societal consequences in all countries (WHO 2018a, b, c, d).

Z. Sako
Deakin University, Kew East, VIC, Australia

S. Adibi
School of Information Technology, Deakin University, East Kew, VIC, Australia

N. Wickramasinghe (✉)
Epworth HealthCare, Richmond, VIC, Australia

Swinburne University of Technology, Hawthorn, VIC, Australia



N. Wickramasinghe, F. Bodendorf (eds.), *Delivering Superior Health and Wellness Management with IoT and Analytics*, Healthcare Delivery in the Information Age, https://doi.org/10.1007/978-3-030-17347-0_16

Chronic diseases along with change in demographics, increasing costs of medical services, ongoing quality and safety issues in healthcare, are all major challenges to the delivery of healthcare services (Armstrong et al. 2007). These healthcare challenges are pushing the boundaries of traditional methods of healthcare delivery by seeking new available alternative methods of delivering faster and reliable healthcare services. Given today's digital economy, it appears logical to look for technology-enabled solutions such as mobile phones. The number of mobile phone subscriptions as per the 2015 statistics released by the International Telecommunication Union (ITU), is 7 billion worldwide (International Telecommunication Union 2015). This presents an opportunity for mobile phones to be used as an intervention in the rising number of chronic diseases and for health management. As half of smartphone owners frequently browse for health information online and monitor their health using mobile health applications (Fox and Duggan 2012), this gives mobile phones a new capacity to be used as mobile health. The definition of mobile health (mHealth) is the use of portable devices such as smartphones and tablets to improve health (Hamel et al. 2014). While smartphones have a new role to play in the effective management of health and diseases, the technology must be clear of any medical errors. A medical error has been defined as a preventable adverse outcome that results from improper medical management (a mistake of commission) rather from the progression of an illness due to lack of care (a mistake of omission) (Van Den Bos et al. 2011). Errors in the medical field belong to a number of domains such as development and use of technologies, ergonomics, administration, management, politics and economics (Vincent 2010). A common root cause of medical errors is human error, where errors are of omission (forgetting to do something) and commission (intentionally doing something that is not meant to be done) (Health Informatics: improving patient care 2012). However, medical errors have progressed from human to technological errors. Jenicek (2010) defines technological error in medicine as errors that relate to data and information recording, processing, and retrieval caused by information technology and its uses (information technology inadequacy and failure). Using mobile phone technology as mHealth devices, has its own set of challenges. These challenges relate to data (accuracy, integrity, privacy, security and confidentiality) and information integrity. To break through these challenges and benefit from such promising technology, techniques such as Machine Learning, which apply probabilistic reasoning after the analysis of data, can help deliver robust and accurate mHealth solutions.

## 2 Literature Review

The literature review explores a number of technical areas related to mHealth technology including the elements of data quality, particularly the element of data accuracy, information integrity and Machine Learning.

## *2.1 Data Quality and the Element of Data Accuracy*

The term data itself can be defined as information in the form of facts or figures obtained from experiments or surveys and used as basis for making calculations or drawing conclusion, as defined by Dumas (2013). Data in the healthcare domain is important as health data is used for multiple purposes including health sector reviews, planning, programme monitoring, quality improvement and reporting (WHO 2018a, b, c, d, ref. 2). With data serving different areas of healthcare, the quality of data becomes critical, and high quality must be present at all times (WHO 2018a, b, c, d, ref. 2). Changes in the delivery of healthcare are being accompanied by technological solutions (Marconi and Lehmann 2014), and mHealth is one of the technologies. mHealth apps and devices like smartwatches, smart scales and insulin pumps are generating large amount of data (Greene et al. 2018). This data may be helpful for patient care, enabling healthcare providers to monitor patients more effectively and tailor their treatments accordingly (Greene et al. 2018). However, data consists of multiple elements that make up data quality (see Tables 1 and 2). The elements of data quality are completeness, conformity, validity, accuracy, timeliness, duplicate avoidance, integrity, consistency and synchronization (Jugulum and Gray 2014). Table 1 lists the main four core data quality dimensions, yet, there are other dimensions (see Table 2) also used by some of the industries.

The quality of the data can sometimes suffer from bad quality, and this leads to a defect in the quality of data where data does not meet the domain constraints and business rules (Jugulum and Gray 2014). Errors in data quality can lead to inaccurate data. Data accuracy according to WHO, is the original source of data and it is an element of data quality that is intended to achieve desirable objectives using legitimate means (WHO 2003). Wang (1996) also defined accuracy as 'data are certified error-free, accurate, correct, flawless, reliable, errors can be easily identified, the integrity of the data, precise'. The quality of the data helps in evaluating

**Table 1** Four core data quality dimensions

| Dimension | Definition |
|---|---|
| Completeness | Completeness is defined as a measure of the presence of core source data elements that, exclusive of derived fields, must be present in order to complete a given business process |
| Conformity | Conformity is defined as a measure of a data element's adherence to required formats (data types, field lengths, value masks, field composition, etc.) as specified in either metadata documentation or external or internal data standards |
| Validity | Validity is defined as the extent to which data corresponds to reference tables, lists of values from gold sources documented in metadata, value ranges, etc. |
| Accuracy | Accuracy is defined as a measure of whether the value of a given data element is correct and reflects the real world as viewed by a valid real-world source (SME, customer, hard-copy record, etc.) |

Adapted from Jugulum and Gray (2014, p. 77)

**Table 2** Other data quality dimensions used by industry

| Dimension | Definition |
|---|---|
| Timeliness | A measure of current data available for business use as defined by established service level agreements (SLAs) for delivery/receipt |
| Duplicate avoidance | A measure of erroneous duplicated records and data elements across or within systems |
| Integrity | An entity-level measure of the existence of a unique primary key field, as well as a measure of whether foreign keys in one table reference a valid primary key in the respective parent table |
| Consistency and synchronization | A measure of data elements or records being equivalent across systems and sources, to include continuity of the data elements and records through its life cycle |
| Data decay | A measure of how current the data is, to include the frequency at which the data is refreshed/updated |

Adapted from Jugulum and Gray (2014, p. 77)

health, assess effectiveness of interventions, monitor trends, inform health policy and set priorities (Van Velthoven et al. 2013). When data lacks accuracy, currency or certainty, it can have catastrophic results (Sadiq 2013), and data of poor quality can result in lack of trust among users (WHO 2018a, b, c, d, ref. 2).

For mHealth solutions to be effective, the data collected from mHealth devices, wearables, and applications must be accurate and secure (Mottl 2014). Accurate data ensures proper assessment and treatment of patients. Unlike electronic medical records (EMRs), where data collection only occurs during in-person appointments, mHealth devices collect data continuously and throughout normal life activities, giving providers a more complete picture of the patient's health (Greene et al. 2018). Some of the traditional methods of assessing patients provide inaccurate data (Lin 2013). The common standard for data collection in the medical field is direct observation (Flocke and Stange 2004). Direct observation is the observation of patients and the different patient characteristics at the clinic (Flocke and Stange 2004). This allows for the collection of accurate data by directly observing the patients and their symptoms (Eisele et al. 2013). While the absence of a direct observation method in mHealth is a challenge, capturing of data also poses another challenge in mHealth.

With the collection of data in mHealth, errors in the quality of the data can be harmful. Some of the ways that lead to inaccurate data are classified into four categories. These categories are initial data entry, data decay, moving and restructuring and using data (Olson 2003).

1. Initial data entry: mistakes, data entry process, deliberate, system errors
2. Data decay: accuracy of data when originally created over time
3. Moving and restructuring: extracting, cleansing, transformation, loading, integration
4. Using: faulty reporting, lack of understanding

In addition to the four (4) categories described above, intentional and unintentional wrong data entry and the speed at which data is collected can be misleading. Misleading data results in misallocating resources or interventions when needed for the patients (Patnaik et al. 2009). Inaccurate readings, insufficient amount of data, movement and physical activities also contribute to inaccurate data provided through the mHealth devices (Mena et al. 2013). Another factor that affects the quality of the data is security breaches, where unauthorized modification or alteration is made to patients' data that compromise their confidentiality and privacy (Mena et al. 2013). Concerns associated with data accuracy and validity are persistent and can become a risk to patients' safety (Linda 2012). mHealth solutions must deliver accurate data. For data to be accurate, they must always consist of completeness, consistency, currency, relevance and accuracy (Narman et al. 2011). In mHealth, these elements of data quality can be compromised as data goes through five different stages. These are (1) collection, (2) transmission, (3) analysis, (4) storage and (5) presentation (Klonoff 2013). This means data must be accurate and consistent over its entire life cycle in order to conform to data integrity (Cucoranu et al. 2013).

Accurate data plays an important role in providing physicians, nurses and administrators with good information that they can rely on (Sayles and American Health Information Management 2013), whereas bad data have the opposite effect and can create errors in medicine (Sayles and Health Information Management 2013). The final product after data is analysed is information. Information is another major part of mHealth that must conform to integrity and achieve high level of quality in order to provide patients and medical professionals with knowledge about health management and the different aspects of the solution.

## 2.2 *Information Integrity*

Information at the very basic level is raw data that is processed and transformed into information, from which then knowledge is extracted (Dumas 2013). In mHealth, information must conform to integrity. The integrity of information is about having the right properties of information including sensitivity in which information is used, as well as encompassing accuracy, consistency and reliability of the information content, process and system (Fadlalla and Wickramasinghe 2004). For information to be useful, the integrity (accuracy and completeness) must be maintained (Taylor n.d.). Information integrity and quality are few of the building blocks of enterprise information management where information adds a value to the business information when data and a content are accurate, up to date, reliable and fit for the purpose (Oachs et al. 2010).

mHealth is used in several ways for the treatment of patients and delivery of healthcare services. It is vital that the information generated is accurate in order to avoid misdiagnosis, delayed care seeking, incorrect self-treatment, conflict over appropriate care or non-adherence to treatment and medication (Kahn et al. 2010).

The shift from clinician care towards patient-centred model is encouraging patients to actively self-manage and make decisions concerning their health (Boulos et al. 2011). To sustain self-management using mHealth, patients must be provided with accurate information that are of high integrity. The integrity of information produced as a result of shift in the dynamics of technology has been getting more focus as the interaction experience has changed (Cunningham 2012). What causes information to lack integrity is errors in healthcare systems due to data loss, incorrect data entry, displayed or transmitted data (Bowman 2013).

To treat patients correctly using mHealth and ensure information integrity, then data governance, information workflow management, internal controls, confidentiality and data privacy processes must exist (Flowerday and Solms 2010). These processes along with information technology can improve the quality of care by decreasing medical errors due to inaccurate and untimely information (Mahmood et al. 2012). Using a semantic tool when processing data and transforming it into information, can prove critical in detecting errors in data and ensuring information are of relevance to the patients and treatments. One common and publicly available semantic tool is the Omaha System.

The Omaha System 'is a complex, multi-axial, hierarchical, relational standardized health services taxonomy' as explained by (Monsen et al. 2009). The Omaha System has been integrated into software programs, recognized by nursing associations, and is in agreement with the International Organization for Standardization (ISO) (Monsen et al. 2009). The three (3) components of the Omaha System (See Fig. 1) are the Problem Classification Scheme, the Intervention Scheme, and the Problem Rating Scale for outcomes. The first component of the

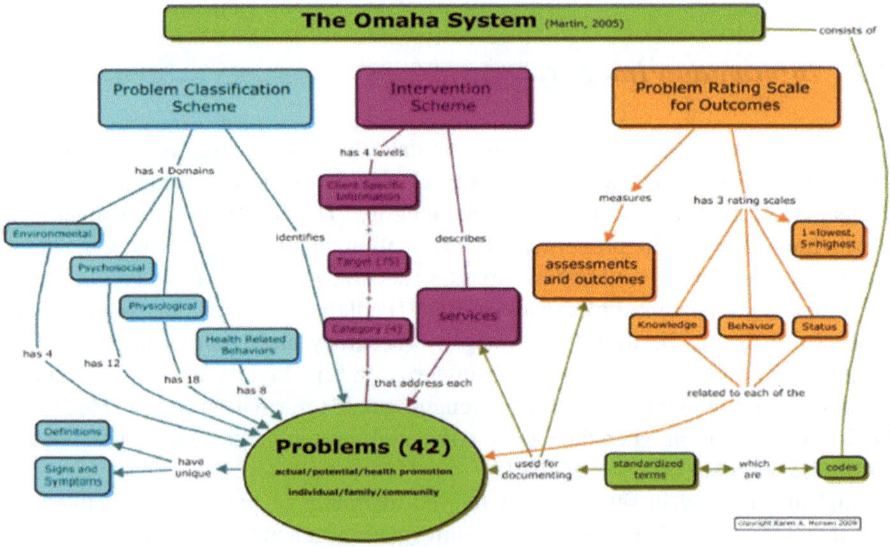


**Fig. 1** The Omaha System 2005 Version. (Adapted from The Omaha System Chart 2018)

Omaha System enables healthcare professionals to collect assessment data such as signs and symptoms, intervention scheme to design intervention and it is driven by the provider, and lastly is an outcome measurement scale for evaluating the interventions and the care process (Topaz et al. 2014).

Using the Omaha System in an accurate and consistent way, would establish an effective basis for documentation, communication, coordination of care and outcome measurement (Garvin et al. 2008). Incorporating Omaha System in mHealth, can potentially increase the accuracy of data and information. Elements from the Omaha System have also been incorporated into Machine Learning Algorithm studies (Monsen et al. 2009). This offers a role for Machine Learning to be adapted in mHealth technology to improve the detection of inaccurate data using the standardized taxonomy, that would enhance the quality and delivery of information of high quality.

## 2.3 Application of Machine Learning in the Health Domain

Machine Learning, according to Mohammed et al. (2016, p. 4), is 'a branch of artificial intelligence that aims at enabling machines to perform their jobs skilfully by using intelligent software'. Machine Learning has enabled smarter use of data in health by shifting from curing diseases to anticipating and preventing them before they occur through real time data analysis (Kumar et al. 2013). Machine Learning has been applied in a number of areas of healthcare including Computer-Aided Diagnosis, whereby machines assist radiologists in analysing the images (Mohammed et al. 2016).

Machine Learning is a continuously growing in the field of computer science and Health Informatics (Holzinger 2016). The growth is due to the probabilistic, uncertain, unknown, incomplete, heterogeneous, noisy, dirty, unwanted and missing datasets that could endanger the modelling of artefacts (Holzinger 2016). Despite the opportunities and benefits mHealth brings, the risk of medical errors occurring in mHealth must be constrained. Varshney (2009) describes common medical errors as those found during investigation, diagnosis, treatment, communication and office administrations errors. Constraining these errors during those stages can be achieved by learning about the collected data and applying analysis techniques to find sources of inaccurate data. The analysis performed by Machine Learning, extracts new knowledge when there is great amount of data (Lambin et al. 2013).

The concept of Machine Learning is learning that improves with experience at some task. That is (Bell 2014):

- Improve over task, T
- With respect to performance measure, P
- Based on experience, E

Machine Learning algorithms can play a pivotal role in acquiring accurate data through pre-trained algorithms that can be deployed in mHealth solutions. Support vector machines (SVM) algorithm was deployed in a blood pressure measurement application on an android tablet that detected the patient's arm and ensured stability, in order to acquire accurate reading of the data performed by the cuffs (Murthy and Kotz 2014). In a fall detection scenario, recorded data were used from a database which contained 95 instances of recorded falls, from which then four types of Machine Learning algorithms were applied to accurately detect a fall (Sannino et al. 2014). The role of Machine Learning in detecting inaccurate data through reasoning could prove crucial in enhancing the quality of the data collection stage of mHealth, as the accuracy aspect of data is a major challenge in itself. Removing inaccuracy and assuring high data quality would result in a deluge of solutions that can be developed to help manage diseases to reduce healthcare costs. With Machine Learning having a role in the delivery of mHealth, the proposed study is to investigate data accuracy and information integrity in the context of mHealth solution by addressing the research question:

*How Can Machine Learning Be Applied in mHealth Solutions to Address Data Accuracy* and Information Integrity?

## 3 Conceptual Model

To help address the accuracy problem in mHealth, the conceptual model (see Fig. 2) has been developed to facilitate the detection of data inaccuracy in mHealth and draws on providing high-quality information using multiple agents. These agents are intended to check for accuracy of the data at different levels in

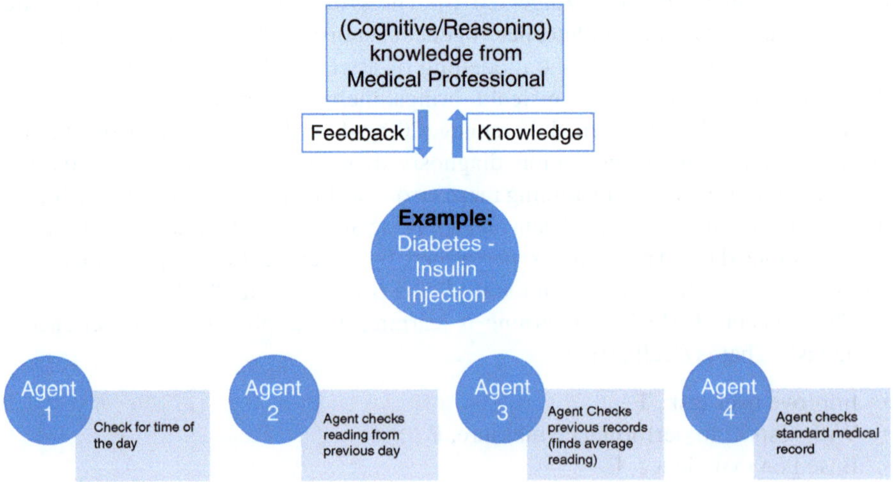


**Fig. 2** Conceptual model

order for the final dataset to be accurate. The conceptual model is built using the elements of data integrity that prove critical to generating high-quality information that is free of errors.

The four (4) agents described in the conceptual model can perform smart functions which help detect and assess the accuracy of the value that is received from a patient during the use of mHealth:

1. *Time of the day*: The function of this agent is to check for data decay, currency and timeliness, which makes the treatment and actions to be relevant and provide information in a timely manner. Thus, ensuring no delay in seeking treatment and allows for monitoring of the patient to be more relevant.
2. *Reading from previous day*: The function of this agent can detect mistakes by comparing the current value against what was provided previously. Where there's a significant difference, it will notify the medical professional of such event to raise awareness about the change in the value.
3. *Average value from previous reading*: The advice given to a patient during mHealth treatment is often based on the current value and does not take into an account the history of the patient. This agent performs calculations that finds the average value as well as providing a better insight of the patient's behaviour by providing a trend using the available historical data.
4. *Medical standards*: This agent checks the current value against the standard, acceptable medical reading that is of the right range and conforms to medical data definition related to the disease.

## 4 Research Methodology

In addressing this study's research question, a qualitative research method is applied using an exploratory case study. The following sections illustrates the different components of the study (the case study, data collection, data sampling, data triangulation and data analysis).

### *4.1 Single Case Study*

Yin (2014) defines case study as 'an empirical inquiry that investigates a contemporary phenomenon (the "case") in depth and within its real-world context, especially when the boundaries between the phenomenon and context may not be clearly evident'. Case studies are not considered a methodology but rather a choice of what is to be studied (Denzin and Lincoln 2011), and they are for studying a single group, event or person (Donley 2012). The selected case study is a mHealth solution for diabetes, with the case being patients' data. The selection of the case is guided by two principles. First is the form of question posed in this research where the form is ‘How’, requires no control over behavioural events (no control over how the data is produced) and focuses on contemporary events as the case (patients’ data) is studied in its real-world context. The second principle in selecting such case study is the single-case study rationale where the case is critical to the theory (Yin 2014) and relevant to the research question. Treating patients via mHealth rather than at the clinic could allow a gap for errors. Accessing such case study enables the research question to be addressed by examining patients’ data and exploring the characteristics of the data, the intended meaning when the data was produced and how it contributes towards the treatment of the patient.

## 4.2 Data Collection

The type of data collected for this study is qualitative secondary, de-identified data of patients with diabetes. Secondary, de-identified data is data that is used for research purposes and does not identify or represent a person (McGraw 2012). The de-identified data will be of records of patients who have diabetes and contain information such as time and date of measurement, glucose reading and a description of the reading. The chosen method of data collection seeks data that presents a chronic disease that is relevant to the case study; it is produced by people in real world and is authentic.

## 4.3 Data Sampling

With the proposed method of data collection being Secondary data, the sampling technique employed in this study is convenience sampling. The selection of this sampling technique is due to the readily available and accessible secondary data that is used for this study and conveniently recruited (Gideon 2012) through two sources. The first sample is diabetes data from clinical solution for the treatment of diabetes, while the second source is data from mHealth solution. The sample represents one of the many developed mHealth solutions and the data characterize the type of data created when using mHealth.

## 4.4 Data Triangulation

The data is triangulated using triangulation of different data sources of information by separating the secondary data into different datasets to build coherent justification for themes (Creswell 2009). The datasets will be numbered to represent different patients and for triangulation to confirm the accuracy of the findings.

## *4.5 Data Analysis*

The data analysis is performed using thematic and hermeneutics techniques. Thematic analysis is will be applied to aid in the interpretation of the texts by coding the data into organized segments of texts before bringing meaning to information (Creswell 2009), and later underlining them for generating themes that describe passages in the data (Cohen et al. 2000). In analysing the themes, Hermeneutics analysis is used to provide a detailed description of the text to capture and communicate the meaning of the lived experience (patients using mHealth) being studied (Cohen et al. 2000). This is to seek interpretation of the mHealth data and understand the meaning of it, accuracy of the values and what the producers of the text initially intended it for (Flick ct al. 2004).

# 5 Research Limitations

A key challenge for this research that requires mentioning is the use of secondary data. Using secondary data does not allow this research to observe the patients nor their behaviour during the use of the mHealth solution, specifically when the patient enters the data. Thus, this research does not take into consideration the human factors that can affect the accuracy of the data. Despite this difficulty, this is a major challenge in mHealth as there is no direct observation of the patient or their behaviour when the data is collected. Using the secondary data helps establish methods that can overcome this challenge and ensure data accuracy and information integrity in mHealth through the use of Machine Learning. Another limitation is the study's focus on a single chronic disease, diabetes. Diabetes is one of the many chronic diseases listed by World Health Organization. However, treatment of diabetes through mHealth is achieved through the transmission of text data that contain diabetes-related information, which allows for the testing of Machine Learning algorithms to be done.

# 6 Discussion and Conclusion

The preceding serves to present a research in progress study that focuses on trying to optimize data assets for mHealth contexts. In particular, it focuses on critical considerations regarding data accuracy and information integrity. While still at an early stage, the research should provide important implications for theory and practice. From the perspective of theory, the study will assist in developing a new area of knowledge that establishes methods similar to direct observation in mHealth using Machine Learning as a step to validate the accuracy of the data. As mHealth grows and the domain of consumer health informatics matures, we will see more and more mobile solutions being embraced to support health and wellness. Central to the success of these solutions is that they provide accurate data and information to consumers who in turn make decisions with far reaching implications and consequences based on the data and information received. The findings from this study will clearly be significant in ensuring optimal value from such mHealth solution. Upon the completion of the study, it will contribute to the hermeneutics field in information systems and a reference for researchers to use in analysing future empirical mHealth-related studies and assisting in the interpretation of their analysis.

Given today's digital economy, findings from this study are relevant to not just for healthcare but transferable to other industries also concerned about accuracy of data input and information integrity. Finally, the practical implication also extend to the field of Data Science as the findings will help bring new information that will assist data scientists in detecting inaccurate data within their datasets. This will enable data scientists to narrow down the data selection so that they can produce truthful answers to their questions when analysing data using Machine Learning.

## 7 Future Research Direction

This research sets a new area of focus for mHealth technologies and that is data accuracy. This research establishes early methods of detecting data inaccuracy within a single context being diabetes data. Future research should review the different mHealth technologies (wearable, ingestible, text-messaging, mHealth Apps, etc.) and study how data inaccuracy can occur within these solutions. This will then help build a framework for standardizing ways of collecting mHealth data to ensure accuracy in health information becomes a standard for delivering safe solutions to patients.

Future studies should also investigate and explore the privacy of users and security of the data due to the potential of Machine Learning being able to reveal more information than what was originally planned for. Machine Learning is a powerful tool, and if designed correctly, the information extraction mechanism can bring a whole new knowledge to the providers that did not exist before.

## References

Armstrong, B. K., Gillespie, J. A., Leeder, S. R., Rubin, G. L., & Russell, L. M. (2007). Challenges in health and health care for Australia. *Medical Journal of Australia, 187*(9), 485–489.

Bell, J. (2014). *Machine learning: Hands-on for developers and technical professionals*. New York: Wiley, ISBN: 978-1-118-88906-0. 408 Pages.

Boulos, M. N. K., Wheeler, S., Tavares, C., & Jones, R. (2011). How smartphones are changing the face of mobile and participatory healthcare: An overview, with example from eCAALYX. *Biomedical Engineering Online, 10*, 24–24. https://doi.org/10.1186/1475-925X-10-24.

Bovell-Benjamin, A. (2016). *Chronic diseases: The escalating dilemma in developing countries*. New York: Nova Science Publishers, Inc.

Bowman, S. (2013). Impact of electronic health record systems on information integrity: Quality and safety implications. *Perspectives in Health Information Management*, 1–19. 19p.

Cohen, M. Z., Steeves, R. H., & Kahn, D. L. (2000). *Hermeneutic phenomenological research: A practical guide for nurse researchers*. Thousand Oaks: SAGE Publications, Inc.

Creswell, J. W. (2009). *Research design: Qualitative, quantitative, and mixed methods approaches* (3rd ed., p. c2009). Thousand Oaks: Sage Publications.

Cucoranu, I. C., Parwani, A. V., West, A. J., Romero-Lauro, G., Nauman, K., Carter, A. B., et al. (2013). Privacy and security of patient data in the pathology laboratory. *Journal of Pathology Informatics, 4*(1), 23–39. https://doi.org/10.4103/2153-3539.108542.

Cunningham, P. (2012). It's most important role: Ensuring information integrity. *Information Management Journal., 3*, 20.

Denzin, N. K., & Lincoln, Y. S. (2011). *The Sage handbook of qualitative research* (4th ed., p. c2011). Thousand Oaks: Sage.

Donley, A. M. (2012). *Research methods*. New York: Infobase Publishing.

Dumas, M. B. (2013). *Diving into the bitstream: Information technology meets society in a Digital World*. New York: Routledge.

Eisele, T. P., Silumbe, K., Yukich, J., Hamainza, B., Keating, J., Bennett, A., & Miller, J. M. (2013). Measuring coverage in MNCH: Accuracy of measuring diagnosis and treatment of childhood malaria from household surveys in Zambia. *PLoS Medicine, 10*(5), e1001417. https://doi.org/10.1371/journal.pmed.1001417.

Fadlalla, A., & Wickramasinghe, N. (2004). An integrative framework for HIPAA-compliant $I_*IQ$ healthcare information systems. *International Journal of Health Care Quality Assurance Incorporating Leadership in Health Services, 17*(2–3), 65–74.

Flick, U., Kardorff, E. v., & Steinke, I. (2004). *A companion to qualitative research*. London: SAGE.

Flocke, S. A., & Stange, K. C. (2004). Direct observation and patient recall of health behavior advice. *Preventive Medicine, 38*(3), 343–349. https://doi.org/10.1016/j.ypmed.2003.11.004.

Flowerday, S., & Solms, R. V. (2010). What constitutes information integrity? *South African Journal of Information Management, 2*, 1–9.

Fox, S., & Duggan, M. (2012). *Mobile health 2012*. Washington, DC: Pew Internet & American Life Project.

Garvin, J. H., Martin, K. S., Stassen, D. L., & Bowles, K. H. (2008). The Omaha system. *Journal of AHIMA, 79*(3), 44–49.

Gideon, L. (2012). *Handbook of survey methodology for the social sciences*. New York: Springer New York.

Greene, E., Proctor, P., & Kotz, D. (2018). Secure sharing of mHealth data streams through cryptographically-enforced access control. *Smart Health*. https://doi.org/10.1016/j.smhl.2018.01.003.

Hamel, M. B., Cortez, N. G., Cohen, I. G., & Kesselheim, A. S. (2014). FDA regulation of mobile health technologies. *The New England Journal of Medicine, 371*(4), 372–379.

Health informatics: Improving patient care. (2012). Swindon: British Informatics Society Ltd.

Holzinger, A. (2016). Interactive machine learning for health informatics: When do we need the human-in-the-loop? *Brain Inform, 3*(2), 119–131.

International Telecommunication Union. (2015). Key ICT indicators for developed and developing countries and the world (totals and penetration rates). Retrieved February 2016, from http://www.itu.int/en/ITUD/Statistics/Documents/statistics/2015/ITU_Key_2005-2015_ICT_data.xls

Jenicek, M. (2010). *Medical error and harm understanding, prevention, and control*. Hoboken: Taylor and Francis.

Jugulum, R., & Gray, D. H. (2014). *Competing with high quality data: Concepts, tools, and techniques for building a successful approach to data quality*. Somerset: Wiley.

Kahn, J. G., Yang, J. S., & Kahn, J. S. (2010). 'Mobile' health needs and opportunities in developing countries. *Health Affairs, 29*(2), 252–258.

Klonoff, D. C. (2013). The current status of mHealth for diabetes: Will it be the next big thing? *Journal of Diabetes Science and Technology, 7*(3), 749–758.

Kumar, S., Nilsen, W. J., Abernethy, A., Atienza, A., Patrick, K., Pavel, M., et al. (2013). Mobile health technology evaluation: The mhealth evidence workshop. *American Journal of Preventive Medicine, 45*(2), 228–236. https://doi.org/10.1016/j.amepre.2013.03.017.

Lambin, P., Roelofs, E., Reymen, B., Velazquez, E. R., Buijsen, J., Zegers, C. M., et al. (2013). 'Rapid learning health care in oncology' – An approach towards decision support systems enabling customised radiotherapy. *Radiotherapy and Oncology, 109*(1), 159–164. https://doi.org/10.1016/j.radonc.2013.07.007.

Lin, J. Y. (2013). *Mobile health tracking of sleep bruxism for clinical, research, and personal reflection.*https://escholarship.org/uc/item/5wr4q4xn

Linda, L. K. (2012). Information integrity: A high risk, high cost vulnerability proper information governance includes paying attention to some key building blocks.(GOVERNANCE). *Health Data Management, 20*(4), 44.

Mahmood, N., Burney, A., Abbas, Z., & Rizwan, K. (2012). Data and knowledge management in designing healthcare information systems. *Growth, 9*(10), 11.

Marconi, K., & Lehmann, H. (2014). *Big data and health analytics*. Philadelphia: Auerbach Publications.

McGraw, D. (2012). Building public trust in uses of health insurance portability and accountability act de-identified data. *Journal of the American Medical Informatics Association: JAMIA, 20*(1), 29. https://doi.org/10.1136/amiajnl-2012-000936.

Mena, L. J., Felix, V. G., Ostos, R., Gonzalez, J. A., Cervantes, A., Ochoa, A., et al. (2013). Mobile personal health system for ambulatory blood pressure monitoring. *Computational and Mathematical Methods in Medicine, 2013*, 13. https://doi.org/10.1155/2013/598196.

Mohammed, M., Khan, M. B., & Bashier, E. B. M. (2016). *Machine learning: Algorithms and applications*. Milton: Chapman and Hall/CRC.

Monsen, K. A., Martin, K. S., Christensen, J. R., & Westra, B. L. (2009). Omaha system data: Methods for research and program evaluation. *Studies in Health Technology and Informatics, 146*, 783–784.

Mottl, J. (2014) *The imperative of safety in mHealth and why it can't be ignored.* http://www.fiercemobilehealthcare.com/story/imperative-safety-mhealth-and-why-itcant-be-ignored/2014-05-26.

Murthy, R., & Kotz, D. (2014). *Assessing blood-pressure measurement in tablet-based mHealth apps*. Paper presented at the COMSNETS.

Närman, P., Holm, H., Johnson, P., König, J., Chenine, M., & Ekstedt, M. (2011). Data accuracy assessment using enterprise architecture. *Enterprise Information Systems, 5*(1), 37–58. https://doi.org/10.1080/17517575.2010.507878.

Oachs, P. K., Eichenwald, S., LaTour, K. M., & American Health Information Management. (2010). *Health information management: Concepts, principles, and practice* (4th ed.). Chicago: AHIMA Press.

Olson, J. E. (2003). Chapter 3 – Sources of inaccurate data. In J. E. Olson (Ed.), *Data quality* (pp. 43–64). San Francisco: Morgan Kaufmann.

Patnaik, S., Brunskill, E., & Thies, W. (2009). *Evaluating the accuracy of data collection on mobile phones: A study of forms, SMS, and voice*. Paper presented at the Information and Communication Technologies and Development (ICTD), 2009 international conference on.

Sadiq, S. E. (2013). *Handbook of data quality research and practice*. Berlin/Heidelberg: Springer.

Sannino, G., De Falco, I., & De Pietro, G. (2014). *A general-purpose mHealth system relying on knowledge acquisition through artificial intelligence ambient intelligence-software and applications* (pp. 107–115). New York: Springer.

Sayles, N. B., & American Health Information Management. (2013). *Health information management technology: An applied approach* (4th ed.). Chicago: AHIMA Press.

Taylor, A. (n.d.). *1.1.4.1 Threat. A potential cause of an incident that may result in harm to a system or organisation (ISO 27002) information security management principles* (2nd ed.). BCS The Chartered Institute for IT. https://www.saiglobal.com/PDFTemp/Previews/OSH/AS/AS20000/27000/27002-2006(+A1).pdf

The Omaha system. (2018). http://www.omahasystem.org/overview.html

Topaz, M., Golfenshtein, N., & Bowles, K. H. (2014). The Omaha system: A systematic review of the recent literature. *Journal of the American Medical Informatics Association, 21*(1), 163–170. 168p. https://doi.org/10.1136/amiajnl-2012-001491.

Van Den Bos, J., Rustagi, K., Gray, T., Halford, M., Ziemkiewicz, E., & Shreve, J. (2011). The $17.1 billion problem: The annual cost of measurable medical errors. *Health Affairs, 30*(4), 596–603.

van Velthoven, M. H., Car, J., Zhang, Y., & Marušić, A. (2013). mHealth series: New ideas for mHealth data collection implementation in low– And middle–income countries. *Journal of Global Health, 3*(2), 020101. https://doi.org/10.7189/jogh.03.020101.

Varshney, U. (2009). *Pervasive healthcare computing*. Dordrecht: Springer.

Vincent, C. (2010). *Patient safety* (2nd ed.). Hoboken: Wiley.

Wang, R. Y. (1996). Beyond accuracy: What data quality means to data consumers. *Journal of Management Information Systems, 12*(4), 5–34.

World Health Organization. (2003). https://books.google.com.au/books?id=Vv-rOQZs_e0C&printsec=frontcover&dq=world+health+organisation+2003&hl=en&sa=X&ved=0ahUKEwiqueX-6snkAhXA8XMBHXiKDTcQ6AEIKjAA#v=onepage&q=world%20health%20organisation%202003&f=false

World Health Organization, Mendis, S. (2014). QR code for global status report on noncommunicable diseases 2014. In: S. Mendis. Global Status Report on Noncommunicable Diseases 2014. World Health Organization, p. 280.

World Health Organization. (2018a). *Data quality review*. Retrieved 20 July 2018. http://apps.who.int/iris/bitstream/handle/10665/259224/9789241512725-eng.pdf?sequence=1

World Health Organization. (2018b).*Global status report on noncommunicable diseases*. Retrieved 27 Apr 2018. http://apps.who.int/iris/bitstream/10665/148114/1/9789241564854_eng.pdf?ua=1

World Health Organization. (2018c). *Noncommunicable diseases*. Retrieved 22 July 2018. http://www.who.int/en/news-room/fact-sheets/detail/noncommunicable-diseases

World Health Organization. (2018d). *Time to deliver*. Retrieved 16 July 2018. http://apps.who.int/iris/bitstream/handle/10665/272710/9789241514163-eng.pdf?ua=1

Yin, R. K. (2014). *Case study research: Design and methods* (5th ed.). Los Angeles: SAGE.

# Enabling Value-Based Health Care with Business Analytics and Intelligence

**Nilmini Wickramasinghe**

## 1 Introduction

Although healthcare organizations have a wealth of data available through several data sources, much of this data is challenging to access, discontinuous, incomplete, lacking standardization or even incorrect (Moghimi et al. 2016; Porter and Teisberg 2006; Shafqat et al. 2018). Today, healthcare organizations globally are being challenged by the pressures to reduce costs, improve care coordination and outcomes, provide greater value and focus on a more patient-centric approach (Ball 2000; Raghupathi and Raghupathi 2014; Wang et al. 2018).

To address these challenges and thereby provide optimal patient care outcomes, the application of business analytics and business intelligence (BA/BI) is recommended (Burke and Menachemi 2004; Das et al. 2010; Levinson 2010; Rouse 2010; Frimpong et al. 2013; Miller 2012; Proctor and Compton 2010; Rajiv-Kohli 2008; Ryan 2009; Silvia Piai 2008; Wang et al. 2018; Ward et al. 2014) .

Our findings to date (Moghimi et al. 2016) indicate the existence of an opportunity to leverage the multispectral data collected through the application of strategic BA and BI solutions which will in turn improve knowledge sharing between clinicians as well as between clinicians and administrators and clinicians and patients (Gagnon et al. 2003; Kelley et al. 2011). In so doing it will also be possible to reduce costs, by focusing on key metrics including length of stay, unplanned readmissions and clinical risk mitigation (Shafqat et al. 2018). The research question is:

> *How can business analytics/business intelligence contribute to the provision of value-based care delivery?*

N. Wickramasinghe (✉)
Epworth HealthCare, Richmond, VIC, Australia

Swinburne University of Technology, Hawthorn, VIC, Australia



N. Wickramasinghe, F. Bodendorf (eds.), *Delivering Superior Health and Wellness Management with IoT and Analytics*, Healthcare Delivery in the Information Age, https://doi.org/10.1007/978-3-030-17347-0_17

To fully explore the potential for leveraging BA/BI, we examine a large healthcare data set for 10 years in a tertiary not-for-profit healthcare hospital to identify critical areas to address that will enable high-value patient-centred superior care delivery to ensue. The primary aim of this study is to develop a framework for BA/BI methodologies, tools and techniques to drive operational efficiencies, reduction of costs and improvements in the quality of care and thereby enhance care efficiency.

## 2 Background

Business analytics (BA) consists of a collection of information technology (IT) tools that enable users to transform data into informed actions (Krumholz 2014; Perer 2012). In contrast, business intelligence (BI) consists of a set of techniques and tools for the acquisition and transformation of raw data into meaningful and useful information for business analysis purposes (Azma and Mostafapour 2012; Eaton et al. 2007). Hence, while the two terms are often used together as one necessitates the other, they are indeed distinct.

Data warehouses and BA/BI technologies have been used in healthcare settings to improve workflow efficiency (Borlawsky et al. 2007; Golfarelli et al. 2004), monitor quality and improve outcomes (Resetar et al. 2006), develop best practices (Dhaval et al. 2006), optimize insurance procedures (Ostrander et al. 2007) and uncover patterns of increased expenditures (Chen et al. 2007). Many health systems are also using 'scorecard' (Eaton et al. 2007) and 'dashboard' (Grant et al. 2006) methodologies and developing web-based query and reporting tools (Grant et al. 2006; Roohan 2006) to optimize delivery of services as well as improve their own data warehouse projects (Eaton et al. 2007) (Fig. 1).

### *2.1 Business Analytics*

Business analytics is generally viewed from three major perspectives: descriptive, predictive, and prescriptive (Fig. 2). Descriptive analytics (Evans and Lindner 2012; Elsworthy et al. 2013) refers to the use of data to understand past and current performance and make informed decisions. Descriptive analytic techniques are the most commonly used and most well-understood type of analytics. These techniques categorize, characterize, consolidate and classify data to convert it into useful information for the purposes of understanding and analysing business performance. Descriptive analytics summarize data into meaningful charts and reports, for example, about processes, clinical outcomes, complications, risk factors, budgets, revenues or cost. Typical questions that descriptive analytics help answer are: How many patients are coming from specific geographic regions? What are patient outcomes following specific clinical treatment? How much of our revenue was

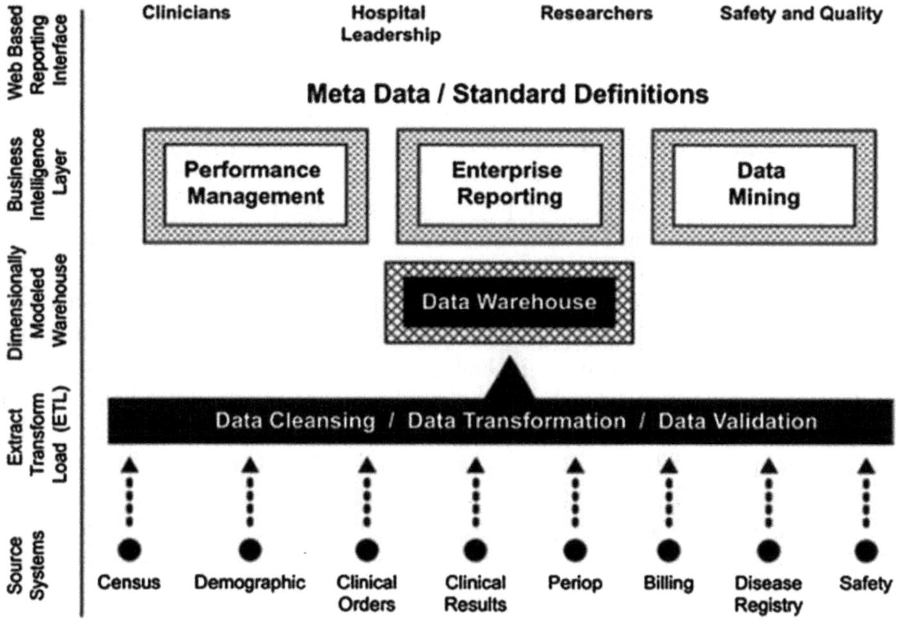


**Fig. 1** Key functionalities of a business intelligence application in the healthcare environment. (Adapted from Ferranti et al. 2010)

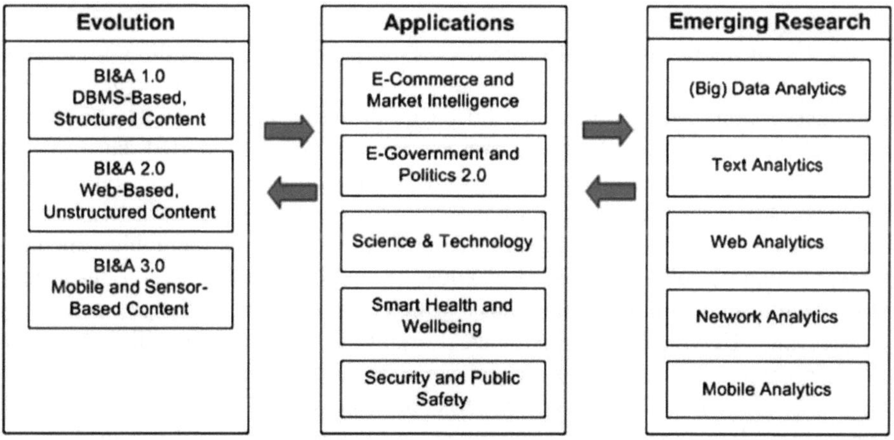


**Fig. 2** BI and BA overview. (Adapted from HIMMS Analytics Database 2009)

generated from different geographic regions? What was our revenue and profit last quarter? How many and what types of complaints did we resolve? Which production unit has the highest or lowest productivity? Descriptive analytics also helps to classify customers into different segments, further enabling the development of specific targeted campaigns and strategies.

In contrast, predictive analytics analyses past performance in order to predict the future by examining historical data, detecting patterns or relationships in these data and then extrapolating these relationships forward in time. Using advanced techniques, predictive analytics can help to detect hidden patterns in large quantities of data to segment and group data into coherent sets in order to predict behaviour and detect trends. Predictive analytics assist in answering questions such as: What will happen if demand falls by ten percent or if supplier prices go up five percent? What do we expect to pay for fuel over the next several months? What is the risk of losing money in a new business venture?

Finally, prescriptive analytics uses optimization to identify the best alternatives to minimize or maximize some objective. Prescriptive analytics are used in many functional areas of organizations, including operations, marketing and finance. The mathematical and statistical techniques of predictive analytics can also be combined with optimization to make decisions that take into account the uncertainty in the data. Prescriptive analytics addresses questions like: What is the best treatment to use for patients with certain disease characteristics? Figure 2 presents an overview of BI and BA evolution, applications, and emerging research.

## 2.2 BA Benefits to Healthcare

Today, healthcare organizations must contend with large volumes of multispectral data generated from numerous patient care points of contact, sophisticated medical instruments and monitoring devices and web-based health communities. Two main sources of health big data are genomics-driven big data (genotyping, gene expression, sequencing data) and payer–provider big data (electronic health records, claims records, pharmacy prescription and patient-reported data) (Miller 2012). The expected raw sequencing data from each person is approximately 4 terabytes. From the payer–provider aspect, a data matrix might have hundreds of thousands of patients with many records and parameters (demographics, medications, outcomes) collected over an extended period of time. Extracting knowledge from health big data poses significant research and practical challenges, especially considering HIPAA (Health Insurance Portability and Accountability Act) and IRB (Institutional Review Board) requirements for building a privacy-preserving and trustworthy health infrastructure and conducting ethical health-related research (Institute of Medicine (U.S.). Committee on Quality of Health Care in America 2001). Healthcare big data analytics, in general, lags behind e-commerce BI and BA applications because it has rarely taken advantage of scalable analytical methods or computational platforms (Miller 2012). Recent efforts have just begun to scratch the surface of these challenges (Shafqat et al. 2018).

Over the past decade, electronic medical records (EMRs) have been widely adopted in hospitals and clinics worldwide (Services, U. S. D. o. H. a. H 2010). Significant clinical knowledge and a deeper understanding of patient disease patterns can be gleamed from such collections (HIMMS Analytics Database

2009), for example, used a large-scale, longitudinal EHR to research associations in medical diagnoses and consider temporal relations between events to better elucidate patterns of disease progression (Hanauer et al. 2011). Lin et al. (2011) used symptom–disease–treatment (SDT) association rule mining on a comprehensive EHR of approximately 2.1 million records from a major hospital.

In addition to EMR data, health social media sites such as Daily Strength and Patients LikeMe provide unique research opportunities in healthcare decision support and patient empowerment (Miller 2012), especially for chronic diseases such as diabetes, Parkinson's, Alzheimer's and cancer. Association rule mining and clustering, health social media monitoring and analysis, health text analytics, health ontologies, patient network analysis, and adverse drug side-effect analysis are promising areas of research in health-related BI and BA (Gagnon et al. 2003).

## 3 Research Design and Method

Our previous work (Moghimi et al. 2016; Wickramasinghe and Schaffer 2010) has enabled us to establish a systematic approach. Specifically, we have applied the process perspective of Boyd's OODA loop (Fig. 3), which highlights four critical steps of observation of the space, orientation within the space to enable full appreciation of the scenario, determination of a solution set to move forward in the best possible fashion and action to delineate the necessary steps required to move forward with the chosen decision, to our intelligence continuum model (Wickramasinghe and Schaffer 2010; Wickramasinghe et al. 2009).

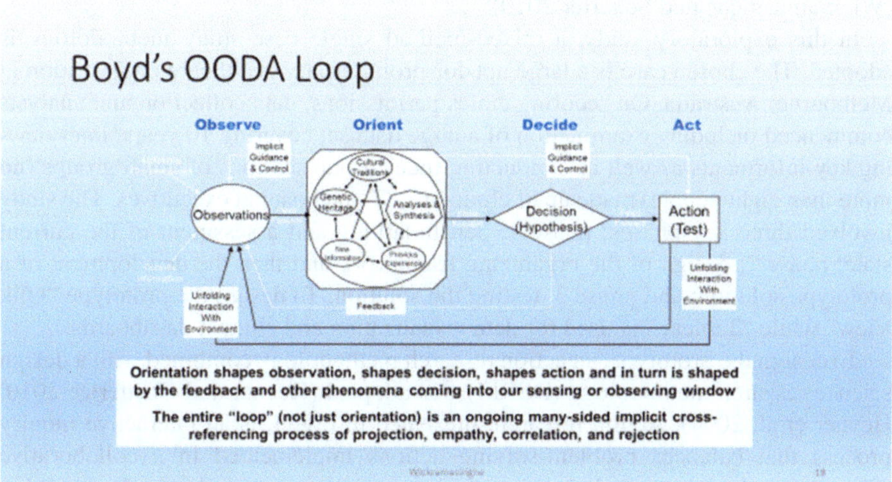


**Fig. 3** Boy's OODA loop

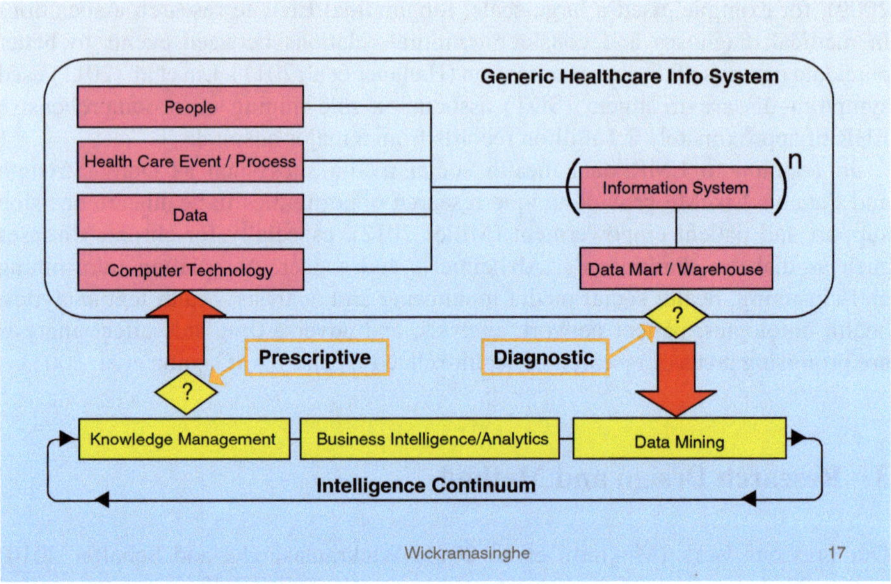


**Fig. 4** The intelligence continuum

The intelligence continuum (Fig. 4) provides a systematic approach to apply to any generic healthcare system consisting of people, healthcare processes, data and IT, the tools and techniques of knowledge management, business intelligence and business analytics to both diagnose (potential) problems and then be able to efficiently and effectively prescribe appropriate solutions in a timely fashion (Wickramasinghe and Schaffer 2010).

In this exploratory study, a mixed-method single case study methodology is adopted. The chosen case is a large not-for-profit tertiary healthcare organization in Melbourne, Australia. On securing ethics permissions, data collection and analysis commenced including examination of a large data set covering 10 years, interviewing key informants as well as conducting focus group sessions of small groups (no more than eight people) made up of clinicians, IT and hospital executives. The study involved three key phases: phase 1, benchmarking and assessment of the current state; phase 2, design of the organizing framework and then the development of a prototype solution; and phase 3, testing the solution. To design the prototype "Qlik View" while "Tableau" is used for data visualization and clinical dashboards.

To design the prototype, an action research methodology combined with a design science research methodology (DSRM) was adopted (Hevner and Chatterjee 2010; Hevner et al. 2004). Action research, presented in Fig. 5, is an interactive inquiry process that balances problem-solving actions implemented in a collaborative context with data-driven collaborative analysis or research to understand underlying causes enabling future predictions about personal and organizational change (Burns 2007). The DSRM is adopted to guide the cycles of problem/opportunity determina-

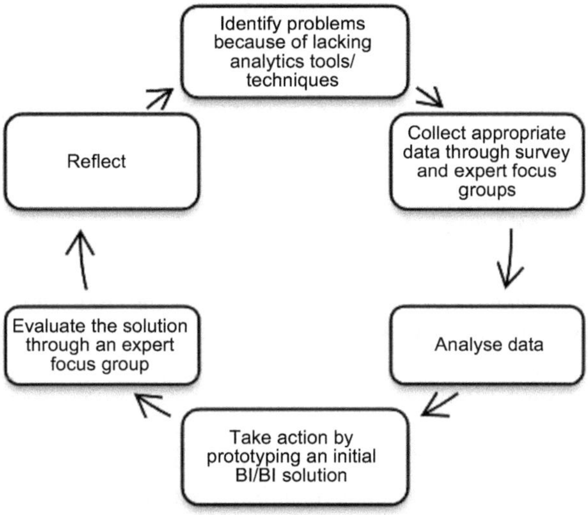


**Fig. 5** Action research steps

tion (relevance cycle), solution design and evaluation examine (design cycle), using the knowledge base to inform design (rigor cycle) (Hevner and Chatterjee 2010; Hevner et al. 2004).

## 3.1 Data Collection Plan and Data Analysis

Thematic analysis (Boyatzis 1998) is used in conjunction with NVivo or Dedoose to perform the qualitative analysis aspects of the data analysis. In addition, descriptive statistical analysis including frequency analysis will be used. Further, the information from the focus group discussions is also analysed. Specifically, the main objectives from the focus groups include:

1. Evaluating and assessing the prototype by clinicians and also executives
2. Identification of any ideal future state for the solution, how this might be addressed and what is required to realize this state
3. Other relevant key points to be considered in the proposed solution

Findings from the focus group are then applied to the prototype in order to improve patients' outcomes.

The findings from the thematic analysis are used to assess the usability and acceptability of the prototype and also to further explore the specific benefits of the proposed solution. Based on the feedback from the focus group, the final prototype is tweaked.

As all data collection and analytical methods have limitations, collecting patients' data during their cancer care has limitations given the high mutability of patients' conditions during their treatment course. In this case, collecting longitudinal data is essential for drawing conclusions about changes occurring within the individual patients, but because they are observational rather than experimental, it is important to recognize that even longitudinal data are limited with respect to data access and integrity issues in the study site.

## 4 Discussion

Outcomes of this study are essential for the study site as it addresses issues regarding big data and analytic capabilities so that raw data assets currently residing in various databases can be harnessed and optimally utilized and leveraged to ensure superior clinical outcomes. Given the importance of oncology to the study site, the results from this exploratory study can enable this hospital to further differentiate its operations in this clinical domain. Finally, knowledge and understanding of the BA content and techniques regarding oncology contexts can also be leveraged to other clinical domains across the study site and then on the national level.

To date, realizing the full potential of business analytics (BA) and business intelligence (BI) for healthcare contexts is significantly underutilized. BA/BI can play a pivotal role to assist in designing and planning efficient policies and programmes, improving service delivery and operations, enhancing sustainability, detecting and mitigating clinical and non-clinical risks and providing a means for measuring and evaluating critical organizational data.

Figure 6 succinctly presents the conceptual model which highlights the main components of a possible BA/BI organizing framework as well as highlighting key aspects of the proposed study. We have been using an array of offline analytical platforms such as RapidMiner and SPSS for the mining of data. Also

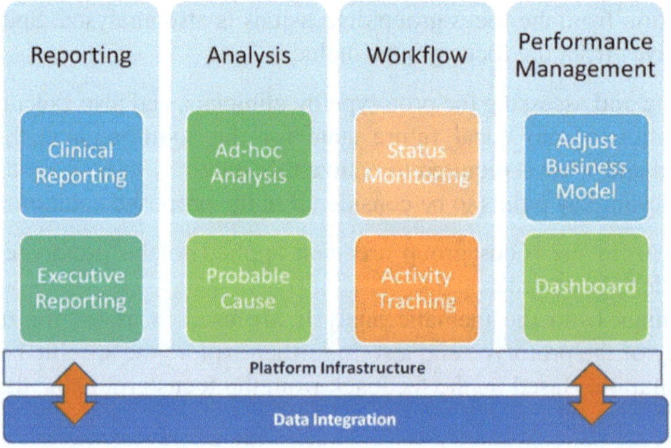


**Fig. 6** Conceptual model to present key aspects of the study

big data platforms like Hadoop will be used for large volumes of data. We will create applications of analytics to build models able to predict risks in cancer treatment processes. Patients' patterns captured through historical data sets may allow clinicians to make evidence-based clinical decision-making in order to treat different types of cancer by comparing data from similar patients in the past with the current observed patient.

At this exploratory study, it will be possible to develop an appropriate organizing framework to apply the necessary BA/BI technologies, tools and techniques in the specific context. This is an essential first step to ensure that the current and future BA/BI assets will be optimally and strategically used.

Thus, a key benefit of this study for the study site is that it will enable the design and development of the most appropriate technology platform and rubrics for BA/BI applications in clinical contexts in order to make more efficient, effective and analytical approaches to ensure a high quality of patient outcomes, care efficiency as well as a high patient experience. From this a secondary end point is the development of a prototype to demonstrate the bespoke uses of BA for the study case.

## 5 Conclusion

This research in progress is set to develop an appropriate organizing framework and a bespoke prototype to apply the necessary BA/BI tools and techniques in the specific cancer treatment context. This is an essential first step to ensure that the current and future critical data assets will be optimally and strategically used.

In order to develop this framework, this study has focused on cancer care across three study sites at one of the biggest private hospitals in Melbourne, as a single case study. An Action Research Methodology coupled with DSRM is applied to design and develop an appropriate strategy to adapt BA/BI to this context as well as to develop a prototype solution.

Thus, a key benefit of this study is designing and developing the most appropriate BA/BI application in cancer context in order to make more efficient, effective and analytical approaches to ensure a high quality of patient outcomes and care efficiency as well as a high patient experience in Australian healthcare sectors.

## References

Azma, F., & Mostafapour, M. A. (2012). Business intelligence as a key strategy for development organizations. *Procedia Technology, 1*, 102–106.

Ball, M. J. (2000). *Nursing informatics: Where caring and technology meet* (3rd ed.. [with] forewords by Sue Karen Donaldson and Ulla Gerdin. ed). New York: Springer Verlag.

Borlawsky, T., LaFountain, J., Petty, L., Saltz, J., & Payne, P. (2007). *Leveraging an existing data warehouse to annotate workflow models for operations research and optimization.*

Paper presented at the AMIA... Annual Symposium proceedings/AMIA Symposium. AMIA Symposium.
Boyatzis, R. (1998). *Transforming qualitative research*. Thousand Oaks: Sage.
Burke, D. E., & Menachemi, N. (2004). Opening the black box: Measuring hospital information technology capability. *Health Care Management Review, 29*(3), 207–217.
Burns, D. (2007). *Systemic action research: A strategy for whole system change*. Bristol: Policy Press.
Chen, Y., Matsumura, Y., Nakagawa, K., Ji, S., Nakano, H., Teratani, T., et al. (2007). Analysis of yearly variations in drug expenditure for one patient using data warehouse in a hospital. *Journal of Medical Systems, 31*(1), 17–24.
Das, S., Yaylacicegi, U., & Menon, N. M. (2010). The effect of information technology investments in healthcare: A longitudinal study of its lag, duration, and economic value. *IEEE Transactions on Engineering Management, 58*(1), 124–140. https://doi.org/10.1109/TEM.2010.2048906.
Dhaval, R., Buskirk, J., Backer, J., Sen, C., Gordillo, G., & Kamal, J. (2006). *Leveraging an information warehouse to create translational research environment for wound care center.* Paper presented at the AMIA... Annual Symposium proceedings/AMIA Symposium. AMIA Symposium.
Eaton, S., Ostrander, M., Santangelo, J., & Kamal, J. (2007). *Managing data quality in an existing medical data warehouse using business intelligence technologies*. Paper presented at the AMIA... Annual Symposium proceedings/AMIA Symposium. AMIA Symposium.
Elsworthy, A. M., Claessen, S. M., Graham, B., Guo, Y., Innes, K. C., Loggie, C. L., et al.. (2013). ICD-10-AM: The international statistical classification of diseases and related health problems, 10th revision, Australian modification: Tabular list of diseases.
Evans, J. R., & Lindner, C. H. (2012). Business analytics: The next frontier for decision sciences. *Decision Line, 43*(2), 4–6.
Ferranti, J. M., Langman, M. K., Tanaka, D., McCall, J., & Ahmad, A. (2010). Bridging the gap: Leveraging business intelligence tools in support of patient safety and financial effectiveness. *Journal of the American Medical Informatics Association, 17*(2), 136–143.
Frimpong, J. A., Jackson, B. E., Stewart, L. M., Singh, K. P., Rivers, P. A., & Bae, S. (2013). Health information technology capacity at federally qualified health centers: A mechanism for improving quality of care. *BMC Health Services Research, 13*(1), 1.
Gagnon, M. P., Godin, G., Gagne, C., Fortin, J. P., Lamothe, L., Reinharz, D., & Cloutier, A. (2003). An adaptation of the theory of interpersonal behaviour to the study of telemedicine adoption by physicians. *International Journal of Medical Informatics, 71*(2–3), 103–115. https://doi.org/10.1016/S1386-5056(03)00094-7.
Golfarelli, M., Rizzi, S., & Cella, I. (2004). *Beyond data warehousing: What's next in business intelligence?* Paper presented at the Proceedings of the 7th ACM international workshop on Data warehousing and OLAP.
Grant, A., Moshyk, A., Diab, H., Caron, P., Lorenzi, F. d., Bisson, G., et al. (2006). Integrating feedback from a clinical data warehouse into practice organisation. *International Journal of Medical Informatics, 75*(3), 232–239.
Hanauer, D., Zheng, K., Ramakrishnan, N., & Keller, B. (2011). Opportunities and challenges in association and episode discovery from electronic health records. *IEEE Intelligent Systems, 26*(5), 83–87.
Hevner, A., & Chatterjee, S. (2010). *Design research in information systems*. New York: Springer Science+Business Media.
Hevner, A., March, S., Park, J., & Ram, S. (2004). Design science in information systems research. *MIS Quarterly, 28*(1), 75–105.
HIMMS Analytics Database. (2009). Available from HIMSS Retrieved 5/03/2012 http://www.himssanalytics.org/hcproider/emradoption.asp
Institute of Medicine (U.S.). Committee on Quality of Health Care in America. (2001). *Crossing the quality chasm: A new health system for the 21st century*. Washington, D.C.: The National Academies Press.

Kelley, T. F., Brandon, D. H., & Docherty, S. L. (2011). Electronic nursing documentation as a strategy to improve quality of patient care. *Journal of Nursing Scholarship, 43*(2), 154–162. https://doi.org/10.1111/j.1547-5069.2011.01397.x.

Krumholz, H. M. (2014). Big data and new knowledge in medicine: The thinking, training, and tools needed for a learning health system. *Health Affairs (Millwood), 33*(7), 1163–1170.

Levinson, D. R. (2010). *Adverse events in Hospitals: National incidence among medicare beneficiaries*. Washington, DC: US Department of Health and Human Services, Office of the Inspector General.

Lin, Y., Brown, R., Yang, H., Li, S., Lu, H., & Chen, H. (2011). Data mining large-scale electronic health records for clinical support. *IEEE Intelligent Systems, 26*(5), 87–90.

Miller, K. (2012). Big data analytics in biomedical research. *Biomedical Computation Review, 2*, 14–21.

Moghimi, H., Schaffer, J. L., & Wickramasinghe, N. (2016). Exploring the possibilities for intelligent risk detection in healthcare contexts. *International Journal of Networking and Virtual Organisations, 16*(2), 171–190.

Ostrander, M., Parikh, D., & Tennant, M. (2007). *Improving daily workflow to improve insurance rejection handling by leveraging an existing data warehouse.* Paper presented at the AMIA... Annual Symposium proceedings/AMIA Symposium. AMIA Symposium.

Perer, A. (2012). Healthcare analytics for clinical and non-clinical settings. Proceedings of CHI Conference.

Porter, M. E., & Teisberg, E. O. (2006). *Redefining health care: Creating value-based competition on results*. Boston: Harvard Business School Press.

Proctor, P. R., & Compton, W. D. (2010). Engineering and the healthcare delivery system. In *Handbook of healthcare delivery systems* (p. 2-1-2-8). New York: CRC Press.

Raghupathi, W., & Raghupathi, V. (2014). Big data analytics in healthcare: Promise and potential. *Health Information Science and Systems, 2*(1), 3.

Rajiv-Kohli, V. G. (2008). Business value of IT: An essay on expanding research directions to keep up with the times. *Journal of the Association for Information Systems, 9*(1), 1.

Resetar, E., Noirot, L., Reichley, R., Storey, P., Skiles, A., Traynor, P., et al. (2006). *Using business intelligence to monitor clinical quality metrics.* Paper presented at the AMIA... Annual Symposium proceedings/AMIA Symposium. AMIA Symposium.

Roohan, P. J. (2006). Integration of data and management tools into the New York state medicaid managed care encounter data system. *The Journal of Ambulatory Care Management, 29*(4), 291–299.

Rouse, W. B. (2010). *Engineering the system of healthcare delivery*. Amsterdam: IOS Press.

Ryan, L. (2009). System Overhaul. Retrieved 1/09/2014, 2014, from http://www.the-hospitalist.org/details/article/184492/System_Overhaul.html

Services, U. S. D. o. H. a. H. (2010). Electronic health records and meaningful use.

Shafqat, S., Kishwer, S., Rasool, R. U., et al. (2018). Big data analytics enhanced healthcare systems: A review. *The Journal of Supercomputing*. https://doi.org/10.1007/s11227-017-2222-4.

Silvia Piai, J. D. (2008). Western Europe, Healthcare Sector, IT Spending Forecast Update, 2007–2011.

Wang, Y., Kung, L., & Byrd, T. (2018). Big data analytics: Understanding its capabilities and potential benefits for healthcare organizations technology. *Forecasting and Social Change, 126*, 3–13. https://doi.org/10.1016/j.techfore.2015.12.019.

Ward, M. J., Marsolo, K. A., & Froehle, C. M. (2014). Applications of business analytics in healthcare. *Business Horizons, 57*(5), 571–582.

Wickramasinghe, N., & Schaffer, J. (2010) Realizing value driven ehealth solutions IBM Center for the Business of Government http://www.businessofgovernment.org/report/realizing-value-driven-e-health-solutions

Wickramasinghe, N., Bali, R. K., Lehaney, B., Schaffer, J., & Gibbons, M. C. (2009). *Healthcare knowledge management primer*. Hoboken: Taylor and Francis.

# Part III
# Critical Issues Around Aspects of IoT in Healthcare Contexts

The technologies of the Internet of Things (IoT) are disrupting the traditional approaches to deliver healthcare that have remained relatively unchanged since the time of Hippocrates. With such changes, many good things are clearly possible such as providing anytime, anywhere care to many people; however, there is also a dark side and often unintended consequences that follow with the embracement of technology into healthcare delivery. This third section, thus, serves to highlight some of the critical issues that have arisen due to the incorporation of the technologies of the IoT. Specifically the following five chapters address some important aspects as follows:

Chapter 18: A Review of Mixed Reality in Healthcare by John and Wickramasinghe
Chapter 19: Implementing Lean Principles in the Healthcare Industry: A Theoretical and Practical Perspective by Pakdil et al
Chapter 20: Data, Denial and Disparity: Is This a New Digital Divide by Wickramasinghe et al.
Chapter 21: The Enabling Role for Technology in the Support of Care Coordination In Healthcare by Gibbings and Wickramasinghe.
Chapter 22: Managing the Risks of Emerging IoT Devices by Paxton and Branca.

What is evident from these chapters is that issues around equity, disparities, privacy, security, and cost of care are important to keep in mind as we design and develop technology solutions. Moreover, by having a broad appreciation for all the players in the healthcare web and all stakeholders it is possible to understand how changes effect each and every group. In addition, there is a need for policy, regulations, and laws to be revised and modified given the new solutions and possibilities afforded by technology. Clearly, it is not feasible to address all these issues in one section, but we feel it is important to note these issues so that they are kept in mind.

# A Review of Mixed Reality in Health Care

**Blooma John and Nilmini Wickramasinghe**

## 1 Introduction

Mixed reality (MR) is an evolving form of experience in which the real world is enhanced by an overlay of computer graphics-based interaction which is tangled to an activity (Azuma 1997). MR seamlessly overlay 2D and 3D objects such as audio files, videos and textual content into the real world (Azuma et al. 2001). Thus, jointly, MR makes its users see the real world, along with augmented data, as an integrated and improved environment. In an MR interface, the user views the real world through a handheld or head-mounted device by coating illustrations on the surrounding environment (Harborth 2017). The idea of MR stems from Sutherland (1968), when head-mounted three-dimensional display was introduced. MR system is defined by Azuma et al. (2001) to have the following properties: "combines real and virtual objects in a real environment, runs interactively, and in real time, registers (aligns) real and virtual objects with each other". Milgram and Kishino (1994) presented a continuum of real-to-virtual environments as given in Fig. 1.

Today, the terms 'augmented' reality (AR) and 'mixed' reality (MR) are used interchangeably (Farshid et al. 2018). AR refers to the overlay of data onto the visible world, while MR technologies display virtual objects over the real-world background (Cuendet et al. 2013). These objects are often 'locked' in space as though they were a natural part of the otherwise real environment. *Microsoft's*

B. John
University of Canberra, Bruce, ACT, Australia
e-mail: blooma.john@canberra.edu.au

N. Wickramasinghe (✉)
Epworth HealthCare, Richmond, VIC, Australia

Swinburne University of Technology, Hawthorn, VIC, Australia



N. Wickramasinghe, F. Bodendorf (eds.), *Delivering Superior Health and Wellness Management with IoT and Analytics*, Healthcare Delivery in the Information Age, https://doi.org/10.1007/978-3-030-17347-0_18

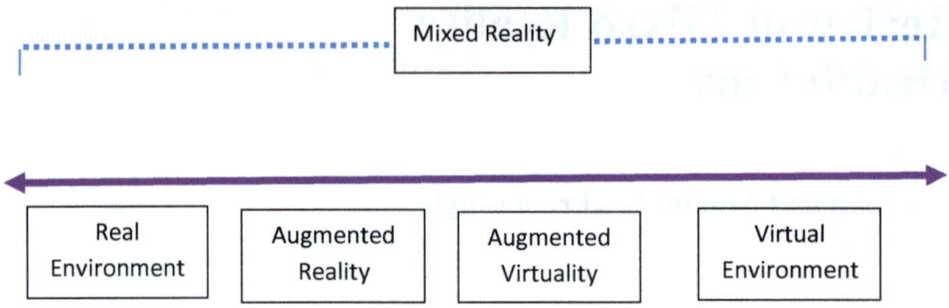


**Fig. 1** Reality–virtuality continuum. (Adapted from Milgram and Kishino 1994)

**Table 1** The actual reality/virtual reality continuum

| Reality | Augmented reality | Virtual reality | Mixed reality | Augmented virtuality | Virtuality |
|---|---|---|---|---|---|
| The actual world that we experience with all our senses | Information and data overlaid on top of the actual world | A complete digital representation of the actual world | The introduction of possible elements into an actual world | The introduction of actual elements into a possible world | An imaginary world that mostly follows the rules of the actual world |
| *An example* | | | | | |
| An actual hospital | A reality app provides details of an actual hospital | A 3D image of actual furniture. A virtual tour of an actual hospital | Simulation of different furniture, virtual or new, in an actual hospital | Staging of actual furniture in a new hospital room | A 3D model for a new hospital or of new furniture |
| *Key concept* | | | | | |
| Physical co-presence of people and objects | Add utility to physical co-presence | Enable perceived presence and full immersion | Adaptation of actual scenarios | Participation in possible scenarios | Vision of a completely different world |

Adapted from Farshid et al. (2018)

*HoloLens* is an example of an MR device, while *Pokémon Go* is an example of an MR application (Harborth 2017). Table 1 adapted from Farshid et al. (2018) gives a clear distinction between actual reality and virtual reality continuum.

MR applications are used in various fields like health, sports, business and more, and it is very much applicable to education (Yuen et al. 2011). The educational value of using MR in learning design is to provide exceptional student experience, by helping them to 'see the unseen' through the capacity of MR and to visualize and interact with complex and abstract concepts (Billinghurst 2002). Students in information technology (IT) field can have a work-integrated learning experience with MR by programming augmented reality applications (Jee et al. 2014; Chong et al. 2009). MR is also found to be very significant in health as well as health education domain by using MR application for conducting virtual bowel cancer surgery or communicating with virtual patients (Nicolau et al. 2011; Wu et al. 2013). However, very few studies presented a review of AR specifically for healthcare.

AR/MR overlay virtual objects over the real visible world and enhance our sensory-motor engagement with the world (Lindgren et al. 2016). A systematic literature review by Harborth (2017) highlights a shortage of information systems technology papers in developing or reviewing AR/MR technologies and outlines many promising areas for future work on augmented and mixed reality. A recent study used MR to help students learn the anatomy of the human body mediastinum, and in this case, MR was found to strengthen the students' self-efficacy and motivation, improve learning and provide a good learning experience (Nørgaard et al. 2018).

The differences between VR, AR and MR have important practical and theoretical implications for learning design, and there is a strong argument available to set aside the technical similarities of the technologies and to treat them separately (Hugues et al. 2011). In many respects, the affordances of virtual reality have been well explored in the literature on the educational use of video games (Waddington 2015), although the immersive nature of more advanced VR technologies does appear to enhance these effects (Clark et al. 2016).

## 2 Methodology

To analyse the literature on AR in healthcare, we identified the top journals in healthcare domain as listed in Serenko et al. Among the management-centred journals, the researchers ranked *Journal of the American Medical Informatics Association* and *Journal of Medical Internet Research* as A+ journals, and among the clinical-focused journals, they ranked *BMC Medical Informatics and Decision Making* and *IEEE Journal of Biomedical and Health Informatics* as A+ journals. We selected only the top five journals from both e-health management-focused and e-health clinical-focused as listed in Tables 2 and 3, respectively.

**Table 2** Ranking of e-health management-focused journals

| Tier | Rank | Title | Number of articles found |
|---|---|---|---|
| A+ | 1 | *Journal of the American medical informatics association* | 4 |
| A+ | 2 | *Journal of medical internet research* | 24 |
| A | 3 | *JMIR medical informatics* | 3 |
| A | 4 | *Methods of information in medicine* | 0 |
| A | 5 | *Journal of Telemedicine & Telecare* | 2 |

**Table 3** Ranking of e-health clinical-focused journals

| Tier | Rank | Title | Number of articles found |
|---|---|---|---|
| A+ | 1 | *BMC medical informatics and decision making* | 2 |
| A+ | 2 | *IEEE journal of biomedical and health informatics* | 2 |
| A | 3 | *International journal of medical informatics* | 0 |
| A | 4 | *Journal of biomedical informatics* | 2 |
| A | 5 | *Applied clinical informatics* | 4 |

We conducted the literature search in August 2018. We used the terms 'virtual reality', 'augmented reality' and 'mixed reality' to search in the respective databases. All publications issued through the keyword search were regarded as 'hits'. The hits were evaluated based on their titles, abstract and content. The number of relevant articles found is also listed in Tables 2 and 3. From the tables, it is evident that the *Journal of Medical Internet Research* had the maximum number of articles related to augmented reality. We went through the papers and classified the papers based on the categories discussed. The findings of this study are discussed in the next section.

## 3 Findings

This section discusses the results and findings of the literature review. Among the four articles from the journal of the American Medical Informatics Association, three were review of the studies and literature, and only one paper was about the annotations to physical workspaces such as signs and notes using MR technology. The paper emphasized on the fact that MR can be achieved using personal head-up displays, digital spectacles or handheld devices like camera-enabled smartphones. They emphasize on the fact that MR appears as an ideal vehicle to support the need to annotate space.

In the *Journal of Medical Internet Research*, 19 papers were focused on AR/MR or VR technology with 16 papers on VR and 3 papers covered AR/MR. The topics ranged from playing Pokémon Go as an emerging health-risk behaviour, the use of second life for clinical psychology, the use of Google Glass for documentation of medical findings and the use of MR glasses for assisted paediatric cardiopulmonary resuscitation. The three papers in the *Journal of Medical Internet Research* – Medical Informatics – were all focused on VR for chronic pain management, clinical dentistry and posturography.

The relevant articles in *Applied Clinical Informatics* covered the effect of VR on the doctor patient relationship. Both articles from *BMC Medical Informatics and Decision Making* were focused on VR. Overall, the papers were found to be moving towards MR from VR in the recent years. The topics covered other than the VR and AR technology are health, medicine, informatics, training, user behaviour, surgery, anaesthetics, anxiety and pre-operation. These topics came up by analysing the words used in the abstract.

On further analysis, papers that covered VR were analysed. The first paper sought to explore interactions with the neuropsychologist's avatar in virtual locations using a VR social network (Bernard et al. 2018). The second paper dealt with the creation of a hybrid augmented experience merging physical and virtual worlds for immersive e-therapy (Gorini et al. 2008). The study concluded that 'the interaction between real and 3-D virtual worlds may convey greater feelings of presence, facilitate the clinical communication process, positively influence group processes and cohesiveness in group-based therapies as well as foster higher levels of interpersonal trust between therapists and patients'. Another study tested the use of VR for enhancing the cognitive behavioural treatment of obesity with binge eating disorder (Cesa et al. 2013). They found that the VR-based treatment could better prevent weight regain in comparison with the standard cognitive behaviour therapy approach. Yet another study found that the patient hit by the van was playing Pokémon Go on his mobile phone while crossing a street, despite red traffic lights, which he did not notice due to the distraction induced by the game (Barbieri et al. 2017). The study emphasized the need for comprehensive, multilevel interventions, to reduce accidents caused by distraction, and to stress findings on the positive and negative effects of video games, which were becoming a source of public health concern. In these studies, VR has played a vital role as a device that is integrated into health information and treatment.

Studies using MR are very promising. One recent paper aimed to determine whether adapting American Heart Association guidelines for MR glasses increased adherence by reducing deviation and time to initiation of critical life-saving manoeuvres during paediatric cardiopulmonary resuscitation when compared with the use of Pediatric Advanced Life Support pocket reference cards (Siebert et al. 2017). They found that MR glasses did not decrease time to first defibrillation attempt and other critical recovery endpoints when compared with PALS pocket cards. However, they improved adherence and performance among residents in terms of administering the defibrillation doses set by AHA. Yet another study empirically attempted to determine the feasibility of deploying Google Glass in forensics setting (Albrecht et al. 2014). It was found that the Google Glass was efficient for acquiring images for documentation in forensic medicine, but the image quality was inferior compared to a digital single-lens reflex camera. Albrecht et al. (2013) compared the impact of the heightened realism of a self-developed mAR-blended learning environment (mARble) on learners to textbook material and concluded that mARble group performed considerably better regarding learning efficiency. Thus, mixed reality has a legitimate status for further exploration as a tool for teaching, monitoring and recording for medical use.

## 4 Discussion

By synthesizing the review, we suggest that there is a need to conduct more user studies in medical domain with MR and VR. Opportunities in this direction arise with VR in handheld mobile phone and MR by using Microsoft HoloLens and other devices for experiments as well as for diffusion of such studies in the mass market. This will help in conducting large-scale user studies which are missing currently. The second area that needs focus is in the medical education and training. There is a lack of studies that focus on the use of new technologies in the medical classes. Education and learning is a potential area that can teach and learn the various perspectives of the use of MR, AR and VR in medical education. A possible approach to provide learning opportunities is the use of mixed reality, where virtual learning experiences can be embedded in a real physical context.

Another area that needs emphasis is in psychology. Using these technologies, users can experience their problematic situations and be taught to handle mental health problems. The capability of VR, AR and MR to simulate reality could greatly increase access to psychological therapies, while treatment outcomes could be enhanced by the technology's ability to create new realities. VR and AR need to be more widely applied in the medical fields like medical education and training, surgical simulation, neurological rehabilitation psychotherapy and telemedicine. It is evident that the use of AR in healthcare improves the inconvenience of traditional medical care, reduces medical misconduct caused by untrained operation and lowers the cost of medical education and training.

## 5 Conclusion

In summary, we found that using MR in healthcare is promising. We analysed the top ten journals in healthcare domain. We found that VR, AR and MR technology were mostly covered in the *Journal of Medical Internet Research* when compared to other journals. This study was limited in its focus on the use of VR, AR and MR words in the ten top medical journals. As this technology is new and evolving, there is a need to expand the literature search into more medical informatics domain so that we can get more practical studies with respect to the technology.

MR is still considered as a novelty in the literature while VR is more significant. As technology advances, mobile devices have gradually turned into wearable devices, and VR, AR and MR have been applied more and more widely. It is evident from the early studies that the application of MR-related technology in medical field can enhance effectiveness of medical education and training, raise the level of diagnosis and treatment, improve the doctor-patient relationship and boost efficiency of medical implementation.

## References

Albrecht, U. V., Folta-Schoofs, K., Behrends, M., & Von Jan, U. (2013). Effects of mobile augmented reality learning compared to textbook learning on medical students: Randomized controlled pilot study. *Journal of Medical Internet Research, 15*(8), e182.

Albrecht, U. V., Von Jan, U., Kuebler, J., Zoeller, C., Lacher, M., Muensterer, O. J., et al. (2014). Google glass for documentation of medical findings: Evaluation in forensic medicine. *Journal of Medical Internet Research, 16*(2).

Azuma, R. T. (1997). A survey of augmented reality. *Presence, 6*(4), 355–385.

Azuma, R. T., Baillot, Y., Feiner, S., Julier, S., Behringer, R., & Macintyre, B. (2001). Recent advances in augmented reality. *IEEE Computer Graphics and Applications, 21*, 34–47.

Barbieri, S., Vettore, G., Pietrantonio, V., Snenghi, R., Tredese, A., Bergamini, M., et al. (2017). Pedestrian inattention blindness while playing Pokémon Go as an emerging health-risk behavior: A case report. *Journal of Medical Internet Research, 19*(4), e86.

Bernard, F., Lemée, J. M., Aubin, G., Ter Minassian, A., & Menei, P. (2018). Using a virtual reality social network during awake craniotomy to map social cognition: Prospective trial. *Journal of Medical Internet Research, 20*(6), e10332.

Billinghurst, M. (2002). Augmented reality in education. *New Horizons for Learning, 12*(5), 1–5.

Cesa, G. L., Manzoni, G. M., Bacchetta, M., Castelnuovo, G., Conti, S., Gaggioli, A., et al. (2013). Virtual reality for enhancing the cognitive behavioral treatment of obesity with binge eating disorder: Randomized controlled study with one-year follow-up. *Journal of Medical Internet Research, 15*(6), e113.

Chong, J. W. S., Ong, S., Nee, A. Y., & Youcef-Youmi, K. (2009). Robot programming using augmented reality: An interactive method for planning collision-free paths. *Robotics and Computer-Integrated Manufacturing, 25*(3), 689–701.

Clark, D. B., Tanner-Smith, E. E., & Killingsworth, S. S. (2016). Digital games, design, and learning: A systematic review and meta-analysis. *Review of Educational Research, 86*(1), 79–122.

Cuendet, S., Bonnard, Q., Do-Lenh, S., & Dillenbourg, P. (2013). Designing augmented reality for the classroom. *Computers & Education, 68*, 557–569. https://doi.org/10.1016/j.compedu.2013.02.015.

Farshid, M., Paschen, J., Eriksson, T., & Kietzmann, J. (2018). Go boldly!: Explore augmented reality (AR), virtual reality (VR), and mixed reality (MR) for business. *Business Horizons, 61*(5), 657–663.

Gorini, A., Gaggioli, A., Vigna, C., & Riva, G. (2008). A second life for eHealth: Prospects for the use of 3-D virtual worlds in clinical psychology. *Journal of Medical Internet Research, 10*(3).

Harborth, D. (2017). Augmented reality in information systems research: a systematic literature review. Twenty-third Americas Conference on Information Systems, Boston, 2017.

Hugues, O., Fuchs, P., & Nannipieri, O. (2011). New augmented reality taxonomy: Technologies and features of augmented environment. In *Handbook of augmented reality* (pp. 47–63). New York: Springer.

Jee, H. K., Lim, S., Youn, J., & Lee, J. (2014). An augmented reality-based authoring tool for E-learning applications. *Multimedia Tools and Applications, 68*(2), 225–235.

Lindgren, R., Tscholl, M., Wang, S., & Johnson, E. (2016). Enhancing learning and engagement through embodied interaction within a mixed reality simulation. *Computers & Education, 95*, 174–187.

Milgram, P., & Kishino, F. (1994). A taxonomy of mixed reality visual displays. *IEICE Transactions on Information and Systems, 77*(12), 1321–1329.

Nicolau, S., Soler, L., Mutter, D., & Marescaux, J. (2011). Augmented reality in laparoscopic surgical oncology. *Surgical Oncology, 20*(3), 189–201.

Nørgaard, C., O'neill, L. D., Nielsen, K. G., Juul, S., & Chemnitz, J. (2018). Learning anatomy with augmented reality. In 10th Annual International Conference on Education and New Learning Technologies.

Siebert, J. N., Ehrler, F., Gervaix, A., Haddad, K., Lacroix, L., Schrurs, P., et al. (2017). Adherence to AHA guidelines when adapted for augmented reality glasses for assisted pediatric cardiopulmonary resuscitation: A randomized controlled trial. *Journal of Medical Internet Research, 19*(5).

Sutherland, I. E. (1968). A head-mounted three-dimensional display. In *Proceedings of the December 9-11, 1968, fall joint computer conference, part I* (pp. 757–764). ACM.

Waddington, D. I. (2015). Dewey and video games: From education through occupations to educations through simulations. *Educational Theory, 65*(1), 21.

Wu, H. K., Lee, S. W. Y., Chang, H. Y., & Liang, J. C. (2013). Current status, opportunities and challenges of augmented reality in education. *Computers & Education, 62*, 41–49.

Yuen, S. C. Y., Yaoyuneyong, G., & Johnson, E. (2011). Augmented reality: An overview and five directions for AR in education. *Journal of Educational Technology Development and Exchange (JETDE), 4*(1), 11.

# Implementing Lean Principles in the Healthcare Industry: A Theoretical and Practical Overview

**Fatma Pakdil, Timothy N. Harwood, and Feride Bahar Isin**

## 1 Introduction

Healthcare industry has a relatively high proportion in the developed countries' gross domestic products (GDP) in the last six decades, and the USA has been shown to have the highest level of health spending in the world (c.f. Borger et al. 2006; Congressional Budget Office). As stated by McLaughlin and Olson (2012), the share of healthcare expenditures in GDP is expected to be 19.6% by 2019 in the USA (CMS 2010). Taking into consideration the rising demand for improved healthcare and decreasing costs, both healthcare providers and managers in the micro level, as well as healthcare policymakers in the macro level, strive to figure out how to improve healthcare outcomes for all stakeholders. In this regard, lean principles have been shown as one of the most future-promising means by several parties in the healthcare industry (c.f. PCAST Report 2014; Womack et al. 2005; Caldwell et al. 2005; Chalice 2005). President Obama's healthcare plans strongly mentioned the necessity of implementing lean in the nationwide healthcare system (PCAST Report 2014).

Lean principles were successfully implemented in other industries after Toyota originated lean in the automobile industry (c.f. Laursen et al. 2003; De Souza 2009; Liker 2004). A great variety of industry has benefited from lean by reducing

F. Pakdil (✉)
Eastern Connecticut State University, Willimantic, CT, USA
e-mail: pakdilf@easternct.edu

T. N. Harwood
Wake Forest University, Winston-Salem, NC, USA

F. B. Isin
Baskent University, Ankara, Turkey



N. Wickramasinghe, F. Bodendorf (eds.), *Delivering Superior Health and Wellness Management with IoT and Analytics*, Healthcare Delivery in the Information Age, https://doi.org/10.1007/978-3-030-17347-0_19

waste and increasing reliability, quality, and overall performance (Chalice 2005; Womack and Jones 1996; Womack et al. 2005; PCAST Report 2014). Additionally, Collins et al. (2015: 906) state that "despite the differences in organisation and technology, the challenges of achieving quality and cost advantages are quite similar in both manufacturing and healthcare." Similarly, many theoretical and practical studies have shown that lean can be implemented in healthcare to increase the overall performance, particularly by reducing and eliminating waste in various services (c.f. Manos et al. 2006; Miller 2005; Shahrabi 2015; Berwick et al. 1991; Zidel 2006). However, no systematic effort to provide an approach to integrate 14 lean principles in healthcare service delivery processes has been made so far. Instead, with misunderstandings and lack of knowledge of preconditions for lean implementation (Dahlgaard et al. 2011), lean practitioners in the healthcare industry adopted lean tools in several processes which results in very limited outcomes. Gomez et al. (2010) pointed out that by optimizing only one process in the entire value chain, sub-optimizing components would be at risk, and eventually the entire value chain. Implementing lean in an isolated process would be detrimental for the sake of the lean efforts in healthcare organizations. Based on these facts, the main objective of this chapter is to present how 14 principles of lean can be implemented in healthcare delivery processes. De Souza (2009: 121) points out that "it remains a challenge for academics and practitioners to evaluate lean healthcare under a more critical perspective." Based on the outcomes of this study, healthcare providers, managers, and decision-makers can focus on how to implement lean in healthcare organizations more effectively and efficiently.

This chapter flows as follows. The next section details lean principles and analyzes how to interpret lean in healthcare organizations. The limitations are followed by the conclusion.

## 2 Lean Principles in Healthcare Processes

While Rosmulder (2011) addresses that healthcare delivery does not equal industrial mass production, Allen (2009) expresses that the hospital industry stayed behind other industries in terms of adopting the concepts of operations management. Besides, service chain management in the healthcare setting is built on particular components that make it hard to adopt what other industries accomplished (Rust 2013). Literature also emphasizes the fact that healthcare processes are not designed as to meet higher efficiency and effectiveness expectation, which results in prolonged waiting and processing times, duplication of services, transportations, and operations, and eventually delays and increased adverse events and medical errors (c.f. Manos et al. 2006). According to Frings and Grant (2005: 314), "defects in patient care processes can run the spectrum from minor dietary issues to patient morbidity and fatality" or from transfusion mistakes to incorrect amputations (Kumar and Steinebach 2008). In a general spectrum, errors and mistakes in healthcare can be detrimental and devastating to all parties since human life is at risk (Taner et al. 2007). As reported by Kohn et al. (2000), nearly 98,000 patients admitted to hospitals lost their life because of preventable medical errors and nosocomial infections. Wagner (2004) identifies in-hospital errors as one of the leading killers in the USA. These facts make healthcare organizations as a candidate to integrate operations management to improve overall performance. In this sense, lean philosophy is presented as one of the effective ways in healthcare organizations (c.f. Womack et al. 2005).

Due to the nature of prioritization of the healthcare services that are based on (1) not harming patient and (2) improving patient's health conditions at any costs, ineffective managerial outcomes have not been attractive to the managers, as long as the patient's health conditions are improved. Similarly, healthcare processes are characterized to save human lives in a service delivery environment where many unpredictable risks are high (Tu et al. 2009). However, increasing costs, preventable medical errors, demanding healthcare providers and customers, and raising quality and performance expectations resulted in searching how to improve the outcomes of processes. In a great variety of industry, it is expected to decrease cost while increasing quality and efficiency. The increasing cost of healthcare has been correlated with the inefficient use of resources and non-optimized processes (Arcidiacono and Pieroni 2018). Based on that, lean was perceived as a remedy for the aforementioned concerns in the healthcare industry. The roots of the lean are built upon creating *value* for customers by eliminating waste and variation (c.f. Womack and Jones 1996; Gomez et al. 2010; Liker 2004). In a broader point of view, Radnor et al. (2012: 5) identify lean as:

> a practice based on the philosophy of continuously improving processes by either increasing customer value or reducing non-value adding activities (Muda), process variation (mura), and poor work conditions (muri).

Wickramasinghe et al. (2014) identify the quality of care as meeting the physical, psychological, and social expectations of patients who search for care. Institute of Medicine (IOM) focused on six aims such as safe, effective, patient-centered, timely, efficient, and equitable care. From IOM's perspective, each aim is likely to be achieved based on lean. Similarly, Husby (2012) addresses that if enough resources, discipline, and long-term focus are devoted, lean would be successfully implemented in healthcare, where there is high variability in the processes. According to Black and Miller (2008) and Barnas (2011), lean has already been utilized in the US healthcare system over the last decade. Lean principles are manifested that create a new way of management thinking. On one hand, healthcare is often identified as badly managed (Machado and Leitner 2010) and the ultimate negative outcome is considered the preventable death of a patient (Black and Miller 2008). On the other hand, lean seems more adaptable to healthcare settings based on examples (De Souza 2009) to eliminate the roots of bad management. As a starting point, De Souza (2009: 132) states that "lean healthcare is still in an early stage of development if compared to the same process in the auto industry." Jonsson and Randefelt (2013: 3) point out that "lean is regarded as a solution for many issues connected to healthcare." Lean in healthcare has also been gaining popularity among practitioners and decision-makers not because it is a "management fad," but because it produces sustainable results in healthcare.

The lean philosophy and norms/values in healthcare have many crucial similarities. As a general rule, it is aimed at "doing the right thing at the first time" in lean (Liker 2004). This fundamental rule is centered at the heart of medicine. Due to the fact that healthcare processes cannot tolerate mistakes, delays, and errors, it naturally fits in healthcare. Lean in healthcare is expected to "remove duplicates and unnecessary procedures such as recording patient details in multiple places; excessive waiting for staff; and uncoordinated discharge processes resulting in a longer length of stay" (NHSIII 2007). Specifically, lean in healthcare is commonly used to minimize delays in the emergency rooms, reduce the number of return visits, eliminate medical errors, and prevent inappropriate procedures (Rust 2013: 25).

In the healthcare world, there are good examples of lean. For example, De Souza (2009) and Burgess (2012) addressed that USA, UK, and Australia are countries in which healthcare successfully implemented lean with an expectation of improving overall performance. Similarly, Kollberg et al. (2006) and Jonsson and Randefelt (2013) presented how the Swedish healthcare system developed a measurement system, the "flow model," to reduce lead and waiting times. Kollberg et al. (2006), Karlsson et al. (1995), and Womack and Jones (2003) mentioned how lean is implemented in healthcare, emphasizing that patients should be the first priority, and time and comfort are the key performance indicators. Focusing on identifying "customer" in healthcare, Young et al. (2004) state that lean may benefit healthcare processes. Additionally, Womack et al. (2005), Graban (2009), Black and Miller (2008), Tachibana and Nelson-Peterson (2007), and Breyfogle and Salveker (2004) support the implementation of lean in healthcare.

Translating lean into healthcare has been challenging (Rust 2013: 85). Before starting the lean journey in healthcare, a comprehensive definition of "customer" should be made. Kollberg et al. (2006), Machado and Leitner (2010), and Bushell and Shelest (2002) suggest that the primary customers of the healthcare outcomes are patients. Frings and Grant (2005) consider patients and healthcare service providers as the primary customers. From the financial aspect of healthcare, Kollberg et al. (2006) also consider patients' families, payers, the society in general, and medical students.

Attempts in the use of lean in the healthcare arena remains primarily in smaller, lower-level processes. A literature search for the clinical implementation of the lean indicates sparse publication of its use in the provision of healthcare (Raja et al. 2015; Sugianto et al. 2015; Lunardini et al. 2014). A review of hospitals in England likewise demonstrated that implementation tended to be isolated rather than system-wide (Burgess and Radnor 2013). However, their review of a second dataset projecting lean implementation trajectory across time signaled lean's increasing use by English hospitals and shows progression toward an increasingly systemic approach. In one of the clinical studies of lean in a multicenter examination comparing a variety of approaches to lean, Van Vliet et al. (2011) examined the entire process of moving a patient with a cataract from diagnosis to recover after phacoemulsification. Their publication provides an excellent example of the

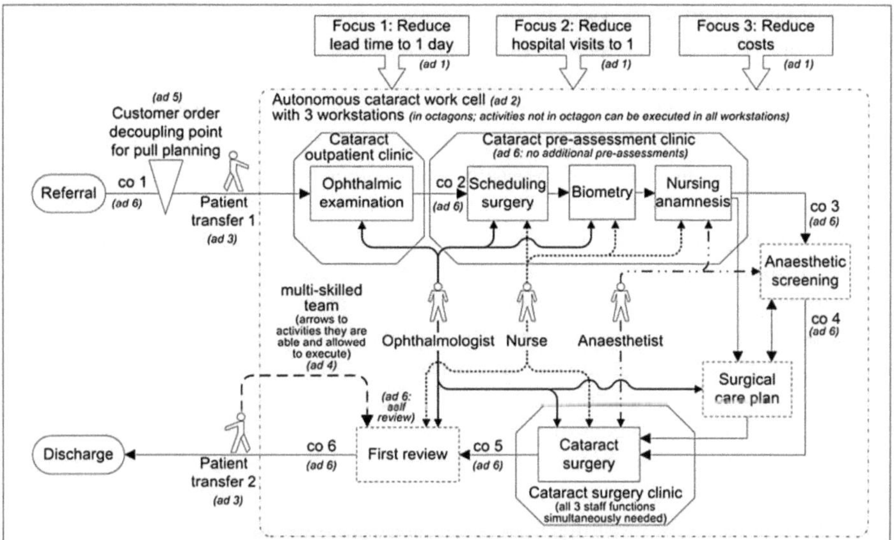


**Operational aspects of lean thinking to design a care process:**

*Define value (focus)*

1. Operational focus: Reduce lead time, hospital visits and costs.

*Make value flow (delegation of activities)*

2. Create an autonomous work cell that consists of a close arrangement of staff, workstations and equipment,
3. in a processing sequence to facilitate flow and minimize the number of physical patient transfers,
4. with 'capable' team members, that is the care team is cross trained in every task, as far as competence allows.

*But make it only flow when pulled by the next step (coordination of activities)*

5. Locate one customer order decoupling point in the care process, where the patient is referred to the hospital for cataract treatment and customer specifications are known, and immediately schedule all activities in the pathway following this customer order (i.e., make-to-order or pull planning). Extra decoupling points in the cataract pathway after the customer order decoupling point (i.e. make-to-stock or push planning) are impediments for flow.

*Eliminate waste (design)*

6. Avoid overprocessing: Waste is (among others) generated when providing higher-quality products than is necessary, such as providing additional pre-assessments (e.g., electrocardiograms, blood tests and internist consultations) and in-hospital reviews conducted by ophthalmologists, or when unneeded steps are taken to coordinate the patient's pathway.

**Fig. 1** Example of lean framework for a cataract removal pathway (Van Vliet et al. 2011)

schematic formulation that is inherent in lean process exploration (Fig. 1). The authors concluded that reducing non-value-added parts of the process appeared to have a stronger influence on improving efficiency than other lean thinking principles.

Some limited studies indicate that concerted lean effort in a limited scope can improve performance. Sugianto et al. (2015) applied lean in processing gastrointestinal biopsy and reduced total time from tissue input to results output by 63%. Lunardini et al. (2014) implemented lean principles to optimize instrument utilization for spine surgery in an effort to standardize instruments and reduce costs. Based on the data and stakeholder consensus, about 40% of the instruments were removed, resulting in a significant instrument set weight reduction and consolidation of two instrument sets into one, with $41,000 cost savings annually. Warner et al. (2013) brought lean to a more multifactorial process, namely, on-time first case vascular surgery starts, and employing DMAIC methodology and a Rapid Process

Improvement Workshop, rapidly improved the proportion of patients arriving on time to the OR. Their results were sustained at 1 year. However, improvement in opportunity costs reached only $12,582 in the first 9 weeks.

In addition to positive outcomes, published results reveal several difficulties in implementation of lean in clinical areas, due to the lack of exposure to these specific principles in the education and experience of managers (Raja et al. 2015). Raja et al. (2015) stated that in a study of implementing an automated process issue reporting system, the implementation of a QI reporting system from psychiatry residents met several limitations. Given the high volume of issues raised and limited resources, this reporting system informally promoted "lean" principles of waste identification and continuous improvement. The authors categorized residents' reports as closely as possible into lean categories of operational waste. The resident's submitted issues were completely addressed with the requested outcome partially or fully implemented or with successful clarification of existing policies in over 80% of cases. The authors concluded that a real-time, voluntary reporting system can effectively capture trainee observations of waste in healthcare and training processes and help contribute to meaningful operations improvement.

Although somewhat pessimistic in tone, but at least displaying some realism, Waring and Bishop (2010) examined and commented on three domains regarding the lean process: Rhetoric (platitudes and slogans), Ritual (the practice), and Resistance. They concluded that some of the inconsistencies found in current healthcare improvement efforts might result, in part, from the poor translation of models and methodologies developed in other settings (e.g., automobile manufacturing), that gives inadequate attention to current practices and potential unintended consequences. In addition, a lack of social and cultural research explores lean implementation and interaction with existing clinical practices. They suggested that making healthcare services lean is likely to be a process fraught with resistance and battles, as it becomes reinterpreted and reshaped by different social actors to ensure that it fits with their prevailing vision or aspirations for clinical practice. As such, it enters into an environment rife with conflict and disagreement, and unlikely to migrate to practice fully intact.

In continuous quality improvement efforts, the impact of IoT (Internet of Things) on lean in healthcare processes is another factor for healthcare providers and designers to be considered. With all algorithms and intelligent systems, IoT will take a role as a game player in the healthcare industry, since its capability of reducing cost and improving quality of care. According to Scarpato et al. (2017), IoT has become a fundamental technology in the medical environment. In the same vein, wired and wireless components of IoT can be expected to function as critical actors in lean, particularly in increasing the value of processes and detecting and decreasing adverse events, errors, failures, and abnormalities. As a paradigm shift, IoT takes a place in Industry 4.0 revolution (Löffler and Tschiesner 2013). The fact that IoT is expected to link all related activities in the manufacturing setting, such as logistics, supply chain, and manufacturing, the same positive effect of IoT in healthcare processes is likely to be active. As stated by Bhatt and Bhatt (2017), "the reason for integrating healthcare with the Internet of Things features into medical devices improves the quality and effectiveness of service, bringing especially high value for elderly, patients with chronic conditions and those that require consistent supervision." The overall positive effects of reduced inventory throughout the healthcare industry, including hospitals, clinics, suppliers, pharmaceutical, and biomedical device manufacturers, may be strengthened by IoT practices. As the main two components of IoT, value creation and cost reduction (Löffler and Tschiesner 2013) take critical roles in lean implementation in healthcare organizations. Parallel to IoT, Big Data should also be considered in the healthcare industry to increase lean performance, specifically on data collection and analysis needs in lean implementation.

Hines et al. (2004) and Radnor et al. (2012) stated that lean in healthcare would be of a limited impact and largely confined to the application of specific tools to local optimization. According to them, lean in healthcare will go through a similar evolution to lean in manufacturing: from shop-floor-based tools to a process view and a holistic understanding of pathways. In Liker and Convis' (2011) book, the lean philosophy is constructed on 4P model, which includes *philosophy*, *process*, *people and partners*, *and problem-solving* (Table 1). Each level in 4P model includes several items of 14 principles (Liker and Convis 2011; Womack and Jones 1996). In light of the 14 principles, the next subsections analyze how to incorporate and interpret them in healthcare.

**Table 1** Liker's (2004) 14 principles

| Group | Principle |
|---|---|
| *Philosophy: Long Term* | 1. Base your management decisions on a long-term philosophy, even at the expense of short-term financial goals |
| *Process: Promote Flow* – create a pull production system that has continuous flow and balanced workload | 2. Create a continuous process flow to bring problems to the surface<br>3. Use pull systems to avoid overproduction<br>4. Level out the workload (*heijunka*)<br>5. Build a culture of stopping to fix problems, to get quality right the first time<br>6. Standardized tasks are the foundation for continuous improvement and employee empowerment<br>7. Use visual control so no problems are hidden<br>8. Use only reliable, thoroughly tested technology that serves your people and processes |
| *People: Respect and Development* | 9. Growing leaders who thoroughly understand the work, live the philosophy, and teach it to others<br>10. Develop exceptional people and teams who follow your company's philosophy<br>11. Respect your extended network of partners and suppliers by challenging them and helping them improve |
| *Problem-Solving: Continuous Improvement*– organize their continuous improvement activities | 12. Go and see for yourself to thoroughly understand the situation (*genchi genbutsu*),<br>13. Make decisions slowly by consensus, thoroughly considering all options, and implement decisions rapidly<br>14. Become a learning organization through relentless reflection (*hansei*) and continuous improvement (*kaizen*). |

## 2.1 Philosophy

**Principle 1: Base your management decisions on a long-term philosophy, even at the expense of short-term financial goals**
In lean, Liker (2004: 37 & 73) focuses on generating value for the customer, society, and the economy by stating that, "*do the right thing for the company, its employees, the customer, and society as a whole.*" Jonsson and Randefelt (2013) propose that the value formulation is perceived as a guide to carrying the organization toward the desired direction. Collins et al. (2015) suggest that the concept of converting raw materials to a product creating value to the customers is equally significant in healthcare.

Womack and Jones (1996: 141) express that "... failure to specify value correctly before applying lean techniques can easily result in providing the wrong product/service in a highly efficient way – pure *muda*." Radnor et al. (2012) point out the importance of identifying "value" and "waste" from a customer's point of view in the process. Liker (2004: 77) points out that "Toyota believes this is what drives profit in the long run." According to Gomez et al. (2010) and Rust (2013), "lean is centered on creating value for the customer with less work." Manos et al. (2006) address that value adding for patients is different for customers of the production systems since healthcare processes are focused on prevention or cure. While literature (c.f. Rosmulder 2011; Rust 2013) focuses on the diagnostic question "is the customer willing to pay for the product/service?" lean in healthcare should start with the definition of value (Gomez et al. 2010). While answering this question, complexity in healthcare processes increasing risk and inefficient use of resources (Rust 2013) should be taken into consideration. However, there are many points in which creating value depends on other processes that do not have a value at the first glance. For example, the value-added processing time may be around 30 minutes while turnover time takes at least 45 minutes or longer in an ENT case in the operating room (OR) (Pakdil and Harwood 2005). From a value-added perspective, the turnover time might be perceived as a non-value-added time even if significant tasks are performed in this sub-process. "The value produced is more or less untouchable, difficult to specify and unclearly priced" (Rosmulder 2011: 2). Value is created indirect interaction with the patients (Rosmulder 2011: 19). Similarly, Radnor et al. (2012) address that the lack of customer and value definition blocks quantifying the productivity and quality improvement efforts.

The lack of value definition affects diagnosing potential wastes and how to implement lean in healthcare. Liker (2004) points out that most business processes are 90% waste and 10% value-added work. Graban (2009) states that healthcare organizations implementing lean focus only on the elimination of waste by ignoring the core of lean, smoothing flow, and improving workplace environments and conditions. In a similar vein, Robinson et al. (2012) state that healthcare organizations focus only on eliminating waste and cost reduction by neglecting system-wide variation reduction efforts.

By integrating a management system focusing on value identification, committed workforce, having the constancy of purpose, and doing the right thing for the customers, healthcare organizations get ready to implement long-term orientation in the first principle's requirements. Robert McCurry states that "the most important factors for success are patience, a focus on long-term rather than short-term results, reinvestment in people, product, and plant, and an unforgiving commitment to quality" (Liker 2004).

## 2.2 *Process*

Liker (2004: 87) expresses that "Toyota leaders truly believe that if they create the right *process* the results will follow." A process is identified "as a set of actions or steps each of which must be accomplished properly in the proper sequence at the proper time to create value for the patient" (Miller 2005). The following principles, principle 2 through 8, are comprised of the details of the "process" segment of lean.

**Principle 2: Create a continuous process flow to bring problems to the surface** Liker (2004) states that "flow" is at the heart of the lean and that decreasing the elapsed time leads to best performance based on highest quality, lowest cost, and shortest delivery time. Liker (2004) also addresses that lean journey should start with creating continuous flow wherever applicable in the core processes. More importantly, flow efficiency is utilized as an indicator measuring the ratio of how much a unit is processed from when a need is determined to when the need is met (Modig and Åhlström 2012). However, Jonsson and Randefelt (2013) address that it is harder to figure out how the flow is built in non-production businesses. Jonsson and Randefelt (2013) see the point of allocating activities in separate departments as problematic to maximize effectiveness, which results in increasing waste throughout the system. This point is also seen as a contradiction with one-piece flow rule in lean.

The understanding of continuous process flow implemented in manufacturing settings cannot be directly adopted in healthcare due to the nature of uncertainty and unexpected conditions of healthcare services. In the manufacturing setting, when the customer places an order, the process starts obtaining raw material and operationalize just for meeting that customer order. In this sense, uncertain or unplanned patient visits may result in deficiencies in continuous flow Kollberg et al. (2006). The identified customer demands in various sections may vary based on test results and diagnoses, affecting continuous flow and smoothing of the flow. As an example, Frings and Grant (2005) point out that inefficient hospital discharges result in a great variety of low performance in terms of the increased bottleneck, capacity issues, and dissatisfied parties in healthcare delivery processes. In this sense, continuous discharge process is capable of bringing bottlenecks and issues to the surface in inpatient clinics.

In the manufacturing setting, lean is built upon U-shaped production lines with all required equipment and multiskilled employees in order to catch the *takt* time.

In healthcare, patient rooms in clinics or ORs can be considered U-shaped service delivery line. Looking into services and care in this perspective, it can be perceived that patients, biomedical equipment, information, etc. are automatically supposed to be transferred from one point to another within designated cycle time.

This principle is most easily applied to processes in which repetitive pathways occur. As mentioned earlier, healthcare administrators and clinicians have generally not thought of the care to patients they provide as analogous to automobile manufacturing. The argument used has frequently centered on the concept that a "human is not a car," and too many variables are inherent in clinical care to analyze it systematically. However, due to the rising costs of healthcare (>20% of GDP) and better education needs in efficiency, it is seen that many care processes are repetitive. To accurately map a system, one must obtain high-fidelity and reliable data about the flow of information. Accurately timing steps and determining multi-departmental teams, if necessary, is essential to obtain a true picture of the processes. To map a patient's path to treatment (the value chain) and identify areas for improvement, a current state map can be created in a tool such as a value stream map (VSM). For example, let's use a process in which the first step a patient takes is to visit his primary care physician about a lump in his groin (Fig. 2). The time the patient spends at any step can be broken down into value-added (VA) and non-value-added (NVA) cycle times. VA is the time the customer is willing to pay for: for example, the 20 minutes spent consulting with the PCP. NVA is the time the customer is not willing to pay for. Items such as spending 45 minutes in the waiting room (see the yellow-colored process) or having blood work drawn for a chemistry panel the patient already had a month ago (see the red-colored process) would be considered

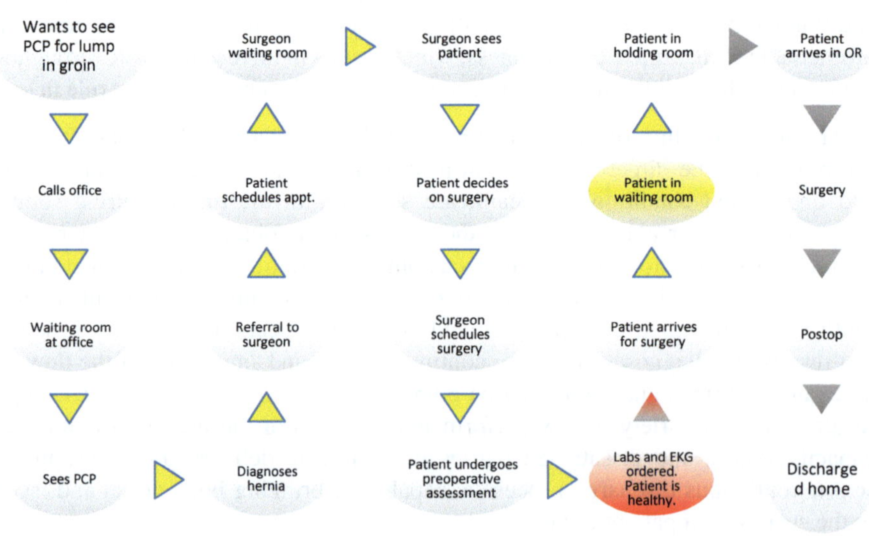


**Fig. 2** A healthcare delivery process flow chart

NVA resource utilization. The pointer between process steps is called a "push" arrow. This shows that once a patient completes a step, they are "pushed" to the next step. This is inefficient, and a more efficient process can be designed by changing push steps to continuous flow or "pull" steps. The yellow triangles indicate the time a patient spends waiting for the next process. These steps are non-value-added actions for the patient.

Damle et al. (2016) determined that VSM could work in a clinical setting such as an endoscopy unit (Fig. 3). With no higher utilization of resources, a single lean process improvement cycle increased their productivity and capacity of their colonoscopy unit by 10%. The authors expected this to result in increased patient access and revenue while maintaining patient satisfaction.

By identifying all the steps, one can begin mapping the entire process, moving from left to right. Once the system is mapped out, the prototypical future state map is created, with the possibility that one could add more detailed microstates within larger streams (Fig. 2). Stream mapping in healthcare can become very complex and detailed. These maps can identify areas for improvement, and once implemented, users can update them, producing more "current state" maps as part of an iterative quality improvement process. At its most basic use, VSM can guide users in strategic directions. Potentially, with the increasing use of artificial intelligence, we can work toward employing these complex maps for use in daily tactical planning

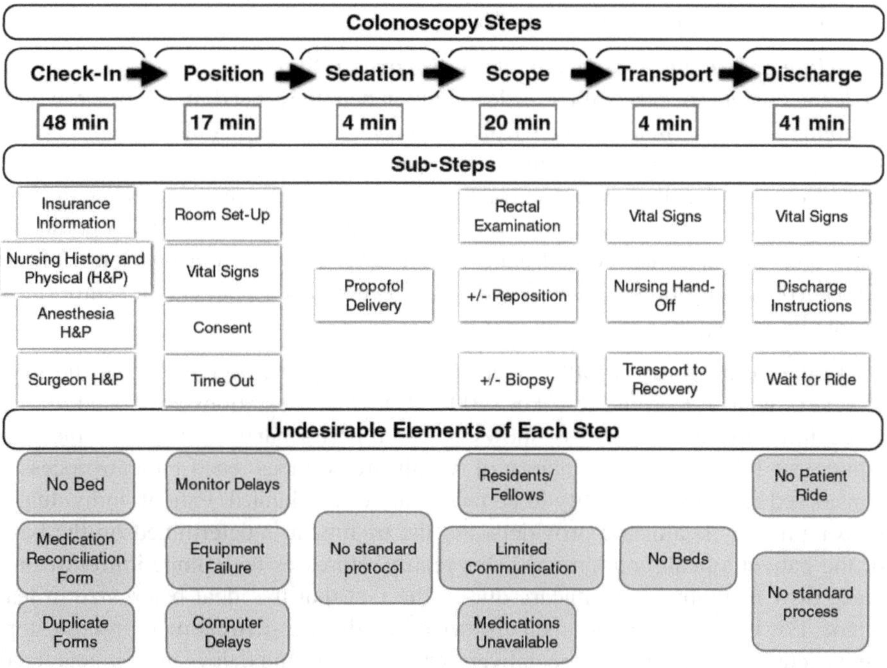


**Fig. 3** Value stream map for colonoscopy (Damle et al. 2016)

and implementation of a variety of healthcare processes (Malcolm and Milstein 2016).

Lean is well suited for application to clinical laboratory settings (Riebling and Tria 2005; Elder 2008; Hurley et al. 2008; Condel et al. 2004). These environments deal with numerical precision and quality control in laboratory testing and measurement activities. It is also of a highly repetitive nature. Thus, mapping the value stream for areas such as laboratories appears readily amenable.

Another clinical arena similar to the laboratory is the Radiology Department. Radiology is a patient care area that employs repetitive steps in a complex system where many opportunities exist for employing lean approach. Process items from appointment scheduling to the final interpretation allow one to reduce clinical and technical errors and mistakes, diminish patient and report waiting times, improve patient outcomes through faster and more accurate reporting, increase the productivity of staff, improve customer (physician/patient) and employee satisfaction, and finally, decrease costs (Seltzer et al. 1997; Kruskal et al. 2012).

Along with all these efforts to create a continuous process flow in healthcare delivery services, IoT can be utilized as a promising platform to smoothly run the processes by reducing traditional hospital visits and waiting lines (Bhatt and Bhatt 2017) as well as reduced inventory and medical errors. The collaboration between Big Data and IoT on managing patient's individual electronic health records (EHR) and overall systematic data in healthcare has a potential for building a continuous process as well. This collaboration may even contribute to providing diagnosis and treatment for patients (Jeong et al. 2016).

**Principle 3: Use pull systems to avoid overproduction**

In lean, customers place their orders only when they need it. In other words, customers *pull* thc final product from the production lines. Pull systems include several important essential requirements such as lowering inventory level, single-piece flow, and Kanban. Toyota Production System (TPS) specifically emphasizes the importance of inventory reduction (Monden 1998). Liker (2004: 99) points out that "inventory enables the bad habit of not having to confront problems." Liker (2004: 105) also addresses that "Toyota Way is not about managing inventory; it is about eliminating it."

As the purest form of pull (Liker 2004), single-piece flow is one of the main drivers in lean. According to Liker (2004: 99), when operations are linked to one-piece flow, the entire process shows a lower performance if any one piece of equipment fails. Due to the nature of healthcare services, each patient/process is considered an individual entity to be taken care of as planned. Patient, individually, is examined by healthcare providers and the treatment is determined on the basis of the patient's health conditions. This reality addresses the compatibility of one-piece flow naturally in healthcare due to the fact that the ideal batch size in lean is one. For instance, if a biomedical device in a diagnostic/treatment process stops unexpectedly, the entire service delivery process stops and delays until it gets fixed. In this example, very differently from the manufacturing setting, the probability of putting the patient's life at risk is another negative outcome that should be taken into consideration by the decision-makers of the process.

There are particular studies in the literature pointing out that lean requirements are not compatible with some certain industries such as healthcare. For example, Christopher and Towill (2001, 2002) state that lean cannot be implemented in industries where demand is highly variable and customization is high. However, due to the fact that each patient has unique symptoms, health conditions, and treatment ways, one-piece flow rule is compatible with the healthcare industry. Similar to production lines that are based on mix-model production systems, admitting each patient individually is a way of implementing one-piece flow rule in healthcare processes. Based on the physician orders, the rest of the diagnostic and treatment processes are shaped as a way of one-piece flow combined with mix-model service delivery. Additionally, healthcare services do not start operating before a patient demand appears; which means that healthcare delivery systems naturally run based on "pull."

The philosophy of just-in-time (JIT) requires the provision of what is needed when they are needed at the right amount, at the right place, and at the right time. Starting from this motto, JIT concept and Kanban can be easily implemented in healthcare (c.f. Rust 2013). For instance, medical inventory should not be stored on the shelves, instead, in lean, the inventory is required to be provided by the supplier when they are needed (c.f. Toussaint and Berry 2013; Hintzen et al. 2009). Similarly, patient transportations between departments should be scheduled in such a way that none of the service providers or patients are required to wait. From another point of view, patients are expected to show up at the designated places at the right time. For example, reducing turnover time in ORs relatively depends on the delay of patients taken to the OR.

From a different perspective, researchers have been examining input into the healthcare system (patient load) more critically since recognizing the input into certain systems may not be appropriate. The US Congressional Budget Office estimates that 30% of healthcare provided is unnecessary, defined as services that do not improve the patient's health. With that in mind, some have challenged specialty societies to develop a list of relatively expensive procedures that do not provide meaningful benefit to at least some categories of patients for whom they are commonly used (Brody 2010). Among others, examination of true need for intervention for health problems such as breast cancer (Pezzin et al. 2016), laboratory orders (Johnson et al. 2016), psychiatric admissions (Shaffer et al. 2015), and prostate cancer screening (Wilt and Dahm 2015) are being scrutinized in an effort to reduce needless input into the system.

Another attempt to reduce overproduction and decrease "pull" targets the consumer side in healthcare. Matching demand to appropriate resources is being attempted by some, with examples ranging from nursing home ratings (Werner et al. 2016) to crowdsourcing costs of care (Meisel et al. 2016). Others have attempted to reduce loading by identifying higher-risk individuals that "burden" healthcare systems with the inability to control the time and quantity of workload available in the system to address these individuals. One study at Cleveland Clinic calculated that 10% of their patients cost 60% of total expenditures (Lee et al. 2017). The authors concluded that higher cost patients are heterogeneous and most are not easily identified patients with frequent admissions. They also concluded that effective interventions to reduce costs by matching resources to workload would require a more multifaceted approach to this population that requires higher resources. However, behavioral changes may be more difficult to implement after obtaining evidence for leaner input and establishing recognition of high value (Vegas et al. 2015; Wilensky 2016). Wilensky (2016) has stated that "it is hard to ignore that there is little evidence that we've made much measurable progress in addressing these problems." Even in the popular press such as the New York Times, lamentations exist that metrics that attempt to alter practice behaviors (Wachter 2016). We are beginning to see more cautionary tales such as these as we attempt to impose more "quality"-related and leaner processes on the human aspect of care delivery.

From an individual clinical practice standpoint, pull reduction may be difficult to implement. However, with large-scale public policy initiatives, e.g., smoking cessation, motivating the populace to improve health habits, while slowly reducing interventions of low value, may be more prudent. Tobacco smoking cessation efforts, while requiring an "extra" expenditure of money, have been proven to reduce healthcare costs. In 2004, the WHO estimated that over 10 years (2006–2015), 13.8 million deaths could be averted by the implementation of smoking cessation interventions. Population-based intervention strategies could therefore substantially reduce mortality from chronic diseases while reducing patient loading into the healthcare systems. Extra evidence that load is reduced is that tobacco smoking is associated with higher hospital lengths of stay (Rezaei et al. 2016) and overall increased costs of care to an estimated total economic cost in the USA at more than $300 billion a year (Hall and Doran 2016).

**Principle 4: Level out the workload (*heijunka*)**

"Operationally, lean synchronizes supply with demand by leveling demand inside production processes and working with suppliers to reduce lead-times toward JIT supply delivery. This goal, called *heijunka* (load-leveling), is the foundation of the TPS and all lean manufacturing" (Rust 2013: 37). Aronsson et al. (2011) express that *heijunka* is linked with process redesign in lean. From this perspective, Christopher (2000) addresses that lean is the best way to follow when load-leveling is possible, demand is predictable, the requirement for variety is low, and volume is high (Christopher 2000). Healthcare industry may rely on "load-leveling" and "high volume" components of lean in order to improve overall performance. However, predictable demand and low variety requirement may not be applicable in all healthcare processes, particularly in trauma centers and ERs where demand is uncertain and unplanned. As stated by Rust (2013: 86), lean aims at stabilizing uneven demand. In addition to this, outpatient centers, ORs, or ambulatory care units run based on planned scheduling where demand is certain and apparent. Appointment-based planning systems may help implement load-leveling in healthcare in addition to the fact that all types of healthcare demands may not be met by planned systems. At that point, Liker (2004) points out that schedules in service operations are leveled by fitting customer demand into a standardized schedule and establishing standard times for delivering different types of service.

According to Naylor et al. (1999), lean aims to ensure level schedule in the manufacturing setting. In service setting, leveling schedule focuses on eliminating variation and uncertainty. In this manner, Rust (2013) links level schedule and high-capacity utilization as the main expectation in healthcare. *Heijunka* is suggested by Weber (2006) to smooth the service delivery pace in healthcare, while continuous flow and *takt* time are suggested by Liker (2004) to be applied in relatively high-volume and repetitive manufacturing and service operations. In this sense, some departments such as test laboratories and treatment and diagnostic procedures in outpatient and inpatient clinics are considered the best places where continuous flow and *takt* time may apply in the healthcare setting.

**Principle 5: Build a culture of stopping to fix problems, to get quality right the first time**

Solving problems at their source results in saving resources (Liker 2004). Along with this understanding, lean suggests stopping the lines in order to prevent defects and errors. Mistake-proofing and standardization are important components of this understanding. Monden (1998: 227) states that "there are two ways to stop the line when abnormalities occur: by relying on human judgment and by means on automatic devices." The first alternative has already been implemented in healthcare processes for many centuries. Since the beginning of the science of medicine, physicians have taken a leadership role in designing the care processes. "Don't harm and don't hurt the patient – *primum non nocere*" is the first motto taught at schools of medicine as a universal rule. The second alternative has been implemented in healthcare along with the integration of various engineering disciplines in medicine. A great variety of engineering-based discoveries and inventions have changed and improved the diagnostic and treatment processes in healthcare. In a general perspective, because of these two distinctive points, healthcare processes have already accommodated the essentials of the culture of stopping to fix the problems. Additionally, in today's healthcare industry, IoT-related systems, devices, and applications, such as medical sensor devices, aim to improve medical decisions made by care providers and minimize and prevent potential abnormalities. As more IoT devices are placed in healthcare processes, the importance of health data collected through these devices cannot be ignored in the future (Anumala et al. 2015), especially in order to determine where, when, and in what condition to stop the healthcare.

Stopping the processes is named as *jidoka* or *autonomation* in lean, as one of the two pillars of lean. Grout and Toussaint (2010) ask this question for healthcare professionals: "When in doubt, is the best action no action at all?" This question takes the conversation to the point where it is more acceptable to stop services rather than let them harm relevant parties in the process. In the manufacturing setting, *jidoka* empowers employees to stop the production lines. In a similar approach, *jidoka*, which means "react at first defect," is naturally used in healthcare processes. For example, if any abnormalities occur in a blood transfusion, the procedure is immediately halted. In any medical procedures, if the vital signs of the patient get worse, the procedure is stopped. Medical devices and biomedical equipment, giving colorful or sound signs that are similar to *andon* in the manufacturing setting, are utilized in these procedures. Burgess (2012) and Machado and Leitner (2010) emphasize that patient safety alert systems are utilized for care providers in case a potential error is about to occur. Furman (2005) perceives each staff at hospitals as an inspector who can stop the line to improve patient safety. IoT-related systems may also help healthcare providers improve patient safety and identify potential areas where the care delivery process needs to be stopped. Grout and Toussaint (2010) reported that Seattle's Virginia Mason Medical Center had 8000 stoppages based on this safety inspector concept.

As a crucial component of *jidoka*, mistake-proofing, namely, *poka-yoke*, is integrated into lean efforts. *Poka-yoke* means avoidance of inadvertent errors (Shingo 1986). In a comprehensive approach, Grout (2007) describes mistake-proofing as "the use of process or design features to prevent the creation of non-conformances." From this perspective, Grout and Toussaint (2010) address that even basic warning labels are considered mistake-proofing tool since these labels are capable of avoiding mistakes and nonconformities. The main idea in *poka-yoke* is to make it impossible for operators and employees to make an error. Implementing poka-yoke in healthcare may also help decrease malpractice lawsuits (Grout et al. 2013). *Poka-yoke* devices are assembled in the processes and equipment and operators do not even recognize that such devices are active to prevent them make an error. *Poka-yoke* including two segments such as (1) detecting the defect and (2) correcting the error is likely to reduce adverse events and medical errors in the healthcare setting. Considering that deaths caused by preventable medical errors in US hospitals have been a rising issue in patient safety (Kumar and Steinebach 2008; McLaughlin and Olson 2012), *poka-yoke* should be utilized as a remedy in healthcare. As an example of effective *poka-yoke* ways in these two segments, Grout and Toussaint (2010) point out how easier to detect surgical sponges left in patients than discovering them after surgery finalized. Standardized double checkpoints on sponges used in the surgery processes should be perceived as an example of procedural *poka-yoke* approach. Similarly, the single-use plastic locks used in blood transfusion, automatic wheelchair brakes, marked floors, wristbands, and central line kits are given as the other examples of *poka-yoke* in healthcare. In a more primitive approach, at the beginning of the care process, patients are involved in the process by orally confirmed their names and the type of medical intervention in the ORs to prevent mixing-up patients. At that point, *timeouts*, *sign-your-site*, and *read-back* are other mistake-proofing examples systematically embedded in healthcare processes (Grout and Toussaint 2010). As given in Grout and Toussaint's (2010) study, ThedaCare showed a high performance on patient safety and healthcare quality especially by focusing on *jidoka* and *poka-yoke* in the USA. According to Manos et al. (2006: 27), "the key is to set up the system so there is no chance for error." They also give bar-coded or computerized physician order entry systems as examples for mistake-proofing practices. Manos et al. (2006) also give pulse oximeters as another example of *autonomation*. When pulse oximeters become loose or disconnected from the patient, the reading sets off an alarm for healthcare providers to intervene the condition. As seen in those examples, the culture of stopping is already an integrated part of the healthcare systems.

The potential impact of IoT should be taken into consideration in this principle. As stated by Anumala et al. (2015), IoT hubs connecting health devices and smart appliances may generate new data for caregivers and decision-makers. Connecting patients with doctors, nurses, and other direct and indirect caregivers via devices and tools embedded in IoT may improve the quality of patient monitoring processes and decrease the likelihood of adverse events, errors, and failures.

**Principle 6: Standardized tasks are the foundation for continuous improvement and employee empowerment**

Standardization was developed and utilized particularly in manufacturing work settings. Standardized work is likely to improve quality since employees will be trained to follow standardized work in lean (Manos et al. 2006). Gomez et al. (2010: 85) stated that "standard work can be thought of as the best combination of activities that will limit activities that don't add value" with the goal of highest quality healthcare. Similarly, standardization is an indispensable component of healthcare. Kumar and Steinebach (2008) and Kohn et al. (2000) report that experts suggest "standardization" as the remedy to reducing medical errors since standardization is likely to reduce the variation which increases complexity, risk, and error. The error rate in the US healthcare system is estimated approximately at sigma level four (Kendall 2003), which turns into 6210 medical errors per million tasks. Lowe et al. (2012) state that standardized work allows doctors and nurses to perform at their licensure level more often because of effectively running processes. As reported by McLaughlin and Olson (2012), Park Nicollet Healthcare System in Minnesota implemented lean tools to help decrease the number and impact of the issues with the delivery of the anticoagulants, focusing on international normalized ratio time (INR) in the desired range. They achieved their goals by *standardized* policies and dosing models related to the administration of the medicine. As the main outcomes of lean, the organization was able to reduce the admission rate from 15.9% to 11.2% and cost from $1300 to $442 patient/year.

Even if some processes have to be patient-centric and can be changeable, the majority of the process flows are standardized. For example, standardized color-coded triage systems help physicians prioritize the patients in ERs. Other examples are standardized patient transfer procedures helping other care providers to figure out what to do in other departments and standardized electroconvulsive therapy (ECT). Jonsson and Randefelt's (2013) study focuses on how these standardized procedures eliminate waste in healthcare. In a similar vein, standardized care paths, treatment protocols, and worksheets are also integrated into healthcare and administrative processes. With regard to standardized care paths, ERAS (Enhanced Recovery After Surgery) can be given as an advanced example of standardization in healthcare. Particularly, standardized worksheets are developed especially for employees who perform in repetitive processes such as in-room patient preparation and patient discharge processes. Similarly, ORs, X-ray, MRI, or CT laboratories operate based on standardized procedures which result in reduced potential wastes. Gomez et al.'s (2010) study is an exemplar for showing the effectiveness of standardized work instruction sheets. Grout and Toussaint (2010) mention the consequences of the violence of hand-washing protocols as a root cause of stopping operations in ORs. The benefit of implementing standardization in healthcare is to easily detect deviations and have a chance to correct them (Jonsson and Randefelt 2013). Furthermore, IoT-oriented standardization has a huge potential to analyze the root causes of potential abnormalities in healthcare processes.

Spear and Bowen (1999) addressed that establishing process stability through standardization is utilized as a way to create a workforce capable of identifying problems and solutions. Even though "standardization" in lean is considered a vital base, variability and uncertainty associated with treatment and patients' health conditions result in problematic (doubtful) areas in healthcare. According to Rust (2013: 17), "the growing interdependence of healthcare delivery, coupled with pressure to reduce costs and serve greater numbers of patients, makes these delivery chains increasingly difficult to manage and coordinate." Rust (2013) also emphasizes that new service management strategies should be implemented in healthcare to manage changes, variability, and adverse effects of variability in patient demand. In addition, preventable medical errors resulting in 98,000 deaths each year were correlated with the lack of standardization and increasing variability in the healthcare industry (Kohn et al. 2000; McLaughlin and Olson 2012). In this manner, standardized care is more likely to contribute to patient safety as well as improved lean outcomes (Gomez et al. 2010).

Rust (2013) calls attention to the point where increasing demand variability causes revenue losses, increasing cost and stress on employees in the healthcare industry. Additionally, high complexity and uncertainty are seen two obstacles to implementing standardization and lean in healthcare (Rust 2013). That high complexity increases risks and makes the processes and healthcare providers harder to manage (Rust 2013).

**Principle 7: Use visual control so no problems are hidden**
Problems in the workplace must be visible to every employee (Monden 1998). In implementing *autonomation*, various visual controls, such as mistake-proofing tools, digital display panels, *kanban* cards, *andon*, call lights, and lines, monitor the flow of the production (Monden 1998; Liker 2004; Manos et al. 2006). Fillingham (2007) states that visual management helps determine whether the process is operated correctly and if problems or errors occur.

Even if it is not directly recognized, visual control approach takes place in a great variety of services in healthcare processes. There are certain departments and processes in which the concept of visual control is naturally implemented to help decision-makers, particularly clinically, in the healthcare industry. For example, visual timetables or schedules used in radiology (Gomez et al. 2010), labels and signs used in ORs, ERs, and clinics, patient monitors used to observe patient's vital signs, etc. As another visual control tool, *kanban* is a card usually placed in production lines to show the type and quantity of units needed in certain production areas (Monden 1998). Kanban system is also identified as "an information system that harmoniously controls the production of the necessary products in the necessary quantities at the necessary time in every process of a factory" (Monden 1998: 15). Rust (2013: 7) expresses that "demand variability may be the most pressing problem facing healthcare delivery today." Unplanned walk-in patients and visits and unpredictable healthcare demand should be considered in lean while focusing on how to manage Kanban systems. Daily-basis supply order systems in JIT have been implemented in some healthcare organizations. For example, Ballé and Régnier (2007) express that daily delivery in healthcare solves the space problems in inventory since it may result in less stock on hand. However, government-driven healthcare systems rather prefer to order for a period of time instead of putting orders on a daily basis. The main benefit of JIT for healthcare organizations is to save storage space and surface the problems hidden in excessive inventory systems. Manos et al. (2006) point out that inventory and supplies are the major costs in healthcare.

As a crucial component, visual control in manufacturing is basically based on 5S. Machado and Leitner (2010: 387) describe 5S as "a place for everything and everything in its place." 5S refers to "organize, set in order, cleanliness, standardize, and discipline." Ikuma and Nahmens (2014) identify 5S as a systematic approach to transfer a company's culture. The main idea behind 5S is to increase effectiveness and efficiency by eliminating unorganized and dirty workplaces. This idea complies with the healthcare processes since the work environment in healthcare cannot tolerate disorganized and contaminated conditions. Machado and Leitner (2010) express that frequently used items should be placed closer to the workplace and easier to attain than rarely used items. In one case, Laing and Baumgartner (2005) reported that an endoscopy unit saved 17 minutes in cycle time by implementing 5S methodology in a community hospital. Gomez et al. (2010) performed first-step 5S efforts during the weekends by offering overtime to healthcare providers. Additionally, Ikuma and Nahmens (2014) showed how 5S could be integrated into safety-oriented efforts in acute care facilities. Kim et al. (2009) showed that nurses could identify quality and safety issues by using 5S supply carts. Esain et al. (2008) reported that the National Health Service (UK) benefited from 5S projects.

Based on the nature of healthcare processes, labeling, coding, sorting, and classification of the drugs, medicines, material, medical equipment, tool and supplies, spaces and even patients have been used for a long time, as identified in the first two components of 5S. The requirements of cleanliness component of 5S are also in the same line with the cleanliness, sterilization, and medical hygiene requirements. Furthermore, the sterilization procedures in healthcare may be beyond the regular cleanliness expectations in lean.

**Principle 8: Use only reliable, thoroughly tested technology that serves your people and processes**
In lean, new technology is implemented after approved by all relevant parties. The main question in this technology vetting process is whether new technology is capable of improving and adding value to the existing system. While analyzing and testing new alternative technology very slowly and safely, implementing new technology is done very quickly. Liker (2004) states that:

> Toyota has tended to lag behind its competitors in acquiring all types of new technology . . . The Toyota Way is to move slowly because more than one technology has failed to meet their "acid" test of supporting people, process, and values and has been yanked in favor of simpler, manual systems. Toyota is still following this policy in the age of digital technology . . . Throughout this analysis and planning, Toyota will broadly involve all key stakeholders in a consensus-building process.

The requirements of this principle have been neglected by lean practitioners in various industries. It may be seen as one of the root causes of failing lean implementation. Even if hospitals in the USA implement the most advanced technology in the processes, the association between the technological investment and patient outcomes is seen weak and not significant by the researchers (c.f. Himmelstein et al. 2010). Additionally, technology is considered one of the primary contributors of rising healthcare costs (Health Care Costs Report 2007) even if the expected impact on overall performance is lower than the actual one (Husby 2012). By looking into this concern, healthcare organizations should (1) analyze, (2) test, and (3) approve the potential risks, benefits, opportunities, and values of improving technology, before transferring it (Liker 2004). This is compatible with the cultural change integrated into lean transformation in healthcare. As stated by Liker (2004), the entire technology transformation process should be analyzed to see if it conflicts with the organization's philosophies and operating principles. Considering that today's healthcare industry talks about IoT, Big Data analytics, medical sensor devices, telemedicine, and e-health, technology adaptation and export requires a huge amount of time and effort devoted to lean implementation in healthcare systems. As a highly regulated industry, any potential or current change in technology implemented in healthcare processes should be comprehensively reviewed.

## *2.3 People: Respect and Development*

As addressed by several parties (c.f. Zidel 2006; Liker 2004; Womack and Jones 1996), a lean transformation requires an organization-wide paradigm shift. That transformation process can be constructed on the following three principles given under *People* principle. Emiliani (2006: 169) states that the "respect for people" has long been unrecognized, ignored, or misunderstood by most senior managers outside Toyota and its affiliated suppliers. Even though the majority of the publications focus on how to implement lean in healthcare, limited studies worked on the *soft* part of lean. From this point of view, Husby (2012) articulated that *people*, process, and technology are the three crucial areas of focus for US healthcare organizations.

**Principle 9: Growing leaders who thoroughly understand the work, living the philosophy, and teaching it to others**
Human assets are one of the crucial resources in lean success. Toyota leaders are raised up and promoted in the hierarchy of the firm in lifelong employment to become effective leaders. Liker (2004) emphasizes that in Toyota's history, key leaders have been found within the company. Toyota expects leaders in all level to introduce, educate, and teach their subordinates the Toyota culture (Liker 2004; Liker and Convis 2011) and encourage them for lean transformation (Atkinson 2004; Boyer and Sovilla 2003). Culturally, these traditions are not shared by Western business organizations. However, healthcare organizations should implement such a strategy to raise their own leaders and train them about lean philosophy. At that point, transformational leadership (Bass 1999) and servant leadership (Greenleaf and Spears 2002) are linked with lean leaders.

McLaughlin and Olson (2012: 4) emphasize that "excellence in healthcare derives from three major areas of expertise: clinical care, *leadership*, and operations." Essentially, top management support is considered a crucial success factor in lean implementation regardless of the type of the industry (Boyer and Sovilla 2003; Womack and Jones 1996). The role of leadership is accepted as an essential factor for improved quality and safety in healthcare (Sollecito and Johnson 2013). For example, Gomez et al. (2010) address that strong leadership support in their study helped in lean transformation. Liker and Convis (2011) and Husby (2012) perceive leadership function as a critical success factor at hospitals for the long-term success in lean. Fine et al. (2009) reported that when the top leaders of the hospital were fully engaged in lean initiatives, the effort spread across the organization more quickly. Within this understanding, leaders in healthcare organizations should be raised up and trained in accordance with lean in order to increase the likelihood of lean achievement.

**Principle 10: Develop exceptional people and teams who follow your company's philosophy**
"Toyota invests in people, in return, it gets committed associated who show up to work every day on time and are continually improving their operations" (Liker 2004: 198). While Zidel (2006) states that lean organizations value their employees and encourage them to involve in organizational initiatives, the healthcare industry has experienced many challenges such as decreasing reimbursement, staffing shortages, and increasing costs over the last decades (Husby 2012). All these challenges limit investments in staff and teams who work on process improvement and problem-solving. Instead, by looking from a lean perspective, it is seen that healthcare organizations invest in employees' and teams' training and education to increase the likelihood of lean achievement in the long run.

Manos et al. (2006) point out the importance of relying on the employees' knowledge and skills especially in problem-solving processes. According to Manos et al. (2006: 25), “the people closest to the work know it best. They are the process experts and they just have to be trained in problem-solving and lean techniques. One of the advantages of the lean technique is that the staff involved directly in the process are the ones who work to improve it.” De Souza (2009) points out “staff empowerment” as one of the key aspects making lean more acceptable in healthcare in addition to gradual and continuous improvement. In the same line, the respondents in Jonsson and Randefelt’s (2013: 31) study state that “it is important that all the employees feel as a part of the organization and all employees are encouraged to suggest improvements.”

Liker (2004) also emphasizes on the necessity of focusing on the interaction of teams and employees and perceiving teams as the focal point for problem-solving. The teams in TPS are located at the top of the hierarchy. Liker (2004: 185) states that:

> All systems are there to support the team doing value-added work. The teams coordinate the work, motive, and learn from each other. Teams suggest innovative ideas, even control through peer pressure… Excellent individual performers are required to make up teams that excel.

In the healthcare industry, patient-centered multispecialty teams including physicians, nurses, social workers, etc. manage the entire patient care process, clinically and administratively (Sollecito and Johnson 2013). Very similar to TPS, since patient care is managed and given within naturally built teams, the teams should be put at the highest hierarchical level and the rest of the organization should be managed to maximize the patient care team performance.

**Principle 11: Respect your extended network of partners and suppliers by challenging them and helping them improve**

As expressed by Liker (2004), TPS establishes high standards and expect all their partners to raise those standards. According to Deming (1986), the fewer the number of suppliers, the higher the organizational performance. Deming (1986) and Bicheno (1999) also recommend working with suppliers very closely in a long-term orientation and supporting them to improve the quality of their outcomes. Similarly, establishing long-lasting collaboration between healthcare providers and reducing the number of suppliers are seen two important requirements in the healthcare industry (Aronsson et al. 2011). Depending on reliable lean suppliers and partners is likely to increase the success of lean implementation (MacDuffie and Helper 1997; Liker 2004; Womack and Jones 1996). MacDuffie and Helper (1997) state that:

> …customers are likely to demand that suppliers assume substantial responsibility during product development, accommodate customer requests for engineering changes in their product or manufacturing process; be highly reliable with respect to quality and delivery; and have the ability to respond quickly in case of problems. These requirements are difficult for a supplier to meet unless they have adopted lean production practices themselves. Thus a lean customer is likely to find it more productive to work with lean suppliers.

In the same vein, healthcare organizations are supplied by a great variety of organizations such as pharmaceutical companies, biomedical device producers, and R&D firms. As the leading industry in lean, the automobile industry has trained its partners and helped them become a lean supplier (MacDuffie and Helper 1997). However, the healthcare industry has not arrived at such a matured customer-partner relationship level yet. Aronsson et al. (2011) pointed out that the healthcare industry is in need of great system-wide changes in the supply management function. Instead, Husby (2012) points out that physicians are separated from lean efforts, which blocks the potential positive outcomes of lean efforts that might be managed by the physicians. It is obvious that the integration of physicians into lean from top to bottom, as the main director of patient care in one-piece flow, is more likely to increase the effectiveness of healthcare. Inspired by best practices, the healthcare industry with its partners should focus on how to implement lean principles to decrease *muda* in the industry.

## 2.4 Problem-Solving: Continuous Improvement

**Principle 12: Go and see for yourself to thoroughly understand the situation (Genchi Genbutsu)**
Processes and performance issues should be analyzed by the service providers, designers, and managers at firsthand to understand the root causes of high variability in healthcare. Performance-related issues should be monitored in the place where they occur, *gemba*, to figure out assignable and common causes resulting in variability (Kruskal et al. 2012). Seasonal or random variability add complexity to healthcare processes (Husby 2012). In order to decrease response time to problems and increase the probability of solving the problems, problem solvers in healthcare processes should be in *gemba* (c.f. Toussaint and Berry 2013; Fine et al. 2009).

**Principle 13: Make decisions slowly by consensus, thoroughly considering all options, implement rapidly**
In lean, it is expected to run decision-making processes collaboratively based on data throughout the hierarchical structure (Liker 2004). Rosmulder (2011) reported that lean efforts would be useless if there was no system-wide participation in lean projects. At the heart of healthcare, routine diagnostic and treatment processes are compatible with this principle due to the fact that diagnostic processes take relatively longer to consider all potential possibilities. Compared to diagnostic processes, treatment ways are determined by physicians based on protocols and standardized procedures. In addition to explicit knowledge, Husby (2012) mentions the importance of using tacit knowledge developed in healthcare processes to improve patient care-related outcomes. In this manner, tacit knowledge, which is seen as a barrier to transfer evidence-based medical knowledge into actual care (Husby 2012), produced with the help of PDCA may be used as a base for decision-making processes in lean practices in healthcare. Parallel to advancements seen in data management, particular topics such as Big Data and IoT have emerged as the promising areas for healthcare organizations. Considering the excessive

need for better and timely decisions in the healthcare industry, all these advanced technologies may help decision-makers function in this principle more effectively.

**Principle 14: Become a learning organization through relentless reflection (hansei) & Continuous Improvement (Kaizen)**
Ballé and Régnier (2007: 33) point out that "lean is a learning method more than anything else." Based on knowledge generated in lean, healthcare organizations become capable of utilizing the current knowledge for providing better care and making better decisions. A continuous learning process is built based on lean pillars such as continuous improvement, standardization, and stability. Taichi Ohno (1988) states that without standards, there is no *kaizen*. At that point, using both explicit and tacit knowledge in lean effort takes the healthcare organizations to a higher performance level, especially in terms of sustaining lean efforts and becoming a learning organization. The learning environment should be created not only for a process but also for the entire healthcare delivery system.

## 3 The Limitations of the Study

The primary limitation of our research and the resulting manuscript is that, although much has been discussed and implemented in the lean process as it relates to manufacturing and service processes, clinical application of the lean method into healthcare processes is relatively scarce. Based on our review of published examples of lean in healthcare, the most consistent feature was the difficulty encountered in implementing the intended training and improvement projects. Most studies demonstrated multiple barriers to progress inherent in the very systems and culture they were attempting to improve. There are clearly important questions here that need to be addressed. The organizational structure of healthcare systems appears to most outside observers as highly complex with multiple lines of accountability. It is important to understand why this structure has evolved and persisted, despite the obvious disadvantages. The assumption that a clearer and simpler structure with stronger accountability would be more effective and safer for patients may need to be tested empirically at some stage, but this would clearly involve a very large-scale experiment (essentially a whole trust, but with implications that would certainly attract comment and concern from national professional bodies, unions, and others).

The business literature demonstrates that dealing with the theory of organizational change and an important component of any investigation of organizational barriers should be an attempt to comprehend these through this perspective. We recommend observational and analytical studies of the relationships between culture, organizational structure, and resistance to change in healthcare organizations, in collaboration with experts in change management. An attempt to create a large and practical trial of lean interventions will need to be designed using an approach framed in an appropriately adapted theory of organizational change.

If we accept that major changes to organizational structures are likely to be difficult at best to achieve, future qualitative research into the attitudes to safety interventions and system improvement expressed by different staff groups will be useful in understanding how these might be changed or circumvented when they appear unhelpful. Research such as this may require the use of sophisticated theory-based questionnaires to differentiate between attitudes and motivations, which are often engendered by the expectations of the organization and professional peer groups. These are deep, and perhaps, subconscious drivers which show themselves in decision-making choices rather than public statements made for political purposes. Using previous research on organizational culture and organizational change theory, it may be possible to create a predictive model that will help to identify how a given organization will react to attempts at systems change and test this against the outcome of real-life improvement projects.

## 4 Conclusion

In this chapter, we have discussed the lean process as one management approach that we can use to examine operational aspects. We also have discussed the social and technical dynamics associated with these improvement systems. We have also focused on how one can start to apply these insights to healthcare.

In lean implementation, one must take into account the variety of issues surrounding the application of lean thinking to healthcare. One difficulty regards the social and technical issues that occur when implementing lean thinking. Research on sociotechnical dynamics in lean organizations, especially in healthcare, is virtually absent.

Operational aspects of lean thinking and their relationship to performance have been looked at thoroughly, but application to healthcare has been limited. The same goes for cumulative capabilities, where research in and out of healthcare is scarce. More attention is needed to verify these key propositions of lean thinking in healthcare. Even if these problems are addressed and lean delivers on its promises, the challenges to increasing the role of lean thinking are very future-promising. The value produced is more or less ephemeral, difficult to specify, and unclearly priced into the equation.

A lean transformation requires an organization-wide paradigm shift. It requires no less than the redesign of the healthcare system as we now see it. Perseverance, high-quality leadership, dedicated professionals, high quality and timely data management and knowledge creation, and patience are surely needed. Doubt and the resulting resistance will be high. Organizations may take a long period of time before embarking on such a journey. Worse, they may superficially implement lean thinking, adding to existing resistance; thus, making it more difficult to improve healthcare in the long term.

We believe lean thinking has the potential to improve healthcare delivery. At the same time, there are methodological and practical considerations that need to be taken into account. Otherwise, lean implementation will be superficial and fail, adding to existing resistance and making it more difficult to improve healthcare in the long term.

## References

Allen, S. (2009). No waiting: A simple prescription that could dramatically improve hospitals and American health care. *The Boston Globe, 276*, C1–C4.

Anumala, H., Busetty, S. M., & Bharti, V. (2015). Leveraging IoT device data for emotional health. In *International internet of things summit* (pp. 487–501). Cham: Springer.

Arcidiacono, G., & Pieroni, A. (2018). The revolution lean Six Sigma 4.0. *International Journal on Advanced Science, Engineering and Information Technology, 8*(1), 141–149.

Aronsson, H., Abrahamsson, M., & Spens, K. (2011). Developing lean and agile healthcare supply chains. *Supply Chain Management: An International Journal, 16*(3), 176–183.

Atkinson, P. (2004). Creating and implementing lean strategies. *Management Services, 48*(2), 18.

Ballé, M., & Régnier, A. (2007). Lean as a learning system in a hospital ward. *Leadership in Health Services, 20*(1), 33–41.

Barnas, K. (2011). ThedaCare's business performance system: Sustaining continuous daily improvement through hospital management in a lean environment. *The Joint Commission Journal on Quality and Patient Safety, 37*(9), 387–399.

Bass, B. M. (1999). Two decades of research and development in transformational leadership. *European Journal of Work and Organizational Psychology, 8*(1), 9–32.

Berwick, D. M., Godfrey, B. A., & Roessner, J. (1991). Curing health care: New strategies for quality improvement. *Journal for Healthcare Quality, 13*(5), 65–66.

Bhatt, Y., & Bhatt, C. (2017). Internet of things in healthcare. In *Internet of things and big data technologies for next generation HealthCare* (pp. 13–33). Cham: Springer.

Bicheno, J. (1999). *The new lean toolbox*. London: Picsie.

Black, J., & Miller, D. (2008). *The Toyota way to healthcare excellence: Increase efficiency and improve quality with lean*. Chicago: Health Administration Press.

Borger, C., Smith, S., Truffer, C., Keehan, S., Sisko, A., Poisal, J., & Clemens, M. K. (2006). Health spending projections through 2015: Changes on the horizon. *Health Affairs, 25*(2), w61–w73.

Boyer, M., & Sovilla, L. (2003). How to identify and remove the barriers for a successful lean implementation. *Journal of Ship Production, 19*(2), 116–120.

Breyfogle, F., & Salveker, A. (2004). *Lean six sigma in sickness and in health*. Austin: Smarter Solutions.

Brody, H. (2010). Medicine's ethical responsibility for health care reform—The top five list. *New England Journal of Medicine, 362*(4), 283–285.

Burgess, N. J. (2012) *Evaluating lean in healthcare* (Unpublished Ph.D. thesis). University of Warwick, Warwick Business School.

Burgess, N., & Radnor, Z. (2013). Evaluating lean in healthcare. *International Journal of Health Care Quality Assurance, 26*(3), 220–235.

Bushell, S., & Shelest, B. (2002). Discovering lean thinking at progressive healthcare. *The Journal for Quality and Participation, 25*, 20–25.

Caldwell, C., Brexler, J., & Gillem, T. (2005). Engaging physicians in lean six sigma. *Quality Progress, 38*(11), 42.

Chalice, R. (2005). *Stop rising healthcare costs using Toyota lean production methods: 38 steps for improvement*. Milwaukee: ASQ Quality Press.

Christopher, M. (2000). The agile supply chain: Competing in volatile markets. *Industrial Marketing Management, 29*(1), 37–44.

Christopher, M., & Towill, D. (2001). An integrated model for the design of agile supply chains. *International Journal of Physical Distribution & Logistics Management, 31*(4), 235–246.
Christopher, M., & Towill, D. R. (2002). Developing market specific supply chain strategies. *International Journal of Logistics Management, 13*(1), 1–14.
CMS. (2010). Accessed on 31st Oct 2016. https://www.cms.gov/Research-Statistics-Data-and-Systems/Statistics-Trends-and-Reports/MedicareMedicaidStatSupp/2010.html
Collins, K. F., Muthusamy, S. K., & Carr, A. (2015). Toyota production system for healthcare organisations: Prospects and implementation challenges. *Total Quality Management & Business Excellence, 26*(7–8), 905–918.
Condel, J. L., Sharbaugh, D. T., & Raab, S. S. (2004). Error-free pathology: Applying lean production methods to anatomic pathology. *Clinics in Laboratory Medicine, 24*(4), 865–899.
Congressional Budget Office. Increasing the value of federal spending on health care. Accessed at http://www.cbo.gov/publication/41717
Dahlgaard, J. J., Pettersen, J., & Dahlgaard-Park, S. M. (2011). Quality and lean health care: A system for assessing and improving the health of healthcare organisations. *Total Quality Management & Business Excellence, 22*(6), 673–689.
Damle, A., et al. (2016). Elimination of waste: Creation of a successful lean colonoscopy program at an academic medical center. *Surgical Endoscopy, 30*(7), 3071–3076.
De Souza, L. B. (2009). Trends and approaches in lean healthcare. *Leadership in Health Services, 22*(2), 121–139.
Deming, W. E. (1986). *Out of the crisis*. Cambridge, MA: MIT Press.
Elder, B. L. (2008). Six Sigma in the microbiology laboratory. *Clinical Microbiology Newsletter, 30*(19), 143–147.
Emiliani, M. L. (2006). Origins of lean management in America: The role of Connecticut businesses. *Journal of Management History, 12*(2), 167–184.
Esain, A., Williams, S., & Massey, L. (2008). Combining planned and emergent change in a healthcare lean transformation. *Public Money and Management, 28*(1), 21–26.
Fillingham, D. (2007). Can lean save lives? *Leadership in Health Services, 20*(4), 231–241.
Fine, B. A., Golden, B., Hannam, R., & Morra, D. (2009). Leading lean: A Canadian healthcare leader's guide. *Healthcare Quarterly, 12*(3), 32–41.
Frings, G. W., & Grant, L. (2005). Who moved my Sigma... effective implementation of the Six Sigma methodology to hospitals. *Quality and Reliability Engineering International, 21*(3), 311–328.
Furman, C. (2005). Implementing a patient safety alert system. *Nursing Economics, 23*, 42–45.
Gomez, M. A., Hirsch, J. A., Stingley, P., Byers, E., & Sheridan, R. M. (2010). Applying the Lean management philosophy to NeuroInterventional radiology. *Journal of Neurointerventional surgery, 2*(1), 83–86.
Graban, M. (2009). *Lean hospitals: Improving quality, patient safety, and employee satisfaction*. Boca Raton: CRC Press.
Greenleaf, R. K., & Spears, L. C. (2002). *Servant leadership: A journey into the nature of legitimate power and greatness*. New York: Paulist Press.
Grout, J. R. (2007). *Mistake-proofing the design of healthcare processes (AHRQ Pub. No. 07-0020)*. Rockville: U.S. Department of Health and Human Services.
Grout, J. R., & Toussaint, J. S. (2010). Mistake-proofing healthcare: Why stopping processes may be a good start. *Business Horizons, 53*(2), 149–156.
Grout, J. R., Hill, J. W., & Langvardt, A. W. (2013). Mistake-proofing medicine: Legal considerations and healthcare quality implications. *Minnesota Journal of Law, Science & Technology, 14*, 387–438.
Hall, W., & Doran, C. (2016). How much can the USA reduce health care costs by reducing smoking? *PLoS Medicine, 13*(5), e1002021.
Health Care Costs, A Primer: Key Information on Health Care Costs and Their IMpact. (2007). Available: http://www.kff.org/insurance/upload/7670.pdf
Himmelstein, D. U., Wright, A., & Woolhandler, S. (2010). Hospital computing and the costs and quality of care: A national study. *The American Journal of Medicine, 123*(1), 40–46.

Hines, P., Holweg, M., & Rich, N. (2004). Learning to evolve: A review of contemporary lean thinking. *International Journal of Operations & Production Management, 24*(10), 994–1011.

Hintzen, B. L., Knoer, S. J., Van Dyke, C. J., & Milavitz, B. S. (2009). Effect of lean process improvement techniques on a university hospital inpatient pharmacy. *American Journal of Health-System Pharmacy, 66*(22), 2042–2047.

Hurley, B., Taylor, T., Levett, J., Huber, C., & Hahn, E. (2008). Implementation of Six Sigma and lean methodology into the anticoagulation management process. *Journal of Thrombosis and Thrombolysis, 25*(1), 106–106.

Husby, B. C. (2012). *Integrating people, process, and technology in lean healthcare* (Ph.D. Dissertation). The University of Michigan.

Ikuma, L. H., & Nahmens, I. (2014). Making safety an integral part of 5S in healthcare. *Work, 47*(2), 243–251.

Jeong, J. S., Han, O., & You, Y. Y. (2016). A design characteristics of smart healthcare system as the IoT application. *Indian Journal of Science and Technology, 9*(37), 52.

Johnson, D. P., Lind, C., Parker, S. E. S., Beuschel, C., VanVliet, S., Nichols, J., et al. (2016). Toward high-value care: A quality improvement initiative to reduce unnecessary repeat complete blood counts and basic metabolic panels on a pediatric hospitalist service. *Hospital Pediatrics, 6*(1), 1–8.

Jonsson, C., & Randefelt, A. (2013). What is Lean?: A case study of how Akademiska Hospital's departments work with Lean, Uppsala University. Unpublished case study report.

Karlsson, C., Rognes, J., & Nordgren, H. (1995). *Model for lean production*. Goteborg: Institute for Management of Innovation and Technology.

Kendall, E. M. (2003). Improving patient care with collaborative rounds. *American Journal of Health-System Pharmacy, 60*(2), 132.

Kim, C. S., Spahlinger, D. A., Kin, J. M., & Coffey, R. J. (2009). Implementation of lean thinking: One health system's journey. *The Joint Commission Journal on Quality and Patient Safety, 35*(8), 406–413.

Kohn, L. T., Corrigan, J., & Donaldson, M. S. (2000). *To err is human: Building a safer health system*. Washington, DC: National Academy Press.

Kollberg, B., Dahlgaard, J. J., & Brehmer, P. O. (2006). Measuring lean initiatives in health care services: Issues and findings. *International Journal of Productivity and Performance Management, 56*(1), 7–24.

Kruskal, J. B., Reedy, A., Pascal, L., Rosen, M. P., & Boiselle, P. M. (2012). Quality initiatives: Lean approach to improving performance and efficiency in a radiology department. *Radiographics, 32*(2), 573–587.

Kumar, S., & Steinebach, M. (2008). Eliminating US hospital medical errors. *International Journal of Health Care Quality Assurance, 21*(5), 444–471.

Laing, K., & Baumgartner, K. (2005). Implementing "lean" principles to improve the efficiency of the endoscopy department of a community hospital: A case study. *Gastroenterology Nursing, 28*(3), 210–215.

Laursen, M., Gertsen, F., & Johanson, J. (2003). *Applying lean thinking in hospitals—Exploring implementation difficulties*. Denmark: Center for Industrial Production.

Lee, N. S., Whitman, N., Vakharia, N., & Rothberg, M. B. (2017). High-cost patients: Hot-spotters don't explain the half of it. *Journal of general internal medicine, 32*(1), 28–34.

Liker, J. (2004). *The Toyota way: 14 managements principles from the world's greatest manufacturer*. New York: McGraw-Hill.

Liker, J. K., & Convis, G. L. (2011). *The Toyota way to lean leadership: Achieving and sustaining excellence through leadership development*. New York: McGraw-Hill.

Löffler, M., & Tschiesner, A. (2013). *The Internet of Things and the future of manufacturing*. (pp. 1–6). McKinsey & Company.

Lowe, G., Plummer, V., O'Brien, A. P., & Boyd, L. (2012). Time to clarify–the value of advanced practice nursing roles in health care. *Journal of Advanced Nursing, 68*(3), 677–685.

Lunardini, D., Arington, R., Canacari, E. G., Gamboa, K., Wagner, K., & McGuire, K. J. (2014). Lean principles to optimize instrument utilization for spine surgery in an academic medical

center: An opportunity to standardize, cut costs, and build a culture of improvement. *Spine, 39*(20), 1714–1717.
MacDuffie, J. P., & Helper, S. (1997). Creating lean suppliers: Diffusing lean production through the supply chain. *California Management Review, 39*(4), 118–151.
Machado, V. C., & Leitner, U. (2010). Lean tools and lean transformation process in health care. *International Journal of Management Science and Engineering Management, 5*(5), 383–392.
Malcolm, E., & Milstein, A. (2016). Achieving higher quality and lower costs via innovations in healthcare delivery design. *America's Healthcare Transformation: Strategies and Innovations*, 105–112.
Manos, A., Sattler, M., & Alukal, G. (2006). Make healthcare lean. *Quality Progress, 39*(7), 24.
McLaughlin, D. B., & Olson, J. R. (2012). *Healthcare operations management*. Chicago: Health Administration Press.
Meisel, Z. F., VonHoltz, L. A. H., & Merchant, R. M. (2016). Crowdsourcing healthcare costs: Opportunities and challenges for patient centered price transparency. *Healthcare (Amsterdam, Netherlands), 4*(1), 3–5.
Miller, D. (2005). *Going lean in health care*. Cambridge, MA: Institute for Healthcare Improvement.
Modig, N., & Åhlström, P. (2012). *This is lean-resolving the efficiency paradox*. Halmstad: Bulls Graphics AB.
Monden, Y. (1998). *Toyota production system, an integrated approach to just-in-time*. Norcross: Engineering and Management Press.
Naylor, J. B., Naim, M. M., & Berry, D. (1999). Leagility: Integrating the lean and agile manufacturing paradigms in the total supply chain. *International Journal of Production Economics, 62*(1), 107–118.
NHSIII. (2007). *Going lean in the NHS*. Warwick: NHS Institute for Innovation and Improvement.
Ohno, T. (1988). *Toyota production system: Beyond large-scale production*. New York: CRC Press.
Pakdil, F., & Harwood, T. N. (2005). Factors that influence efficiency in performing ENT cases: A qualitative and quantitative analysis. *Journal of Medical Systems, 29*(3), 285–301.
PCAST Report. (2014). *Report to the president better health care and lower costs: Accelerating improvement through systems engineering*. Washington, DC: Executive Office of the President, President's Council of Advisors on Science and Technology.
Pezzin, L. E., Laud, P., Neuner, J., Yen, T. W. F., & Nattinger, A. B. (2016). A statewide controlled trial intervention to reduce use of unproven or ineffective breast cancer care. *Contemporary Clinical Trials, 50*, 150–156.
Radnor, Z. J., Holweg, M., & Waring, J. (2012). Lean in healthcare: The unfilled promise? *Social Science & Medicine, 74*(3), 364–371.
Raja, P. V., Davis, M. C., Bales, A., & Afsarmanesh, N. (2015). NPITxt, a 21st-century reporting system: Engaging residents in a lean-inspired process. *American Journal of Medical Quality, 30*(3), 255–262.
Rezaei, S., Akbari Sari, A., Arab, M., Majdzadeh, R., Shaahmadi, F., & Mohammadpoorasl, A. (2016). The association between smoking status and hospital length of stay: Evidence from a hospital-based cohort. *Hospital Practice, 44*(3), 129–132.
Riebling, N., & Tria, L. (2005). Six sigma project reduces analytical errors in an automated lab. *MLO: Medical Laboratory Observer, 37*(6), 20, 22–23.
Robinson, S., Radnor, Z., et al. (2012). SimLean: Utilizing simulation in the implementation of lean in healthcare. *European Journal of Operational Research, 219*, 188–197.
Rosmulder, R. W. (2011). *Improving healthcare delivery with lean thinking: Action research in an emergency department*. Enschede: University of Twente.
Rust, T. C. (2013). *Dynamic analysis of healthcare service delivery: Application of lean and agile concepts* (Doctoral dissertation). Worcester Polytechnic Institute.
Scarpato, N., Pieroni, A., Di Nunzio, L., & Fallucchi, F. (2017). E-health-IoT universe: A review. *International Journal on Advanced Science Engineering Information Technology, 7*(6), 2328–2336.

Seltzer, S. E., Kelly, P., Adams, D. F., Chiango, B. F., Viera, M. A., Fener, E., Hooton, S., Bannon-Rohrbach, S., Healy, C. D., Doubilet, P. M., & Holman, B. L. (1997). Expediting the turnaround of radiology reports in a teaching hospital setting. *AJR: American Journal of Roentgenology, 168*(4), 889–893.

Shaffer, S. L., Hutchison, S. L., Ayers, A. M., Goldberg, R. W., Herman, D., Duch, D. A., et al. (2015). Brief critical time intervention to reduce psychiatric rehospitalization. *Psychiatric Services (Washington, DC), 66*(11), 1155–1161.

Shahrabi, M. S. (2015). Reducing errors by RFID technology to achieve lean healthcare. *International Journal of Hospital Research, 4*(2), 95–100.

Shingo, S. (1986). *Zero quality control: Source inspection and the poka-yoke system*. Cambridge, MA: Productivity Press.

Sollecito, W., & Johnson, J. (2013). *Continuous quality improvement in health care: Theory, implementations, and applications*. Burlington: Jones & Bartlett Learning.

Spear, S., & Bowen, K. H. (1999, September–October). Decoding the DNA of the Toyota production system. *Harvard Business Review, 77*, 96–106.

Sugianto, J. Z., Stewart, B., Ambruzs, J. M., Arista, A., Park, J. Y., Cope-Yokoyama, S., & Luu, H. S. (2015). Applying the principles of lean production to gastrointestinal biopsy handling: From the factory floor to the anatomic pathology laboratory. *Laboratory Medicine, 46*(3), 259–264.

Tachibana, C., & Nelson-Peterson, D. L. (2007). Implementing the clinical nurse leader role using the Virginia Mason Production System. *Journal of Nursing Administration, 37*(11), 477–479.

Taner, T. M., Sezen, B., & Antony, J. (2007). An overview of Six Sigma applications in healthcare industry. *International Journal of Health Care Quality Assurance, 20*(4), 329–340.

Toussaint, J. S., & Berry, L. L. (2013). The promise of lean in health care. *Mayo Clinic Proceedings, 88*(1), 74–82.

Tu, Y. J., Zhou, W., & Piramuthu, S. (2009). Identifying RFID-embedded objects in pervasive healthcare applications. *Decision Support Systems, 46*(2), 586–593.

Van Vliet, E. J., Bredenhoff, E., Sermeus, W., Kop, L. M., Sol, J. C. A., & Van Harten, W. H. (2011). Exploring the relation between process design and efficiency in high-volume cataract pathways from a lean thinking perspective. *International Journal for Quality in Health Care, 23*(1), 83–93.

Vegas, D. B., Levinson, W., Norman, G., Monteiro, S., & You, J. J. (2015). Readiness of hospital-based internists to embrace and discuss high-value care with patients and family members: A single-centre cross-sectional survey study. *CMAJ Open, 3*(4), E382.

Wachter B. (2016). How measurement fails doctors and teachers. *New York Times*. January 16, 2016.

Wagner, J. (2004). Medical errors still plague US hospitals, WebMD, Inc., available at: www.webmd.com/content/article/91/101128.htm

Waring, J. J., & Bishop, S. (2010). Lean healthcare: Rhetoric, ritual and resistance. *Social Science & Medicine, 71*(7), 1332–1340.

Warner, C. J., Walsh, D. B., Horvath, A. J., Walsh, T. R., Herrick, D. P., Prentiss, S. J., & Powell, R. J. (2013). Lean principles optimize on-time vascular surgery operating room starts and decrease resident work hours. *Journal of Vascular Surgery, 58*(5), 1417–1422.

Weber, D. (2006). Toyota-style management drives Virginia-Mason. *Physician Executive, 32*(1), 12–17.

Werner, R. M., Konetzka, R. T., & Polsky, D. (2016). Changes in consumer demand following public reporting of summary quality ratings: An evaluation in nursing homes. *Health Services Research, 51*(Suppl 2), 1291–1309.

Wickramasinghe, N., Al-Hakim, L., Gonzalez, C., & Tan, J. (Eds.). (2014). *Lean thinking for healthcare*. New York: Springer.

Wilensky, G. (2016). Changing physician behavior is harder than we thought. *JAMA, 316*(1), 21–22.

Wilt, T. J., & Dahm, P. (2015). PSA screening for prostate cancer: Why saying no is a high-value health care choice. *Journal of the National Comprehensive Cancer Network, 13*(12), 1566–1574.

Womack, J. P., & Jones, D. T. (1996). Beyond Toyota: How to root out waste and pursue perfection. *Harvard Business Review, 74*(5), 140–158.

Womack, J. P., & Jones, D. T. (2003). *Lean thinking*. London: Simon & Schuster.

Womack, J. P., Byrne, A. P., Fiume, O. J., Kaplan, G. S., & Toussaint, J. (2005). *Going lean in health care*. Cambridge, MA: Institute for Healthcare Improvement.

Young, T., Brailsford, S., Connell, C., Davies, R., Harper, P., & Klein, J. H. (2004). Using industrial processes to improve patient care. *British Medical Journal, 328*(7432), 162–164.

Zidel, T. (2006). *A lean guide to transforming healthcare: How to implement lean principles in hospitals, medical offices, clinics, and other healthcare organizations*. Milwaukee: Quality Press.

# Data, Denial, and Disparity: Is This a New Digital Divide?

**Nilmini Wickramasinghe, J. B. Silvers, Elliot B. Sloane, Michael J. McCoy, and Jonathan L. Schaffer**

## 1 Introduction

Healthcare delivery in the USA is moving to a value-based system (Lynn 2016; Zuvekas and Cohen 2016; Weeks et al. 2013; Bozic 2017), where key components include bundled payments and well-defined care paths that dictate care processes (Weeks et al. 2013; Bozic 2017; Kaplan and Porter 2011; Porter and Teisberg 2009; Porter and Lee 2013; Porter 2010; Betbeze 2016; Bracy 2016). This value-based model has been enabled by the introduction, adoption and increased penetration of electronic health records (EHRs) into the care process providing the necessary data and information to facilitate the design of appropriate care paths and calculation of bundled payments for specific care processes (Hero et al. 2016; Porter and Lee 2013). On the surface, this appears to be a logical and rational approach to stem consistently increasing costs of healthcare delivery in the USA (Porter and Teisberg 2009; Porter and Lee 2013). On deeper examination, however, there appears to be

N. Wickramasinghe (✉)
Epworth HealthCare, Richmond, VIC, Australia

Swinburne University of Technology, Hawthorn, VIC, Australia

J. B. Silvers
Case Western Reserve University, Cleveland, OH, USA

E. B. Sloane
Villanova University, Villanova, PA, USA

M. J. McCoy
Physician Technology Services, Lawrenceville, GA, USA

J. L. Schaffer
Deakin University, Geelong, VIC, Australia

Cleveland Clinic, Cleveland, OH, USA



N. Wickramasinghe, F. Bodendorf (eds.), *Delivering Superior Health and Wellness Management with IoT and Analytics*, Healthcare Delivery in the Information Age, https://doi.org/10.1007/978-3-030-17347-0_20

a serious and troubling potential for unintended consequences leading to the failure of systems to intervene to care for the patient (Weeks et al. 2013; Pearlstein 2005; Economists View 2005; Bracy 2016).

## 2 What Is the Issue?

Over the last 10 years in the USA, the adoption, diffusion and use of EHRs have increased significantly, assisted largely by federal policies and initiatives, including President George W. Bush's 2004 executive order enabling e-health and the subsequent ARRA/HITECH legislation and Meaningful Use programme (Hsu et al. 2005). This rapid uptake has led to the availability and use of significant amounts of captured and generated clinical data. These data are being used by clinicians to provide appropriate care to their patients, which are in turn anticipated to lead to optimal outcomes while minimizing complications and adverse events. However, these data are also being used to scrutinize cost and quality and calculate risk assessments at a specific point in time (Lynn 2016; Zuvekas and Cohen 2016; Porter and Lee 2013; Bracy 2016). In the context of a healthcare system that is facing substantially increasing costs, many encourage the move from a focus on volume to value (Kaplan and Porter 2011; Porter and Teisberg 2009; Porter and Lee 2013; Porter 2010; Betbeze 2016). This shift in focus is defining appropriate care paths and narrow definitions of risk and then applying them broadly to patients presenting for specific treatments, such as a knee arthroplasty (Porter 2010; Betbeze 2016; Bracy 2016). These decisions are often relegated to sole specialists rather than a clinical team dealing with an integrated holistic view of the patient. Taken together, however, this approach may translate into patients being told that their proposed knee surgery is not appropriate at this time and should be delayed until such time it is safe such as when their haemoglobin A1c falls within a specific acceptable range or when their body mass index is below a specific level or they have completed smoking cessation. Many care paths and value-based care systems strongly recommend to the point of dictating care parameters such that patients who smoke or have elevated BMI or HbA1C may not get the elective procedure such as knee arthroplasty when it would alleviate their pain and improve their function. Moreover, the physicians who would recommend surgery are often penalized or not provided with maximal payment for patients who have complications from one of these "out of range" parameters. Patients do not always have the luxury or privilege of deciding which disease or diagnosis it is that challenges their health or is reimbursable. This conundrum may serve to deny the surgery they may need or directly address the issue for which they present, or may not necessarily provide them with the treatment for their abnormal lab values, or allow a satisfactory solution for the pain control they may need. Further, it may also not provide patients and/or surgeons with any sense of urgency to perform the surgery in a timely manner. There is a significant failure of the healthcare system to intervene. In many situations, the patients are left hanging and not able to mitigate their risks.

Admittedly, a knee surgeon is not the one to deal with the higher haemoglobin A1c, and isn't compensated for any coordination efforts, but the reality is the patient now must face new challenges in navigating the healthcare delivery system. This is leading to what is best described as a new digital divide which is really a digital disparity that is exacerbating any pre-existing conditions (Hero et al. 2016; Yamin et al. 2001; Sarkar et al. 2011; Chang et al. 2004).

## 3 Background

Increasing costs, an aging population and subsequent rise of chronic disease have been plaguing the healthcare delivery system for decades, rendering unsustainable the traditional fee for service system (Kaplan and Porter 2011; Moghimi et al. 2013; Haddad et al. 2015). In the fee for service model, financial incentives tend to reward providers for the quantity and complexity of services, regardless of outcome (Zuvekas and Cohen 2016). Since the 1990s, the USA and other countries shifted to diagnostic-related group (DRG) reimbursements in an attempt to limit payments based on each patient's most serious diagnosis, but other unintended consequences emerged, such as treating patients' illnesses sequentially rather than simultaneously, manipulating diagnostic codes to reach the highest-reimbursed DRG categories regardless of evidence-based scientific guidelines and/or providing treatments without regard for subsequent complications or actual outcomes (Zuvekas and Cohen 2016).

In 2016, CMS has begun a shift in payment from volume to value, aligning 85% of all payments it makes linked to quality or value by the end of this calendar year, with 30% of payments tied to quality or value through alternate payment models (APM). In 2016, the USA began shifting to "value-based reimbursement" in the MACRA[1] and MIPS[2] programmes with the goals to overcome waste, benchmark all providers' Medicare payments and adjust them based on comparison with peers' clinical outcomes for the same procedures. In addition, the US government is shifting to fixed-price "bundling" for many known treatments, e.g. knee and hip replacements (Bracy 2016; Bumpass and Samora 2016). These reimbursement changes are intended to guide providers towards evidence-based medical decisions and procedures. Such evidence-based processes are supposed to be unbiased and scientifically optimized formulations that describe the best way to treat patients for the best overall outcomes.

The US Veterans Administration continuously publishes and updates a compendium of best practice guidelines for many chronic diseases.[3] Most hospitals and payers either adopt or create similar guidelines for their own decisions, and those

---

[1]**M**edicare **A**ccess and **C**HIP **Rea**uthorization **A**ct of 2015.

[2]**M**erit-based **I**ncentive **P**ayment **S**ystem.

[3]http://www.healthquality.va.gov/guidelines/

formulas are typically embedded in hospital/physician EHRs and financials software to reinforce "positive" behaviours, compliance and, ultimately, outcomes.

Any clinical "centre of excellence" (CoE) will naturally apply or customize the best evidence-based guidelines with its own expertise, with the expectation and intention that such decisions will result in their outcomes and reimbursement being among the most highly ranked, respected and, hopefully, profitable. Hospitals or physicians who deliver care at lower levels may have to settle for average outcomes reimbursement, and poor performers may be motivated to abandon services they cannot deliver well.

An unintended outcome that can – and does – emerge, however, is that many centres including CoEs will very naturally, and quite scientifically correctly, decline to provide clinical services to patients who have underlying illnesses that are statistically proven to have poor prognosis and outcomes. While that strategy will help push the facilities outcome statistics to the top of their fields, patients with underlying chronic illnesses may be left with few if any alternatives, including, perhaps, doing without, simply opting for care from a US provider with a mediocre or poor outcome, or self-payment for similar services from a foreign medical tourism provider. In some cases, there is a failure to intervene and mitigate the risks that are preventing the elective procedure that may reduce the patient's pain or improve their function and quality of life.

Many scholars, notably Porter (Kaplan and Porter 2011; Porter and Teisberg 2009; Porter and Lee 2013; Porter 2010), have proffered that the only appropriate strategy is to focus on value not volume. Porter and colleagues contend that a key contributor to the current state of escalating costs is that healthcare delivery has been incorrectly defined (ibid). Specifically, they proffer a model focused on value creation around the customer (Kaplan and Porter 2011; Porter and Teisberg 2009), i.e. the patient. To provide the patient with the appropriate value, they further identify six key enablers: (1) organizing around an integrated practice unit, (2) measuring outcomes and costs for every patient, (3) moving to a bundled payment system for care cycles, (4) integrating care delivery across separate facilities, (5) expanding excellent services geographically and (6) having an enabling IT platform (Kaplan and Porter 2011; Porter and Teisberg 2009; Porter and Lee 2013; Porter 2010).

A major contributor to this has been the general embracement of EHR, which, in turn, enables generation of significant data now being used by healthcare administrators and clinicians to analyse care delivery and then design and develop appropriate care delivery paths to standardize care delivery, streamline processes and ensure best practice patterns ensue. Connected with this is the ability to use the generated data to identify key risk factors (Porter and Lee 2013) (e.g. BMI, haemoglobin A1C and smoking) developed very narrowly but may be applied more generally and broadly.

Existing quality measures also look at the care (and cost) provided but fail to identify those who are denied services or care. How can one take part in a patient-reported outcomes survey if denied the opportunity to have the surgery or care? How can the failure of inclusion of an individual in overall statistics be measured to ensure that appropriate exclusions have been applied? These questions address the access to care that is as important in vulnerable populations as the care delivery.

## 4 The Conundrum

The proposed value approach as outlined by Porter (Kaplan and Porter 2011; Porter and Teisberg 2009; Porter and Lee 2013; Porter 2010), while serving to address some issues that have been shown to fuel increasing healthcare costs, neglects to address a key aspect: access to care. Simply stated, if one only focuses on outcomes, quality, cost and compliance with "evidence-based guidelines" to increase value, and defines risk factors very narrowly (e.g. smoking, BMI and HA1C) but applies those risk factors broadly, what we will see is "cherry picking" (Weeks et al. 2013; Economists View 2005); i.e. sicker patients will be denied access to essential care they require.[4] In the long run, we might expect an increase in total healthcare costs as patients who were denied services and did not have a health issue addressed may ultimately re-present themselves in the emergency room incurring a higher cost of care.

We suggest that consideration of *value* with a focus on *quality* and *cost* only is not sufficient even if it is necessary. In particular, it is essential to consider value in terms of *quality*, *cost* and *access*. By not addressing individual access, we ignore a critical component of healthcare; it is a service that should be available to all members of a community. In addition, we also find that the discussion of value to date has failed to include a consideration of the opportunity cost of *not* receiving care. Should a patient not be immediately eligible for an elective procedure, then the factors responsible should be mitigated by a coordinated multidisciplinary care team with appropriately aligned incentives.

## 5 The Solution Set

The current approaches to value typically are based on a simplistic formula, $V = Q/C$, which may be easy to understand, but grossly misrepresents what value (V) is in the real world. Value is not linear where more quality (Q) or lower cost (C) results directly in equivalent units of value, and it omits the critical indicator of access (A) – the widely recognized third leg of health policy.

We must recognize all three indicators and at least consider the non-linear nature of how value changes in healthcare with each. To be general, the formula should be $V = \mathrm{f}(Q, C, A)$. To be simplistic again, one might use $V = QxA/C$ although this begs the question of how the combination of these variables and the metrics used to

[4]For example the VA ones can be found at http://www.healthquality.va.gov/guidelines/

represent them may actually result in value. The relationship is illustrated (Fig. 1) in a more generalized way as the impact of different approaches to try to increase quality, reduce cost and thereby increase value. If we look at point A in Fig. 1, what managed care did was to create strategies that targeted cost structures in an attempt to lower cost and simultaneously increase value. The result, though, also meant that quality was not improved. Thus, to address quality, various quality initiatives have been introduced so that it is possible to move to point C, but in doing so cost may be increased. The most recent initiatives of bundled payments, for instance, attempt to increase value by increasing quality and simultaneously reducing cost, i.e. moving to point D. However, in doing this without considering the other important dimension of access means rather than moving to D in Fig. 1, we actually may move to point $D_L$ in Fig. 2, i.e. lower value due to the opportunity cost of limited or no access to care.

Another problem we believe that will arise with the bundled payment approach is that translating any set of product attributes for a service or treatment into value is not the same for each recipient. In the commercial world, this variation is accepted as difference in taste or need which vary among individuals. Thus, markets are segmented. In health insurance, we recognize that some may be willing to bear more risk and will buy bronze plans, while others desire more complete coverage and are willing to pay much more for platinum. Concierge service also recognizes the segment of patients who are willing to pay for special coverage. The point is that value is contextual not absolute. It is only in a competitive market where many buyers and sellers establish an equilibrium price that we can have any degree of consensus about what constitutes marginal value – and that is meaningful only for the last set of purchasers and sellers. Economists translate this into "willingness to

**Fig. 1** Policy actions and anticipated results

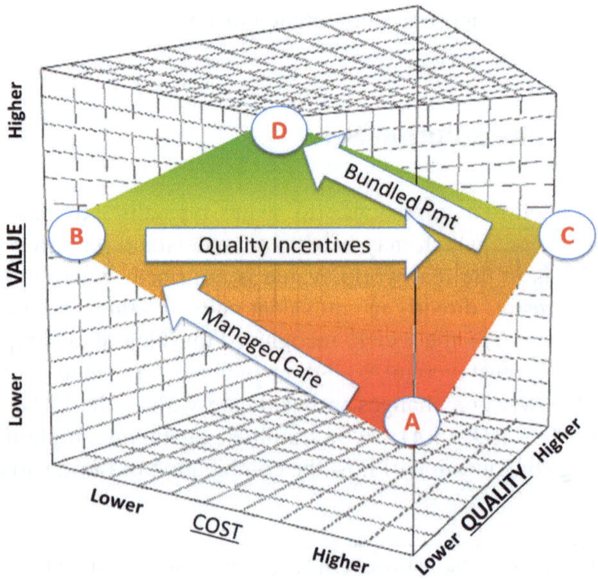

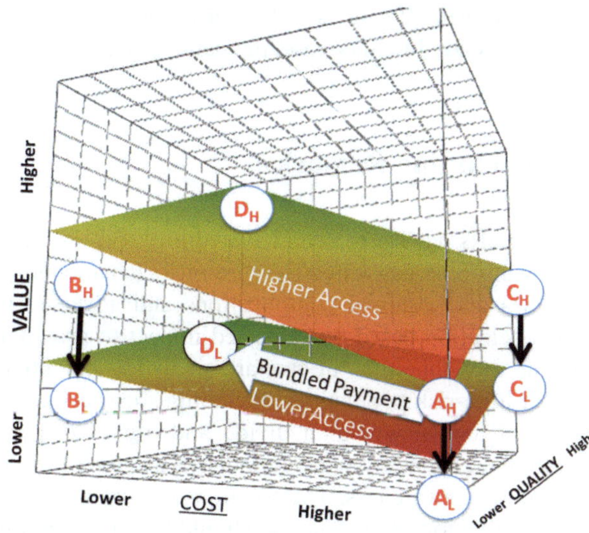


**Fig. 2** Policy actions and likely results with lower access to marginal patients

pay" as a measure of the value of a year of life or the worth of a safety device. Yet we are loath to take this variation in value too far in healthcare in the name of equity. The converse is the high deductible health plan that will pay for surgery such as a knee arthroplasty but effectively denies appropriate postoperative care by imposing an exorbitant co-payment for postoperative physical therapy. Many surgeons and many patients are not aware preoperatively of this potential postoperative trap and do not have an opportunity to mitigate the issues and may not have sufficient services to achieve an optimal postoperative outcome.

Thus, value-based payment and the loss of access discussed above makes the assignment of one measure of the worth of a procedure for all people very questionable. While the value of a knee replacement for the overweight diabetic in the initial scenario may be less, it clearly is not zero – unless the patient is denied care. The resulting loss of access has a very real cost, albeit an indirect "opportunity cost" rather than a direct cost of care. Figure 2 illustrates this loss of value by depicting two possible scenarios: one lower access and the other high access and for each point, respectively, $A_H$ and $A_L$, $B_H$ and $B_L$, $C_H$ and $C_L$ and $D_H$ and $D_L$. The difference in value between the levels is a very real cost to the patient who would have benefited from the new knee even if the "quality" as indicated by the provider's overall outcome metric would have suggested a lower value given the likelihood of a less favourable outcome.

A further nuance is that access has another dimension – timeliness. Care that would be of a given value if given when first recommended declines in worth if delayed due to these concerns. Ultimately with many patients, the delay is accompanied by further deterioration in the underlying conditions until the procedure cannot be done at all. Thus, one could consider Fig. 2 to be representative of the timeliness of care as well as direct access, i.e. the opportunity cost for loss of immediate care. In either case, the patient loses.

## 6 Concluding Remarks

Use of EHRs has led to actual availability and use of data by clinicians to provide high-quality and cost-effective care of their patients, hopefully leading to optimal outcomes and minimal complications and adverse events. Specific examples of patient selection based on projected impact to provider quality and outcomes that can lead to withholding of care are cited. The potential for denial or diminution of care for high-risk populations, including socio-economic and marginalized groups, may lead to a new digital divide and digital disparity if not accounted for in operationalizing new payment models. The goal of this paper has been to outline the issue to raise awareness. The solutions require awareness and definitions including the identification of criteria that should be considered in an attempt to mitigate the dilemma.

While increasing the value provided in the delivery of healthcare, the preceding discussion is intended to call attention to logical flaws and serious concerns regarding the new approach of focusing solely on value as currently promulgated by many advocates. Specifically, we have identified that the often used definition of risks is very narrowly defined but extremely broadly and generally applied. Further, we highlight that no consideration to opportunity costs or access has been given. The likely consequence of the current move from volume to value, designed to stem escalating healthcare costs, may instead paradoxically increase healthcare costs, if those denied service currently due to high-risk factors do not have their health issue resolved and re-present later at the emergency room. Further, by not addressing access issues, a healthcare chasm between the "haves" and the "have nots" may only widen the chasm and create a substantially larger digital divide.

We thus urge meaningful debate and discourse to develop more appropriate and successful solutions, which not only address the healthcare cost issue but are also patient-centred, socially responsible and equitable.

## References

Betbeze, P. (2016). Get comfortable with bundles. www.healthleadersmedia.com. Accessed 040416.

Bozic, K. (2017). Value-based healthcare: A novel transitional care service strives to improve patient experience and outcomes. *Clinical Orthopaedics and Related Research., 475*(11), 2638–2642. https://link.springer.com/article/10.1007/s11999-017-5481-8.

Bracy, C. (2016). The healthcare value creation equation. http://www.aaos.org/AAOSNow/2016/Apr/Managing/managing5/. Accessed Apr 1 2016.

Bumpass, D., & Samora, J. (2016). Price transparency in health care: The other half of the value equation. http://www.aaos.org/AAOSNow/2015/Jan/advocay/advocay1/?ssopc=1. Accessed Apr 11 2016.
Chang, B., Bakken, S., Brown, S., Houston, T., Kreps, G., Kukafka, R., Sfran, C., & Stavri, Z. (2004). Bridging the digital divide: Reaching vulnerable populations. *White Paper JAMIA, 11*, 448–457.
Economists View (2005). Cherry picking profitable operations from general service hospitals.
Haddad, P., Schaffer, J. L., & Wickramasinghe, N. (2015). Evaluating Business Value of IT in Healthcare: Three Clinical Practices from Australia and the US. In MEDINFO 2015 August 19–23 Sao Paulo, Brazil: EHealth-enabled Health: Proceedings of the 15th World Congress on Health and Biomedical Informatics (Vol. 216, p. 183). IOS Press. 19–23 August 2015.
Hero, J., Blendon, R., Zaslavsky, A., & Campbell, A. (2016). Understanding what makes Americans Dissatisfied with their health care system: An international comparison. *Health Affairs, 35*(3), 502–509.
Hsu, J., Huang, J., Kinsman, J., Fireman, B., Miller, R., Selby, J., & Ortiz, E. (2005). Use of e-health services between 1999 and 2002 a growing digital divide. *JAMIA, 12*, 164–171.
Kaplan, R., & Porter, E. (2011). How to solve the cost crisis in health care. *Harvard Business Review, 89*(9), 46–52. Reprint R1109B.
Lynn, J. (2016). Value based care hurting most vulnerable hospitals. http://www.hospitalemrandehr.com/2016/03/25/value-based-care hurting-most-vulnerable-hospitals/
Moghimi, H., de Steiger, R., Schaffer, J., & Wickramasinghe, N. (2013). The benefits of adopting e performance management techniques. *Health & Technology, 3*(3), 237–247. ISSN: 2190–7188.
Pearlstein, S. Free Market Philosophy doesn't always work for health care. *Washington Post*, June 8 2005.
Porter, M. (2010). What is value in health care? *The New England Journal of Medicine, 363*(26), 2477–2481.
Porter, M., & Lee, T. (2013). The strategy that will fix health care. *Harvard Business Review, 91*, 50–70. Reprint R1310B.
Porter, M., & Teisberg, E. (2009). *Redefining health care*. Boston: Harvard Business Review Press.
Sarkar, U., Karter, A., Liu, J., Adler, E., Nguyen, R., Lopez, A., & Schilling, D. J. (2011). Social disparities in internet patient portal use in diabetes: Evidence that the digital divide extends beyond access. *Journal of the American Medical Informatics Association, 18*, 318–321.
Weeks, W., Rauh, S., Wadsworth, E., & Weinstein, H. (2013). The unintended consequences of bundled payments. *Annals of Internal Medicine, 158*, 62–64.
Yamin, C., Emani, S., Williams, D., Lipsitz, S., Karson, A., Wald, J., & Bates, D. (2001). The digital divide in adoption and use of a personal health record. *Archives of Internal Medicine, 171*(6 Mar 28), 568–574.
Zuvekas, S., & Cohen, J. (2016). Fee-for-service, while much maligned, remains the dominant payment method for physician visits. *Health Affairs, 35*(3), 411–414.

# The Enabling Role for Technology in the Support of Care Coordination in Health Care

**Rima Gibbings and Nilmini Wickramasinghe**

## 1 Introduction

Current advancements in technology in the healthcare sector provide promising tools for the improvement of care coordination. As such, there has been an increased awareness and focus on patient-centered care. Technology in this sector contributes greatly to the improvement of the overall quality of the sector (Tricco et al. 2014). However, although ICT in the healthcare provides positive hope for the numerous stakeholders in this sector, its implementation is still difficult and normally involves changes at different levels. These include the patients, healthcare providers as well as healthcare organizations. This paper seeks to provide a systematic literature review as pertained to how Information Technology facilitates care coordination in healthcare.

Care coordination refers to the process that ensures that patients' health services and information sharing requirements are met in the most effective manner. It is therefore a critical component that emphasizes on continuity and accountability on the part of the caregivers. Moreover, care coordination calls for effective collaboration between the various organizations and stakeholders taking care of each individual patient.

Much of the existing evidence suggests that healthcare is still uncoordinated. For instance, a look at the United States shows that care is uncoordinated although other pieces of evidence tend to suggest that quality improvement strategies within

R. Gibbings
University of Illinois Chicago (UIC), Chicago, IL, USA

N. Wickramasinghe (✉)
Epworth HealthCare, Richmond, VIC, Australia

Swinburne University of Technology, Hawthorn, VIC, Australia

 
N. Wickramasinghe, F. Bodendorf (eds.), *Delivering Superior Health and Wellness Management with IoT and Analytics*, Healthcare Delivery in the Information Age, https://doi.org/10.1007/978-3-030-17347-0_21

the sector can improve performance (Davis et al. 2015). In the United States, the average Medicare beneficiary has seen many different providers in a given year and network characteristic. It is worth noting that the Health Information Technology for Economic and Clinical Health (HITECH) Act has resulted in the rapid adoption of electronic health records in many hospitals, nursing homes, post-acute settings, and home care which were not eligible for the HITECH incentives. Moreover, the information that was vital for the improvement of the situation (clinical data exchange) has stayed behind in the adoption of the electronic health records.

A review by Ovretveit that was published in 2011 in Sweden contains a summary of the evidence that determines whether clinical coordination can help improve quality and in turn help in saving money. Some of the conclusions and deductions seem useful to clinicians and the healthcare systems. They include:

- Some of the coordination improvements will tend to reduce wastage in the sector and in turn improve the quality. However, the fact that this will save money or allow increased income will mainly depend on the commitment of healthcare providers. Additionally, this review concluded that the most cost-effective approaches are those that use reliable data in the identification of patients that are at high risk of deterioration. The best intervention method (coordinated care) is then employed to the specific patients.
- Changes in the payment systems, professional education, regulations, and codes of practice are needed. Additionally, savings will depend on the implementation of any improvements and their timescale.

  Kane et al. completed a technical report that touched on this topic for the United Agency for Healthcare Research and Quality. The main objectives of this project included developing a working definition of care coordination, conduct a systemic review of care coordination through 2006 as well as identifying the theoretical frameworks that could end up explaining how care coordination mechanisms are influenced by various factors in the healthcare setting. Additionally, this relates to how to connect to the patient outcomes and healthcare costs. In this, more than 40 definitions were touched on. As such, the one that was arrived at stated that care coordination is the deliberate organization of the patient care activities between one or more providers of healthcare in the effort to facilitate efficient service delivery.

The AHRQ report, on the other hand, identified 75 good systemic reviews. Out of these, 43 had their full focus on care coordination. However, only half covered explicitly the multiple settings. To that effect, the authors were able to come up with the following conclusions:

- A gap exists for the need of consensus definitions and conceptual models. However, they were able to identify four frameworks that were believed to be helpful in the development and study of coordination intervention factors. As such, a wide array of factors have been tested in the care coordination rubric. However, the evidence about the key intervention components is lacking. Moreover, the strongest evidence only shows benefits for patients that have

ailments such as diabetes, heart failure, a recent stroke or even depression. Through this, the evidence is not clear for the other patient populations that will obviously have coordination wants. These include the patients that have complex medical issues, the elderly that live independently and the physically disabled people.
- Another conclusion was that there was insufficient evidence that is supposed to aid in the drawing of the conclusions especially on the costs that are associated with the care coordination interventions.

It is evident that the two reports almost have the same information and recommendations. However, they seem to be specifically aimed at the policymakers and researchers. On the models of care coordination, Gensichen et al. (2006) conducted an interview in which 23 family doctors were involved in a study of practice-based healthcare assistants for patients that had major depression in the country of Germany. As such, they perceived it to be beneficial largely for patient care. However, they involved different approaches to the implementation of the model. This is a variation that was not focused on but it could have been more beneficial for future implementers. Additionally, Hall et al. conducted a modeling exercise for the shared coordinated care of cancer survivors between their consultants and the Gps. In this, it was found that patients perceived potential value in such kind of arrangements. However, the patients wanted a reassurance that their Gps had received extra training and that they were supported fully by the specialists.

Still on the care coordination, Melby et al. illustrate on the unintended consequences of health information technology to improve care coordination (Steichen and Gregg 2015). As such, the previously reported effects of health information technology were mainly negative. However, Melby aimed at illustrating on some of the unintended consequences of health information technology. It is specific on the general practitioners and the home care nurses. For instance, the use of electronic messaging between the home care nurses and the general practitioners allowed for more and efficient inter-professional communication. This is the intended consequence. However, electronic messaging reduced face-to-face communication and consequently the interpersonal relationships were discouraged. This is the unintended negative effect. Additionally, the documentation of the nurses' messages in the electronic health records ended up giving weight to the requests that the nurses made and in turn enhanced their relationship with the general practitioners. This is an unintended positive consequence. In this study, it is evident that this study underscores that organizational changes are more important to the outcome and the acceptance of the health information technology tools. Since many of these changes are not able to be anticipated, the evaluation of the impact of the health information technology tools must at least be evaluated from the onset to allow corrective actions.

Stephan et al. studied how medical informatics can be vital in the study of patient-centered care coordination (Stephan et al. 2015). The authors of this paper screened a large number of nurse notes using natural language processing techniques in the identification of the activities that are associated with care coordination. They organized the activities into an ontology which then was used to quantify the global amount and the type of care coordination per patient (these were mainly communication and management activities). On the other hand, other authors parsed event logs in gynecologic oncology departments to picture what was done to patients and in which order (Caron et al. 2014). As such the analysis of this low-level administrative data by these researchers were able to generate interesting insights and conclusions into prototypic or sometimes problematic patient pathways.

There are numerous other works that were published in 2014 that are worth noting. An example involves the direct association of patients in the communication and generation of their personal health records. This helps increase the level of patient engagement and in turn gives appropriate and timely information to the caregivers. In order to illustrate this effect, there are two studies that support this. The first one involved parents that were offered the possibility to interact with an automated voice response system in the input of data in the EHR of their child or to express concerns to be addressed in the upcoming visit (Adams et al. 2014). On the other hand, the second study involved patients from the Department of Veterans Affairs who used the opportunity to download a comprehensive extract of their HER. These were able to better understand their own medical history and to share it efficiently with caregivers outside the various organizations that offer care coordination (Turvey et al. 2014). However, there is still no adequate evidence that supports these beneficial effects of patient involvement that in turn translates into better health outcomes. However, it is worth noting that patient satisfaction may be a better measure of patient-centeredness than health outcomes (Scholl et al. 2014).

It is also vital to note that medical homes represent an important effort to try to better manage and coordinate care for populations. A study by O'Malley et al. explored how electronic records were working in medical homes. This was with respect to care coordination in the healthcare sector (O'malley et al. 2010). The team that was involved in the study identified four especially important domains for electronic health records in care coordination. These are reconciling medications, tracking laboratory tests, communicating across settings, and mediating care plans between one discipline and another. Additionally, the management of referrals, consultations, and care transitions were also important in the study. In present times, electronic health records tend to have substantial deficiencies in all these sectors.

Another study was conducted on the national evaluation of how health information technology was being used in the support care transitions across settings (Samal et al. 2013). Some of the expectations, in this case, were to find some organizations that had sophisticated tools for the management of the health of their patients. In this, it was evident that some organizations appeared to be performing better than others, but in every instance, this was occurring largely with human involvement and without numerous technological tools. However, it was found that some health information technology tools were being used within one healthcare setting in the preparation for the transition of care.

In care coordination, it is clear that there is a need to build and test interventions to improve this sector. An example of an intervention that targets this area is a toolkit that was built for the PROSPECT trial (Bates 2015). This intervention attempted to improve outcomes in intensive care and also in general care units. Some of the tools built include a patient-centered toolkit that the patient or care partner can access. Additionally, it includes information about the plan of care and goals of care visible to the entire care team. Moreover, a microblog function was developed so that anyone on the team including the patient or the care partner is able to ask questions, responses can be made, and additionally everyone in the care team is able to see what the responses of the others have been (Bates 2015).

Morton et al. (2015) state that patient-centered medical home practices still depend on nontechnological methods in the management of care coordination. As such, the use of computerized systems in the support of healthcare and specifically on care coordination was limited. It was not consistent with the stakeholders' priorities. Many state that the computerized systems do not support the care coordination activities that they value most. However, it was noted that in this study, assistance or participation of consultants in a collaborative setting had a positive impact on the overall performance of care coordination activities in the health sector. In order to improve the uptake of health IT for care coordination that in turn meets the clinician's needs, the various practices will need technical assistance in redesigning workflows. Additionally, this will work toward ensuring that technological capabilities are enhanced. For the other activities that are not valued as much by the clinicians, greater technical knowledge will be needed. Moreover, education on the same ought to be passed down to all the stakeholders so that they are able to identify efficiently the many benefits of having a well-framed care coordination. The technical assistance can also work toward ensuring that there is increased use of change management strategies. This will influence positively both the level of care coordination activity and the electronic support for the activity.

A factor that Morton et al. (2015) considered to be a remedy, in this case, was to have a non-clinician being responsible for care coordination. However, dedicating staff to care coordination will end up requiring additional resources for the numerous practices. Clinicians who are concerned about their practices' financial situation should work in practices that perform fewer care coordination activities. Additionally, this study identified some of the barriers to the implementation of care coordination activities. It showed that time and money were the greatest barriers to coordinating patient care with the other practices. However, reimbursement approaches that usually support non-visit care are vital in the improvement of care coordination.

Rudin and Bates identified four areas in their study in a care coordination framework that would likely benefit from the information technology solutions in the sector. These include the ability to identify a patient's care team both within and across settings or disciplines (this includes primary care providers, specialists, social worker, case managers, and substance use counselors), the ability to collaborate with different care teams in a quick and efficient manner, the ability to collaborate in the sharing and formulation of care plans, and the ability to monitor and track task responsibilities in care coordination. However, it is worth noting that electronic health record (EHR) systems still lag in several areas that include information exchange across settings, development, storing, and sharing of major care plans, tracking of referrals and improvement of team-based communication.

Claborn et al. (2017) stated that mobile health technology presents a potential platform for efficient communication among the care coordination providers. Additionally, it will come in handy in the tracking of patient service utilization and treatment approaches. Moreover, this has the potential of improving team-based approaches that are consistent with the collaborative models. This will work toward promoting more effective care in hard-to-retain populations. For instance, in the fight against HIV, there are currently over 200 applications. A survey conducted in 2013 showed that only 55 of these applications promoted HIV prevention and care services. However, a majority of these apps only focused on providing disease-specific educational information. On the other hand, there were no identified applications that focused on care coordination for the healthcare providers.

Lindberg et al. (2013) state that a wide variety of terms were used in the reviewed studies in the definition of ICT. The most used definitions, in this case, were telehealth and telemedicine. This is similar to Koch's review of the current state and future trends in the home telehealth sector. In this case, the term telehealth has mainly been defined as the use of telecommunication and information technologies in the provision of healthcare to individuals at a geographical distance. Additionally, telehealth will involve a wide variety of specific modalities that will include telephone-based interactions, internet-based information, live imaging, personal digital assistants, and interactive audio-video communication or television (Sato et al. 2009). Moreover, eHealth is basically defined as the overall umbrella that includes ICT and telehealth. In this, it combines the use of electronic communication and information technology in healthcare. Consequently, this may explain the results of this review with many different terms used to define the technology that is in turn linked to care coordination.

Lindberg et al. (2013) also describe how ICT is used for communication in home care, and a key result found was that the most frequent type of communication was between patients and healthcare professionals. It, therefore, means that user focus needs to be shifting from tools for professionals to tools for patients and family members. The review by Lindberg et al. is in accordance with Koch, describing trends toward tools and services not only for professionals but also for patients and citizens. However, from a nursing perspective, it would be ideal to note that there is lack of knowledge about how to use ICT solutions to meet the needs of numerous groups of patients such as those with chronic illnesses. Specifically, by performing qualitative studies, people's needs related to living with chronic illness can be elucidated. A major challenge in home care will be to use existing ICT tools in order to meet the caring needs of people with chronic illness but based on the specific experiences.

Lindberg et al. add that ICT applications used in home care should ensure that they take into consideration the roles that technology should play in the use of patient and healthcare professionals. This is more specifically in care coordination between the various stakeholders. Neglecting this aspect may lead to technology that not provides the needed support for communication. As such, it leads to a scenario in which ICT does not facilitate efficiency in care coordination and in turn maims the healthcare of individual patients.

Owens states that ICT in the healthcare sector should aim at providing proper care coordination amid the shift toward value-based care. As such, value-based care creates a new paradigm, one that ensures that care is no longer delivered by only doctors and nurses but by a community of providers that focus on the "whole" patient rather than just treating the disease. Moreover, with the advancements in technology in the healthcare sector, care coordination has been moving on the positive as concepts such as communication are not confined to the four walls of a doctor's room. Through efficient integration of the different technological tools in care coordination, value-based healthcare should be achieved in both the short and long runs.

However, care coordination has been coupled with numerous challenges and barriers. Carayon et al. state that some of these barriers to effective care coordination include poor communication and tracking patient populations. In this, most healthcare frameworks globally are disjointed. As such, most patients will never comprehend why they are being referred. Others will lack the knowledge on how to create appointments. On the other hand, specialists from higher levels lack adequate information as to why patients are being referred to them. Consequently, the primary care providers end up not getting feedback from the specialists. This creates confusion in care coordination.

Additionally, although digitalization of healthcare has proved to be beneficial to the sector, it is not yet sufficient for care coordination. Seroussi et al. state that creating information availability does not necessarily mean that the information will be accessed. This is the case with electronic health records. They have made operations easier in the healthcare sector. However, not many specialists will take the task of accessing the EHRs. Moreover, the multiple IT systems that have been created do not talk to each other. This creates a scenario where information is present but is not synchronized to best serve the interests of the stakeholders such as patients. At the end of the day, the healthcare sector remains stagnated.

## 1.1 Barriers to Care Coordination Table

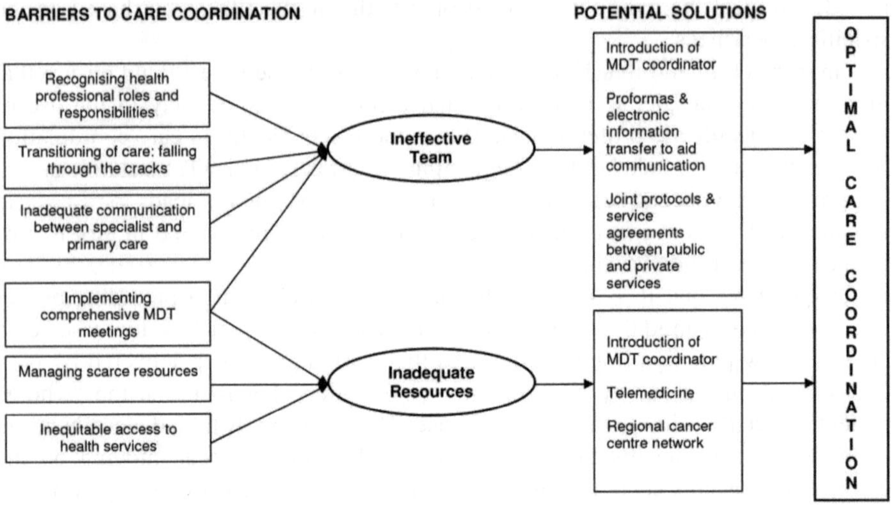


Source: https://media.springernature.com/full/springer-static/image/art%3A10.1186%2F1472-6963-10-132/MediaObjects/12913_2009_Article_1269_Fig1_HTML.jpg

In the above table, it is evident that the barriers to care coordination are mainly linked to inadequate resources and ineffective teams. These are exhaustively discussed below:

- Ineffective team—In this, different teams do not recognize their roles and responsibilities. As a result, care coordination is not conducted in the most efficient manner. In addition, there is a lack of adequate communication between the various stakeholder teams. Without communication, it becomes difficult to push forward the agenda of care coordination. Moreover, the transitioning of care becomes hard and limited. This is the reason communication becomes inefficient. These could be solved through collaborations between the various stakeholders, the use of electronic information transfer to aid in the effective communication between different parties.
- Inadequate resources—In this, inadequate resources lead to a scenario in which care coordination is unable to be implemented. A perfect framework might be present but without the obliged resources, this concept cannot be adequately pushed forward. It is important to note that in some instances, managers are faced with the difficulty of managing scarce resources. Moreover, the lack of well-framed meetings between different stakeholders leads to a situation in which funds can be embezzled or mismanaged. Through this, care coordination becomes difficult to propagate positively. The possible remedies to deal with these challenges include the introduction of telemedicine in this sector which will enable efficient management of the available resources. Additionally, each

stakeholder should be accountable in their areas of play. If each performs their part well, the goals of care coordination will be highly achieved.

## *1.2 Review on Value-Based Care*

It is worth noting that as the healthcare sector propels toward the transition toward value-based care, the various discrepancies in technology advancements (telemedicine) continue to deter the progress. As such, it is vital to align the available IT frameworks for value-based care and specifically on coordinated care. Currently, most clinicians feel that reporting regulations have changed roles from patient care to data entry. In this, it would be ideal to focus on the physician well-being, industry initiative process, as well as making sure that patient care is steadfast. Focusing on these three pillars will ensure that telemedicine is able to create a link with the current health frameworks globally and this will facilitate value-based care. Additionally, it will aid in a well-driven coordinated care that in the long run enhances value-based care.

The value-based care seeks to transform the healthcare sector since the model mainly concentrates on the wellness of the patients first as opposed to the traditional fee-for-service models that were paid for the amount of services they offered. This had led to a scenario where the healthcare providers could align numerous tests (some unnecessary to the clients) with the aim of getting more money at the expense of the patients' wellness. However, value-based care is based on real-time data that demonstrates improvement in the quality of the healthcare sector. Evidence-based methods are largely applied in the value-based models, and this helps in fast achievement of healthcare objectives such as healthy populations globally. Some models that push forward the value-based healthcare include bundled payments, accountable care organizations, and patient-centered medical homes (Friedman et al. 2016).

## *1.3 Impact of Electronic Health Records and Their Role in Care Coordination*

Globally, the number of people living with at least one chronic condition is increasing. For instance, taking a case of the United States, almost half of all Americans had at least one chronic condition in 2005 (Kvedar et al. 2014). Additionally, one person in four had multiple chronic conditions. This indicates that these patients will tend to visit multiple healthcare providers yearly and will have numerous prescription drugs. As a result, care for these kinds of patients is becoming complex and requires high levels of coordination so as to ensure quality healthcare is provided. Other patients in the healthcare sector are too in need of a well-coordinated framework. In this, the use of the electronic health records System is a vital concept in facilitating the transfer of information. Additionally, it is due to improve the coordination of patient care across different healthcare providers. The EHR systems have constantly been promoted as a policy priority in many nations that could improve the quality and efficiency of the various healthcare sectors.

Current evidence tends to indicate that healthcare providers will rarely have access to complete medical information especially when patient care is transferred from one provider to another. It is during such transitions in healthcare that the safety of the patient is put at stake (Krist et al. 2014). Moreover, the lack of timely information in the current healthcare frameworks usually has led to inadequate patient monitoring, medical errors, redundant care, and even use of wrong tools. As such, it is worth noting that any model of coordinated care must heavily rely on the timely availability of health information of specific individuals. This could be efficiently acquired through the use of an integrated EHR system. This will help in compiling a comprehensive patient clinical record and in turn will assist with coordination of clinical care delivery through improvement in the availability and timeliness of medical information regarding individuals. Focusing on the impact of implementing an integrated EHR system based on access to timely and complete information and agreement on roles or the responsibilities among clinicians, the following points prove that EHR systems are the way to go.

- Improved clinical decision-making—Appropriate clinical decision-making is vital in coordinated care. This is especially in the case of first responders. In case of an emergency, these care providers must strive to keep the lives of individuals in stable conditions until they are handed over ton hospital emergency units. However, the introduction of the electronic health records has created a scenario in which these first responders are able to access complex patient health records (King et al. 2014). This aids in choosing the right first aid medication which is crucial in such situations. Additionally, in patients with chronic conditions, different healthcare providers will look at what their counterparts prescribed to the patient. This way, they will tend to use alternative means in treating the patients. Additionally, the availability of the health records with ease facilitates fast decision-making since one party is able to understand the cause of a certain ailment and the treatments that an individual has passed through.

  Moreover, in certain situations such as during emergencies, patients will tend to forget their health information. This could be due to shock and trauma. However, due to the adoption of electronic health records, decisions concerning the treatment of such patients become easier. Their health information can be accessed and through this, their lives can be saved.
- Improved collaboration and care management for specialty care physicians—EHR use enables improved care coordination between primary care physicians and behavioral healthcare providers. A research team found out that integrated primary and behavioral healthcare is increasing in popularity due to its numerous benefits especially in dealing with behavioral health disorders. In this, the

integrated EHR systems end up standardizing and streamlining collaboration and care delivery between various providers.

- Less paperwork and fewer storage issues—In the healthcare sector, it is worth noting that administrative duties usually represent a significant amount of time and costs. As such, clinicians will tend to spend lots of time filling in data. However, the use of electronic health records tends to streamline a number of routine tasks. Due to this, the amount of paperwork is highly reduced. As the amount of paperwork reduces, the required storage space also declines. As a result, with instant storage and retrieval of digital EHRs, healthcare providers may help health offices become less cluttered and efficiency increases in the health sector (King et al. 2014).
- Financial incentives impact on health providers—The process of procuring the necessary equipment, hiring dedicated personnel, and training staff on new procedures can be costly to healthcare providers. However, it is worth noting that various incentives are available in order to help organizations recoup their investment. Through Medicare incentive programs, eligible providers can earn incentives for the adoption and meaningful use of technology. In turn, this leads to improved care coordination and overall provision of quality healthcare.

It is therefore evident that integrated EHR systems have a great positive impact on healthcare and specifically on coordinated care. Through its adoption, various stakeholders in this sector are due to achieve numerous benefits that are currently not available in the traditional models. However, there are barriers to the widespread adoption of electronic health records. In this, four of the core barriers to the adoption of EHR systems that would in turn allow clinicians to share information easily include the lack of a common format for recording the information, high costs that are involved with the installation and implementation of these systems, privacy issues that patients are concerned with regard to their personal information being shared, and physicians' concerns about legal liability (Nguyen et al. 2014). These barriers prevent various stakeholders from achieving the benefits of electronic health records. However, these are issues that can be dealt with by the various stakeholders. In this, the governments of specific countries and bodies such as the World Health Organization should ensure that the barriers to the adoption and implementation of HER systems are minimized. This will facilitate efficient care coordination that in turn leads to quality healthcare for individuals globally.

## *1.4 Role of Care Coordination in Chronic Disease Management*

Chronic ailments have remained to be a persistent burden on both care providers and patients globally. In this, the chronically ill patients will often have multiple conditions that oblige a well-coordinated system of care in different settings across multiple providers. To support these facts, Maeng et al. (2012) have found out that in the United States alone, a typical Medicare beneficiary will tend to see two primary care physicians and five specialists per year. Moreover, the patients with complex conditions may end up seeing up to 16 specialists per year. With the aging population and advances in the treatment of chronic diseases, teamwork with regard to care coordination is critical. A basic team composition in care coordination for effective chronic disease management consists of the nurse case managers, medical specialists, clinical pharmacists, social workers, and lay health workers. As such, successful chronic disease interventions will usually involve a well-coordinated multidisciplinary team.

Coordination is a multifaceted activity that in most cases will require effective participation by different professionals, service organizations, and the patients. In this, most successful interventions involve the delegation of responsibilities by the primary care doctor to team members who in turn ensure that patients receive efficient clinical and self-management support services. In chronic ailment programs, effective care coordination tends to exploit the varied skills of the team using the following set of strategies according to Wagner (2000):

- Population-based care—This is an approach in the planning and delivering care to defined patient populations that ensure that interventions are able to reach all the patients that need them. This strategy begins with a guideline that defines various components (assessments and treatments) of high-quality care with all the stakeholders in mind. The various steps that are required in order to deliver effective interventions to the chronically ill patients are outlined and in turn delegated to members of the team.
- Treatment planning—Care coordination in chronic disease management will require treatment planning for each of the patients. In this, plans (especially the formal ones) will help in the organization of the work of the various teams and additionally help the patients navigate easily the complexities of this multidisciplinary care. It is worth noting that the treatment plans that include the patients' treatment preferences are the ones that are more likely to yield, resulting in both satisfied and compliant patients.
- Self-management support—Effective care coordination in the management of chronic ailments and conditions will usually involve educational yet supportive interventions. These are aimed at helping patients change certain risky behaviors. Additionally, these interventions in most cases will tend to encourage patients to become better self-managers and this improves outcomes in a wide range of chronic ailments. Moreover, in care coordination, interventions emphasize on the acquisition of skills rather than just knowledge and systematically will boost the patients' motivation and confidence in the management of their conditions. This in turn discourages dependency on the part of the patients. However, currently, most doctors and health professionals do not have the training or the time to engage their patients in counseling on behavior change. This creates a scenario where self-management support is not accorded to the patients, lagging behind the benefits of care coordination in the chronically ill.

- More effective consultations—The chronically ill will tend to have multiple needs. However, there are the limitations of brief consultations which leave patients partially attended to. However, care coordination seeks to turn this around. This is because the health professionals will have access to a patient's detailed information before they even arrive for the consultation. This is bound to be made possible by the availability of electronic health records. Additionally, some models of care coordination will talk about implementing group consultations. Zwar et al. (2017) analyzed group consultations for the elderly, and the results were that the patients ended up being more satisfied, informed on their preventive care, and in turn used health services less often than the comparison group. Consultations could also be made for patients with similar needs. For instance, in the United Kingdom, asthma or diabetic clinics are part of medical practice. This shows that group consultations may provide an efficient tool for the complementary functions in line with care coordination.
- Sustained follow-up—Close follow-up will always ensure that there is an early detection of the adverse effects, failure to respond to treatment, or even the recurrence of the symptoms. In care coordination for the chronic ailments, follow-up provides an opportunity to solve problems while at the same time demonstrating the undivided concern of the care teams. In this, some of the methods that have been found to be effective in the provision of a sustained follow-up include the use of telephones by the nurses in chronic illness care.

Intervention studies that have been conducted in the last few decades have started to clarify the role and advantages of care coordination to the chronically ill. It is worth noting that the involvement and leadership by appropriately trained nurses or other staff who complement the doctor in critical care functions (assessment, treatment management, self-management support, and follow-up) have been found to repeatedly improve professionals' adherence to guidelines and patients' satisfaction, clinical and health status, and use of health services. Moreover, the chronically ill patients are bound to benefit from a care team that includes skilled clinicians and educators who have efficient clinical skills and self-management support skills. Additionally, there are population managers who understand team function and public health principles and approaches. As such, care coordination has the potential to improve the quality of care for the chronically ill on the condition that the roles of the team members are clearly outlined, delegated and adequate training conducted to the team (Fromer 2011).

## 2 Care Coordination Conceptual Framework

Rapid and continuous advancements in the healthcare industry in care delivery, patient-clinician interaction, and clinical information exchanges have introduced a new level of challenges. If these challenges can be addressed we could realize positive impacts for both care outcome quality and cost. Currently, however, a lack of comprehensiveness and efficiency in a wide range of care delivery systems tends to promote the wasteful consumption of resources (Bush 2007). Health information technologies have the potential to be utilized to enhance clinical communication and collaboration and address many of these challenges. A large amount of patient health information is stored and retrieved during different clinical processes; however, if these processes are not synchronized to communicate with one another in a timely manner and to include all active stakeholders' needs, health information effectiveness will be futile (Kersten 2013).

An essential component in healthcare systems is the integration of care services that relies on collaboration, communication, and information exchange between clinicians, patients, and caregivers. Healthcare integration highlights care coordination principles that seek to improve care efficiency and value by connecting/matching all stakeholders and their needs appropriately. To optimize the integration process, a team approach must be developed that emphasizes on the empowerment of care coordination strategies that are embedded within EHR systems (Burton et al. 2004). This approach will extend the use of commonly adopted EHR systems to cover clinical communication and improve efficiency. As patients transition through different care points, the need for accurate transfer of information between practitioners and also patient clinical communication plays an important role in reducing redundancy and improving care outcomes. The availability of standardized patient summary records and clinical information exchange forms (for referral and test results) could pose genuine challenges in addition to the untimely inclusion of such information/data in the treatment/clinical process.

There are many established definitions for care coordination with some focused on a certain portion of the patient population or certain disease categories. Care coordination in its simplest form means delivering appropriate care, at the right time, in the right setting (McDonald et al. 2007). To this end, care coordination is identifying all contributing stakeholders in the process of care; determining their needs, roles, and expectations; and establishing clear communication methods between stakeholders that distribute care-relevant information to improve care task impact. Prioritizing healthcare tasks requires a systematic effort to track patient treatment cycles. Care-relevant information must be validated, authenticated, organized, and delivered as actionable data (steps) to their proper destinations (Raghupathi and Raghupathi 2014). In order to structure a reliable care coordination system successfully, stakeholders and their assigned tasks must be appropriately outlined in the care delivery system to assess their impact on each other and their influence on the care outcome. Identifying clear distinctions between stakeholder contributions and their interdependencies will extend the applicability of system feedback.

Today a key void in the literature with regard to care coordination is the lack of any robust and systematic frameworks to assist with addressing this critical issue for better healthcare delivery. To address this void, we proffer a suitable care coordination framework. The framework presented below is based on the principles of systems thinking (Gibbings 2017). The framework partitions care coordination system into three main categories that regularly interact and impact each other. Primary stakeholders in this model include patients, providers, payors, and potentially regulatory bodies and healthcare agencies. The communications or interdependencies layer is mainly embodied by the EHR/EMR systems utilized by a large number of clinics and hospitals. And the system goals section is assessed in the three main areas of healthcare system outputs: population health, patient experience, and cost (Berwick et al. 2008). Care coordination measures are attributes that impact the progress and performance of the framework's interdependence component and hence they have an undeniable impact on the entire framework performance. A care coordination system is similar to the medical care provided for a human body. Similar to medical care, care coordination must be comprehensive and it must provide an overview of the entire care process to support a healthy body. The credit or lifeline of a care coordination system is similar to the blood flow in a human body, as it must transfer the information accurately and in a timely manner to all sections of the system. Analyzing the interdependencies in a healthcare system will assist in the evaluation of the collaboration impact and identify possible impediments (stakeholders) that impact the care/process outcome. The care coordination measures (accountability, information exchange, communication, patient-clinician communication, etc.) that characterize the interdependence intricacies will provide an appropriate context for evaluating the stakeholder collaboration in a given care setting (Fig. 1).

## 3 Current Lack of Health Quality Issues in the United States

In the effort toward improving the US healthcare system, many goals and objectives have been introduced and experimented. The increase in the demand for healthcare services due to the aging population (Mulero 2017) combined with a sharp incline of advanced and often costly care services' expenditure (Callahan 2008) is putting a substantial pressure on the US economy. These strains are urging policymakers and healthcare leaders to provide potential solutions to control the cost and improve overall healthcare outcome indicators by improving the quality of services rendered (Hussey et al. 2013). These solutions are aiming at care quality evaluation and promoting value-based care by improving the adoption of health information technology (Wheatley 2013).

The uncontrolled rising cost of the US healthcare system has produced a perplexing array of expectations, discussions, and competing resolutions. With the National Health Expenditure accounting for 17.8% of the GDP in 2015, and the largest share of this cost (28.7%) being sponsored by the federal government, projections are indicating that between 2017 and 2022, this trend will continue to rise (5.6%) at a faster rate (CMS.gov 2017) than the GDP growth (2.0–2.9%). It is a basic disproportionate equation of spending at a higher rate than producing without provisions to implement any correcting measures that could fix such imbalance. These numbers combined with aging population, innovative and often more expensive medical treatments, and a jump in the chronic patient population

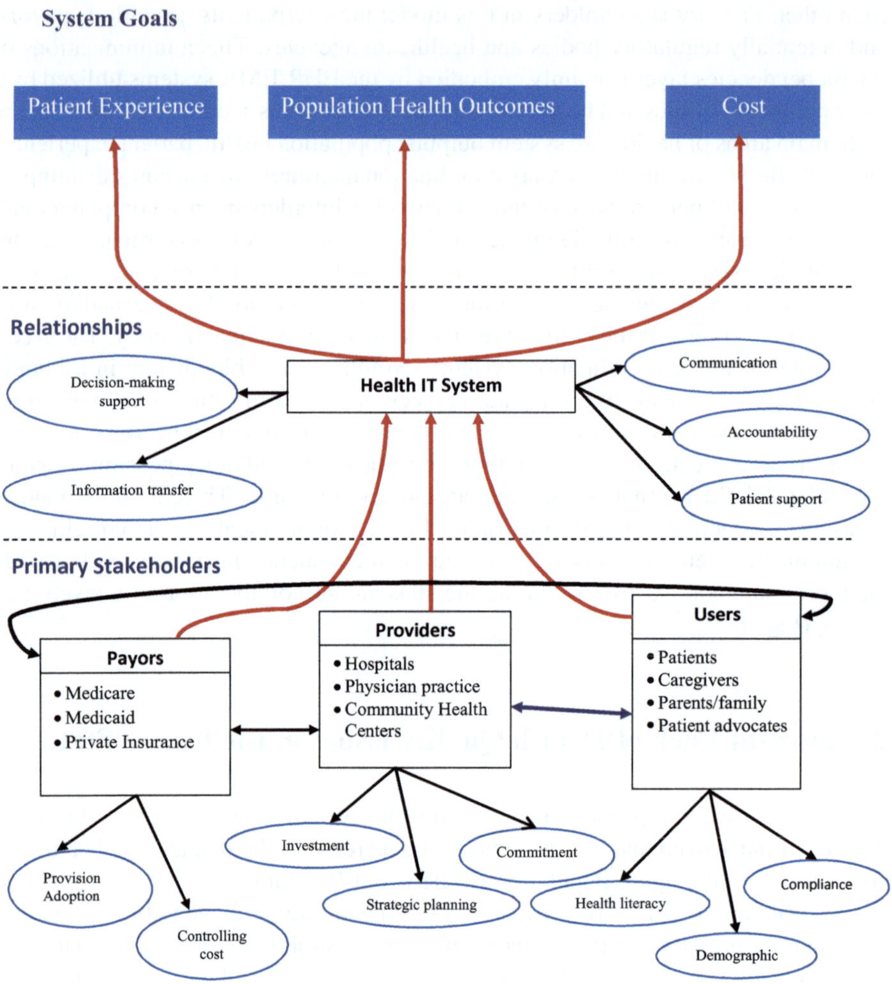


**Fig. 1** Care coordination framework

numbers (half of Americans having one or more chronic conditions and one fourth of adults having two or more chronic conditions) (Ward et al. 2014) are posing serious questions and inquiries for healthcare policymakers and leaders.

Unfortunately, rising healthcare cost is not entirely due to modern/progressive developments in the care services provided to patients. In many cases, growing healthcare expenditure is caused by the lack of synchronization between a large spectrum of entities that are included in the care processes (Berwick et al. 2008). These entities are also identified as stakeholders in the care coordination field as they are both impactful and influenced by the care processes. Furthermore, lack of active (and monitored) integration between stakeholders can critically impact the process of delivering care and hence the quality and cost of services provided. To improve care quality and manage resources (cost) effectively, healthcare organizations must systematically monitor and assess stakeholders' participation to generate actionable feedback.

Replacing fee-for-service (FFS) with fee-for-value (FFV) or pay-for-performance (value-based purchasing, VBP) has been a focus point for healthcare organizations. FFS creates the potential of overusing costly care services in addition to duplicating test/diagnostic services. Lack of accountable and integrated models that promote efficiency in utilizing care delivery services could also increase hospitalization and cause harm to patients. Value-based purchasing however relies on measuring the value produced or achieved in relation to the expenditure spent on producing care delivery services (Porter 2010). VBP programs could be classified into several categories: improving quality measurement science, strengthening both the size and design of incentives, minimizing health disparities, establishing broad outcome measurement, choosing appropriate comparison targets, and determining the optimal role of VBP relative to alternative payment models (Chee et al. 2017). Two main concepts emerge in VBP models: measuring the quality of care processes and identifying the standard of measurement.

Allocating resources and controlling care delivery expenditure efficiently drives healthcare organizations and regulatory bodies to impose spending cuts that are tied to care outcomes or are predetermined by payor (discounted) rates. Although these cost reductions might be beneficial in impeding the sharp rise in healthcare expenses for a period of time or for targeted patient populations. However, these cuts could introduce inadvertent consequences such as diminishing the quality of care and raising cost of care in the long term, especially, if they are implemented unconditionally and without adequate consideration to all care participants. Conservative estimates indicate that controlling six main categories of wasteful spending, overtreatment, failures of care coordination, failures in execution of care processes, administrative complexity, pricing failures, and fraud and abuse, could reduce healthcare expenditure by 20% of total cost (Berwick and Hackbarth 2012). Detecting the waste caused by these six categories must be detected, corrected, and reported at the earliest stages of care delivery services. An overall tracking structure (applied as an API to EHRs) that can monitor, report, and alert care participant contributions and potential shortcomings will be able to essentially produce actionable steps that positively influence the six abovementioned wasteful factors in healthcare. Generally, when products and services are wasteful to any industry or field, they are identified, studied/monitored, and eventually eliminated from production/service line to improve profits and enhance performance. In healthcare, eliminating (inefficient) services must be implemented with high precision to prevent any type of patient injury or harm in addition to being replaced by more effectual care delivery services. As care services in many instances include a large number of clinical processes and staff, identifying inefficiencies in the delivery system must include accurate tracking and reporting mechanism.

## 4 Policy Recommendations

Healthcare organizations and care entities are actively investing in improving their involvement in value-based purchasing, its potential feasibility, and its long-term influence on the care delivery systems. Alternative payment models that are introduced in healthcare systems emphasize on concepts such as care process management and stakeholder-contribution alignment which consequently expands the impact variability in the care delivery system determinants. This expansion not only requires extension in the breadth/scope of information systems for identifying such factors but it also relies largely on the systematic data collection methods and pragmatic knowledge base application approaches. In simple terms, improving care quality means improving the utilization of current care systems in general and IT systems in particular to identify and enhance care value. Healthcare value is defined as the care outcomes achieved per cost (Porter 2010). The main goal of increasing value is enhancing efficiency to achieve the best outcome relative to cost instead of merely cost reduction without any notable improvement (maintaining) of care outcome measurements. The prevailing care payment system is (predominantly) based on the volume of the services delivered with little or no consideration to the value they produce.

Improving the quality of care must be focused on the outcome of the care services delivered, divided by the cost or expenditure spent per these services (Porter 2010). Both of these essential facets of care delivery system rely on several factors and components that must be empowered and enhanced in order for the outcome and cost to be improved. Considering the outcome and cost association in a simple division equation, it becomes clear that in order to increase the equation outcome (quality of care), we must either increase the outcome value per a group of services delivered or reduce their cost. Main components that will impact the care delivery outcome measures are comprehensiveness, accountability, and efficiency.

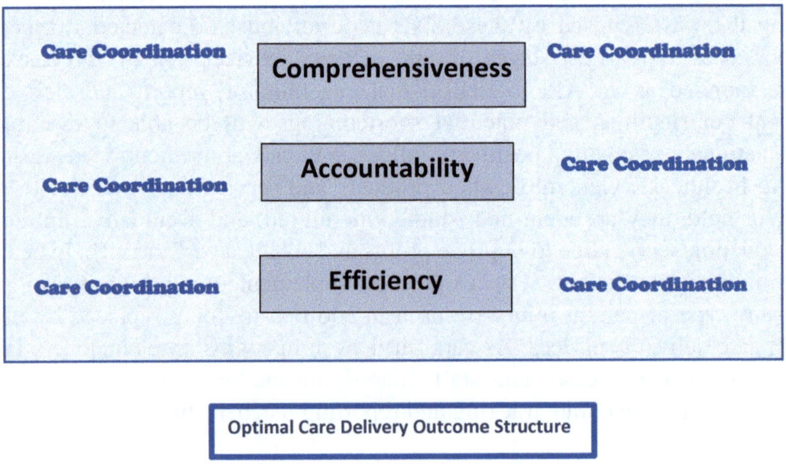

These components require a solid and integrated framework that must be supported by care coordination measures.

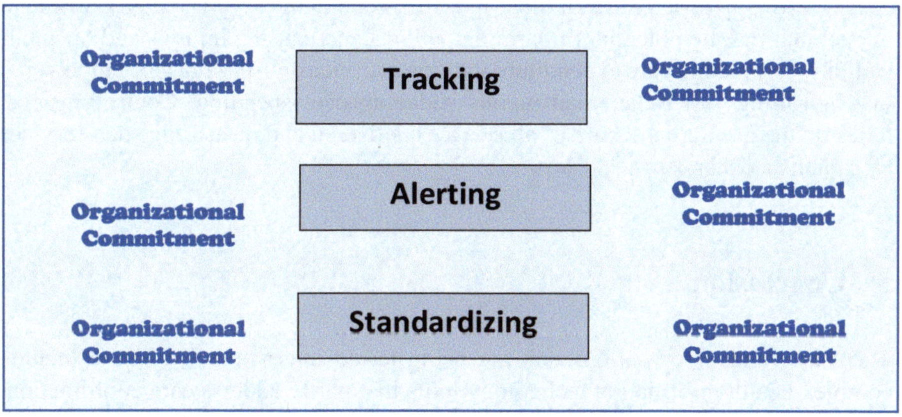


**Care Cost Control Structure**

The main components in the care cost structure are tracking, alerting, and implementing standardization in the care cost evaluation approaches. These components must be considered in an organization structure that is fully committed to unifying these components in a single formation that can be parallelly assessed with care quality.

Care quality improvement must consider the integrated enhancement of care delivery outcome structure components. Incremental improvements that are designed based on a constricted outlook of processes, payors, providers, and patient populations or any restricted combination of these categories will fail to form a complete image that depicts/assesses care outcome based on stakeholder participation and input. The comprehensive overview of care outcome processes must be considered by holding all participating contributors accountable to their share of care process/task while implementing efficient methods of care delivery services. In order for accountability and efficiency measures to positively impact care quality, they must be evaluated based on the desired (or previously systematically proven) outcomes. To this end and based on the severity and resource availability of care processes (in healthcare organizations) that are included in a given episode (treatment period/measuring unit) of care, outcomes must be identified and recorded. Distinct care processes could be included in different care episodes but each episode must be associated with a continuous measure of outcome and eventually be stored with such outcome once the episode is complete. Such tracking data could be later used in standardizing care processes based on stakeholder input and identifying more achievable episode outcomes. Efficiency will be considered in both sharing the care process to provide optimal care and in assessing individual care task performance.

Healthcare cost is another imperative factor in determining care quality, it is a driving force for providers that must be regulated with high accuracy. Healthcare cost is distinctly different from other industries due to the severity of any oversight impact that has the potential of harming patients and causing injuries and hospitalizations. Correlating care expenditure with care service delivery must be empowered by controlling cost rather than merely reducing care spending. Controlling cost relies on the accurate tracking of all services delivered and measuring such services by established standards.

## 5 Conclusion

There are other interventions that are being tested internationally. These include complex health information technology tools that partly address care coordination. An example is two linked trials that aim at improving care for patients with depression and cardiovascular disease risk in the United Kingdom (Bates 2015). Another perfect example is the Finnish Geriatric Intervention Study that is aimed at preventing cognitive impairment (Bates 2015). However, the issues involved in this sector vary substantially by specific settings. These include outside the hospital, access to information about community resources like substantial abuse treatment and mental healthcare represents a particular challenge. Such interventions need to be designed and tested, in a variety of settings. If the healthcare organizations seek to succeed in the improvement of quality and at the same time reduce costs, they should be ready to provide better care coordination. As such, the electronic health records of today do not yet fully support care coordination.

## References

Adams, W. G., Phillips, B. D., Bacic, J. D., Walsh, K. E., Shanahan, C. W., & Paasche-Orlow, M. K. (2014). Automated conversation system before pediatric primary care visits: A randomized trial. *Pediatrics, 134*, e691–e699, peds-2013.

Bates, D. W. (2015). Health information technology and care coordination: The next big opportunity for informatics? *Yearbook of Medical Informatics, 10*(1), 11.

Berwick, D. M., & Hackbarth, A. D. (2012). Eliminating waste in US health care. *JAMA*, 1513–1516. https://doi.org/10.1001/jama.2012.362.

Berwick, D. M., Nolan, T. W., & Whittington, J. (2008). The triple aim: Care, health, and cost. *Health Affairs, 27*. Retrieved from https://www.healthaffairs.org/doi/full/10.1377/hlthaff.27.3.759.

Burton, L. C., Anderson, G. F., & Kues, I. W. (2004). Using electronic health records to help coordinate care. *Milbank Quarterly, 82*, 457–481.

Bush, R. (2007). Reducing waste in US health care systems. *JAMA Network, 297*, 871–874.

Callahan, D. (2008). Health care costs and medical technology. *The Hastings Center*. Retrieved from https://www.thehastingscenter.org/briefingbook/health-care-costs-and-medical-technology/

Caron, F., Vanthienen, J., Vanhaecht, K., Van Limbergen, E., De Weerdt, J., & Baesens, B. (2014). Monitoring care processes in the gynecologic oncology department. *Computers in Biology and Medicine, 44*, 88–96.

Chee, T. T., Tyan, A. M., Wasfy, J. H., & Borden, W. B. (2017). Current state of value-based purchasing programs. *HHS Author Manuscripts*, 2197–2205. Retrieved from https://www.ncbi.nlm.nih.gov/pmc/articles/PMC5378385/

Claborn, K., Becker, S., Ramsey, S., Rich, J., & Friedmann, P. D. (2017). Mobile technology intervention to improve care coordination between HIV and substance use treatment providers: Development, training, and evaluation protocol. *Addiction Science & Clinical Practice, 12*(1), 8.

CMS.gov. (2017, June 14). NHE fact sheet. Retrieved from Centers for Medicare & Medicaid Services: https://www.cms.gov/research-statistics-data-and-systems/statistics-trends-and-reports/nationalhealthexpenddata/nhe-fact-sheet.html

Davis, K., Stremikis, K., Schoen, C., & Squires, D. (2015). Mirror, mirror on the wall, 2014 update: How the US health care system compares internationally. The Commonwealth Fund, June 2014. *Google Scholar*.

Friedman, A., Howard, J., Shaw, E. K., Cohen, D. J., Shahidi, L., & Ferrante, J. M. (2016). Facilitators and barriers to care coordination in patient-centered medical homes (PCMHs) from coordinators' perspectives. *The Journal of the American Board of Family Medicine, 29*(1), 90–101.

Fromer, L. (2011). Implementing chronic care for COPD: Planned visits, care coordination, and patient empowerment for improved outcomes. *International Journal of Chronic Obstructive Pulmonary Disease, 6*, 605.

Gensichen, J., Beyer, M., Muth, C., Gerlach, F. M., Von Korff, M., & Ormel, J. (2006). Case management to improve major depression in primary health care: A systematic review. *Psychological Medicine, 36*(1), 7–14.

Gibbings, R. A. (2017). *The impact of care coordination strategies in health information systems on health care outcomes*. Chicago: University of Illinois.

Hussey, P. S., Wetheimer, S., & Mehrota, A. (2013). *The association between health care quality and cost: A systematic review*. American College of Physicians. https://www.ncbi.nlm.nih.gov/pubmed/23277898

Kersten, S. (2013). Moving the needle toward a data-driven health care system: Optimizing the EHR through information governance. *Health Information Management*. https://www.coursehero.com/file/27194990/Article-Review-3pptx/

King, J., Patel, V., Jamoom, E. W., & Furukawa, M. F. (2014). Clinical benefits of electronic health record use: National findings. *Health Services Research, 49*(1pt2), 392–404.

Krist, A. H., Beasley, J. W., Crosson, J. C., Kibbe, D. C., Klinkman, M. S., Lehmann, C. U., et al. (2014). Electronic health record functionality needed to better support primary care. *Journal of the American Medical Informatics Association, 21*(5), 764–771.

Kvedar, J., Coye, M. J., & Everett, W. (2014). Connected health: A review of technologies and strategies to improve patient care with telemedicine and telehealth. *Health Affairs, 33*(2), 194–199.

Lindberg, B., Nilsson, C., Zotterman, D., Söderberg, S., & Skär, L. (2013). Using information and communication technology in home care for communication between patients, family members, and healthcare professionals: A systematic review. *International Journal of Telemedicine and Applications, 2013*, 1–31.

Maeng, D. D., Martsolf, G. R., Scanlon, D. P., & Christianson, J. B. (2012). Care coordination for the chronically ill: Understanding the patient's perspective. *Health Services Research, 47*(5), 1960–1979.

McDonald, K., Sundaram, V., Bravata, D., Lewis, R., Lin, N., Kraft, S., et al. (2007). *Closing the quality gap: A critical analysis of quality improvement strategies (Vol. 7: Care coordination).* NCBI. Retrieved from https://www.ncbi.nlm.nih.gov/pubmed/20734531

Morton, S., Shih, S. C., Winther, C. H., Tinoco, A., Kessler, R. S., & Scholle, S. H. (2015). Health IT–enabled care coordination: A national survey of patient-centered medical home clinicians. *The Annals of Family Medicine, 13*(3), 250–256.

Mulero, A. (2017, February 15). CMS: Medical prices, aging population to speed up health spending growth. Retrieved May 14, 2018, from healthcaredive.com: https://www.healthcaredive.com/news/cms-medical-prices-aging-population-to-speed-up-health-spending-growth/436285/

Nguyen, L., Bellucci, E., & Nguyen, L. T. (2014). Electronic health records implementation: An evaluation of information system impact and contingency factors. *International Journal of Medical Informatics, 83*(11), 779–796.

O'malley, A. S., Grossman, J. M., Cohen, G. R., Kemper, N. M., & Pham, H. H. (2010). Are electronic medical records helpful for care coordination? Experiences of physician practices. *Journal of General Internal Medicine, 25*(3), 177–185.

Porter, M. E. (2010). What is value in health care? *The New England Journal of Medicine*, 2477–2481. Retrieved from http://www.nejm.org/doi/full/10.1056/NEJMp1011024.

Raghupathi, W., & Raghupathi, V. (2014). Big data analytics in healthcare: Promise and potential. *Health Information Science and Systems*. Retrieved from https://www.ncbi.nlm.nih.gov/pmc/articles/PMC4341817/

Samal, L., Dykes, P. C., Greenberg, J., Hasan, O., Venkatesh, A. K., Volk, L. A., & Bates, D. W. (2013). The current capabilities of health information technology to support care transitions. In *AMIA... Annual symposium proceedings. AMIA symposium* (Vol. 2013, pp. 1231–1231). American Medical Informatics Association.

Sato, A. F., Clifford, L. M., Silverman, A. H., & Davies, W. H. (2009). Cognitive-behavioral interventions via telehealth: Applications to pediatric functional abdominal pain. *Children's Health Care., 38*(1), 1–22.

Scholl, I., Zill, J. M., Härter, M., & Dirmaier, J. (2014). An integrative model of patient-centeredness–a systematic review and concept analysis. *PLoS One, 9*(9), e107828.

Steichen, O., & Gregg, W. (2015). Health information technology coordination to support patient-centered care coordination. *Yearbook of Medical Informatics, 10*(1), 34.

Stephan, C., Popejoy, L. L., Mehr, D. R., Rahmani, E., Wilkinson, K., & Simoes, E. (2015). Using the consolidated framework for implementation research (CFIR) to evaluate EHR population health management tools. In AMIA. http://toc.proceedings.com/28699webtoc.pdf

Tricco, A. C., Antony, J., Ivers, N. M., Ashoor, H. M., Khan, P. A., Blondal, E., et al. (2014). Effectiveness of quality improvement strategies for coordination of care to reduce use of health care services: A systematic review and meta-analysis. *Canadian Medical Association Journal, 186*(15), E568–E578.

Turvey, C., Klein, D., Fix, G., Hogan, T. P., Woods, S., Simon, S. R., et al. (2014). Blue Button use by patients to access and share health record information using the Department of Veterans Affairs' online patient portal. *Journal of the American Medical Informatics Association, 21*(4), 657–663.

Wagner, E. H. (2000). The role of patient care teams in chronic disease management. *BMJ, 320*(7234), 569–572.

Ward, B., Schiller, J., & Goodman, R. (2014). Multiple chronic conditions among US adults: A 2012 update. *Preventing Chronic Disease*. Retrieved from https://www.ncbi.nlm.nih.gov/pubmed/24742395

Wheatley, B. (2013). Transforming care delivery through health information technology. *The Permanente Journal*. Retrieved from https://www.ncbi.nlm.nih.gov/pmc/articles/PMC3627785/

Zwar, N., Harris, M., Griffiths, R., Roland, M., Dennis, S., Powell Davies, G., & Hasan, I. (2017). A systematic review of chronic disease management. https://www.ncbi.nlm.nih.gov/pmc/articles/PMC5759778/

# Managing the Risks of Emerging IoT Devices

**Napoleon C. Paxton and David L. Branca**

## 1 Introduction

Connecting medical monitoring devices to the Internet of Things (IoT) ecosystem gives trained professionals or additional medical systems immediate access to the health state of a monitored patient. Access to this data can significantly reduce reaction times in the event of a medical emergency and improve monitoring of a patient's medical status. Unfortunately, connecting these devices to the Internet is a risky endeavor due to persistent threats from cybercriminals. In this chapter, we discuss the current state of IoT from a cybersecurity perspective. We also present possible solutions to improve medical device cybersecurity; including a network science-based method to identify and analyze related threats in near real time.

This chapter is structured as follows: Sect. 2 provides a brief history of IoT followed by a discussion in Sect. 3 of the risks involved with incorporating medical devices to the IoT ecosystem. Section 4 discusses how design decisions of the Internet have led to the vulnerable state of anything that connects to it and Sect. 5 builds on the content from Sect. 4 by discussing unsecure design decisions specifically related to IoT architectures. Section 6 presents viable solutions to reduce IoT-related cybersecurity issues and Sect. 7 explores advanced IoT security, specifically the application of a network science-based analysis suite. Section 8 concludes this chapter.

N. C. Paxton (✉) · D. L. Branca
IOMAXIS, LLC, Lorton, VA, USA
e-mail: npaxton@iomaxis.com

 
N. Wickramasinghe, F. Bodendorf (eds.), *Delivering Superior Health and Wellness Management with IoT and Analytics*, Healthcare Delivery in the Information Age, https://doi.org/10.1007/978-3-030-17347-0_22

## 2 Brief History of IoT

For the purpose of this chapter, we define IoT as a network of physical devices (e.g., computers, sensors, items with embedded electronics) that are interconnected on the Internet and can exchange data. The origins of IoT can be traced back to the 1930s and 1940s with the discovery of radar and the creation of radio-frequency identification (RFID) (https://iot.ieee.org/images/files/pdf/IEEE_IoT_Towards_Definition_Internet_of_Things_Revision1_27MAY15.pdf). Incremental improvements to RFID were made over the years until the late 1990s into the early 2000s where the term "Internet of Things" was first published in a report from the International Telecommunication Unit (ITU) (https://iot.ieee.org/images/files/pdf/IEEE_IoT_Towards_Definition_Internet_of_Things_Revision1_27MAY15.pdf). The term IoT became more prominent soon after, when the Auto-ID Center began promoting it. Recent advancement in IoT can be tracked to 2008 when Amazon fully released Amazon Elastic Cloud Computing (EC2) (Amazon Elastic Compute Cloud (Amazon EC2), Cloud Computing Servers 2014). Shortly thereafter, a multitude of conferences were held promoting the theory, development, and application IoT. Recent forecasts predict that over 20 billion "things" will be connected to the Internet by 2020 (Gartner-forecast) (https://www.gartner.com/newsroom/id/3598917) (Fig. 1 shows an IoT timeline).

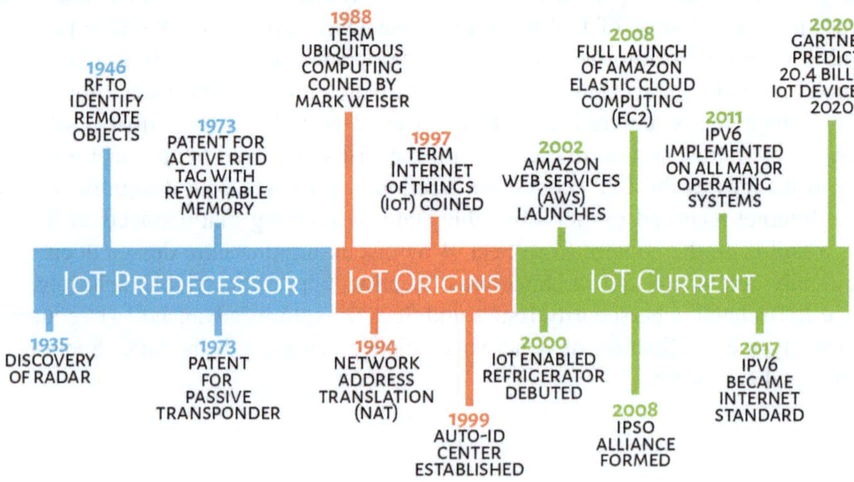


**Fig. 1** IoT timeline

## 2.1 IoT Predecessor

In the 1950s and 1960s, scientists in the United States, Europe, and Japan began exploring how radio frequency (RF) could be used to remotely identify objects. The early efforts of RF object identification led to the development of radio-frequency identification (RFID) projects in 1970s. In particular, Los Alamos National Laboratory developed a system to track nuclear materials by installing a transponder in a truck and readers at the gates of secure facilities. This system used an antenna to trigger the truck's transponder, which then broadcasted an ID and other data programmed in the transponder. This system was commercialized in the 1980s after proving its utility for tracking tagged items. More successful RFID projects followed and as technological advancements were made, RFID developers were able to transition their technology to higher radio-frequency bands which increased the distance the devices could communicate with one another.

In 1999, the Auto-ID Center at the Massachusetts Institute of Technology (MIT) was established (http://vs.inf.ethz.ch/publ/papers/Internet-of-things.pdf). The research conducted at this center paved the way for IoT as we know it today. Expanding the technologies utility, two professors at MIT, David Brock and Sanjay Sarma, conducted research on using RFID tags to track the movement of all products through a supply chain. Brock and Sarma developed a method composed of a central database which contained the information of all tagged data and a barcode installed on the devices which linked back to location of that data. Before their research, people using RFID technology maintained a mobile database of information from tagged devices, which they physically carried with them everywhere they went (https://iot.ieee.org/images/files/pdf/IEEE_IoT_Towards_Definition_Internet_of_Things_Revision1_27MAY15.pdf). This new method enabled businesses that were using RFID to be more efficient, leading to widespread adoption of RFID among large companies.

## 2.2 Birth of IoT

As support for RFID grew, the Auto-ID Center created important interface components which shifted focus from connecting the product supply chain for manufacturing companies, to the possibility of connecting any device to the Internet. These components included: Air Interface Protocols, the Electronic Product Code (EPC), and a network architecture for referencing data associated with an RFID tag across the Internet. This technology was then licensed to the Uniform Code Council, which then created EPCglobal to commercialize the EPC technology. The creation of the EPC technology prompted the Auto-ID Center (which later became the Auto-ID Labs) to promote the idea of an "Internet of Things" using the EPC technology to connect to the Internet. Although the work to create the initial IoT concept was largely done at the Auto-ID Center and Labs, it has been noted that the term actually originated from a workshop conducted by the International Telecommunication Union (ITU) in 1997 (https://iot.ieee.org/images/files/pdf/IEEE_IoT_Towards_Definition_Internet_of_Things_Revision1_27MAY15.pdf).

## 2.3 The Current IoT Platform

In the 1990s, the growth of connected IoT devices was constrained by limits in technology. Cellular networks were not fully IP-based at the time and there was a finite and ever-decreasing amount of available IP addresses to use. Technologies to address IP address availability such as network address translation (NAT) was in its infancy and IPv6 had not yet been implemented. The limitation of available IP addresses prevented a large number of devices from connecting to the Internet since they simply could not be addressed over a network.

Today, several technological breakthroughs have paved the way for mass implementation of IoT devices. These breakthroughs include advanced networking capabilities, artificial intelligence (AI)-based large-scale analytics tools, and cloud computing technologies.

### 2.3.1 Advanced Networking Capabilities

In 1994, network address translation (NAT) was created. This networking methodology enables devices that do not have Internet addressable IP addresses to be accessible by enabling communication between them and an intermediate device that can be addressed and accessed through the Internet. This technology exponentially increased the number of devices which could now access the Internet and significantly reduced the connectivity barrier initially faced by IoT devices. Another communication medium which will significantly improve IoT deployment is fifth generation (5G) cellular connections. The standard for this technology was published late in 2017 and devices have already begun to hit the marketplace. 5G technology brings three major advantages of fourth generation (4G) technology; The ability to move more data through greater speed, the ability to provide lower latency for greater response time, and the ability to connect a great number of devices concurrently (Akpakwu et al. 2018). All these developments were added with IoT in mind. Once this technology matures, IoT devices will explode on to the market place.

### 2.3.2 AI-Based Large-Scaled Analytics Tools

Incorporating data-bearing devices to the Internet also faces the barrier of making sense of all the data which is being added. Over the past decade, significant progress has been made in AI-based analytics tools, particularly in the area of convolutional neural network (CNN) machine learning (ML). Machine learning can be implemented in two ways, supervised or unsupervised. In a supervised implementation, the analyst creating the application define categories and explicitly program what the data should look like for each category. When the data is analyzed, each instance of the data that fits the defined category model will be grouped under that category. In an unsupervised implementation, the algorithm creates the categories based on the structure and perceived symantics of the data. This is the area which has seen the largest growth over the past decade. Unsupervised CNN ML is a branch of AI which mimics the way a brain learns. CNN ML identifies the context of data by breaking it down into multiple layers. For instance, instead of trying to identify a dog in a group of pictures, CNN ML first identifies, with high confidence, the components that make up a dog such as the dogs' ears, the nose, the head, tail, body, and so on. The end result is a highly reasonable conclusion that the image processed contained a dog. With unsupervised CNN ML, accuracy is greatly improved as the dataset increases. This is important since the volume of data will continue to increase as new devices are added to the IoT ecosystem. It is also important to note that supervised ML methods are a less viable option for big data since human resource limitations make it is infeasible for a human analyst to create and maintain accurate definitions for new categories as new data are analyzed.

### *2.4 Cloud Technologies*

Arguably, the most important technological breakthrough that has enabled mass implementation of IoT is the proliferation of cloud computing. Cloud computing refers to off-loading of computer processing and storage to remotely accessible and centralized computer servers. Cloud computing has enabled innovative researchers, hobbyists, and entrepreneurs with low budgets, access to massive amounts of on-demand compute resources and storage. The cloud has also supported the networking component of the IoT ecosystem, enabling IoT data to be remotely processed. Readily available access to the cloud platform has sparked a rapid increase in innovation for big data analytics as well as ML and has provided IoT a platform to be connected to and monitored with.

## 3 Risk Involved with Applying IoT to Medical Devices

The marketplace has exploded with IoT-enabled devices recently. In 2017, Gartner estimated that by 2018, there will be 8.3 billion devices connected to the Internet and predict that by 2020 there will be 20.4 billion devices connected (https://www.gartner.com/newsroom/id/3598917). Another study has estimated that linking physical devices to the Internet will generate between 4 and 11 trillion dollars of economic value by 2025 (https://www.mckinsey.com/industries/semiconductors/

our-insights/whats-new-with-the-internet-of-things), meaning most industries will be looking to capitalize off this growing trend and capability. Appliances such as IoT-enabled refrigerators, security cameras, thermostats, vacuum cleaners, and cars are already becoming commonplace in our daily lives. It will not be long before the proliferation of IoT-enabled medical devices follow suit.

Heart monitors as well as other medical devices are starting to enter the IoT ecosystem, but wide-spread adoption has not yet been begun. In the near future, it is likely that the majority health care providers will utilize IoT-enabled medical devices, enabling immediate patient diagnosis, optimized and personalized preventive care, emergency dispatch during patient distress, and increased accuracy of patient treatment due to cross-correlated, on-demand, and medical device patient data analysis. Similar to other industries, the added capabilities provided by IoT can enhance the lives of those that use them. Unfortunately, as other industries have experienced, there are significant cybersecurity risks involved with incorporating Internet access to medical devices (https://hitrustalliance.net/documents/content/Securing-Connected-Hospitals.pdf). These risks can often be magnified by the sensitivity of the data contained on the devices. Examples of the risks for medical devices include: compromised confidentiality, compromised integrity, and compromised availability.

## 3.1 Compromising the Confidentiality of Medical Devices

Doctors and patients who utilize IoT-enabled medical devices to monitor the patient's state of health will likely mandate that the data on the devices remain confidential and private to a select set of people and medical devices. Leakage of this data could cause irreparable harm. Additionally, malware, such as a worm or ransomeware, can gain access to an IoT medical device connected to the Internet and steal or destroy sensitive information stored on the device (https://hitrustalliance.net/documents/content/Securing-Connected-Hospitals.pdf).

The reproductions for loss of IoT medical device data can be huge. Take, for instance, this scenario: a highly popular and influential CEO of a fortune 500 company has a health monitoring IoT device installed and has just been diagnosed with an incurable disease. Prior to the company devising and announcing a succession plan, one of the IoT medical sensors used by the CEO leaks data and a media outlet obtains the leaked data. Shortly thereafter, the media outlet breaks a story of the CEOs health and the company is caught off guard and not prepared to respond. As a result, the company's valuation goes down due to the uncertainty of the company's future caused by the leaked information. This example is a highly likely possibility due to the current state of IoT device security.

### 3.2 *Compromising the Integrity of Medical Devices*

In general, medical device integrity is especially important. The integrity of a device's output could dictate if a proactive or reactive medical action should be taken. A malfunctioning or nefariously exploited IoT medical device that outputs incorrect data can cause a patient to apply the incorrect treatment. This could directly result in worsening of the condition or even death. Similarly, a device that does not function as intended could also worsen the condition or cause death. Traditional man-in-the-middle (MITM) attacks have been used to target Internet connected devices to intercept communications, change the content of the message, and transmit altered content.

### 3.3 *Compromising the Availability of Medical Devices*

The availability of IoT medical device data and services can externally be affected. Once doctors and patients begin to depend on IoT-enabled medical devices to perform services, denied access to the device services and data could be extremely detrimental. Several threats, such as denial-of-service (DOS) attacks are prevalent on the Internet and are designed to cause this type of effect. Put in context, this can adversely impact a patient's medical treatment. For example, if obtaining a drug prescription requires information from a patient's IoT medical device, but access to the device has been denied, they will not be able to receive their prescription.

## 4 The Role of the Internet in Risk of Attacks

A major reason attacks on the Internet are so prevalent is due to design decisions made during its inception and throughout its evolution. This section discusses prevalent design flaws of the Internet to provide an educated understanding of the risks involved with incorporating medical devices into the IoT ecosystem.

### 4.1 *Origins of the Internet*

The philosophies and characteristics of today's Internet are similar to the philosophies and characteristics of its predecessors. Therefore, to understand the inherent flaws of the Internet, one should first understand how the Internet was conceived and evolved. The progenitor of modern-day Internet is the Advanced Research Projects Agency Network (ARPANET). ARPANET was a research project funded by the Federal Government of the United States Advanced Research Projects

Agency (ARPA) – renamed to the Defense Advanced Research Projects Agency (DARPA) in 1972 (A Selected History of DARPA Innovation 2018). In the 1960s, ARPANET was devised to address the need for a system and method that could easily connect researchers with geographically distributed compute resources. As Charles M. Herzfeld, the director of DARPA (1965–1967) stated "[ARPANET] came out of our frustration that there were only a limited number of large, powerful research computers in the country and that many research investigators who should have access were geographically separated from them" (Deffree 2017).

Officially deployed in 1969 with the first network connection between UCLA and Stanford Research Institute (Bryner 2009), ARPANET started as a small experimental network used exclusively for research and academic purposes only. Overtime, the network expanded and the project evolved from an experimental research project to a proven solution. In 1975, DARPA transferred management and operation of ARPANET to the Defense Communications Agency (Stewart 2000) – now known as Defense Information Systems Agency (DISA). In 1983, the US military enclave of ARPANET split off to form Military Network (MILNET) – latter becoming the DoD Defense Data Network (DDN). The civilian portion of ARPANET continued its operations supporting research and academia.

In 1985, the National Science Foundation (NSF) established the NSF Network (NSFNET) (NSFNET 2017). NSFNET was a network backbone service used primarily by academic and educational institutions and linked five super computers in America (Goldstein 2016). Shortly after NSFNET was established, ARPANET was interlinked with it. Subsequently ARPANET was decommissioned in 1990 (Deffree 2017). In 1991, NSF allowed restricted access of the NSFNET backbone to commercial networks (Internet service providers) (NSFNET 2017). In 1995, the NSFNET backbone was decommissioned as privatization of network backbones occurred (A Brief History of NSF and the Internet 2003) – resulting in a system that is considered the modern-day Internet.

Technologically, the Internet is based on the same concepts used in ARPANET and NSFNET for connecting devices and enabling communications. At its start, ARPANET used a communications protocol called Network Control Program (NCP) to send data between hosts of a single network. Inherently, the NCP capabilities were limited to communications between two devices on the same network – the NCP protocol could not function in an open-architecture network. To overcome this limitation, Robert Kahn and Vint Cerf designed a new protocol called the Transmission Control Protocol (TCP) (Cerf and Kahn 1974). Their 1974 paper established the TCP/IP framework that current day network communication protocols (e.g., IPv4, IPv6) are constructed from. In 1983, ARPANET permanently transitioned from the NCP to TCP/IP (Postel 1981). The transition to the TCP/IP suite made ARPANET the first network to route TCP/IP using packet switching (Robert Kahn and the Internet Protocol: http://scihi.org/robert-kahn-internet-protocol/Sack 2016). ARPANETs adoption of TCP/IP enabled the network to communications between heterogeneous devices connected to local area networks (LAN) and wide area networks (WAN).

Like the ARPANET project, NSFNET also utilized the TCP/IP stack (MERIT Networks, Inc 1995) making the integration of the two networks seamless. Concurrent to the absorption of ARPANET into NSFNET, NSF promoted commercial entities to not only access NSFNET, but to build their own networks and exchanges. Thus, the Internet backbone service providers' business model was in part based on the federally funded operational model of NSFNET (Cerf 1990), and NSF promotion of such commercial models. As the result of the ideas and technologies developed for APRANET, the expansive adoption of them, and the exemplar model of federated networks services realized by NSFNET, today's Internet is similar to the communications networks of 30 years ago.

## 4.2 *Inherent Flaws of the Internet*

As exemplified by its history, the Internet was not built from one single idea or technology. The Internet has been evolving for over 50 years, addressing issues of the moment. The foundational systems, methods, technologies, and protocols of the internet trace their beginnings to the ARPANET project. The Internet was not built considering today's requirements, nor was security built-in from the start. Additionally, the Internet was purposefully designed and built to be an open platform – enabling any device that can connect, the ability to communicate with all other devices using at the least the TCP/IP protocol stack. It is the complex and unique development history of the Internet that has enabled it to transform from a simple communications technology to an inherently complex and flawed system.

Throughout its existence, the Internet has provided the functionality to easily connect heterogeneous devices that are geographically separated. This was the requirement from which ARPANET was conceived nearly half a century ago. It is by design that the Internet is distributed and network connections and communications are standardized. These characteristics have enabled the Internet to provide readily available network access to devices. The complex evolution of the Internet continues today as the utility of the Internet expands with the proliferation of IoT devices.

The inherent flaws of the Internet are concerning when innovation pushes the limits of the Internet, or when issues manifest themselves as a result of reactionary and incremental changes. Within the scope of IoT, the most concerning Internet flaws are the ones that manifest themselves as privacy, security, and safety issues. The initial design the Internet naïvely omitted privacy, security, and safety considerations. It was not until the late 1980s that the need for network security became evident and the US government established the Computer Emergency Response Team (CERT). Currently, privacy has become a main concern as large data breaches (Armerding 2018) and user data misuse scandals (Lomas 2018) became prevalent. Unfortunately, no strong technological solutions or government policies have been developed and implemented to address this matter at the time of this writing.

The privacy, security, and safety of devices and data on the Internet are exacerbated by the fact that the Internet has evolved from a patchwork of assumptions and fragmented technologies that span over a half century. As the hardware and software fields continue to advance, innovative technologies continue to stress and illuminate flaws of the Internet. In many cases, new innovations will, and already have, outpaced the ability of the Internet to effectively implement and maintain privacy, security, and safety of devices and data. Consider the current state of the Internet with the recent and growing rise of IoT devices, the limitations of the Internet from a privacy, security, and safety perspective is extremely concerning and overwhelming.

Efforts over the years have been made to improve Internet security which indirectly improves privacy and safety. Yet, most improvements are reactive and after the fact – such as how most Internet advancements occur. Mechanisms such as implementing network device firewalls, device access/authentication standards (e.g., 802.1X), and new cryptographic protocols (i.e., Signal Protocol) have been developed and implemented. These solutions do improve security when implemented correctly but are still patches to the inherent design flaws of the Internet. These solutions are not de facto standards like the TCP/IP protocol and therefore are not an intrinsic component of the Internet.

The inherent flaws of the Internet are complex because of its development history. The Internet was designed to be open and support interoperability. Over the years, it has evolved by a patchwork of changes and has become technically fragmented. The foundational protocols of the Internet were primarily developed in the 1970s and the current Internet lags in incorporating standardized, effective, and efficient security.

## 5 Design Flaws of IoT Networks

Couple the inherent Internet flaws with the introduction of IoT device ecosystem and the inherent Internet flaws are exponentially magnified. Therefore, great consideration and precautions should be taken prior to incorporating new technologies into the Internet. Unfortunately, that has not been the case so far. This section discusses some of the security flaws caused by the design of IoT ecosystem.

### *5.1 Exploiting the Weakest Link*

Most IoT deployments involve the devices, middleware, and a gateway. In order to protect IoT devices from having their confidentiality, integrity, or availability compromised, security needs to be applied across every segment of the network. If any network segment has a vulnerable vector, a cybercriminal can potentially gain access to the data from the device and cause irreparable harm and/or loss.

## 5.2 *Insecure Message Protocols*

The excitement, opportunity, and potential for making money in selling IoT solutions has prompted many vendors to release their products to market as fast as possible at the cost of security. When products are rushed to market, it is often at the detriment of the security mechanisms. The following illustrate common issues encountered when security procedures are poorly implemented or omitted entirely.

### 5.2.1 MQTT Protocol

The message queue telemetry transport (MQTT) protocol is an Oasis standard for IoT communication. Within the standard, an option exists to use transport layer security (TLS) to encrypt communications between devices and the Internet. This option is recommended by the Cloud Security Alliance, but many vendors do not follow the recommendation. As a result, all communication between devices that do not follow the recommendations is easily available for anyone to intercept. Another issue with the MQTT protocol is that the standard does not allow devices to authenticate to servers. Because of this, a cybercriminal impersonating a management server can intercept all traffic from a MQTT-enabled device. In a 2017 study, a researcher discovered over 60,000 IoT message servers that did not require authentication. In the study, the researcher demonstrated that using a MITM attack, he could quickly intercept and modify data destined for hospitals, prisons, and satellite control systems due to the lack of security to the MQTT protocol (MQTT).

### 5.2.2 Telnet Protocol

Telnet is a protocol which has been around since ARPANET. It is a simple management protocol that gives administrators easy remote access to devices. Telnet does not encrypt its communications and it is typically configured to be enabled by default in most IoT devices. Some Telnet implementations do require authentication, but since all communication is transferred openly (no data encryption), a cybercriminal need to only discover a way to observe the network traffic to and from the Telnet port in order to identify sensitive information such as device username and password. Other methodologies to gain access to devices using Telnet include brute force access (submitting common device users/password combinations such as user/password) or trying default user login credentials that can easily be obtained from searching the Internet (most common access method). In many cases, the username and password are hard coded in the devices, allowing cybercriminals unfettered access to these devices most times without the knowledge of the owner and eliminating the owner's ability to change the hard-coded login credentials. Hard-coded device passwords is what the Mirai botnet used with the

Telnet protocol to gain access to millions of IoT devices worldwide. This botnet attacked and shutdown Twitter, as well as some other well-known services.

# 6 Reducing the Risk of IoT-Related Threats

As chapter "Converting Disability into Ability Using IT/IS and Smart Textiles" illuminated, any device or system that connects to the Internet will also inherit the risks of the Internet. In addition to that, chapter "A Mobile Nursing Solution" pointed out that the IoT ecosystem inherits risks that originate from the IoT device themselves. Fortunately, there are steps that can be taken to reduce and mitigate the risks as discussed in this section.

## *6.1 Demand Secure Devices*

Section 3 highlighted the fact that the type of data medical devices generate is high value – it is sensitive in nature and can literally mean life or death for a patient that depends on it. Because of this, manufacturers should take a zero-trust mentality to designing the software and hardware of the devices. A zero-trust-based approach assumes not to trust a device by default. This approach is the exact opposite of the design of the Internet, which was initially engineered assuming all devices can be trusted. In addition, the most vulnerable ports and services, such as Telnet, should not be enabled or in IoT devices. The utilization of vulnerable ports and services such as Telnet are not necessary with modern-day protocols. Also, encryption of all device stored data and communications must be mandatory to further reduce the possibility for cybercriminals to access and eavesdrop on communications.

## *6.2 User Cybersecurity Training*

Throughout the history of the Internet, holistically, the weakest link has always been the end user. Studies show that most users use unsecure passwords (if any), do not update their system software, and always leave their systems on and connected to a network. Compounded with the fact that cybercriminals are getting increasingly better at phishing attacks and malware exploitation, it is easy to understand how cyber-attacks of every kind are on the increase. While there is no way to completely stop these attacks, studies show that user training is the most effective way to reduce the amount of attacks in an organization. In particular, users that have been trained to recognize the importance of updating device software reduce the threat vector with every system update. Such measures minimize the cybercriminals attack surface for breaking into a device.

### 6.3 Use Network Security Tools

Tools such as anti-virus software and firewalls are still mandatory to protect against attacks to IoT devices. Unfortunately, these devices depend on signatures and models which can take a significant amount of time to identify and develop. Because of this, protection against cyber threats is highly reactionary and poses no real immediate threat to the cybercriminals. To address this deficiency, a more proactive approach is needed. Such an approach would effect precursors to attack activity through system-wide monitoring. An example is shown in Fig. 2. Tests using this framework have been able to detect the interactions of botnet agents within a network. This could be useful in protecting IoT devices. For example if multiple IoT devices are compromised and placed in the same compromise network, the collective use of these devices now become part of the botnet. The approach shown in Fig. 2 could detect botnet activity locally. Additionally, locally compromised devices are often joined with other compromised devices from diverse and unrelated locations to comprise a a larger botnet. The monitoring approach could also identify key characteristics of those devices as well. More details concerning this approach are given in Sect. 7.3.

## 7 Advanced Security Through Network Science

Network Science is the study of complex networks such as social, biological, or communication networks, where networks are represented by graphs. Elements of the graph are referred to as vertices or nodes and connections between the elements are referred to as links or edges. Identifying the structural properties of networks can have many advantages. Community structure is a property which has proven to be effective at analyzing complex networks.

### 7.1 Community Detection

Nodes in a graph that are tightly correlated through edges have community structure as opposed to more disparate nodes. Any correlation between nodes can be considered an edge. Examples of edges between nodes are: scientists collaborating on research papers, members within a karate club, or computer messages transmitted from a sender to a receiver. The latter is of particular interest to IoT devices since network traffic is transferred from sending "things" to receiving "things." It is important to note that community discovering algorithms are different from clustering algorithms. This difference is because community detection algorithms group nodes based on context and not just distance, which is the case in clustering algorithms. There are several methods used to construct the communities. The most

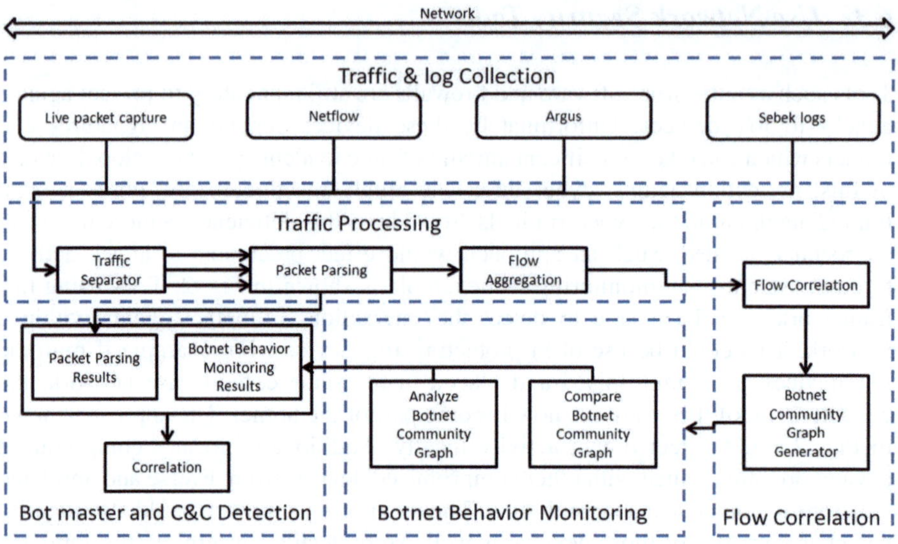


**Fig. 2** Network science based analysis framework

prevalent methods are hierarchy-based (Wang et al. 2011), modularity (null model)-based (Perry and Wolfe 2012), information theory-based (Rosvall and Bergstrom 2007), and clique-based (Palla et al. 2005). Each model has its drawbacks and advantages. The clique-based method forms communities by the percolation of fully connected adjacent subgraphs (Palla et al. 2005). Subgraphs are said to be adjacent when they share *k-1* nodes. The major drawback to this approach is it is not considered the most efficient community detection approach, although recent tests show that it functions well for networks up to several million (Palla et al. 2005) (Fig. 2).

## *7.2 Identifying Hidden Correlations*

Correlating nodes based on context allows nodes to be members of multiple groups. This property is called node overlap. Allowing nodes to be members of multiple communities allows cross-correlation between communication streams, which sheds light on the underlying meaning for the network as a whole. This includes correlations between nodes which go beyond the intended the community structure. An example of this can be found in the widely used Karate club dataset used to test community detection algorithms. In this dataset, there was a split in the Karate club due to a popular instructor starting his own club after a dispute. The intended catalyst of the community development are the two instructors. Members are either grouped in one class or the other, but while evaluating the metadata of the members several members also belong to the same social groups beyond the karate club which makes it more likely that they will stay in a class with their friend.

## *7.3 Network Science-Based Analysis Framework*

The lack of security in the current IoT communication protocols of choice coupled with the inability of current tools to identify rapid changes in malicious code, the characteristics of malicious code use have inspired the creation of a flexible approach to monitoring and analysis. Community structure describes data by evaluating trends within data which are not directly related. To demonstrate how community structure analysis can be used to enhance the security of Internet-enabled devices, we present a framework which was designed to identify threat actors involved in operating a peer-to-peer (P2P) botnet. Botnets are currently the attack platform of choice for cybercriminals targeting and compromising IoT devices. In this example, traffic was captured and analyzed using a honeynet. The framework has five components; traffic and log collector, traffic processor, flow correlation, bot master, and C&C detector, botnet behavior monitor, and flow correlator.

### 7.3.1 P2P Botnet

A P2P botnet is a network of compromised devices connected to the Internet. Typically, the owner is unaware that the device(s) have been compromised. The term P2P refers to the command and control structure, where control of the botnet is peer-based and not centralized to one command and control host. This approach is flexible and robust since any device in the network architecture can be used to manage the botnet. Utilization of this command structure has traditionally been hard to analyze and prevent due to the unstructured nature of P2P.

### 7.3.2 Honeynets

A honeynet is a network of Internet-enabled devices which are designed to attract and study Internet threats. To attract threats, a honeypot (a single device in the honeynet) is designed to emulate the services of a user device. For example, as mentioned earlier, Telnet is a protocol which is used extensively to gain access to many IoT devices. In an attempt to study and analyze how cybercriminals are exploiting Telnet, a honeypot is set up with the Telnet port open. When attackers connect to the honeypot, all interactions are recorded and visualized using specialized analysis tools. Adding more than one honeypot creates the honeynet.

### 7.3.3 Traffic and Log Collection Module

The traffic and log collection module is the entry point to the network science-based analysis framework. This module captures and monitors network traffic at the edge of networks and at each sensor of installed honeypots within the honeynet. This module has two modes: live traffic capture from a network in the form of packets using the packet capture (PCAP) library, and loading captured files stored on a file system. Currently, the types of stored traffic that can be read with this module are PCAP, Netflow, Argus, and Sebek. Other forms of log data can be formatted to be included in the framework. Once collected, data is forwarded to the traffic-processing module.

### 7.3.4 Traffic-Processing Module

The traffic-processing module preprocesses the traffic for flow correlation. The traffic separator module within the component separates the network traffic into 10-minute time windows in order to illustrate the changes in the behavior over time. The time window in this experiment was set to 10 minutes because it is the default time programmed into the malware for connection intervals. Future work will modify the time window and evaluate whether or not 10 minutes is the optimal value for community detection. The packet parsing module extracts header level information from the network traffic such as source IP (SrcIP), destination IP (DstIP), source port, destination port, TCP/UDP payload size, source to destination packet count and data size, destination to source packet count and data size, and session interval. This module also generates plain text and comma separated value (CSV) files based on above parameters. The flow aggregation module generates an input file for flow correlation.

### 7.3.5 Flow Correlation Module

The flow correlation module creates communities from the normalized data inputted from the flow aggregation module located in the traffic-processing component. A clique-based method is used to generate communities. In order to discover communities using cliques, the nodes and edges must be identified. The original *k*-clique percolation algorithm (Xie et al. 2012) is modified to perform the analysis. Instead of identifying one entity to be a node, secondary node is added. For the analysis, each IP address was equal to a “primary” node, and the message size was equal to the “secondary” node. The link between nodes is a tuple of {Time,SrcIP,DstIP,MsgLen}. For the purpose of constructing the communities, direction was not considered. Once the communities were formed the tuple {SrcIP, DstIP} was used to add direction in order to identify the role of each primary node. The purpose of the secondary node is to identify all relevant communications. For instance, many botmaster communications only include the bot master and the command and control node. By including the message size as a secondary node, we are able to separate each meaningful communication into a community even if it only contains two nodes.

### 7.3.6 Botnet Behavior Monitoring Module

In the botnet behavior monitoring module, a custom python script is used to analyze the current community graph and then compare it to previous graphs to determine changes that have occurred over time. Here, the evolution of nodes and fluid relationships within the same community and across multiple communities can be observed.

### 7.3.7 Bot Master and C&C Detection Module

The bot master and C&C detection module is where the results from the other modules are combined and correlated to produce pre-imminant cyber security intelligence. The botnet behavior modeling results component gives a display of what has changed over time within the community graphs.

The packet parsing results module captures and displays parsing results from each data source that was used in the analysis. This module is included to provide the analyst with a lens into the lower level data in the event that there is a question regarding the validity of the monitoring results. For instance, the full packet from PCAP data is available here. By including the payload of the monitoring results, which is based on flows, context is added to the analysis by being able to inspect the payload content. The analyst can also manipulate the data before correlation is facilitated.

The correlation module adjusts the output of the behavior monitoring results depending on what has changed in the packet parsing results. If no change has been made, the correlation results will be identical to the input for the behavior monitoring results. If a change has been made, the results will change accordingly. An example change would be an inaccuracy that the analyst notices. The significance of this module is, if there are minor issues with the analysis, they can be corrected at this point instead of having to render the entire analysis useless.

### 7.3.8 Framework Evaluation Results

The honeynet used captured live PCAP traffic to and from the command and control servers. In the initial experiment, connections between nodes were identical for each of the collection components. Because of this, the results will only be discussed in terms of PCAP traffic. After capturing and processing the data in the *traffic and log collection module* and the *log-processing module*, respectively, the data were

**Table 1** Community graph metric values

| Metric | Value |
|---|---|
| Analysis period (hours) | 168 |
| Community graphs | 1008 |
| Communities | 3120 |

**Table 2** Community graph 168

| Community | Nodes | Overlap |
|---|---|---|
| 128.1 | 3 | {128.1.1:128.2.1;128.4.1},{128.1.2:128.2.2;128.3.2;128.4.2} |
| 128.2 | 3 | {128.2.1:128.1.1;128.4.1, {128.2.2:128.1.2;128.3.2;128.4.2} |
| 128.3 | 24 | {128.3.2:128.1.2;128.2.2;128.4.2} |
| 128.4 | 3 | {128.4.1:128.1.1;128.4.1, {128.4.2:128.1.2;128.3.2;128.2.2} |

analyzed by the *flow correlation module*, and full PCAP data were sent to the *Bot master and C & C detection module.*

As a component of the *flow correlation module*, the *botnet community graph generator* runs the community detection algorithm to discover graphs of communities within the data. Table 1 contains the details of the results.

As shown in Table 1, the analysis period was approximately 168 hours, which is 7 days of analysis. Since the flow correlation module generates community graphs in 10-minute increments, 1008 community graphs were generated. Within those graphs 3120 communities were observed.

The *botnet community graph module* then analyzed the details of the communities in each graph using the *analyze community graph component*. Table 2 shows the details of one of the community graphs. Communities and each node within every community are delineated using a period. Table 2 shows the details of community graph 128. There are 4 communities within the 10-minute community graph. Communities 128.1, 128.2, and 128.4 each have 3 nodes within them. Community 128.3 has 24 nodes within it.

The overlap of communities 128.1, 128.2, and 128.4 show that they are closely related. For instance, node 128.1.1 is the same node as 128.2.1 and node 128.4.1. The second component within the botnet community graph module is the *compare botnet community graph component*. In this component, the current incoming community graph is compared with all the previous community graphs. The overall node overlap is significant within the botnet (84%). This was an expected finding since most attacks and other communications between bots and the command and control server are coordinated. Previous research found that most connections on the Internet have a very low overlap rate among nodes (Perry and Wolfe 2012). Since it was discovered that botnets have a high overlap rate, the findings could prove to be significant if the presence the high overlap is applied as a detection tool.

The *botmaster and C&C detection module* attempts to discover the role of each node within the communities. In the *botnet behavior monitoring results component*,

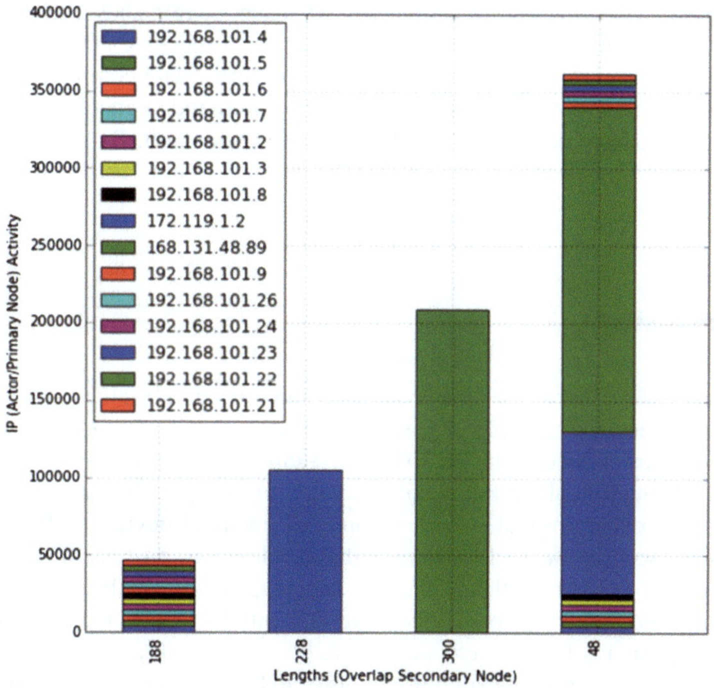


**Fig. 3** G-test of top 15 actors based on aggregate message sizes

relationships in the flow data that were collected were observed. Each node is grouped based on overlap. The packet payloads within the packet parsing results component were then used to discover the relationships between overlapping nodes.

To evaluate the significance of community overlap, the g-test is applied to the 15 most active actors in the botnet that sent messages to a command and control server. A g-test is a measure of fitness to a distribution of data. The g-test was used to discover correlations between actors and the overlap nodes. In the experiment, active actors refer to aggregated IP length sizes. Figure 3 displays the g-test results. The results showed that the 15 most active actors only sent 4 message types of sizes (188, 228, 300, 48). The varying activity of the actors highlight their roles. The IP length size 228 was only sent by actor (192.168.101.4) and size 300 was only sent by actor (192.168.101.5). This suggests a lack of coordination, which leads to the belief that these two actors are bot masters. IP length size 188 was used by the other 13 actors equally. This suggests that the activity was coordinated which leads to the belief that these actors are bots. Length size 48 was correlated with each of the actors, but actors (192.168.101.5 and 192.168.101.6) had uneven distributions while the other 13 actors had an equal amount of data sent. This is further evidence that the 2 uncoordinated actors are bot masters and the 13 coordinated actors are bots.

To evaluate the ability to discover the overall purpose of the botnet using this framework, the content of the overlap node messages was investigated. Message 48 was the basic "GET // HTTP" message sent when a node refreshes its command and control administration information. Message 228 and message 300 are distributed denial-of-service commands being sent to the command and control server. When the bots check back with the server, they get the command to attack a target. These four overlap nodes represent ~85% of the data in the dataset. By correlating these nodes it can be reasonably assumed that this botnet is being used specifically for DDoS attacks.

## 8 Conclusion

In this chapter, we discuss security concerns which need to be addressed before the wide-spread adoption of deploying medical devices on IoT networks. In particular, medical devices connected to the IoT ecosystem will undoubtedly contain highly sensitive information about the people they are monitoring. This data is extremely valuable to a cybercriminal, which means IoT medical devices will need to be created with security at the forefront of the development process in order to have any chance of protecting them. We present suggestions, such as implementing a zero-trust methodology for device development. In reality, many medical IoT devices have already been developed and are being sold today, and as mentioned in Sect. 5, the typical communication protocols of IoT devices are insecure. Because of this, new monitoring and analysis tools are required to complement IoT ecosystem security. To address this deficiency, we discuss a novel approach to identify components in a botnet attack, which has become the attack structure of choice for cybercriminals that compromise IoT devices. This approach is based on network science and provides a highly accurate way to detect malicious activities without using the standard signature or behavioral-based tools which can be subverted by changing the malicious code base or model of the attack. We believe that further development of this method will improve attack attribution, which will serve as another deterrent for would be cybercriminals.

## References

A Brief History of NSF and the Internet. (2003). Retrieved from https://www.nsf.gov/news/news_summ.jsp?cntn_id=103050

A Selected History of DARPA Innovation. (2018). Retrieved from https://www.darpa.mil/Timeline/index.html

Akpakwu, G. A., Silva, B. J., Hancke, G. P., & Abu-Mahfouz, A. M. (2018). A survey on 5G networks for the internet of things: Communication technologies and challenges. *Browse Journals & Magazines, IEEE Access, 6*, 3619–3647.

Amazon Elastic Compute Cloud (Amazon EC2), Cloud Computing Servers. Aws.amazon.com (2014-07-01). Retrieved on 2014-07-01.

Armerding, T. (2018). The 17 biggest data breaches of the 21st century. Retrieved from https://www.csoonline.com/article/2130877/data-breach/the-biggest-data-breaches-of-the-21st-century.html

Bryner, J. (2009). 40 years ago: The message that conceived the Internet. Retrieved from https://www.livescience.com/5839-40-years-message-conceived-internet.html

Cerf, V. (1990). Thoughts on the national research and education network. Retrieved from https://tools.ietf.org/html/rfc1167

Cerf, V. G., & Kahn, R. E. (1974). A protocol for packet network intercommunication. *IEEE Transactions on Communications, 22*(5), 637–648.

Deffree, S. (2017). ARPANET establishes 1st computer-to-computer link, October 29, 1969. Retrieved from https://www.edn.com/electronics-blogs/edn-moments/4399541/ARPANET-establishes-1st-computer-to-computer-link%2D%2DOctober-29%2D%2D1969

Goldstein, P. (2016). NSFNET served as a precursor to the Internet, helped spur scientific research. Retrieved from https://fedtechmagazine.com/article/2016/11/nsfnet-served-precursor-internet-helped-spur-scientific-research

http://vs.inf.ethz.ch/publ/papers/Internet-of-things.pdf

https://hitrustalliance.net/documents/content/Securing-Connected-Hospitals.pdf

https://iot.ieee.org/images/files/pdf/IEEE_IoT_Towards_Definition_Internet_of_Things_Revision1_27MAY15.pdf

https://www.gartner.com/newsroom/id/3598917

https://www.mckinsey.com/industries/semiconductors/our-insights/whats-new-with-the-internet-of-things

Lomas, N. (2018). Facebook data misuse scandal affects "substantially" more than 50M, claims Wylie. Retrieved from https://techcrunch.com/2018/03/27/facebook-data-misuse-scandal-affects-substantially-more-than-50m-claims-wylie/

MERIT Networks, Inc. (1995). NSFNET: A partnership for high-speed networking final report. Retrieved from https://www.nsf.gov/news/special_reports/nsf-net/textonly/related.jsp

NSFNET. (2017). Retrieved from http://www.internet-guide.co.uk/NSFNET.html

Palla, G., Derenyi, I., Farkas, I., & Vicsek, T. (2005, June). Uncovering the overlapping community structure of complex networks in nature and society. *Nature, 435*, 814.

Perry, P., & Wolfe, P. (2012). Null models for network data. Available at http://arxiv.org/abs/1201.5871v1

Postel, J. (1981). NCP/TCP transition plan. Retrieved from https://tools.ietf.org/html/rfc801#page-2

Rosvall, M., & Bergstrom, C. T. (2007). An information-theoretic framework for resolving community structure in complex networks. *Proceedings of the National Academy of Sciences of the United States of America, 18*, 7327–7331.

Stewart, W. (2000). The living Internet. Retrieved from https://www.livinginternet.com/i/ii_arpanet.htm

Wang, J., Li, M., Chen, J., & Pan, Y. (2011). A fast hierarchical clustering algorithm for functional modules discovery in protein interaction networks. *IEEE/ACM Transactions Computational Biology and Bioinformatics, 8*, 607–620.

Xie, J., Kelley, S., & Szymanski, B. (2012). Overlapping community detection in networks: the state of the art and comparative study. In *Social and information Networks*. ACM.

# Mosquitoes and Public Health: Improving Data Validation of Citizen Science Contributions Using Computer Vision

**J. Pablo Muñoz, Rebecca Boger, Scott Dexter, and Russanne Low**

## 1 Introduction

Billions of people around the world live in areas where mosquito-borne diseases such as malaria, dengue fever, and West Nile virus are endemic. Each year, over a billion people are infected, and more than one million people die from these diseases (WHO 2014). While many species in several genera can carry diseases, there are three genera that are the most common disease bearers, *Anopheles*, *Culex*, and *Aedes*. *Anopheles* species spread malaria, while *Culex* species are important vectors for West Nile virus (WNV). One species in particular, *Aedes aegypti*, spreads several diseases that include dengue fever, Zika, and chikungunya. A close relative, *Aedes albopictus*, also spreads these diseases but to a lesser degree than *A. aegypti*. These two species are found in almost all the countries in the world (Leta et al. 2018).

Given the immensity of the public health problem, a variety of methods are being used to control the mosquito populations and spread of diseases. The challenges are difficult to overcome. Climate change, land-use patterns, and the large numbers of people and goods moving from place to place over short and long distances all contribute to exposing people to mosquito-borne diseases. Furthermore, there are many species with different habitat patterns and behaviors. For instance, some are dusk and dawn biters while others prefer their blood meal during the day. While all species require standing water for the early life stages of eggs, larvae, and pupae,

---

J. P. Muñoz (✉) · R. Boger · S. Dexter
Brooklyn College, CUNY, Brooklyn, NY, USA
e-mail: rboger@brooklyn.cuny.edu; sdexter@brooklyn.cuny.edu

R. Low
Institute for Global Environmental Strategies, Arlington, VA, USA
e-mail: rusty_low@strategies.org



N. Wickramasinghe, F. Bodendorf (eds.), *Delivering Superior Health and Wellness Management with IoT and Analytics*, Healthcare Delivery in the Information Age, https://doi.org/10.1007/978-3-030-17347-0_23

different mosquito species prefer different habitats to lay their eggs. Some like wetlands, ponds, and other natural water bodies. *A. aegypti* likes small patches of water created by container trash and other artificial artifacts of human activities (e.g., tires, bases of flower pots, and water tanks).

Public health officials, entomologists, epidemiologists and other researchers modeling disease are always in need of data to better predict potential disease outbreaks and identify communities at risk. Public health officials do not have the resources to fully monitor and track the mosquitoes that transmit the diseases. When an outbreak does occur, public health officials may not have the capabilities to effectively react, mitigate and warn the public.

One way to increase data collection, increase communication among the public, mosquito scientists and public health officials in order to avoid or mitigate disease outbreaks is to involve citizens in addressing the problem and to make them part of the solution. Participation in citizen science projects is rapidly growing both in the number of people participating and the different types of projects offered. For mosquito monitoring, citizen scientists can collect data on adult mosquitoes or the early life stages (egg, larva, pupa), where they are found, and the type of water habit for the early stages and perhaps more importantly report or remove the standing water. As part of the data collections, photos are submitted of the mosquito larvae.

State-of-the-art image understanding techniques can assist communities in detecting dangerous specimens. We describe a system and the implementation of a prototype that combines image classification techniques, citizen science, and mobile applications to respond to mosquito-borne epidemics. Furthermore, the architecture of our system can support a variety of applications that involve a combination of image understanding and citizen science.

Citizen science and crowdsourcing are important means of deterring the spread of disease. With the state-of-the-art technology, these approaches can greatly increase and improve their positive contributions. One of our goals is to provide citizen scientists with access to the latest technologies for mobile communication, image classification, and object recognition, in addition to information about their potential uses, so they can better help their communities. The optical recognition system described here is an important element for streamlining the data quality/data assurance process. This is particularly important for a citizen science project that is global in scope with the potential to have thousands of people submitting photo data.

The prototype of our system described later in this chapter is an innovative solution that follows the premises of the *Design Science* paradigm (Hevener https://aisel.aisnet.org/misq/vol28/iss1/6/), in which novel artifacts help improve *Information Systems (IS)*. In brief, our prototype allows anyone with a smartphone and an Internet connection to submit images of mosquito larvae that are then identified using artificial intelligence, creating conditions for an arbovirus outbreak to be recognized very early. With additional improvements, this prototype system will have obvious utility in real-world applications. The ongoing improvement of the system can also benefit from *Action Design Research (ADR)* (Sein et al. 2011) in which iterations of concurrent development and user evaluation of the system will help us identify emergent behavior of the system that may not have been anticipated during the design of our prototype.

This chapter is organized as follows: Sect. 1 provides an overview of trends in citizen science; Sect. 2 discusses applications of computer vision techniques in support of citizen science; Sect. 3 introduces the particular computer vision technique (image classification) used in our proposed system; Sect. 4 describes the general structure of our proposed system; Sect. 5 describes our first proof-of-concept implementation of the system; Sect. 6 shows the initial results of our experiments with the first prototype of the system; and Sect. 7 presents some possible directions for improving our prototype. This chapter is an extension of the conference paper *Image Recognition of Disease-Carrying Insects: A System for Combating Infectious Diseases Using Image Classification Techniques and Citizen Science* presented at HICSS-51 (Munoz et al. 2018).

## 2 The Growth of Citizen Science and Data

Over the past 20 years, citizen science has become increasingly popular, with growth in the number of citizen science projects, the number of citizen participants, and the number of formal institutions supporting citizen science specifically. As a result, the number of publications about or using citizen science data has risen dramatically (Bonney et al. 2014; McKinley et al. 2015). The phrase "citizen science" now has its own dictionary entry: *Oxford Dictionaries* defines citizen science as "[t]he collection and analysis of data relating to the natural world by members of the general public, typically as part of a collaborative project with professional scientists".[1] Worldwide, three professional organizations have been established: The Citizen Science Association (CSA) in the USA, the European Citizen Science Association (ESCA) in Europe, and the Australian Citizen Science Association (ACSA) in Australia. Nascent regional citizen science organizations are forming in Asia and Africa.[2] A new journal, *The Theory and Practice of Citizen Science*, was launched in 2016. In 2017, the United Nations in collaboration with the Wilson Center and other organizations launched the *Earth Challenge 2020 Citizen Science Initiative* whereby millions of people around the world are challenged to collect one billion data points for air and water quality, pollution, and human health.[3]

This rapid growth of citizen science has been enabled by the use of the Internet and smartphones (Land-Zandstra et al. 2016; Mazumdar et al. 2017), but also indicates that millions of people around the world want to participate in some

[1] https://en.oxforddictionaries.com/definition/citizen_science

[2] Co-author Boger participates in Global Mosquito Alert, a UN sponsored consortium.

[3] https://www.earthday.org/earth-challenge-2020-a-citizen-science-initiative/

manner. Various websites such as SciStarter[4] have been created to help people learn about the different citizen science opportunities available, connect people with projects, and even help people design new projects. A US government website[5] provides federal practitioner information about citizen science projects across the country in order to identify ways to leverage the citizen science movement.

Citizen science projects can be classified by different typologies (Wiggins and Crowston 2011). Projects can be defined by the level of engagement of the participants in the science process (Bonney et al. 2009). At the less involved end of the spectrum, citizen scientists may contribute to data collection of a scientist-sponsored project. In "extreme citizen science," participants collaborate in the science process from problem definition, data collection, to analysis (Haklay 2013). Subject matter spans many different scientific disciplines such as astronomy, mapping the human brain, bird counts, rain data, and many more. The majority of natural science citizen science projects are biological in nature, such as Cornell's long-standing eBirds project,[6] although the types of projects are expanding in health and virtual projects such as Galaxy Zoo (Follett and Strezov 2015).

Participation in citizen science projects can be an effective vehicle for enhanced science literacy and knowledge; strengthened connections between people, nature and place; increased community-building, social capital, social learning and trust; and motivation by citizen scientists to influence policy and/or improve living environments (Haywood et al. 2016; Lawrence 2006; Jordan et al. 2011), However, the types of impacts vary with the way the citizen scientists participate. Bonney et al. (2016) use a typology that includes four citizen science project categories based on the broader public participation science framework – data collection, data processing, curriculum-based, and community-based. Within this framework, Bonney et al. (2016) review the literature on impacts on people participation and conclude that there is limited but growing evidence of participant gains in science knowledge and process, participant attitudes toward science, along with other measures. As regards the data collection category, in which participants are primarily data gatherers, small to no change is evidenced in content knowledge, understanding of science process, and change in attitude. This lack of change may be attributed to the fact that citizen scientists who self-select to participate in science activities already have positive attitudes to science. Data processing projects show even less evidence for science impacts on participants despite the fact that there are many scientific publications based on the data. Furthermore, although there is a large amount of data processed by citizen scientists, the bulk of this is done by relatively few contributors (Sauermann and Franzoni 2015). Curriculum-based projects like GLOBE and BirdSleuth generally do show increased understanding of science content and process, although not always increased interest in science. The last category that (Bonney et al. 2016) discuss is the community, or co-created, science.

[4]https://scistarter.com

[5]https://www.citizenscience.gov/

[6]https://ebird.org/home

This type of citizen science is thought to have the strongest impact on participants' understanding of the science process since participants are intimately involved in the entire process from framing questions, designing methods, collecting, analyzing, to presenting results. The longest record of projects of this type is in the biological sciences.

The benefits for science and community-based researchers are well established, including the potential to gain spatial and temporal data that would not be collected otherwise (Chandler et al. 2017; McKinley et al. 2017). Many scientists/researchers, though, have viewed citizen science collected data with skepticism, believing that the data do not have the scientific rigor necessary for research, resource management, or policy decisions. Rather, it has been treated as a means for education and outreach, as well as to make people better stewards for their communities, wildlife, or biomes.

This attitude toward the use of citizen science data is changing. Over the past 1015 years, there has been a sharp rise in the number of peer-reviewed articles that have used or studied citizen science (Duncan https://onlinelibrary.wiley.com/doi/full/10.1111/iej.12590). The types of articles vary with an increasing number of them addressing methodology and validation indicating that scientists are addressing concerns over data quality and being confident that the data can be used for research (Follett and Strezov 2015). There is currently a focus in the literature on ensuring data quality and the steps that need to be taken within the project design to assure sufficient data quality for project needs (Bonney et al. 2014; Hunter et al. 2013). Citizen science has shown to help improve conservation science, natural resource management, and environmental protection and can be indistinguishable from conventional science (McKinley et al. 2017). Steps taken for Quality Assurance/Quality Control (QA/QC) vary and depend on the types of questions being asked, methods used, and the accuracy and precision of the data collected. These are the same considerations for all research, whether or not citizen scientists are involved. Organizations and individual scientists need to design their research thoughtfully to clearly identify their research needs and goals as well as the ways the public can contribute and are motivated to be involved (McKinley et al. 2015; Shirk et al. 2012).

The design of our project follows best practices in QA/QC for citizen science. Most if not all projects that involve the public will provide instructions on how to collect the data including types of equipment used to make measurements or observations, when and where to collect, and other information deemed important for QA/QC. Some citizen science organizations require or encourage training before collecting data. The curriculum-based GLOBE program, for instance, requires teachers to participate in either face-to-face or online training before implementing citizen science activities in their classroom. The new GO platform does not require certification through training, but does provide background information and ways to practice before collecting and submitting data on the mobile-based app.

When designing a citizen science project, one needs to consider the geographic scope, number of participants, and frequency of data collection. GO and GLOBE are global, while others are national, regional, or local in scope. The number of data observations could be large, and if expert validation is required, could quickly become overwhelming for the researchers leading the project. With the rise of technology, there are various ways that large amounts of data can be checked after submission to the cloud, and the exact ways are dependent on the types of data. One way is to establish minimum and maximum values to be included in the database. For instance, if air temperature is being collected and degrees Celsius are required and a participant submits a value of 90, it might be rejected or flagged as a questionable value; follow-up to the participant may be needed to establish correctness. It may be a correct value for degrees Fahrenheit and the participant could revise the submission.

## 3 Computer Vision and Citizen Science

Many citizen science projects use photos for identification of objects. iNaturalist[7] and Leafsnap,[8] for instance, ask participants to collect and submit photos of wildlife and trees respectively. Other projects use photos collected by other methods; Snapshot Serengeti[9] collects photos of wildlife using a network of over 200 automated camera traps while Galaxy Zoo[10] provides photos of galaxies taken by the Hubble Space Telescope and the Sloan Digital Sky Survey. Citizen scientists are asked to identify objects in these photo collections. In projects such as Snapshot Serengeti and Galaxy Zoo, thousands and even millions of photos have been collected while new ones are being collected on an ongoing basis (Willi et al. 2018). Projects that involve citizen scientists in collecting new data through photos may not have millions of photos, but in the hundreds and thousands and could reach millions in the near future. Photos submitted by citizen scientists can be used as part of the QA/QC process. GO Mosquito Habitat Mapper (GO MHM) photos are used as validation for the type of mosquito habitat, identification of mosquito larvae, and lastly for identifying the genus or species of the mosquito larvae. At present, experts have to manually verify the photos to confirm or not the citizen scientist data submitted. Since GO MHM is a global effort, this means that there could quickly be thousands of photos submitted that need expert validation. Without a team of experts who either devote their time and the lack of funds to maintain a team of experts, the project can quickly become overwhelmed, and the data submitted could be excellent, but this may not be known or certain. Scientists and public health officials who may want to use the data will be hesitant to use the data. Image recognition, thus, can be a means to greatly reduce the need for experts to validate

[7]https://www.inaturalist.org/

[8]http://leafsnap.com/

[9]https://scistarter.com/project/646-Snapshot-Serengeti

[10]http://zoo1.galaxyzoo.org

every photo by removing those photos that are obviously incorrect when examined by eyesight.

Having expert validation is seen as an important step for data usage. A related project, Mosquito Alert[11], has three or four independent entomologists examine submitted photos of adult *Aedes albopictus* (Tiger Mosquito), an aggressive invasive species that can carry many diseases like dengue fever, Zika, and chikungunya. Mosquito Alert started in Spain and this helps to maintain the quality of the dataset through the network of entomologists in Spain. Other countries, such as Pakistan and France are adopting the Spanish app.[12]

With the rapid developments in computer vision algorithms based on artificial neural networks, many citizen science organizations have developed or are in the process of developing computer vision systems to increase the rate of photo identification as part of the QA/QC. Many are for wildlife identification (Horns et al. 2018; Willi et al. 2018; Kress et al. 2018; Mattos et al. 2014; Von Alan et al. 2004; Berger-Wolf et al. 2017), although there are several other foci including galaxies (Dieleman et al. 2015; Dhami et al. 2017) and neuroimaging (Keshavan et al. 2018). For the project Leafsnap, citizen scientists use the camera on a smartphone to collect photos of leaves. A shape-matching algorithm matches the shape of the leaf in the photo with reference photos in a database housed in a remote server. Within 5–20 seconds, the top matches are identified, and the app shows high-resolution images and background information of the species matched. It is up to the citizen scientist to identify the tree species. The high-resolution photo repository contains over 9000 photos taken by botanists using high-quality photographic equipment (Kress et al. 2018).

Our proposed system relies on a technique – image classification – which is slowly being incorporated in image-based citizen science projects. Image classification refers to the general problem of having a computer accurately assign a label to a provided image. Image classification techniques have reached very high accuracy in the past few years. Until 2011, the error rate in the *ImageNet Large-Scale Visual Recognition Challenge* – one of the most popular international image recognition competitions – was about 25%. In 2012, Krizhevsky et al. won this challenge using an approach known as *Deep Learning* (LeCun et al. 2015), obtaining an outstanding error rate of 16% (Krizhevsky et al. 2012), significantly outperforming traditional methods for image classification. Since that time, more researchers using Deep Learning approaches have entered the competition, reducing the error rate to the single digits (Russakovsky et al. 2015).

Broadly speaking, Deep Learning refers to the use of artificial neural networks (ANN) with multiple nonlinear layers. Each layer in the network is composed of neurons, i.e., the nodes of the network. The values computed at each of these neurons are transformed using nonlinear activation functions (e.g., sigmoid, rectified linear unit), and are weighted when passed to the next layer. Deep Learning

[11] https://www.gbif.org/dataset/1fef1ead-3d02-495e-8ff1-6aeb01123408#geographicCoverages

[12] Personal communication with the Global Mosquito Alert Consortium.

algorithms for image classification follow a supervised learning approach, that is, they "train" on a large number of labeled images, slowly adjusting the classification model's parameters using a method known as backpropagation until the algorithm correctly classifies most of the images in the training set. The particular kind of network used for image classification is called a Convolutional Neural Network (CNN).

Deep Learning has been applied to many problems over the last few decades: in one early example from the late 1990s, Deep Learning was applied to document recognition (LeCun et al. 1998). The revival and relatively recent success of Deep Learning algorithms is due mostly to the exponential increase in computing power known as Moore's Law (Schaller 1997), and the widespread availability of computers with powerful Graphics Processing Units (GPU). In addition to the improvements in GPUs, ongoing refinement of *artificial neural network* implementations has resulted in better classification schemes (Srivastava et al. 2014; Donahue et al. 2014).

The need for a large dataset of training images – often training requires thousands of images, or more – is one of the most challenging aspects of these state-of-the-art image classification algorithms. In some cases, public datasets are available to researchers and developers that want to train their algorithms. In others, such as the prototype for larvae recognition described below, there is no public dataset that is adequate for this kind of training. In the past few years, several projects, e.g., iNaturalist mentioned above, have created a variety of new datasets with the help of citizen scientists. For this project, we collected hundreds of images of larvae, and will make this available to the public for future applications and research.

Several other projects have or are developing computer vision systems using state-of-the-art techniques for object identification. A project similar to Leafsnap, LeafNet, is developing a Deep Learning system using CNNs to identify plant species in photos (Barré et al. 2017). The code used in LeafNet (Flavia[13]) has been made available to the public. Datasets, e.g., Foliage[14] and Leafsnap, are also publicly available to further develop and evaluate identification systems. These repositories of plant photos are two of several others around the world that have been developed for various leaf identification techniques (e.g., shape, texture, and color) using Deep Learning methods.

Zooniverse and iNaturalist have developed automated identification with CNNs with varying levels of success (Willi et al. 2018; Horns et al. 2018), and this is largely due to the large datasets used to train the CNNs. Hundreds of thousands and millions of photos for 40–55 species were used in the development of CNNs for the different citizen science projects (e.g., Snapshot Serengeti and Snapshot Wisconsin) in the Zooniverse platform (Willi et al. 2018). While iNaturalist has a dataset of over 850,000 photos, there are over 5000 species of plants and animals with large differences in the number of photos per species. The species with fewer

[13]http://flavia.sourceforge.net/

[14]https://www.researchgate.net/publication/235189451_Dataset_Foliage_-_Introduction

photos had poorer identification results. While promising, the results for iNaturalist illustrate many challenges in adopting Deep Learning methods to other citizen science projects with user generated photos and hence less controlled and consistent photo quality, and species with a limited number of photos. Here, we describe the prototype for image recognition for mosquito larvae that, like iNaturalist, has relatively few photos.

## 4 System Overview

The core of our system is a computational process that "recognizes" objects of interest, e.g., mosquito larvae. That is, users of our system provide an image and ask the system to assign one of a set of predetermined labels (such as mosquito genera) to that image.

The system is composed of a suite of applications. Each application plays a crucial role in the different phases of the system, as illustrated in Fig. 1. In Phase 1, we train an image classifier using images collected by our collaborators – trained entomologists – who carefully label the images. These researchers use a mobile application, the *Image Collector*, to label their images. We use these images to train a classification model. In Phase 2, citizen scientists use a mobile *Recognition Client* to upload images captured in the field and receive recognition results. Finally, in Phase 3, the system presents inconclusive images submitted by citizen scientists to entomologists who review and verify the presence of larvae in these images. The system sorts these inconclusive images in ascendant order based on their confidence

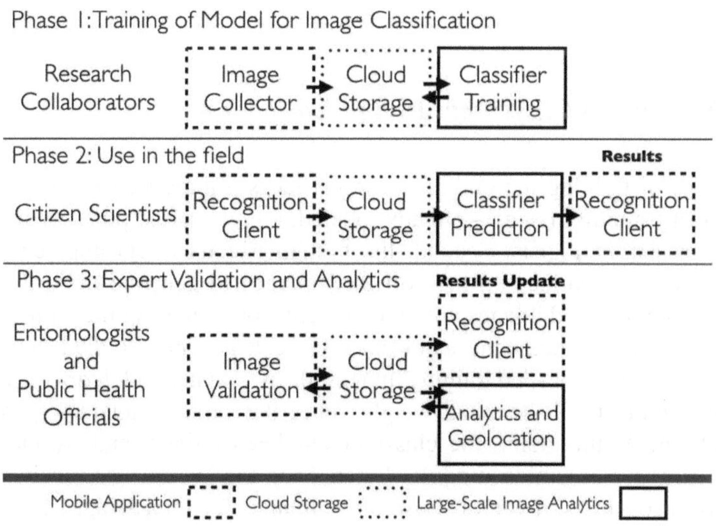


**Fig. 1** Phases of the system

value assigned by the classification algorithm. Once inconclusive images have been validated, the system can be used to alert experts in infectious diseases about geographic concentrations of dangerous larvae. Expert users, such as public health authorities, have access to all of the results produced by the recognition module, which allows them to confirm and oversee the whole recognition process.

At the end of Phase 3, our working prototype produces data that can be used by external analytics tools. Currently, we are visualizing this data using Google Maps. However, future iterations of this software, intended for actual field work, will include a fourth phase, in which a more sophisticated statistical analysis of user-submitted images will be used to provide public health experts with real-time information about developing or ongoing outbreaks of dangerous mosquitoes.

The different applications that compose the system streamline the processes of collecting and labeling the images; training classifiers with the collected data; and using these models for recognition of images submitted by citizen scientists. Raw image data as well as results of image analysis are stored in the cloud and are thus available, through the Internet, to all modules of the system. This data takes the form of text files, or small binary files that store recognition results.

## 5 Working Prototype

In this section, we discuss the details of the prototype implemented as a proof-of-concept of the system. We begin our discussion with the process of training the classifier for the image recognition approach currently used by the system, and then discuss the set of applications that we have implemented for our first working prototype.

### *5.1 Training of Classifier: Image Collection*

The process of training the classification model is a three-step process. First, we must collect training images – already labeled, e.g., as larva, pupa, or mosquito – from expert entomologists. Second, the system executes data augmentation and image processing routines on the collected images to generate multiple different versions of the received images with the objective of increasing the variety and size of the training data. Finally, the set of images is given as input to the Deep Learning network. Using these labeled training images, the neural network *learns* to recognize the images, adjusting its internal weights until it produces satisfactory results on a test dataset. At this point, the classifier has been trained, and we can make it available to the recognition module. Later, once the system has received images from citizen scientists, it can also use those images for improving, i.e., retraining, the classification model.

**Fig. 2** 60× Clip light-emitting diode (LED) microscope attached to a collaborator's smartphone

*Image Collection:* Deep Learning approaches require an immense amount of training data. For instance, in the *ImageNet* challenge mentioned in Sect. 3, 1.2 million hand-labeled images are used for 1000 different classes. We have developed a standalone mobile application, *Image Collector,* through which experts can feed the system with training data. First, the user identifies a training image, either retrieved from the device's storage or taken with the device's camera. For photographing mosquito larvae and pupae, users can attach a mobile microscope, such as the 60X Clip light-emitting diode (LED) microscope shown in Fig. 2. Then, the user labels the image. Currently, this application has been tailored to those genera known to transmit diseases to people, so the user can choose among *Aedes*, *Culex*, *Anopheles*, and Unknown. Each image must also be labeled with its metamorphic phase – *Egg*, *Larvae*, or *Pupa* – and whether the image contains a *Head*, *Tail*, or *Whole body*, or if it is a *Side view* of the insect. Close-up photos of the head and tail are necessary to correctly identify the genus of the larvae.

The application uploads the labeled image to our cloud storage, and that image is later used to train the recognition model.

Figure 3 shows a screenshot of the data collection mobile application, *Image Collector*. If the expert submitting the image is not sure about the correct label, she can simply use the last option, "Unknown," allowing other experts to address the labeling of the image. This application also collects the user name of our collaborator and a timestamp of the image, allowing the system to keep track of each collaborator's contributions. In Phase 1, the goal is to collect images from our collaborators to train the classification model for the first time. Once citizen

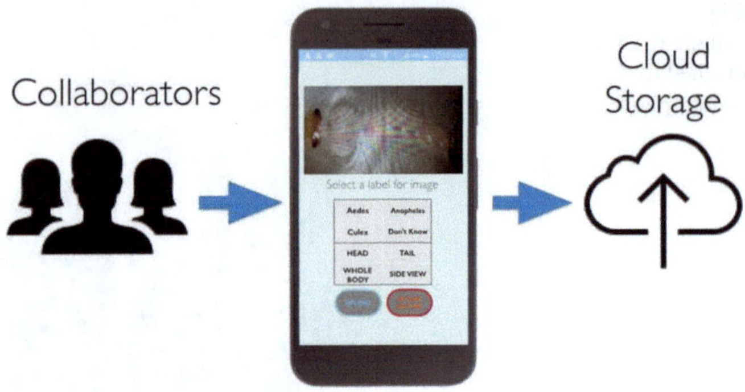


**Fig. 3** *Image Collector*: a standalone mobile application for image collection, labeling, and uploading to our cloud storage

scientists use their mobile application (Phase 2), their images can also be used to retrain the model with the expectation of making it more robust and accurate.

*Image Processing and Data Augmentation* To increase the variety in the data received by our collaborators, and to increase the size of the training dataset, we execute a data augmentation process on the collected images. First, each image received by our collaborators is manually processed to remove extraneous additional information contained in the image. For instance, borders are cropped since they do not usually contain any useful information for the classifier. The resulting cropped image is subject to the automatic data augmentation procedure that creates new images at different scales and orientations. Figure 4 shows an example of the data augmentation procedure. The transformed copies of the original image are generated using:

- Random rotation of the original image in a range of 0–60°
- Random horizontal and vertical shift in a range of 0–1/10 of the size of the image
- Random shear in a range of 0.20 radians
- Random flip either vertically or horizontally

In addition, as shown in Fig. 5, we perform histogram equalization to enhance the contrast of the images.

*Model training* Once we have increased the size of our training set of images, we proceed to train the Deep Learning model. We currently use the *CaffeNet* network from the *Caffe Deep Learning Framework* (Jia et al. 2014), which is an implementation of the *AlexNet* neural network, with some minor variations. *AlexNet* has historic importance in image classification, as it was the first *Deep Neural Network* to win the *ImageNet* classification competition in 2012. Details about the inner workings of our current selected neural network architecture can be found in (Krizhevsky et al. 2012). *AlexNet* was developed to classify images based on the 1000 classes of the *ImageNet* competition. This means that this network is over-

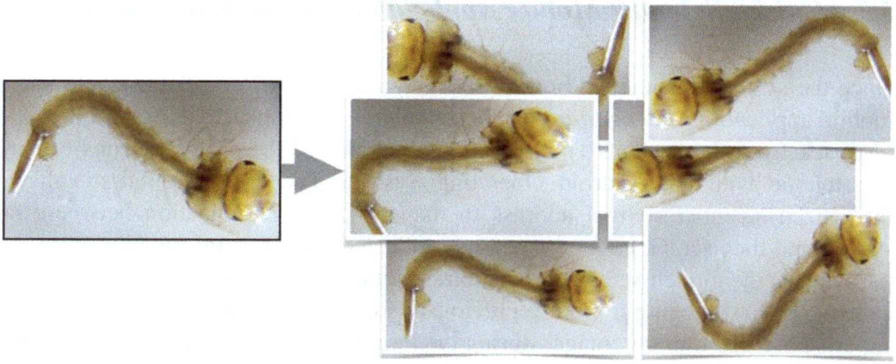

**Fig. 4** Data Augmentation Step. Each processed image is transformed to create multiple different versions of the original image. On the left is the original image, and on the right are samples of the new images generated from transforming the original image

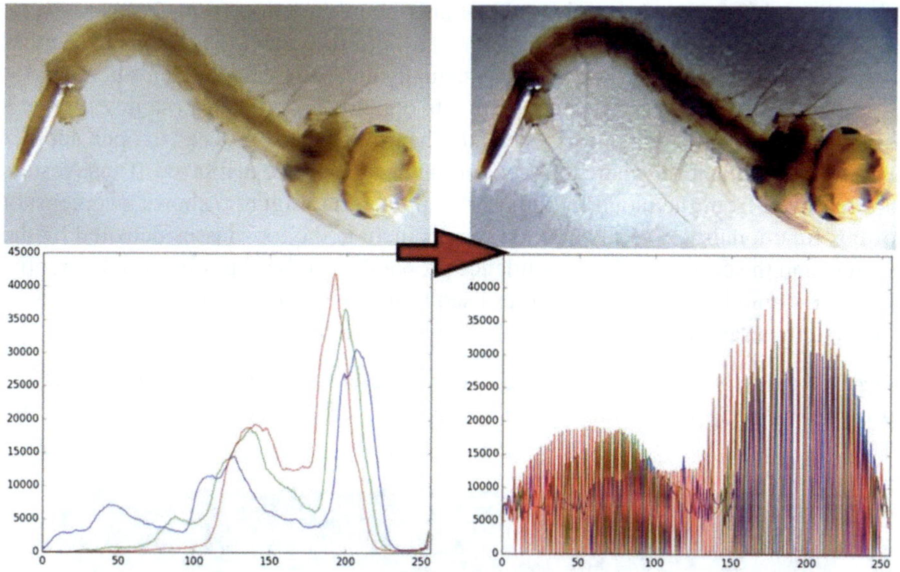


**Fig. 5** Histogram equalization

specified for our problem, since, for now, we only need to discriminate among 48 different classes. In the future, we plan to simplify the network to improve performance without losing accuracy.

We trained the model using 80% of the augmented dataset (original plus transformed images). The remaining 20% of the images are used for validation of the model. Once we have a trained model, we proceed to deploy it in the system's recognition module, which is explained later on in this paper.

## 5.2 *Citizen Scientist Mobile Application: Recognition Client*

Once the system has learned how to recognize larvae, citizen scientists can use a mobile application, *Recognition Client*, to submit their images and get immediate feedback from the system. The mobile application for the prototype runs on any smartphone with the Android Operating System. The application first collects the citizen scientist's email address to manage all communication between the system and the user. Once the system has collected this basic information from the participant, the citizen scientist can submit pictures, either by using a smartphone camera or by retrieving a picture previously stored on the device, similar to how experts provided training images during the training phase. In both cases, the image is sent to the system's cloud storage, along with a text file containing metadata such as the geolocation of the image. Separating this metadata from the image file can be used in the future to prevent unauthorized actors to identify the user that provided the image. This process is helpful for implementing user privacy settings, but it will also allow the system to receive other kinds of information from the user – e.g., additional descriptions of the environment where the image was collected, either collected from the device's sensors or manually entered by the citizen scientist.

After the user has submitted her picture to our cloud server, the system will send a confirmation that the image was received. The user can now proceed to open another view in the application where the recognition results will be displayed. These results are retrieved from the cloud by the user's application. Figure 6 shows a screenshot of this functionality. The recognition results include the top classes detected by the system and the corresponding confidence probability of the classification. Currently, the user's email is the only identifier used by the system, but in future versions of the system, a random unique ID might be generated for each new user.

*Overlay guidelines to assist the end user in taking a picture* To further assist the user when taking the picture of the specimen, we have included an overlay with

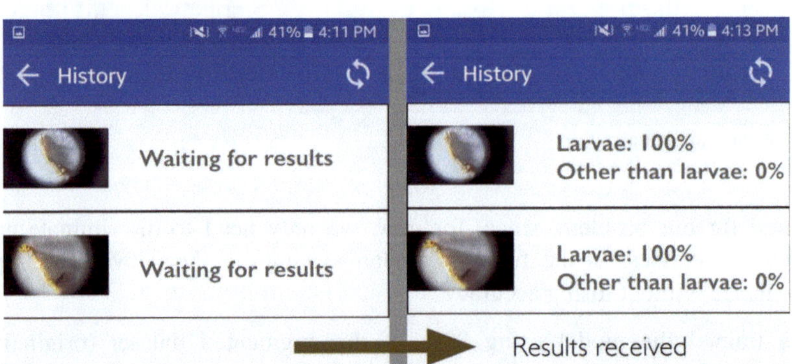


**Fig. 6** Screenshot of the list of images submitted by the user with their respective recognition result

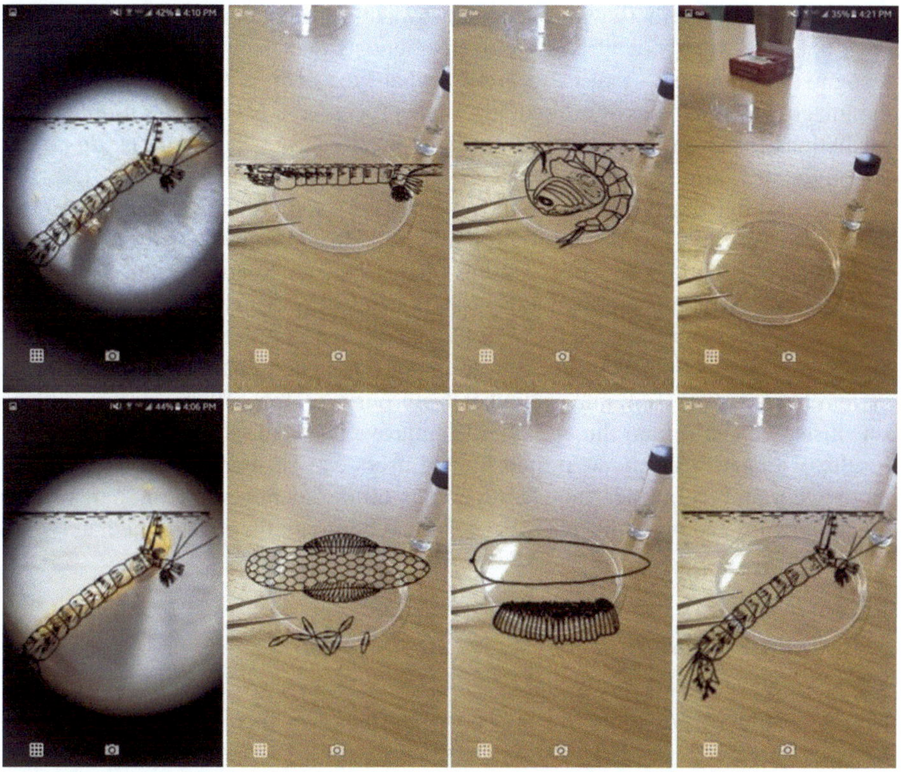

**Fig. 7** Example of the overlays used to assist the user when taking a picture

transparencies that guide the user and assist her positioning the mobile device. In the particular case of mosquito-borne diseases, and as shown in Fig. 7, the overlays include pictures of eggs, larvae, and pupa obtained from the *Centers for Disease Control and Prevention (CDC)* website. The purpose of the overlay feature is to improve the quality of the image taken by the user, and also to improve the positioning of the specimen to be recognized. The overlays implemented in our prototype range from a simple *water line* to help the user align the specimen if hanging on water and on side view, to more complex overlays, such as mosquito species in different stages (*egg, larvae, or pupa*). (The overlays are *not* part of the image, and are not transmitted to cloud storage.)

## 5.3 *Recognizing New Images: Classifier Prediction*

*Classifier Prediction* is at the heart of the running system. Once the classifier has been trained, the system can be deployed in area at risk for an outbreak.

This component interacts with our cloud storage, retrieving new images that have been uploaded for recognition by citizen scientists and using the trained CNN to recognize the images – that is, to predict what type of mosquito is represented in the image. Once the classifier/recognizer has processed a new image, it uploads its predictions to the cloud so the end user's application, i.e., *Recognition Client*, can access the prediction results.

The classifier has been conceived to be portable and efficient. We envision this module to run on a workstation by public health authorities. In the case of an outbreak, this module can be used anywhere with an Internet connection. This module also maintains a copy of the results locally so that they can be used later for analytics without the need for an Internet connection. Centralizing the classifier on one workstation makes it very easy to upgrade the classification model. Once the system has received new images, either from our research collaborators or citizen scientists, we can retrain the classifier to improve its prediction accuracy. Existing results of images already submitted by end users can be reprocessed to provide a more accurate label. Since the classifier runs on the manager's computer, we can quickly deploy the upgraded classifier to obtain better predictions immediately.

### *5.4 Image Validation by Expert Entomologists*

Despite all the recent successes in image classification, there is always the possibility of an erroneous classification (Nguyen et al. 2015). To confront this issue, we anticipate the use of human experts to monitor the recognition results of the system. We have created a mobile application that we term *Expert Validator* for image validation (Phase 3 in Fig. 1). This application connects to the cloud storage and allows experts to review the classification results. Images are presented to the expert, sorted by the confidence value of the classification. The user can click on the image to get a complete set of results if the list of results is too large. Figure 8 shows a screenshot of the image verification process. First, the images submitted by end users are marked as *pending validation*. Second, experts confirm that the label assigned to the images is correct. Finally, the end user is notified that an expert has validated the image.

On the end user's application, images that have not undergone validation show a *Pending Validation* label next to the prediction results. Those images that have been validated by an expert show either the *Accepted* or *Rejected* label next to the prediction results.

## 6 Pilot Evaluation of the First Prototype

We tested the system to determine its recognition performance on random images obtained online. We also tested the whole application pipeline simulating an

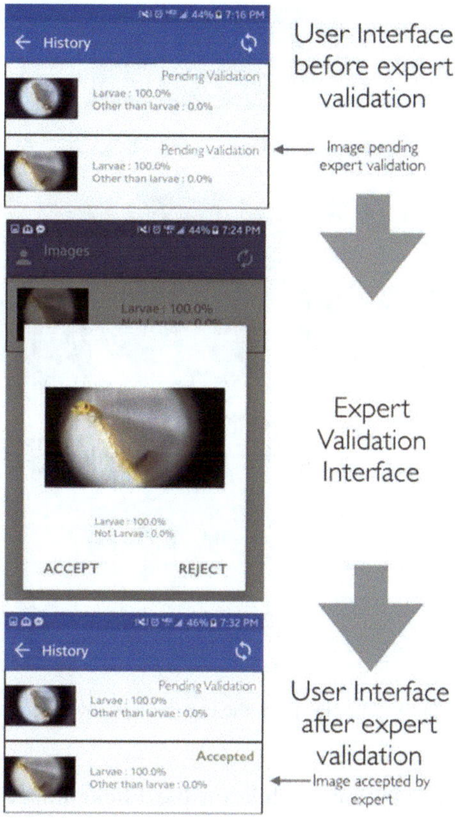


**Fig. 8** Top: End-user interface showing two images that have not been verified by an expert. Middle: The *Expert Validator* application is used by experts to verify images that have been flagged by the system. Bottom: The user is notified whether a submitted image has been accepted or rejected by an expert, despite of its recognition result

outbreak in the vicinity of our campus. We used a Dell Precision 5510 laptop computer with 8GiB of RAM, an Intel Core i7-6700HQ processor and an Nvidia Quadro M1000 M GPU. The recognition module runs on Ubuntu 14.04. The mobile components of the system were tested using three Android smartphones: a Samsung Galaxy S5, a Samsung Galaxy S6, and a Google Pixel.

We first tested the recognition performance of our classifier to discriminate between larvae and non-larvae. We used a random set of 550 images from the website *Flickr*.[15] These images were obtained by implementing a web crawler that downloaded random images of anything other than larvae, and 100 random images with the words "mosquito larvae" in their description. Figure 9 shows some examples of the images in this dataset. The images were then labeled to indicate whether they were considered to be *Larvae* or *Other than larvae* based on the label assigned by each *Flickr* user. We trained a classifier using the images received by our research collaborators in other labs and universities. We then ran the trained

[15] www.flickr.com

**Fig. 9** Samples from the test dataset downloaded from the Internet

recognition module on the Flickr images. The classifier produced the following results: Precision 100% and Recall 47.4%.

One hundred percent precision means that every single prediction that an image *does* mosquito larvae was correct, i.e., the system's prediction did not include any false positives. The system did, however, have a few false negatives, where the system failed to detect some images containing larvae. Of the set of 550 images collected randomly from the web, 57 images contained mosquito larvae. The recognition module recognized 27 of these 57 images or 47.4 percent. This relatively low recall rate is primarily due to the fact that the online dataset contained features that didn't appear in the training data, such as images containing multiple, sometimes overlapping, specimens. Figures 10 and 11 show examples of these images. The recognition process was unable to recognize the arrangement and quantity of the specimens in the picture as containing larvae. We are currently exploring the use of object proposal algorithms (e.g., Dollár and Zitnick 2013; Zitnick and Dollár 2014) to help the system improve its performance in these kinds of images. Additionally, we expect citizen scientists will use overlay guidelines to take better pictures than the images gathered randomly from the Internet to test our first prototype. In any case, the precision and recall values obtained from this dataset are interesting when looked at from the perspective that they reflect the behavior of the recognition module in the "wild." In future iterations of the system, with further training, the recall value will be greatly improved.

**Fig. 10** Examples of false negatives. Larvae are present in the images, but the classifier does not recognize them as containing larvae, likely due to the arrangement and quantity of the specimens

**Table 1** Execution time of procedures in the recognition module

| Execution time | |
|---|---|
| Procedure | Milliseconds |
| *One time* | |
| Setup classifier | 3940 |
| *Per image* | |
| Load image | 0.94 |
| Resize image | 1.30 |
| Histogram equalization | 3.24 |
| Prediction | 6.47 |
| – (Forward Pass – GPU enabled) | |

Table 1 lists the execution time of procedures executed in the recognition component. Once the classifier is running, it can process approximately 83 images per second in GPU-enabled mode.

## 6.1 System Response to Low-Quality Images

Based on our tests, we expect that if a citizen scientist uploads low-quality images, the recognition module, which has been trained using high-quality labeled images, will reject them by giving them a low probability value for the class in question.

**Fig. 11** True positives. Images correctly identified as mosquito larvae by the system

In such cases, the second line of defense is in the expert validation phase, in which experts could reject low-quality images that have been flagged by the system.

## 6.2 *Outbreak Simulation and Sample Visualization of Geographical Regions of Interest*

In the event of an outbreak, we expect citizen scientists to upload their observations and receive recognition results in return. The recognition module will be continuously updated with every batch of images received from citizen scientists. Epidemiologists will get precise information about geographic regions where relevant specimens have been spotted. This information will help them organize

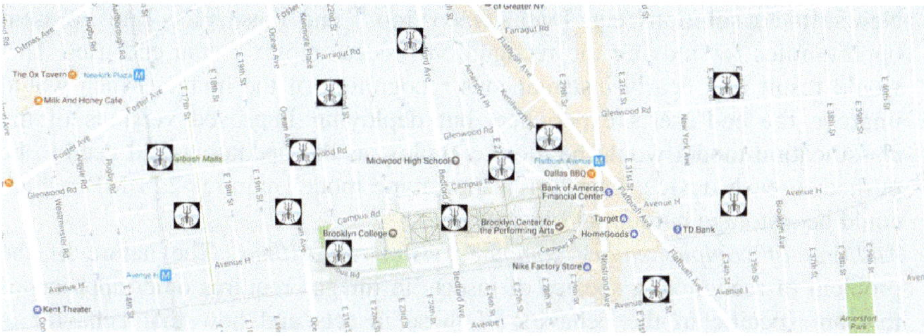


**Fig. 12** Testing the system and visualizing the locations of the positive recognition results

a response to the crisis. The system will help experts filter information received from citizen scientists, and this filtering will help them efficiently process more information than if they had to manually classify the images received from the users.

We tested the system in the vicinity of Brooklyn College. Volunteers walked around campus with the end user application installed on their smartphones. These volunteers chose places where mosquito larvae could be present in warmer months. Instead of using pictures taken from the camera on the device, they used images that were previously stored on the devices. The end-user applications uploaded the images to the cloud server and these images were later classified by the recognition module. The system correctly identified the images that contained larvae and passed the information to a Google Maps visualization script. Figure 12 shows a visualization produced from the exploration of the volunteers. Further analytics can be performed with the data collected (geolocation and classification results). The modularity of the system makes it easier to streamline the resulting data to other visualization tools, e.g., Google Maps. In the event of a real outbreak, experts could use these types of visualizations to identify regions that may require urgent attention from public health experts.

## 7 Discussion and Future Work

As exciting as we find this prototype, we believe that the system can further be improved with contributions from external developers. We have made our software public so others can continue building on the work we have done. Below are aspects of the system where we would like to see further improvements.

- *Deploying the model to the end user's mobile device*. Our current prototype carries out image recognition on a central device, communicating results back to the citizen scientist. Deploying the model and executing the classification algorithms on the citizen scientist's device (e.g., a smartphone) may be possible.

New software solutions, e.g., Intel's OpenVino,[16] and TensorFlow Lite[17] present opportunities for moving the recognition process closer to the end user. This would result in a nearly instantaneous recognition of the image, which would improve the end-user's experience, but deploying improved versions of the classification model would be more complex, as the updates would need to be pushed to each device. Our current prototype model requires 228 MB, which could be a storage burden for older devices.

- *Addition of complementary computer vision algorithms.* The nature of the problem of recognizing species of insects in images requires other approaches that are specific to the behavior of these insects and how that behavior is captured by the images. We believe that using other complementary approaches to image segmentation and recognition can refine the suggested classifications obtained from the *Deep Learning* algorithms. Another idea is to include video capturing functionality in the citizen scientists' applications, which can allow us to determine the species of mosquito based on their movements. We have considered including an *object proposal* algorithm (e.g., Dollár and Zitnick 2013; Zitnick and Dollár 2014) in future iterations of the system to facilitate the selection of the specimens on the training images. These algorithms analyze the image and suggest possible locations of objects. These approaches use features in the images, e.g., edges, saliency, or color variation, to make predictions about the location of objects in images.
- *Simplification of current Deep Learning network.* The current architecture used by our recognition module, *Caffenet/AlexNet*, was developed to classify images based on the 1000 classes of the ImageNet competition. This means that this network is over-specified for our problem since, for now, we would only need to discriminate between a few different classes. Future work will include the simplification of the network to meet the needs of our problem. This reduction in the complexity of the network will improve performance without sacrificing accuracy.
- *Multiplatform end-user application.* Future work includes expanding the supported software platforms for the system. Even though the Android Operating system is used by a large number of end users, we believe that our prototype must also include versions that run on other operating systems, including iOS, the operating system used by Apple devices.
- *Privacy considerations.* It is important to mention that the goal at this stage of our research is to use the prototype as a proof-of-concept. We must develop a more secure and robust system in future development iterations, particularly if the proposed system is going to be deployed on a large scale. A system deployed to the public must make sure that any information that can possibly link images to the user must be very carefully handled, particularly, when geolocation is included. Future work must include the generation of secure user

[16] https://software.intel.com/en-us/openvino-toolkit

[17] https://www.tensorflow.org/lite

IDs. Actions must be taken to prevent malicious people from gathering data on citizen scientists.

- *Providing additional information to the user.* In addition to recognition results, citizen scientists could receive other information that is tailored to the recognition results and the disease that the system is helping to combat in the future on their mobile applications.
- *Training the system with images that the system may recognize as false positives.* To make the recognition module more accurate, we are currently collecting images of Chironomidae Larvae, Oligochaeta, and Planaria. Since these specimens can sometimes be confused for mosquito larvae, we plan to train the system using those images as negative samples. We hope to make the system more accurate and reduce misclassifications.
- *Real-time analytics and management applications.* In the case of an outbreak, the system could benefit from new applications that display real-time analytics of the state of the system. More sophisticated Graphical User Interfaces (GUIs) can also be developed to better assist the users of the system.
- *Large-scale simulation of an outbreak.* Future iterations of the system must be evaluated to determine its capacity to handle a large number of submissions from citizen scientists. This evaluation will allow us to better estimate the necessary amount of computing power required and the level of organizational and governmental commitment in case of an outbreak. This evaluation must also include varying configurations for the number of image uploads per user with the goal of making the system more robust.

## 8 Ending Remarks

Current state-of-the-art technology allows for high-accuracy classification of images. In this paper, we have presented a modular system and its first prototype that is capable of leveraging computer vision techniques to empower communities affected by an arbovirus epidemic. The proposed system can provide valuable information to experts in charge of coordinating solutions. Our modular system provides a suite of applications that are capable of recognizing mosquito larvae in images with great accuracy. Future iterations of the system can be deployed to the citizen science community and help in the fight to control potential epidemics.

## References

Barré, P., Stöver, B. C., Müller, K. F., & Steinhage, V. (2017). LeafNet: A computer vision system for automatic plant species identification. *Ecological Informatics, 40*, 50–56.

Berger-Wolf, T. Y., Rubenstein, D. I., Stewart, C. V., Holmberg, J. A., Parham, J., Menon, S., Crall, J., Van Oast, J., Kiciman, E., & Joppa, L. (2017). Wildbook: Crowdsourcing, computer vision, and data science for conservation. *arXiv preprint arXiv:1710.08880.*

Bonney, R., Cooper, C. B., Dickinson, J., Kelling, S., Phillips, T., Rosenberg, K. V., & Shirk, J. (2009). Citizen science: A developing tool for expanding science knowledge and scientific literacy. *Bioscience, 59*(11), 977–984.
Bonney, R., Shirk, J. L., Phillips, T. B., Wiggins, A., Ballard, H. L., Miller-Rushing, A. J., & Parrish, J. K. (2014). Next steps for citizen science. *Science, 343*(6178), 1436–1437.
Bonney, R., Phillips, T. B., Ballard, H. L., & Enck, J. W. (2016). Can citizen science enhance public understanding of science? *Public Understanding of Science, 25*(1), 2–16.
Chandler, M., See, L., Copas, K., Bonde, A. M., López, B. C., Danielsen, F., Legind, J. K., Masinde, S., Miller-Rushing, A. J., Newman, G., & Rosemartin, A. (2017). Contribution of citizen science towards international biodiversity monitoring. *Biological Conservation, 213*, 280–294.
Dhami, D. S., Leake, D., & Natarajan, S. (2017, March). Knowledge-based morphological classification of galaxies from vision features. In *Workshops at the thirty-first AAAI conference on artificial intelligence*.
Dieleman, S., Willett, K. W., & Dambre, J. (2015). Rotation-invariant convolutional neural networks for galaxy morphology prediction. *Monthly Notices of the Royal Astronomical Society, 450*(2), 1441–1459.
Dollár, P., & Zitnick, C. L. (2013). Structured forests for fast edge detection. In *Proceedings of the IEEE international conference on computer vision* (pp. 1841–1848).
Donahue, J., Jia, Y., Vinyals, O., Hoffman, J., Zhang, N., Tzeng, E., & Darrell, T. (2014, January). Decaf: A deep convolutional activation feature for generic visual recognition. In *International conference on machine learning* (pp. 647–655).
Follett, R., & Strezov, V. (2015). An analysis of citizen science based research: usage and publication patterns. *PLoS One, 10*(11), e0143687.
Haklay, M. (2013). *Citizen Science and volunteered geographic information: Overview and typology of participation* (pp. 105–122). Dordrecht: Springer Netherlands.
Haywood, B. K., Parrish, J. K., & Dolliver, J. (2016). Place-based and data-rich citizen science as a precursor for conservation action. *Conservation Biology, 30*(3), 476–486.
Horns, J. J., Adler, F. R., & Şekercioğlu, Ç. H. (2018). Using opportunistic citizen science data to estimate avian population trends. *Biological Conservation, 221*, 151–159.
Hunter, J., Alabri, A., & van Ingen, C. (2013). Assessing the quality and trustworthiness of citizen science data. *Concurrency and Computation: Practice and Experience, 25*(4), 454–466.
Jia, Y., Shelhamer, E., Donahue, J., Karayev, S., Long, J., Girshick, R., Guadarrama, S., & Darrell, T. (2014, November). Caffe: Convolutional architecture for fast feature embedding. In *Proceedings of the 22nd ACM international conference on Multimedia* (pp. 675–678). ACM.
Jordan, R. C., Gray, S. A., Howe, D. V., Brooks, W. R., & Ehrenfeld, J. G. (2011). Knowledge gain and behavioral change in citizen-science programs. *Conservation Biology, 25*(6), 1148–1154.
Keshavan, A., Yeatman, J., & Rokem, A. (2018). Combining citizen science and deep learning to amplify expertise in neuroimaging. *BioRxiv*, 363382.
Kress, W. J., Garcia-Robledo, C., Soares, J. V., Jacobs, D., Wilson, K., Lopez, I. C., & Belhumeur, P. N. (2018). Citizen science and climate change: Mapping the range expansions of native and exotic plants with the mobile app leafsnap. *Bioscience, 68*(5), 348–358.
Krizhevsky, A., Sutskever, I., & Hinton, G. E. (2012). Imagenet classification with deep convolutional neural networks. In *Advances in neural information processing systems* (pp. 1097–1105).
Land-Zandstra, A. M., Devilee, J. L., Snik, F., Buurmeijer, F., & van den Broek, J. M. (2016). Citizen science on a smartphone: Participants' motivations and learning. *Public Understanding of Science, 25*(1), 45–60.
Lawrence, A. (2006). 'No personal motive?' Volunteers, biodiversity, and false dichotomies of participation. *Ethics, Place & Environment: A Journal of Philosophy & Geography, 9*, 279–298.
LeCun, Y., Bottou, L., Bengio, Y., & Haffner, P. (1998). Gradient-based learning applied to document recognition. *Proceedings of the IEEE, 86*(11), 2278–2324.
LeCun, Y., Bengio, Y., & Hinton, G. (2015). Deep learning. *Nature, 521*(7553), 436.

Leta, S., Beyene, T. J., De Clercq, E. M., Amenu, K., Kraemer, M. U., & Revie, C. W. (2018). Global risk mapping for major diseases transmitted by Aedes aegypti and Aedes albopictus. *International Journal of Infectious Diseases, 67*, 25–35.

Mattos, A. B., Herrmann, R., Shigeno, K. K., & Feris, R. S. (2014). A mission-oriented citizen science platform for efficient flower classification based on combination of feature descriptors. In *EMR@ ICMR* (pp. 45–52).

Mazumdar, S., Wrigley, S., & Ciravegna, F. (2017, January 19). Citizen science and crowdsourcing for earth observations: An analysis of stakeholder opinions on the present and future. *Remote Sensing, 9*(1), 87.

McKinley, D. C., Miller-Rushing, A. J., Ballard, H., Bonney, R., Brown, H., Evans, D. M., French, R. A., Parrish, J. K., Phillips, T. B., Ryan, S. F., & Shanley, L. A. (2015). Investing in citizen science can improve natural resource management and environmental protection. *Issues in Ecology, 2015*(19), 1–27.

McKinley, D. C., Miller-Rushing, A. J., Ballard, H. L., Bonney, R., Brown, H., Cook-Patton, S. C., Evans, D. M., French, R. A., Parrish, J. K., Phillips, T. B., & Ryan, S. F. (2017). Citizen science can improve conservation science, natural resource management, and environmental protection. *Biological Conservation, 208*, 15–28.

Munoz, J. P., Boger, R., Dexter, S. Low, R., & Li, J. (2018). Image recognition of disease-carrying insects: A system for combating infectious diseases using image classification techniques and citizen science. In *HICSS*.

Nguyen, A., Yosinski, J., & Clune, J. (2015). Deep neural networks are easily fooled: High confidence predictions for unrecognizable images. In *Proceedings of the IEEE conference on computer vision and pattern recognition* (pp. 427–436).

Russakovsky, O., Deng, J., Su, H., Krause, J., Satheesh, S., Ma, S., Huang, Z., Karpathy, A., Khosla, A., Bernstein, M., & Berg, A. C. (2015). Imagenet large scale visual recognition challenge. *International Journal of Computer Vision, 115*(3), 211–252.

Sauermann, H., & Franzoni, C. (2015). Crowd science user contribution patterns and their implications. *Proceedings of the National Academy of Sciences, 112*(3), 679–684.

Schaller, R. R. (1997). Moore's law: Past, present and future. *IEEE Spectrum, 34*(6), 52–59.

Sein, M., Henfridsson, O., Purao, S., Rossi, M., & Lindgren, R. (2011). Action design research. https://aisel.aisnet.org/misq/vol35/iss1/5/

Shirk, J. L., Ballard, H. L., Wilderman, C. C., Phillips, T., Wiggins, A., Jordan, R., McCallie, E., Minarchek, M., Lewenstein, B. V., Krasny, M. E., & Bonney, R. (2012). Public participation in scientific research: A framework for deliberate design. *Ecology and Society, 17*(2), 29.

Srivastava, N., Hinton, G., Krizhevsky, A., Sutskever, I., & Salakhutdinov, R. (2014). Dropout: a simple way to prevent neural networks from overfitting. *The Journal of Machine Learning Research, 15*(1), 1929–1958.

Von Alan, R. H., March, S. T., Park, J., & Ram, S. (2004). Design science in information systems research. *MIS Quarterly, 28*(1), 75–105.

Wiggins, A., & Crowston, K. (2011, January). From conservation to crowdsourcing: A typology of citizen science. In *2011 44th Hawaii international conference on system sciences* (pp. 1–10). IEEE.

Willi, M., Pitman, R. T., Cardoso, A. W., Locke, C., Swanson, A., Boyer, A., Veldthuis, M., & Fortson, L. (2018). Identifying animal species in camera trap images using deep learning and citizen science. *Methods in Ecology and Evolution*. https://besjournals.onlinelibrary.wiley.com/doi/10.1111/2041-210X.13099

World Health Organization. (2014). *A global brief on vector-borne diseases* (No. WHO/DCO/WHD/2014.1). World Health Organization.

Zitnick, C. L., & Dollár, P.. (2014, September). Edge boxes: Locating object proposals from edges. In *European conference on computer vision* (pp. 391–405). Cham: Springer.

# Part IV
# Applications of Portals in Healthcare Contexts

As with the developments of mobile and sensors, the adoption of portals to support healthcare delivery has grown rapidly in the last 5 years. While it is not possible to cover all areas, what we have tried to achieve in this section is to present a representation of key areas where portals are being embraced with success as the following eight chapters illustrate.

Chapter 23: Using Responsive Web Design to Enhance the User Experience of Chronic Disease Management Portals for Clinical Uses by Gunawardane and Wickramasinghe

Chapter 24: Older Adults Empowerment Through Training and Support and Its Implication on Proactive Self-Monitoring, Patient Engagement, and Connected Health by Bozan and Mooney

Chapter 25: An Evaluation of a Point-of-Care System Implementation and Adoption in a Multi-Campus Private Hospital in Melbourne by Muhammad and Wickramasinghe

Chapter 26: Leveraging the IOT to Enable the Guided Self-Determination Method by Wickramasinghe et al.

Chapter 27: Determining Missing Key Elements in OIS to Improve the Patient Experience and Clinical Care by Shaukat et al.

Chapter 28: Toward Actionable Knowledge: A Systematic Analysis of Mobile Patient Portal Use by Noteboom and Abdel-Rahman

Chapter 29: A Lazy User Perspective to Patient Adoption and Use of Personal Health Records by Kunene

Chapter 30: The Australian PCEHR or My Health Record: The Journey Around a Large-Scale Nationwide Digital Health Solution by Wickramasinghe and Zelcer

On reading these chapters, what should be evident is that portals can be used to support micro-level, meso–level, and macro-level health and wellness issues both within an acute care context, to facilitate better and on-going chronic condition support or even to provide a national e-Health solution. Thus, the versatility and adaptability of portals is evident. Once again however, as noted with the presentation of the chapters around mobile and sensors, it is important to note that successful solutions depend on technical and clinical outcomes being appropriate as well as high levels of clinical and patient user satisfaction.

# Using Responsive Web Design to Enhance the User Experience of Chronic Disease Management Portals for Clinical Users

**Ranganath Gunawardane and Nilmini Wickramasinghe**

## 1 Introduction

In general, four major disease groups, cardiovascular diseases, cancer, chronic obstructive pulmonary disease and diabetes, are discussed as chronic diseases (Australian Institute of Health and Welfare 2015a, b).

### *1.1 Disease Burden*

In a global study of the Australasia region in 2010 on the burden of disease, a measure that simultaneously compares the nonfatal burden (impact of ill health) and fatal burden (impact of premature death) of a comprehensive list of diseases, chronic diseases accounted for close to 85% of the total burden (Australian Institute of Health and Welfare 2014) (Fig. 1).

Cancer tops the list of the largest contributors to the total burden, while cardiovascular diseases come at a close third. These numbers are worse when considering only the fatal burden. Cancer contributed 33% and cardiovascular diseases 26% amounting to more than half of premature deaths in the region (Australian Institute of Health and Welfare 2014) (Fig. 2).

In addition, chronic diseases are a significant economic burden due to the combined effects of healthcare costs and lost productivity from illness and death.

R. Gunawardane
Epworth HealthCare, Deakin University, Geelong, VIC, Australia

N. Wickramasinghe (✉)
Epworth HealthCare, Richmond, VIC, Australia

Swinburne University of Technology, Hawthorn, VIC, Australia

 
N. Wickramasinghe, F. Bodendorf (eds.), *Delivering Superior Health and Wellness Management with IoT and Analytics*, Healthcare Delivery in the Information Age, https://doi.org/10.1007/978-3-030-17347-0_24

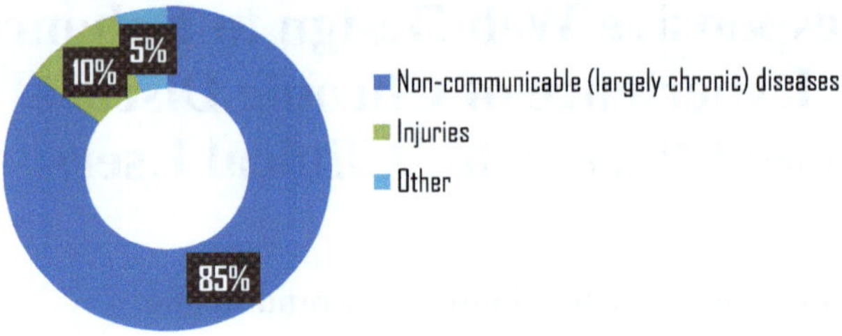


**Fig. 1** Contributors to the burden of disease in Australasia 2010 (Australian Institute of Health and Welfare 2014)

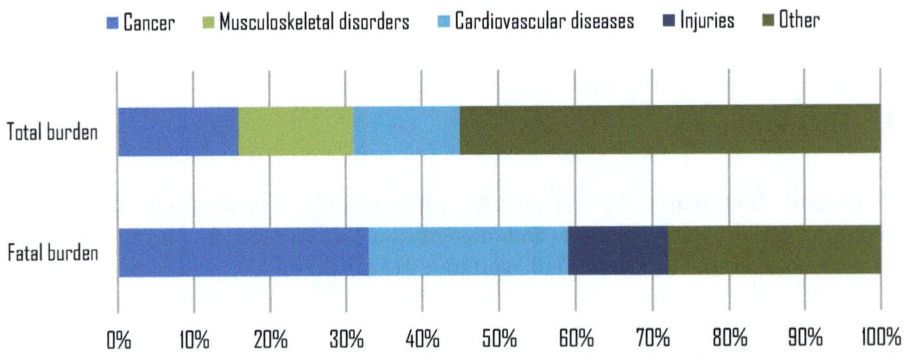


**Fig. 2** Largest contributors to the total burden and fatal burden of disease in 2010 (Australian Institute of Health and Welfare 2014)

Cancer is responsible for about one in every ten hospital admissions in Australia (Willcox 2014). Chronic diseases consume billions of dollars in avoidable health expenditure every year. The Business Council of Australia has estimated that if chronic diseases were totally eliminated, workforce productivity could increase by 10% (Willcox 2014).

As highlighted above, the burden of chronic diseases is multifaceted and massive. Moreover, these diseases are comparatively complex, and traditional methods of healthcare delivery, where the care is frequently fragmented, disorganized and duplicated, are inapt in managing them (Norris et al. 2003). Also, health systems that maintain current disease management practices cannot afford to continue caring for the ever-increasing numbers of people with chronic diseases (Singh 2008), which beseeches a new and improved disease management methodology.

## *1.2 Chronic Disease Management*

Chronic disease management is an organized, proactive, multicomponent, patient-centred approach to healthcare delivery (Norris et al. 2003; Barr et al. 2003; Lorig et al. 2001; McCorkle et al. 2011; Pincus et al. 2001; Thompson 1986). Such an approach also facilitates a strong coordination across care settings and providers, and there is mounting evidence to suggest that it is more effective than single or uncoordinated interventions (Singh 2008).

Essential components of a chronic disease management system include the following (Norris et al. 2003):

1. Identification of the population
2. Implementation of clinical practice guidelines or other decision-making tools
3. Implementation of additional patient-, provider- or healthcare system-focussed interventions
4. Use of clinical information systems to measure and manage the outcomes

### 1.2.1 Importance

Compared to other illnesses, the disease management plays an important role in chronic diseases due to a number of reasons.

#### Chronic Diseases Last Longer

Chronic diseases have no cure, and also, with improved treatment methods and facilities, people with chronic diseases have a better chance of survivorship (Australian Institute of Health and Welfare 2015a, b). Patients have to live with, through and beyond the disease, making the chronic disease management an ongoing process, not a one-off event.

It is important to have a capable disease management to lead survivors to adjust to physical and emotional changes after treatment, to help them cope with fear of recurrence, to supply follow-up medical care and to guide them through post-illness socioeconomic hardships (National Cancer Institute 2014).

#### Chronic Diseases Are Preventable

Although chronic diseases are among the most common and costly health problems, they are also among the most preventable and most effectively controllable (Center for Managing Chronic Disease - University of Michigan 2011). This is largely due to the fact that many chronic diseases not only share common risk factors but can also be risk factors for each other.

One worthy example is cancer, which currently places the largest and a growing burden on patients, families and the health system in Australia. It is also potentially one of the most preventable and treatable of all diseases. Almost one-third of all cancers may be avoidable, with more than a quarter attributable to just three risk factors: smoking, alcohol misuse and obesity (Department of Health, Australia 2015a, b).

Hence, when correctly managed, risk of many chronic diseases will decrease, healthcare costs will go down and overall quality of life will improve. This again highlights the importance of having a proficient chronic disease management.

### 1.2.2 Role of Technology

Today, we are witnessing more and more adoption of technology to support healthcare delivery especially technologies which make up the Internet of Things (Georgeff 2014).

Web Portals

As discussed earlier, one of the key features of chronic disease management methodology is the use of information technology – Internet portals, in particular, to facilitate tracking and monitoring interventions and outcomes and also enable regular measurement and collection of patient information, allowing clinicians to revise the treatments proactively. Known as chronic disease management portals, these portals allow patients to interact and communicate with and receive coaching from their healthcare providers, empowering the patient and increasing the patient contribution in the disease management.

Chronic disease management portals have come a long way from the early stand-alone computer applications that were confined to an intranet, often a supplement to an electronic health record system (Goldberg et al. 2003). Over time, with the rapid advancement of the World Wide Web, almost all these portals have moved onto become online web solutions. Web-based disease management portals have since become the de facto standard allowing clinicians and patients to communicate and share health information remotely, increasing the patient involvement in care planning and facilitating an ongoing doctor–patient dialogue.

This transition from stand-alone applications to the web has increased the reach of disease management portals.

One of the prominent features these portals provide is the patient self-management. Self-management is defined as a person's ability to manage the symptoms and the consequences of living with a chronic condition, including treatment, physical, social and lifestyle changes (Barlow et al. 2002). This with a collaborative and interactive relationship between patients and healthcare professionals empowers patients to take on responsibility for their condition with the appropriate clinical support (Davies and Batehup 2010; Kruse et al. 2015; Gardener 2011).

## Mobile Web

In the past decade, there has been an explosive growth in the use of mobile phones. The number of people using mobile phones is steadily increasing. In 2015, a record 4.8 billion mobile phones (out of which close to 2 billion are smartphones) are in use worldwide.

Australians, in particular, have embraced this digital phenomenon. An estimated 12.5 million adult Australians (70%) used a mobile phone to access the Internet for personal use in the 6 months to May 2014, up eight percentage points on the previous year. This proportion is larger than those who specified a laptop (68%) or desktop computer (62%) as a way to access the Internet for personal use. Tablets are not far behind either – 50% of Australians used a tablet to go online in the 6 months to May 2014, up eight percentage points from the previous year. Notable growth occurred in the volume of data downloaded to mobile devices, which almost doubled from the year before (Australian comminucations and media authority 2015).

The rise of wireless Internet has played a major role in mobile revolution.

## Responsive Web Design

The popularization of handheld communication devices and the rise of wireless Internet access have created a new paradigm in the way the web is accessed. For the past few years, mobile digital media time is fast overtaking static desktop-based media time. In fact, in 2015, in the USA, the mobile digital media time is now significantly higher at 51% compared to desktop (42%) (Meeker 2015).

All these numbers point to one clear fact; the future of web is through mobile devices. As more and more people start browsing the web through their smartphones and tablets, website publishers are now challenged to focus not only on better content and high discoverability but also on better presentation of that content over multitude of devices. The obvious answer to this is the responsive web, which enables a user to consume content on a website through the device of his/her choice and preference irrespective of the size of the screen or the method of input of that device.

While this transformation has made the web immensely popular, the growing use of mobile devices has posed quite unique problems to the web. The small form factor of these devices has shrunken the screen real estate and rendered the usual input devices such as keyboards and mice unusable. So in order to maintain and possibly improve the user experience, web designers improved their approach to web design, to provide an optimal viewing and interaction experience – easy reading and navigation with a minimum of resizing, panning and scrolling – across the full range of devices from desktop computers to mobile phones. This design approach is known as responsive web.

Known sometimes as Web 3.0, responsive web design is now an integral part of any decent website (Natda 2013). When designed to be responsive, the user experience across different devices of varying form factors becomes seamless and predictable. Responsive websites are agnostic to devices and their operating systems (ibid). A responsive web design ensures that users get the best and consistent experience of a website on any device of the user's choice and preference – be that the iPhone, the iPad, the smartphones running the Android OS or the Windows OS and several others (ibid). As a result, website owners and content publishers can need not exercise the option to build versions of their website for every popular device platform which they expect their audience might be using.

Even though the advantages of being a responsive website are glaringly obvious, most of the websites today are still not mobile friendly. A research testing the top 100,000 websites for responsive indicators shows that only about 12% of them are actually responsive (Podjarny 2015).

These figures are far worse for specialized fields such as health web portals. Even though there is a lack of such quantitative research specifically for health portals, a cursory look at the existing disease management portals, and disease management portals in general, suggests that they have not enjoyed the same amount of attention compared with other market segments such as social media – which have thrived on the ability to provide a responsive website to its users.

It must be noted that there has been a considerable amount of attention and research around the application of earlier web trends. Among them, a more prevalent topic is the use of Web 2.0, which represents the transition of the static web into a more interactive user experience. Web 2.0 is widely discussed as a way to engage a broader audience towards disease management portals and to empower the patient for a better self-management of the conditions (Haddad 2013).

The underlying idea of the responsive web design is that a website should gracefully 'respond to' every size of screen or display of devices, such as smartphone, tablet, notebook or any size desktop monitor (Natda 2013).

Studies have identified three major components of a responsive web design (Marcotte 2011).

1. A fluid layout to ensure a website can scale to the form factor of the browser
2. Images that work in a flexible context
3. Media queries, which optimize the design for different viewing contexts

## 2 Literature Review

This section starts by examining the prominent role the information technology plays in chronic disease management. Then, it examines the emergence and application of the responsive web design. Finally, it argues for the significance of the proposed research – to enhance the user experience of chronic disease management portals for patients and clinicians by using responsive web design.

## 2.1 Information Technology in Chronic Care

Information technology features prominently in every aspect of healthcare today, and chronic disease management is no exception. A clinical information system is one such element, which is established as one of the core components of an effective chronic disease management.

Every chronic disease management model, from the well-established Wagner care model (Thomas Bodenheimer et al. 2002) to more recent and refined models such as integrated chronic disease management (ICDM) model (Department of Health & Human Services 2015), emphasizes the importance of such clinical information systems.

Clinical information systems are often a combination of subsystems (Ahmed et al. 2011). One of the major components among them is disease management web portal, providing electronic patient self-assessment and management.

## 2.2 Web Portals in Chronic Disease Management

Disease management web portals have now become a central part of healthcare facilities, especially in the treatment of chronic diseases. The ability to engage patients and clinicians in a more customized, dynamic and longitudinal manner – sending a reminder saying 'It has been one year since surgery, you should have a PSA test' instead of mentioning at baseline that 'A PSA test is recommended one year after surgery', for example (Vickers et al. 2010) – is what makes these web portals so effective.

The disease management portals in chronic care are designed to solve three main problems which are more critical in the context of chronic diseases (Vickers et al. 2010).

1. Patients typically do not know how they are doing.
2. Healthcare providers frequently do not know how their patients are doing.
3. Patients often do not receive the services that they should.

To solve the above problems, these chronic disease management portals attempt to implement the key elements of chronic care models such as (Georgeff 2014):

1. Better integrated and coordinated care
2. Collaboration across a multidisciplinary team of care providers
3. Planned care with regular follow-up and review
4. Support for patient self-management

There are a number of studies on the effective use of such portals in chronic disease management, especially related to high impact diseases, for example, cancer or diabetes. These studies focus on supplementing chronic care with web-based portals and measuring the impact.

Aforementioned researches in this area dissect in detail how the traditional one-way passage of information – a typical website dealing with recovery after cancer surgery, for example – provides little more information than an online brochure and how effective the portals can be by facilitating interactivity in real time, personalized information, longitudinal follow-up or data analysis (Vickers et al. 2010).

Most of these studies, in addition, have developed a working prototype of a chronic disease management web portal based on the theoretical mode. And the findings on both the theoretical and the practical points of view are undeniably encouraging.

One such study involves the development of a novel multimodal diabetes model, which then implements a smart Web 2.0 portal based on the proposed multimodal diabetes model. The study has found that on average 87% of the patients and 98% of doctors agree on the usefulness of such web portals (Haddad 2013).

Another trial has subjected a set of type 2 diabetes patients to a web-based portal in addition to the usual care. The portal included patient access to electronic medical records, secure e-mail with providers, feedback on blood glucose readings, an educational website and an interactive online diary for entering information about exercise, diet and medication. The results have indicated that the GHb levels declined by 0.7% (95% CI 0.2–1.3) on average among intervention patients compared with usual-care patients (Ralston et al. 2009).

Another such clinical trial measuring the effectiveness of Internet-based asthma management has concluded that by using web portals to provide feasible electronic monitoring, easily accessible information, e-mail communication and use of an electronic action plan improves the ability to react to change and to optimize asthma control (Victor van der Meer et al. 2007).

## 2.3 The Gap

As highlighted in the previous subsections, a considerable amount of research has been conducted to discuss the importance and the effectiveness of the use of web portals in chronic disease management. They have developed both theoretical and working models, and the results are significantly positive.

One aspect that has not been received much attention though is the user experience of these web portals. Especially, there is no inclusive study to date on the use of responsive web design to improve the user experience of chronic disease management. With the huge transformation in the way we access web caused by the development of mobile technologies and wireless revolution, responsive web has become one of the buzz words of the past decade. Other ecosystems, social media in particular, have embraced the change and in the process have attested the importance of using the responsive web design to provide a better user experience.

The proposed research aims to address this gap, by proposing an improved approach to developing disease management web portals using responsive web design. In addition, the research attempts to develop a proof of concept working model based on the discussed theoretical framework.

## 3 Theoretical Framework

This section introduces the initial theoretical framework (Fig. 3) upon which the proposed research is built. The proposed research which focusses on enhancing the user experience of chronic disease management portals using responsive web design incorporates the interactions of three major entities: the web portal, its user interfaces and the potential users.

Based on the review of the exiting literature and around the gap identified, the development of the research is phased out as follows:

1. Identify the key features presented by relevant care models.

   Critically review the key features presented by each care model and discuss how they are already used and can be used in web portals. This review will help set the scope of the research.
2. Review existing chronic disease management models and their characteristics.

   Discuss both the well-established chronic care models such as Wagner chronic care model (Thomas Bodenheimer et al. 2002) and more modern and locally adopted models such as the integrated chronic disease management (ICDM) model (Department of Health & Human Services 2015).
3. Study the principals of responsive web design.

   Review the basics of responsive web design and its use in portals. The purpose of this study is to draw comparisons and learn lessons from other portal ecosystems with rich user experience, especially social media.
4. Construct a model to utilize the responsive web design in chronic disease management portals.

   Apply the reviewed body of knowledge to construct the proposed model.
5. Develop a proof of concept oncology management web portal with enhanced user interfaces for mobile web.

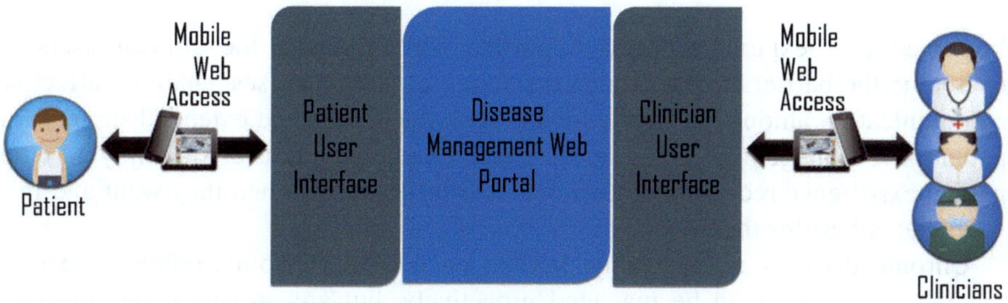


**Fig. 3** Overview of disease management web portals

As a proof of concept, design and develop a disease management web portal for oncology treatment management following the proposed model. This prototype portal will be used to evaluate the validity of the proposed model.

### 3.1 Technology Brief

The key technology component of the research is the responsive web design. The underlying idea of responsive web design is that a website should gracefully respond to every device, i.e. smartphone or tablet. This study employs responsive web design to enhance user experience, with special focus on usability and findability – two key features of the Morville's user experience honeycomb.

### 3.2 Implementation Processes

In developing the theoretical component, the study employs grounded theory (Charmaz 2006) to analyse data collected using single-case study strategy. To construct the prototype technology solution iteratively in a build-and-evaluate loop, design science research principles are used (Hevner and Chatterjee 2010).

An initial prototype of the web portal is developed after a systematic synthesis of the extant literature is completed. This portal is then demonstrated to the clinical users of the target data site to collect feedback. This user input, combined with other data collected from interviews and the literature, is analysed to develop the theoretical model. These results and the responsive indices as reported by responsive web analytical tools are also fed back to update the portal. This process will iterate a several times as necessitated by design science research strategies until no significant improvements are requested by users.

## 4 Discussion and Conclusions

Improved user experience of web portals will empower the clinical users to overcome the barriers in effectively managing chronic diseases, enabling effective communication among the care team and with the patient and extended support for patient self-management. In addition, it will enable patient users to enjoy a better patient experience receiving critical data and information when they want and in a form that is best for them.

Chronic diseases are a leading burden on health, economy, culture and productivity. They need to be managed proactively, but are proven to be complex, mainly due to difficulties in care collaboration and communication. Chronic disease management web portals have emerged as a prospective solution to these problems, facilitating a collaborative and interactive relationship between patients and healthcare professionals to assist better patient-centred care. This chapter has served to proffer a framework to ensure maximum benefit of web portals by enabling them to be as usable and beneficial whether accessed on a PC, mobile device or tablet. Given the trend for more and more portals to support healthcare delivery especially in the domain of chronic and non-communicable diseases, this becomes an important issue and we urge for more research in this area. The proposed research aims to enhance the user experience of these portals on mobile devices for clinical users and patients using responsive web design, thus maximizing the reach and the overall effectiveness.

## Bibliography

Ahmed, S., Bartlett, S. J., Ernst, P., Paré, G., Kanter, M., Perreault, R., Grad, R., Taylor, L., & Tamblyn, R. (2011). Effect of a web-based chronic disease management system on asthma control and health-related quality of life: Study protocol for a randomized controlled trial. *Trials, 12*, 260.

Australian communications and media authority. (2015). *Australians embrace a digital life | ACMA*. Retrieved from ACMA: http://www.acma.gov.au/~/media/Research%20and%20Analysis/Research/Word%20Document/Australians%20digital%20livesFinal%20docx.docx

Australian Institute of Health and Welfare. (2014). *Australia's health: Burden of disease*. Retrieved from AIHW: http://www.aihw.gov.au/WorkArea/DownloadAsset.aspx?id=60129547724

Australian Institute of Health and Welfare. (2015a). *Chronic diseases (AIHW)*. Retrieved from AIHW: http://www.aihw.gov.au/chronic-diseases/

Australian Institute of Health and Welfare. (2015b). *Leading types of ill health (AIHW)*. Retrieved from AIHW: http://www.aihw.gov.au/australias-health/2014/ill-health/#t2

Barlow, J., Wright, C., Sheasby, J., Turner, A., & Hainsworth, J. (2002). Self-management approaches for people with chronic conditions: A review. *Patient Education and Counseling, 48*, 177–187.

Barr, V., Robinson, S., Marin-Link, B., Underhill, L., Dotts, A., Ravensdale, D., & Salivaras, S. (2003). The expanded chronic care model. *Hospital Quarterly, 7*, 73–82.

Center for Managing Chronic Disease - University of Michigan. (2011). *What is chronic disease? | Center for Managing Chronic Disease - Putting People at the Center of Solutions*. Retrieved from CMCD University of Michigan: http://cmcd.sph.umich.edu/what-is-chronic-disease.html

Charmaz, K. (2006). *Constructing grounded theory: A practical guide through qualitative analysis*. Los Angeles: Sage Publications.

Davies, N. J., & Batehup, L. (2010, March). Retrieved from NCSI UK: http://www.ncsi.org.uk/wp-content/uploads/Guidance-for-Developing-Cancer-Specific-Self-Management-Programmes.pdf

Department of Health & Human Services. (2015). *Integrated chronic disease management in Victoria - health.vic*. Retrieved from health.vic: https://www2.health.vic.gov.au/primary-and-community-health/primary-care/integrated-care/integrated-chronic-disease-management/icdm-in-victoria

Department of Health, Australia. (2015a). *Department of Health | Cancer*. Retrieved from Department of Health: http://www.health.gov.au/internet/main/publishing.nsf/Content/cancer

Department of Health, Australia. (2015b). *Department of Health | Chronic Disease*. Retrieved from Australian Government, Department of Health: http://www.health.gov.au/internet/main/publishing.nsf/Content/chronic-disease

Gardener, B. S. (2011). Responsive web design: Enriching the user experience. *Sigma Journal: Inside the Digital Ecosystem*, 13–19. Retrieved from Smashing Magazine: http://www.noblis.org/media/2dd575c1-2de9-4d92-9bdb-f72ad9fb9a19/docs/SigmaDigEco2011_pdf#page=15

Georgeff, M. (2014, December). Digital technologies and chronic disease management. *AUSTRALIAN FAMILY PHYSICIAN VOL. 43*, pp. 842–846. Retrieved from RACGP - Digital technologies and chronic disease management: http://www.racgp.org.au/download/Documents/AFP/2014/December/AFP-2014-12-Focus-Georgeff.pdf

Goldberg, H. I., Ralston, J. D., Hirsch, I. B., Hoath, J. I., & Ahmed, K. I. (2003). Using an Internet comanagement module to improve the quality of chronic disease care. *Joint Commission Journal on Quality and Patient Safety, 29*, 443–451.

Haddad, M. (2013, November). Development of a smart e-health portal for chronic disease (Diabetes) management. Canberra, Capital Territory, Australia.

Hevner, A., & Chatterjee, S. (2010). *Design science research in information systems*. New York: Springer.

Kruse, C. S., Argueta, D. A., Lopez, L., & Nair, A. (2015). Patient and provider attitudes toward the use of patient portals for the management of chronic disease: A systematic review. *Journal of Medical Internet Research, 17*, 1438–8871.

Lorig, K., Sobel, D., Ritter, P., Laurent, D., & Hobbs, M. (2001). Effect of a self-management program on patients with chronic disease. *Europe PMC*. Retrieved from http://europepmc.org/abstract/med/11769298

Marcotte, E. (2011). *Responsive web design* (2nd ed.). New York: A Book Apart.

McCorkle, R., Ercolano, E., Lazenby, M., Schulman-Green, D., Schilling, L., Lorig, K., & Wagner, E. (2011). Self-management: Enabling and empowering patients living with cancer as a chronic illness. *CA: A Cancer Journal for Clinicians, 61*, 50.

Meeker, M. (2015, May 25). *2015 Internet trends*. Retrieved from KPCB: http://www.kpcb.com/blog/2015-internet-trends

Morville, P. User experience design. http://semanticstudios.com. [Online] 21 June 2004. http://semanticstudios.com/user_experience_design/.

Natda, K. V. (2013). Responsive web design. *Eduvantage, 1*, 1.

National Cancer Institute. (2014, December 2). *Coping after treatment - National Cancer Institute*. Retrieved from National Cancer Institute - U.S. Department of Health and Human Services: http://www.cancer.gov/about-cancer/coping/survivorship

Norris, S. L., Glasgow, R. E., Engelgau, M. M., O'Connor, P. J., & McCulloch, D. (2003). Chronic disease management: A definition and systematic approach to component interventions. *Disease Management and Health Outcomes, 11*, 477.

Pincus, T., Esther, R., DeWalt, D., & Callahan, L. (2001). Social conditions and self-management are more powerful determinants of health than access to care. *American College of Physicians–American Society of Internal Medicine*.

Podjarny, G. (2015). *RWD Ratio in Top 100,000 websites - refined*. Retrieved from Guy's Pod: http://www.guypo.com/rwd-ratio-in-top-100000-websites-refined/

Ralston, J., Hirsch, I., Hoath, J., Mullen, M., Cheadle, A., & Goldberg, H. (2009). Web-based collaborative care for type 2 diabetes: A pilot randomized trial. *Diabetes Care, 32*, 234–239.

Singh, D. (2008). How can chronic disease management programmes operate across care settings and providers? *WHO European ministerial conference on health systems*. Tallinn, Estonia: World Health Organization.

Thomas Bodenheimer, M., Wagner, E. H., & Kevin Grumbach, M. (2002). Improving primary care for patients with chronic illness: The chronic care model, part 2. *JAMA, 288*, 1909–1914.

Thompson, C. (1986). Flexibility of single-subject experimental designs. Part III: Using flexibility to design or modify experiments. *The Journal of Speech and Hearing Disorders, 51*, 214–225.

van der Meer, V., van Stel, H. F., Detmar, S. B., Otten, W., Sterk, P. J., & Sont, J. K. (2007). Internet-based self-management offers an opportunity to achieve better asthma control in adolescents. *Chest, 132*, 112–119.

Vickers, A. J., Salz, T., Basch, E., Cooperberg, M. R., Carroll, P. R., Tighe, F., et al. (2010). Electronic patient self-assessment and management (SAM): a novel framework for cancer survivorship. *BMC Medical Informatics and Decision Making, 10*, 34.

Willcox, S. (2014, October). *Chronic diseases in Australia: The case for changing course.* Retrieved from Victoria University: https://www.vu.edu.au/sites/default/files/AHPC/pdfs/Chronic-diseases-in-Australia-the-case-for-changing-course-sharon-willcox.pdf

# Older Adults Empowerment Through Training and Support and Its Implication on Proactive Self-Monitoring, Patient Engagement, and Connected Health

**Karoly Bozan and David Mooney**

## 1 Introduction

In the age of digitization and government regulations, patient health information that was once stored on paper is now stored digitally. Many studies have investigated the advantages and challenges of electronic medical and health information and have found that efficient access to relevant health information is one of the major benefits of electronic health information (Chrischilles et al. 2014).

Communicating with healthcare providers is a concern of patients, as they often feel that their communication is restricted to their face-to-face appointments, during which they often feel rushed and sometimes intimidated (Chrischilles et al. 2014; Kronish et al. 2012). Importantly, electronic health records (EHR) software, specifically, patient portals, includes the capability to provide patients with access to their health records and laboratory and diagnostic results and to schedule appointments, request medication refills, make payments, download and complete forms, view educational material, and send secure messages to providers (Krist et al. 2012). Portals are defined as a "single door to [health] information" (Davey and Tatnall 2007, p. 690). They increase administrative efficiency and accessibility as well as communication between patients and providers. These portals are also viewed as a tool to increase patient satisfaction and to involve patients, especially elderly patients, in their own care (Bierman 2012).

K. Bozan (✉)
Duquesne University, Pittsburgh, PA, USA
e-mail: bozank@duq.edu

D. Mooney
Ohio State University, Columbus, OH, USA
e-mail: mooney.122@osu.edu

 
N. Wickramasinghe, F. Bodendorf (eds.), *Delivering Superior Health and Wellness Management with IoT and Analytics*, Healthcare Delivery in the Information Age, https://doi.org/10.1007/978-3-030-17347-0_25

While patient portals are mainly viewed as a display of patient records, portals are also becoming more widely available and viewed and are expected to promote patient engagement. There is also a growing evidence that patient engagement improves health outcomes and reduces healthcare costs (Forbat et al. 2009; Ferman 2010). As patient portals are considered to increase patient engagement, numerous large-scale surveys found that older adults, those of over the age of 55, are less likely to use the patient portals even though this demographic utilizes most of the healthcare resources (Sarkar et al. 2011; Graetz et al. 2016). Some studies investigated the driving forces that promote the use of portal by the older generation. For example, Bozan et al. (2016) found that the healthcare providers' directions are the most powerful drivers for the elderly's patient portal "use behavior" and the older users' peers also have an influence on their behavior intention.

The low adoption rates among the elderly is found the most pronounced among those with low health literacy, less access and experience with technology, and less education (Gordon and Hornbrook 2016; Nahm et al. 2016). Numerous studies investigated the possible barriers to adoption, and the technical support and initial training were two of the main reasons the elderly have a positive experience and attitude toward the technology use for healthcare and portal adoption (Irizarry et al. 2017).

Elderly people have varied levels of experience with computers and the Internet. The Pew Research Center (Anderson and Perrin 2017) found that, in the United States, individuals 65 years and older have shown an increase in regular Internet use from 12% in 2000 to 67% in 2017. Nevertheless, approximately half of these elderly users indicate that they need help with using technology. This "barrier to adoption ranges from physical challenges to lack of comfort and familiarity with technology" (Anderson and Perrin 2017, p. 3). Only 26% of Internet users aged 65 and over state that they are very confident about using computers or other devices to do what they need to online. Older people are more likely to say that they need someone to initially show them how to use devices or the Internet. Three-quarters of seniors who go online use the Internet daily, indicating that after initial help, older users have the interest to use the technology regularly after they got familiar with it.

This intervention study investigates the effect of training and support for older people in regard to their perceived ease of use of a patient portal. The training was based on Bandura's self-efficacy model. Computer anxiety, confidence, and self-efficacy were measured, using corresponding subscales from the Computer Attitude Scale and Computer Self-Efficacy Scale. The questionnaire was administered before and immediately after the training and then 6 weeks after the training. We used a pre-post repeated measures design that involved 108 participants who were 65 years and older in a training-and-control group setting, with randomized assignment of participants. This study expands the understanding of older IS users' training needs for the use of patient portals and adds to the literature on the various technology acceptance models.

Furthermore, this study aims to enhance the understanding of the determinants of patient empowerment among the elderly by technology use. We also intend to better understand the precursors of connected care, which is recognized as a technology enabled model of healthcare delivery.

## 2 Background and Literature Review

Hospitals and providers' offices are digitizing their patients' health records and information to enhance quality of care and reduce risks, as provided through timely access to relevant health information. In the United States, certified EHR systems provide patients with access to their personal health information. Studies suggest that patient engagement increases patient satisfaction and health outcomes and reduces cost (Forbat et al. 2009) and that providers offer patient portals to enhance communication between themselves and patients (GPO 2016). Studies also indicate that older adults are less likely to use patient portals (Smith et al. 2015; Sarkar et al. 2011; Graetz et al. 2016). The characteristics of older adults who are less likely to use patient portals included having less education, low health literacy, less experience with technology, and low computer confidence and self-efficacy (Gordon and Hornbrook 2016; Nahm et al. 2016). Recent multiple focus group-based research (Irizarry et al. 2017) found the following themes among older people in regard to technology use for healthcare, specifically, patient portal adoption:

- Having limited or poor relationship with technology: "Difficulty troubleshooting without having access to live technical support," "no prior experience and training using computers," "afraid of making a mistake," and "lack of knowledge, which caused avoidance of portals or reliance on family members"
- Experiencing fear and frustration with technology and portals: "not understanding the security risks but afraid of them"
- Being willing to adopt portal with support: "Lack of confidence in their own abilities but interested in someone's accessing the portal on their behalf," "indicating that task-specific training would help them to learn the purpose of each function"
- Viewing portals as beneficial: "Interested in finding personal information and results and communicating with providers"

The above themes clearly indicate that older people have an interest in portals but that their lack of confidence and self-efficacy as well as computer anxiety hinder their perceptions of the effort that it takes to use a patient portal. It is clear that there is demand for training in the use of patient portals, and accessible support may increase confidence and reduce anxiety, resulting in older individuals' finding portal use to be easier. The literature includes several empirical studies in which technical and user support is given to healthcare stakeholders, rather than patients, in the use of electronic patient records (Michel-Verkerke and Hoogeboom 2012; Yu et al. 2008). Thus, the role of technical support on the perceived ease of use of a patient portal, especially for older uses, warrants research.

Although much attention has been paid to the characteristics of older adults who do not adopt patient portals, little has been given to how older people's perceptions of ease of use of patient portals changes when they receive customized training and on-demand support. To this end, our research was guided by the following research question: To what extent do computer self-efficacy, computer anxiety, and computer confidence change due to training the elderly, and how does this change affect the perceived ease of use of patient portals?

Furthermore, we investigate the motivation of self-monitoring and the sharing self-generated health information with providers among those who received training and support and compare this data with those in the treatment group. We seek to understand the possible antecedents of self-monitoring and willingness to sharing that information with the providers with the intent that patients better engage with their own care through connected health. It is found that empowered patients are more likely to participate in a connected health system (Caulfield and Donnelly 2013), which supports an enhanced preventative and predictive health decision-based care. It is mainly driven by the pervasive and participatory effect of patients to drive a more personalized care (Poon and Zhang 2008).

Caulfield and Donnelly (2013, p. 704) defined connected care as "Connected Health encompasses terms such as wireless, digital, electronic, mobile, and tele-health and refers to a conceptual model for health management where devices, services, or interventions are designed around the patient's needs, and health related data is shared, in such a way that the patient can receive care in the most proactive and efficient manner possible. All stakeholders in the process are 'connected' by means of timely sharing and presentation of accurate and pertinent information regarding patient status through smarter use of data, devices, communication platforms and people." This definition highlights the shift from reactive to proactive care, and our focus is on the elderly patients sharing their self-generated health information with their physicians using advanced information technology. Currently, manually entering information is a more feasible approach, but the focus must shift toward smart devices, which automatically upload measurements of their users. As a first step, we need to find a determinant of the willingness of elderly users to share this information using a more advanced technology that they are comfortable with. We propose that computer training and provided support will increase the confidence in self-ability among the elderly to a level which will enable them to share their information with their physicians using current pervasive technology, which they can use. This may also enable elderly patients to adopt smart devices, which will automatically upload their information. But that is a farfetched goal, and this study hypothesizes that a targeted training to increase comfort and confidence with computers and reduce anxiety for the purpose of accessing health records through a patient portal is the first step in this process. Once the elderly users are comfortable with the concept and purpose of accessing health information through a portal and have a better understanding of the way technology used to display and capture health information, they will better engaged through the use of technology.

## 2.1 Theoretical Background

Organizational software users usually receive training on how to use a new information system (IS) effectively and efficiently to increase productivity (Lamb and Kling 2003). Consumers, in contrast, are less likely to receive training when considering to use a new IS for personal use and are motivated by a balance of utilitarian and hedonic factors (Agarwal and Karahanna 2000; Shiv and Fedorikhin 1999). Utilitarian factors are objective in nature and refer to the usefulness or cost benefit of the IS. Researchers have found that technology acceptance is greatly influenced by these objective factors, which, in technology acceptance models, comprise *perceived usefulness* (Davis 1989; Venkatesh and Bala 2008). Hedonic factors concern the enjoyment and ease of use of the technology; more recent acceptance models refer to this as *perceived ease of use* (Agarwal and Karahanna 2000; Dabholkar 1996; van der Heijden 2004; Venkatesh 2000; Venkatesh et al. 2012).

A variety of acceptance models have evolved in recent decades, but all grew out of the Technology Acceptance Model (TAM), introduced by Davis (Davis 1986), which evolved from the Theory of Reasoned Action (TRA) by Ajzen and Fishbein (1980). Both of these powerful and parsimonious theories (Lucas and Spitler 1999; Venkatesh and Davis 2000) assume that an individual's acceptance of IS is determined by the following two major variables:

- Perceived usefulness (PU)—utilitarian factor: "The degree to which a person believes that using a particular system would enhance his or her job performance" (Davis 1989, p. 320).
- Perceived ease of use (PEU)—hedonic factor: "The degree to which a person believes that using an IT will be free of effort" (Davis 1989, p. 320).

This study explores the PEU of patient portals by the elderly through the lens of the PEU construct in acceptance models. It has been shown empirically that PEU has a direct positive effect on behavioral intention, which, in turn, has a direct effect on use behavior. Therefore, a better prediction of the driving forces of PEU will increase understanding of the intention to use and actual use of patient portals.

More recent acceptance models have the same underlying constructs, which have been proven to affect intention and use. Social influence processes and cognitive instrumental processes, such as job relevance and output quality, were incorporated into later acceptance models, such as TAM2 (Venkatesh and Davis 2000). Venkatesh and Bala (2008) further identified PEU determinants in TAM3, namely, anchor determinants (computer self-efficacy, perception of external control, computer anxiety, computer playfulness) and adjustment determinants (perceived enjoyment, objective utility). Venkatesh (2000) suggested that perceived enjoyment, computer self-efficacy, and perception of external control play a strong role in PEU even after individuals gain experience with a new system. In contrast, computer anxiety and playfulness were found to diminish over time.

PEU of an information system is supported by Bandura's self-efficacy research (Li and Kishore 2006) and defined as "judgement on how well one can execute courses of action required to deal with prospective situation" (Li and Kishore 2006, p. 122). Davis (1989) drew the association between self-efficacy and PEU as proximal determinants of behavior. Our intervention study is based on Bandura's self-efficacy theory as applied to computer training.

Bandura's self-efficacy theory (Bandura 1977) is a major part of his social learning theory. According to Bandura's self-efficacy theory, combining the four indicators of efficacy expectations in a training program will increase computer self-efficacy, and based on the literature, we argue that it also will increase computer confidence and lower computer anxiety. These determinants, coupled with accessible support, will increase PEU.

Infrastructural and user support after training is comparable to the facilitating conditions in the unified theory of acceptance and use of technology (UTAUT) model and has been shown to affect technology use and trust in users' own abilities (Venkatesh and Bala 2008; Venkatesh and Davis 2000; Lu et al. 2005; Ratnasingam et al. 2005; Rensel et al. 2006; Winter and Chudoba 1998).

## 3 Conceptual Framework and Hypotheses

We empirically measured the effect of training on these constructs and on elderly people's PEU of their patient portal. We measured training effectiveness, using computer self-efficacy, computer anxiety, and computer confidence scales. Self-efficacy, anxiety, and confidence were found to be strong determinants of PEU. The conceptual model is shown in Fig. 1, which indicates the theoretical model of PEU and its constructs that are affected by computer and patient portal training.

Computer anxiety is defined as the degree of "an individual's apprehension, or even fear, when she/he is faced with the possibility of using computers" (Venkatesh 2000 p. 349, drawn on Simonson et al. 1987). Empirical studies have found that computer anxiety is negatively related to confidence and comfort with computer use

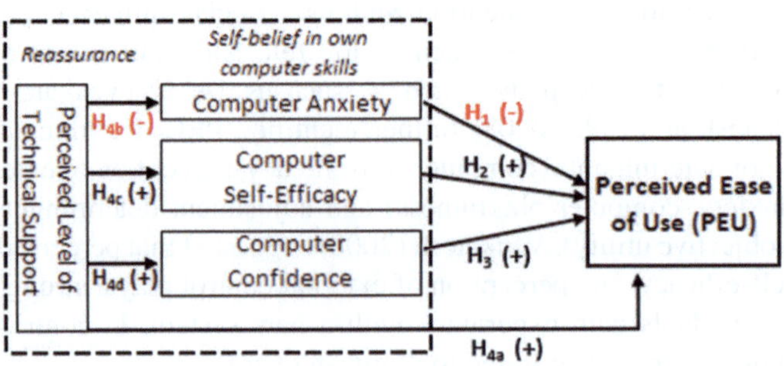


**Fig. 1** Conceptual framework

(Harrington et al. 1990) and positively related to poor computer use and resisting the adoption and use of computer software (Torkzadeh and Angula 1992; Heinssen et al. 1987). Therefore, we posit:

*Hypothesis 1 Training decreases computer anxiety and will increase the PEU of patient portals by the elderly.*

Computer self-efficacy is defined as the degree to which an individual believes that he or she has the ability to successfully perform a specific task/job, using the computer (Compeau and Higgins 1995a, b; Doll and Torkzadeh 1989). Bandura's self-efficacy concept has been used in the computer use context in many studies and empirically shown to be positively related to users' perception of software ease of use. Therefore, we propose:

*Hypothesis 2 Training increases computer self-efficacy and will increase the PEU of patient portals by the elderly.*

Computer confidence is defined as "one's ability to do well in activities that involve the computer" (Gardner et al. 1993, p. 429). Confidence has been measured in several empirical works to evaluate user characteristics that influence computer use (Rovai and Childress 2002; Yildirim 2000) and perception of effort it takes to use computers (Davis 1989). Hence, we posit:

*Hypothesis 3 Training increases computer confidence and will increase the PEU of patient portals by the elderly.*

Technical support is used directly as a predictor of IS ease of use and is part of a group of perceived resources (Ngai et al. 2007). Perceived resources are similar to perceived behavioral control construct in the theory of planned behavior: "... the presence or absence of requisite resources and opportunities" (Ajzen and Madden 1986, p. 457).

Ralph defined technical support as involving "people trained to help users in solving problems related to computer hardware and software" (Ralph 1991, p. 57). It can be in a variety of forms, such as in person, over the phone, through instant messaging, and so forth. Numerous empirical studies have identified technical support as one of the most important factors in technology acceptance and user satisfaction with ISs in multiple industries (Hofmann 2002; Sumner and Hostetler 1999; Mirani and King 1994). Commercial or consumer IS support, including technical support, results in more favorable attitudes that should lead to greater acceptance and the success of personal computing systems (Igbaria 1990). Lack of technical support, however, has been shown to be a barrier to adopting an IS in the field of healthcare (Igbaria 1990; Jha et al. 2009) as is a lack of help from peers (Randeree 2007; Vishwanath and Scamurra 2007) Hence, we propose:

*Hypothesis 4a An increase in the perceived level of computer technical support will increase the PEU of patient portals by the elderly.*

Adult and elderly learners are often characterized as having a lack of confidence when learning the use of new software and require support (Russell

1995). Infrastructure and technical support were found to positively influence IT usage perceptions and increase user confidence and comfort with the software (Bhattacherjee and Hikmet 2008). Researchers have studied technical support as a subconstruct of other variables (Bhattacherjee and Hikmet 2008; Yiong et al. 2008) or as a construct of its own that directly affects technology ease of use (Abbad et al. 2009; Gazinoory and Afshari-Mofrad 2011). To enhance technology learning outcomes and satisfaction, technical support should be provided, as it increases confidence in users' abilities to perform the tasks learned in training (Kidd 2010). Thus, we hypothesize that technical support is also a subconstruct of computer anxiety, computer self-efficacy, and computer confidence:

*Hypothesis 4b An increase in perceived level of computer technical support will decrease computer anxiety among the elderly.*

*Hypothesis 4c An increase in perceived level of computer technical support will increase computer self-efficacy among the elderly.*

*Hypothesis 4d An increase in perceived level of computer technical support will increase computer confidence among the elderly.*

# 4 Methodology

This randomized, control-group, repeated-measures intervention study involves the measurement of context- and learner-specific training's effect on comfort of using a computer and, ultimately, on PEU of a patient portal. Computer confidence, self-efficacy, and anxiety were measured, and the results between the training and control groups were compared three times: (1) before training, (2) immediately after training, and (3) 6 weeks after training. The control group received training after the final assessment was taken.

## *4.1 Participants and Setting*

Computer training was offered to a group of volunteers aged 65 years or older who identified themselves as uncomfortable with computers yet interested in accessing their own personal health records through a patient portal. We recruited participants from health fairs in the United States and retirement homes as well as asked volunteers to encourage their qualified acquaintances to sign up. An informed consent form was provided to all participants.

The training also included a detailed printed manual that consisted of screenshots of five patient portals from the large providers in the area and short explanations of what each field and button represent. Registration, appointment scheduling/modifying, messaging doctors, and prescription refill requests were detailed step by step for each portal. This manual and the purpose of the patient portal were covered after a basic computer and web browser skills review. The training was delivered face to face by qualified instructors, who covered pre-established material in an agreed way that followed Bandura's self-efficacy expectations. Participants engaged in training until they felt that they could comfortably log into a test account on a portal and perform basic transactions. Most participants needed only one session.

## 4.2 Instrumentation

Subscales from Gressard's Computer Attitude Scale (Gressard and Loyd 1986) were used to measure computer anxiety (CA) and computer confidence (CC). The 10-item Likert scale answers for CA ranged from "strongly agree" to "strongly disagree," with higher scores' indicating lower levels of anxiety. Similarly, CC also was measured by 10 items, with higher scores' indicating higher computer confidence. The computer self-efficacy (SE) construct was measured through Compeau and Higgins' 10-item instrument (Compeau and Higgins 1995a, b), with questions that started with "I could complete the job, using the software package ..." and 10 response items that included statements such as "... if there was no one around to tell me what to do as I go along." Answers were *yes* or *no*, and respondents chose from a 1 to 10 scale that ranged from "not at all confident" (1) to "totally confident" (10). Only *yes* answers and their corresponding answers were counted for scoring, and, again, higher scores indicated higher SE.

Content validity was reviewed by experts who worked with older adults in a validity survey and were based on six responses. The content validity indexes ranged from 0.89 to 0.98 for representativeness and 0.85 to 0.94 for clarity. Reliability coefficients were 0.92, 0.90, and 0.89 for CA, CC, and SE, respectively. Other studies also found high internal consistency scores for these scales (Nash and Moroz 1997).

## 4.3 Data Collection

Eligible participants were randomly assigned to the training (intervention) group or the control group, which did not receive training until the study was concluded. The survey, including the CA, CC, and SE items, was administered prior to training, immediately after training, and 6 weeks after training.

# 5 Results

Initially, there were 126 participants, but multiple people discontinued their involvement during the study. A total of 109 participants of a variety of ethnicities finished all three surveys. The mean age was 70.4 years, 68% of the participants were female, and all participants had some kind of formal education. In addition, 48% of the participants defined themselves as unfamiliar with computers, 32% defined themselves as somewhat familiar, and 20% indicated that they are fairly familiar with computers. Further, 62% of participants revealed that they have no Internet access, and 82% of participants indicated that they feel that patient portals are good idea even though they are accessed through computers.

## *5.1 Analysis of Variance*

We analyzed the repeated measures analysis of variance (ANOVA) of the mean scores across participants at the three time intervals (before training, immediately after training, and 6 weeks after training). A total of 48 participants completed the training, and 61 were in the control group. We chose a repeated-measures ANOVA versus an independent (between-subject) ANOVA because the repeated-measures ANOVA further divides the within-group variability (that is, the error term in an independent ANOVA) into subject variability and a smaller error term. We used SPSS statistical software for the pairwise, repeated-measures ANOVA analysis.

Table 1 summarizes the mean scores of the CC, CA, and SE questionnaire across participants in the treatment group and their F-score across the three measurement points: before training (T1), right after training (T2), and 6 weeks after training (T3). The control group maintained a relatively similar level to the baseline across the measurement points; their score is decreased but not significantly.

### 5.1.1 Computer Anxiety

The CA scores increased compared to the baseline scores. This indicates reduced anxiety related to using a computer and to accessing the patient portal after training.

**Table 1** Mean scores of treatment group

| | T1 | T2 | T3 |
|---|---|---|---|
| CA (Mean/SD) | 24.68/4.67 | 36.18/6.13 | 32.81/5.91 |
| CC (Mean/SD) | 27.11/5.41 | 36.18/6.13 | 32.81/5.91 |
| SE (Mean/SD) | 38.81/6.93 | 67.43/8.07 | 61.83/7.63 |
| F ($df_{time}/df_{error}$) | 34.71*** | 19.95*** | 26.42*** |

***$p < 0.001$

Baseline CA score mean ($M$) = 24.68 ($SD$ = 4.67);
After training, $M$ = 36.18 ($SD$ = 6.13);
6 weeks after training, $M$ = 32.81 ($SD$ = 5.91).

The change in CA level was statistically significant over time, $F\ (2, 94) = 34.71$, $p < 0.001$.

### 5.1.2 Computer Confidence

The CC scores also increased compared to the baseline scores. This indicates increased confidence in computer and patient portal use after training.

Baseline CC score mean ($M$) = 27.11 ($SD$ = 5.41).
After training, $M$ = 41.93 ($SD$ = 7.03).
6 weeks after training, $M$ = 38.39 ($SD$ = 6.61).

The change in CC level was statistically significant over time, $F(2, 94) = 19.95$, $p < 0.001$.

### 5.1.3 Computer Self-Efficacy

The SE scores increased compared to the baseline scores. This indicates increased computer self-efficacy after training.

Baseline SE score mean ($M$) = 38.81 ($SD$ = 6.93).
After training, $M$ = 67.43 ($SD$ = 8.07).
6 weeks after training, $M$ = 61.83.81 ($SD$ = 7.63).

The change in SE level was statistically significant over time, $F(2, 94) = 26.42$, $p < 0.001$.

## *5.2 Partial Least Squares*

Partial least squares (PLS) is a second-generation regression model that combines a factor analysis with linear regression, making only minimal distribution assumptions (Gefen et al. 2000). We chose to measure the path coefficients in the conceptual model to see the determinants of PEU of patient portals before training (T1), in immediate response to training (T2), and 6 weeks after training (T3). Venkatesh and Davis used stepwise regression analysis when empirically testing their TAM2 model (Venkatesh and Davis 2000); they opted against structural equation modeling due to the two-item scales in the measurement model that can cause instability in the parameter estimates. Our conceptual model has far fewer constructs, and the number of measurement items is sufficient for the model; hence, PLS appears to be a reasonable choice for path analysis. Gefen et al. (2000) tested TAM using four different methods, and PLS was found to provide strong predictive power in construct relationships. Several researchers regard PLS to be superior to more rigorous covariance-based structural equation modeling (SEM) (Thompson et al. 1995; Chin 1998a, b).

### 5.2.1 Instrumentation and Measurement Model

The same measurement scales were used for CA, CC, and SE across the three time periods, and we used Venkatesh's scales for PEU (Venkatesh and Davis 2000). The Cronbach alpha for internal consistency across the three time periods were 0.82–0.92 for CA, 0.84–0.93 for CC, 0.81–0.91 for SE, and 0.86–0.97 for ease of use; therefore, high reliability could be assumed for all measurement scales. Principal component analysis, with direct oblimin rotation exhibited, strongly supported construct validity and showed low cross-loadings. Measurement items came from the literature (Davis 1989; Ngai et al. 2007; Davis and Venkatesh 1996; Mathieson 1991; Taylor and Todd 1995), and the high reliability and validity are consistent with earlier empirical studies that use these items.

The measurement was created in structural equating software, SmartPLS, to assess the properties of the latent constructs. Sample covariance matrices were utilized to test the explanatory power and overall fit of the research model and, ultimately, the relative strengths of the causal paths between the variables described in the model. Common model-fit measures were used to evaluate the model's goodness-of-fit using pooled data, and all measures were within the tolerance limits found in the literature. Non-normed fit index (NNFI) $= 0.927$ ($>0.90$), comparative fit index (CFI) $= 0.931$ ($>0.90$), root mean square error of approximation (RMSEA) $= 0.077$ ($<0.10$), normed chi-square $= 2.014$ ($<3.0$), and GFI $= 0.935$ ($>0.90$).

### 5.2.2 Explaining Perceived Level of Patient Portal Ease of Use

Table 2 provides a summary of the effect of CA, CC, and SE and perceived level of technical support (TS) on perceived level of patient portal perceived ease of use. SE was shown to be a strong determinant of perceived ease of use, which is consistent with the literature (Davis and Venkatesh 1996). CA was a significant secondary determinant. Perceived level of TS had a somewhat strong effect on perceived level of patient portal ease of use among the elderly, but its effect weakened as time passed after training, also consistent with longitudinal technology acceptance literature (Venkatesh and Davis 2000).

**Table 2** Regression results explaining PEU

| Time of measurement | $R^2$ | *ß* |
|---|---|---|
| *Before training (T1)* | 0.46 | |
| Computer anxiety | | 0.39* |
| Computer self-efficacy | | 0.48** |
| Computer confidence | | 0.12* |
| PL of technical support | | 0.24* |
| *Right after training (T2)* | 0.56 | |
| CA | | 0.34** |
| SE | | 0.45* |
| CC | | 0.19** |
| TS | | 0.33** |
| *6 weeks after training (T3)* | 0.48 | |
| CA | | 0.41* |
| SE | | 0.45** |
| CC | | 0.23* |
| TS | | 0.18* |

*$P < 0.05$, **$P < 0.01$

## 6 Factors Affecting Willingness to Share Self-Generated Data

In this study, patient portals were mainly viewed as display of health information generated at the time of healthcare visit. While this is important to enhance patient's awareness of the information, it can also enhance adherence (McInnes et al. 2013; Bozan 2017), trust (Lyles et al. 2013), and satisfaction with care among the elderly (Otte-Trojel et al. 2014). However, it is not well understood whether patients are willing to share their own, self-generated data with the providers and what factors affect this willingness. In this regard, we had an interview-live discussion with each participant after the training and asked few open-ended questions in regard to their routine on self-monitoring, such as vital signs, medication adherence, etc., and their willingness to share this information electronically with their provider. In order to better understand the determinants for patient engagement, we categorized the motivating and contributing factors and their proposed effect on willingness to share self-generated data with providers.

The interviews were relatively short during the discussion after training with the treatment group and before the actual training of the treatment group. Only those participants who stayed after training were asked, and out of the 48 who received training, 12 discussed their perceptions of sharing self-generated health information through patient portal or other application with the their providers. Out of the 61 people in the control group, 14 discussed the same before they received the actual training. The discussions were recorded and analyzed using qualitative clustering method, where we grouped and conceptualized the excerpts with similar characteristics (Miles and Huberman 1994). The object of this investigation was to see the willingness and determinants that enhance or hinder the sharing of self-generated health information with health professionals using technology among the elderly. As no developed methods were demonstrated, the discussions concerned the adoption intention of such process, considering the identified determinants.

## *6.1 Results*

The average experience with computers and the Internet was similar among the respondents, except the fact that those who went through patient portal training exhibited increased computer confidence and self-efficacy and reduced anxiety. This was not evident at the time of data collection as the analysis of the survey following the training took place later in time.

### 6.1.1 Perceived Advantage

The perceived advantage of self-monitoring and sharing the information with healthcare providers included keeping the physicians "up-to-date" or "in the loop" as some respondents described it. Generally, all respondents agreed that the appointment-based face-to-face meeting is the preferred method of their healthcare, but it limits the interaction and health data collection to a small segment of the possibility. As one respondent puts it:

> I have been measuring my own blood pressure for many years and noticing some irregular readings, which makes me wonder. When I see my doctor and they take my measurements there, it never shows those irregularities that concern me.

Other said:

> I feel more comfortable if my doctor sees my glucose numbers even before I have an appointment. I don't always know what measurement is alarming and which one is OK.

The idea of adding notes was important to some respondents:

> I don't think my measurements always give the full picture. I have the best blood pressure numbers but I feel dizzy and fatigued. I would like the doctor know these episodes rather than just have occasional readings and have a quick recap of my past few months. I may forget to mention specific situations.

The interviewers mentioned the possibility of manual data entry to an application that is connected to the personal or electronic health records systems and also described a hypothetical scenario, where the self-monitoring device would automatically send the reading information to the physicians' system. The difference between confidences in the ability to perform such submission was clear in the manual type of data entry, while the automatic data upload scenario seemed equally attractive to respondents from both the treatment group, which just finished the training and the control group, which was about to receive the training.

### 6.1.2 Learnability

The main concern from respondents was clustered together in the effort it takes to learn a system that would be used to capture and transmit self-generated health information. When the patient portal was mentioned as a possible medium, respondents from the treatment group felt much more confident in their responses about their perceived ability to learn and use that platform.

> I like this portal and if I am shown where to enter my glucose numbers and blood pressure measures, I am sure I could get used to it quickly.

Respondents from the treatment group showed some extent of apprehension when they discussed their ability to enter information into an application.

> I am confused about where to write what and what to click. Calling my doctor is just much easier.

It is clear that those who went through the training had increased confidence to use the portal and they also felt that other applications would be easy to learn as long as they get some training in that as well.

### 6.1.3 Security

We received a harmonious response from both groups in regard to the perceived security risk of entering and sending their health information through an application to their physicians. It is understandable as the training was not covering such topics.

> I don't know where my data will end up and who will read it. How do I know it goes directly to my doctor? I feel more comfortable telling certain issues in person.

> What if the system breaks and never send my message?

This signals that not only the user interface usability but also the medium used and network provider contribute to the perceived security formation.

### 6.1.4 Trust

Trust can be a barrier to sharing electronic data; existing research found that it is not always supported when patients voluntarily share their heath information with their physicians using electronic medium (Goel et al. 2011). This is especially holding to older patients. The application needs to be affiliated with the hospital in order to gain older patients' trust entering their health information.

> If I see my doctor's name or her office on the webpage, I would feel better about entering my information.

> ...(the application) needs to show me that it is associated with the hospital I usually go to. I don't think I would enter anything into a website, who knows where my information would end up.

This apprehension is lightened by those respondents who went through the training.

> I like how I can see my information on the portal and if I could just send in my blood pressure and glucose readings, it would be great.

Some respondents expressed their unease with the personnel who may have access to the information that they send to their physicians. The concern was around if other physicians, nurses, or support staff has access to the information.

> I wouldn't want my information and notes to my doctor accessed and reacted by others. I need to make sure it is read, understood and responded on by my doctor who has been seeing me for quite some time.

### 6.1.5 Effort

Most websites and patient portals are created for "digital natives," those who have extensive experience using the Internet and websites, and the "digital immigrants," those elderly who have not used the Internet and computers during their work years. Many cognitive and physchomotor functions make it challenging to the elderly physically perform the tasks to manually enter information into applications for sharing with their physicians. Accessing information seems less on an effort but the responses touched on entering information into fields and the effort it takes to learn what to enter where and how to access the particular application. Overall, it was obvious that the effort it takes may outweigh the perceived benefit to some respondents.

> I was just fine talking to my doctor and going to appointments and why should I change that?

> I will just bring in my blood pressure numbers on paper and my doctor will review it during my appointment.

Those who just finished training found the perceived effort to capture their self-generated health information more manageable but counted on technical support.

> I think it is not much more difficult than accessing my information on the patient portal. With some training or support, I am sure I can learn it and I would want see if it helps my doctor better diagnose me with that information.

## 7 Discussion

Intervention studies have found training to be an important factor in technology acceptance. Nevertheless, the design of the training and its specific psychological effect on users' perceptions have not been studied. Based on Bandura's self-efficacy theory, context- and trainee-specific training have a significant effect on computer comfort among the elderly.

The limited understanding of meaningful training design to enhance acceptance is evident in literature (Davis and Venkatesh 1996). Although computer self-efficacy is empirically proven to enhance perceived ease of use, as is objective usability, other psychological factors have not received attention. Following the literature, this study hypothesized that computer anxiety and computer confidence, along with perceived level of technical support, have positive effects on perceived ease of use of patient portals among the elderly.

The repeated-measures ANOVA showed statistically significant changes in the mean scores of computer anxiety, computer confidence, and computer self-efficacy across the temporal measurements. In addition, the results indicated that the control group did not experience an increase in CA, CC, or SE. The training was shown to increase computer self-efficacy and computer confidence and to reduce computer anxiety in the context of patient portal use among the elderly. The significant changes in the above scores were maintained 6 weeks after training, even though no additional training was provided.

PLS found SE to be the strongest determinant across all three measurement points. This can be attributed to the fact that elderly people find it important to acquire skills that enhance their ability to perform tasks and actions using computers. Elderly users are enthusiastic learners, and we found that the training, based on Bandura's self-efficacy model, worked well with their learning style. The sense of accomplishment, coupled with a positive emotional response and peer-related hands-on training, increased older individuals' perceptions of self-efficacy in regard to patient portal ease of use.

The increasing coefficients of SE and CC may be due to the realization of newly acquired skills and the enjoyment that they generate. Older computer users appreciate the confidence, self-esteem (Gatto and Tak 2008), and accomplishments that they gain through training (Reed et al. 2005).

The perceived level of TS showed an interesting result. TS, as a determinant of PEU of the patient portal by the elderly, was shown to be less significant prior to training. This may be due to the fact that elderly people do not know what to expect from TS and initially perceive it as not beneficial. As the training took place, however, elderly users realized that technical support would benefit them, as they may need only minor assistance to move forward. Hence, the strength of TS increased after training, and the increase in the other constructs (SE, CC, and CA) may be attributed to having TS available, providing peace of mind. Although the literature often uses TS as a precursor of self-efficacy and computer confidence, we opted to test its effect on ease of use directly as well. First, the perceived level of TS is considered when the ease of use of a software is consciously or unconsciously assessed. Having a resource to depend on should have a direct effect on the level of PEU. This is especially the case among the elderly, who may be less resourceful and creative with software use during their initial phase of learning it (Roskos-Ewoldsen et al. 2008; Salthouse 2004). In addition, the large amount of new information may leave elderly users confused, and they are comforted by knowing that TS is available. The decline of the perceived importance of TS at 6 weeks after the training indicated that TS seemed more important immediately after training but that, once the regular tasks that involve the use a patient portal became a habit, the importance declined.

Understanding the key constructs of patient portal PEU for the elderly can help with effective training design and intervention (Davis and Venkatesh 1996). The constructs that this study theorized and empirically tested proved to be strong indicators of patient portal PEU and the effect that training had on them.

This study contributes to existing acceptance theories by including a learner- and software-specific training intervention. The existing literature has examined mainly software acceptance in an organizational setting in which software use is mandated. Therefore, PEU has not been properly investigated, as the focus was mainly on perceived usefulness. The voluntariness of software use is greatly dependent on the usability of the software, and this study addressed this issue and identified the constructs of PEU and the effect of training among the elderly.

Future research should investigate how peers influence elderly people to use patient portals. Feedback from this training indicated that some attendees took the training because they were "jealous" that some of their friends were able to log in to their own health information. Others said that their doctors recommended the patient portal or that they had received mail that indicated the availability of their information online. They indicated that their PEU of the portal and their lack of skills prevented them from exploring all that the patient portals had to offer.

The purpose of the interviews was to identify the willingness of the elderly people to share their self-generated health information, such as vital signs, glucose, etc., with their physicians using an information system application. The respondents from the group, which just finished the training showed an increased confidence in the technology and in their own ability to enter such information into an application. They also showed increased trust in technology.

The other group has not yet received the training and their responses showed lower level of confidence and increased computer anxiety.

Respondents from both groups found the sharing of self-generated data advantageous and to their benefit and felt that the doctors would be better informed this way to provide better diagnosis. The answers were grouped into five major themes based on the qualitative clustering method and identified as the determinants of willingness to share user-generated data. We labeled them as perceived advantage, learnability, security risk, trust, and effort. Table 3 summarizes the contributing aspects of these determinants and the proposed effect of these determinants on the willingness to share user-generated data. Positive indicates the aspects impact on the willingness to share user-generated data, while negative effect indicates a possible barrier. We also noted the difference we found in the responses from the respondents between the treatment and control group.

The study is aimed to increase elderly users' confidence with technology in order to access patient portal through a training and technical support. We also qualitatively measured the training's impact on the willingness to share self-generated data in order to shift the reactive care toward a proactive care. While the feedback is encouraging and we found empirical evidence that the training helped the elderly users to better understand the purpose of sharing their self-generated

**Table 3** Summary of qualitative clustering of willingness to share self-generated health information by the elderly using information technology application

| Determinant of willingness to share user-generated data | Contributing aspect | Proposed effect on willingness to share user-generated data | Difference evident between treatment and control groups |
|---|---|---|---|
| Perceived advantage | Keeping doctor up-to-date<br>Peace of mind<br>Irregular readings are noticed by physician real time<br>Counting on physician to let them know when appointment is required based on data reading<br>Having a "second set of eyes" on the results<br>Have a chance to share how patients feel and not only the actual measurement | Positive<br>Positive<br>Positive<br>Positive<br>Positive | Treatment group was comfortable with entering their own data manually<br>Control group liked the idea but did not think it is possible unless data is uploaded automatically |
| Learnability | Enter data in patient portal<br>Logging into a new application<br>User interface complexity<br>Support provided<br>Easy to follow training material | Negative<br>Negative<br>Negative<br>Positive<br>Positive | Clear difference between control and treatment group |
| Security risks | Unauthorized access to data<br>Privacy<br>Device reliability<br>Error in transition | Negative<br>Negative<br>Negative<br>Negative | Both groups had the same concerns |
| Trust | In physicians<br>In other healthcare stuff<br>In technology<br>Worries about information is stolen | Positive<br>Negative<br>Positive/negative<br>Negative | Treatment group had much higher confidence in technology |
| Effort | To learn<br>To access<br>To enter information | Negative<br>Negative<br>Negative | Treatment group found it less of a perceived effort |

health information that empowers them and be a part of connected health care to receive proactive and personalized care.

## 8 Limitations

We designed one training session to reduce the chance of participants' not coming back and completing the survey; however, some participants did not show up or did not complete the follow-up survey at 6 weeks after training. This reduced the power of significance. Nevertheless, because we trained the control group after the study concluded and collected the same longitudinal samples, the major findings in this study were confirmed by their responses as well.

We did not include actual use to control for variations of SE, CC, CA, and TS between those who used the patient portals during the 6 weeks between the second and third measurement points and those who did not log in after training. Self-reported confidence and actual skills to perform an action in a patient portal may yield different results, and self-reported measures have been debated in IS research (Straub et al. 1995).

This study did not control for the actual level of TS, which may have resulted in a more granular elicitation of the phenomenon if the study had considered whether the comfort of knowing that help is available is sufficient to increase SE, CC, and CA or whether the actual use of such support has a different effect.

The comfort that users developed in a patient portal during training may not be the same when they are faced with a different user interface or menu layout. The perceived level of technical support may comfort users to rely on and users may lose confidence without the presence of support. Although being resourceful was outside of the scope of this study, it would be worth to measure older users' self-efficacy and creativity when faced with an unfamiliar portal.

## References

Abbad, M., Morris, D., & De Nahlik, C. (2009). Looking under the bonnet: Factors affecting student adoption of e-learning systems in Jordan. *The International Review of Research in Open and Distance Learning, 10*(2), 1–25.

Agarwal, R., & Karahanna, E. (2000). Time flies when you're having fun: Cognitive absorption and beliefs about information technology usage. *MIS Quarterly, 24*(4), 665–694.

Ajzen, I., & Fishbein, M. (1980). *Understanding attitudes and predicting social behavior.* Englewood Cliffs: Prentice Hall.

Ajzen, I., & Madden, T. J. (1986). Prediction of goal-directed behavior: Attitudes, intentions and perceived behavioral control. *Journal of Experimental Social Psychology, 22*, 453–474.

Anderson, M., & Perrin, A. (2017). Tech adoption climbs among older adults. Pew Research Center. Retrieved From http://www.pewinternet.org/2017/05/17/tech-adoption-climbs-among-older-adults/ Accessed 20 May 2017.

Bandura, A. (1977). *Social learning theory*. Englewood Cliffs: Prentice-Hall.

Bhattacherjee, A., & Hikmet, N. (2008). Reconceptualizing organizational support and its effect on information technology usage: Evidence from the health care sector. *The Journal of Computer Information Systems, 48*(4), 69–76.

Bierman, A.S. (2012) Averting an impending storm: Can we reengineer health systems to meet the needs of aging populations? *PLoS Medicine, 9*(7), Retrieved from https://doi.org/10.1371/journal.pmed.1001267 Accessed 03 May 2017.

Bozan, K. (2017). Can information availability increase patient compliance? Mitigating uncertainty perceptions in the provider-patient relationship. In *50th Hawaii international conference on system sciences (HICSS)* (pp. 3105–3114). Waikoloa. https://doi.org/10.1109/HICSS.2016.391.

Bozan, K., Parker, K., & Davey, B. (2016). A closer look at the social influence construct in the UTAUT model: An institutional theory based approach to investigate health IT adoption patterns of the elderly. In *49th Hawaii international conference on system sciences (HICSS)* (pp. 3347–3356), Koloa. https://doi.org/10.24251/HICSS.2017.406

Caulfield, B., & Donnelly, S. (2013). What is connected health and why will it change your practice? *QJM: Monthly Journal of the Association of Physicians, 106*. https://doi.org/10.1093/qjmed/hct114.

Chin, W. W. (1998a). Issues and opinion on structural equation modeling. *MIS Quarterly, 22*(1), vii–xvi.

Chin, W. W. (1998b) The partial least squares approach to structural equation modeling. In G. A. Marcoulides (Ed.) *Modern methods for business research* (pp. 295–336). London: CRC Press.

Chrischilles, E. A., Hourcade, J. P., Doucette, W., et al. (2014). Personal health records: A randomized trial of effects on elder medication safety. *Journal of the American Medical Informatics Association, 21*(4), 679–686.

Compeau, D. R., & Higgins, C. A. (1995a). Application of social cognitive theory to training for computer skills. *Information Systems Research, 6*(2), 118–143.

Compeau, D. R., & Higgins, C. A. (1995b). Computer self-efficacy: Development of a measure and initial test. *MIS Quarterly, 19*(2), 189–211.

Dabholkar, P. A. (1996). Consumer evaluations of new technology-based self-service options: An investigation of alternative models of service quality. *International Journal of Research in Marketing, 13*(1), 29–51.

Davey, B., & Tatnall, A. (2007). The paradox of social portals. In A. Tatnall (Ed.), *Encyclopaedia of portal technology and applications* (pp. 689–693). Hershey: Information Science Reference.

Davis, F. D. (1986). *Technology acceptance model for empirically testing new end-user information systems theory and results* (Unpublished Doctoral Dissertation). MIT.

Davis, F. D. (1989). Perceived usefulness, perceived ease of use, and user acceptance of information technology. *MIS Quarterly, 13*(3), 319–340.

Davis, F., & Venkatesh, V. (1996). A critical assessment of potential measurement biases in the technology acceptance model: Three experiments. *International Journal of Human-Computer Studies, 45*(1), 19–45.

Doll, W., & Torkzadeh, G. (1989). A discrepancy model of end-user computing involvement. *Management Science, 35*(10), 1151–1171.

Ferman, J. H. (2010). Healthcare reform: Medicare & Medicaid Innovation Center. CMS provision will test new payment and service delivery models. *Healthcare Executive, 25*(4), 58–60.

Forbat, L., Cayless, S., Knighting, K., Cornwell, J., & Kearney, N. (2009). Engaging patients in health care: An empirical study of the role of engagement on attitudes and action. *Patient Education and Counseling, 74*(1), 84–90. https://doi.org/10.1016/j.pec.2008.07.055.

Gardner, D. G., Dukes, R. L., & Discenza, R. (1993). Computer use, self-confidence and attitudes: A causal analysis. *Computers in Human Behaviour, 9*(4), 427–440.

Gatto, S. L., & Tak, S. H. (2008). Computer, Internet, and E-mail use among older adults: Benefits and barriers. *Educational Gerontology, 34*(9), 800–811.

Gazinoory, S., & Afshari-Mofrad, M. (2011). Ranking different factors which affect E-learning outcomes. In *International conference on information and computer applications* (pp. 400–403). Dubai.

Gefen, D., Straub, D. W., & Boudreau, M. C. (2000). Structural equation modeling and regression: Guidelines for research and practice. *Communications of the Association for Information Systems, 4*(7), 1–79.

Goel, M. S., Brown, T. L., Williams, A., et al. (2011). Patient reported barriers to enrolling in a patient portal. *Journal of the American Medical Informatics Association, 18*, i8–i12.

Gordon, N. P., & Hornbrook, M. C. (2016). Differences in access to and preferences for using patient portals and other eHealth technologies based on race, ethnicity, and age: A database and survey study of seniors in a large health plan. *Journal of Medical Internet Research, 8*(3), e50. https://doi.org/10.2196/jmir.5105.

GPO. (2016, November 26). Medicare and Medicaid programs; electronic health record incentive program. Retrieved from https://www.gpo.gov/fdsys/pkg/FR-2010-07-28/pdf/2010-17207.pdf. Federal Register/Vol. 75, No. 144/Wednesday, July 28, 2010/Rules and Regulations Section.

Graetz, I., Gordon, N., Fung, V., Hamity, C., & Reed, M. E. (2016). The digital divide and patient portals: Internet access explained differences in patient portal use for secure messaging by age, race, and income. *Medical Care, 54*(8), 772–779. https://doi.org/10.1097/MLR.0000000000000560.

Gressard, C. P., & Loyd, B. H. (1986). Validation studies of a new computer attitude scale. *Association for Educational Data Systems Journal, 18*(4), 295–301.

Harrington, K. V., McElroy, J. C., & Morrow, P. C. (1990). Computer anxiety and computer-based training: A laboratory experiment. *Journal of Educational Computing Research, 6*(3), 343–358.

Heinssen, R. K., Glass, C. R., & Knoght, L. A. (1987). Assessing computer anxiety: Development and validation of the computer anxiety rating scale. *Computers in Human Behavior, 3*(1), 49–59.

Hofmann, D. W. (2002). Internet-based distance learning in higher education. *Tech Directions, 62*(1), 28–32.

Igbaria, M. (1990). End-user computing effectiveness: A structural equation model. *Omega, 18*(6), 637–652.

Irizarry, T., Shoemake, J., Nilsen, M. L., Czaja, S., Beach, S., & DeVito, D. A. (2017). Patient portals as a tool for health care engagement: A mixed-method study of older adults with varying levels of health literacy and prior patient portal use. *Journal of Medical Internet Research, 19*(3), e99. https://doi.org/10.2196/jmir.7099.

Jha, A. K., Bates, D. W., Jenter, C., Orav, E. J., Zheng, J., Cleary, P., & Simon, R. (2009). Electronic health records: Use, barriers and satisfaction among physicians who care for black and Hispanic patients. *Journal of Evaluation in Clinical Practice, 15*(1), 158–163.

Kidd, T. T. (2010). *Online education and adult learning: New frontiers for teaching practices*. Hershey: IGI Global.

Krist, A. H., Woolf, S. H., Rothemich, S. F., et al. (2012). Interactive preventive health record to enhance delivery of recommended care: A randomized trial. *Annals of Family Medicine, 10*(4), 312–319.

Kronish, I. M., Leventhal, H., & Horowitz, C. R. (2012). Understanding minority patients' beliefs about hypertension to reduce gaps in communication between patients and clinicians. *The Journal of Clinical Hypertension, 14*(1), 38–44.

Lamb, R., & Kling, R. (2003). Reconceptualizing users as social actors in information systems research. *MIS Quarterly, 27*(2), 197–235.

Li, J. P., & Kishore, R. (2006). How robust is the UTAUT instrument? A multigroup invariance analysis in the context of acceptance and use of online community weblog systems. In *Proceedings of the 2006 ACM SIGMIS CPR conference* (pp. 183–189).

Lu, J., Yu, C., & Liu, C. (2005). Facilitating conditions, wireless trust and adoption intention. *Journal of Computer Information Systems, 46*(1), 17–24.

Lucas, H. C. J., & Spitler, V. K. (1999). Technology use and performance: A field study of broker workstations. *Decision Sciences, 30*(2), 291–311.

Lyles, C. R., Sarkar, U., Ralston, J. D., Adler, N., Schillinger, D., Moffet, H. H., Huang, E. S., & Karter, A. J. (2013). Patient-provider communication and trust in relation to use of an online patient portal among diabetes patients: The Diabetes and Aging Study. *Journal of the American Medical Informatics Association, 20*(6), 1128–1131. https://doi.org/10.1136/amiajnl-2012-001567.

Mathieson, K. (1991). Predicting user intentions: Comparing the technology acceptance model with the theory of planned behavior. *Information Systems Research, 2*(3), 173–191.

McInnes, K., Shimada, S. L., Rao, S. R., Quill, A., Duggal, M., Gifford, A. L., Brandt, C. A., Houston, T. K., Ohl, M. E., Gordon, K. S., Mattocks, K. M., Kazis, L. E., & Justice, A. C. (2013). Personal health record use and its association with antiretroviral adherence: Survey and medical record data from 1871 US veterans infected with HIV. *AIDS and Behavior, 17*(9), 3091–3100. https://doi.org/10.1007/s10461-012-0399-3.

Michel-Verkerke, M. B., & Hoogeboom, A. M. G. M. (2012). Evaluation of an electronic patient record in a nursing home: One size fits all? In *The 45th annual Hawaii international conference on system sciences*. Maui: Conference Publishing Services, IEEE.

Miles, M. B., & Huberman, A. M. (1994). *Qualitative data analysis: An expanded sourcebook* (2nd ed.). Thousand Oaks: Sage.

Mirani, R., & King, W. R. (1994). Impacts of end-user and information center characteristics on end-user computing support. *Journal of Management Information Systems, 11*(1), 141–166.

Nahm, E., Sagherian, K., & Zhu, S. (2016). Use of patient portals in older adults: A comparison of three samples. *Studies in Health Technology and Informatics, 225*, 354–358.

Nash, J. B., & Moroz, P. (1997). *Computer attitudes among professional educators: The role of gender and experience*. Annual meeting of the Southwest Educational Research Association. Austin, TX.

Ngai, E., Poon, J., & Chan, Y. (2007). Empirical examination of the adoption of WebCT using TAM. *Computers & Education, 48*(2), 250–267.

Otte-Trojel, T., de Bont, A., Rundall, T. G., & van de Klundert, J. (2014). How outcomes are achieved through patient portals: A realist review. *Journal of the American Medical Informatics Association, 21*(4), 751–757. https://doi.org/10.1136/amiajnl-2013-002501.

Poon, C. C. Y., & Zhang, Y. T. (2008). Some perspectives on high technologies for low-cost healthcare—The Chinese scenario. *IEEE Engineering in Medicine and Biology Magazine, 27*, 42–47.

Ralph, W. (1991). *Help! The art of computer technical support*. Berkeley: Peachpit Press.

Randeree, E. (2007). Exploring physician adoption of EMRs: A multi-case analysis. *Journal of Medical System, 31*(6), 489–496.

Ratnasingam, P., Gefen, D., & Pavlou, P. (2005). The role of facilitating conditions and institutional trust in electronic marketplaces. *Journal of Electronic Commerce in Organizations, 3*(3), 69–82.

Reed, K., Harold, D., & May, D. R. (2005). The impact of aging on self-efficacy and computer skill acquisition. *Journal of Managerial Issues, 17*(2), 212–228.

Rensel, A., Abbas, J., & Rao, H. (2006). Private transactions in public places: An exploration of the impact of the computer environment on public transactional web site use. *Journal of the Association for Information Systems, 7*(1), 19–51.

Roskos-Ewoldsen, B., Black, S. R., & Mccown, S. M. (2008). Age-related changes in creative thinking. *Aging and Creativity, 42*(1), 33–59.

Rovai, A. P., & Childress, M. D. (2002). Explaining and predicting resistance to computer anxiety reduction among teacher education students. *Journal of Research on Technology in Education, 35*(2), 226–235.

Russell, A. L. (1995). Stages in learning new technology: Naïve adult email users. *Computers in Education, 25*(4), 173–178.

Salthouse, T. A. (2004). Localizing age-related individual differences in a hierarchical structure. *Intelligence, 32*(6), 541–561.

Sarkar, U., Karter, A. J., Liu, J. Y., Adler, N. E., Nguyen, R., López, A., & Schillinger, D. (2011). Social disparities in internet patient portal use in diabetes: Evidence that the digital divide extends beyond access. *Journal of the American Medical Informatics Association, 18*(3), 318–321. https://doi.org/10.1136/jamia.2010.006015.

Shiv, B., & Fedorikhin, A. (1999). Heart and mind in conflict: The interplay of affect and cognition in consumer decision making. *Journal of Consumer Research, 26*(3), 278–292.

Simonson, M. R., Maurer, M., Montag-Torardi, M., & Whitaker, M. (1987). Development of a standardized test of computer literacy and a computer anxiety index. *Journal of Educational Computing Research, 3*(2), 231–247.

Smith, S. G., O'Conor, R., Aitken, W., Curtis, L. M., Wolf, M. S., & Goel, M. S. (2015). Disparities in registration and use of an online patient portal among older adults: Findings from the LitCog cohort. *Journal of the American Medical Informatics Association, 22*(4), 888–895. https://doi.org/10.1093/jamia/ocv025.

Straub, D., Limayem, M., & Karahanna-Evaristo, E. (1995). Measuring system usage: Implications for IS theory testing. *Management Science, 41*(8), 1328–1342.

Sumner, M., & Hostetler, D. (1999). Factors influence the adoption of technology in teaching. *Journal of Computer Information Systems, 40*(1), 81–87.

Taylor, S., & Todd, P. A. (1995). Understanding information technology usage: A test of competing models. *Information Systems Research, 6*(2), 144–176.

Thompson, R., Barclay, D. W., & Higgins, C. A. (1995). The partial least squares approach to causal modeling: Personal computer adoption and use as an illustration. *Technology Studies: Special Issue on Research Methodology, 2*(2), 284–324.

Torkzadeh, G., & Angula, I. E. (1992). The concept and correlates of computer anxiety. *Behavior & Information Technology, 11*(2), 99–108.

van der Heijden, H. (2004). User acceptance of hedonic information systems. *MIS Quarterly, 28*(4), 695–704.

Venkatesh, V. (2000). Determinants of perceived ease of use: Integrating control, intrinsic motivation, and emotion into the technology acceptance model. *Information Systems Research, 11*(4), 342–365.

Venkatesh, V., & Bala, H. (2008). Technology acceptance model 3 and a research agenda on interventions. *Decision Sciences, 39*(2), 273–315.

Venkatesh, V., & Davis, F. D. (2000). A theoretical extension of the technology acceptance model: Four longitudinal field studies. *Management Science, 46*(2), 186–204.

Venkatesh, V., Thong, J. Y. L., & Xu, X. (2012). Consumer acceptance and use of information technology: Extending the unified theory of acceptance and use of technology. *MIS Quarterly, 36*(1), 157–178.

Vishwanath, A., & Scamurra, S. D. (2007). Barriers to the adoption of electronic health records: Using concept mapping to develop a comprehensive empirical model. *Health Informatics Journal, 13*(2), 119–134.

Winter, S., & Chudoba, K. (1998). Attitudes toward computers: When do they predict computer use? *Information & Management, 34*(5), 275–284.

Yildirim, S. (2000). Effects of an educational computing course on pre-service and inservice teachers: A discussion and analysis of attitudes and use. *Journal of Research on Computing in Education, 32*(4), 479–495.

Yiong, B. L. C., Sam, H. K., & Wah, T. K. (2008). Acceptance of e-learning among distance learners: A Malaysian perspective. In *Proceedings of ASCILITE – Australian Society for Computers in learning in tertiary education annual conference* (pp. 541–555).

Yu, P., Hailey, D., & Li, H. C. (2008). Caregivers' acceptance of electronic documentation in nursing homes. *Journal of Telemedicine and Telecare, 14*(5), 261–265.

# An Evaluation of the Point-of-Care (PoC) System Implementation and Adoption in a Multi-Campus Private Hospital in Melbourne

**Imran Muhammad and Nilmini Wickramasinghe**

## 1 Introduction

Most if not all countries today are investing heavily in health information technology in general and e-health in particular. One of the key reasons for this is the immense pressures currently facing healthcare delivery caused by a confluence of factors including changing patient demographics, financial implications, work force shortages, advancements in medical technologies and the need to provide efficient and effective patient-centric healthcare delivery. Due to an ever-increasing need for effective and efficient healthcare services and delivery, healthcare organizations are now trying to find integrated point-of-care health IT solutions to gain a strategic advantage (Ives and Jarvenpaa 1991).

The complex nature of healthcare services is compelling healthcare organizations to adopt best available technologies; in doing so healthcare organizations are facing many problems. Thus, healthcare organizations need to adopt technologies according to their requirements and best fit having considered their environment, infrastructure, government regulations, and scope of their business, availability of funds, and local culture and norms (Ignatiadis and Nandhakumar 2007). Healthcare organizations are inclusive organizations, involving different stakeholders, partners, customers, and suppliers from different cultures and different systems; they need more sophisticated means of communication and interaction.

I. Muhammad
Deakin University, Burwood, VIC, Australia

N. Wickramasinghe (✉)
Epworth HealthCare, Richmond, VIC, Australia

Swinburne University of Technology, Hawthorn, VIC, Australia



N. Wickramasinghe, F. Bodendorf (eds.), *Delivering Superior Health and Wellness Management with IoT and Analytics*, Healthcare Delivery in the Information Age, https://doi.org/10.1007/978-3-030-17347-0_26

Healthcare globally is at a crossroads. E-health is being seen as a vital necessity to address escalating costs and yet provide quality healthcare. This in turn means that the successful assimilation of these e-health solutions into their respective healthcare contexts is vital so that the technology-mediated collaborations in healthcare which serve to provide quality patient care and life-saving treatments must not only be understood but also be facilitated and supported by these e-health solutions.

Health information technology implementation is not limited to just installing software; i.e. it is much more than a technology adoption. Rather, it involves people issues more than technological issues (Cresswell et al. 2011). Research indicates that people issues are more to blame for the unsuccessful efforts of e-health implementations (Mukesh and Betsy 2009). The effect of sociotechnical issues such as macro level or external factors including political, social, economic, environmental infrastructure and technology, and laws and regulations; meso level or organizational factors such as leadership, management style, policies, and structure; and micro level or tactical factors such as information sharing, training and learning, technical staff or user behaviour has been less widely studied (Nguyen et al. 2015). Yet, it is precisely these issues that separately or in combination derail numerous HIS implementations (Nguyen et al. 2015). To examine this dilemma, we proffer a unique application of the fit-viability model (FVM) (Liang et al. 2007) to facilitate a better understanding of key issues pertaining the implementation and adoption of a point-of-care (PoC) system at one of the not-for-profit private hospitals in Australia. In so doing, we answer the research question: 'How can an FVM assist in unpacking the varied sociotechnical issues in the adoption and implementation of PoC system?'. This will help the decision-makers in hospital to understand how the new system fits within the different departments and also if it is a viable option to install such a new system. This study deals with just two departments of the hospitals, namely, Food Services and Environment Services. An exploratory single case study methodology is adopted. From such an analysis, we believe it will be possible to identify optimal PoC solutions and opportunities to add value. The remainder of this paper is structured as follows: the following section presents a brief literature on point-of-care solutions followed by the theoretical underpinning, methodology, and research design including data collection and data analysis phases, as well as describing the examined point-of-care system and its main elements. Section 5 presents and discusses the results, followed by a conclusion that summarizes the research, its implications and limitations, and future research directions.

## 2 Literature Review

Today, there exist various studies that investigate the impact of IS/IT on specific aspects of healthcare delivery such as patient outcomes (Goldzweig et al. 2013; Tucker et al. 2014), patient safety (Farley et al. 2013; Middleton et al. 2013), quality of care (Kellermann and Jones 2013), the efficiency of healthcare delivery operations (Goldzweig et al. 2013; Lee 2013), and the cost of these operations (Balabanova et al. 2013). That being said, the majority of the reviewed studies, however, share two types of limitations: (1) they tend to study specific systems and their impacts on a particular output, and (2) most of these studies lack adopting sociotechnical aspects to cover the different levels of healthcare delivery, which makes their findings questionable, especially when issues around patient outcomes and safety are of concern (ibid). Moreover, they focus narrowly on cost containment rather than how to support value-based care delivery. In addition, the current literature shows contradictory results from different studies on the impact of IS/IT on different outputs such as quality of care, patient safety, patient outcomes, cost, and efficiency (Chaudhry et al. 2006; Jones et al. 2014) which calls for a deeper examination of IS/IT in healthcare to investigate the impact of IS/IT on the organizational performance of healthcare providers.

Patient-centric care, an emerging key success factor for healthcare delivery, provides customized precision care to each patient's individual needs and requirements (Reynolds 2009). The literature identifies several components of patient-centric care delivery, including; patient participation and involvement and the context where care is delivered (Kitson et al. 2012). Hence, investigating the role of IS/IT in delivering patient-centric care is vital for today's healthcare systems and structures, and to date is not well studied.

Many healthcare information systems have been implemented around the globe with mixed results, despite the claims that HIS can play a significant role in efficiency and effectiveness of healthcare service delivery (Muhammad and Wickramasinghe 2014) The literature provides evidence of failed clinical systems and lack of adoption by users (DesRoches et al. 2008; Protti et al. 2009). Challenges and barriers to implementation and adoption of bedside PoC systems in hospital wards have been extensively debated (Brailer 2005; Choi et al. 2004; Yao et al. 2005; Nguyen et al. 2015).

Researchers have divided these barriers into different categories ranging from environmental, social, technical, and organizational (André et al. 2008). These factors can play a very crucial role in the decision-making process of technology adoption (Huang and Palvia 2001). In a healthcare service context, where organizations are now required to work as a networked framework, health information technology implementation and adoption would be a more complex and challenging endeavour because of the different business processes, the available infrastructure, compatibility issues, decision centres, authorization mechanisms and hierarchies, enterprise systems, and data semantics (Avgerou 2008; Liu et al. 2011; Trudel 2010). IT implementations can cause serious disruptions in service deliveries, and in result, at productivity and healthcare services are one of the very critical areas of services that cannot afford disruptions (Kralewski et al. 2010; Scott et al. 2005).

There are many organizational barriers to the implementation and the adoption of e-health technologies, for example, poor governance, organizational culture, and proper management of the change process, that could harm the flow of transformation (Greenhalgh and Stones 2010; Kennedy 2011). These issues can aggravate the resistance to the change process and complicate the dissemination of the e-health technology. Technological issues can also exacerbate the resistance to the adoption of health information technology (Muhammad and Wickramasinghe 2014). The lack of infrastructure and standards results in a fragmentation of healthcare information systems and this contributes to creating a very complex situation for coordination (Kennedy 2011; Trudel 2010). Pre-implementation and post-implementation vendor support is another key concern for organizations (Kennedy 2011; Liu et al. 2011). Lack of technical resources and experience with information technology implementation within healthcare settings are other problems faced by many (Trudel 2010; Liu et al. 2011; Kennedy 2011).

People issues, ranging from user acceptance (Trudel 2010), perceived ease of use (Al-Azmi et al. 2009), lack of knowledge about the system (André et al. 2008; Liu et al. 2011), lack of training, lack of stakeholder consultation (Showell 2011), lack of willingness to assimilate the technology in to daily routines and processes (Greenhalgh and Stones 2010), conflict between system and user-embedded values (Greenhalgh and Stones 2010), complex and complicated user interfaces, conflict between physician activities and training schedules (André et al. 2008), and complications in patient-provider communications are some of the major concerns. Further, it is paramount that the systems are user-centric and have a good fit with user values as well as existing healthcare systems (Liang et al. 2007).

## *2.1 The Point-of-Care (PoC) System*

The point-of-care system is an integrated care coordination platform that sits at the patient's bedside consisting of various modules which all work in concert to provide integrated care and support functions in an acute care context. A point-of-care solution is distinct from an electronic medical record as it does not support all medical and clinical input and primarily is tailored around nursing and patient care activities. Many healthcare organizations are turning to point-of-care solutions as they enable more tailoring and are not as expensive and disruptive as large- scale EMR adoptions. As a system, PoC consists of 4 tiers as described in Table 1.

## *2.2 Fit-Viability Model*

Tjan (2001) proposed fit-viability dimensions for evaluating Internet initiative projects. Liang and Wei (2004), by taking these two dimensions and adding task-technology fit (TTF) theory, proposed a fit-viability model to study m-commerce applications as shown in Fig. 1.

In this framework, viability measures the readiness of the organization for the technology adoption and implementation, and fit measures capabilities of the systems to optimally perform the required tasks. These two dimensions make a simple matrix with fit on horizontal and viability on vertical axis. By using the

**Table 1** The four tiers of the point-of-care system

| Tier | Description |
|---|---|
| Core platform | This is the core foundation of all services and applications that may run on top of the system. The main features of this core platform are:<br>1. Centralized content management<br>2. Hospital systems integration through HL7 to provide real-time information at the point of care<br>3. Highly secure environment with 256-bit encryption for all patients' information<br>4. Multimedia for patients, including IPTV, internet access, and video streaming on any connected terminal<br>5. Customized branding to provide a consistent experience as possible for both patients and clinicians<br>6. Proactive management for patient satisfaction through real-time surveys about various aspects of their care, stay, or treatment |
| Feature packs | This tier addresses three areas:<br>1. Enhancing patient experience through personalized treatment plans, better communications with their clinicians, as well as providing a wide range of entertainment choices<br>2. Improving clinical workflow by assuring the access to valid information and/or valid real-time data<br>3. Engaging patients in their care plans, education programs, and communicating with their clinicians through efficient and easy to use patient portals |
| Add-on items/ modules | This tier aims to provide patients with help and support for their orientation and introduction to available services. This is enabled through providing digital signage, way finding, accessibility options (screen readers, eye-tracking, speech recognition and so on), and presence (Who's done what) |
| Third party content | This tier is designed to provide third-party contents for both patient entertainment and patient education. While the former is based on personal choices, the latter is based on a patient's profile, age, or cognitive ability. The care team can also assign personalized/customized content based on their judgement/ assessment for patients' individual health conditions. |

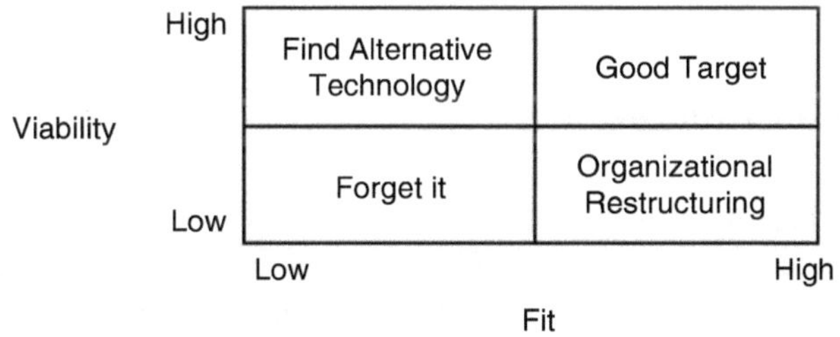


**Fig. 1** The fit-viability framework. (Adopted from: Liang and Wei 2004)

four corners of the matrix, organizations can understand the feasibility of system implementation based on its viability and fit and thus make an informed decision for technology adoption and implementation.

## 2.3 *Task-Technology Fit*

The theoretical basis of the fit construct is derived from the task-technology fit model which according to Goodhue (https://pdfs.semanticscholar.org/668e/58d4e3479317257a41ce66c688a8aa663399.pdf) argues that a fit between task characteristics and system features need to be high for the better performance and success, and this will have effect on the decision-making process of an organization. Research (Soh et al. 2000) has indicated that if a system is more aligned with the requirements of the users there are greater chances of system success which leads to better performance. It means that if the features offered by the system fit with the task requirements the users will be more incline to use it.

## 2.4 *Viability*

Viability refers to the degree of impact of environment and organizational factors on a system adoption and implementation decision. These factors at the macro level include political and social, economic, environmental, as well as infrastructure/technology factors. At the organizational level, literature has proposed many factors at the strategic and tactical levels (Umble et al. 2003). These factors include leadership, management style, polices, information sharing, training and learning, technical staff, and user behaviour. Taking the example of PoC, economic and technological factors are crucial factors in HIS implementations; and ignoring these factors could lead to unsuccessful projects. Management support, physical and IT infrastructure create stronger desire of system implementation and innovation adoption, that positively impact viability of the system.

# 3 Research Framework

The research framework shown in Fig. 2 is used to identify the key constructs and factors affecting point-of-care (PoC) system implementations. The PoC is a patient bedside solution that can be accessed across the hospital by health service providers and also has a patient portal component. It is not an EMR (electronic medical record) but has many features and capabilities similar to an EMR. It also has a patient entertainment component. The PoC system was implemented into the not-for-private tertiary hospital system (the chosen case study) gradually starting in late 2015. Primary objectives for its implementation included to enhance the patient experience and provide value-based patient-centred care. The PoC system is consist of many integrated modules, but this paper will only focus on two modules, namely, catering/food services and environmental services.

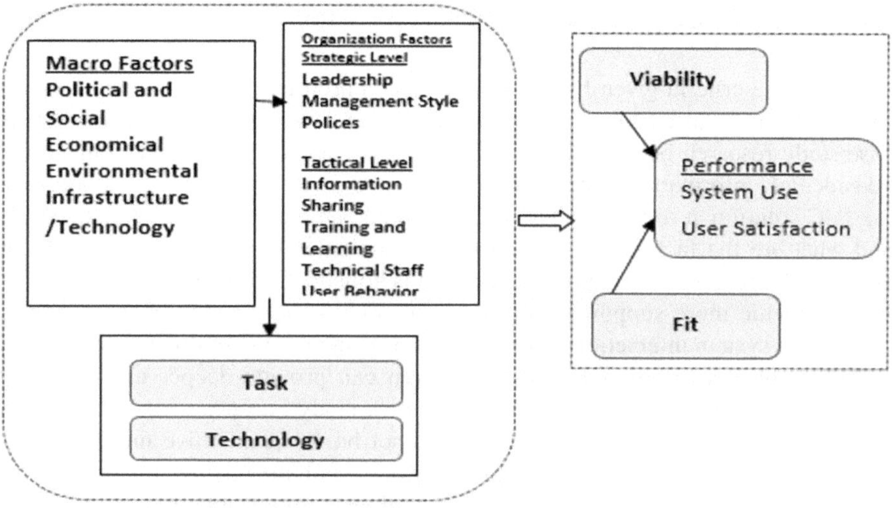


**Fig. 2** FVM

Fit is measured by matching the requirements of the organization with the functionalities offered by the PoC system e.g. data format, operating procedures, and output format while viability is measured by assessing the impact of national and organizational factors on the adoption decision of the organization and individual user adoption.

Considering that the PoC has many similar factors to other e-health solutions such as political and sociotechnical factors identified by Muhammad and Wickramasinghe 2017 in evaluation of MyHealth record in Australia (the Australian national e-health solution) and smart card solution in Germany (the German national e-health solution), thus, it is logical to use these factors as the basis of the model. This conceptual model serves to capture the important aspects of the barriers and facilitators for the prediction of the successful adoption and implementation of the PoC. The proposed model identifies a network of different actors interconnected to each other. It further illustrates that a central issue with the evaluation of IT based healthcare is influenced by the complexity of the evaluation objects and includes both social and technical considerations (Greenhalgh and Stones 2010). For instance, the nature of the integration of healthcare information systems with the culture and business processes of healthcare organizations puts more emphasis on the evaluation methods and goes beyond the technology aspects of hardware and software; furthermore, external and internal environmental factors as well as an understanding of the diverse nature of system effects in the healthcare settings are required (Greenhalgh and Stones 2010). This emphasis is on creating a better fit between human, contextual, and technological factors for the successful implementation and adoption of health information systems (Yusof et al. 2008).

## 4 Methodology

Based on the criteria given by Yin (2014); the appropriate choice of methodology to test the use and usability of the proposed framework (Fig. 1) is a qualitative case study research because this is an exploratory study of a new phenomenon of bedside PoC information system's implementation. Further, we wish to explore how the PoC solution at different sites of the hospital can be implemented successfully and what are the factors that impact on the implementation and adoption of this HIS based intervention. Qualitative research is holistic, humanistic, and interactive; it can provide more support to focus on the study of a complex phenomenon of human and system interaction and relationship, as in our case multisite bedside PoC system implementations. Qualitative research can provide deeper understanding of the phenomenon as compared to quantitative study because of the exploratory nature of the study and focus which would not be on quantitative measures (Yin 2014). For this study, several archival records and documents relating to the health information and communication technologies implementation and adoption in healthcare service delivery settings along with hospital and OneView reports and evaluations were critically analysed. These documents were of great value in developing an understanding of the need for a PoC system and factors important for the implementation and adoption of this system. This analysis assisted in developing the theoretical re-search framework and in planning the primary data collection strategies for the larger study. A priori themes were developed through a pilot study and then literature was analysed using thematic analysis and hermeneutic analysis (Boyatzis 1998; Kvale 1996), then we performed a gap analysis. The analysis led us towards the development of FVM for this study.

## 5 Analysis and Results

Data for this piolet study was collected using unstructured interview and analysis of archival material. As far as possible, the multiple stakeholder views (i.e. service providers, regulator, payer, food and environment services, and hospital and patient perspectives) in healthcare were captured. Data analysis included standard qualitative techniques such as thematic analysis where a priori themes were derived from the components in the conceptual model. The case studies all exemplify various aspects of the proffered conceptual model in this way serving to validate the model and demonstrate its usefulness in unpacking critical aspects with HIS implementation.

## *5.1 The Context Before and After OneView PoC System*

This section presents a process map for two departments of the hospital before and after the implementation of the OneView PoC system. It must be noted here that prior to the PoC system, there was another system known as Infotainment. This system had many problems concerning poor technical support, frequent hardware and software failures, smaller screens, and usability issues especially with the user interface and the biometrics registries. That system really did not fit with in the hospital quality services environment; thus, the change was impending. This section will explain the process before and after OneView PoC system within two departments, namely, Food Services and Environmental Services. These departments are chosen as they are fully transitioned to the new PoC system now.

### 5.1.1 The Context of Food Services

Prior to the current PoC system, the processes in the Food Services were facilitated using an application called Delegate. This application had been in use since the opening of Epworth Eastern and has been replaced by OneView PoC system. Hence, the old PoC system was not used by the staff from Food Services.

### 5.1.2 The Context of Food Services Prior to OneView PoC System

Managing patients' meal orders was handled by the Department of Food Services. The contact point was the menu monitors, who take the meal orders 24 h prior to the actual delivery of meal. A computerized system called Delegate Prior to OneView PoC system was used for the order and delivery services. Delegate system was installed on a number of computers on wheels (mainly laptops). Taking orders and delivery meals to the patient includes discussing the options and personal preferences for the breakfast, lunch, and dinner for the next day. This process would usually consume nearly 70% of the Menu Monitor's work time with a rate of 14 patients per hour. That is 3–4 min per patient. Patients can choose three full courses for the next day from a menu that changes every 3 years hospital wide. The gathered information about the meal preferences was then entered by the menu monitors into Delegate, and then spread sheets were printed off this system to circulate to kitchens. These sheets were then assigned to different chefs as 'Production Lists'. Throughout the day, three production lists were prepared, one in the morning (7:00 am), the second is after breakfast (10:00 am), and the third is at 3:30 PM for the evening as Fig. 3 below illustrates.

Two main issues were found in this way of handling patients' meals. The first relates to the information collection, and the second issue about Delegate performance and functionality.

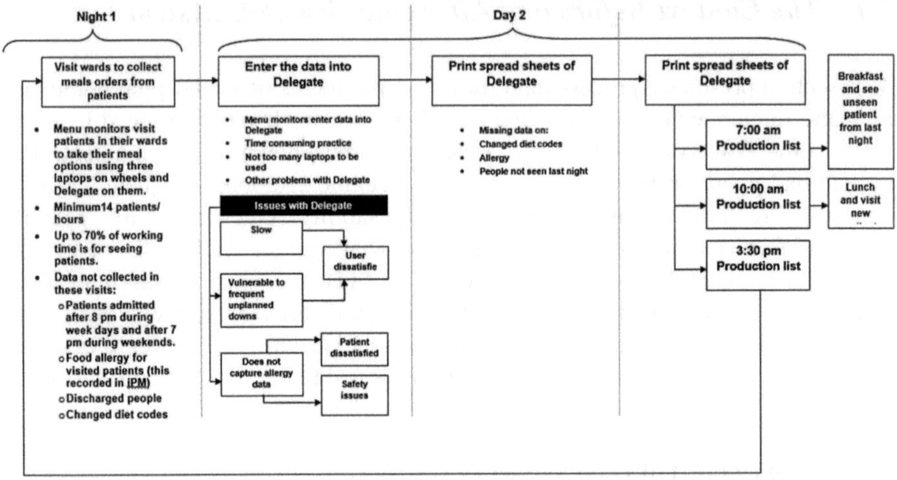


**Fig. 3** A flow chart for ordering meals before PoC

Although preparing the production lists was time consuming and required loads of time for the interaction between patients and menu monitors, still four types of information were not possible to capture using Delegate and the process around it. Those are:

I. *Late orders*: Meal orders for patients admitted after 8 pm during week days and after 7 pm during weekends were not attended. This group of patients did not have a choice of food for the first day of their admission because they could not see the menu monitors. Rather, they would have the default meal for the day.
II. *Food allergy for visited patients*: Delegate did not have any capability to record any food allergy patients may have. This piece of information normally comes from iPM. In many cases this information was then not passed on to the chefs, so they make meals without taking that into consideration.
III. *Discharged people*: Depending on the time patients are discharged, many cases reportedly happened where meal orders were made, but patients had been discharged. Again, this piece is coming from iPM, and not passed on to the Food Services staff at the right time.
IV. *People with changed diet codes*: As treatment plans progress, patients may change their diet codes, such as changing from 'not eating' to 'eating' and from 'soft' to 'hard' food. These changes were also managed by nurses using iPM; however, the cooperation between nurses and Food Services was not maintained at all times, which resulted in many cases meals were not made according to these changes.

These issues had direct and indirect impacts on the cost and quality of provided services as Table 2 summarizes.

**Table 2** Issues resulting from the conventional way to handle meal orders prior to OneView PoC system

| Issue | Impact on cost | Impact on quality |
|---|---|---|
| Late orders | None, as patients receive the default meal for the next day | Negative impact on patient satisfaction |
| Food allergy for visited patients | Wasted food, unplanned care for the resulting allergy which may result in implications on patients' insurance cover | Negative impact on patient satisfaction and safety as well as trust in the hospital |
| Discharged patient | Wasted food | More pressure on chefs |
| Patients with changed diet codes | Wasted food | Negative impact on patient satisfaction |

Delegate has a number of issues as the interviewees from the Food Services agreed. These issues include:

I. It was perceived to be a slow system.
II. System was not really fit for the task.
III. Long-term viability of the system was really doubtful even though infrastructure and management support were available.
IV. The system had many technical problems, and fixing these problems took long times as described by this interviewee: 'If there is some technical problems with Delegate this would take long time to fix. I mean when the system goes down it really goes down and takes a while to fix'.
V. As mentioned before, Delegate did not have the capability to capture food allergy data, which forced the staff at the Food Services to look up these data from iPM. This was not possible all the time, which resulted in a number of cases where patients had food that they were allergic to; as a consequence, this had negative impact on patient satisfaction and healthcare services and delivery of the hospital.

While some issues with Delegate had negative impacts on the users, thus they became dissatisfied about the system; but few problems were really serious problem with real consequences on patient's health and safety. Not only did it negatively affect patient satisfaction and experience, but it also represented a real risk factor and caused safety issues to the patients and the hospital at the same time.

### 5.1.3 The Context of Food Services After OneView PoC System

Using OneView PoC system, patients can place their orders of meals through their user interface. The arrival of this function to the PoC system has partially solved the issues faced by the conventional system, namely, late patients (after 8 pm weekdays and after 7:00 pm weekends), allergy data, and patients with changed diet codes.

Late patients can order their favourite meals for the next day if they want to, they can state their allergy status through the admission form, and nurses can change patients' diet codes right from OneView PoC system.

Currently, no more than 10% of the patients are using the PoC system to order their meals. On asking on the reasons behind that, two main reasons were identified. The first is some issues with the user interface, especially with elderly patients, as patients need to scroll down to the bottom of the screen to reach the meal ordering function. During the scrolling down, a number of pop-ups will appear and may disrupt their endeavour: 'We go up and introduce the system to them. When you go into the system at the minute, the way they implemented it, it is hard to use. You've got to scroll down to find the ordering' (Fig. 4).

The other reason relates to the conceptual resistance by some patients to rely on a 'machine' to order their meals, preferring human-human interaction more than human-machine interaction. This was agreed upon by both interviewees from the Food Services.

As a result of being in a hybrid environment, i.e. a minority of patients are using the PoC system to order their meals, and the majority are still using conventional way to order their meals, Delegate now has coloured dots to indicate the patients who used the PoC system to order their meals. Hence, the menu monitors do not need to visit them to organize their meals.

With the expected increase of the uptake of this function in the PoC system, more patients will adopt this function, and more time required to see patients by their menu monitors will be freed-up. This is expected to have positive impacts on the hospital and its patients. In addition, it is expected that there will be a significant saving regarding food wastage.

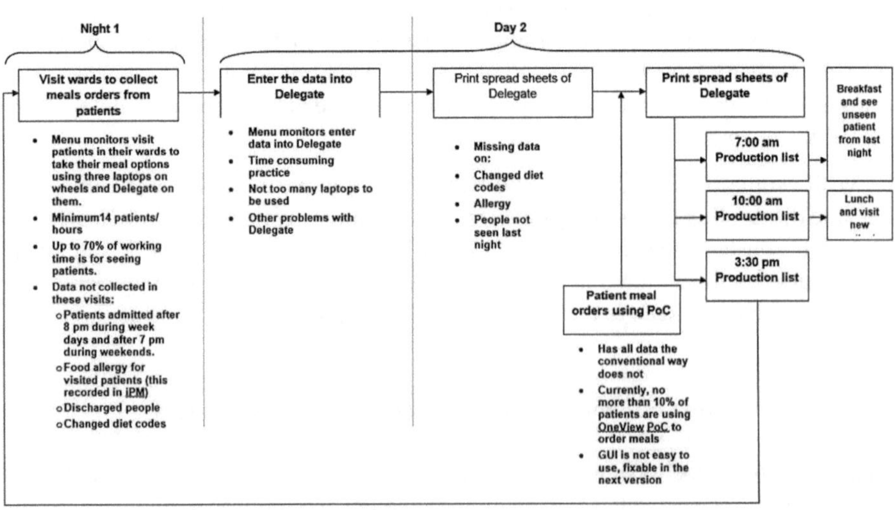


**Fig. 4** A flow chart for ordering meals after PoC

## 5.2 The Context of Environmental Services

The Environmental Services is a dedicated team whose two main responsibilities are (1) providing all types of cleaning (steam cleaning, buffer cleaning, advanced cleaning, curtains cleaning, etc.), and (2) patient transportation. In terms on human resources, the Environmental Services team comprises about 60 staff. Of this figure, about 40 work in cleaning, and the reminder work in patient transportation.

Unlike the Food Services, which had Delegate as a computerized system to facilitate food-related processes prior to the OneView PoC system, the operations of the Environmental Services at the hospital were based on over-the-phone and face-to-face communications. Introducing the OneView PoC system and integrating the Room Ready Module into it have made considerable change in the processes of this vital department. The following is a summary of the process map of the Environmental Services at the hospital before and after OneView PoC system.

### 5.2.1 The Context of Environmental Services Before OneView PoC System

The process of performing jobs by the Environmental Services before the PoC had three main steps: (1) initiating the job by nurses and specific cleaners; (2) receiving job orders by the supervisors within the Environmental Services; and (3) assigning tasks to cleaners as Fig. 5 depicts.

Nurses normally initiate job orders when needed. This includes preparing rooms before admitting new patients to these rooms, cleaning rooms after patients have been discharged, and as needed if a patient had an incident such as bleeding or

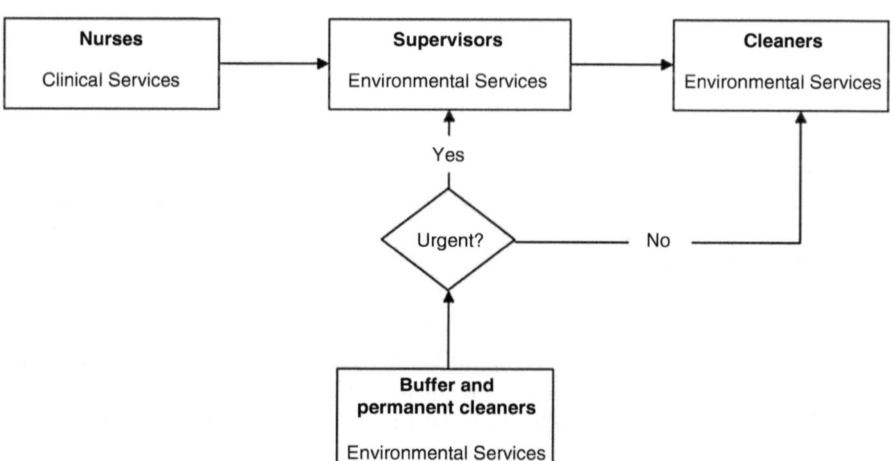


**Fig. 5** A map for the Environmental Services before OneView PoC

vomiting. These job orders go to the supervisors from the Environmental Services using face-to-face or phone-to-phone communications tools. The supervisors in turn convey these orders to the cleaners across the hospitals using same communication means, i.e. phone and face-to-face. Apart from nurses, buffer cleaners and permanent cleaners can initiate job orders if need be. This normally happens when one of these cleaners realizes, while doing their jobs, there exist some curtains or carpets that need to be cleaned, and they do not have the required equipment to do so. Once the need of a cleaner has been established, the path of this order in dependent upon its urgency. If the job was of a higher urgency, then these cleaners inform their supervisors either by phone calls or by handwritten notes. Then the supervisors assign the tasks to different cleaners on floor accordingly. If the job was not considered urgent by these cleaners, then they will wait till the next shift of cleaners has come to do the job based on hand written notes, which caused extended times to do specific jobs. The level of urgency was left to the cleaners to decide.

This system had caused many problems, which can be summarized as follows:

1. Over-reliance on human factors: As can be seen, the whole processes of the Environmental Services team are centred on human communications and judgement. This resulted in many cases where jobs were not done or took longer time to be addressed. This is particularly acute in the cases where patients had to wait outside their rooms while the room is being cleaned, which resulted in unsatisfied patients and nurses.
2. Lack of coordination in the multi levels of communications: Given that nurses, supervisors, and permanent and buffer cleaners could initiate job orders, cleaners in many cases were confused about their tasks and what tasks had higher priorities. This lack of coordination was due to adopting phone calls, handwritten notes, and face-to-face means.
3. Inability to address language barriers: Given that a considerable portion of the cleaners had language barriers, their understanding of their assigned tasks over the phone or handwritten notes was reportedly limited in many cases, which caused many jobs not done properly.
4. More effort by cleaners: As the job orders did not have enough information about their jobs (locations, level of urgency, and required equipment), the cleaners had to go to the site to manually collect all of this information and come back to their workplace to collect the right equipment for that specific job. This caused them to walk back and forth many times, which reduced their productivity and the quality of their jobs, and increased their fatigue.
5. Lack of accountability: As the majority of the needed jobs were verbally conveyed from one stakeholder to another, the possibility of creating accountability and tracking the performance of different units and individuals were almost impossible. This is especially acute in the case of cleaners with limited literacy, as well as the communications between nurses (Clinical Services) and the cleaners and their supervisors as Fig. 4 depicts.

These issues had direct and indirect impacts on the cost and quality of provided services as Table 3 below summarizes.

**Table 3** Issues resulting from the conventional way to handle environmental services prior to PoC system

| Problems | Impact on cost | Impact on quality |
|---|---|---|
| Over-reliance on human factors | Double work resulted in many cases, which implied extra cost and less quality | |
| Lack of coordination: | | |
| Multi levels of communications | | |
| Language barriers | As many jobs needed to be repeated, double works resulted in many cases, which implied extra cost and less quality | |
| More effort by cleaners | Given the missing information on the nature of their tasks, cleaners had to survey the location of their jobs in person, which resulted in them walking for extended times/distances | Due to the unneeded increased workload, the quality of cleaning services was negatively affected |
| Lack of accountability | Tracking the performance of individuals and different unit was almost impossible | |

### 5.2.2 The Context of Environmental Services After OneView PoC System

Introducing the OneView PoC system, particularly the Room Ready Module, has notably streamlined the cleaning-related processes of the Environmental Services, while the other vital role of the Environmental Services, namely, patient transportation, is still conducted using the conventional way, with a vision to integrate this function into the OneView PoC in later enhancements.

The Room Ready Module enables nurses, permanent cleaners, and buffer cleaners to log into the system and place cleaning orders with enough details about the job, its location, requirements and level of urgency. This information is then conveyed to the cleaners on floor as short text messages on their PoC phones. Based on the nature of the jobs, cleaners can choose the jobs of higher urgency, closer to their geographic location, and/or achievable using their current equipment. This has resulted in saving cleaners' times and efforts, which in turn has shown faster responses to the cleaning needs initiated by different wards, units, and individuals. Not only has the Room Ready module enhanced quality and productively of cleaners, but it has also resulted in a simpler map of cleaning processes performed by the Environmental Services as Fig. 6 depicts.

The initial findings from the interviews show that the use of OneView PoC system to support the cleaning processes at Epworth Eastern has addressed most of the problems faced by the conventional way to manage the cleaning needs for the hospital. Table 4 summarized these findings:

The initial results suggested that the PoC system for food and environment services is fit for task and very much viable to implement. Almost all sites of hospital were ready for implementation. IT Infrastructure was fit for the purpose while

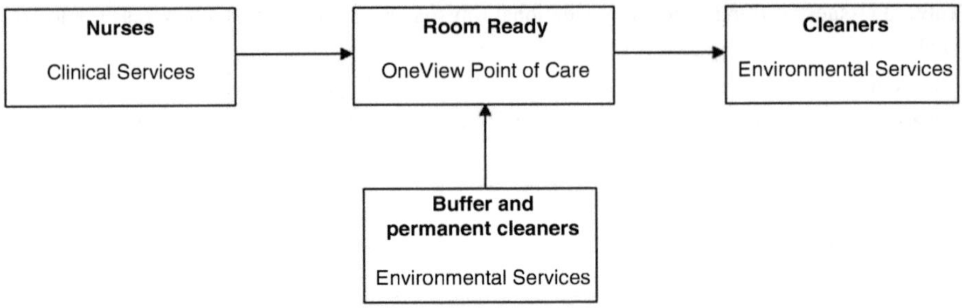


**Fig. 6** A map for the Environmental Services after OneView PoC

**Table 4** The impact of Room Ready Module from the OneView PoC system

| Problems | How has OneView PoC addressed the problems | Impact of OneView PoC on | |
|---|---|---|---|
| | | Quality | Cost |
| Over-reliance on Human Factors | This problem has been partially solved, as human still need to log in and place job orders. The existence of Room Ready though has increased the ability to place jobs and track them. | Positive M | Positive M |
| Lack of coordination: | The introduction on Room Ready Module has eliminated one layer of communications in the process map of cleaning services. That is the supervisors, which has freed up their time, and has rebuilt their roles around coordinating different tasks and following up with different stakeholders. | Positive H | Positive M |
| Multi levels of communications | | | |
| Language barriers | As job orders come to cleaners in a form of short text messages on their PoC phones, this problem has been partially solved. | Positive L | Positive L |
| More effort by cleaners | As the jobs orders come with a relatively comprehensive set of information, cleaners don't need to go and assess the job before actually doing the job, which resulted less effort from them, and more tasks performed every day than before | Positive H | Positive M |
| Lack of accountability | All job orders are now documented and stored in the system. Hence, tracking different jobs and their progress and the responses from different stakeholders is always possible. | Positive H | Positive H |

*Legends*: *L* low, *M* medium, *H* high

physical infrastructure needs changes for one site. Environment was favourable, and project has full support from top management. Appropriate budget was allocated for the project. Our initial analysis suggests that the use of system will depend on the fitness of the system for the tasks. Users were very positive in its very early stages of deployment. We contend that a large-scale study is prudent to further understand the capacities of implementation and use of this system for clinical services.

## 6 Conclusion

The purpose of this research paper is to conceptualize a framework to investigate the implementation and use viability of a bedside PoC health information and entertainment system for food and environment services for patients in private hospital wards, since we recognize that with e-innovations not only the technology solution is necessary but it is also critical to look at organizational and societal aspects concurrently especially in today's global business environment. E-health system implementations comprise of technology as well as human involvement. In view of this, it is important to investigate the impact of political and social, economic, environmental and infrastructure/technology factors on the organizational decision-making. We underscore the importance of studying the system viability and fit before making any decision about system adoption and deployment. Technology needs to have a good fit to perform the required tasks, only then can system viability and fit have positive effects on the organization's performance and that can be measured by system use and user satisfaction and thus ensure the full potential of a solution is realized.

The theoretical contribution of the paper is the use of fit-viability model for the evaluation of health information technology (HIT) implementation in private hospital settings. This model has never been used of HIT implementations, thus we believe the model will present more informed lens for decision-makers to understand HIT implementation. The practical implication of the fit-viability model presented can be realized by using the developed model to measure the fitness and viability of implementing health information technology in hospital settings. This should be able to envisage the possibility of success.

There are also limitations of this study. First, the theoretical generalizability of the FVM needs critical testing in future studies. Second, the findings of this paper are based on just two segments of the hospital, namely, Food Services and Environmental Services; thus, more extensive studies including more segments of hospital may be necessary.

## References

Al-Azmi, S. F., Al-Enezi, N., & Chowdhury, R. I. (2009). Users' attitudes to an electronic medical record system and its correlates: A multivariate analysis. *The HIM Journal, 38*(2), 33–40.

André, B., Ringdal, G. I., Loge, J. H., Rannestad, T., Laerum, H., & Kaasa, S. (2008). Experiences with the implementation of computerized tools in health care units: A review article. *International Journal of Human–Computer Interaction, 24*(8), 753–775.

Avgerou, C. (2008). Information systems in developing countries: A critical research review. *Journal of Information Technology, 23*, 133–146. https://doi.org/10.1057/palgrave.jit.2000136.

Balabanova, D., Mills, A., Conteh, L., Akkazieva, B., Banteyerga, H., Dash, U., Gilson, L., Harmer, A., Ibraimova, A., Islam, Z., Kidanu, A., Koehlmoos, T. P., Limwattananon, S., Muraleedharan, V. R., Murzalieva, G., Palafox, B., Panichkriangkrai, W., Patcharanarumol, W., Penn-Kekana, L., Powell-Jackson, T., Tangcharoensathien, V., & McKee, M. (2013). Good health at low cost 25 years on: Lessons for the future of health systems strengthening. *Lancet (London, England), 381*(9883), 2118–2133. https://doi.org/10.1016/S0140-6736(12)62000-5.

Boyatzis, R. E. (1998). *Transforming qualitative information: Thematic analysis and code development* (1st ed.). Thousand Oaks: Sage Publications, Inc.

Brailer, D. J. (2005). Interoperability: The key to the future health care system. *Health Affairs, 24*, W5.

Chaudhry, B., Wang, J., Wu, S., Maglione, M., Mojica, W., Roth, E., Morton, S. C., & Shekelle, P. (2006). Systematic review: Impact of health information technology on quality, efficiency, and costs of medical care. *Annals of Internal Medicine, 144*(10), 742.

Choi, J., Chun, J., Lee, K., Lee, S., Shin, D., Hyun, S., Kim, D., & Kim, D. (2004). MobileNurse: Hand-held information system for point of nursing care. *Computer Methods and Programs in Biomedicine, 74*(3), 245–254.

Cresswell, K., Worth, A., & Sheikh, A. (2011). Implementing and adopting electronic health record systems: How actor-network theory can support evaluation. *Clinical Governance, 16*(4), 320–336. https://doi.org/10.1108/14777271111175369.

DesRoches, C. M., Campbell, E. G., Rao, S. R., Donelan, K., Ferris, T. G., Jha, A., Kaushal, R., Levy, D. E., Rosenbaum, S., Shields, A. E., & Blumenthal, D. (2008). Electronic health records in ambulatory care – a national survey of physicians. *New England Journal of Medicine, 359*, 50–60.

Farley, H. L., Baumlin, K. M., Hamedani, A. G., Cheung, D. S., Edwards, M. R., Fuller, D. C., Genes, N., Griffey, R. T., Kelly, J. J., McClay, J. C., Nielson, J., Phelan, M. P., Shapiro, J. S., Stone-Griffith, S., & Pines, J. M. (2013). Quality and safety implications of emergency department information systems. *Annals of Emergency Medicine, 62*(4), 399–407. https://doi.org/10.1016/j.annemergmed.2013.05.019.

Goldzweig, C. L., Orshansky, G., Paige, N. M., Towfigh, A. A., Haggstrom, D. A., Miake-Lye, I., Beroes, J. M., & Shekelle, P. G. (2013). Electronic patient portals: Evidence on health outcomes, satisfaction, efficiency, and attitudes: A systematic review. *Annals of Internal Medicine, 159*(10), 677–687. https://doi.org/10.7326/0003-4819-159-10-201311190-00006.

Greenhalgh, T., & Stones, R. (2010). Theorising big IT programmes in healthcare: Strong structuration theory meets actor-network theory. *Social Science & Medicine (1982), 70*(9), 1285–1294. https://doi.org/10.1016/j.socscimed.2009.12.034.

Huang, Z., & Palvia, P. (2001). ERP implementation issues in advanced and developing countries. *Business Process Management Journal, 7*(3), 276–284. https://doi.org/10.1108/14637150110392773.

Ignatiadis, I., & Nandhakumar, j. (2007). The impact of enterprise systems on organizational resilience. *Journal of Information Technology, 22*, 36–43.

Ives, B., & Jarvenpaa, S. L. (1991). Applications of global information technology: Key issues for management. *MIS Quarterly, 15*(1), 33–49.

Jones, S. S., Rudin, R. S., Perry, T., & Shekelle, P. G. (2014). Health information technology: An updated systematic review with a focus on meaningful use. *Annals of Internal Medicine, 160*(1), 48–54. https://doi.org/10.7326/M13-1531.

Kellermann, A. L., & Jones, S. S. (2013). What it will take to achieve the as-yet-unfulfilled promises of health information technology. *Health Affairs, 32*(1), 63–68. https://doi.org/10.1377/hlthaff.2012.0693.

Kennedy, B. L. (2011). *Exploring the sustainment of health information technology: Successful practices for addressing human factors* (pp. 7–17). New York, Northcentral University.

Kitson, A., Marshall, A., Bassett, K., & Zeitz, K. (2012). What are the core elements of patient-centred care? A narrative review and synthesis of the literature from health policy. *Journal of Advanced Nursing, 69*. https://doi.org/10.1111/j.1365-2648.2012.06064.x.

Kralewski, J., Dowd, B. E., Zink, T., & Gans, D. N. (2010). Preparing your practice for the adoption and implementation of electronic health records. *Physician Executive, 36*(2), 30–33.

Kvale, S. (1996). *InterViews: An introduction to qualitative research interviewing* (1st ed.). Thousand Oaks: Sage Publications, Inc.

Lee, T.-F. (2013). An efficient chaotic maps-based authentication and key agreement scheme using smartcards for telecare medicine information systems. *Journal of Medical Systems, 37*(6), 9985. https://doi.org/10.1007/s10916-013-9985-9.

Liang, T. P., & Wei, C. P. (2004). Introduction to the special issue: A framework for mobile commerce applications. *International Journal of Electronic Commerce, 8*(3), 7–17.

Liang, T.-P., Huang, C.-W., Yeh, Y.-H., & Lin, B. (2007). Adoption of mobile technology in business: A fit-viability model. *Industrial Management & Data Systems, 107*(8), 1154–1169. https://doi.org/10.1108/02635570710822796.

Liu, L. S., Shih, P. C., & Hayes, G. R. (2011). *Barriers to the adoption and use of personal health record systems*. In Proceedings of the 2011 iConference (pp. 363–370).

Middleton, B., Bloomrosen, M., Dente, M. A., Hashmat, B., Koppel, R., Overhage, J. M., Payne, T. H., Rosenbloom, S. T., Weaver, C., Zhang, J., & American Medical Informatics Association. (2013). Enhancing patient safety and quality of care by improving the usability of electronic health record systems: Recommendations from AMIA. *Journal of the American Medical Informatics Association, 20*(e1), e2–e8. https://doi.org/10.1136/amiajnl-2012-001458.

Muhammad, I., & Wickramasinghe, N. (2014). How an Actor Network Theory (ANT) analysis can help us to understand the Personally Controlled Electronic Health Record (PCEHR) in Australia. In *Technological advancements and the impact of actor-network theory* (pp. 15–34). Hershey: IGI Global.

Muhammad, I., & Wickramasinghe, N. (2017). My health record: The journey to date. Heidelberg Press, VIC. ISBN 978-1-920889-69-2.

Mukesh, S., & Betsy, J. G. (2009). Chinese cultural implications for ERP implementation. *Journal of Technology Management & Innovation, 4*(1), 105–113.

Nguyen, L., Bakewell, L., Wickramasinghe, N., Haddad, P., Muhammad, I., Moghimi, H., Wilde, D., Redley, B., Considine, J., & Botti, M. (2015). Transition from paper to electronic nursing documentation in residential aged care: An actor network theory analysis. *Electronic Journal of Health Informatics, 9*(1), 4.

Protti, D., Johansen, I., & Perez-Torres, F. (2009). Comparing the application of health information technology in primary care in Denmark and Andalucía, Spain. *International Journal of Medical Informatics, 78*(4), 270–283. https://doi.org/10.1016/j.ijmedinf.2008.08.002.

Reynolds, A. (2009). Patient-centered care. *Radiologic Technology, 81*(2), 133–147.

Scott, J. T., Rundall, T. G., Vogt, T. M., & Hsu, J. (2005). Kaiser Permanente's experience of implementing an electronic medical record: A qualitative study. *BMJ (Clinical Research Ed.), 331*(7528), 1313–1316.

Showell, C. M. (2011). Citizens, patients and policy: A challenge for Australia's national electronic health record. *Health Information Management Journal, 40*(2), 39–43.

Soh, C., Kien, S. S., & Tay-Yap, J. (2000). Enterprise resource planning: Cultural fits and misfits: Is ERP a universal solution? *Communications of the ACM, 43*(4), 47–51.

Tjan, A. K. (2001). Finally, a way to put your internet portfolio in order. *Harvard Business Review, 79*(2), 76–85.

Trudel, M. C. (2010). *Challenges to personal information sharing in interorganizational settings: Learning from the Quebec Health Smart Card project*. New York: The University of Western Ontario.

Tucker, C. A., Cieza, A., Riley, A. W., Stucki, G., Lai, J. S., Bedirhan Ustun, T., Kostanjsek, N., Riley, W., Cella, D., & Forrest, C. B. (2014). Concept analysis of the patient reported outcomes measurement information system (PROMIS(®)) and the international classification of functioning, disability and health (ICF). *Quality of Life Research, 23*(6), 1677–1686. https://doi.org/10.1007/s11136-014-0622-y.

Umble, E., Haft, R., & Umble, M. (2003). Enterprise resource planning: Implementation procedures and critical success factors. *European Journal of Operational Research, 146*(2), 241–257.

Yao, J., Schmitz, R., & Warren, S. (2005). A wearable point-of-care system for home use that incorporates plug-and-play and wireless standards. *IEEE Transactions on Information Technology in Biomedicine, 9*(3), 363–371.

Yin. (2014). https://www.researchgate.net/publication/308385754_Robert_K_Yin_2014_Case_Study_Research_Design_and_Methods_5th_ed_Thousand_Oaks_CA_Sage_282_pages.

Yusof, M. M., Papazafeiropoulou, A., Paul, R. J., & Stergioulas, L. K. (2008). Investigating evaluation frameworks for health information systems. *International Journal of Medical Informatics, 77*(6), 377–385. https://doi.org/10.1016/j.ijmedinf.2007.08.004.

# Leveraging the IoT to Enable the Guided Self-Determination Method

**Nilmini Wickramasinghe, Bodil Rasmussen, Freimut Bodendorf, Vibeke Zoffmann, Lars Kayser, and Judy Currey**

## 1 Introduction

Chronic conditions also referred to as non-communicable diseases (NCD) are the largest causes of death in the developed world (AIHW 2017; WHO 2014). Moreover, due to their very nature, individuals who develop one or more chronic conditions must continue to live their lives being challenged by both the burden of their condition and treatment. The burden of living with a chronic condition is well established (AIHW 2016, 2018a; WHO 2014). According the WHO Global

N. Wickramasinghe (✉)
Epworth HealthCare, Richmond, VIC, Australia

Swinburne University of Technology, Hawthorn, VIC, Australia

B. Rasmussen
Deakin University & Western Health, Melbourne, VIC, Australia
e-mail: bodil.rasmussen@deakin.edu.au

F. Bodendorf
Friedrich-Alexander-University Erlangen-Nuremberg, Erlangen, Germany
e-mail: freimut.bodendorf@fau.de

V. Zoffmann
Juliane Marie Centret, Rigshospitalet, Women's and Children's Health and University of Copenhagen, Copenhagen, Denmark
e-mail: vibeke.zoffmann@regionh.dk

L. Kayser
University of Copenhagen, Copenhagen, Denmark
e-mail: lk@sund.ku.dk

J. Currey
Deakin University, Melbourne, VIC, Australia
e-mail: judy.currey@deakin.edu.au



N. Wickramasinghe, F. Bodendorf (eds.), *Delivering Superior Health and Wellness Management with IoT and Analytics*, Healthcare Delivery in the Information Age, https://doi.org/10.1007/978-3-030-17347-0_27

status report on non-communicable diseases report 2014, of the 38 million deaths due to NCDs in 2012, more than 40% were premature, affecting people under 70 years of age. The majority of premature NCD deaths are preventable. Four major NCDs (cardiovascular diseases, cancer, chronic respiratory diseases and diabetes) are responsible for 82% of NCD deaths (WHO 2014). In Australia, 50% of individuals have one chronic condition such as cancer, obesity, diabetes or depression; 20% have at least two chronic diseases (AIHW 2017). An estimated 1.2 million (6%) Australian adults aged 18 years and over have diabetes, with the prevalence tripled between 1989 and 2015 (AIHW 2018b). Given the increasing numbers of individuals presenting with one or multiple chronic diseases today, there is an imperative to develop superior approaches for supporting people in managing chronic conditions so that they enjoy a high quality of life. Correspondingly, there is also an imperative to stem the rising economic burden to healthcare systems and society (WHO 2014).

People living with one or more chronic conditions, their families and their extended social networks experiencing burden of symptoms with severe social and economic consequences that can impact on peoples' quality of life (AIHW 2018a). However, they are also experiencing a burden of treatment related to having to drastically change their behaviours to accommodate recommended lifestyle changes; monitor and manage symptoms at home; and accessing, coping and navigating complex health services. For many people this level of accommodation is overwhelming and time consuming, and it requires a high level of health literacy and often technical skills (Mair and May 2014). The burden of treatment dramatically changes the relationship between people with chronic conditions and health professionals (Mair and May 2014) and demands on health services. The healthcare changes are associated with a higher number of primary care consultations, hospital outpatient visits and hospital admissions (Mclean et al. 2015).

These demands or health service changes required in order to provide safe quality care and treatment inversely vary with the need in the population it serves. This phenomenon is known as the inverse care law which states that the availability of good medical care tends to vary inversely with the need for it in the population served (Hart 1971). Despite the importance of ensuring that resources are allocated in line with the health needs of practice populations, there is evidence that people with chronic conditions are not well served as the system is not geared into the required changes (O'Connor et al. 2016; May 2006; Paterson 2001). Innovative, cost-effective, patient-centred care models are called for to meet these changes.

In this chapter, a research stream that is embarking on identifying a new approach to chronic condition management by focusing on developing the life skills and ability to co-manage their condition, using Guided Self Determination (GSD) is presented. Supported by positive results from Danish studies, the GSD method can alleviate the burden of treatment through creation of provision of insights into one's own challenges and problems and, by empowering citizens/patients, will contribute to the ongoing paradigm shift towards co-production of health and well-being for people living with one or more chronic conditions (WHO 2016).

Guided self-determination may contribute to this ongoing paradigm shift in how chronic disease management is addressed (Rasmussen et al. 2017a, b). Zoffmann and Kirkevold (2012) explain how GSD can overcome barriers to empowerment which were seldom overcome in usual diabetes care. Zoffmann argues that to realise empowerment, health professionals need detailed knowledge of the barriers, their own roles in these barriers, ways to overcome them, and recognisable evidence of having succeeded (ibid.). Further, through theory-driven, qualitative evaluation, the GSD method helps health professionals to recognise changes consistent with empowerment in dyads of health professionals and people with chronic condition. By completing GSD reflection sheets, clients remarkably improve their ability to identify, prioritise, express, and share unique and unexpected difficulties related to living with diabetes (Zoffmann et al. 2015; Rasmussen et al. 2017a, b). As signs of empowerment, clients and health professionals accomplished shared decision-making, resolved life–disease conflicts, and established meaningful and effective relationships (ibid.).

The GSD method is universal and can be used for both distress in relation to non-chronic conditions and to support treatment/monitoring of conditions. However, the GSD is mainly intended for people living with one or more chronic conditions who are having problems coping and managing their overall situation with respect to physical, emotional or social wellbeing. To illustrate how the method works, we applied the GSD method in a diabetes population although evidence highlights the method can be applied to a palette of chronic conditions, even in multi-morbidity, and can work in most age groups (Zoffmann et al. 2008; Olesen et al. 2016; Jørgensen et al. 2015).

In the following section, the GSD method in the context of diabetes is presented. The role for IS/IT (information systems/information technology) is described, and then key aspects of the diabetes pilot to be facilitated in Australia are highlighted to assess the potential of this approach for managing chronic conditions in Australia.

## 2 Background: The GSD Method in Diabetes Context

Living with diabetes requires constant discipline and benefits significantly from access to timely and targeted information. On average, a person with diabetes will have contact with health professionals for 10 hours of care per year. For the rest of the year, people with diabetes manage their condition on their own; hence learning to self–manage diabetes is vital for good health outcomes.

Less than optimal self-management can result in serious complications in people with type 1 diabetes. In Australia, 28,036 people with T1DM aged 21–39 years are registered with 5962 in the state of Victoria in Australia. Many of these live in regional and rural areas, often with little access to appropriate health services. A recent Australian study found this target group reported current health services in rural and regional Victoria do not meet their information and support needs, placing them at a heightened risk for developing serious diabetes complications.

Current educational interventions to achieve glycaemic control are not always successful, particularly in those people who have lost motivation for self-management. People with low motivation are often also put in the "too hard basket" by health professionals which exacerbates the risks for negative and severe physical, emotional and social impacts on people's lives. Novel strategies are therefore required to assist those with suboptimal glycaemic outcomes to prevent complications and adverse events.

The guided self-determination (GSD) program was developed on basic grounded theory research specifically for people with diabetes by Dr. Vibeke Zoffmann. The GSD program aims to improve the life skills of people with diabetes, i.e., those personal, social, cognitive, and physical skills that enable people to control and direct their lives, and to develop the capacity to live with and produce change in their environment. It is also a method to enhance communication and problems solving skill in health professionals as a part of their professional development (Rasmussen et al. 2017a, b).

Thus the GSD method is designed to guide both people with diabetes having persistent suboptimal glycaemic control and their health professionals through mutual reflection, drawing on a number of semi-structured reflection sheets (ibid.). Participants are prompted to systematically explore and express their personal difficulties and experiences with diabetes through words, photos and drawings. Reflections are recorded on worksheets designed to increase participants' ability to clarify and express their views and prepare them for active participation in the care process (Zoffmann and Lauritzen 2006).

Therefore, GSD is a problem-solving and decision-making method designed to develop life skills and overcome the barriers often occurring in the traditional dialogue between citizens/patients and health professionals, thereby increasing the co-production of health and co-management of the condition (Zoffmann and Kirkevold 2007).

As such, GSD is underpinned by life-skills theory, empowerment theory and motivational theory of self-determination. The program has been tested and proven to reduce HbA1c (up to 0.6%) and to improve life skills in adults with persistently suboptimal glycaemic outcomes of type 1 diabetes (Zoffmann and Lauritzen 2006; Zoffmann et al. 2015; Husted et al. 2011, 2014). The original method comprises eight sessions. An overview of the sessions is outlined in the next section. However, recent studies have shown that GSD can be dosed differently, and even an intervention providing two to four conversations was shown to be effective among cancer survivors (Olesen et al. 2016).

## 3 Overview of the GSD Course Activities Between Health Professionals and Client/Patient Eight Meetings with Activities

**Meeting 1:** Collaborating about life with diabetes – Conversation 1: One-to-one conversation, sheet sent to participants prior to meeting, at the meeting discuss three sheets: (1) two ways to use the long-term test (HbA1c), (2) important events and periods in your life and (3) what is currently most challenging or difficult for you living with diabetes?

*Activity:* Three sheets must be available to health professionals (HP) and patient (P) at the same time so they can work on them together.

**Meeting 2:** The collaboration process about life with diabetes – Conversation 2: One-to-one conversation, working on two sheets simultaneously around (1) Unfinished sentences – about values, experiences and needs (2) Your plans for making changes in your lifestyle.

*Activity:* Two sheets must be available to HP and P so they can work on them simultaneously, however, focus will be on the conversation prompted by questions on work sheet 1.

**Meeting 3:** Your life with diabetes – Conversation 3: One-to-one conversation, working with two sheets simultaneously (1) Pictures, metaphors or automatic thoughts about diabetes and (2) Space for diabetes in your life.

*Activity:* HP and P work on the sheets (drawing, showing metaphors etc.) and must be able to see and work on the sheets at same time.

**Meeting 4:** Conversation about agreement to make changes – Conversation 4: One-to-one conversation, working on two sheets simultaneously (1) Agreed description of main challenge or problem in your life with diabetes and (2) "For and against".

*Activity:* HP and P work on the sheets (drawing, showing metaphors etc.) and must be able to see and work on the sheets at same time).

**Meeting 5:** Work to make changes – Conversation 5: One-to-one conversation, working on five sheets simultaneously: (1) ways the described challenge/problem has been solved until now (four pages) and (2) dynamic problem-solving (one page) which is a mutual reflection exercise to how patients can expand their current problem-solving abilities.

*Activity:* HP and P work on the sheets together and need to be able to read and write on the sheets at same time.

**Meeting 6:** Status: your goals in relation to blood glucose and HbA1c – Conversation 6: some of the sheets will have been filled out prior to the meeting. The conversation will focus on (1) blood glucose checks and your reasons for doing them, (2) your ideal and actual daily blood glucose levels, (3) advances and disadvantages of monitoring blood glucose levels, and (4) your plan for achieving blood glucose control in the short and long term.

*Activity:* HP and P work on the sheets together and need be able to see and work (possible write and draw) on the sheets at same time. The HP will teach (talk and explain sheets).

**Meeting 7:** New strategies and long-term plan – Conversation 7: One-to-one conversation, working on one sheet simultaneously.

*Activity:* HP and P need to be able to read and write on same work sheet.

**Meeting 8:** Hand-over after consultation being integrated into the course and the long-term plan – Conversation 8 between P and HP and the third person (HP).

*Activity:* Conservation and working on one sheet with three communication points (at least).

Results of a feasibility study (Rasmussen et al. 2017a, b) indicate that the transferability of the GSD method to an online platform. Further, the GSD online program provided demotivated and geographically isolated clients with a novel, inexpensive and readily accessible therapeutic intervention to improve their self-management of diabetes. The GSD online program improved quality of life of clients and the communication between diabetes educators and young adults regarding better diabetes management. Further development and inclusion of new technologies was required and has now been integrated in a newly developed GSD platform in order to fully realise its capacity to engage clients in self-management to improve health outcomes. As with the face-to-face GSD method, diabetes educators using the GSD online program require education and training in GSD methods to optimise their professional role in the program and realise the self-management potential of clients.

The results of the proposed study will be of paramount importance to the future implementation of the online interactive GSD program in Australia. The use of this program in other countries has yielded positive results. The systematic, inclusive and rigorous process of development, modification, refinement and advancement of the online interactive GSD program is pivotal for its future impact on the quality of healthcare services. It is expected that this program will be a substantial positive influence on optimal health outcomes for young adults affected by type 1 diabetes and better enable them to fulfil their potential in work, study and daily life.

## 4 The Enabling Role for Technology

Today, information and communication technologies (ICTs) are enablers in many aspects of healthcare delivery (Wickramasinghe and Schaffer 2018). However, it has become increasingly recognized as being critical that end users identify, understand, and apply information provided to them through the growing number and range of electronic resources becoming available these systems must be usable and the content must be understandable (Monkman and Kushniruk 2015). It is also increasingly recognised that the success or failure of health information systems and applications designed for laypeople, patients, and consumers depends on factors related to the match between the demands a system (or application) places on a user and the end user's level of eHealth literacy (Monkman and Kushniruk 2015; Kayser et al. 2015). Norman and Skinner (2006) defined eHealth literacy as "the ability to seek, find, understand, and appraise health information from electronic sources and apply the knowledge gained to addressing or solving a health problem" (p. 3). Identified a new concept for eHealth literacy: a model based on systematic and inductive methods that sought to identify the full range of elements relevant to individuals attempting to understand and use eHealth technologies and digital services. This model, the eHealth Literacy Framework (eHLF), consists of seven dimensions that describe the attributes of the users (information and knowledge about their health), the intersection between users and the technologies (their feeling of being safe and in control and their motivation) and user experience of systems (they work and are accessible and suit users' needs). The eHLF provides a comprehensive map of the individual technology user's health literacy that covers his or her knowledge and skills, the system, and how the individual interacts with the system (Nørgaard et al. 2015) and can be assessed using the eHealth Literacy Questionnaire (Kayser et al. 2018).

These factors associated with eHealth literacy are important to consider in development of next step of GSD platform. The study participants will be aside to fill out a survey comprising validated questionnaires concerning their competence, self-management and communication with HP. Focus group interviews will add qualitative data and in-depth information about the participants' experiences.

The GSD methods aligns with the principle of empowerment. Hence, GSD has a tool that, through reflection exercises, interacts with HP, the system and potentially can enhance patients' eHealth literacy to cope better with their diabetes. Data collection and analysis, as well as communication within the GSD method, is strongly based on documents called reflection sheets. A first step to support GSD by IT (information technology) is to use electronic documents instead of conversation-based hard copy sheets. As a consequence, HPs as well as patients (Ps) would edit and handle those documents on digital devices like smart phones, tablet computers or notebooks. For processing the documents as well as storing, retrieving and analysing the data contained, there is a document management system as a backend application which also comprises security and privacy mechanisms. Document upload and download is done via cellular networks or the internet. There is a large variety of added applications for document management systems which can support GSD-course activities. A prominent one is Co-Authoring which is required in conversations 2, 3, 4, 5, 6, and 7 (HP and P working on the sheets together/simultaneously/at the same time). Guided Self Determination online is thereby an enabling role to support the health professionals need to learn which would be dependent on their specific context but always enabling key lessons and reflections to be stored for later access by them as required. In addition, social media functions can be helpful to comment or discuss on sheets, either synchronously or independent of time and place. Theory-driven qualitative evaluations have confirmed GSD's ability to realise empowerment in practice in individual conversations for adults with type 1 diabetes (Zoffmann et al. 2015). The online version of the GSD would increase the flexibility of the GSD process, taking into account that HP and patient might be geographically distributed and would like to meet virtually to collaborate. This will make it possible to reach out to otherwise difficult to reach populations.

Given the breadth and depth of the tools and technologies afforded to us by IoT, we anticipate that in the future it would be possible to also include key developments in AR and VR (augmented reality and virtual reality), sensors, mobile, analytic capabilities around prediction and prevention as well as advances in AI (artificial intelligence) and Bots. Clearly, it is important to first establish the approach and delivery. However, the existence of all these technology developments makes the future bright and exciting with possibilities that ultimately will ensure that solutions and strategies can be highly tailored to each individual's needs, level of health and technology literacy so that the end result is a program that enables each individual to manage with critical life skills their chronic condition to ensure they have the highest possible quality of life.

## 5 Research Methodology and Design

To assess the likely benefits of the GSD method for the Australian environment a two state approach is embarked upon. First, an Australian feasibility study to transfer the GSD method to online was conducted with a small number (11) of individuals with type 1 diabetes and 7 diabetes educators in 2016. Results demonstrated that the transferability of the GSD method to an online platform provided demotivated and geographically isolated participants with a novel, inexpensive and readily accessible therapeutic intervention to improve their diabetes self-management (Rasmussen et al. 2017a, b). In addition, the study found that translating the GSD program to an online platform clearly helped these young adults with type 1 diabetes to improve their capacity to find solution to their pertinent health issues as well as improving communication processes between diabetes educators and patients (Rasmussen et al. 2017a, b). It was however clear that further development and inclusion of new technology were required for the GSD online platform to fully realise its capacity to engage participants to improve diabetes self-management and health outcomes (Rasmussen et al. 2017a, b).

Thus, although results of the feasibility study highlighted potential strengths, a larger-scale proof of concept study was subsequently required. In order to establish full proof of concept of the GSD method in the Australian healthcare environment, a larger scale study was designed. To ensure consistency with the feasibility study outcomes, and building on the feedback from the participants, we chose the same target population of young adults with type 1 diabetes. A pre- and post-interventional study will use a sequential, two-phase design. The first phase is the delivery of the training program in GSD conversations for diabetes educators to enable facilitation of GSD online with young adults with type 1 diabetes. The second phase is client participation in, and completion of, the GSD online program with a diabetes educator. A strength of this proposed study is the inclusion of patient and diabetes educator feedback informing the design of the new GSD online platform. The proposed two-phase study involves training diabetes educators in the GSD program, and patient completion of the GSD program was conducted over a 6-month period. Surveys will be used to measure young adults' self-management skills before and 3 months after completing the GSD program with their specifically trained diabetes educator. Findings of this study will inform future implementation and scalability of GSD online as an official educational tool in Australia for Diabetes Educators to motivate and assist young adults in self-management of type 1 diabetes, as well as the provision of the GSD online program with other client and patient groups.

Clients or patients will have access to the GSD solution via an online portal. All necessary security and privacy standards are being satisfied and research will commence after ethical approvals are secured.

The results of the proposed study will be of paramount importance to the future implementation of the online interactive GSD program in Australia. The use of this program in other countries has yielded positive results. The systematic, inclusive and rigorous process of development, modification, refinement and advancement of the online interactive GSD program is pivotal for its future impact on the quality of healthcare services. It is expected that this program will be a substantial positive influence on optimal health outcomes for young adults affected by type 1 diabetes and better enable them to fulfil their potential in work, study and daily life.

## 6 Discussion and Conclusion

The preceding has reported on the early stages of a research stream focussed at designing and developing a new approach to provide superior chronic care management. A central aspect of this GSD method is to empower and engage the client so that he/she can actively participate in their treatment and wellness management. Further, through GSD, clients develop necessary life skills to navigate their life and ensure they can enjoy a high-quality lifestyle irrespective of their chronic conditions. Our initial discussions with key clinicians and consumer groups have been very positive. Based on the strong results from the Danish studies to date, we are confident that the pilot studies when completed will result in similarly strong results.

The burden of the growing prevalence of people with chronic and multi-comorbidities is an enormous burden on our health systems. There is little doubt that current approaches to address the treatment and management of chronic diseases is not adequate and that new approaches desperately need to be developed, designed and trialled. We contend that GSD is such a method that offers a paradigm shift in the management of chronic care. Moreover, we contend that GSD will enable a high-value, client-/patient-centred superior experience to chronic care management. In this chapter, the enabling role for a web-based platform has been outlined.

Finally, this study has implications for theory and practice, most notably the following: (1) for theory, designing and developing methods to provide superior client-/patient-centred high-value care, extending design science research methodology and user-centred design to IoT technologies in healthcare and, (2) for practice, collaborating across countries to design and develop healthcare technology-enabled solutions and provide clients/patients with a better quality of life, necessary life skills and empowerment to take an active role in their chronic care management. In today's twenty-first century, the tools and technologies that make up the IoT offer us a great opportunity to design and develop solutions to provide all individuals with the appropriate level of care, data and information they require to best manage their health situations and thereby ensure they can enjoy a high quality of life. It is thus incumbent on us to work together and provide the needed solutions so that chronic conditions are no longer a limitation to enjoying a high quality life.

## References


Australian Institute of Health and Welfare (AIHW). (2016). *Australian Burden of Disease Study: Impact and causes of illness and death in Australia 2011* (Australian Burden of Disease Study series no. 3. BOD 4). Canberra: AIHW.

Australian Institute of Health and Welfare (AIHW). (2017). Health conditions, disabilities and deaths. https://www.aihw.gov.au/reports-statistics/health-conditions-disability-deaths.

Australian Institute of Health and Welfare (AIHW). (2018a). Burden of disease. Last updated Jan 2018. https://www.aihw.gov.au/reports-statistics/health-conditions-disability-deaths/chronic-disease/overview.

Australian Institute of Health and Welfare (AIHW). (2018b). Diabetes snapshot web report. https://www.aihw.gov.au/reports/diabetes/diabetes-snapshot/contents/how-many-australians-have-diabetes.

Hart, J. T. (1971). The inverse care law. *Lancet, 1*(7696), 405–412.

Husted, G., Thorsteinsson, B., Esbensen, B. A., Hommel, E., & Zoffmann, V. (2011). Improving glycaemic control and life skills in adolescents with type 1 diabetes: A randomised, controlled intervention study using the Guided Self-Determination-Young method in triads of adolescents, parents and health care providers integrated into routine paediatric outpatient clinics. *BMC Pediatrics, 11*(55), 1–12.

Husted, G. R., Esbensen, B. A., Hommel, E., Thorsteinsson, B., & Zoffmann, V. (2014). Adolescents developing life skills for managing type 1 diabetes: A qualitative, realistic evaluation of a guided self-determination-youth intervention. *Journal of Advanced Nursing, 70*, 2634–2650.

Jørgensen, R., Licht, R. W., Lysaker, R. H., Munk-Jørgensen, P., Buck, K. D., Jensen, S. O. W., Hansson, L., & Zoffmann, V. (2015). Effects on cognitive and clinical insight with the use of Guided Self-Determination in outpatients with schizophrenia: A randomized open trial. *European Psychiatry, 30*(5), 655–663. https://doi.org/10.1016/j.eurpsy.2014.12.007.

Kayser, L., Kushniruk, A., Osborne, R. H., Norgaard, O., & Turner, P. (2015). Enhancing the effectiveness of consumer-focused health information technology systems through eHealth literacy: A framework for understanding users' needs. *JMIR Human Factors, 2*(1), e9.

Kayser, L., Karnoe, A., Furstrand, D., Batterham, R., Christensen, K. B., Elsworth, G., & Osborne, R. H. A. (2018). Multidimensional tool based on the eHealth literacy framework: Development and initial validity testing of the eHealth literacy questionnaire (eHLQ). *Journal of Medical Internet Research, 20*(2), e36. https://doi.org/10.2196/jmir.837.

Mair, F. S., & May, C. R. (2014). Thinking about the burden of treatment: Should it be regarded as an indicator of the quality of care? *BMJ, 349*, g6680. https://doi.org/10.1136/bmj.g6680.

May, C. (2006). Self-management of chronic conditions: Re-engineering patient-hood. *Chronic Illness, 2*, 15–16.

McLean, G., Guthrie, B., Mercer, S. W., & Watt, G. C. (2015). General practice funding underpins the persistence of the inverse care law: Cross-sectional study in Scotland. *British Journal of General Practice, 65*, e799–e805.

Monkman, H., & Kushniruk, A. W. (2015). eHealth literacy issues, constructs, models, and methods for health information technology design and evaluation. *Knowledge Management & E-Learning, 7*(4), 541–549.

Norgaard, O., Furstrand, D., Klokker, L., Karnoe, A., Batterham, R., Kayser, L., & Osborne, R. H. (2015). The e-health literacy framework: A conceptual framework for characterizing e-health users and their interaction with e-health systems. *Knowledge Management & E-Learning, 7*(4), 522–540.

Norman, C. D., & Skinner, H. A. (2006). eHealth literacy: Essential skills for consumer health in a networked world. *Journal of Medical Internet Research, 8*(2), e9.

O'Connor, S., et al. (2016). Understanding factors affecting patient and public engagement and recruitment to digital health interventions: A systematic review of qualitative studies. *BMC Medical Informatics and Decision Making, 16*(1), 120.

Olesen, M. L., Duun-Henriksen, A. K., Hansson, H., Ottesen, B., Andersen, K. K., & Zoffmann, V. (2016). A person-centered intervention targeting the psychosocial needs of gynecological cancer survivors: A randomized clinical trial. *Journal of Cancer Survivorship, 10*(5), 832–841. https://doi.org/10.1007/s11764-016-0528-5.

Paterson, B. L. (2001). The shifting perspectives model of chronic illness. *Journal of Nursing Scholarship, 33*, 21–26.

Rasmussen, B., Currey, J., Haigh, B., Dunning, T., & Zoffmann, V. (2017a, September). An online self-management intervention for young adults with type 1 diabetes: Guided Self-Determination program. *Australian Diabetes Educator Association, 3*(20), 1–9.

Rasmussen, B., Currey, J., Dunning, T., & Zoffmann, V. (2017b, June). Major changes in a young woman's diabetes self-management by using the Guided Self-determination online program. *Australian Diabetes Educator Association, 2*(20), 1–5.

Wickramasinghe, N., & Schaffer, L. J. (2018). *Theories in health informatics management*. New York: Springer.

World Health Organization (WHO). (2014). Status report of non communicable diseases. http://www.who.int/nmh/publications/ncd-status-report-2014/en/.

World Health Organization (WHO). (2016). WHO framework on integrated people-centred health services. http://www.who.int/servicedeliverysafety/areas/people-centred-care/en/.

Zoffmann, V., & Kirkevold, M. (2007). Relationships and their potential for change developed in difficult type 1 diabetes. *Qualitative Health Research, 17*(5), 625–638. https://doi.org/10.1177/1049732307301230.

Zoffmann, V., & Kirkevold, M. (2012). Realizing empowerment in difficult diabetes care: A guided self-determination intervention. *Qualitative Health Research, 22*, 103–118.

Zoffmann, V., & Lauritzen, T. (2006). Guided self-determination improves life skills with type 1 diabetes and A1C in randomized controlled trial. *Patient Education and Counseling, 64*(1–3), 78–86. https://doi.org/10.1016/j.pec.2005.11.017.

Zoffmann, V., Harder, I., & Kirkevold, M. (2008). A person-centered communication and reflection model: Sharing decision-making in chronic care. *Qualitative Health Research, 18*, 670–685.

Zoffmann, V., Vistisen, D., & Due-Christensen, M. (2015). Flexible guided self-determination intervention for younger adults with poorly controlled type 1 diabetes, decreased HbA1c and psychosocial distress in women but not in men: A real-life RCT. *Diabetic Medicine, 32*(9), 1239–1246. https://doi.org/10.1111/dme.12698.

# Determining Missing Key Elements in Oncology Information System to Improve Patient Experience and Clinical Care

**Muhammad Nadeem Shaukat, Stephen Vaughan, and Nilmini Wickramasinghe**

## 1 Introduction/Background

In 2016, the total number of deaths registered in Australia was 158,504 consisting of 81,867 males and 76,637 females (Australian Bureau of Statistics (ABS) 2017a), with cancer being one of the leading causes (Australian Bureau of Statistics (ABS) 2016). By the age of 85, one in two Australians will be diagnosed with cancer and one in five will die from cancer (AIHW 2018a). Males (around 54%) are more likely to suffer from cancer than females (around 46%), or in other words, one in three males whereas one in four females will be identified for cancer by the age of 75 (AIHW 2017a). It is estimated that in 2018 around 138,321 people will be identified as cancer sufferers (Cancer Australia 2018), and the number of cancer cases is expected to rise to 150,000 till 2020. Cancer-related treatments and medicines are extremely expensive; in Australian healthcare system, this is more than $4.5 billion (6.9%) (AIHW 2018b) and likely to increase at a rate much greater than inflation over the next 5–10 years because of the availability of new drugs (Armstrong et al. 2007). Treatment of cancer involves initially the obtaining of a tissue diagnosis followed by staging to determine the extent of the cancer and more recently the performance of genomic tests to indicate more effective targeted treatments (Ramaswamy et al. 2001; Balch et al. 2009). Physical treatment modalities include surgery, radiotherapy and systemic treatment which might be chemotherapy, hormone therapy or immunotherapy (Solomon et al. 2014;

M. N. Shaukat · S. Vaughan
Epworth HealthCare, Melbourne, VIC, Australia

N. Wickramasinghe (✉)
Epworth HealthCare, Richmond, VIC, Australia

Swinburne University of Technology, Hawthorn, VIC, Australia



N. Wickramasinghe, F. Bodendorf (eds.), *Delivering Superior Health and Wellness Management with IoT and Analytics*, Healthcare Delivery in the Information Age, https://doi.org/10.1007/978-3-030-17347-0_28

Goldhirsch et al. 2013; Dempke et al. 2017). Analgesia and other drugs are used throughout the cancer journey particularly in palliative care (Vogelaar et al. 2015; Cleary 2000). Psychosocial care is also a critical part of the whole cancer journey (Jacobsen and Jim 2008; Surbone et al. 2010). Given the number of treatment modalities and number of persons involved in making treatment recommendations, multidisciplinary meetings are the usual vehicle by which cancer treatment plans are formulated (Patkar et al. 2011; Kesson et al. 2012).

The total number of cancer cases includes ICD-10 cancer codes C00–C97 (malignant neoplasms of specific sites), D45 (polycythaemia), D46 (myelodysplastic syndromes) and D47.1, D47.3, D47.4 and D47.5 (myeloproliferative diseases); however, it does not include basal cell carcinoma (BCC) and squamous cell carcinoma (SCC) of the skin. Both BCC and SCC are the most common skin cancers, thus not considered as notifiable diseases in Australia and are not registered in the Australian Cancer Database (Australian Bureau of Statistics (ABS) 2017a; Cancer Australia 2018b).

Since 2015, along with other diseases cancer remains in the top five leading causes of death (Australian Bureau of Statistics (ABS) 2017a; Australian Bureau of Statistics (ABS) 2016). The Australian Bureau of Statistics reported that half of the deaths (49.4%) for young Australians (aged under 40) was due to suicide or road accidents while more than half of deaths (50.6%) were because of natural reasons. It is alarming to note that cancer caused the largest proportion of natural deaths (12.8%) in 2016 for young Australians (Australian Bureau of Statistics (ABS) 2018). Even though young Australians only constitute overall a very small fraction (1.7%) of the total cancer toll, the inclusive impact of social, emotional and economic is very large and entrenched deep in Australian society leaving a big footprint (Australian Bureau of Statistics (ABS) 2018). The top 5 cancer types affecting young Australians (0–39 years) include brain cancer, cancer of the lymphoid, haematopoietic and related tissues (leukaemia), colorectal cancer, breast cancer and skin cancer.

Today, there is an increase in general chronic diseases (cancer, coronary heart disease and diabetes) in Australia which is due to an ageing population, quality of air and changes in lifestyle, yet the advancements in diagnostic, imaging, medicine and technology in medical care available today have helped us in living longer with illnesses and diseases (AIHW 2016). In general, people living in Australia have a better chance of survival than people living in other parts of the world. This is because of better awareness and improved facilities for testing and treatment (AIHW 2018c). Persons diagnosed with cancer in 2009–2013 had 68% chances of survival (males 68% and females 69%). By the end of 2012, there were 106,340 persons living with cancer, while the number of survivors in the previous 5 years (2012–2008) was 410,530 and the number of people still living with cancer since 1982 (31 years of survival) was 994,605 (Cancer Australia 2018b). By and large, there has been a declining trend in cancer in general population (Fig. 1); this can be attributed to early detection and intervention due to decreased number of cases for breast cancer and prostate cancer. This is owing to increased perception about breast and prostate cancer and availability of PSA (prostate specific antigen) testing for males (Rycaj and Tang 2015) and breast screen for females (AIHW 2017b).

**Table 1** Estimated incidence of all cancers combined, by sex, 2018 (Cancer Australia 2018b; AIHW 2017b)

| | Males | Females | Persons |
|---|---|---|---|
| Number of cases | 74,644[a] | 63,676[a] | 138,321[a] |
| Age-standardized incidence rate per 100,000 persons (2014) | 555 | 424 | 484 |
| Age-standardized rate incidence per 100,000 persons (2018) | 529 | 424 | 472 |
| Number of deaths due to cancer (2016) | 45,782 | 25,910 | 19,872 |
| Number of deaths due to cancer (2018) | 48,586 | 27,552 | 21,034 |
| Per cent of all cancer cases | 53.8 | 46.2 | 100.0 |
| Risk to diagnose for cancer by age 85 | 1 in 2 | 1 in 2 | 1 in 2 |
| Risk to death due to cancer by age 85 | 1 in 5 | 1 in 5 | 1 in 6 |

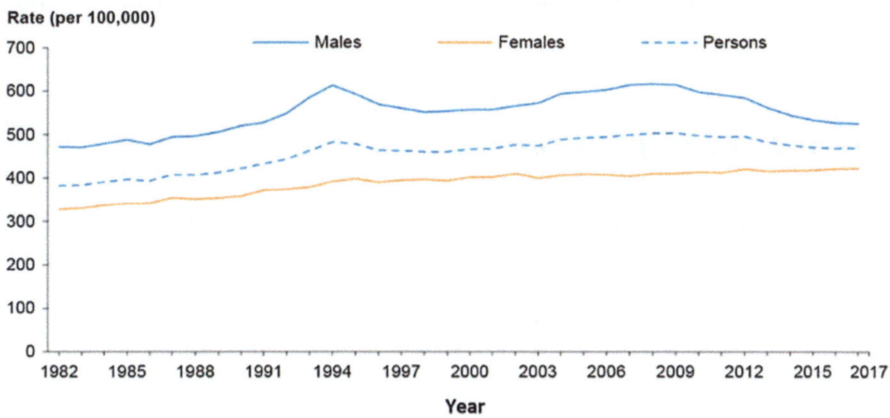


**Fig. 1** Incidence and mortality trends, 1982–2017 (AIHW 2018c)

## 2 Burden of Disease

Apart from direct treatment, emotional and psychological burden, there is another aspect of disease burden which is calculated in terms of disability-adjusted life years (DALY). This is a technique for determining and comparing the overall effect on a population for a variety of diseases, conditions/injuries and risk factors. This includes calculating fatal (dying from chronic disease) and nonfatal (living with chronic disease) (Goodwin 2017). It was reported that cancer is still the cause of highest disease burden in Australian healthcare system (19% of total disease burden) (Cancer Australia 2017). In simple words, DALY estimates the number of Years of Life Lost (YLL) because of premature death by disease and number of Years Lived with Disability (YLD) because of disease.

## 2.1 DALY (Disability-Adjusted Life Years)

DALY is a measure (in years) of healthy life lost, either through premature death defined as dying before the ideal life span (YLL) or, equivalently, through living with disability due to illness or injury (YLD) (Goodwin 2017).

## 2.2 YLD (Years Lived with Disability)

YLD is a measure of the years of what could have been a healthy life but were instead spent in states of less than full health. YLD represents nonfatal burden (Goodwin 2017).

## 2.3 YLL (Years of Life Lost)

Years of life lost due to premature death is defined as dying before the ideal life span. YLL represents fatal burden (Goodwin 2017).

$$\mathrm{DALY} = \mathrm{YLD} + \mathrm{YLL}$$

## 2.4 Burden of Cancer in Australia

It was reported that even though cancer survival rates have improved yet, the burden caused by cancer is mainly due to premature death due to cancer. Therefore, cancer has been described as the major contributor for the fatal burden in Australia (around 1.5 times that of cardiovascular diseases and twice that of injuries) (Goodwin 2017). It was noted in 2011 that cancer was presented the highest disease burden in Australia (Cancer Australia 2017). In general, Australians lost 833,250 disability-adjusted life years (DALY) owing to premature mortality by cancer or while suffering from cancer (19% of total DALY) (Goodwin 2017). In the male population, lung cancer (3.9%) has been identified as the highest proportion of the cancer burden, followed by after coronary heart disease (9.4%) (AIHW 2016; AIHW 2018d). Even though the survival rate from cancer is improving, yet almost all of the disease burden was due to early deaths (94%) while burden of living with caner is comparatively very small (6%) (AIHW 2017a).

# 3 Oncology Information System

## *3.1 Oncology*

According to the *Free Dictionary*, oncology is 'the branch of medicine that deals with tumours' (Saletti et al. 2018). This encompasses prevention, screening or early diagnosis, treatment and follow-up either in survivorship programmes for palliative care (Crowther 1982; Hesketh et al. 2017). Given the rapid development of new drugs research is becoming increasingly integrated into cancer treatment (Amaravadi et al. 2011; O'Connell et al. 2000). An additional important consideration is that most cancers occurs in elderly people who often have significant comorbidities that influence their cancer and its treatment particularly their ability to tolerate intensive treatment (Staples et al. 2006; Chamberlain et al. 2002). This paper will restrict its scope to the diagnosis and treatment of cancer but be mindful of the broader healthcare system in which these aspects of cancer treatment operate.

## *3.2 Oncology Information System*

Oncology is an attractive specialty for IT implementation as it is more standardized than most other medical specialties particularly in diagnosis, staging and treatment specification with well-recognized protocols (Herre and Heller 2004; Enterline et al. 2012; Lenhard et al. 1983; Siochi et al. 2009). This often led to the development of oncology information systems in isolation from larger enterprise-wide systems (Yu et al. 2010).

With increase in and ageing of the general population along with migration from rural to metropolitan areas for various reasons, the number of patients presenting themselves at hospitals is increasing day by day (AIHW 2011). To facilitate the patients, clinicians and healthcare providers, the use of ICT (information and communication technology) has become a vital tool in managing day-to-day task (Centre for Clinical Governance Research 2011; Ochs and Casagrande 2008; Shortliffe et al. 1984). Although it takes a long time for any system to be adopted or diffused into any healthcare system (Berwick 2003), there is growing evidence that ICT is necessary to avoid errors and wastage in medicine (Fasola et al. 2008) and increase overall effectiveness and efficiency (Fasola et al. 2014).

As cancer treatment often involves multiple modalities of treatment that interact and the difference between a therapeutic dose and a toxic dose is small, meticulous attention to the details of treatment is critical, and relatively small errors can have catastrophic consequences. Oncology information systems (OISs) are solutions providing such benefits for these complicated tasks and facilitate their management through one system (Ando 2014).

Although oncology information systems do address many of the errors of manual systems which they replace, they do introduce the potential for new types of errors which in some respects are more serious as once the error occurs it is disseminated widely (Klein et al. 2005; Surbone and Rowe 2015). Traditionally, radiotherapy has had its own oncology information system developed by manufacturers of the radiotherapy equipment in part related to the IT requirements for tight control of radiotherapy doses.

Chemotherapy oncology information systems have tended to be developed separately from broader drug prescription systems in part related to the complexity of chemotherapy administration (Bourret et al. 1996). Integration of chemotherapy prescribing with more general drug prescribing is usually only part of large enterprise-wide IT systems (Moen 2011).

## *3.3 Need for OIS (Oncology Information System)*

Owing to new advancements in diagnostic and treatment therapies, today we are in a better position to identify early and better understand diseases and their management, resulting in longer and healthier life outcomes (AIHW 2017b). This era of knowledge and easy access to information through Internet, print and electronic media has made us more demanding of every aspect of health and well-being (Duckett and Willcox 2015). Overall, it has put an enormous burden on the healthcare systems (Nutbeam 2008). To cope with these errors and get up-to-date information about patient history, medication and treatments, electronic health records and data systems have been introduced in many advanced countries (Sulaiman and Wickramasinghe 2014; Wickramasinghe and Schaffer 2010; Bernstein 1983; Boonstra and Broekhuis 2010). These electronic healthcare records (EHR) can help in managing a single record instead of multiple records for same patient, reducing multiple testing/duplication of treatment or medication (Duncan et al. 2010; Hillestad et al. 2005). Especially, when it comes to treating of cancer patients, the need to avoid every possible error becomes even more crucial due to the toxicity and cost of drugs and radiation therapies (Fasola et al. 2008; Evans et al. 2014). Therefore, the need to synchronize all treatments, provide instant access to updated records and information on latest state of tumour/disease and informed decision making through cancer registry becomes very crucial (Duncan et al. 2010). No two cancer patients are the same, and even though there might be two persons having similar conditions and symptoms of cancer, their treatment plan and recovery will vary depending on a variety of different factors (MesotheliomaGroup 2018; Štambuk et al. 2010). Therefore, their treatment needs to be very personal and entirely customized to help them recover better and faster (Total Health 2018), thus arising a necessity for incorporating a variety of treatment plans and multidisciplinary departments within or across the hospitals into one single system (Herre and Heller 2004; Crawford 2013). The utilization of an oncology information system (OIS) helps in the integration of radiation oncology, particle therapy and medical oncology patient data into a single database with multiuser and multiple site access (Janssen et al. 2017). The OIS seamlessly connects to any linear accelerator and treatment planning system (Wickramasinghe et al. 2015). It also helps in the evaluation and comparison of patients under treatment to existing data in cancer registries. Development of disease-specific clinical protocols is made easy by OIS, and it helps in generating standard and consistent care (Krayenbuehl et al. 2015). By utilizing OIS, clinicians can monitor the amount of dose and review treatment images in one single system to determine if they need to change the dosing plan (Cheng et al. 2011).

### *3.4 Current and Expected Market Share of OIS*

At present, the global oncology information system market is valued at US $ 2247 million, and it is expected to attain value of US $ 3755 million (Sajeev and Sumant 2017). The market is expected to continue growing at cumulative annual growth rate (CAGR) of 7.5% since 2016–2022 (Sajeev and Sumant 2017). There are a number of elements which contribute in such rapid growth in the usage of OIS, e.g., an ageing population and thus the increasing rate of cancer, increasing rates of obesity associated with increased cancer risk (Calle et al. 2003; Calle and Kaaks 2004; Calle and Thun 2004) and the presence of certain infections that predispose to cancer advancement in research and technology, adoption of electronic health record systems and funding of governments and corporate sector (Insight Pharma Reports 2018). North America possesses the greatest share of global oncology information system industry due to advancement in technology and expertise in IT along with pioneer cancer treatment centres situated in this region (TMR 2018) and incentives provided by various legislative programmes. The European Union is second in OIS market share, while Asia-Pacific is still lagging behind but is expected to at highest rate (11.5% CAGR) (Markets and Markets 2015). The main factors for such a high growth are large patient population size, involvement of governments for cancer prevention and treatment, and rising expertise in IT and healthcare sectors (Markets and Markets 2015).

## 4 Methodology of Study

The primary focus of current fact-finding study was to compare and identify missing key elements from the Australian healthcare context with respect to OIS. The point of view of patients and their care providers, clinicians and healthcare providers was sought to accommodate leading oncology information systems already developed and used in the USA, Europe or other parts of the world. Initially, we selected three leading OIS solutions to perform a hermeneutic analysis on publicly available data. In addition, group discussions with patients and their care providers, clinicians and healthcare providers (in private and public sectors) were conducted to gain insight and useful information about OIS under consideration. It is expected that these missing key factors described by different users will help in designing better OIS for Australian end users in the healthcare industry. The preliminary comparison and key missing/suggestive elements are detailed in tabular form in the following sections.

## 5 Comparing OIS with Australian Point of View

Three different leading international oncology information systems were analysed. Specifically, their main features as relevant and beneficial for the Australian healthcare system for cancer care were examined. The summary of the comparison is presented in Table 4.

Though the clinical process of cancer care is essentially the same in all jurisdictions, the organizational arrangements differ widely, in particular how care is funded. Given that the initial primary purpose of many patient information systems including oncology systems was for administrative purposes including medical billing, there are substantial difficulties in adapting these systems to jurisdictions other than the ones in which they were developed (Lenhard et al. 1983; Blum 1983). Oncology information systems also tend to reflect the interests of the particular cancer treatment modality craft group of the several involved in cancer care that was involved in their development, i.e., pharmacy, radiotherapy etc. (Miller and Phillips 2005; Islam et al. 2018).

After completing the comparison for these systems, a focus group discussion with oncology patients, clinicians and healthcare management administrator was conducted to further gain important insights. From a comparative point of view, most of the systems have been designed either in USA, Europe or other advanced countries, which have different healthcare system to Australia.

The Australian healthcare system has both a public and private component and ambulatory and inpatient care (Duckett and Willcox 2015; Wilson et al. 1995). The majority of cancer care is delivered in the ambulatory outpatient setting with about half in the public sector and half in the private sector.

The system is funded by both the state and federal governments with the states funding most inpatient public care and the federal government funding most ambulatory care including drugs either by block grants or on a fee-for-service basis (Duckett and Willcox 2015; Wilson et al. 1995; Newman et al. 2006). Diagnostic services are critical to cancer services including medical imaging and pathology are mostly in the private sector even when located within public sector hospitals with significant market concentration (Duckett 1999; Forbes et al. 2010). They tend to have their own in-house IT systems and are usually national or international organizations. Thus, any system designed overseas needs to be customized to the Australian healthcare context if its full potential is to be realized.

Almost all the systems have in common inclusion of billing system, image visualization and management, paperless communication and fast data transfer, while all of them lack the inclusion of PBS/MBS and insurance share calculation. The

**Table 2** Most common cancers to cause death in younger Australians (0–39 years), 1997 and 2016 (Australian Bureau of Statistics (ABS) 2018)

| | 1997 | | | | 2016 | | | |
|---|---|---|---|---|---|---|---|---|
| Type of cancer | Rank | No. of deaths | ASDR | % of total cancers | Rank | No. of deaths | ASDR | % of total cancers |
| Brain cancer | 2 | 134 | 1.2 | 13.9 | 1 | 133 | 1.0 | 17.5 |
| Malignant neoplasms of lymphoid, haematopoietic and related tissue | 1 | 233 | 2.1 | 24.2 | 2 | 112 | 0.9 | 14.7 |
| Leukaemia | – | 149 | 1.4 | 15.5 | – | 83 | 0.6 | 10.9 |
| Colorectal cancer | 5 | 55 | 0.5 | 5.7 | 3 | 94 | 0.7 | 12.4 |
| Breast cancer | 3 | 128 | 2.4 | 13.3 | 4 | 89 | 1.4 | 11.7 |
| Skin cancer | 4 | 74 | 0.7 | 7.7 | 5 | 39 | 0.3 | 5.1 |
| Melanoma | – | 73 | 0.7 | 7.6 | – | 37 | 0.3 | 4.9 |
| Total cancers | – | 962 | 8.9 | 100.0 | – | 760 | 5.9 | 100.0 |
| Total deaths | – | 8272 | 76.1 | NA | – | 5918 | 45.6 | NA |

**Table 3** Overall burden of all cancers by sex and as persons, 2011 (AIHW 2017b)

| | Males | | Females | | Persons | |
|---|---|---|---|---|---|---|
| | Number | % | Number | % | Number | % |
| Fatal burden (YLL) | 442,228 | 94.1 | 340,121 | 93.7 | 782,349 | 93.9 |
| Nonfatal burden (YLD) | 27,882 | 5.9 | 23,019 | 6.3 | 50,901 | 6.1 |
| Total burden (DALY) | 470,110 | 100.0 | 363,140 | 100.0 | 833,250 | 100.0 |

OIS 1 has some features which distinguish it from others including review of dosimetric images for IMRT, ability of scheduling appointment, remote image review, customization of forms and questionnaire, ability to attach images/files/photographs to documents and continuous data protection while ensuring minimal data loss and immediate data recovery through HARRP system. Whereas prominent features for OIS 2 are 3D viewer, RTP, XVI, iView GT integration, stereoscopic and volume registration and elimination of error-prone work through tight integration that enables transition of seamless data between different tools, in OIS 3, along with other features, the most intriguing feature is estimation and comparison of cost, complications, toxic analysis and even analysis between different practice centres and clinics.

Table 4 gives comparative details of the OISs under consideration. On initial inspection, it is obvious that these OISs have been designed to provide a unique solution for oncology clinicians and healthcare providers to access data regarding patient history, treatment plan, amount of dose, tracking improvement, image visualization, medication prescription and other relevant information but with little emphasis on the patient, carers and patient-reported outcome (Saletti et al. 2018).

In the modern era, the focus has shifted from clinicians and hospitals to patient, family members and care givers. The patients were desirous of more empowerment and access to information and data through psychological surveys, awareness/training, visualization of their treatment and booking/cancelling/rescheduling their appointments as mentioned in Table 5. Further, having the capacity to participate in patient-reported outcomes and self-reporting pain and progress of recovery on a daily and hourly basis was key for them to improve their patient experience. Many patients also wanted the ability to log their medication, e.g., missing and on-time or ordering for repeats.

Australia is one of the most culturally and linguistically diverse countries around the world. There is a variety within communities in terms of origin of birth, spoken language, religion and ethnicity (Aboriginal or Torres Strait Islanders descent) (Australian Bureau of Statistics (ABS) 2017b). According to 2016 census, almost half (49%) of the Australian population was either born overseas or one of their parents was born overseas. It was also estimated that over 300 different languages were spoken in Australia (Australian Bureau of Statistics (ABS) 2017b). This highlights the need to incorporate major languages of Australia in OIS for patient informational materials regarding diagnosis, treatment plan, survivor support and stories and videos. The absence of material in an ethnic language does affect early

**Table 4** Comparison of different leading international oncology information systems (OISs) in Australian context

| | OIS 1 | OIS 2 | OIS 3 |
|---|---|---|---|
| Therapies (chemo and radiation) | Yes | Yes | Yes |
| Payment | Billing<br>Bill adjustment of missed procedure<br>Export billed items to billing software | Billing and other third-party systems | Claims and billing data from financial systems |
| Image management | Comprehensive review of clinical images<br>Optimized image-guided treatment techniques<br>Review images remotely and send set-up instructions to the treatment machine<br>Review dosimetric images for IMRT pre-treatment quality assurance<br>Compare images using automatic, manual or fiducial marker-matching algorithms<br>Attach images, files or patient photographs to documents | Sophisticated image visualization and distribution<br>3D viewers: RTP, XVI and iView GT™ integration, adding and synchronizing tools, 2D image registration, volume image viewing registration, volume image viewing<br>External image registration<br>Stereoscopic and volume image registration<br>Setup details, immobilization devices and reference images entered into treatment chart | Present complex, personalized analytics on individual patients based on clinical, molecular and other relevant data in a clinically actionable format |
| Clinician led and patient centric | Appointment scheduling, review images remotely, customizable data collection forms and questionnaires<br>Attach images, files or patient photographs to documents | Electronic patient management allows users to achieve a paperless, filmless working environment; treatment setup workflow management | Configure CARE reports to contain personalized analytic content and historical patient content formatted to your institution's and clinician's specific needs |
| Information and analytics | Pain scoring to identify trends | Trend analysis | Integration, aggregation and personalized analytics on highly complex, disparate, multi-institutional data sources |

(continued)

**Table 4** (continued)

| | OIS 1 | OIS 2 | OIS 3 |
|---|---|---|---|
| Adherence to agreed standards and interoperability and leveraging existing assets and capabilities | Automates patient-data transfer with external hospital systems using standard communication protocols | Seamless connectivity to virtually any linear accelerator and treatment planning system from any vendor, providing unmatched integration, freedom and flexibility | IT team, data and network security resulting in very fast time-to-value |
| PBS/MBS | – | – | – |
| Robust privacy, security and data protection | Built-in EDI (electronic data interchange) for secure online claim submissions High Availability and Rapid Recovery Protection (HARRP) continuous data protection and ensures minimal data loss and immediate recovery | Eliminate error-prone work through tight integration that enables the seamless transition of data between tools | – |
| Safety and quality | Facilitates compliance with the Work Health and Safety (WHS) Act | Manual, barcode and biometric patient identification and verification options provide multilevel safety checkpoints | Quality and safety reporting |
| Comparison of cost, complications, toxicity and practice | – | – | Comparison of cost, complications, toxicity and even comparison between different practice centres or clinics |

**Table 5** Data comparison of feedback for missing elements from different user's perspective

| OIS providers | Healthcare management | Patient and carers | Clinician |
|---|---|---|---|
| * Value | * Value for OIS and services provided, efficieny, adoptibility in existing system and integration for future planning. | * Value for treatment, procedures and services provided, ease of date/time/place for treatment. | * Value for performing treatment, procedures and services, increasing efficieny by reducing paper work, accessing images and prescibing medicine and amount of dose for radiation. |
| * Payment and share calcuation | * Seamless calculation and transfering data for VMOs, nurses, PBS, MBS and insurance company share. | * Seamless payment transfer through PBS, MBS and insurance company. | * Calcuation of share for payment by healthcare providers, medicare and insurance company. |
| * Standardization | * Adherence to different Standards e.g., international, Austraian, regional systems. The access of OIS for VMOs without requiring any further licensing. | * Standardization for different organizations e.g., international, Austraian, regional systems to move across in different systems in case patient has to go abroad or interstate for procedure / treatment. | * Compatible with different Standards e.g., international, Austraian, regional systems to accept and send patient overseas or interstate. |

| OIS providers | Healthcare management | Patient and carers | Clinician |
|---|---|---|---|
| * Appointment | * Real time update on vacant spot to fill in for another patient in case of rescheduling or cancelation of an appointment due to patient / doctor. | * Easy access and realtime for confirmation, cancellation, and rescheduling of appointments, treatment, procedure place, time and date | * Change of plan in case of cancellation or resheduling of an appointment due to patient/ equipment failure. |
| * Medication logging | – | * Registering missing, late or overdosing of medication and ability to order online for repeats. | * Ability to check missing, late or overdosing of medication and ordered repeats. |
| * User interface | * Easy access to data for finance, safety and quality control, legislation, transfer between existing and future equipment. | * Easy graphical user interface, accesible from mobile / tablet and desktop computres with easy login option to keep privacy and data security. | * Easy graphical user interface, with access to from mobile / tablet and desktop with easy login option to keep privacy and data security |
| * Encryption | * Encrypted data transfer within same site and other sites, secure from cyber attacks and malwares. | * Security and privacy for personal data | * Secure access to data keeping privacy for personal data for patient |

(continued)

**Table 5** (continued)

| OIS providers | Healthcare management | Patient and carers | Clinician |
|---|---|---|---|
| * Language | *Ability to intergrate appointment booking/relevant disease or treatment literature in different langauges. Possiblity of adding tag for easy identification of multi-lingual staff, nurses and doctros. | * Access for booking/relevant disease or treatment literature in patient's own langague. | * Options to tag in as multi-lingual capabilities. |
| * Videos | * Easy access to data for finance, safety and quality control, legislation, transfer between existing and future equipment. | * Ability to watch videos in patient's own language to learn about prognosis, possible treatments and outcomes. | * In case of multi-lingual ability to store guidance in thier language to help thier community better understand the procedures, course of treatment and possible outcome. |
| * Survivor supports | * Connect survivor to the current patients to give them emotional, psychological and social support. | * Connect survivor to the current patients to give them emotional, psychological and social support. | * Connect survivor to the current patients to give them emotional, psychological and social support. |

| OIS providers | Healthcare management | Patient and carers | Clinician |
|---|---|---|---|
| * Consistency | * Consistency around treatments protocols, timelines and cost forecasting | * Consistency of treatments protocols, timelines and cost. | * Consistency for treatments plans, procedures and payments of shares for VMOs and nurses |
| * 24 / 7 access | * Easy access to data for finance, safety and quality control, legislation, transfer between one location to another | * Easy access to financial matters for billing and treatment procedure. | * Easy access to data for finance, safety and quality control, legislation, transfer between one location to another |
| * Available online | * Always online for prompt decision and procedure in case of any emergency. | * Online availability through portal or mobile app for booking / treatment details. | * Always online for prompt decision and procedure in case of any emergency. |
| * Technical support | *Around the clock technical support for intergration / software malfunctioning | *24/7 technical support for potral / app malfunctioning | *24/7 technical support for potral / app malfunctioning |

detection, treatment and decision making for specific treatment. The GP (general practitioners) from the same ethnic group and material published in the targeted ethnic language will have definitely a significant impact over early detection and treatment, hence reducing overall economic, emotional and psychological burden in the society (Tavasoli et al. 2016; Morrell et al. 2010; McDonald and Kennedy 2007; Byles et al. 1994; Henderson et al. 2012).

From clinician and healthcare provider's point of view, it was noted that most of the providers (public and private hospitals) have had different central systems of patient data recording and maintaining (as detailed in Table 5). This makes it challenging for clinicians who must work with multiple systems. Different hospitals are now collaborating or amalgamating to become a big single healthcare provider through acquisitions. Therefore, it is becoming increasingly important that OIS can communicate with existing equipment and infrastructure systems planned for future within and across different sites. Otherwise, there will be inadequate incorporation of new systems with old one creating disruption in smooth work flow. Consequently, variety in user interfaces will result in unwillingness in staff and management for adoption of any new OISs. The discussions with oncology clinicians and managers at different healthcare organizations were very fruitful and highlighted a number of key missing elements for any OIS to be successful and improve overall performance of OIS and make it acceptable in Australian healthcare context. Healthcare management pointed out that the central existing equipment, assets and solutions are equally important as it involves huge investment and had great willingness for utilization/integration with OIS and future proofing the planning to make OIS more desirable. Adherence and compliance with regional, national and international standards was also highlighted as another aspect of the concern.

To increase the overall efficiency of the system especially the oncology, it is very highly expected that system should be automated to create alert in case of cancellation of appointment by patient, to fill in the next patient. Systematic alert on feedback or complaint system would be another element highly important for healthcare providers' point of view to achieve higher patient satisfaction. Healthcare providers also had desire to have a registry for safety and quality control forms and protocols to avoid any legal or ethical issues. In Australia, a unique concept of independent visiting medical officer exists, and this warrants the need for web-based login, to enable specialist to work in different organizations and have access to patient data/images and treatment equipment. In short, for a system developed for another country/medical system (as in Fig. 2), it will need to consider these missing elements to make it attractive and seamlessly incorporate into Australian healthcare. The point of views of patients and their care providers, clinicians, healthcare management and finally OIS developers all will have to be considered to get most appropriate system.

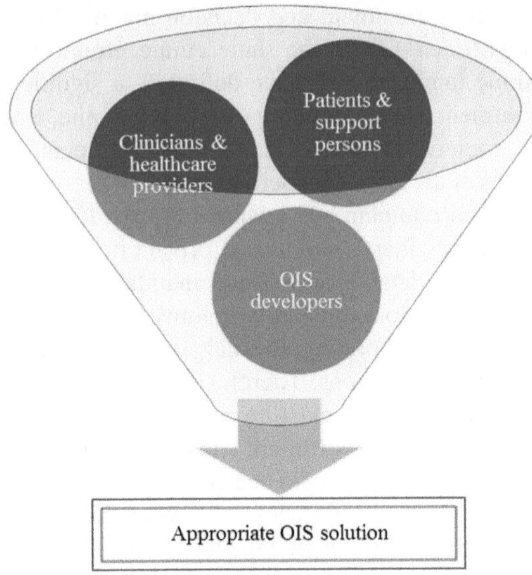


**Fig. 2** Filter to develop appropriate OIS soluation using missing elements

## 6 Conclusion

Cancer is one of the leading chronic diseases throughout the world. Owing to increasingly costly treatments, the funding of cancer care is coming under increasing scrutiny by government and other third-party payers which means OIS has to easily provide the necessary justification for care decisions. OISs have been designed to integrate imaging, radiation, surgery, recovery and progress data all in one place and access for multiusers from multiple sites. These systems reduce paper work, give seamless connectivity to linear accelerator and planning system and attachment of images to files and prescribe medicine remotely to avoid any system or human mistake. However, to maximize their true potential, they need to be tailored to the specific healthcare environment. This study compares three leading oncology systems along with focus group discussion with patients, patient carers, clinicians and healthcare management to identify missing key elements in Australian context. Though this study is in its early stages, it is expected that it will serve as a key building block for any future study focused on oncology information system targeting Australian healthcare market. The key missing elements described by different user participants will also make any oncology system work better for Australian patients, their care providers and overall healthcare industry.

## References

AIHW. (2011). *$137 billion in welfare spending driven by 'changing, complex and varied society'*. AIHW media releases. 31 Oct 2017, v5.0 [cited 13 Sept 2019]. Available from: https://www.aihw.gov.au/news-media/media-releases/2011/november/137-billion-in-welfare-spending-driven-by-changi.

AIHW. (2016). *Australia's health 2016 (Australia's health series no. 15)*. Canberra: Australian Institute of Health and Welfare.

AIHW. (2017a). *Cancer in Australia 2017 (Cancer series)*. Canberra: Australian Institute of Health and Welfare.

AIHW. (2017b). *Cancer in Australia* [19 Feb 2018]. Available from: https://www.cancer.org.au/about-cancer/what-is-cancer/facts-and-figures.html.

AIHW. (2018a). *Twitter graphic cancer in Aust 2017*. Can in Brief [cited 19 Feb 2018]. Available from: https://www.aihw.gov.au/getmedia/96aa3854-ccb0-40c5-8d25-27d4b1098490/twitter-graphic-cancer-in-aust-2017.pdf.aspx.

AIHW. (2018b). *Cancer in Australia*. What is cancer? 3 Sept 2018 [cited 13 Sept 2018]. Available from: https://www.cancer.org.au/about-cancer/what-is-cancer/facts-and-figures.html.

AIHW. (2018c). *Cancer in Australia in brief 2017* [19 Feb 2018]. Available from: https://www.aihw.gov.au/getmedia/a7ef1bb6-c61a-4046-97f3-8dea2fd315c0/cancer-in-australia-in-brief-2017-slideshow.pptx.aspx.

AIHW. (2018d). *Australia's health 2018 (Australian Institute of Health and Welfare 2018)*. Canberra: AIHW.

Amaravadi, R. K., et al. (2011). Principles and current strategies for targeting autophagy for cancer treatment. *Clinical Cancer Research, 17*(4), 654–666.

Ando, Y. (2014). Oncology information system. In H. Tsujii et al. (Eds.), *Carbon-ion radiotherapy: Principles, practices, and treatment planning* (pp. 113–117). Tokyo: Springer Japan.

Armstrong, B. K., et al. (2007). Challenges in health and health care for Australia. *Medical Journal of Australia, 187*(9), 485.

Australian Bureau of Statistics (ABS). (2016). *Cancer within Australia's leading causes of death*. 3303.0—Causes of Death, Australia, 2015. 26 Sept 2017 [cited 5 Sept 2018]. Available from: http://www.abs.gov.au/ausstats/abs@.nsf/Lookup/by%20Subject/3303.0~2015~Main%20Features~Cancer%20within%20Australia's%20leading%20causes%20of%20death~6.

Australian Bureau of Statistics (ABS). (2017a). 3303.0—Causes of Death, Australia, 2016. 20 Mar 2018 [cited 5 Sept 2018]. Available from: http://www.abs.gov.au/ausstats/abs@.nsf/mf/3303.0.

Australian Bureau of Statistics (ABS). (2017b). *Census reveals a fast changing, culturally diverse nation*. 2016 census: Multicultural. 11 Jan 2018 [cited 4 Sept 2018]. Available from: http://www.abs.gov.au/ausstats/abs%40.nsf/lookup/Media%20Release3.

Australian Bureau of Statistics (ABS). (2018). *Cancer deaths in younger Australians—changes over 20 years*. 3303.0—Causes of Death, Australia, 2016. 20 Mar 2018 [cited 5 Sept 2018]. Available from: http://www.abs.gov.au/ausstats/abs@.nsf/Lookup/by%20Subject/3303.0~2016~Main%20Features~Cancer%20deaths%20in%20younger%20Australians%20-%20changes%20over%2020%20years~10000.

Balch, C. M., et al. (2009). Final version of 2009 AJCC melanoma staging and classification. *Journal of Clinical Oncology, 27*(36), 6199.

Bernstein, R. J. (1983). *Beyond objectivism and relativism: Science, hermeneutics, and praxis*. Oxford: B. Blackwell.

Berwick, D. M. (2003). Disseminating innovations in health care. *JAMA, 289*(15), 1969–1975.

Blum, B. I. (1983). An information system for developing information systems. In *Proceedings of the May 16–19, 1983, national computer conference*. New York: ACM.

Boonstra, A., & Broekhuis, M. (2010). Barriers to the acceptance of electronic medical records by physicians from systematic review to taxonomy and interventions. *BMC Health Services Research, 10*(1), 1.

Bourret, J., et al. (1996). Medication use review process and information systems utilized for oncology chemotherapy quality improvement. *Pharmacy Practice Management Quarterly, 16*(1), 1–17.

Byles, J. E., et al. (1994). Effectiveness of three community based strategies to promote screening for cervical cancer. *Journal of Medical Screening, 1*(3), 150–158.

Calle, E. E., & Kaaks, R. (2004). Overweight, obesity and cancer: Epidemiological evidence and proposed mechanisms. *Nature Reviews Cancer, 4*(8), 579.

Calle, E. E., & Thun, M. J. (2004). Obesity and cancer. *Oncogene, 23*(38), 6365.

Calle, E. E., et al. (2003). Overweight, obesity, and mortality from cancer in a prospectively studied cohort of US adults. *New England Journal of Medicine, 348*(17), 1625–1638.

Cancer Australia. (2017). *Cancer biggest burden on Australia's health*. 7 Sept 2017–4:05 pm [cited 13 Sept 2018]. Available from: https://canceraustralia.gov.au/about-us/news/cancer-biggest-burden-australias-health.

Cancer Australia. (2018). *All cancers in Australia* [cited 5 Sept 2018]. Available from: https://canceraustralia.gov.au/affected-cancer/what-cancer/cancer-australia-statistics.

Centre for Clinical Governance Research. (2011). *Medical electronic systems in oncology: A review of the literature*. Sydney: University of New South Wales.

Chamberlain, A. J., et al. (2002). Nodular type and older age as the most significant associations of thick melanoma in Victoria, Australia. *Archives of Dermatology, 138*(5), 609–614.

Cheng, H. C., et al. (2011). Evaluation of radiation dose and image quality for the Varian cone beam computed tomography system. *International Journal of Radiation Oncology ·Biology·Physics, 80*(1), 291–300.

Cleary, J. F. (2000). Cancer pain management. *Cancer Control, 7*(2), 120–131.

Crawford, S. (2013). Is it time for a new paradigm for systemic cancer treatment? Lessons from a century of cancer chemotherapy. *Frontiers in Pharmacology, 4*, 68.

Crowther, D. (1982). Cancer—principles and practice of oncology. *British Journal of Cancer, 46*(2), 308.

Dempke, W. C., et al. (2017). Second-and third-generation drugs for immuno-oncology treatment—the more the better? *European Journal of Cancer, 74*, 55–72.

Duckett, S. (1999). Policy challenges for the Australian health care system. *Australian Health Review, 22*(2), 130–147.

Duckett, S., & Willcox, S. (2015). *The Australian health care system*. South Melbourne: Oxford University Press.

Duncan, L. D., et al. (2010). Clinical integration of picture archiving and communication systems with pathology and hospital information system in oncology. *The American Surgeon, 76*(9), 982–986.

Enterline, J. P., Raymond, E., Jr., & Blum, B. I. (2012). *A clinical information system for oncology*. New York: Springer Science & Business Media.

Evans, W., et al. (2014). Implementing a regional oncology information system: Approach and lessons learned. *Current Oncology, 21*(5), 224.

Fasola, G., et al. (2008). Drug waste minimisation and cost-containment in medical oncology: Two-year results of a feasibility study. *BMC Health Services Research, 8*(1), 70.

Fasola, G., et al. (2014). Health information technology in oncology practice: A literature review. *Cancer Informatics, 13*, CIN. S12417.

Forbes, M., et al. (2010). Quality of care in Australian public and private hospitals. In *Australian Conference of Health Economists, Sidney*.

Goldhirsch, A., et al. (2013). Personalizing the treatment of women with early breast cancer: Highlights of the St Gallen International Expert Consensus on the Primary Therapy of Early Breast Cancer 2013. *Annals of Oncology, 24*(9), 2206–2223.

Goodwin, M. (2017). *Burden of cancer in Australia: Australian Burden of Disease Study 2011*. Canberra: AIHW.

Henderson, B. E., et al. (2012). The influence of race and ethnicity on the biology of cancer. *Nature Reviews Cancer, 12*(9), 648.

Herre, H., & Heller, B. (2004). Ontological foundations of medical information systems. In H. Fujita & V. Gruhn (Eds.), *New trends in software methodologies, tools and techniques: Proceedings of the third SoMeT_W04, volume 111 of Frontiers in Artificial Intelligence and Applications*. Amsterdam: IOS Press.

Hesketh, P. J., et al. (2017). Antiemetics: American Society of Clinical Oncology clinical practice guideline update. *Journal of Clinical Oncology, 35*(28), 3240–3261.

Hillestad, R., et al. (2005). Can electronic medical record systems transform health care? Potential health benefits, savings, and costs. *Health Affairs, 24*(5), 1103–1117.

Insight Pharma Reports. (2018). *Oncology information system market by software, application, end user—global forecasts to 2019* [cited 30 Aug 2018]. Available from: http://www.insightpharmareports.com/Affiliated-Reports/Markets-and-Markets/Oncology-Information-System-Market-by-Software,-Application,-End-User%2D%2D-Global-Forecasts-to-2019/.

Islam, M. M., Poly, T. N., & Li, Y.-C. J. (2018). Recent advancement of clinical information systems: Opportunities and challenges. *Yearbook of Medical Informatics, 27*(01), 083–090.

Jacobsen, P. B., & Jim, H. S. (2008). Psychosocial interventions for anxiety and depression in adult cancer patients: Achievements and challenges. *CA: A Cancer Journal for Clinicians, 58*(4), 214–230.

Janssen, A., et al. (2017). Interdisciplinary eHealth practice in cancer care: A review of the literature. *International Journal of Environmental Research and Public Health, 14*(11), 1289.

Kesson, E. M., et al. (2012). Effects of multidisciplinary team working on breast cancer survival: Retrospective, comparative, interventional cohort study of 13 722 women. *BMJ, 344*, e2718.

Klein, E. E., et al. (2005). Errors in radiation oncology: A study in pathways and dosimetric impact. *Journal of Applied Clinical Medical Physics, 6*(3), 81–94.

Krayenbuehl, J., et al. (2015). Evaluation of an automated knowledge based treatment planning system for head and neck. *Radiation Oncology, 10*(1), 226.

Lenhard, R. E., et al. (1983). The Johns Hopkins oncology clinical information system. *Journal of Medical Systems, 7*(2), 147–174.

Markets and Markets. (2015). *Oncology information system market by software (patient information system & TPS), & professional service), application (medical, radiation, & surgical oncology), end user (hospital, government institution, & research center)—global forecasts to 2019* [cited 30 Aug 2018]. Available from: https://www.marketsandmarkets.com/Market-Reports/oncology-information-systems-market-1721676.html.

McDonald, J. T., & Kennedy, S. (2007). Cervical cancer screening by immigrant and minority women in Canada. *Journal of Immigrant and Minority Health, 9*(4), 323–334.

MesotheliomaGroup. (2018). *Understanding cancer* [cited 5 Mar 2018]. Available from: https://www.mesotheliomagroup.com/our-mission/resources/understanding-cancer/.

Miller, A. A., & Phillips, A. K. (2005). A contemporary case study illustrating the integration of health information technologies into the organisation and clinical practice of radiation oncology. *Health Information Management, 34*(4), 136–145.

Moen, A. (2011). What makes an information system more preferable for clinicians? a qualitative comparison of two systems. *User Centred Networked Health Care: Proceedings of MIE 2011, 169*, 392.

Morrell, S., et al. (2010). Outcomes from a mass media campaign to promote cervical screening in NSW, Australia. *Journal of Epidemiology & Community Health, 64*(9), 777–783.

Newman, L., Baum, F., & Harris, E. (2006). Federal, state and territory government responses to health inequities and the social determinants of health in Australia. *Health Promotion Journal of Australia, 17*(3), 217–225.

Nutbeam, D. (2008). The evolving concept of health literacy. *Social Science & Medicine, 67*(12), 2072–2078.

O'Connell, B., Kristjanson, L., & Orb, A. (2000). Models of integrated cancer care: A critique of the literature. *Australian Health Review, 23*(1), 163–178.

Ochs, M. F., & Casagrande, J. T. (2008). Information systems for cancer research. *Cancer Investigation, 26*(10), 1060–1067.

Patkar, V., et al. (2011). Cancer multidisciplinary team meetings: Evidence, challenges, and the role of clinical decision support technology. *International Journal of Breast Cancer, 2011*, 1–7.

Ramaswamy, S., et al. (2001). Multiclass cancer diagnosis using tumor gene expression signatures. *Proceedings of the National Academy of Sciences, 98*(26), 15149–15154.

Rycaj, K., & Tang, D. G. (2015). Cell-of-origin of cancer versus cancer stem cells: Assays and interpretations. *Cancer Research, 75*(19), 4003–4011.

Sajeev, S., & Sumant, O. (2017). *Oncology information system market overview* [cited 5 Sept 2018]. Available from: https://www.alliedmarketresearch.com/oncology-information-system-market.

Saletti, P., et al. (2018). Choosing wisely in oncology: Necessity and obstacles. *ESMO Open, 3*(5), e000382.

Shortliffe, E. H., et al. (1984). An expert system for oncology protocol management. In B. G. Buchanan & E. H. Shortiffe (Eds.), *Rule-based expert systems* (pp. 653–665). New York: Springer.

Siochi, R. A., et al. (2009). Information technology resource management in radiation oncology. *Journal of Applied Clinical Medical Physics, 10*(4), 16–35.

Solomon, B. J., et al. (2014). First-line crizotinib versus chemotherapy in ALK-positive lung cancer. *New England Journal of Medicine, 371*(23), 2167–2177.

Štambuk, S., et al. (2010). Future perspectives of personalized oncology. *Collegium Antropologicum, 34*(2), 763–769.

Staples, M. P., et al. (2006). Non-melanoma skin cancer in Australia: The 2002 national survey and trends since 1985. *Medical Journal of Australia, 184*(1), 6–10.

Sulaiman, H., & Wickramasinghe, N. (2014). Assimilating healthcare information systems in a Malaysian hospital. *CAIS, 34*, 77.

Surbone, A., & Rowe, M. (2015). Introduction to oncology and medical errors. In *Clinical oncology and error reduction* (pp. 1–13). Hoboken: John Wiley & Sons, Inc.

Surbone, A., et al. (2010). Psychosocial care for patients and their families is integral to supportive care in cancer: MASCC position statement. *Supportive Care in Cancer, 18*(2), 255.

Tavasoli, S. M., et al. (2016). Impact of invitation and reminder letters on cervical cancer screening participation rates in an organized screening program. *Preventive Medicine, 88*, 230–236.

TMR. (2018). *Global oncology information system market: Snapshot* [cited 30 Aug 2018]. Available from: https://www.transparencymarketresearch.com/pressrelease/oncology-information-system-market.htm.

Total Health. (2018). *No two breast cancer patients are the same and treatment should be tailored* [cited 5 Mar 2018]. Available from: http://www.totalhealth.co.uk/blog/no-two-breast-cancer-patients-are-same-and-treatment-should-be-tailored.

Vogelaar, F., et al. (2015). Epidural analgesia associated with better survival in colon cancer. *International Journal of Colorectal Disease, 30*(8), 1103–1107.

Wickramasinghe, N., & Schaffer, J. (2010). *Realizing value driven e-health solutions*. Washington, DC: Report for IBM.

Wickramasinghe, N., et al. (2015). Identifying key success factors for the adoption and implementation of a chemotherapy ordering system: A case study from the Australian private healthcare sector. In *21st Americas Conference on Information Systems, AMCIS 2015*.

Wilson, R. M., et al. (1995). The quality in Australian health care study. *Medical Journal of Australia, 163*(9), 458–471.

Yu, P., Gandhidasan, S., & Miller, A. A. (2010). Different usage of the same oncology information system in two hospitals in Sydney—lessons go beyond the initial introduction. *International Journal of Medical Informatics, 79*(6), 422–429.

# Toward Actionable Knowledge: A Systematic Analysis of Mobile Patient Portal Use

**Cherie Noteboom and Mohammad Abdel-Rahman (Al-Ramahi)**

## 1 Introduction

The American elderly population aged 65 and older is expected to continue to grow rapidly. This population segment is expected to reach 89 million by 2015 (Dall et al. 2013). The population reports increasing health complexities, due to increases in chronic disease, and many report suffering from multiple chronic disease. Estimates predict that by 2030, there will be an additional 27 million Americans with hypertension, eight million with coronary heart disease, and three million with heart failure (Heidenreich et al. 2011). Cancer cases are expected to increase to 27 million (Suzman and Beard 2011). Alzheimer's is estimated to grow to 7.1 million, a 40% increase, by 2025 (Alzheimer's Association 2013).

Healthcare in America costs 2.5 trillion a year and is expected to grow to 4.5 trillion in six years (Clifton 2012). According to the Agency for Healthcare Quality and Research, automation can improve the quality and safety of care delivered by healthcare facilities by enabling collaboration among physicians, medical personnel and patients (Clifton 2012). Understanding the healthcare context is key to understanding the integration of information systems (IS) into the fabric of their organizations. According to Fichman, Kohli, and Krishnan (2011):

> *at the most general level, a striking feature of healthcare industry is the level of diversity that characterizes patients (e.g., physical traits and medical history), professional disciplines*

C. Noteboom (✉)
Dakota State University, Madison, SD, USA
e-mail: Cherie.Noteboom@dsu.edu

M. Abdel-Rahman (Al-Ramahi)
Texas A&M, San Antonio, San Antonio, TX, USA
e-mail: Mohammad.Abdel@tamusa.edu



N. Wickramasinghe, F. Bodendorf (eds.), *Delivering Superior Health and Wellness Management with IoT and Analytics*, Healthcare Delivery in the Information Age, https://doi.org/10.1007/978-3-030-17347-0_29

*(e.g., doctors, nurses, administrators and insurers), treatment options, healthcare delivery processes and interests of various stakeholder groups. (p. 419)*

Patient-centered care is seen to be a natural progression toward greater efficiency and effectiveness in healthcare provision. This form of care is one in which a patient actively participates in his or her care, delivery of care takes place from a patient's point of view, there is greater communication with the patient, and therapy is tailored to the needs of the patient (Murphy 2011; Sacristán 2013; Stewart et al. 2000). The implementation of health information technology (HIT) appears to have enabled greater patient-centered care through better access to patient data, shorter recovery through targeted care, and lower cost through fewer tests (Blumenthal and Tavenner 2010; Cliff 2012; Cohen et al. 2010; Fichman et al. 2011).

Patient-centered care implies a paradigm shift in the relationship between doctors and patients, but also requires the development of self-management practices (Sacristán 2013). Kane and Labianca (2011, p. 510) note that, "if patients fail to manage their chronic diseases adequately, escalating conditions can become extremely expensive to treat and can significantly compromise the patient's quality of life." In order to help patients self-manage their diseases, information is made available from HIT products, such as home health devices and patient portals. Patient health records (PHR) technology, often known as patient portal, provides patients with online access to their health records, which in turn enables better disease management through tracking of comprehensive health indicators and lower the cost of care (Cliff 2012; Cohen et al. 2010). Through the potential to provide continuity of service and better care, and the potential to change the physician-patient relationship and enable chronic disease self-management, patient portals are positioned as a central component of patient engagement (McAlearney et al. 2016).

Healthcare organizations in the United States are investing in information technology (IT) to reduce the associated cost of services and improve the quality of patient care in a move toward population health initiatives. IT systems in healthcare organizations must meet requirements, as they positively impact patients. Many of these initiatives focus on education and the engagement of the patient population. Internet of Things (IoT)-supported devices and applications, which continue to experience great growth, are considered a key IT strategy to engage and educate the healthcare population (Allan and Wang 2017; Sacristán 2013).

Healthcare continues to integrate IT solutions to transform the methods of patient interaction to support patient engagement and education. These solutions are transforming how patients participate in their individual care. Seventy-eight percent of healthcare customers either wear or are willing to utilize wearable technology solutions to track their lifestyle choices and vital signs. Mobile medical technology is advocated by 75.5% of physicians who feel that the technology simplifies access and is one of the greatest benefits of mobile medical technology. Nearly half of hospitals provide applications (apps) for patient education and engagement; 58% of hospitals have patient portal solutions (Stewart et al. 2000). The number of health apps exceeds 165,000 (Microsoft n.d.).

Previous technology research (e.g., Qureshi and Keen 2005; Qureshi et al. 2005; Qureshi and Noteboom 2006) has investigated collaboration effects and provides insight to inform the patient portal research in the areas of collaboration, coordination, communication, and adaptation. In addition, the adaptation insights at the work, social, and technology levels inform this research. However, up to now, very few efforts have been made to extract knowledge from online user reviews from actual use of mobile patient portals to help understand patients' concerns that deter effective use of these health technology. Therefore, this research answers the call for the development of patient-oriented research by investigating the key challenges relating to the use of mobile patient portals (i.e., determining the gaps in the design of patient portals) via extracting insights from negative users' reviews of patient portal mobile apps. Such insights can improve the partnership and collaboration between patients and healthcare providers.

The advances of Web 2.0 technologies have enabled consumers to easily and freely exchange opinions on products and services on an unprecedented scale (volume) and in real time (velocity). Online user review systems provide us with one of the most powerful channels for extracting user feedback that can help enhance patient portals design. In the e-commerce domain, user reviews have long been widely recognized as a crucial factor that influences product sales (e.g., Chevalier and Mayzlin 2006) and shapes consumers' purchase intention (e.g., Yang et al. 2016). In the domain of patient portal systems, analyzing users' reviews has the potential to greatly inform developers of patients' preferences and how they engage with health portals, and provide opportunities for further enhancing their efficacy.

In this study, we systematically analyze users' reviews to identify design gaps from the actual use of mobile patient portal. The question investigated in this paper is: What are the gaps in mobile patient portal service to enable patient-centered care? To answer this question, we use an implementation of EPIC's mychart as an instance of mobile patient portal and discover design gaps based on a systematic analysis of users' negative reviews. Instead of manually analyzing the data, which is time-consuming, we utilize a text-mining technique, specifically topic modeling, to analyze the contents of user reviews and identify design gaps for mobile patient portal systems. The key contribution of this research is in discovering the gaps that may exist in current mobile patient portal solutions and identifying the opportunities for mobile patient portal enhancement to achieve improved patient-centered care.

## 2 Related Work and Background

### *2.1 Patient-Centered Care*

Effective patient-centered care centers on the identification of the best intervention for every individual patient using personalized medicine and tailored therapeutics (Sacristán 2013). To provide patient-centered care, physicians will have to collaborate. Collaboration is a purposeful joint action through the construction of relevant meanings that are shared among members. Collaboration is needed to (1) determine what action is required and relevant, (2) identify knowledge to carry out a required action, and (3) demand for action. To support collaboration, it is necessary to have a media with which to communicate and a social network or "community of minds."

In this regard, patient portals have the potential to better inform and engage patients in their care. Patient portals, now commonly used in ambulatory settings, leverage integration with electronic health record (EHR) efforts to inform and engage patients. Healthcare providers feel the information provided by the portal helps to facilitate patient engagement in care and identification of errors (O'Leary et al. 2016).

## *2.2 Patient Health Records (Patient Portal)*

With the exponential growth of the communications technologies that allow us to potentially reach more individuals regardless of their locations, new types of health intervention have emerged. Smartphone or mobile-based patient portals can enhance patients' engagement at a very low cost. Due to the promising influence of these smartphone-based technologies in supporting healthy lifestyle and self-care practices, researchers have been inspired to explore the impact and use of mobile applications. For example, the fact that women widely used mobile apps for health information during pregnancy, but reported apps as unavailable or invaluable postpartum, highlights the need for the development of more mobile apps with postpartum content (Guerra-Reyes et al. 2016). With this respect, Zhang, Ho, Cassin, Hawa, and Sockalingam's (2015) study is one of the first few studies to describe the methodology of developing an online and smartphone compatible cognitive behavioral therapy intervention program for bariatric surgery patients.

While the results of HIT use by providers are mixed, it appears that motivated patients can achieve significant improvements in their health outcomes when they use mobile applications (García-Gómez et al. 2014). There are currently over 3000 mobile applications available through Google Play Store and Apple Store to support lifestyle changes such as fitness, calorie counter, and body mass index calculators, some of which are used to control diabetes (García-Gómez et al. 2014; Qureshi et al. 2015).

Little research, however, has been done to connect the growing mobile application use by patients to accessing their healthcare data. From a public health perspective, patient-centered care requires "a partnership among practitioners, patients, and their families (when appropriate) to ensure that decisions respect patient's wants, needs, and preferences and that patients have the education and support they need to make decisions and participate in their own care" (IOM

2001, p. 7). Robinson, Callister, Berry, and Dearing (2008) also offer an economic view of the patient as the informed consumer who makes decisions based on cost and quality of care. They also identify care from a patient's perspective to include "respect, courtesy, competence, efficiency, patient involvement in decisions, time for care, availability/accessibility, information, exploring patient's needs, and communication" (Robinson et al. 2008, p. 602). To address these views, patient health portals will need to be customized for patient-centered care.

The focus of previous studies included providing access to the patient record and information on the care team through a mobile phone app (e.g., Pfeifer Vardoulakis et al. 2012), a tablet computer app to view care team profiles and hospital medication records, and a tablet app with the plan of care, diet and safety information (Dykes et al. 2012). Providing patients real-time access to health information has been demonstrated as a positive force for change in the way care is provided (McAlearney et al. 2016). In this regard, Lu et al. (2017) develop an app to inspect controlled substances in patient care units. Using a web-enabled smartphone, pharmacist inspection can be performed on site and the inspection results can be directly recorded into the HIS through the Internet, so human error of data translation can be minimized and the work efficiency and data processing can be improved.

While previous studies reported positive findings, including patient reports of enhanced engagement in the care process and satisfaction with care, none included *patient-centered functionality*, such as the ability to send messages to the care team, allowing patients to input information or record notes—elements that have been demonstrated to further enhance patients' engagement (McAlearney et al. 2016). This is especially true with the proliferation of wearable devices, where data about an individual's health state can be collected by real-time sampling and analysis of a few parameters using noninvasive, inexpensive, and portable devices (Pierleoni et al. 2014). With this respect, Neubeck et al. (2016) adopted a collaborative user-centered design process to develop a patient-centered care tool. O'Leary et al. (2016) concluded that optimizing a hospital-based patient portal will require attention to type, timing and format of information provided, as well as the impact on patient-provider communication and workflow. Patients can identify areas of improvement that could enhance the design of portals. For example, patients suggested inclusion of a test result feature (O'Leary et al. 2016). Therefore, further research is needed to work in concert with patients to explore patient-centered functionalities that help develop a patient-centric portal to increase patients' engagement in their care.

Leveraging user feedback from the actual use of mobile patient portal, this research contributes to an understanding of how the technology architecture can enable patients to interact with patient portal functionality, which is technological adaptation, to work (work adaptation) together with their physicians and care providers (social adaptation) using the content available to them and using the collaboration media to provide patient-centered care.

### 2.3 *The Impacts of User-Generated Content*

Several researchers in the areas of social media and e-commerce have studied the effects of user-generated content, such as online users' reviews and rating systems, on product sales and consumers' purchase intention. The findings of the existing research have demonstrated that analyzing and measuring these electronic word-of-mouth (eWOM) messages is quite valuable in product design, sales prediction, marketing strategy, and other decision-making tasks (e.g., Al-Ramahi et al. 2015; Phillips et al. 2016; Yu et al. 2013). In this regard, Guo et al. (2017) adopted latent Dirichlet allocation (LDA) as topic modeling technique to discover key dimensions from online user reviews for hotels located in 16 countries.

Recently, few researchers have been attracted to explore the impact of user-generated content in healthcare domain. For example, Jung et al. (2015) proposed a text mining approach to identify hospital service quality factors and overtime trends automatically from user-generated content from online health communities. Xu et al. (2016) examined the impact of online information on patient choice of outpatient care doctors. Al-Ramahi et al. (2016, 2017) use topic modeling, LDA algorithm, to discover design principles for Health Behavioral Change Support Systems (HBCSSs) from online user reviews of mobile diabetes applications. However, to the best of our knowledge, no research to date has looked at online users' reviews in the context of mobile patient portals. User reviews implicitly communicate satisfaction/dissatisfaction based on actual usage experience and may provide a good opportunity for extracting design dimensions that can strongly influence users' satisfaction and then informing the design of these systems.

## 3 Method

This section describes the methodology used to investigate: What are the gaps in mobile patient portal service to enable patient-centered care? Figure 1 shows the framework of the text mining-based method, which Al-Ramahi et al. (2017) adopted. We propose to use an unsupervised topic model, LDA, to extract latent dimensions (i.e., design gaps) from user-generated data. Below, we first discuss the data collection and preparation process. We then explain the topic modeling technique used to extract design gaps from online low ratings users' reviews.

### 3.1 *Data Collection and Preparation*

In this study, our target population is mobile patient portal users. The patient portal selected as empirical setting of this research is EPIC's mychart. We selected this patient portal for study as EPIC is replacing other vendors in the EHR market

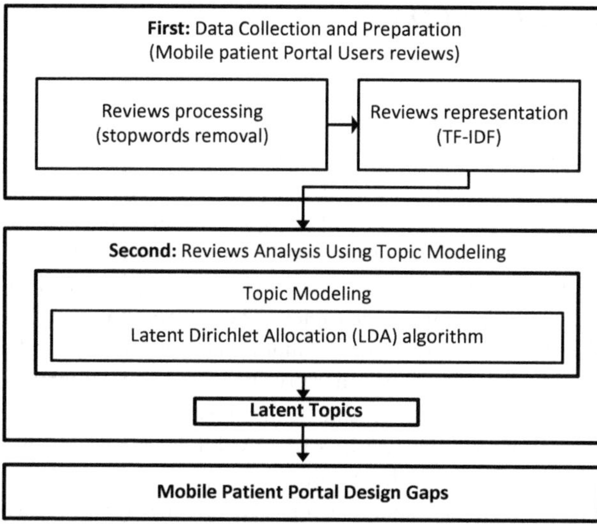


**Fig. 1** Architecture of our text mining-based method (Al-Ramahi et al. 2017)

and is beginning to establish a single vendor landscape. Reportedly, EPIC has at least partial health information for over 51% of the U.S. population (Koppel and Lehmann 2014). MyChart mobile app is available for Apple and Android devices. The data were collected from Apple iTunes store, where the online reviews posted by the users were gathered using the Apple store API. We developed a web crawler to automatically collect data. Through this process, we obtained our dataset consisting of 500 reviews. Since the main objective of this research is to identify design gaps of mobile patient portal, we focused on users' complaints contained in 1- or 2-star reviews. In comparison with the high-rated reviews, these low-rated reviews are more likely to reflect users' concerns and shed light on gaps that should be considered, but that, unfortunately, have been ignored in current research and practice. In total, we analyzed 258 1- and 2-star reviews. When preprocessing the data, we removed stop words and represented each document using the well-known term frequency inverse document frequency (TF-IDF) weighting scheme (Haddi et al. 2013). In order to adopt this weighting scheme, we treated each user review in the dataset as a document. Specifically, TF-IDF weight of a word *i* in a user review *j* is given by

$$F_{i,j} * \log(N/\mathrm{DF})$$

Where $F_{i,j}$ is the frequency of the word *i* in the user review *j*, *N* indicates the number of user reviews in the corpus, and DF is the number of user reviews that contains word *i*.

### 3.2 Topic Modeling: LDA

Topic models are statistical-based algorithms for discovering the main themes (i.e., set of topics) that describe a large and unstructured collection of documents. Topic models allow us to summarize textual data at a scale that is impossible to be tackled by human annotation. We selected the LDA model, the most common topic model currently in use, due to its conceptual advantage over other latent topic models (Blei et al. 2003). The model generates automatic summaries of topics in terms of a discrete probability distribution over words for each topic, and it also infers per-document discrete distributions over topics. The interaction between the observed documents and hidden topic structure is manifested in the probabilistic generative process associated with LDA. This generative process can be thought of as a random process that is assumed to have produced the observed document (Bao and Datta 2014). In order to illustrate the results of LDA, let $M$, $K$, $N$, and $V$ be the number of documents in a collection, the number of topics, the number of words in a document, and the vocabulary size, respectively. The first result is an $M \times K$ matrix, where the weight $w_{m,k}$ is the association between a document $d_m$ and a topic $t_k$. In our case, the documents are user reviews for patient portal mychart app (i.e., we integrated the reviews of the app in a data file and treated each user review as a single document) ($M = 258$). The second result is an $N \times K$ matrix, where the weight $w_{n,k}$ is the association between a word $w_n$ and a topic $t_k$. The notations *Dirichlet*(·) and *Multinomial*(·) represent Dirichlet and multinomial distribution with parameter (·), respectively. The graphical representation of LDA is shown in Fig. 2, and the corresponding generative process is shown below:

(1) For each topic $t \in \{1, \ldots, K\}$,
    (a) draw a distribution over vocabulary words $\beta_t \sim Dirichlet(\eta)$.
(2) For each document $d$,
    (a) draw a vector of topic proportions $\theta_d \sim Dirichlet(\alpha)$.
    (b) For each word $w_n$ in document $d$, where

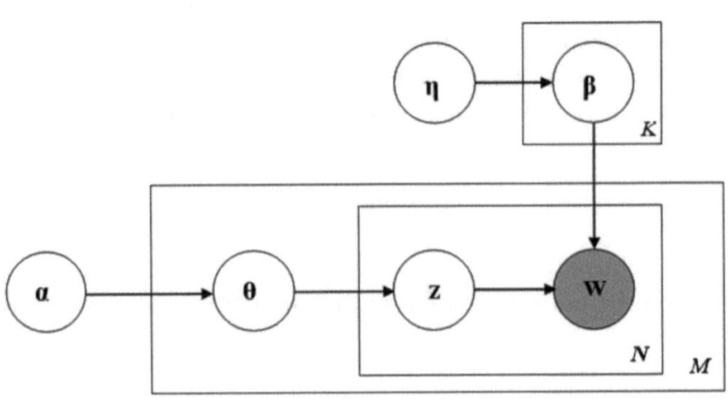


**Fig. 2** Graphical model of LDA

```
n∈ {1, ..., N},
(i) draw a topic assignment
    z_n ~ Multinomial(θ_d);
(ii) draw a word w_n ~ Multinomial(β_zn).
```

The notation $\beta_t$ is the $V$-dimensional word distribution for topic $t$, and $\theta_d$ is the $K$-dimensional topic proportion for document $d$. The notations $\eta$ and $\alpha$ represent the hyperparameters of the corresponding Dirichlet distributions.

# 4 Results

## 4.1 Gaps Discovered

In this section, we summarize and discuss the results of the extraction of the topics of the users' negative experience (i.e., gaps of mobile patient portal). We apply LDA to extract and label the topics of users' concerns across all collected low rating reviews of mychart portal in our sample. The LDA identified 25 topics and, within each topic, showed the top-10 words and their relative weight (i.e., probability). The naming of topics was first conducted by the first author and confirmed by the second author. Naming was initially based on the identification of a logical connection between these 10 most frequent words for a topic. For example, in Table 1, the topic name "Way to update and schedule appointment" is based on the words "Appointments," weighted 1.3%; "update," weighted 0.9%; and "Schedule," weighted 0.7%, which appear at the top 10 words. Once we specified a candidate topic label, we further tested it via investigating the reviews that are highly associated with that topic.

Then, we mapped the topics we obtained into 10 design gaps, which Table 2 shows along with examples from user feedback. The mappings between the topics and the gaps were sometimes one-to-one. For instance, the topic "notifications" was

**Table 1** Examples of identifying topics labels

| Topic 1: Way to update and schedule appointment | Weight (%) | Topic 2: Log in using touch ID | Weight (%) |
|---|---|---|---|
| Needs | 1.5% | ipad | 1.0% |
| Appointments | 1.3% | Beneficial | 0.9% |
| App | 1.0% | Available | 0.8% |
| Info | 0.9% | Log | 0.8% |
| Update | 0.9% | Id | 0.8% |
| Updated | 0.8% | App | 0.8% |
| Way | 0.7% | Touch | 0.8% |
| Unable | 0.7% | Alert | 0.8% |
| Schedule | 0.7% | Longer | 0.7% |
| Ability | 0.7% | Topic | 0.7% |

**Table 2** Identified gaps, supported by examples from user feedback

| Gap | Examples from the user feedback |
|---|---|
| Gap 1: Appointments management<br>This gap refers to patients' inability to manage their appointments over the portal. For examples, they cannot request, schedule, print, track appointments and cannot even view their appointment schedule in an appropriate format | Can't even schedule or request an appointment with it. Just downloaded the app, but could not schedule appointments on iPhone or iPad. You no longer can read your appointment schedule or any information in a normal sentence<br>Appointment layout horrible |
| Gap 2: Notifications/alerts<br>This gap pertains to the lack of providing patients with notifications/alerts when a doctor has new results or a message for them, or when they have an upcoming appointment | Needs notifications/alerts. Useless if not notified of the important messages received within the app. Unable to find any way to enable notifications<br>Please add an email alert option! I can't even get new notifications when my doctor has a new results or a message for me<br>I can't believe that there has been another update and still no push notifications |
| Gap 3: Integration with other health apps<br>This gap relates to not having the portal (Mychart) integrated with other health apps (e.g., FitBit and Apple HealthKit), so that patients can synchronize their health data both ways, from Mychart to health apps (like lab results) and from health apps to Mychart | No sync with Apple Health<br>The MyChart app should integrate with Health on iOS. Ideally, lab results would be sourced from MyChart and feed into Apple's Health. iOS Health data should be read by MyChart and placed in to a message or a report for my physician |
| Gap 4: Communication with health providers<br>This gap refers to patients' inability to communicate well with their health providers via the portal. In this regard, patients neither can exchange messages with health providers nor send them their test results | This app doesn't give you the option to reply when you get a msg from your doctor. You have to send a new msg<br>No way to send test results to doctors<br>I cannot send my doctor any messages<br>Actually makes communicating with your doctor harder<br>Need to add the feature of getting messages from doctors<br>I want to read sent emails |
| Gap 5: Security issues<br>This gap refers to the absence of the "Touch ID" feature, which allows patients to log in using their fingerprint, and to patients' inability to manage their password | Why isn't Touch ID available<br>I would really like Touch ID support for logging into my provider, instead of entering my password every time<br>It also lacks integration with popular password managers, such as 1Password<br>For many of us who have multiple changing passwords this app is a nightmare |

(continued)

**Table 2** (continued)

| Gap | Examples from the user feedback |
|---|---|
| Gap 6: Access and retrieve data<br>This gap refers to patients' inability to access and retrieve health related data like diagnostic or bloodwork info, weight, bp, medical tests or history | I have several health data about my sleep, exercises, and some more that my doctor would love to have access to<br>Can't access any of my diagnostic or bloodwork info<br>Cannot access my weight, bp, etc. record<br>Cannot access to any of the medical tests or history<br>Miss a lot of information especially in the health summary it's not an exact copy of your medical records<br>Cannot retrieve any data since most recent update |
| Gap 7: Informative presentation (data to knowledge presentation)<br>This gap refers to patients' inability to track their health with graphs/charts and to display their health data at multiple levels of aggregations, "drill-down" and "roll-up", on Dashboard | No ability to track health with graphs or charts<br>Instead of displaying data on Dashboard in discussions with doctors, I have to manually download data to a spreadsheet (unneeded waste of effort)<br>Also, the Health Summary: Heath Issues; section is a weird subset of random one liners, none of which are clickable to get more information. I wish it was much more complete with all the notes my doctors have written |
| Gap 8: Update medical data<br>This gap refers to patients' inability to update/correct/upload medical data, such as blood pressure | Upload updated insurance cards and be able to indicate which one is the primary and which one is the second one<br>It be great if I could update my shots and other medical issues<br>Gives no ability to patient/user to correct/update data. Have to request medical personnel to make changes<br>Should be able to upload common health metrics like blood pressure<br>Add support to upload pdf files |
| Gap 9: Install/open the app<br>This gap refers to patients' inability to install and open the app | Haven't been able to open the app for weeks now<br>I can't open or use this app<br>Problems downloading and installing<br>Cannot install on iPhone 7 plus |
| Gap 10: Communicating with server problems<br>This gap refers to patients' inability to connect to the server sometimes | Keeps saying cannot connect to server<br>Down more than a week. Can't communicate with the server<br>Will not let me sign in to my chart from Cleveland clinic, and it keeps saying Problem communicating with the server |

mapped to the gap "Notifications/alerts" and the topic "information summaries" to the gap "Informative presentation". There are also some gaps that correspond to multiple topics. For example, the "Appointments"-related topics (i.e., "way to update and schedule appointment", "Print appointments", "track appointments with providers", and "update appointment") were mapped into the gap "Appointments management."

## 4.2 *Validity of Gaps Discovered*

We examined the validity of the extracted gaps by comparing the results of LDA analysis with that of human analysis (see Table 3). In order to conduct the manual analysis, we adopted open coding technique for data analysis. Two independent researchers read the collected 1- and 2-star reviews and then identified the gaps which were mentioned in these reviews. We compared the gaps derived from the LDA analysis with the ones the two researchers identified, to calculate the reliability of the LDA result. The Jaccard coefficient[1] is 0.71 and 0.6 between the automated analysis and the two researchers, A and B respectively. As Table 3 shows, the manual coding of the data revealed four new gaps: "Export/Import data", that refers to the inability to export and save health data to files, "Support multiple health providers", "Technical support", and "Billing issues".

**Table 3** A comparison of gaps between LDA analysis and human analysis

| Gap | LDA analysis | Researcher A | Researcher B |
|---|---|---|---|
| Appointments management | √ | √ | √ |
| Notifications/alerts | √ | √ | √ |
| Integration with other health apps | √ | √ | √ |
| Communication with health providers | √ | √ | √ |
| Security issues | √ | √ | √ |
| Access and retrieve health data | √ | √ | √ |
| Informative presentation (data to knowledge presentation) | √ | √ | √ |
| Update medical data | √ | √ | x |
| Install/open the app | √ | √ | x |
| Communicating with server problems | √ | √ | √ |
| Export/Import data | X | √ | √ |
| Support multiple health providers | x | √ | x |
| Technical support | x | √ | √ |
| Billing issues | x | √ | √ |

[1]The Jaccard coefficient measures similarity between finite sample sets, and is defined as the size of the intersection divided by the size of the union of the sample sets.

# 5 Conclusion

The rising cost and decreasing quality of healthcare has raised the impetus toward the use of patient portals and EHRs, in order to overcome these issues with the increased transparency and efficiency that the technology allows. However, the challenges of patient portals have tempered efforts to improve efficiency of healthcare through the technology. This paper aims to investigate the gaps in mobile patient portal service to enable patient-centered care by analyzing the actual use of these systems. We adopt a text mining-based approach to leveraging online user reviews as a primary data source. Given the market prominence of EPIC's mychart patient portal, we use it as a problem domain.

The results of this research identified opportunities for improvement from patients' perceptions. Several gaps emerged, which reveal opportunities to enhance the design of portals intended for patient-centered care. These important findings can inform design decisions to promote use and foster engagement. Overall, optimizing a mobile patient portal will require careful attention to particular functionalities to enable collaboration, such as allowing patients to request and schedule an appointment by "Appointments management" and notifying them regarding new results and messages by "Notifications/alerts". It is also paramount to enable patients to "Communicate with health providers", "Export/Import" their health data, and use their fingerprint to log in "Security issues". Also, the portal should be "Integrated with other health apps" (i.e., fitness apps). We found also that patients were interested to having the ability to update their health data (e.g., shots and common metrics such as blood pressure) as well as update insurance information (i.e., insurance cards).

Regarding the portal content, patients described the utility of accessing and retrieving health data (e.g., test results and prescription) and patient's information (e.g., weight and blood pressure). This finding is consistent with O'Leary et al.'s (2016) research, demonstrating that hospitalized patients have a strong interest in all types of test results. It is also crucial that the portal provide patients with readable and informative graphs, reports, and charts of their health-related data, depicting their improvement patterns and historical trends in "Informative presentation". In order to achieve better quality of care, the patient portals can provide the transparency needed, as patients utilize the technology to support patient interaction, and enable patients to access the information they need to make better decisions about their healthcare. These changes will positively influence patient-centered care.

For the future, it is difficult to see anything other than refinements and growth of current healthcare strategies to utilize technology to improve patient engagement and support (Cliff 2012; Kane and Labianca 2011). The expansion of patient portals, chronic disease apps, and educational tools to support patients are expected to grow at increasing rates (Cohen et al. 2010; McAlearney et al. 2016; O'Leary et al. 2016). The use of connected health solutions is becoming standard practice among hospitals in the U.S., as 81% of hospitals leverage this type of IT (IOM 2001).

According to a 2016 HIMSS survey, 47% of respondents emphasized personal technology to influence patient satisfaction, treatment monitoring, patient engagement, and patient education. These individuals planned on continuing to grow in these areas (IOM 2001). Patients want to be engaged in their healthcare decision-making process, and those who are engaged as decision-makers in their care tend to be healthier and have better outcomes. However, the change from episodic-based care to life care will be a long transition. The way we pay for healthcare and the way we deliver healthcare are changing. Technology advancements will play a role in empowering and engaging people before they become patients. The culture change will be similar to other health culture changes like smoking and seatbelts.

This study contributes to the design and development of future technologies (Al-Ramahi et al. 2015). As technological advances continue, healthcare stakeholders agree: "Prepared, engaged patients are a key stepping stone toward high-quality care, lower costs, and better health."

Theoretically, this work contributes to the existing knowledge base of mobile patient portal design by presenting some existing opportunities for design enhancements and inferring new ones. Methodologically, this study exploits users' feedback in form of online reviews. In essence, the design of mobile patient portals requires understanding of users' perceptions and concerns. In this regard, user involvement is key in portal systems design, which can help shift the focus of innovation from pure technology to the context of daily life (Thackara 2001). Hence, we developed and presented opportunities for design enhancements based on users' reviews.

## References

Allan, T. Z., & Wang, Y. (2017). The demand for technical safeguards in the healthcare sector: A historical perspective enlightens deliberations about the future. In *Proceedings of 23rd Americas Conference on Information Systems (AMCIS 2017)*, Boston.

Al-Ramahi, M., Chang, Y., El-Gayar, O., & Liu, J. (2015). Predicting big movers based on online stock forum sentiment analysis. Paper presented at the *Twenty-first Americas Conference on Information Systems (AMCIS 2015)*, Puerto Rico.

Al-Ramahi, M., El-Gayar, O., & Liu, J. (2016). Discovering design principles for persuasive systems: a grounded theory and text mining approach. In *49th Hawaii International Conference on System Sciences (HICSS), 2016* (pp. 3074–3083).

Al-Ramahi, M. A., Liu, J., & El-Gayar, O. F. (2017). Discovering design principles for health behavioral change support systems: A text mining approach. *ACM Transactions on Management Information Systems (TMIS), 8*(2–3), 5.

Alzheimer's Association. (2013). *Alzheimer's disease: Facts and figures*. Chicago: The Association. Retrieved from http://www.alz.org/downloads/facts_figures_2013.pd.

Bao, Y., & Datta, A. (2014). Simultaneously discovering and quantifying risk types from textual risk disclosures. *Management Science, 60*(6), 1371–1391.

Blei, D. M., Ng, A. Y., & Jordan, M. I. (2003). Latent Dirichlet allocation. *The Journal of Machine Learning Research, 3*(Jan), 993–1022.

Blumenthal, D., & Tavenner, M. (2010). The "meaningful use" regulation for electronic health records. *The New England Journal of Medicine, 2010*, 501–504.

Chevalier, J. A., & Mayzlin, D. (2006). The effect of word of mouth on sales: Online book reviews. *Journal of Marketing Research, 43*(3), 345–354.
Cliff, B. (2012). Using technology to enhance patient-centered care. *Journal of Healthcare Management, 57*(5), 301–303.
Clifton, J. (2012). Healthcare is killing us. *Gallup Business Journal.* Retrieved from https://news.gallup.com/businessjournal/151862/Healthcare%20killing.aspx?utm_source =add+this&utm_medium=%20addthis.com&utm_campaign=sharing
Cohen, S. B., Grote, K. D., Pietraszek, W. E., & Laflamme, F. (2010). Increasing consumerism in healthcare through intelligent information technology. *The American Journal of Managed Care, 16*(12 Suppl HIT), SP37–SP43.
Dall, T. M., Gallo, P. D., Chakrabarti, R., West, T., Semilla, A. P., & Storm, M. V. (2013). An aging population and growing disease burden will require a large and specialized health care workforce by 2025. *Health Affairs, 32*(11), 2013–2020.
Dykes, P. C., Carroll, D. L., Hurley, A. C., Benoit, A., Chang, F., Pozzar, R., & Caligtan, C. A. (2012). Building and testing a patient-centric electronic bedside communication center. *Journal of Gerontological Nursing, 39*(1), 15–19.
Fichman, R. G., Kohli, R., & Krishnan, R. (2011). Editorial overview—The role of information systems in healthcare: Current research and future trends. *Information Systems Research, 22*, 419–428.
García-Gómez, J. M., de la Torre-Díez, I., Vicente, J., Robles, M., López-Coronado, M., & Rodrigues, J. J. (2014). Analysis of mobile health applications for a broad spectrum of consumers: A user experience approach. *Health Informatics Journal, 20*, 74–84.
Guerra-Reyes, L., Christie, V. M., Prabhakar, A., Harris, A. L., & Siek, K. A. (2016). Postpartum health information seeking using mobile phones: Experiences of low-income mothers. *Maternal and Child Health Journal, 20*, 13–21.
Guo, Y., Barnes, S. J., & Jia, Q. (2017). Mining meaning from online ratings and reviews: Tourist satisfaction analysis using latent Dirichlet allocation. *Tourism Management, 59*, 467–483.
Haddi, E., Liu, X., & Shi, Y. (2013). The role of text pre-processing in sentiment analysis. *Procedia Computer Science, 17*, 26–32.
Heidenreich, P. A., Trogdon, J. G., Khavjou, O. A., Butler, J., Dracup, K., Ezekowitz, M. D., et al. (2011). Forecasting the future of cardiovascular disease in the United States: A policy statement from the American Heart Association. *Circulation, 123*(8), 933–944.
Institute of Medicine (IOM). (2001). *Envisioning the National Health Care Quality Report.* Washington, DC: The National Academies Press.
Jung, Y., Hur, C., Jung, D., & Kim, M. (2015). Identifying key hospital service quality factors in online health communities. *Journal of Medical Internet Research, 17*(4), e90.
Kane, G. C., & Labianca, G. (2011). IS avoidance in health-care groups: A multilevel investigation. *Information Systems Research, 22*(3), 504–522.
Koppel, R., & Lehmann, C. U. (2014). Implications of an emerging EHR monoculture for hospitals and healthcare systems. *Journal of the American Medical Informatics Association, 22*(2), 465–471.
Lu, Y. H., Lee, L. Y., Chen, Y. L., Cheng, H. I., Tsai, W. T., Kuo, C. C., Chen, C. Y., & Huang, Y. B. (2017). Developing an app by exploiting web-based mobile technology to inspect controlled substances in patient care units. *BioMed Research International, 2017*, 22–34.
McAlearney, A. S., Sieck, C. J., Hefner, J. L., Aldrich, A. M., Walker, D. M., Rizer, M. K., Moffatt-Bruce, S. D., & Huerta, T. R. (2016). High touch and high tech (HT2) proposal: Transforming patient engagement throughout the continuum of care by engaging patients with portal technology at the bedside. *JMIR Research Protocols, 5*(4), e221.
Microsoft. (n.d.). Engage your patients [Infographic]. Retrieved from http://pages.healthcareitnews.com/rs/922-ZLW-292/images/Engage%20your%20patient%20infographic.pdf?alild=678046187
Murphy, J. (2011). Patient as center of the health care universe: A closer look at patient-centered care. *Nursing Economics, 29*(1), 35.

Neubeck, L., Coorey, G., Peiris, D., Mulley, J., Heeley, E., Hersch, F., et al. (2016). Development of an integrated e-health tool for people with, or at high risk of, cardiovascular disease: The Consumer Navigation of Electronic Cardiovascular Tools (CONNECT) web application. *International Journal of Medical Informatics, 96*(December 2016), 24–37.

O'Leary, K. J., Sharma, R. K., Killarney, A., O'Hara, L. S., Lohman, M. E., Culver, E., Liebovitz, D. M., & Cameron, K. A. (2016). Patients' and healthcare providers' perceptions of a mobile portal application for hospitalized patients. *BMC Medical Informatics and Decision Making, 16*, 123.

Pfeifer Vardoulakis, L., Karlson, A., Morris, D., Smith, G., Gatewood, J., & Tan, D. (2012). Using mobile phones to present medical information to hospital patients. In *Proceedings of the SIGCHI Conference on Human Factors in Computing Systems, 2012* (pp. 1411–1420), Austin.

Phillips, P., Barnes, S., Zigan, K., & Schegg, R. (2016). Understanding the impact of online reviews on hotel performance: An empirical analysis. *Journal of Travel Research, 56*(2), 235–249. https://doi.org/10.1177/0047287516636481.

Pierleoni, P., Pernini, L., Belli, A., & Palma, L. (2014). An android-based heart monitoring system for the elderly and for patients with heart disease. *International Journal of Telemedicine and Applications, 2014*, 10.

Qureshi, S., & Keen, P. (2005). Activating knowledge through electronic collaboration: Vanquishing the knowledge paradox. *IEEE Transactions on Professional Communication, 48*(1), 40–54.

Qureshi, S., & Noteboom, C. (2006). Adaptation in distributed projects: Collaborative processes in digital natives and digital immigrants. In *Proceedings of the 39th Annual Hawaii International Conference on System Sciences, 2006, HICSS'06* (pp. 202c–202c), HI, USA.

Qureshi, S., Liu, M., & Vogel, D. (2005). A grounded theory analysis of e-collaboration effects for distributed project management. In *Proceedings of the 38th Annual Hawaii International Conference on System Sciences, 2005, HICSS'05* (pp. 264c–264c), Big Island, HI, USA.

Qureshi, S., Noteboom, C., & Schumaker, A. M. (2015). Mobile access for patient centered care: The challenges of activating knowledge through health information technology. In *48th Hawaii International Conference on System Sciences, 2015, HICSS* (pp. 3227–3236), HI, USA.

Robinson, J. H., Callister, L. C., Berry, J. A., & Dearing, K. A. (2008). Patient-centered care and adherence: Definitions and applications to improve outcomes. *Journal of the American Academy of Nurse Practitioners, 20*(12), 600–607.

Sacristán, J. A. (2013). Patient-centered medicine and patient-oriented research: Improving health outcomes for individual patients. *BMC Medical Informatics and Decision Making, 13*, 6.

Stewart, M., Brown, J., Donner, A., McWhinney, I., Oates, J., Weston, W. W., & Jordan, J. (2000). The impact of patient-centered care on outcomes. *The Journal of Family Practice, 49*(9), 796–804.

Suzman, R., & Beard, J. (2011). *Global health and aging*. National Institute on Aging. Retrieved from https://www.nia.nih.gov/sites/default/files/2017-06/global_health_aging.pdf

Thackara, J. (2001). The design challenge of pervasive computing. *Interactions, 8*(3), 46–52.

Xu, Y., Armony, M., & Ghose, A. (2016). The effect of online reviews on physician demand: A structural model of patient choice. *SSRN Electronic Journal, 6*, 132–135.

Yang, J., Sarathy, R., & Lee, J. (2016). The effect of product review balance and volume on online shoppers' risk perception and purchase intention. *Decision Support Systems, 89*, 66–76.

Yu, Y., Duan, W., & Cao, Q. (2013). The impact of social and conventional media on firm equity value: A sentiment analysis approach. *Decision Support Systems, 55*(4), 919–926.

Zhang, M. W., Ho, R., Cassin, S. E., Hawa, R., & Sockalingam, S. (2015). Online and smartphone based cognitive behavioral therapy for bariatric surgery patients: Initial pilot study. *Technology and Health Care, 23*(6), 737–744.

# A Lazy User Perspective to Patient Adoption and Use of Personal Health Records

K. Niki Kunene

## 1 Introduction

The adoption and use of personal health records is expected to benefit the individual patient, contribute overall to the healthcare system via cost savings and cost avoidances, and contribute to population health from the potential for a large, minable, or query-able data set.

Personal health records (PHRs) in contrast with electronic health records (EHRs) were conceived as patient-controlled records. Personal health records were intended to give the individual consumer the ability to manage, track, share, and participate in his/her own healthcare (Jones et al. 2010; Kaelber et al. 2008; Tang et al. 2006; Vydra et al. 2015). Personal health records, known to many as patient portals, are regarded as the bedrock of patient engagement.

In practice, we have seen two types of implementations: those records that are tethered to healthcare providers' electronic health record (EHR) systems and those that are untethered. Typically, untethered manifestations allow for the entry of patient-generated health data (PGHD). Tethered systems may or may not allow for some patient-generated data. Tethered PHRs are by definition provided and maintained by healthcare covered entities such as hospitals, doctor groups, health insurance companies, and even employers. Untethered PHRs are provided by private vendors who may or may not charge a use fee (Jones et al. 2010).

Policymakers and researchers have written and acted on the understood benefits of using personal health records. Following the HITECH Act of 2009, five objectives of the meaningful use (MU) of electronic healthcare records were articulated in the realization of the act. Three of these objectives (therein numbered 2–4)

K. N. Kunene (✉)
Eastern Connecticut State University, Willimantic, CT, USA
e-mail: kunenek@easternct.edu



N. Wickramasinghe, F. Bodendorf (eds.), *Delivering Superior Health and Wellness Management with IoT and Analytics*, Healthcare Delivery in the Information Age, https://doi.org/10.1007/978-3-030-17347-0_30

are especially pertinent to PHR benefits: (a) engaging patients and families in their health, (b) improving care coordination, and (c) improving population and public health. (The first objective calls for improving quality, safety, efficiency, and reducing health disparities (Stage 3). Objective 5 calls for ensuring adequate privacy and security protection for personal health information.) Wynia and Dunn (2010) also identified the expansion of health education opportunities and strengthening disease prevention as potential PHR benefits. PHRs also offer patients greater access to a wide and customizable array of credible health information, data, and knowledge (Tang et al. 2006). Mobile personal health records (mPHRs) can help in case of emergencies when a patient sees a new provider, or where the patient's primary EHR is not accessible, or interoperable with the new provider's systems (Bouri and Ravi 2014). One design researcher suggested the PHR as a landing solution for overcoming the lack of interoperability between providers' EHRs that serve the same patient.

PHRs have been around for some time, and with beliefs in their various potential benefits holding firmly among policymakers and researchers, efforts to encourage their adoption were well under way by 2014. As a result, patient access to PHR has been growing rapidly. Lester et al. (2016) reported that over 70 million consumers had access to tethered PHRs in the United States. In addition, free access to untethered PHRs, like Microsoft's HealthVault, has existed for over a decade. Legally, the implementation of Meaningful Use Stage 2 mandates providers to use the technical capabilities offered by PHRs, for example, communicate electronically with patients and provide patients the ability to view, download, and transmit their electronic health information within days. Providers have been responding by creating patient portals (tethered PHRs) where EHR health information can be viewed but frequently not modifiable by the patient. Some providers may even show the patient how to log into the portal during patient visits. And yet patient use of personal health records remains low, despite policy efforts to promote their use (Vydra et al. 2015; Wynia and Dunn 2010). Google pulled its untethered platform Google Health in 2012 citing a lack of widespread adoption.

Researchers and policy experts continue to deliberate over whether PHRs will ever gain the necessary adoption or use rates to make the systems effective (Lester et al. 2016; Vydra et al. 2015; Wynia and Dunn 2010). Several barriers to adoption exist. These barriers impact the decision to start using a PHR, the adoption process and the continued use of PHRs (Showell 2017). Patient-centered barriers to adoption include concerns for information privacy (Carrión et al. 2011; Lehnbom et al. 2016; Señor et al. 2012), patient awareness, and/or interest (Vance et al. 2014); patients' ability to understand medical records (Vydra et al. 2015). From a provider-centered perspective, the lack of provider reimbursement for time spent in portal communication (Lester et al. 2016), and the response time required of providers (Vydra et al. 2015). Furthermore, from a social and economic perspective, a digital divide (Yamin et al. 2011), socioeconomic status, race, and education are also important factors in the adoption and diffusion of PHRs (Showell 2017).

A lot of quantitative research has been done on the adoption of PHRs; there are however few qualitative and recent studies on the adoption of PHRs; there are fewer still that have generalized to theory. The purpose of our research project is to get an in-depth qualitative understanding of (1) how individual healthcare users in New England, USA currently manage their personal health information, (2) whether they have adopted an electronic personal health record (PHR) and the factors that influenced their voluntary adoption or non-adoption of a PHR. This study uses grounded theory, an interpretive methodology. The data collection strategy is semi-structured interviews. Two generative open-ended questions formed the basis of the interview: How do consumers currently manage their personal health information? Do they use personal health records and what dissuades or encourages them to use PHRs? Data was collected between spring of 2015 and spring of 2016. Interview transcripts were coded iteratively through open, axial, and selective coding. Kunene et al. (2016) reported on the initial thematic categories emerging from the data. In this study, we use grounded theory to marry axial themes emerging from the data to a universal central theme or theory. We found the lazy user theory (Tétard and Collan 2009) has better explanatory capacity from first-level constructs for the low adoption rates.

The contribution of the study is it presents a new and alternative understanding of the low adoption rates associated with personal health records. Will legislative imperatives driving the adoption of these technologies by providers translate to patient adoption and use? All things being equal, our findings suggest not. Changes will have to target incentives for user laziness.

The rest of the chapter is organized as follows: Section 2, background literature where we summarize related prior research in the literature and the role of legislation particularly the meaningful use of certified electronic health records, and review theories and models of technology acceptance and adoption. Section 3 presents our research questions and methodology. Section 4 provides results and discussion and Sect. 5 presents the conclusion.

## 2 Background Literature

The first articulation of a personal health record (PHR) in an academic journal dates back to 1969 (Kim et al. 2011). Early references to a PHR referred to information about one's health on some form of paper. In the early 1970s, some studies also started to reflect the need for personal records to address community or social needs; for example, the problems associated with data collection in monitoring and managing maternal and child health in developing countries (Wunsch 1982). By the 1990s, we started to see a shift toward emphasizing the personal health record as a patient-centered and patient-controlled record. In other words, a patient record as a patient-held record in contrast to the health record controlled by the healthcare provider. Later, in Australia, for example, we saw the PHR record first introduced as a personally controlled electronic health record (PCEHR). In the United States, personal health records are frequently known as patient portals, and are typically tethered to a provider's electronic health record. Commercially produced untethered

PHRs do exist, e.g., Microsoft's HealthVault; the low adoption and use rates have led to private companies like Google and Microsoft to abandon their platform or reduce their support. On the other hand, belief in their potential has not dimmed among researchers and policymakers; emerging developments in personal health records include the advent of blockchain-secured, patient-controlled health records (Borfitz 2018).

The benefits of using personal health records are based on a clinically well-understood phenomenon that patient participation in their own care is associated with better quality of care and patient outcomes. Information sharing between the clinician and the patient is critical to successful patient engagement (Houston and Ehrenberger 2001). The use of personal health records is seen as the bedrock of patient engagement (Ancker et al. 2015). With personal health records, patients can participate in their own care by tracking, managing, and sharing information about their care (Jones et al. 2010; Kaelber et al. 2008; Tang et al. 2006). The capacity to share information by patients can also improve care coordination among their providers. Further, a widespread use of electronic personal health records is projected to offer opportunities to strengthen disease prevention, improve population health, and expand health education (Department of Health and Human Services 2010). In general, researchers and policymakers seem persuaded about the potential value of PHRs to patients and indeed the healthcare system as a whole. Furthermore when asked, studies show many patients do believe that using PHRs can help them better manage their health and improve care quality (Abramson et al. 2014; Lester et al. (2016) and that this belief is generalizable, including among minority racial/ethnic groups and people at low-income levels (Patel et al. 2011). Yet adoption rates in the United States have remained low even with the now widely distributed access to personal health records.

## 2.1 *Managing Electronic Personal Health Information (ePHI)*

Prior research on PHRs seeks to understand consumer perspectives toward personal health records within communities, in light of the low adoption rates. A majority of these studies are quantitative, survey studies; a few are structured reviews of the literature; fewer still are qualitative studies or experimental trials. The shared understandings found in the literature include privacy and security concerns (Abramson et al. 2014; Carrión et al. 2011; Kaelber et al. 2008; Kavoussi et al. 2014; Patel et al. 2011; Señor et al. 2012) as a barrier to adoption; socioeconomic differences, e.g., the digital, racial/ethnic, and income divide (Yamin et al. 2011; Zapata et al. 2014), which can divide groups into adopters and non-adopters; and the intersection of age and comfort with technology that is a factor among middle-aged and older patients (Taha et al. 2013). Physicians can serve as an adoption influencer, so that the rate of adoption of electronic health records (EHRs) and the variances in physician willingness to use technology in managing patient care influences patient adoption of PHRs (Archer et al. 2011). The interoperability problem, a technical challenge that affects both PHRs and EHRs (Lester et al. 2016) may be a barrier to adoption, where patients are granted access to multiple non-interoperable, tethered PHRs or patient portals.

Young et al. (2014) looked at the barriers to PHR adoption among older adults in home-based care. They interviewed adults in the age range of 46–72 years and found that barriers to adoption were characterized by four factors: (1) privacy and security concerns; (2) general technological discomfort which made participants view the idea of electronically communicating with healthcare providers of "dubious value" (Young et al. 2014); (3) lack of relative advantage, where the use of PHRs or their equivalent were not perceived to be more advantageous to paper; and (4) an undesirable user representation – where participants imagined a user of a PHR as someone infirm or with a chronic condition or someone irresponsible requiring constant reminders, in other words, someone unlike themselves.

Mitchell and Begoray (2010) also show that patients with serious and/or long-term illnesses can benefit the most from these systems (Mitchell and Begoray 2010). Yamin et al. (2011) found patients with comorbidity were more likely to use a patient portal. Wagner et al. (2012), in a cluster-randomized effectiveness trial with PHR and non-PHR groups, investigated the impact of using PHRs among hypertensive patients on multiple outcomes. Wagner et al. (2012) observed no PHR effects on blood pressure (BP). They found that few patients with access to a PHR actually used one with any frequency. They conclude that merely providing a PHR had no impact on the defined outcomes: namely, blood pressure, empowerment, satisfaction with care, or use of healthcare services without additional education, or the clinical interventions targeted at increasing PHR use (Wagner et al. 2012).

Some studies have looked at the barriers to adoption from a physician perspective. Vydra et al. (2015) found there was mismatch of the time physicians spent on portals and the lack of compensation for that time which acted as a barrier to PHR use. In a structured review of existing literature, Lester et al. (2016) also found patients' lack of understanding of medical records, legal liability, as well as the time required for physicians to respond to PHR-related queries to be a concern and burdensome to physicians.

### 2.1.1 The Role of the Meaningful Use (MU) on Certified Electronic Health Records

Following the imperatives of the HITECH Act of 2009, the Centers for Medicare and Medicaid Services (CMS) instituted the EHR Incentive Programs, now called the Promoting Interoperability (PI) Programs, to encourage the so-called eligible professionals (EPs), eligible hospitals, and critical access hospitals (CAHs) to adopt, implement, upgrade, and demonstrate meaningful use of certified EHR technology (CEHRT) in three progressive stages (Services 2018). The first stage establishes requirements for the electronic capture of patient health information. Stage 2 expands on the first stage and focuses on clinical processes and aligns meaningful use (MU) of EHRs with the goals and priorities of the National Quality Strategy

(NQS) of the Patient Protection and Affordable Care Act (PPACA). Stage 3 focuses on using CEHRT to improve health outcomes. Of significance to our research, both MU Stage 2 and the National Quality Strategy call for the patient engagement. From the six NQS priorities, the second priority states, "ensuring that each person and family is engaged as partners in their care" (Agency for Healthcare Research and Quality 2017). As hospitals and eligible professionals started implementing MU Stage 2 requirements, they were explicitly obligated to give patients the ability to view online, download and transmit their health information within days of an encounter as well as the ability to interact with the health system electronically. The adoption of patient portals by providers rose significantly with MU Stage 2. With regulatory compliance driving the adoption of personal health record by providers and near global access for patients, will actual use of the technology by patients follow?

Data from the American Hospital Association (AHA) shows the progressive increase in adoption rates of personal health records between 2012 and 2016/17 (Association 2018). Figure 1 shows the adoption of technologies that allow patients to view their health information, download it, and have the capacity to share it with a designated caregiver (see Fig. 1). What is interesting for our study about Fig. 1 is, around 2014, the adoption rates by hospitals more than doubled; this is consistent with the timelines for penalties and incentives associated with the requirements for Meaningful Use Stage 2. By 2014, even laggards in the adoption of CEHRT were approaching 2–3 years of having met MU Stage 1 requirements, implying they could proceed to Stage 2 requirements.

However, this AHA report speaks to the technological capability offered by hospitals rather than actual use by patients. According to the Office of the National Coordinator (ONC), as of 2017, 52% of individuals have been offered access to their

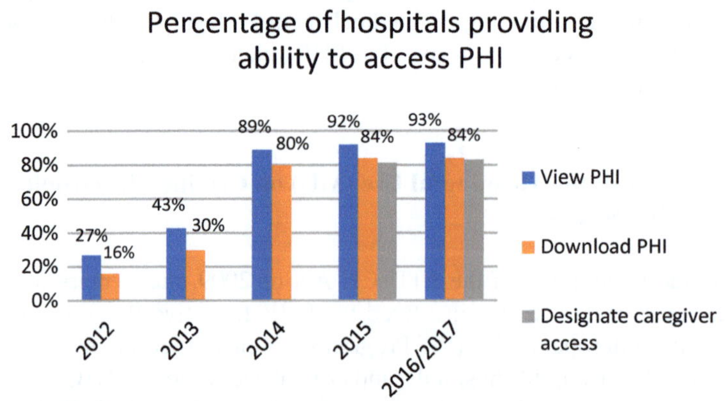


**Fig. 1** Adoption of personal health records by American Hospitals

medical record by a health provider or health insurer. Of those, over half *viewed*[1] their online record within the past year. This represents 28% of individuals in the United States (Patel and Johnson 2018).

Is it therefore the case that if you build a personal health record, they will come? Will the widespread availability of and access to PHRs encourage users to use personal health records and realize their potential benefits? Human-computer interaction (HCI) researchers postulate that the adoption of PHRs may be subject to captology (Saparova 2012). Captology is a theoretical framework in HCI research consistent with build it and they will come. Captology conceives of computers as "persuasive technologies" that can motivate, influence, and persuade users toward the adoption of target behaviors (Saparova 2012) or engineered behaviors (Alter 2010), i.e., patient engagement in their own care. Will MU Stage 2 lead to the captology of PHRs? Or is there something else at play. Tang et al. (2006) argued for education and research to unravel the barriers to adoption and use claiming we "do not know enough about health care consumers' need for and potential use of PHRs" (p. 55).

## 2.2 *Theories of Technology Adoption and Diffusion*

Theoretically, our work is related to technology adoption and diffusion models. Specifically, this study is supported by the Lazy user theory (Tétard and Collan 2009) to technology adoption. There are several technology adoption and diffusion models for studying the adoption, diffusion, and acceptance of technology and technical devices; these include the theory of reasoned action (Ajzen and Fishbein 1975), the theory of planned behavior (Fishbein and Ajzen 1977), the technology acceptance model (Bagozzi et al. 1992; Davis 1989), the unified theory of acceptance and use of technology (UTAUT) (Venkatesh et al. 2003), task-technology fit (Goodhue 1995; Goodhue and Thompson 1995), Roger's diffusion of innovations (Rogers 2003), and cognitive fit theory (Vessey 1991; Vessey and Galletta 1991). The most widely used of these has been the technology acceptance model (TAM) and its variants. More recently, Tétard and Collan (2009) proposed a lazy user theory to technology adoption research.

**Theory of Reasoned Action (TRA)** TRA posits three constructs: behavioral intention, attitude, and subjective norms, where behavioral intention is dependent on attitude toward the behavior and subjective norms (one's beliefs about the normative expectations of others and the motivation to comply with these expectations) (Ajzen 1985). In a meta-analysis, Sheppard et al. (1988) show the limitations of TRA in the

[1]In attempts to comply with the law, physicians and physician groups made patients enroll to the patient portal during a patient visit.

context of choice-making in the presence of multiple alternatives (Sheppard et al. 1988; Tétard and Collan 2009).

**Theory of Planned Behavior** The theory of planned behavior extended the theory of reasoned action adding the construct perceived behavioral control and actual behavior, where intention is the immediate antecedent of behavior. Beliefs about the presence or absence of factors that may facilitate or hinder the performance of the behavior and the perceived power of these factors give rise to perceived behavioral control. In general, "the more favorable the attitude and subjective norm, and the greater the perceived control" (Ajzen 1985), the stronger the intention to perform the behavior in question.

**Technology Acceptance Model** TAM introduced by Davis (1989) is an adaption of TRA tailored for modeling the user acceptance of information systems. A key purpose of TAM was tracing the impact of external factors on internal beliefs, attitudes, and intentions (Davis et al. 1989). TAM posits two distinct beliefs perceived usefulness (U) and perceived ease of use (EOU). Perceived usefulness is defined as the probability that using a specific system will increase one's job performance within an organizational context. PEOU is the degree to which a user expects, using the system to be free of effort. TAM is a frequently used model in information systems research; it has however been criticized for its low predictive value (Tétard and Collan 2009).

**Diffusion of Innovations** Rogers (2003) in diffusion of innovations defines adoption as a decision to fully "use an innovation as the best course of action available" and innovation as "the process by which an innovation is communicated through certain channels over time among members of a social system" (p. 5). The four key components of a diffusion of innovations by this definition are the innovation, communication channel, time, and social system. Further, the innovation decision-decision process is defined as an information-seeking and information-processing activity involving five steps: knowledge, persuasion, decision, implementation, and confirmation, and usually in that sequence over time. A user may adopt or reject an innovation during the decision stage, i.e., after gaining awareness knowledge, how-to knowledge, and principle knowledge about the innovation and developing a positive or negative (affective) attitude about the innovation. Roger's diffusion of innovations is better suited as process theory than a variance model.

**Unified Theory of Acceptance and Use of Technology (UTAUT)** Proposed by Venkatesh et al. (2003), UTAUT is a model that sets out to consolidate eight prior models including TRA, theory of planned behavior, TAM and one of its variants, and social cognitive theory. The authors derive four key constructs that have a direct effect on intention to use: performance expectancy, social influence, effort expectancy, and facilitating conditions. Age, gender, experience, and voluntariness of use are modeled as key moderators.

**Lazy User Theory** Tétard and Collan (2009) proposed the lazy user theory as a technology adoption model in information systems as an alternate model for dealing with choice among alternatives. Important for our study, the model does not presuppose a user as an employee in an organizational context; the model focus is more user-centered. The model assumes users tend to follow a "path of least resistance"; that is, given options, they will exert the least effort or energy to fulfill their need. Though new to information systems, the concept of "path of least resistance" has existed in physics, linguistics, informatics, and information seeking (Tétard and Collan 2009). In linguistics for instance, Zipf (1949) explaining the scaling of human language called his theory the principle of least effort, where speakers seek to minimize articulatory effort, and hearers also want to minimize the effort of understanding. The lazy user model has also been used in information retrieval as one of the principles guiding information seeking behavior and consequently the design of modern libraries (Collan and Tétard 2011). In finance, "lazy banking" is used to explain the decision-making associated with the complexity inherent in evaluating risk connected to credit disbursal when the bank may not have the necessary expertise to screen potential borrowers; thus lazy banking becomes a manifestation of risk averseness (Bhaumik and Piesse 2008).

The lazy user theory posits a user will most often choose the solution that will fulfill their (information) needs with the least effort (lazy user behavior):

> The lazy user theory of solution selection tries to explain how an individual (user) makes her selection of solution to fulfill a need (user need) from a set of possible solutions (that fulfill the need). The set of possible solutions is a subset of universal solutions that is constrained (limited) by the user state (circumstances) ... (Tétard and Collan 2009).

Graphically, the model is shown in Fig. 2.

Figure 2 shows that the set of possible solutions available to the user are defined by a user need and limited by the circumstances, e.g., resources and time, of the user (user state). The selection of a solution is based on assessing a possible solution set and effort required, so that the solution meeting then need with the lowest effort is prioritized.

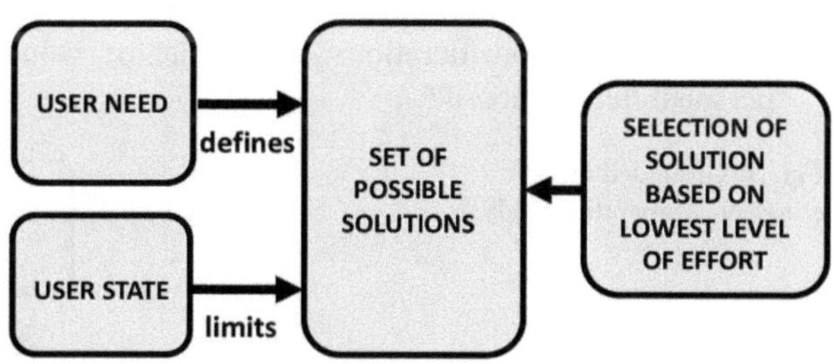


**Fig. 2** The lazy user theory (Tétard and Collan 2009)

## 3 Research Questions and Methodology

This study is a qualitative, interpretive study where study participants do not merely serve as a conduit of information but also participate in meaning-making (DiCicco-Bloom and Crabtree 2006). Data is collected using semi-structured interviews. The research method is grounded theory (Strauss and Corbin 1990; Strauss and Corbin 1994). Grounded theory is an interpretivist methodology; the purpose of using grounded theory is to construct a cohering or universal story by unearthing, from the data, a central theme or category from multiple emergent themes.

A central theme has analytic power: "the ability to pull other categories together to form an explanatory whole" (p. 146) while accounting for variation within categories (Strauss and Corbin 1998). Grounded theory is appropriate for the development of theory grounded in data that is systematically gathered and analyzed where theory then evolves (ibid).

As part of constant comparative analysis, three types of coding are used, i.e., open coding, axial coding, and selective coding (see Fig. 2). Open coding is used for preliminary segmenting of the data, axial coding for generating "codes" or themes, and selective coding for analyzing themes into cohered understandings or theory. Constant comparative analysis allows the researcher to iterate through the interplay between data collection and analysis for sense-making and theory development (Fig. 3).

Because the objective of this study is to understand what incentivizes healthcare consumers to use PHRs, and what dissuades them, we needed to first understand how they were currently managing their personal health data. Following grounded theory, we started with two generative questions as the basis for inquiry during our semi-structured interviews:

i. How are you currently managing all your healthcare data (prescription data; medical bills, test and lab results, historical medical data)?
ii. Do you use a personal health record (PHR)? What considerations incentivize (encourage) or would incentivize (encourage) you to use a personal health record? What considerations discourage or would discourage you to use a personal health record?

**Fig. 3** Grounded theory: constant comparative analysis

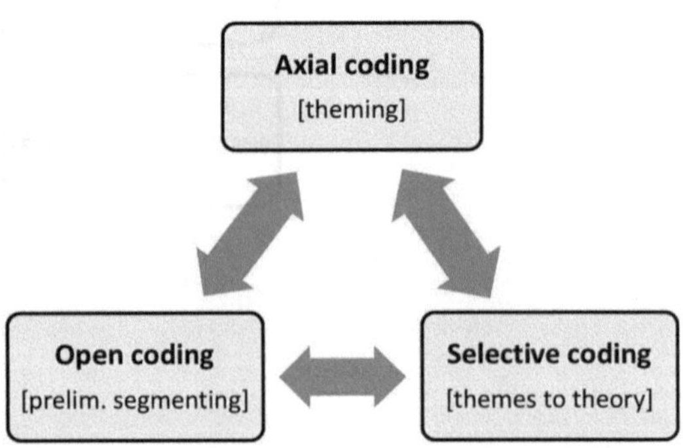

**Table 1** Age and gender descriptives

| Age | | | Gender | | |
|---|---|---|---|---|---|
| age group | N | % | | N | % |
| 20–34 | 40 | 47% | | | |
| 35–49 | 34 | 40% | Female | 46 | 54% |
| 50–65 | 11 | 13% | Male | 39 | 46% |
| Total | 85 | | Total | 85 | |

## 3.1 *Data Collection*

The research was carried out in New England, USA. Data collection interviews were conducted in three periods: spring 2015, fall 2015, and spring 2016. The study was approved by our institutional IRB. The only exclusion criterion was, subjects had to be 21 years or older. An average of 28 subjects are interviewed in each round. Each interview lasted between 20 and 60 minutes and was recorded for transcription.

## 3.2 *Data Analysis*

Two researchers coded each of the interview scripts. For open coding the initial categories were age, gender, education, prior knowledge of PHRs, privacy attitudes, security attitudes, and current state of health. Emerging themes (axial coding) from the abovementioned categories are partially shown in Table 3. With selective coding, we integrate and refined categories to a more abstract or universal concept or theory from the data. We kept an open mind with respect to mapping back to existing technology adoption theory as a central concept.

### 3.2.1 Subjects

Eighty-five subjects have been interviewed over a period of over 18 months. Demographically the gender split was 54% female and 46% male. Nearly half were under 35, 40% between 35 and 49, and 13% in the 50–65 age group (Table 1).

# 4 Results and Discussion

A majority of our respondents considered themselves to be in good health; only five explicitly disclosed a chronic health condition that required ongoing treatment: anxiety, type I and II diabetes, thyroid disease, severe sinusitis, and celiac disease, plus another with an undisclosed serious prior illness. At the same time, only 55% knew what a PHR or patient portal was, and the same number reported being aware

**Table 2** More summary information about our respondents

| | |
|---|---|
| Self-reported relative wellness | Majority report being healthy<br>6 indicated chronic conditions |
| Knowledge of a PHR | 55% knew what a PHR/patient portal was or recognized it when it was described to them |
| Access to and use of a PHR | 55% knew they had access to a tethered one or thought they had access to one<br>7% of the 85 subjects were actually using a PHR in some form |
| Primary care physicians using CEHRT | Nearly 96% reported yes<br>Others could not say definitively<br>Dentists were reported to be using paper |
| Use of health or wellness app | 63.5% had used or were using a health or wellness app for diet and activity tracking |

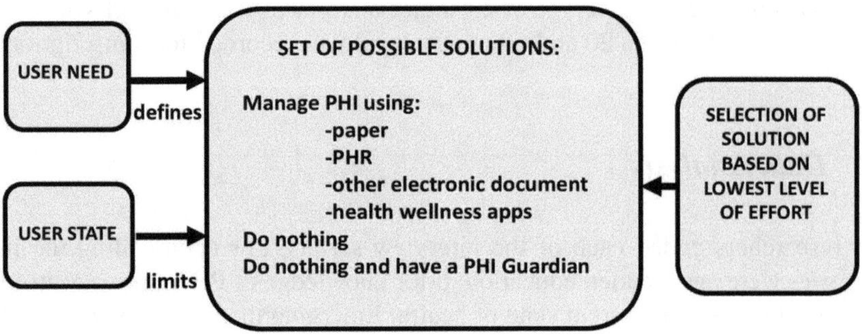


**Fig. 4** Lazy user analysis model for PHR adoption

of having access to a PHR. Nearly all our respondents reported having a primary care physician who had adopted a certified EHR technology (CEHRT). Interestingly, 54 respondents said they were using or have used a health or wellness app at some point (Table 2).

## *4.1 The Lazy Personal Health Information User*

In this study, we focus especially on the first question, "how are you currently managing your personal healthcare data (prescription data; medical bills, test and lab results, historical medical data?" However, Kunene, Zysk, and Diop (2016) discuss some of the emergent themes from both generative questions. The first question explores the array of possibilities available to the respondent for managing personal health information (PHI), and the solutions they choose. Using grounded theory's constant comparative method, we iterated the lazy user theory (Tétard and Collan 2009) as the universal, organizing central theme of our data (Fig. 4).

In Fig. 3, the user need is managing personal health information (PHI) to support one's quality of health and quality of care.

**User Need** Managing personal health information.

*The user state* is a description of the user and their context at the time of need. The context for healthcare consumers is: relative state of health, this includes the presence or absence of chronic diseases particularly; but it also encompasses relative concern for health wellness which in our case may be reflected via reported diet and exercise behaviors and/or concerns. It also includes the healthcare consumer's age. It may include gender; women may feel the need to monitor their menstrual cycle.

**User State** Relative health and wellness; the presence or absence of chronic disease; diet/exercise behaviors and concerns; age.

*The set of possible solutions* for managing personal health information for healthcare consumers ranges from traditional paper solutions where users keep some file or folder related to their personal health information (some of our older respondents do). A second option is, maintaining electronic records of some kind, some of the respondents reported "keeping files on the computer." A third and fourth option is using a personal health record, and here one can distinguish between PHRs tethered to provider EHRs and untethered PHRs like Microsoft's HealthVault. Each option comes with its own set of challenges and conveniences. For instance, in the United States, tethered PHRs open a consumer to the very real possibility of having multiple PHRs each associated with a different healthcare provider (i.e., until the interoperability issue is settled). An untethered PHR comes with its own set of constraints: learning investment; privacy and security challenges, whether real or perceived. A fifth option is a partial solution to managing PHI, but does reflect a degree of patient engagement, that is the use of heath wellness mobile apps and wearables that electronically record and store information, and frequently on ubiquitous cloud storage. There are associated learning investment constraints in general, but these may be perceived as transferable learning investments (Tétard and Collan 2009). Many of our female respondents in particular used at least one health wellness app or wearable. The sixth option is to simply do nothing about managing one's personal health information. Finally, the seventh option is to do nothing and simultaneously have a guardian or agent manage your personal health information for you. The PHI guardian is the healthcare provider, particularly the primary care physician; it may also be the health plan or insurance.

**Possible Solutions** Paper management; electronic documentation; tethered PHR; untethered PHR; health and wellness apps and wearables; do nothing; do nothing and have a guardian or agent acting on your behalf.

## 4.2 Results

Iterating between open codes and axial coding, for the initial question, "how are you currently managing your personal health information?" the following emerges. Notwithstanding the rollout of tethered patient portals as providers started looking toward complying with Meaningful Use Stage 2 requirements; a small minority of patients were actually using personal health records. Instead, a significant majority of our respondents (across age and gender groups) were not in fact using PHRs. See Figs. 5, 6, and 7 in the Appendix for additional responses. An overwhelming number responded to effect, "I don't, my doctors do it for me." A number of respondents equated visiting the primary care physician to monitoring or managing their personal health information. A handful of respondents maintained some form of paper record, and fewer still maintained their records using other electronic documentation. Over half the respondents (47) knew they had access to a patient portal; however actual use of the portal was negligible; only six study patients overall said they used a PHR in some form or another; another admitted to partially registering for a PHR.

We had also expected that patients with chronic conditions would be more encouraged to monitor their own health and therefore had some incentive to use a personal health record. Three of the six participants with a chronic condition actually used a PHR.

Over all three data collection periods (spring 2015, fall 2015, spring 2016), the predominant approach taken by our respondents is to rely on their healthcare providers to manage their personal health information for them. Table 3 shows 47 participants (55%) indicated, without prompting, that they do nothing themselves but rely on their doctors. An additional 9 (11%), without prompting, simply said they don't manage their personal healthcare information, e.g., "I don't" or "I don't manage my personal health data." Thirteen used some form of paper records, two saved some documents electronically on a desktop computer, and six (7%) said they used a PHR from a healthcare provider, health plan (payer), or an employer tethered PHR.

When asked specifically if they used a personal health record and what dissuades them from using a PHR, the predominant themes were "I'm healthy" and/or "I'm lazy" or "I'm too busy."

To illustrate, one participant responded in the following way to the inquiry:

> Ok, actually I have been introduced to those but I haven't even gone on because I haven't had time but I have been directed toward them. Both my gynecologist and my general practitioner through the XXX Network of doctors use them... I don't use it because, it's so new and I just don't have the time to even look at it. I probably should though. (Study participant; female; age, 50–65; spring 2016)

or

> I do not. I guess, probably the biggest one is at this point when I go to the doctor I talk to the doctor. I don't feel like going to my health record to read what he verbally told me. So, I'm

**Table 3** Themes in descending order of prevalence

| Sample subject responses | Theme (N) |
|---|---|
| I rely on my doctor . . .<br>I rely on my health providers . . . when I need it, I ask them to provide it.<br>I rely on my doctor's files . . . .<br>Go to doctor every 6 months . . . | Have a guardian for patient personal health data (47) |
| I don't.<br>I don't currently manage my health information | Do nothing (17) |
| I have a binder.<br>I keep a binder.<br>I have some paper records in a folder.<br>I have a planner with a notes section where I write down important things but besides that not really | Paper records (11) |
| "I keep some files on my computer" | Other electronic documents (2) |
| My employer offers . . .<br>My insurance to an HAS a health savings account I do now track financially my account . . .<br>I also have a program on my computer, which is my MyChart, which is an application . . . | Tethered PHR (6) |
| I have a mobile glucose monitor (DEXCOM) and . . . | Medical devices plus (1), plus guardian |

> assuming there's no additional information on that than when I had my face-to-face visit. (Study participant; male; age, 50–65; fall 2015)

The responses above embody the experience of our participants, in that they had heard of patient portals, had access to them, and had either never logged in or had logged in once to register. And yet, they understood, even if vaguely, the purpose and value of a PHR. Three of the six participants with chronic conditions were similar. There was an overlap between the two categories: healthy and lazy; some participants identified both.

> But I didn't complete the registration. And I only see her (doctor) . . . besides um my once a year. If I even go that much, I'll see her if I'm ill, which *I don't get ill very often*. It's just not right now, um, *it's just not something I think about.* (Study participant; female; age, 50–65; fall 2015)

Table 4 shows sample responses related to the perceived time and effort required to use a PHR and perceived health status. The two themes are related. One could be too healthy to bother with a PHR, or the time effort required isn't worth the bother, I'm healthy.

Therefore, for participants with access to a PHR, their self-assessment of relative good health together with the perceived effort and time required to upkeep a personal health record seems to discourage healthcare consumers from using personal health

**Table 4** Predominant reasons for not using a PHR

| Discouraging considerations | |
|---|---|
| Sample responses | Themes |
| Honestly I don't have time for that<br>I'm lazy, I wouldn't update it<br>I'm not in the habit of keeping journal and record of any kind<br>I'm busy<br>I'm too busy for that | Perceived or requiredTime effort |
| I don't visit the doctor nearly often enough for the hassle<br>I'm healthy<br>Perhaps when I'm older<br>I haven't had any issues, perhaps if I had a chronic disease<br>I don't visit the doctor that often | I'm healthy |

records. Both themes were the predominant response. This is not to suggest users did not have other concerns about PHRs, e.g., information privacy and the ability to control who sees that information. Another user complained about the difficulty with logging in and setting a password. These concerns were rarer as reasons cited for not using a PHR. Yet at the same time, consistent with other research, our respondents believed a PHR is a beneficial tool. For participants who had never heard of PHR or patient portal, some felt they needed "to learn more about these systems."

## 4.3 *Discussion*

The axial codes arising from the first question indicate the choices that healthcare consumers have for managing personal information, whereas one axial code (time/effort) from the second question speaks the disposition participants had toward the effort and time exerted to ostensibly perform a task they view as already being performed by the primary care physician.

> No I don't, it just seems like too much work when the doctor already has all the information he needs in that EHR thing

The combination of themes arising from both questions iterated to the lazy user theory during selective coding. The lazy user theory knits together themes arising from both questions into a coherent explanation for PHR adoption behavior. We believe the lazy user and the busy user have similar needs with respect to expending (the least) effort.

Almost half of our respondents had some knowledge of what is PHR; by spring 2016 almost all participants knew what it was; many knew it as a (tethered) patient portal, and a majority thought using a PHR would positively impact their health. The lazy user theory helps to explain why they are nevertheless not persuaded to use it.

From our findings, it is clear that given the set of options healthcare consumers have for managing PHI: do the perceived work for managing one's personal health information, or a guardian or do the work on one's behalf, the choice for the lazy user is easy. This seems especially true for consumers who perceive themselves healthy users. It may also be true for some consumers with chronic conditions. Three of the six participants with chronic conditions relied on their physicians. In fact, one of them had a thyroid condition and had been explicitly instructed by their primary care physician to use a PHR to keep track of certain information; the respondent reported they didn't do it and were not planning on doing it anytime soon, "perhaps, if it got really bad." In their words, they "kept up with all their scheduled doctor's appointments where all this information is recorded, anyway."

The belief that PHI management a hassle, too much effort. The thyroid patient had a sense that constantly recording this data was a hassle or something that their time did not warrant, and they did not think they were sufficiently unwell to do so. More importantly, they felt they already had the safeguard of the physician already doing it for them. The idea that the provider was already managing this information was not unique to the participant with a thyroid condition. A majority of respondents answered the first question with "I don't, my doctor does," "If I need it, I call the doctor." This is clearly not the same as "I do nothing, my personal health information goes unmanaged." With the lazy user theory, these first-level constructs (Lee and Baskerville 2003; Schutz 1962) (the respondents' own understandings) cohere well with the second-level constructs of the lazy user theory. From our data, the lazy user has greater explanatory capacity in our context than competing theories.

Why we ruled out competing theories? The TRA-informed theories were ruled out during iterative analysis for several reasons. (1) The use of PHRs by patients does not occur within an employee/organization context; there is no direct (job) performance effect that presupposed both short and longer term effects. Patient engagement technologies have projected longer-term effects for the individual. (2) Adoption or use rates remain low. (3) Healthcare patients perceive multiple options for managing their personal health information; personal health records or patient portals happen to be the one option that has broad institutional backing in the domain of healthcare. Rogers' (2003) diffusion of innovations, we recognize, can be used retrospectively to explain the process of PHR adoption when significant rates of adoption or use are reached. On the other hand, based on our data, the decision by consumers to reject the innovation does not appear to be based on evaluating the technology per se but rather that the perceived effort required to engage with the technology is not worthwhile when an agent freely provides the value offered by the technology, particularly when one is not perceptibly ill enough.

Because our methodology is interpretive, grounded theory, where the first-level constructs (Van Maanen 1983) or the "understandings held by the observed people themselves" (p.39) "are the facts of an ethnographic investigation," the interpretations of our study participants are valid data points subject to the second-level constructs of the researcher.

Second-level constructs are the researcher's constructs of first-level constructs (Lee and Baskerville 2003); they are "interpretations of interpretations" (Van Maanen 1983) (p. 40) "The second-order concepts are the 'theories' an analyst uses to organize and explain these [first-level] facts" (Van Maanen 1983) (p. 39). In our findings, the twin notions of perceived effort required to maintain a PHR in the absence of pressing bad health and the presence of an agent freely willing and legally required to keep one's personal health information predominate the data. It is not insignificant that when we asked if a personal health record would be useful or difficult to use, respondents believe in their usefulness and do not believe that the technology is difficult to use. One could argue this is an incomplete understanding of principle knowledge about the role of patients' monitoring of their health information toward their care and wellness. That the rejection decision is grounded in a lack of understanding is consistent with Roger's (2003) diffusion of innovations theory. However, that would hardly capture the more dominant first-level understandings related to the effort expended and the presence of choice. The evidence iterates more readily to the lazy user theory.

## 5 Conclusion

Personal health records as a technology are no longer new. By 2014, every American was poised to have an electronic health record as the incentive phase for the requirements of MU Stage 1 of the Incentives Program ended and penalties would kick in. The next phase, MU Stage 2, effectively mandated tethered patient portals (PHRs) where providers were required to communicate electronically with their patients and allow patients to view and download their health information. So PHR adoption by providers has grown exponentially. Since 2017, as the incentive phase of MU Stage 2 also reached its useful life, providers, health plans, and employers have rolled out personal health records, providers sign up patients during a patient visit, and health plans and employers sometimes offer incentives for use. Will patients actually use these tools?

If primary care physicians are de facto guardians of personal health information, it should also worry anyone concerned about health disparities that those without normal access to primary care also have little to no systemically documented health record and that, as it is, personal health records likely make little to no contribution in this regard.

Several researchers (Kunene et al. 2016; Tang et al. 2006) have advocated for more consumer education about personal health records and their role in engaging in one's care. Some qualitative researchers (Lafky and Horan 2011; Spil and Klein 2014) on the adoption of PHRs, there is an understanding that the design of these tools need to be user-centered and offer strong data security and information privacy protections particularly for untethered systems. One of the focuses of such educational and design efforts likely needs to make the case for the importance of patient-generated data. It is clear from our study that consumers implicitly equate information in a personal health record to provider-generated information. (If the provider is tracking it, what is the value added in me tracking it, as it were.) This implicit equating of personal health information as provider-generated health information by both healthcare consumers and the designers of PHR tools they use seems to weaken the value proposition for consumers and may contribute the least effort choice.

Lastly, designers should be asking, what does design captology for least effort look like in the personal health record space? Could PHR designs allow for more patient-generated data and efficient interfaces for data entry, e.g., an ability to sync it with summary diet, exercise, and menstrual cycle data that are already monitored by healthcare consumers through mobile apps while preserving the user's desired information privacy preferences.

The contribution of our study is bringing a new perspective to low adoption rates of PHRs – a perspective derived from marrying study participants' understandings of their behavior (first-level constructs) with researcher second-level constructs. The implication of our findings is that any design and education efforts must address the effort problem (real or perceived), as well as the perception that provider-maintained health information is sufficient would duplicate the effort of the consumer. Effort must be made to demonstrate the value proposition for maintaining one's own electronic records. Otherwise, why should consumers bother?

There are several limitations to our study, the primary one being that it was conducted in New England where our participants had regular access to care. The other is data was collected over a long period of time at a dynamic time around personal health records. For instance, almost all the participants who had not heard of PHRs were limited to early periods of data collection, and the rate of adoption of health and wellness mobile apps by our participants appears to increase significantly in the final period.

# Appendix

"I really don't have much of a system, **I rely on the doctors** to maintain the information"...

"By **probably asking my doctor how am I doing**?"

"I don't really manage much of my health information right now, **I leave that stuff to my health care provider** and if I ever need access to any of it, then I can just contact them and get the information that I need"

"Uh, **I know it is kept in the hospital** and it is available for me to see probably if I went to through human resources..."

**"I rely on my doctor's** files and I make sure that I have my annual checkups and physicals and then I rely on the doctors to see if I need any follow up blood work or any other tests pertaining to myself."

"As far as I know I am not keeping track of them formally. ...**I have left it up to my doctor** to keep track of my health records."

"No I don't, it just seems like too much work when **the doctor already has all the information he needs in that EHR thing**"

"They are at **my doctors**"

"I'm not really managing it, **my doctor** has things in different areas, although she just moved to electronic record keeping"

"**My doctors** do it in their record-keeping on the on-base"

"Uh verbally with **my doctor**"

"If I'm managing it at all it's by way of **my doctor**"

"**My doctors** do it in their record-keeping on the on-base"

**Fig. 5** Sample responses of "my doctor does it"

"I keep some **paper records**."

"I have a **planner** with a **notes** section where I write down important things but besides that not really"

"I basically remember the information from each visit and in a **journal** write it down."

"I manage my personal health information by keeping all of my important **documents in a safe place**... drawer at home"

"File. So everything is **pretty much paper**, so file" [Has access to tethered PHR, chooses not to use it; reason: "when I go to the doctor I talk to the doctor. I don't feel like going to my health record to read what he verbally told me.

"Just a **file** in cabinet."

"My health information I have all in a **filing cabinet** downstairs. So I have my doctor's appointment and doctor visits in one folder, the dentist appointments in another folder, and if I've gone to the hospital a few times I have that in another folder as well."

"Well I keep a lot of my health records in a **filing cabinet** at home at least the copies that the office at my doctors provides me with but usually if I have a question I go back to the doctors office and ask them about it."

"I have organized it in my home with the different doctors that I see throughout the year and I keep track of all my records there and just, I guess, a hard copy on paper."

"**Primarily paper-based** although my doctor's office does have an electronic **patient portal** where I can access medical records including lab results, etc. **I have logged in occasionally but primarily to view results of recent tests."**

"Umm, I, using an **app**, but I also keep a **folder** which is a **binder** to hold all my information."

File. So everything is pretty much paper, so file.

I have all in a filing cabinet downstairs.

**Fig. 6** Paper record-keeper responses

| |
|---|
| Well, my **employer** offers a way to go online and record everything I do in terms of like doctor's visits, exercising and stuff like that ..." |
| "Yes ...since becoming part of the XXX Health System... both my gynecologist and my general practitioner offer it (patient portal)....I don't use it because it is just so new and **I just don't have the time to even look at it. I probably should though** |
| Yes, I do, most of my health record are from the **army**. Most of them are online- all I have to do is pretty much sign in using my password and user name. They actually keep track and it automatically updates..." |
| "I also have a program on my computer, which is my **MyChart**" |
| Well, currently I use a patient portal for my various doctors. |
| Yes, I started using it when I came to YNH system |
| "**I think** they (physician) offer access to one on their website. I've seen pamphlets around the office but I just never looked into it.". |
| "Well, **I registered for it**... : Okay. But I didn't complete the registration..." |
| "But it's complicated, the log in very difficult for some reason, and the password setting confuses me so I am not the biggest" (note, still uses own filing systems) |

**Fig. 7** Tethered PHR responses

# References


Abramson, E. L., Patel, V., Edwards, A., & Kaushal, R. (2014). Consumer perspectives on personal health records: A 4-community study. *American Journal of Managed Care, 20*, 287–a298.

Agency for Healthcare Research and Quality. (2017). About the national quality strategy. Agency for Healthcare Research and Quality. https://www.ahrq.gov/workingforquality/about/index.html. Accessed 22 Sept 2018.

Ajzen, I. (1985). From intentions to actions: A theory of planned behavior. In *Action control* (pp. 11–39). Berlin, Heidelberg: Springer.

Ajzen, I., & Fishbein, M. (1975). *Belief, attitude, intention and behavior: An introduction to theory and research*. Reading: Addison-Wesley.

Alter, S. (2010). Designing and engineering for emergence: A challenge for HCI practice and research. *AIS Transactions on Human-Computer Interaction, 2*(4), 127–140.

Ancker, J. S., Osorio, S. N., Cheriff, A., Cole, C. L., Silver, M., & Kaushal, R. (2015). Patient activation and use of an electronic patient portal. *Informatics for Health & Social Care, 40*, 254–266.

Archer, N., Fevrier-Thomas, U., Lokker, C., McKibbon, K. A., & Straus, S. E. (2011). Personal health records: A scoping review. *Journal of the American Medical Informatics Association, 18*, 515–522.

Association, A. H. (2018). Expanding electronic patient engagement. Annual Survey, IT Supplement Brief March.

Bagozzi, R. P., Davis, F. D., & Warshaw, P. R. (1992). Development and test of a theory of technological learning and usage. *Human Relations, 45*, 659–686.

Bhaumik, S. K., & Piesse, J. (2008). Does lending behaviour of banks in emerging economies vary by ownership? Evidence from the Indian banking sector. *Economic Systems, 32*, 177–196.

Borfitz, D. (2018). *Blockchain-secured, patient-controlled health records*. https://www.clinicalinformaticsnews.com/2018/05/08/real-world-blockchain-applications-in-healthcare.aspx.

Bouri, N., & Ravi, S. (2014). Going mobile: How mobile personal health records can improve health care during emergencies. *JMIR mHealth and uHealth, 2*, e8.

Carrión, I., Alemán, J. L. F., & Toval, A. (2011). Assessing the HIPAA standard in practice: PHR privacy policies. In: *Engineering in Medicine and Biology Society, EMBC, 2011 Annual International Conference of the IEEE, 2011*. IEEE, pp 2380–2383.

Collan, M., & Tétard, F. (2011). Lazy user model: Solution selection and discussion about switching costs. In *Scandinavian conference on information systems* (pp. 56–68). Berlin, Heidelberg: Springer.

Davis, F. D. (1989). Perceived usefulness, perceived ease of use, and user acceptance of information technology. *MIS Quarterly, 12*, 319–340.

Davis, F. D., Bagozzi, R. P., & Warshaw, P. R. (1989). User acceptance of computer technology: A comparison of two theoretical models. *Management Science, 35*, 982–1003.

Department of Health and Human Services CfMaMS. (2010). 42 CFR parts 412, 413, 422 et al. Medicare and medicaid programs: Electronic health record incentive program; Final Rule.

DiCicco-Bloom, B., & Crabtree, B. F. (2006). The qualitative research interview. *Medical Education, 40*, 314–321.

Fishbein, M., & Ajzen, I. (1977). *Belief, attitude, intention, and behavior: An introduction to theory and research*. Reading: Addison-Wesley.

Goodhue, D. L. (1995). Understanding user evaluations of information systems. *Management Science, 41*, 1827–1844.

Goodhue, D. L., & Thompson, R. L. (1995). Task-technology fit and individual performance. *MIS Quarterly, 7*, 213–236.

Houston, T. K., & Ehrenberger, H. E. (2001). The potential of consumer health informatics. *Seminars in Oncology Nursing, 17*, 41–47. https://doi.org/10.1053/sonu.2001.20418.

Jones, D. A., Shipman, J. P., Plaut, D. A., & Selden, C. R. (2010). Characteristics of personal health records: Findings of the medical library association/national library of medicine joint electronic personal health record task force. *Journal of the Medical Library Association: JMLA, 98*, 243.

Kaelber, D. C., Jha, A. K., Johnston, D., Middleton, B., & Bates, D. W. (2008). A research agenda for personal health records (PHRs). *Journal of the American Medical Informatics Association, 15*, 729–736.

Kavoussi, S., Huang, J., Tsai, J., & Kempton, J. (2014). HIPAA for physicians in the information age. *Connecticut Medicine, 78*, 425–427.

Kim, J., Jung, H., & Bates, D. W. (2011). History and trends of "personal health record" research in PubMed. *Healthcare Informatics Research, 17*, 3–17. https://doi.org/10.4258/hir.2011.17.1.3.

Kunene, K. N., Zysk, K., & Diop, M. -F. (2016). Healthcare consumers' voluntary adoption and non- adoption of electronic personal health records. In: *The 27th Australasian Conference on Information Systems, Wollongong, NSW, Australia*, 2016. University of Wollongong, Faculty of Business.

Lafky, D. B., & Horan, T. A. (2011). Personal health records: Consumer attitudes toward privacy and security of their personal health information. *Health Informatics Journal, 17*, 63–71.

Lee, A. S., & Baskerville, R. L. (2003). Generalizing generalizability in information systems research. *Information Systems Research, 14*, 221–243.

Lehnbom, E., Douglas, H., & Makeham, M. (2016). Positive beliefs and privacy concerns shape the future for the personally controlled electronic health record. *Internal Medicine Journal, 46*, 108–111.

Lester, M., Boateng, S., Studeny, J., & Coustasse, A. (2016). Personal health records: Beneficial or burdensome for patients and healthcare providers? *Perspectives in Health Information Management, 13*, 115–118.

Mitchell, B., & Begoray, D. (2010). Electronic personal health records that promote self-management in chronic illness. *OJIN: The Online Journal of Issues in Nursing, 15*, 1B–10B.

Patel, V., & Johnson, C. (2018). *Individuals' use of online medical records and technology for health needs* (Vol. 40). Washington, DC: Office of the National Coordinator.

Patel, V. N., et al. (2011). Low-income, ethnically diverse consumers' perspective on health information exchange and personal health records. *Informatics for Health & Social Care, 36*, 233–252.

Rogers, E. M. (2003). *Diffusion of innovations* (5th ed.). New York: Free Press.

Saparova, D. (2012). Motivating, influencing, and persuading patients through personal health records: A scoping review. *Perspectives in Health Information Management, 1* 243–249.

Schutz, A. (1962). Concept and theory formation in the social sciences. In *Collected papers I* (pp. 48–66). Dordrecht: Springer.

Señor, I. C., Fernández-Alemán, J. L., & Toval, A. (2012). Are personal health records safe? A review of free web-accessible personal health record privacy policies. *Journal of Medical Internet Research, 14*, e114.

Services CfMaM. (2018). Promoting interoperability. Centers for Medicare and Medicaid Services. https://www.cms.gov/Regulations-and-Guidance/Legislation/EHRIncentivePrograms/index.html?redirect=/EHRIncentivePrograms. Accessed 22 Sept 2018.

Sheppard, B. H., Hartwick, J., & Warshaw, P. R. (1988). The theory of reasoned action: A meta-analysis of past research with recommendations for modifications and future research. *Journal of Consumer Research, 15*, 325–343.

Showell, C. (2017). Barriers to the use of personal health records by patients: A structured review. *PeerJ, 5*, e3268.

Spil, T., & Klein, R. (2014). Personal health records success: why Google health failed and what does that mean for Microsoft HealthVault? In: *2014 47th Hawaii International Conference on System Sciences (HICSS), 2014*. IEEE, pp 2818–2827.

Strauss, A., & Corbin, J. (1990). *Basics of qualitative research* (Vol. 15). Newbury Park: Sage.

Strauss, A., & Corbin, J. (1994). Grounded theory methodology. *Handbook of Qualitative Research, 17*, 273–285.

Strauss, A., & Corbin, J. (1998). *Basics of qualitative research: Techniques and procedures for developing grounded theory. Second edn*. Thousand Oaks: Sage Publications.

Taha, J., Czaja, S. J., Sharit, J., & Morrow, D. G. (2013). Factors affecting usage of a personal health record (PHR) to manage health. *Psychology and Aging, 28*, 1124.

Tang, P. C., Ash, J. S., Bates, D. W., Overhage, J. M., & Sands, D. Z. (2006). Personal health records: Definitions, benefits, and strategies for overcoming barriers to adoption. *Journal of the American Medical Informatics Association, 13*, 121–126.

Tétard, F., & Collan, M. (2009). Lazy user theory: A dynamic model to understand user selection of products and services. In: *System Sciences, 2009. HICSS'09. 42nd Hawaii International Conference on, 2009*. IEEE, pp 1–9.

Van Maanen, J. (1983). Reclaiming qualitative methods for organizational research: A preface. In J. Van Maanen (Ed.), *Qualitative methodology*. Beverly Hills: Sage.

Vance, B., Tomblin, B., Studeny, J., & Coustasse, A. (2014). Personal health records: Benefits and barriers for its adoption. *Insights to a Changing World Journal, 2014*, 48–67.

Venkatesh, V., Morris, M. G., Davis, G. B., & Davis, F. D. (2003). User acceptance of information technology: Toward a unified view. *MIS Quarterly, 12*, 425–478.

Vessey, I. (1991). Cognitive fit: A theory-based analysis of the graphs versus tables literature. *Decision Sciences, 22*, 219–240.

Vessey, I., & Galletta, D. (1991). Cognitive fit: An empirical study of information acquisition. *Information Systems Research, 2*, 63–84.

Vydra, T. P., Cuaresma, E., Kretovics, M., & Bose-Brill, S. (2015). Diffusion and use of tethered personal health records in primary care. *Perspectives in Health Information Management, 12*, 109–114.

Wagner, P. J., Dias, J., Howard, S., Kintziger, K. W., Hudson, M. F., Seol, Y.-H., & Sodomka, P. (2012). Personal health records and hypertension control: A randomized trial. *Journal of the American Medical Informatics Association, 19*, 626–634.

Wunsch, G. (1982). Maternal and child health in the third world: Problems of data collection. *Popul Famille, 53*, 19–33.

Wynia, M., & Dunn, K. (2010). Dreams and nightmares: Practical and ethical issues for patients and physicians using personal health records. *The Journal of Law, Medicine & Ethics, 38*, 64–73.

Yamin, C. K., Emani, S., Williams, D. H., Lipsitz, S. R., Karson, A. S., Wald, J. S., & Bates, D. W. (2011). The digital divide in adoption and use of a personal health record. *Archives of Internal Medicine, 171*, 568–574.

Young, R., Willis, E., Cameron, G., & Geana, M. (2014). "Willing but unwilling": Attitudinal barriers to adoption of home-based health information technology among older adults. *Health Informatics Journal, 20*, 127–135. https://doi.org/10.1177/1460458213486906.

Zapata, B. C., Niñirola, A. H., Idri, A., Fernández-Alemán, J. L., & Toval, A. (2014). Mobile PHRs compliance with android and iOS usability guidelines. *Journal of Medical Systems, 38*, 81.

Zipf, G. (1949). *Human behavior and the principle of least effort*. Reading: Addison-Wesley.

Services (n.d.) [illegible] Centers for Medicare and Medicaid Services. [illegible] Regulations and Guidance/Legislation/EHRIncentivePrograms/[illegible] Accessed [illegible] 2016

[illegible]

[illegible]

[illegible]

[illegible]

[illegible]

[illegible]

[illegible]

[illegible]

[illegible]

[illegible]

[illegible]

[illegible]

Young, R., Willis, E., [illegible], M. (2014). [illegible] Willing but unwilling: [illegible] [illegible]

[illegible]

# The Australian PCEHR or My Health Record: The Journey Around a Large-Scale Nationwide Digital Health Solution

**Nilmini Wickramasinghe and John Zelcer**

## 1 Introduction

Healthcare systems around the globe are facing numerous and substantial challenges (Wickramasinghe and Schaffer 2010). These challenges range from changing demographics of patients to present and developing medical technologies and their implications on the cost of service (Hartman et al. 2009), loss of lives and wastage of resources due to medical errors, and inefficient and inconsistent information systems (Berwick 2003). Demand for better healthcare services is increasing, whilst human and fiscal resources are lessening (Duckett and Willcox 2011). In response to these challenges, e-Health initiatives, particularly the electronic health record (EHR) and various derivatives such as the electronic medical record (EMR) or

---

This case study has been written to capture the developments of a large-scale health technology journey to facilitate discussions, teaching, and learning only. It is not designed to identify good or bad management practices or make any comments about any of the organizations or individuals involved.

This case study is forthcoming as a chapter in a book publication by Springer, New York Eds Wickramasinghe, N and Bodendorf, F. 2019 Delivering superior health and wellness management with IoT and analytics and may only be used for educational purposes with the written permission of the authors.

---

N. Wickramasinghe (✉)
Epworth HealthCare, Richmond, VIC, Australia

Swinburne University of Technology, Hawthorn, VIC, Australia

J. Zelcer
Swinburne University of Technology, Hawthorn, VIC, Australia



N. Wickramasinghe, F. Bodendorf (eds.), *Delivering Superior Health and Wellness Management with IoT and Analytics*, Healthcare Delivery in the Information Age, https://doi.org/10.1007/978-3-030-17347-0_31

personal health record (PHR) are being adopted and implemented around the globe (Wickramasinghe and Schaffer 2010).

Like many other countries, Australia is investing heavily in e-Health initiatives (DoHA & NEHTA 2011). Specifically, in the federal budget of 2010–2011, the government allocated $466.7 million over 2 years to develop and implement a national personally controlled electronic health record (PCEHR later renamed My Health Record) system for all Australians by 2012 (ibid). This was a significant investment aimed at transforming the healthcare delivery system in Australia by enabling the secure sharing of health information between an individual's healthcare providers whilst enabling the individual to control who can access their PCEHR (DoHA & NEHTA 2011). Recently, after 8 years along the journey, questions are naturally being raised regarding the value of the PCEHR and its overall return on investment (ROI) as well as questions concerning (i) policy issues – such as patient privacy, security of information, identification and management of patient's consent for participation and data collection; (ii) technical issues concerning system complexity and user understanding of system, lack of standards and protocols, disparate health information systems and frameworks for their integrations and complex user interfaces; and (iii) clinical issues including benefits to care and quality of care (Currell et al. 2000; DoHA & NEHTA 2011; Foo 2012; Hall 2010; Lehnbom et al. 2012; Liaw and Hannan 2010; McDonald 2012; Muhammad et al. 2012; Naismith 2012; Showell 2011; Spriggs et al. 2012; Westbrook et al. 2009).

## 2 Background

Access to basic healthcare varies in terms of availability and affordability across countries, regions, groups and even individuals, essentially influenced by the socio-economic conditions as well as the healthcare policies of the country concerned (Wickramasinghe and Schaffer 2010). Healthcare is considered a major expenditure across the globe and it consumes a major portion of a given country's budget (ibid). Per the World Health Organization (WHO) data, in 2011, the healthcare sector of OECD countries consumed 9.3 percent of the total GDP across 34 member countries of OECD. In terms of spending, the USA led with 17.7% of GDP followed by the Netherlands 11.9%, France 11.6%, Germany 11.3%, Canada 11.2% and Switzerland 11%, whilst Australia at the same time had healthcare expenditure of 8.9% of GDP (OECD 2013). Moreover, healthcare systems are usually required to achieve multiple objectives including providing improvements in overall health for all people in a responsive, financially affordable and accessible manner by using all available resources and without compromising any efforts to ensure the quality and safety of healthcare services provided (WHO 2000).

There are different types of healthcare systems implemented in different regions of the world (Anonymous 2010). The fundamental building block of the national healthcare system of a nation will always be based on the history, politics, economy and the value system of that nation (ibid). There are four main healthcare models discussed in the healthcare literature, namely (ibid):

(i) The Beveridge model
(ii) The Bismarck model
(iii) The national health insurance model
(iv) The out-of-pocket or market-driven healthcare model

The fundamental philosophy behind these models is to ensure access to basic universal health coverage for all legal residents and citizens of a country by governments of that country regardless of citizens' demographic or financial status (ibid). Furthermore, these models stress a non-profit philosophy, and comprehensive health insurance for everyone (ibid). The models also dictate strict control over the pharmaceutical industry (ibid). The first three models are characterized by social harmony where citizens would pay according to their means and receive care according to their needs; hence more affluent individuals subsidize the poor and sick through taxation or buying their healthcare services out of pocket (ibid). One or more combinations of these models are being used in all the industrialized and democratic countries of the world including Australia, the USA, the UK, Gulf and European Union countries and Scandinavia (ibid). These models are further divided into two categories, namely the single-payer health insurance systems and the multi-payer health insurance systems, with the exception of the market-driven healthcare model (ibid).

## 3 The Australian Healthcare System

Australia is a large country (7.7 m sq km) with a land mass of roughly the same size of Western Europe or the USA with the exemption of the state of Alaska (Muhammad et al. 2012). Much of the Australian population (about 80%) lives in cities or urban areas and the remainder of the population is settled in large scattered areas throughout the country (ABS 2013). Compared to many other developed nations, the Australian healthcare system delivers above average outcomes (DoHA 2010) and the Australian population is ranked one of the best in terms of health status, with an average life expectancy at birth of 81.4 years and there has been a significant decline in infant and youth mortality rates over the period of 1988 to 2007 (AIHW 2012). (refer to Appendix 1).

The healthcare system in Australia is a combination of private and public sector care providers comprising over 1326 hospitals, and the healthcare system services around 22.6 million citizens across different geographic and socio-economic settings (DoHA & NEHTA 2011). At the core of this system is the National Health Act passed in 1953 to regulate the provision of pharmaceutical, medical and hospital benefits (ibid). The federal minister for health administers the health policy of the nation, and some elements of that policy (such as the operation and administration of public hospitals) fall under the state government's responsibility (ibid). Healthcare service delivery and financing are a joint responsibility of federal, state and territory governments through taxation, the Medicare levy and council rates, along with some contribution coming from local governments and private health insurance companies and consumers (Duckett and Willcox 2011; Heslop 2010; Willis et al. 2009). (See Appendix 2).

## *3.1 Issues and Challenges Facing the Australian Healthcare System*

The Australian healthcare system is confronting major healthcare funding and delivery challenges not dissimilar to all OECD countries (DoHA 2010; Muhammad and Wickramasinghe 2014). The rapid developments in medical technologies and treatments have led healthcare systems towards control of communicable diseases and delaying in non-communicable conditions, and that has contributed to the significant decrease in overall mortality and morbidity (OECD 2013). People are living longer which is significantly contributing to the ageing of the population (Armstrong et al. 2007). A report by the Australian Bureau of Statistics (ABS) has predicted that, by 2020, 16% of Australians would be aged 65 or over and this could rise to 27% by 2101 because of low fertility rate and increased life expectancy (ABS 2013). This would mean a greater need for more aged care facilities and an expanding aged care workforce. This can put enormous pressure on healthcare budgets (WHO 2006), as Australia is already spending 9.6% of its GDP on healthcare, and this is expected to rise by 16–20% if current trends for healthcare demands continue (DoHA 2010). Another devastating effect would be a decrease in the working population and a shortage of workforce in all areas, including healthcare that will lead to reduced tax collection which is a major source of healthcare funding (DoHA 2010; Jones 2011; Rhyne 2008). Increased prevalence and diagnosis of chronic disease is another major issue especially with the increasing ageing population, with an estimated 25% of the population of Australia suffering from chronic illness (AIHW 2012). Persistent health inequalities, the rapid pace of inventions into new technologies and medicines and consumer expectations are also putting more pressure for increased healthcare spending (Productivity Commission 2005).

One way of handling these issues is by reorienting the healthcare system towards prevention rather than cure and a community-based self-management care system for chronically ill patients by involving them in their care management process (NHHRC 2009a, b). This would reduce the burden on hospitals and health budgets (ibid). One solution proposed for this problem is implementing and operationalizing health information technology such as the My Health Record, which has capacity to decrease health disparities and improve self-management of healthcare (Shields et al. 2007), and healthcare efficiency, quality and safety (Wise and Bankowitz 2009).

Another major issue with the Australian healthcare delivery system is its fragmented nature (Dwyer and Eagar 2008). The Australian healthcare system operates as an uncoordinated and disparate set of services – there is no systemic coordination or integration of services between primary healthcare service providers and acute healthcare service providers (ibid). This lack of integration even exists within the primary healthcare sector (Duckett and Willcox 2011). The great need for coordination between and across different elements and areas of the healthcare system for fast, cost-effective, accessible and high-quality service delivery presents many challenges (DoHA 2010).

In addition, the inability to collect and share the health information of patients amongst different sectors of healthcare is another major issue (Duckett and Willcox 2011; Heslop 2010; Jones 2011). It is argued that this can pose a serious risk to patients by creating diagnostic and treatment errors, increased waiting times for referrals, increased diagnostic test duplications, delayed hospital discharge and adverse effects on administration staff work efficiency (Heslop 2010; Jones 2011).

Limitations of traditional Paper Medical Records (PMR) are also noted as a major issue in terms of safe and efficient healthcare delivery (Rhyne 2008). Limited availability and accessibility, poor legibility and missing information are a few examples of such shortcomings. Further, the content of PMRs is prone to errors. As illustrated by Maghazil, the medical data from PMR is often found to be missing or incorrectly recorded without proper validation; sometimes handwritten information is not legible. Maghazil further outlined the main factors contributing to incomplete or wrong data in PMRs, including:

- Omission – information may have been omitted due to incomplete documentation, lack of necessary investigations, test results or language hindrance.
- Delays – information might have been delayed in being recorded at the time of collection, with the nurse or GP later recording information from memory.
- Misplacement – information in paper files may have been lost or misplaced in storage or lost due to fire or other natural disasters.

## 4 The Australian E-Health System (My Health Record)

The My Health Record, previously known as the PCEHR, was launched in July 2012 as a key element of the Australian government's healthcare reform agenda (DOHA & NEHTA 2011). The My Health Record is a shared electronic health summary set up by the Australian government with implementation overseen, at that time, by the National Electronic Health Transition Authority (NEHTA), now succeeded by the Australian Digital Health Agency (ADHA) (ibid). It brings together the standards of unique health identifiers, being the Individual Healthcare Identifier (IHI), Healthcare Provider Identifier-Individual (HPI-I) and Healthcare Provider Identifier-Organization (HPI-O), along with encrypted authentication, to provide the foundations for securely exchanging, storing and accessing patients' health-related information (ibid).

### 4.1 The Journey of the My Health Record

In Australia, work on a nationally coordinated electronic health record was initiated in 1993 with the creation of the National Health Information Agreement (NHIA) (ibid). The primary function of NHIA was to develop strategy and tools for better coordination between the Australian government and state and territory governments for the collection and exchange of healthcare data and information (Bartlett et al. 2008).

After a series of government and semi-government committees had identified the need for a nationally coordinated approach to e-Health standards, responsibilities for the development of the national e-Health project were shifted to the new National e-Health Transition Authority (NEHTA) created in 2005 (DOHA & NEHTA 2011). NEHTA's responsibility was to develop the essential foundations, including healthcare identifiers (for individuals, providers and provider organizations), secure messaging, the national security and access framework and national clinical terminologies (e.g. Australian Medicines Terminology) for providers and patients by 2009 (ibid). In 2008, Deloitte, a consulting firm, was engaged to prepare a blueprint for the national strategy for e-Health development and deployment (ibid). In 2009, the federal government, with all state and territory governments, announced the introduction of health identifiers (developed by NEHTA), and in 2010, the introduction of the Health Identifiers Act established their role in law (ibid). Later with the budget of $AUD446.7 million dollars, the government successfully achieved the goal of having Healthcare Identifiers (HI) services for all Australians by July 2012 (ibid). The launch of the personally controlled electronic health record (PCEHR) in July 2012 was underpinned by the HI service which includes 16-digit reference numbers for consumers as the Individual Healthcare Identifier (IHI) and for Healthcare Providers as the Identifier-Individual (HPI-I) as well as the Healthcare Provider Identifier-Organization (HPI-O) (ibid). The HI service provides certainty of identity for digital health information and therefore for exchange between all e-Health information residing in services such as pathology, discharge summaries, referrals and medication management and the PCEHR itself (ibid).

### 4.2 What Is the My Health Record

In 2016 the Australian government made the decision to rename the PCEHR to My Health Record, as a continuation of the development of the PCEHR and its enabling standards and underpinning technologies (Australian Digital Health Agency 2018). The My Health Record is a unique solution that sits between an individually controlled health record and a healthcare provider e-Health record (ibid). In addition, the My Health Record has a shared use and mixed governance model (Bartlett et al. 2008; Muhammad et al. 2012), Fig. 1.

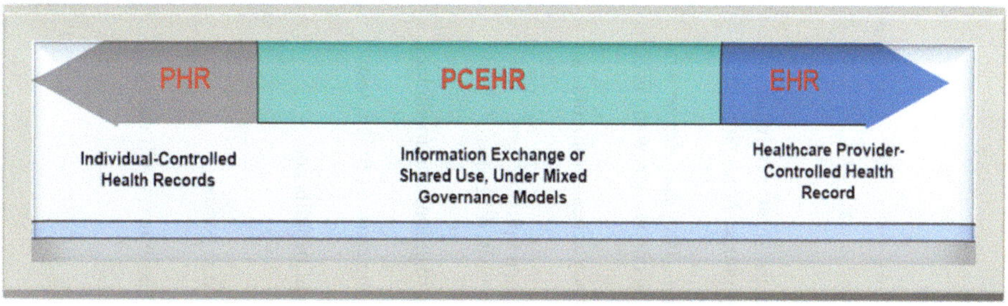


**Fig. 1** The position of the My Health Record. (Adapted from Muhammad et al. 2012)

The My Health Record is a person-centric secure repository of electronic health and medical records of an individual's medical history that can act as a hub for linking hospital, medical and pharmaceutical systems using a patient unique identifier (NHHRC 2009a, b). The My Health Record can capture information from different systems and present information in a single view to consumers and authorized service providers for better decision-making about healthcare and service delivery (DoHA & NEHTA 2011). This is a hybrid health information system that integrates web-based personal health records with clinical electronic health record system and allows shared access to both consumers and providers based on a shared responsibility and mixed governance model. The key features of the My Health Record are described in Table 1.

The MyHR (My Health Record) system provides a number of core services that allow authorized users to search for an individual's MyHR and clinical documents, view clinical documents and access reports (NEHTA 2013, 2014). A key feature of the system is its ability to provide a series of views leveraging different clinical documents in an individual's record (ibid). These views allow users of the system to easily see a consolidated overview of an individual's allergies/adverse reactions, medicines, medical history, immunizations, directives and recent healthcare events from different information sources (ibid). The flow of information into and out of the MyHR (PCEHR) is depicted in Fig. 2.

A report published by Deloitte predicted $AUD11.5 billion in net direct benefits from the use and adoption of the My Health Record system over a period of 15 years from 2010 to 2025. Deloitte categorized this benefit into three categories: the government sector is expected to have $AUD9.5 billion net benefits, whereas $AUD 2.0 billion in direct benefits is expected for private sector. Deloitte's economic modelling was performed at the commencement of the investment in the My Health Record in 2010 and modelled benefits for the forward period of 15 years (ibid). For their modelling, Deloitte considered key areas where the My Health Record could have financial as well as quality and efficiency benefits, including reduced avoidable hospital admissions and GP visits due to the more effective medication management and improved continuity of care (ibid).

**Table 1** Key features of the My Health Record (DoHA & NEHTA 2011)

| Key features | Description |
|---|---|
| Participation model | Opt-in for both consumers and service providers<br>Registration is a prerequisite for the My Health Record use<br>To upload information compliant software installation essential for service providers<br>Registration process for consumer is easy. They can register by phone, Medicare office, by mail or GP practice<br>Consumers will have full control over the healthcare information. They can choose what they want to share and with who they want to share. Further they can withdraw the access any time or even opt-out from the system whenever they desire to<br>The rule of access will be relaxed in emergency, where emergency access protocols can be used to search and use e-Health record |
| Information contribution by service providers | Authorized healthcare professionals will be able to create and then upload event summaries with the detail of GP or organization, reason for the visit, diagnoses, treatments, results and observations<br>Shared healthcare summaries containing medical history, immunizations and medicines, allergies and adverse reactions<br>Discharge summaries from hospital may include detail containing reason of visit, tests ordered and their results, diagnosis, drugs and interventions<br>Medical data about child immunization, organ donation, claim of benefits and PBS-related data<br>Specialist letters<br>The availability of pathology test (deadline was July 2013 which is passed |
| Information contribution by consumers | Patient-provided information about allergies and medicines<br>Notes about diet, exercise, blood glucose and blood pressure. These notes will not be accessible to health professionals |

| | |
|---|---|
| Benefits for consumer and service providers | The summary of patient's health information including medication, medical history referrals, lab test results, prescriptions, discharge summaries, allergies and immunizations will be available for both consumers and health service providers<br>Rigorous governance and privacy procedures will be in place for the security and privacy of both consumers and service providers<br>High-quality data for policy development as well as research and planning to government bodies will be provided<br>All activity history of any actions preformed on the My Health Record by a service provider and consumer will be provided<br>Individuals can make enquires and complaints about the management of their information<br>Easy access to health literacy information via direct link. Can collect information from consumer devices such as blood pressure monitors and blood glucose monitors<br>Fast and reliable accessing of patient's health information for both consumers and service providers remotely via fast Internet connection to enhance decision-making for diagnoses and treatments and preventive actions |
| Participation incentives for service providers including individuals and organizations | An e-Health practice incentive payment to participate in the My Health Record<br>GP can claim a certain amount of time under the MBS for creating or changing the shared health summary |

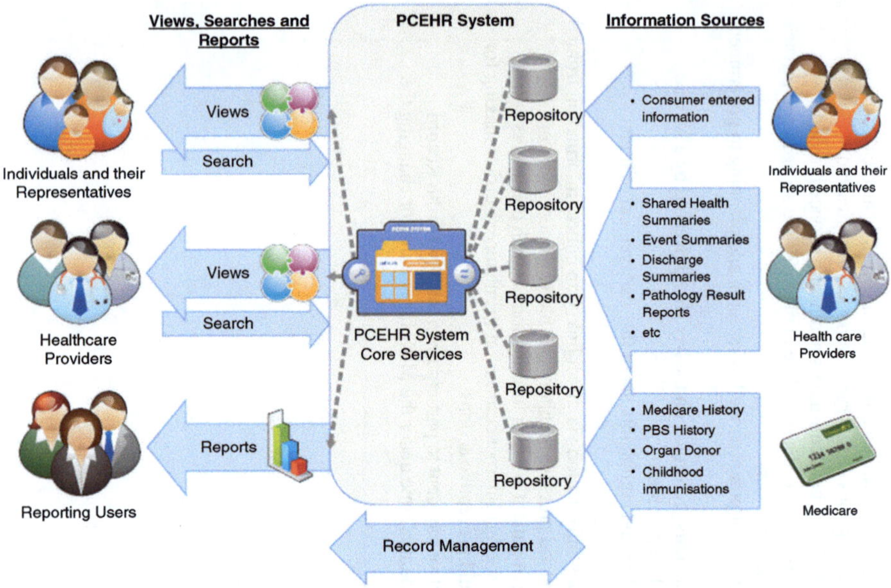


**Fig. 2** The flow of information within the MyHR system. (Adapted from NEHTA 2013, 2014)

# 5 Key Considerations for Large-Scale Information Technology (IT) Projects

For all large-scale IT projects, there are a set of key considerations that should be noted, namely, (i) governance, (ii) risk management and (iii) portfolio and project management (Wickramasinghe and Schaffer 2010); thus we now examine the My Health Record in terms of these considerations.

## *5.1 Governance*

IT governance provides a structure for aligning IT strategy with business strategy. By following a formal framework, organizations can produce measurable results towards achieving their strategies and goals (Lindros 2017). A formal programme also takes stakeholders' interests into account, as well as the needs of staff and the processes they follow. In the big picture, IT governance is an integral part of overall enterprise governance. (ibid).

One definition of IT governance (Gartner 2006) is 'The processes which ensure the effective and efficient use of IT in enabling an organization to achieve its goals'. This definition encapsulates the following key concepts (ibid):

- IT governance is composed of processes with the inputs, outputs, roles and responsibilities that are inherent in a process definition.
- The role of IT governance is identified as 'ensuring' as opposed to having a primary role of 'executing'.
- The goal of IT governance is defined as a business goal, not just IT-related.
- Key performance measures are identified as effectiveness and efficiency, together representing business value.

The purpose of IT governance is to direct IT endeavours, to ensure that IT's performance meets the following objectives (ITGI 2003):

- Alignment of IT with the enterprise and realization of the promised benefits
- Use of IT to enable the enterprise by exploiting opportunities and maximizing benefits
- Responsible use of IT resources
- Appropriate management of IT-related risks

The overall objective of IT governance, therefore, is to understand the issues and the strategic importance of IT, so that the enterprise can sustain its operations and implement the strategies required to extend its activities into the future. IT governance aims at ensuring that expectations for IT are met and IT risks are mitigated (ibid).

A key reference standard for IT governance is ISO/IEC Standard 38,500-2015 (ISO/IEC 2015). The standard presents six *principles*:

1. Responsibility: Individuals and groups within the organization understand and accept their responsibilities in respect of the supply of and the demand for IT. Those with responsibility for actions also have the authority to perform those actions.
2. Strategy: The organization's business strategy takes into account the current and future capabilities of IT; the strategic plans for IT satisfy the current and ongoing needs of the organization's business strategy.
3. Acquisition: IT acquisitions are made for valid reasons, on the basis of appropriate and ongoing analysis, with clear and transparent decision-making. There is an appropriate balance between benefits, opportunities, costs and risks, in both the short term and the long term.
4. Performance: IT is fit for purpose in supporting the organization and in providing the services, the levels of service, and the service quality required to meet current and future business requirements.
5. Conformance: IT complies with all mandatory legislation and regulations. Policies and practices are clearly defined, implemented and enforced.
6. Human behaviour: IT policies, practices and decisions demonstrate respect for human behaviour, including the current and evolving needs of all the 'people in the process'.

The major steps in establishing and implementing IT governance *processes* (Gartner, ibid) are:

1. Strategize: Develop a strategy for IT governance, that is in turn driven by the business strategy, which will ensure participation on a sustained basis by key participants.
2. Plan: Create a comprehensive plan for the implementation of that strategy. This should address the design of the IT governance process steps and support logistics, as well as the culture-change management aspects of ensuring that all participants are educated and motivated to make IT governance a success.
3. Implement: Implement the plan in such a way that the importance of IT governance to the business and to the success key participants is clear and that process responsibilities and accountabilities are understood.
4. Manage: Manage and support the ongoing IT governance processes.
5. Monitor: Monitor the results and effectiveness of IT governance and feedback results to the IT governance strategy and planning cycle to maintain performance. IT governance works best when it is integrated with existing decision-making processes and reflects their style and management culture.

It is critical to develop a common understanding of the role of IT governance between the IT leadership and key members of business management (ibid). This shared understanding is an important prerequisite to developing sustainable efficient and effective governance processes to manage the supply side of IT governance and, in particular, demand IT governance, which is essentially a business investment decision-making and benefits-realization process (ibid).

### *5.2 Risk Management*

Despite decades of experience with formalized project management methodology, widespread awareness of the need to identify and manage risks, and increasingly sophisticated supportive technology, organizations still struggle to execute projects successfully (Deloitte 2015a). Problems such as cost overruns, missed deadlines and failure to meet business requirements have become so frequent they are largely expected and, for the most part, accepted as the norm (ibid).

Risk management – in particular, operational and IT risk management – continues to mature as a discipline and is becoming a more practical approach to improving corporate governance and internal control (Gartner 2018). Practical approaches supported by innovative technology are critical to manage the increasing risks associated with digital business transformation (ibid).

Effective risk management includes elements of formalized risk assessment, controls assessment, risk decision-making, risk tracking and sign-off for residual risk and accountability (Gartner 2014). These elements result in balanced mitigation, acceptance and transfer of residual risk through conscious decision-making (ibid).

In simple terms, risk is uncertainty. From a business perspective, risk is the uncertainty of economic loss, and enterprise risk is the uncertainty of economic loss across the enterprise (ibid). How an enterprise, or an organization within an enterprise, addresses risk is a function of factors such as risk appetite, risk tolerance, perceived severity and frequency of threats, perceived impact of negative events, accuracy of measurement, transparency of controls, risk awareness in the enterprise culture and maturity of methodology (ibid). True risk management addresses risk with a balance of mitigation through the application of controls, transfer through insurance and acceptance through governance mechanisms (ibid). Gartner identifies six required elements for effective risk management (ibid):

1. Risk assessment: Risk assessment is a formal process of analysing threats and their potential business impact. Risk assessment is the cornerstone of good risk management and creates the foundation organizations need to prioritize their risks.
2. Controls assessment: Controls assessment is a formal process of analysing the current and potential controls to address threats. Controls assessment establishes the foundation organizations need to prioritize and develop appropriate treatment plans.
3. Risk decision-making: Risk decision-making is a formal process of choosing what the organization will do and, more importantly, what it will not do to address a risk. The artefacts from risk decision-making include tactical and strategic choices that represent official direction. Risk decision-making and the resulting artefacts provide the basis for accountability.
4. Risk tracking: Risks and risk treatment plans are very dynamic, so they must be tracked. This is typically done through a risk register that provides ongoing visibility into known risks and their current treatment plans. It should also maintain the artefacts from risk/controls assessments and the resulting decisions.
5. Residual risk sign-off: Signing off on residual risk is the formal acceptance of the risk that remains after a treatment plan has been adopted. It should be done by the accountable business unit/process owner. Formal recognition of the remaining risk after the application of controls helps the organization develop a risk-aware culture and establishes the foundation of accountability.
6. Accountability: In a risk-aware culture, organizations work towards making good risk-based decisions, but they understand that things can still go wrong. Accountability should be for following the process and making an informed decision.

Organizations need to have assurance that their risk management processes are at a level of maturity such that the organization is able to have confidence that (ISO 2018):

1. Management has identified, assessed and responded to risks above and below the risk appetite.
2. The responses to risks are effective but not excessive in managing inherent risks within the risk appetite.
3. Where residual risks are not in line with the risk appetite, action is being taken to remedy that.

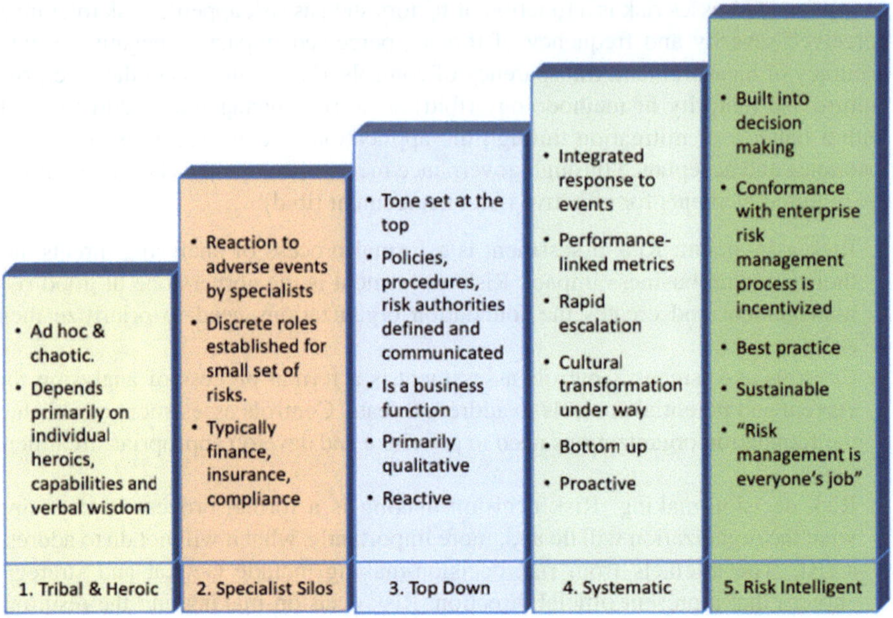


**Fig. 3** Model for evaluating the maturity of an enterprise risk framework. (Adapted from Deloitte 2015b)

4. Risk management processes, including the effectiveness of responses and the completion of actions, are being monitored by management to ensure they continue to operate effectively.
5. Risks, responses and actions are being properly classified and reported.

A model for evaluating the maturity of an enterprise risk framework is presented in Fig. 3.

## *5.3 Portfolio and Project Management*

There are many tools, techniques and frameworks that are used in large-scale portfolio and project management, all of which are well described in reference, training and qualification sources available via the Internet (Smartsheet 2018). These include Agile methods Lean (an Agile method), Scrum (an Agile method), Six Sigma, Waterfall methods (also called traditional) and PRINCE2 (ibid). However, going beyond those methodologies, there are a number of contemporary challenges in managing and successfully delivering large-scale projects (Stead 2010).

The Gartner Program and Portfolio Management (PPM) maturity model elucidates the core requirements designed to help organizations identify relevant best practices in PPM to maximize strategic value delivery (Gartner 2017). The underlying principles for this model assume that organizations progress through a maturity curve and that each level of organizational maturity directly affects the level of investment and types of PPM approaches organizations choose to adopt (ibid).

They describe five interdependent core dimensions of critical importance in (ibid):

1. People: People are the most critical part of any project- or programme-centric endeavour. The interdependencies amongst people in terms of their availability, their skills, their contribution to the work that needs to be done and their career aspirations are of critical importance. At higher levels of maturity, the leadership ability of the individuals involved in supporting PPM activities becomes critical.
2. PPM practices and processes: PPM practices and processes comprise disciplines, such as portfolio management, and programme as well as classic project management processes, such as risk and resource management. One of the most common activities also included in this dimension is the establishment of a PMO, be it a project management office, programme office or portfolio management office.
3. Value and financial management: Systems that might be adequate for monitoring the financials when projects are paid for as part of a lump sum in the budget (a common level 1 practice) rarely cast much light on value and become completely inadequate when forced to support a more detailed look at multiple projects, programmes and products. This dimension focuses on understanding how to ensure that projects, programmes and products offer value for the money spent.
4. Technology: The requirements for technology support tend to evolve in sync with the overall maturity of the approach to PPM. This dimension is designed to help organizations understand which and how much automation technology will yield the greatest return at a given maturity level.
5. Relationships: PPM fundamentally is focused on teams of people working to create a specific outcome. This dimension offers appropriate guidance on the necessary touchpoints. This includes identifying who needs to be informed, who needs to be consulted and whose help is mandatory to ensure that the desired outcome is achieved.

The maturity model stages and attributes are depicted below (Fig. 4):

Gartner emphasizes that the benefits of PPM using advanced practices include:

- Optimizing a complex set of portfolios across products, projects, programmes, streams and teams
- Ensuring you have ongoing transparency to key information that allows you to make better decisions faster and enhance organizational agility
- Ensuring that there is consensus on value and priority for all activities
- Building capability to execute strategy and complex programmes reliably

Gartner further concludes that PPM leaders must reach out to other IT and business leaders to streamline processes, making them faster, more efficient and

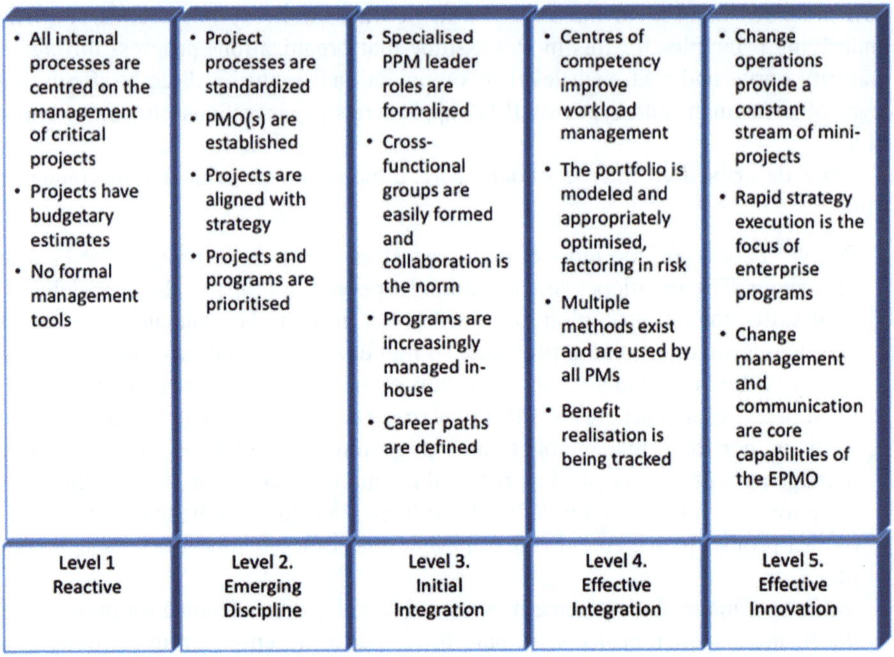


**Fig. 4** Maturity model of Program and Portfolio Management. (Adapted from Gartner 2017)

more collaborative (ibid). They must also build a business agile mindset to enable enterprise agility, including but not exclusive to IT. Gartner recommends that PPM leaders continue to identify new roles within enterprise agility approaches, including collaboration and support (ibid). For an organization to move towards digital technology leadership, it needs to break down the barriers between the IT organization and other technology enablers of the enterprise (ibid). Digital technologies are everyone's business, and digital decisions need to be embedded everywhere (ibid).

As we come to the conclusion of the second decade of the new millennium, it is an appropriate time to reflect on the journey to date with the My Health Record. In particular, given the significant investment as is necessary with any large-scale health IT project, it is useful to explore if the full value has been realized and if not what should now form the focus of the government and other stakeholders. Moreover, in light of recent reports on healthcare quality at various state and federal levels, it is also important to evaluate the My Health Record in terms of policy dimensions as well as privacy and security aspects, technical issues and clinical issues. Specifically, what are the lessons from this initiative that can also be applied to other health IT initiatives and how can the benefits of the My Health Record be fully realized are important areas to explore. Given the journey is ongoing, it is not possible to draw conclusions at this point; rather we invite our readers to reflect on the preceding material and think about the questions we posed. We conclude, by underscoring that this was a massive undertaking and that time will be the true judge of the merits of the My Health Record.

**Acknowledgements** We wish to acknowledge with thanks all the organizations and individuals who assisted us with information to help us compile this case study. In particular, we acknowledge Dr. Imran Muhammad for his inputs on early drafts.

## Appendix 1: Comparative Performance of the Australian Healthcare System

Compared to many other developed nations, the Australian healthcare system delivers above average outcomes (DoHA 2010), and the Australian population is ranked one of the best in terms of health status, with an average life expectancy at birth of 81.4 years, and there has been a significant decline in infant and youth mortality rates over the period of 1988 to 2007 (AIHW 2012).

Australia's population is healthier than the OECD average, considering life expectancy and other general measures of health status (OECD 2013, 2017). Smoking consumption is also low, as is exposure to air pollution. But obesity rates are the fifth highest in the OECD (ibid). Further, despite universal health coverage, a relatively high share of the population reported skipping consultations due to cost (ibid). Quality of care indicators also show mixed results (ibid). The figure below shows how Australia compares across these and other core indicators from *Health at a Glance* (ibid).

This comparison is shown in a Table 2 (from OECD 2013, 2017).

The following are key comparison statistics:

- Health status: life expectancy at birth was 82.5 years in 2015, the fifth highest in the OECD (the OECD average was 80.6). Mortality from ischaemic heart disease and the prevalence of dementia are also both lower than the OECD average.
- Risk factors: results are mixed. The proportion of Australian adults who smoke is 12.4%, amongst the lowest in the OECD. Air pollution is second lowest. Alcohol consumption is slightly above the OECD average. However, the adult overweight and obesity rate is 27.9%, the fifth highest in the OECD and well above the average of 19.4%.
- Access: all Australians are deemed to have access to healthcare. However, 16.2% of adults report that they skip medical consultations due to cost, well above the OECD average of 10.5%.
- Quality of care: Australia has the third lowest 30-day mortality rate following admission for heart attack in the OECD. Just over 70% of people diagnosed with colon cancer survive – the fourth highest rate in the OECD. However, hospital admission rates for asthma and COPD are high (371 per 100,000 people versus an OECD average of 236). Antibiotic prescribing is also higher than the OECD average (23.4 per 1000 people, compared with an OECD average of 20.6).

**Table 2** Australia - Relative performance compared to the OECD average

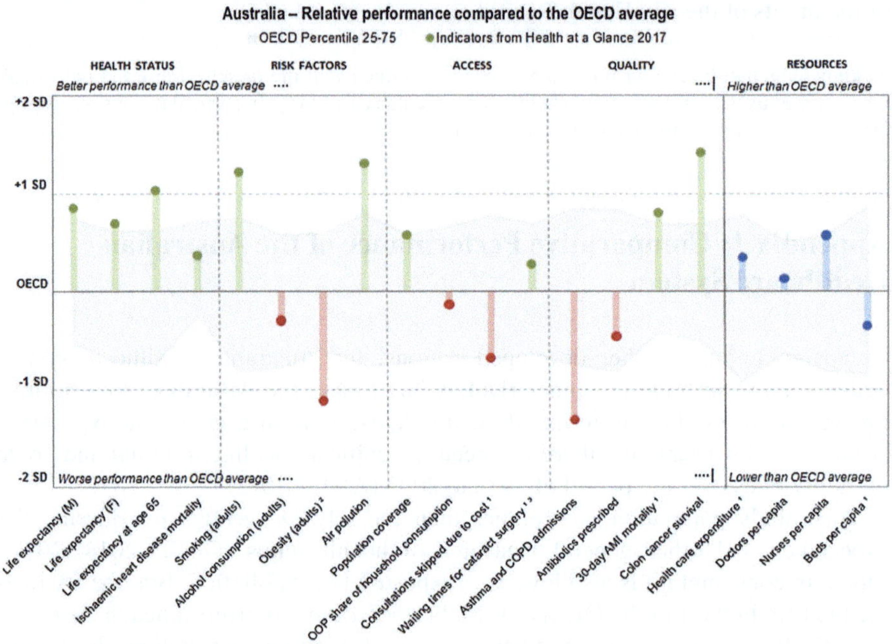


From OECD (2013, 2017)

- Resources: health spending averages \$4708 per person (adjusted for local costs), slightly higher than the OECD average of \$4003. Australia has more nurses and doctors (11.5 and 3.5 per 1000 people) than the OECD average (9.0 and 3.4, respectively), but the number of hospital beds per capita is slightly lower than average.

## Appendix 2 Structure of the Australian Health-Care System

Australia's health landscape has four tiers (Fig. 5; DoHA & NEHTA 2011):

- The first and largest tier is 'determinants of health and other demographic factors'. This includes education, employment, income, family and community, rural and remote and indigenous Australians.
- The second tier is 'health promotion and disease prevention'. This includes immunization, food, physical activity, illicit drug use, tobacco control, alcohol consumption, mental health and cancer screening.
- The third tier is 'primary health and community care'. This includes dental practice, pharmacy, allied health, general practice, primary health networks, community care and aged care.

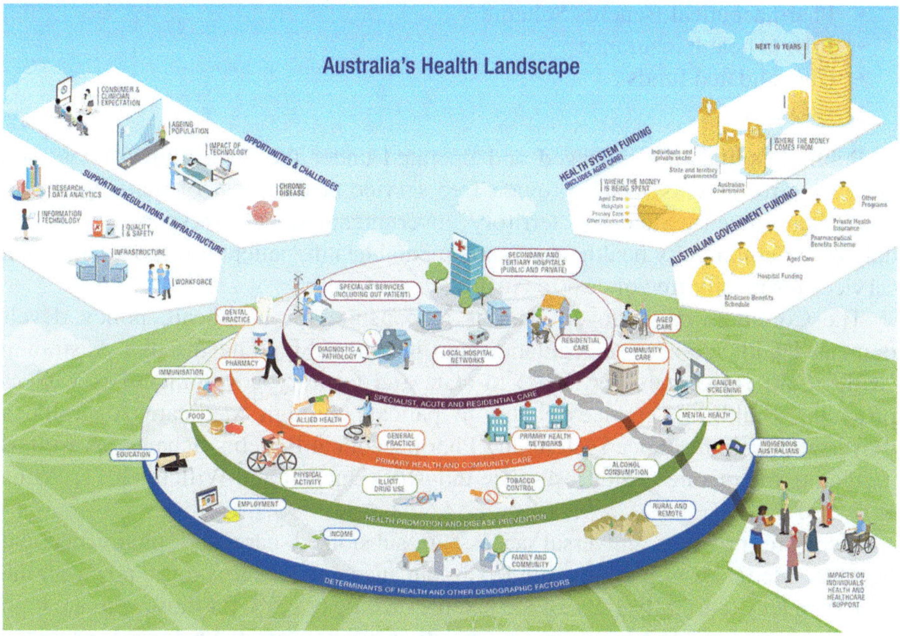


**Fig. 5** The healthcare landscape in Australia. (Adapted from https://beta.health.gov.au/resources/publications/australias-health-landscape-infographic)

- The fourth and smallest tier is 'specialist, acute and residential care'. This includes specialist services (including out patient), diagnostic and pathology, local hospital networks, residential care and secondary and tertiary hospitals (public and private).

  Factors that contribute to the health landscape (ibid):

- Impacts on individuals' health and healthcare support
- Opportunities and challenges, including consumer and clinician expectation, ageing population, impact of technology and chronic disease
- Supporting regulations and infrastructure, including research and data analytics, information technology, quality and safety, infrastructure and workforce
- Health system funding of $2.3 trillion over the next 10 years. The funding in 2014–2015 was $182 billion including aged care. 11.9% was spent on aged care, 36.1% on hospitals, 32.7% on primary care and 19.3% on other recurrent costs. $58 billion came from individuals and the private sector, $43 billion from state and territory governments and $81 billion from the Australian government. The Australian government funding was for (from largest to smallest):

  - Medicare Benefits Schedule
  - Hospital funding
  - Aged care

- Pharmaceutical Benefits Scheme
- Private health insurance
- Consolidated funds
- Research

Public hospitals are managed and operated under the ownership of state and territory governments which provide free service at the point of delivery for all Australian citizens. State and territory governments are also responsible for the delivery of community health, aged care, mental health, patient transport and dental services for mostly free of cost to Australian consumers.

The Commonwealth government is responsible for healthcare policy development, healthcare service regulation and healthcare funding through Australian Health Care Agreements (AHCA) to state and territory governments as well as providing healthcare service rebates to patients through Medicare Australia, a universal health insurance system and Pharmaceutical Benefits Scheme, as well as regulating the private health insurance industry (Duckett and Willcox 2011; Willis et al. 2009).

Medicare provides universal access to subsidized care and pharmaceutical benefits. Medicare is partially funded by an income tax surcharge known as the Medicare levy (currently 1.5% of the taxable income or 2.5% for those who are high income earners and do not have private health insurance) and the balance is provided by government from general revenue (AIHW 2012). The Australian Institute of Health and Welfare has estimated that the total Australian health expenditure is $140.24 billion in 2011–2012 (AIHW 2013). This represents 9.5% of GDP. Another important component of Australian healthcare system financing is private health insurance (ibid). The Australian government encourages citizens to enrol for private insurance by giving them tax benefits for private health insurance premiums paid (ibid).

The flow of money around the healthcare system of Australia is a complex phenomenon and can be controlled by the institutional frameworks in place at both government and non-government levels. The government sector in this respect is comprised of federal, state and territory governments, and in some jurisdictions, local governments are involved (ibid). The non-government sector includes individuals, private health insurers, and other different non-government funding sources (ibid). The other non-government sources include worker's compensation, compulsory third-party motor vehicle insurers, donations for health-related research and miscellaneous non-patient revenue sources for hospitals (ibid). The complexity of these funding arrangements and interaction between different levels of service providers and consumers in healthcare service delivery is explained here (ibid).

The funding model for healthcare services and delivery in Australia has perceived political polarization, with governments being critical in determining nationwide healthcare policy (Behan 2007). This leads to administrative duplications and lack of coordination at national level resulting in slow reactive and inefficient health policy (Wall 2002).

## Appendix 3: System Architecture of the My Health Record

The My Health Record system (previously PCEHR) is situated in an e-Health ecosystem that has been designed to serve a number of communities:

- Consumers – including the individuals themselves, their family members and their carers. This also extends to working with organizations that promote public health and prevention programmes as well as those organizations that support self-managed care.
- Primary/community-based care – organizations and their staff, including, but not limited to, general practices, community pharmacies, allied and complementary healthcare providers, diagnostic imaging and pathology providers, specialists and aged care services.
- Acute/post-acute care – organizations and services, including, but not limited to, admissions, emergency care, outpatient care, surgical and medical care, rehab and subacute care units, hospital pharmacies and hospital diagnostic imaging and pathology providers. [Does this include Medibank Private, MBF, Medicare].
- Vendors – who provide products and services to support primary/community-based care, acute/post-acute care and consumers.
- e-Health services – operated either by public or private sector service providers that deliver online e-Health services, such as secure messaging, directories, prescription services, diagnostic services, registries and other secondary use services.

The e-Health community is also supported by a range of foundations (or national infrastructure) and standards for e-Health solutions (Fig. 6).

Building on a series of stages of development, the PCEHR system provides a number of core services that allow authorized users to search for an individual's PCEHR, clinical documents, view clinical documents and access reports.

A key feature of the PCEHR System is its ability to provide a series of views leveraging different clinical documents in an individual's PCEHR. These views allow users of the system to easily see a consolidated overview of an individual's allergies/adverse reactions, medicines, medical history, immunizations, directives and recent healthcare events from different information sources.

In the context of a national approach to e-Health, local systems at the point of care or in the home will be able to access a range of services, including:

- National infrastructure services, including, but not limited to, the Healthcare Identifier Service, National Authentication Service for Health, etc.
- Public- or private-operated online services, including, but not limited to, pathology services, radiology services, prescription exchange services, etc.

The interfaces for these services are based on national and international standards and other agreed specifications (Fig. 7).

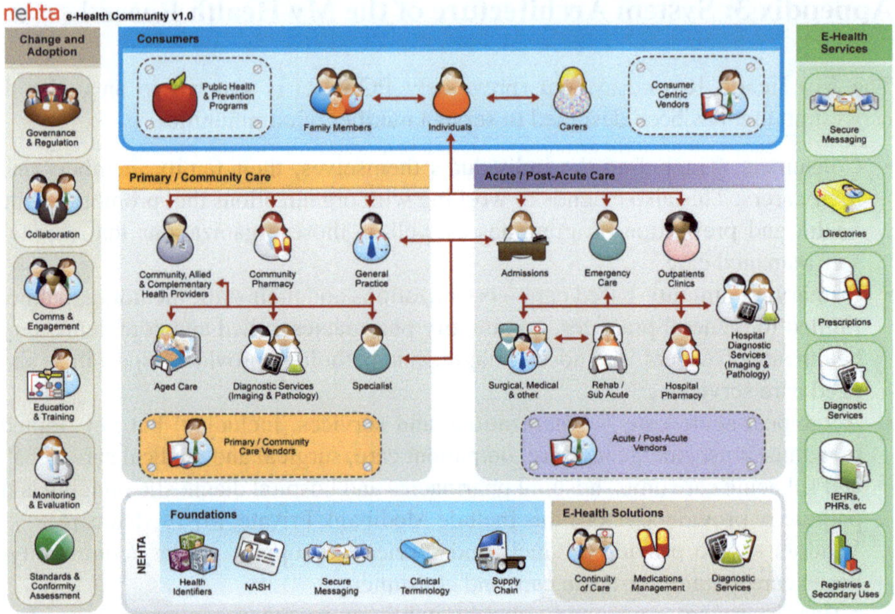


**Fig. 6** The e-Health community architecture. (Adapted from NEHTA 2013, 2014)

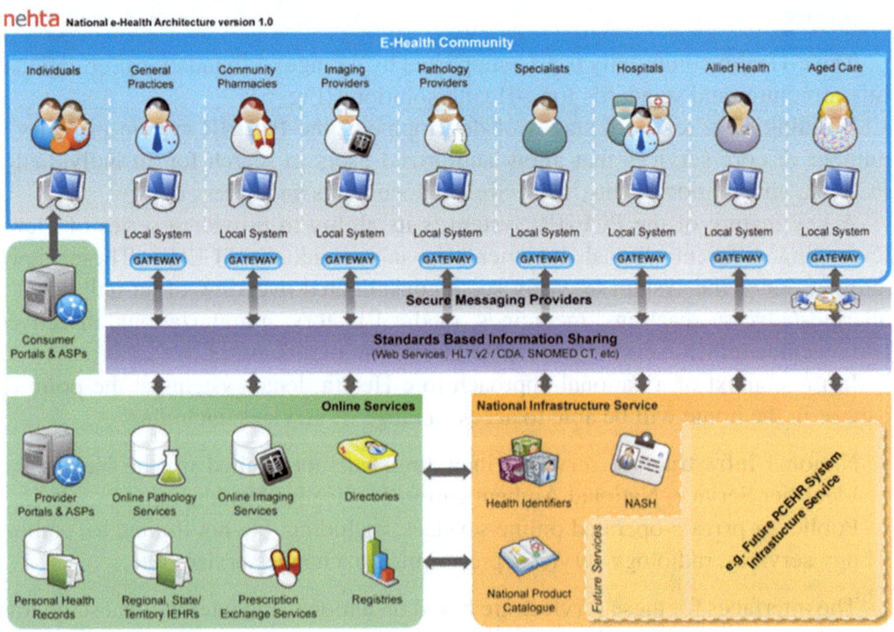


**Fig. 7** The target state architecture of the My Health Record. (Adapted from NEHTA 2013, 2014)

## References


ABS. (2013). Australian demographic statistics, Mar 2013. Retrieved July 26, 2011, from http://www.abs.gov.au/ausstats/abs@.nsf/mf/3101.0

AIHW. (2012). Australia's health 2012: The Thirteenth Biennial Health Report of the Australian Institute of Health and Welfare. AIHW.

AIHW. (2013). Health expenditure Australia 2011–12 (AIHW). Retrieved from http://www.aihw.gov.au/publication-detail/?id=60129544658

Anonymous. (2010). Our current health care system is simple: Don't get sick, PNHP. Retrieved Aug 2018 at http://www.pnhp.org/single_payer_resources/health_care_systems_four_basic_models.php

Armstrong, B. K., Gillespie, J. A., Leeder, S. R., Rubin, G. L., & Russell, L. M. (2007). Challenges in health and health care for Australia. *The Medical Journal of Australia, 187*(9), 485–489.

Australian Digital Health Agency. (2018). What is My Health Record. Accessed Aug 2018 form https://www.myhealthrecord.gov.au/for-you-your-family/what-is-my-health-record

Bartlett, C., Boehncke, K., & Haikerwal, M. (2008). E-health: Enabler for Australia's health reform (p. 66). National Health & Hospitals Reform Commission. Retrieved from http://www.health.gov.au/internet/nhhrc/publishing.nsf/Content/16F7A93D8F578DB4CA2574D7001830E9/$File/E-Health%20-%20Enabler%20for%20Australia%27s%20Health%20Reform,%20Booz%20&%20Company,%20November%202008.pdf

Behan, P. (2007). *Solving the health care problem: How other nations succeeded and why the United States has not*. New York: SUNY Press.

Berwick, D. M. (2003). Disseminating innovations in health care. JAMA. https://doi.org/10.1001/jama.289.15.1969.

Currell, R., Urquhart, C., Wainwright, P., & Lewis, R. (2000). Telemedicine versus face to face patient care: Effects on professional practice and health care outcomes. In The Cochrane Collaboration & R. Currell (Eds.), *Cochrane database of systematic reviews*. Chichester, UK: John Wiley & Sons, Ltd. Retrieved from http://www2.cochrane.org/reviews/en/ab002098.html.

Deloitte. (2015a). IT project risk an abiding CIO challenge. CIO Journal. Retrieved from https://deloitte.wsj.com/cio/2015/05/14/it-project-risk-an-abiding-cio-challenge/

Deloitte. (2015b). Enterprise risk management. A 'risk-intelligent' approach. Deloitte Risk Advisory 2015. Retrieved from https://www2.deloitte.com/content/dam/Deloitte/uk/Documents/audit/deloitte-uk-erm-a-risk-intelligent-approach.pdf

DoHA. (2010). Building a 21st century – Primary health care system – Australia's First National Primary Health Care Strategy. Australian Government Department of Health and Ageing. Retrieved from http://www.yourhealth.gov.au/internet/yourhealth/publishing.nsf/Content/Building-a-21st-Century-Primary-Health-Care-System-TOC

DoHA & NEHTA. (2011). Concept of operations: Relating to the introduction of a personally controlled electronic health record system (Article). p. 159.

Duckett, S., & Willcox, S. (2011). *The Australian health care system*. (4th revised ed.). Melbourne: Oxford University Press.

Dwyer, J., & Eagar, K. (2008). Options for reform of commonwealth and state governance responsibilities for the Australian health system. A Paper Commissioned by the National Health and Hospitals Reform Commission.

Foo, F. (2012). *Canberra admits PCEHR delays COMPLETE ROLLOUT MAY TAKE A FEW MORE MONTHS*. Canberra, ACT, Australia: The Australian.

Gartner. (2006). *Defining IT governance: Roles and relationships*. New York: Gartner, Inc.

Gartner. (2014). *Six required elements of effective risk management*. New York: Gartner, Inc.

Gartner. (2017). *ITScore overview for program and portfolio management*. New York: Gartner, Inc.

Gartner. (2018). *Hype cycle for risk management, 2018*. New York: Gartner, Inc.

Hall, J. (2010). Australian health care reform: Giant leap or small step? *Journal of Health Services Research & Policy, 15*(4), 193–194. https://doi.org/10.1258/jhsrp.2010.010090.

Hartman, M., Martin, A., McDonnell, P., & Catlin, A. (2009). National health spending in 2007: Slower drug spending contributes to lowest rate of overall growth since 1998. *Health Affairs, 28*(1), 246.

Heslop, L. (2010). *Patient and health care delivery systems in the US*. Canada and Australia: A critical ethnographic analysis. LAP LAMBERT Academic Publishing.

ITGI. (2003). *Board briefing on IT governance* (2nd ed.). New York: IT Governance Institute.

ISO/IEC. (2015). Information technology – Governance of IT for the organization. *ISO/IEC, 38500*, 2015.

ISO. (2018). *ISO 31000:2018 – Risk management*. New York: International Organization for Standardization.

Jones, T. (2011). *Developing an E-Health strategy: A commonwealth workbook of methodologies, Content and models*. New York: Commonwealth Secretariat

Lehnbom, E. C., McLachlan, A., & Jo-anne, E. B. (2012). A qualitative study of Australians' opinions about personally controlled electronic health records. *Studies in Health Technology and Informatics, 178*, 105–110.

Liaw, S.-T., & Hannan, T. (2010). Can we trust the PCEHR not to leak? *Australian Family Physician, 39*, 809–810.

Lindros, K. (2017). What is IT governance? A formal way to align IT & business strategy. Retrieved from https://www.cio.com/article/2438931/governance/governanceit-governance-definition-and-solutions.html

McDonald, K. (2012). Use it or lose it—PCEHR for PIP', PULSE+IT. https://www.pulseitmagazine.com.au/news/australian-ehealth/983-use-it-or-lose-it-pcehr-for-pip

Muhammad, I., Teoh, S. Y., & Wickramasinghe, N. (2012). Why using Actor Network Theory (ANT) can help to understand the Personally Controlled Electronic Health Record (PCEHR) in Australia. *International Journal of Actor-Network Theory and Technological Innovation (IJANTTI), 4*(2), 44–60.

Muhammad, I., & Wickramasinghe, N. (2014). How an Actor Network Theory (ANT) analysis can help us to understand the Personally Controlled Electronic Health Record (PCEHR) in Australia. In *Technological advancements and the impact of actor-network theory* (pp. 15–34). Heidelberg: IGI Global.

Naismith, P. (2012). eHorizons: The PCEHR future is here. *AJP: The Australian Journal of Pharmacy, 93*(1106), 62.

NEHTA. (2013). *NEHTA's Annual Report 2012–13*.

NEHTA. (2014). *Secure Message Delivery (SMD) – NEHTA*.

NHHRC. (2009a). *A Healthier future for all Australians: National Health and Hospitals Reform Commission – Final Report June 2009*. New York: Australian Government Department of Health and Ageing.

NHHRC. (2009b). *Future for all Australians: National Health and Hospitals Reform Commission-Final Report June*. New York: Australian Government Department of Health and Ageing.

OECD. (2013). *Health at a glance 2013: OECD indicators*. New York: OECD.

OECD. (2017). *Health at a glance 2017: OECD indicators*. New York: OECD.

Productivity Commission. (2005). *Impacts of advances in medical technology in Australia*. New York: Productivity Commission.

Rhyne, L. (2008). *An explorative case study of the transition to the electronic clinical record in the south Australian acute healthcare context: A diffusion of innovation perspective*. New York: Citeseer.

Shields, A. E., Shin, P., Leu, M. G., Levy, D. E., Betancourt, R. M., Hawkins, D., & Proser, M. (2007). Adoption of health information technology in community health centers: Results of a National Survey. *Health Affairs, 26*(5), 1373–1383. https://doi.org/10.1377/hlthaff.26.5.1373.

Showell, C. M. (2011). Citizens, patients and policy: A challenge for Australia's national electronic health record. *Health Information Management Journal, 40*(2), 39–43.

Smartsheet. (2018). Demystifying the various project management frameworks and methodologies. Retrieved from https://www.smartsheet.com/demystifying-various-project-management-frameworks-and-methodologies

Spriggs, M., Arnold, M. V., Pearce, C. M., & Fry, C. (2012). Ethical questions must be considered for electronic health records. *Journal of Medical Ethics, 38*(9), 535–539. https://doi.org/10.1136/medethics-2011-100413.

Stead, D. (2010). Improving project success – Managing projects in complex environments and project recovery. Retrieved from https://www.pmi.org/learning/library/complex-environments-resolving-challenges-success-6923

Wall, A. (2002). *Health Care Systems in Liberal Democracies*. New York: Taylor & Francis

Westbrook, J. I., Braithwaite, J., Gibson, K., Paoloni, R., Callen, J., Georgiou, A., & Robertson, L. (2009). Use of information and communication technologies to support effective work practice innovation in the health sector: A multi-site study. *BMC Health Services Book, 9*(1), 201. https://doi.org/10.1186/1472-6963-9-201.

WHO. (2000). *The world health report 2000: Health systems: improving performance*. World Health Organization.

WHO. (2006). *The world health report 2006: Working together for health*. World Health Organization.

Wickramasinghe, N., & Schaffer, J. (2010). *Realizing value driven e-health solutions*. Washington, DC: Report for IBM.

Willis, E., Reynolds, L., & Helen, K. (2009). *Understanding the Australian health care system*. Australia: Elsevier.

Wise, G. R., & Bankowitz, R. (2009). Keys to engaging clinicians in clinical IT. *Healthcare Financial Management: Journal of the Healthcare Financial Management Association, 63*(3), 66–72.

# Epilogue

Our book has served to present a miscellany of papers which focus on critical aspects around embracing the technologies of the IoT to enable and support superior healthcare delivery and wellness management. This is still a very nascent domain with many issues around health literacy, policy, privacy and security not to mention the direct and subtle as well as far-reaching implications for the various stakeholders (patients, clinicians, healthcare organisations, regulators, payers and the community at large) which have yet to be fully understood or identified.

These are exciting times, and technology holds the answer to many possibilities often only limited by our imaginations. Clearly the future is bright but challenging, and we can participate to ensure that superior technology solutions can provide high-value, patient-centred quality care. We trust that on the completion of this book researchers, scholars, practitioners, consultants and the general public will all have a better understanding of how the technologies of the IoT can be harnessed to provide superior healthcare delivery and wellness management and will rise to the challenge of starting to build a better health and wellness environment for tomorrow, today.

The Editors
Professors Nilmini Wickramasinghe and Freimut Bodendorf, December 2018.



N. Wickramasinghe, F. Bodendorf (eds.), *Delivering Superior Health and Wellness Management with IoT and Analytics*, Healthcare Delivery in the Information Age, https://doi.org/10.1007/978-3-030-17347-0

# Index




N. Wickramasinghe, F. Bodendorf (eds.), *Delivering Superior Health and Wellness Management with IoT and Analytics*, Healthcare Delivery in the Information Age, https://doi.org/10.1007/978-3-030-17347-0

**B**

## C

**I**

MIX
Papier aus verantwortungsvollen Quellen
Paper from responsible sources
FSC® C105338

If you have any concerns about our products,
you can contact us on
**ProductSafety@springernature.com**

In case Publisher is established outside the EU,
the EU authorized representative is:
**Springer Nature Customer Service Center GmbH**
**Europaplatz 3, 69115 Heidelberg, Germany**

Printed by Libri Plureos GmbH
in Hamburg, Germany